Arthur Cayley

∾ Arthur Cayley ∾

Mathematician Laureate
of the Victorian Age

TONY CRILLY

The Johns Hopkins University Press
Baltimore

The Johns Hopkins University Press
2715 North Charles Street
Baltimore, Maryland 21218-4363
www.press.jhu.edu

Library of Congress Cataloging-in-Publication Data
Crilly, A. J.
Arthur Cayley: mathematician laureate of the Victorian age / Tony Crilly.
    p. cm.
Includes bibliographical references and index.
ISBN 0-8018-8011-4 (acid-free paper)
1. Cayley, Arthur, 1821–1895. 2. Mathematicians—Great Britain—Biography. I. Title.
QA29.C39 C75 2004
510′.92—dc22

                                                                    2004005682

A catalog record for this book is available from the British Library.

The last printed page of the book is an extension of this copyright page.

For my wife Ann

Contents

Acknowledgments   ix
Introduction   xiii
Chronology   xix
Genealogy   xxiii

∾ PART 1 ∾

Growing Up, 1821–1843 | 1

CHAPTER 1   Early Years | 3

CHAPTER 2   A Cambridge Prodigy | 27

CHAPTER 3   Coming of Age | 57

∾ PART 2 ∾

New Vistas, 1844–1849 | 79

CHAPTER 4   A Mathematical Medley | 81

CHAPTER 5   From a Fenland Base | 101

CHAPTER 6   The Pupil Barrister | 120

∾ PART 3 ∾

A Rising Star, 1850–1862 | 155

CHAPTER 7   Barrister-at-Law | 157

CHAPTER 8   A Grand Design | 184

CHAPTER 9   Without Portfolio | 214

CHAPTER 10   The Road to Academe | 235

~ PART 4 ~

The High Plateau, 1863–1882 | 259

CHAPTER 11   The Mathematician Laureate | 261

CHAPTER 12   Years of Challenge | 290

CHAPTER 13   A Representative Man | 312

CHAPTER 14   March On with Step Sublime | 341

~ PART 5 ~

Make One Music as Before, 1882–1895 | 369

CHAPTER 15   "A Tract of Beautiful Country" | 371

CHAPTER 16   The Old Man of Mathematics | 390

CHAPTER 17   Last Years | 417

Appendix A: Arthur Cayley's Social Circle   443
Appendix B: Glossary of Mathematical Terms   468
Abbreviations   477
Notes   483
Bibliography   559
Index   589

Illustrations follow pages 100 and 234

Acknowledgments

While "Cayley" is a household name in mathematics, my first acquaintance with the man was as a postgraduate student in the 1960s. My supervisor, the late Sir W. H. Cockcroft, suggested a research topic that involved "Cayley numbers." For this, combined with his infectious enthusiasm for mathematics, my first debt is to him and also to my other teachers at the University of Hull.

My colleague Ivor Grattan-Guinness has always been encouraging. I am most grateful to him, not least for the transmission of energy he devotes to the history of mathematics. The shift in my interest to biography, in which mathematics is not the only consideration, I owe to Clive Kilmister, who encouraged my thinking of mathematics as a human enterprise. Over the years, Karen Parshall has been generous in contributing historical insights and providing constructive criticism. It has been stimulating to experience her enthusiasm and read her work on Cayley's great friend, James Joseph Sylvester. My friend Colin Fletcher has been instrumental in my keeping a balanced approach to the whole project.

I owe most to my family: my wife Ann and my children, Charlotte, Eleanor, Sophie, and James. For years they showed understanding, as all too frequently I disappeared behind the green baize door that leads to the nineteenth century—and with good humor, they accommodated the extra lodger in their midst.

The focus of research has been in both London and Cambridge. I have always been made welcome in Cambridge, at the University Library and the various college libraries. In the early days of this project, sorting through the cache of Cayley-Sylvester correspondence at St. John's College, Mr. A. G. Lee and the late Mr. N. C. Buck were ever helpful. Later, when my focus moved to Trinity College, I had the good fortune to be guided by Jonathan Smith, who unsparingly gave his time and expertise. I also had the pleasure of visiting the Boole Library at University College, Cork, where I studied the Boole Papers.

It is a pleasant task to publicly thank those who read drafts of chapters as the work proceeded. I would like to thank June Barrow-Green, Colin Fletcher, Desmond MacHale, Philip Maher, Jonathan Williams, Robin Wilson, and the late John Fauvel. James and Peta Abbott, Clive Fleay, Ivor Grattan-Guinness, the late Gordon Smith, and Patricia Wackrill all had the stamina to read a complete draft and offer valuable suggestions. I am most grateful to them all.

For conversations, encouragement, help in the location of sources, illumination of the biographical craft, and not least for taking an interest in the work, I would like to thank, in addition, Anne Barrett, Hugh Barty-King, Harvey Becher, Thomas Bending, Richard Beskeen, the late Neil Bibby, David R. Briggs, W. H. Brock, Tony Carew, Sir Digby Cayley, Karine Chemla, Ellie Clewlow, John Coates (the current holder of the Sadleirian Chair of Pure Mathematics at Cambridge), Alex Craik, the late Vera Crilly, Jim Cross, Richard Crossley, Joe Dauben, Richard Davies, Helmut and Maureen Dhein, Jerry Edwards, C. W. Francis Everitt, Winnie and Malcolm Dudley Ewen, Allan Findlay, Victoria Fox, Frances Gandy, Rod Gow, Jeremy Gray, Theodore Hailperin, Jonathan Harrison, Thomas Hawkins, Guy Holborn, Dale Johnson, Kathleen Jones, Ann Kennedy, Patricia Kenschaft, Elaine Koppelman, Walter Ledermann, Elizabeth Leedham-Green, Albert Lewis, Roger Mallion, David McKitterick, Rita McWilliams-Tullberg, Androulla Michaeloudis, Frank Miles, Stella Mills, Michael Mortell, Peter M. Neumann, Jennifer Newman, John North, Laura Nurzia, Susan Oakes, Maria Panteki, the late E. S. Pearson, Michael Price, Alan Pryor, Neil Rhind, Adrian Rice, John Rollett, Steve Russ, Ron Shaw, the late Frank Smithies, Sofron Sofroniou, Gary Tee, Anne Thomson, Antony R. Unwin, Andrew Warwick, Tom Whiteside, Iolo Wyn Williams, and Peter Williams. I have benefited from membership of the British Society for the History of Mathematics and have enjoyed the collegiality of historically minded people.

For permission to publish quotations from archives under their care, I wish to acknowledge the Master and Fellows of Trinity College, Cambridge; the Master and Fellows of St. John's College, Cambridge; Newnham College Library; Cambridge Philosophical Society; Cambridge University Library; Royal Society Library; Manuscripts and Special Collections, and Information Services, of the University of Nottingham; University of Exeter Library (Special Collections); Department of Special Collections, Glasgow University Library; Boole Library, University College Cork (which owns the copyright of the Boole Papers); Archives of the Royal Geographical Society; London Mathematical Society; Library of the Royal Astronomical Society; the Treasurer and Benchers of Lincoln's Inn; University of Reading Library; Wellcome Library for the History and Understanding of Medicine, London; King's College, London, Archives; Cornwall Record Office, Truro; the late Reverend A. H. M. Kempe and the County Archivist of West Sussex, Chichester, England; Archives de l'École Polytechnique, Paris; Staatsbibliothek Preussischer Kulturbesitz, Berlin; Universitäts und Landesbibliothek, Bonn; Niedersächsische Staats und Universitätsbibliothek, Göttingen; Scuola Normale Superiore, Pisa; David Eugene Smith Papers (Rare Book and Manuscript Library), Columbia University in the

City of New York; Special Collections Department of Brown University (John Hay Library), U.S.A.; Special Collections, Milton S. Eisenhower Library, Johns Hopkins University, U.S.A. In addition to permissions, I would like to thank the staff of other libraries and repositories who have helped me: Middlesex University; British Library; Public Record Office, Kew; Family Records Centre, Islington; Guildhall Library in the City of London; Library of the Royal Institution of Great Britain; Imperial College Archives, London; Library of University College, London; St. Bride Printing Library, London; Cambridge Central Library (Local Studies); Lewisham Local Studies and Archives; Greenwich Local History Library; the libraries of Girton College and of Gonville and Caius College, Cambridge; the Brotherton Library in the University of Leeds; Instituto Matematico, Università di Roma. In particular, I would like to thank the staff of the Science Museum Library in South Kensington for their courteous and efficient service over the years.

For granting me permission to publish illustrations and acknowledgements, I wish to thank the Master and Fellows of Trinity College, Cambridge; the Master and Fellows of St. John's College, Cambridge; Newnham College Library; King's College, London, Archives; National Portrait Gallery, London; the Treasurer and Benchers of Lincoln's Inn; Lewisham Local Studies and Archives; City of Westminster Archives Centre; Cambridge University Press; the Syndics of Cambridge University Library; the London Mathematical Society; the Harry Ransom Humanities Research Center, University of Texas at Austin; Mathematical Institute, University of Oxford; *Encyclopaedia Britannica*; Royal Society; G. Eisenstein, *Mathematische Werke*; and the London University Press. Numerous images are part of my own collection. While every effort has been made to trace all copyright holders, where these are known, if any have been inadvertently overlooked, the publishers will be pleased to make the necessary arrangements at the first opportunity.

I am happy to acknowledge study grants generously given to me by Middlesex University, the British Academy, and the Royal Society; without this support, the research could not have been completed.

The staff of the Johns Hopkins University Press have smoothed my path toward publication. Kathryn Kraynik took on the immense task of copyediting the manuscript, miraculously turning a large draft into a beautifully turned-out printer's copy—and this done with efficiency and a touch of humor.

Introduction

Will our grandchildren live to see . . . an Arthur Cayley Court
in Lincoln' Inn where he once abode?
—James Joseph Sylvester, 1886

To my mind Cayley is the greatest English mathematician
of the last century—& this.
—Herbert Westren Turnbull, 1923

Cayley's work in all parts of algebra and geometry
has been extremely influential.
—Bartels van der Waerden, 1980

It was a hot day. The windows of Garden House were flung open, and one of the last family photographs in which Arthur Cayley would appear was about to be taken. The bent old man with piercing gray eyes sat on a chair waiting, his hands clasped together over his knees. Two housemaids and his wife were standing on the lawn in their heavy Victorian costumes appropriate to their station, and another servant looked on from the window above, almost lost in the ivy that covered the face of the dwelling. Waiting for the photographer's moment, his daughter, sitting in front of the group, smiled in readiness. From this domestic scene, Cayley could count himself a fortunate man. He had lived a full life and done so much.

The professor who resided in Garden House, at the end of Little St. Mary's Lane in Cambridge, was a phenomenon to those who knew him. Within the world of nineteenth-century mathematics, Arthur Cayley (1821–95) was an undisputed leader and one of the most prolific practitioners. He made his mark at the highest level and is now, more than one hundred years later, an indelible figure easily located in the vast panorama of modern mathematics. Here was a man who spent his most productive years as a London barrister and at the same time led his compatriots out of the intellectual hibernation that had existed since the time of Newton.

Yet Cayley's life poses a conundrum: why is he not more widely known? When he threaded his way through the narrow passageways of Cambridge to and from the university, he was recognized, for he was a part of the place. He was a principal member of the Victorian scientific establishment, where to be a mathematician was to be "like Cayley." Thus today we might expect a glimmer of recognition in the wider population for the reputation of this man of attainment. Other scientific leaders have received attention, but Cayley has not. His *Collected Mathematical Papers* aside, there is very little physical evidence to remind us of his existence. Portraits and a marble bust in Trinity College, a brass plaque in its Chapel and the "Cayley Suite" in the modern Garden House International Hotel, a crater on the Moon's surface, accorded to almost all with a connection to astronomy, all exist but one might expect more. There is now no headstone to mark his grave, and the "Arthur Cayley Court" mooted by his great friend J. J. Sylvester has never materialized.

Mathematicians are aware of his individual contributions to areas of current interest, but the passage of his life is chiefly informed by the brief sketch appearing in E. T. Bell's *Men of Mathematics* (1937). At Cayley's memorial service in 1895 it was lamented that no account of his life was in the offing, not even a multivolume "Life and Letters" of the type that might appear in a great man's latter years or closely following his demise. Perhaps contemporaries were persuaded by the old lines written about him: "Story, God bless you, there is none to tell, Sir."[1]

Cayley was no self-publicist. He never sought the limelight and constrained his ambition to the mathematical field. This noble ideal is recognized in contemporary biographical sketches, but today we are interested in a broader perspective in the way he went about research and his views on mathematics itself. While his life as a mathematician of international repute is of primary interest, it was not the only life lived. His years as a lawyer in London have to be woven into the story, as well as the later professor involved with the life of the university. Cambridge, the "mathematical university," with its system of education and its rituals is especially important. Apart from his early youth "cradled amidst the snows of St. Petersburg" and the fourteen years at the Bar, he spent most of his life at Cambridge.[2] There he was a familiar figure: champion student of his day, research professor, active supporter of women's education, an eminence in old age.

Cayley was quintessentially "Victorian." Born when the Regency ended, he progressed from school to Cambridge University in 1838, three months after the coronation of the young Queen. For this label, he is doubly qualified, since his mathematical production, already of voluminous proportions, accelerated at the midcentury, the point which many historians regard as the beginning of

"Victorianism," the era of ambitious public-spirited projects, industrial expansion, and Empire. Importantly for Cayley, communication with the Continent had been activated following the cessation of the Napoleonic Wars; his subject was growing in stature, and young mathematicians of the 1840s, seeing the possibilities of progress with new topics, were able to build on the mathematics of their parents' generation, a generation who had stepped outside Britain's eighteenth-century isolation into the wider mathematical world.

Very few pursue mathematics throughout the years so relentlessly as Cayley. George Boole regarded him, with some amusement, as the "most favored wooer of the Nymph Mathesis."[3] Cayley was certainly contemptuous of the view that mathematics was something for the parlor or a desultory engagement for leisure hours, an uncompromising attitude he made clear in his disparagement of mathematics as a mere avocation. Musing over problems with "one foot on the fender" was not his way and though attracted to recreational mathematics where it had serious mathematical content, he was not given to casual puzzle-solving. Cayley was being "professional" toward mathematics in a period more notable for its armies of amateurs and gentlemen of science. Being professional in the modern sense, not merely being a member of the professional classes, would include awareness of the latest European research and seeing mathematical problems in a broad historical context.

When the reforms of the Royal Society materialized in the 1850s, Cayley was in the vanguard of this reorientation toward professionalism on behalf of pure mathematics. Moving in the same scientific circles as William Thomson, George Gabriel Stokes, and James Clerk Maxwell, he represented the pure mathematician *par excellence*—no less than the Mathematician Laureate of the Victorian Age. To the broad church of Victorian mathematicians, amateur and "professional" alike, he was a leader and if, in his latter years, the conservative side of his nature suggests a rather formal upright Victorian gentleman, in his youth he was instrumental in making a break from the past. He took to new ideas with alacrity; above all else Cayley was an adventurous explorer.

The accessibility of Cayley's work owes much to existence of the *Collected Mathematical Papers*, a rich source for mathematicians and historians alike and yet another monument to Victorian industry.[4] The very fact that one person should write so much mathematics is a fascination in itself. His energy can be compared with Anthony Trollope in the field of literature, who also managed a double life. Trollope's Post Office, like Cayley's legal career, did not prevent him living an artistic existence.

The oft-quoted "967" statistic, the count of articles in the *Collected Mathematical Papers* is indeed impressive, but as a substitute, I offer two others. In the *Cambridge and Dublin Mathematical Journal*, a journal published between 1846 and 1854, the average number of published papers by any author is just two, but Cayley's personal tally is fifty-one. During this period, he also published work further afield. Inquiring into his record in the prestigious *Journal für die reine und angewandte Mathematik (Crelle's Journal)*, for instance, we find he published fifty-nine articles during the course of his life; in the nineteenth century he was runner-up to his hero Carl Gustav Jacob Jacobi, who published one hundred, though Jacobi hardly published elsewhere.[5] There was little deviation in the regularity of his mathematical publications. The year of his marriage was a rare instance when there was brief respite, when the torrent was switched off, but even here the break lasted barely three months.

Of Cayley's myriad contributions, his forays in invariant theory were the most significant. It was modern algebra to British mathematicians and they made it their own. Mathematicians from France, Italy, and Germany enthusiastically took up the theme in the 1850s, but it was the burst of activity from Cayley and Sylvester at the midcentury mark that was portentous. Progress in invariant theory was far from smooth but it mushroomed in the second half of the century. David Hilbert, who made his reputation in the subject in the 1880s and 1890s, when it was a more streamlined subject, divided its history into three epochs—the *naive, formal,* and *critical.*[6] Accordingly, he placed Cayley in the naive period, for Cayley detected the simplest concepts and experienced the joy of first discovery.

In raising the prospect of an "Arthur Cayley Court," Sylvester saved his literary best when thinking of his friend and of their mathematical adventures. Sylvester's admiration was total: He was simply "the central luminary, the Darwin of the English school of mathematicians."[7] Cayley formed a special alliance with this man, the two forming one of the most productive associations in the history of mathematics. It was an unlikely amalgam but one which proved its worth. Cayley's calm demeanor counterbalanced Sylvester's excitable and flamboyant style.

Within this Victorian man of few words, there was an artist constantly engaged in the mathematical struggle; in common with a swathe of the Victorian intelligentsia, the high-minded Platonist sought the eternal truths of mathematics. Indeed, his student Charlotte Scott addressed her sensitive eulogy to "the life of a great artist."[8] On first glimpsing a geometrical model, or a promising mathematical idea, he might pronounce it "wonderfully interesting" or often simply "beautiful," with the same fervor as Tennyson. Of key significance for a proper understanding of Cayley is his love for the classical culture of the ancient Greeks, and central to this was his admiration for their mathematical world.

The young man was seen clearly by Robert Leslie Ellis, an intimate of his at Cambridge. He had Dante's loss of Beatrice in mind when he spoke of Cayley, as "one of the best, as well as ablest men I have known. There is no purer minded man, nor any to whom love has been more or might have been, what it was to Dante such as you find it in the *Vita Nuova.*"[9] His Cambridge colleague

J. W. L. Glaisher wrote of the "beautiful character of the man who so long has lived among us."[10] The inescapable fact that Cayley was England's leading mathematician during the entire nineteenth century is a strong reason for having a biography of him, but Ellis and Glaisher's commendation presents an intriguing one.

Generally speaking, historians of mathematics want to see mathematics in context. How did mathematics evolve? What difficulties were encountered? Success should be recognized, but we might learn much from failure. After all, a finished text duly polished by author and editor often gives the impression that mathematics rolled out this way, a circumstance that is rarely the case. The biographer has to meet added demands. What was Cayley like? Who were his friends? What did he read? What were the conditions of his life, mathematically and otherwise? Account must be taken of people and places he knew and the time and cultural affinities which bound his existence. Historians may legitimately divorce mathematics from the person, but in the biographer's task they must be seen together and the life of mathematics woven into the fabric of their subjects. Personality, stories, and even sentimentality have a place in sculpting a mathematician's life-story. Small details begin to matter; the postscripts to hurriedly written notes, and "off-the-cuff" remarks, which can be better than any printed record in conveying the sense of the creative process with its ups-and-downs, strokes of luck, and wrong turnings.

In this biography, I have attempted to convey the "fish in water," to view Cayley's work in the context of his life, which *was* mathematics. To divorce him from it would run the risk of giving a very lopsided view. In setting out on this writing task, mathematics at first assumed the central role. If this had been pursued to its logical conclusion, what might have appeared—one day—would be a new *Collected Mathematical Papers* ten times the length of the original. This would surely miss the point of a biography. For the purpose of biography, the *Collected Mathematical Papers* can be read differently. Among his works are the five-minute Note and the weighty memoir. They are fine when read as mathematics, but as they appear in the *Collected Mathematical Papers* they do not convey the life. Contributions are not organized chronologically but in journal runs, thus camouflaging his activity, month by month, year by year. Sedate progress in keeping with the stereotype of a Victorian gentleman does not fit the whirlwind activity of a Cayley who constantly diverted from subject to subject as he pursued the mathematical zeitgeist. Mathematicians wanting Cayley's technical expertise can fill in the mathematical details for themselves and interpret his work in a modern setting better than any biographer.

To a large extent, we are now cut off from the world Cayley inhabited. In 1972, the old Garden House Hotel, the setting of his family photograph, the sedate ivy-clad building of pale yellow Cambridgeshire brick on the banks of the Granta, was consumed by fire.[11] As the world of the Victorians gradually crumbles away, our most tangible link with Cayley remains his *Collected Mathematical Papers*. If a portrait of his life helps to read between the lines of his work it will have fulfilled a central object of biography.[12] Cayley's life is still relevant today, and the internal desires and aspirations which drove him as a mathematical creator are practically identical to the ones which motivate mathematicians of all periods.

Chronology

Date	Event in Cayley's Life	Historical Context
1821	Born (16 August) in England at Richmond, in Surrey	Coronation of George IV
1821–28	Spends early life in St. Petersburg	
1835	Enters King's College, London, as a school pupil	
1838	Enters Trinity College, Cambridge	Coronation of Queen Victoria
1840	Scholar of Trinity College	
1841	Publishes first mathematical paper (as an undergraduate)	
1842	BA University of Cambridge, Minor Fellow, Trinity College	"Railway Mania" in Britain
1843	Visits Switzerland and Italy	
1844	Collaborates with George Boole on theory of invariants, recognizes the importance of n-dimensional geometry	Irish potato famine begins
1845	Discovers the algebra of octaves, which become "Cayley numbers"; visits Scandinavia and Germany	
1846	Enters Lincoln's Inn as a pupil barrister, leading light in the expansion of the *Cambridge Mathematical Journal* under William Thomson's editorship	Louis Napoleon escapes to London

1847	Cayley meets his lifelong friend, the mathematician James Joseph Sylvester, about this time	
1848	Visits Dublin and meets the Irish mathematicians	Year of European Revolutions
1849	Called to the Bar, Lincoln's Inn	
1851	Discovers a new basis for invariant theory	Great Exhibition in Hyde Park, London
1852	Elected Fellow of the Royal Society of London	
1854	Commences sequence of ten memoirs on quantics; widens the Galois notion of a group of permutations	British War with Russia in the Crimea
1857	Produces his widely praised first "Report on the Recent Progress of Theoretical Dynamics"	
1858	Develops the notion of a matrix and explores the algebra of matrices	
1859	Awarded Royal Medal of Royal Society of London; defines a notion of distance in projective geometry	Charles Darwin publishes *On the Origin of Species*
1863	Marries Susan Moline; elected to Sadleirian Chair, University of Cambridge	
1864	Introduces non-Euclidean geometry to England	
1868–70	President of the London Mathematical Society	Paul Gordan's fundamental theorem of invariant theory for binary forms
1869–70	President of the Cambridge Philosophical Society	
1870	Birth of son, Henry	Franco-Prussian War, 1870–71
1872	Birth of daughter, Mary	

1872–74	President of the Royal Astronomical Society	Introduction of James Clerk Maxwell's equations of electromagnetism, 1873
1874	Applies theory of mathematical "trees" to organic chemistry	
1876	Publishes *An Elementary Treatise on Elliptic Functions*	
1878	Completes sequence of ten memoirs on quantics; introduces the notion of colorgroups, states "Cayley's theorem" in group theory	Telephone invented by Alexander Graham Bell
1880	Visits Felix Klein in Munich	Time measurement in Britain synchronized to Greenwich Mean Time
1882	Visits the Johns Hopkins University, Baltimore, USA; awarded Copley Medal of the Royal Society of London	
1883	President of the British Association for the Advancement of Science	Robert Louis Stevenson publishes *Treasure Island*
1884	Awarded (first) De Morgan Medal of London Mathematical Society	
1886		David Hilbert's major result in invariant theory
1895	Dies (26 January) at Cambridge, England	Birth of future George VI

Genealogy

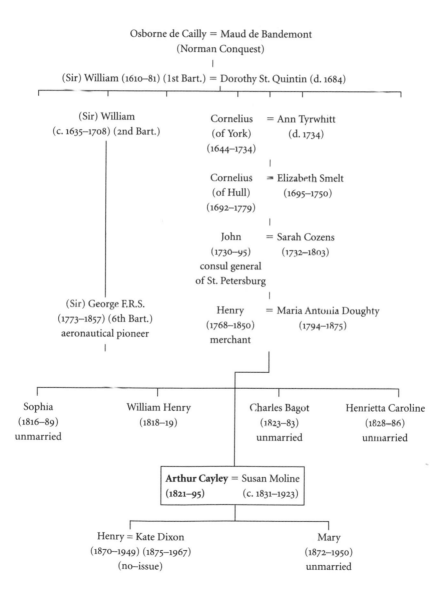

Osborne de Cailly = Maud de Bandemont
(Norman Conquest)

(Sir) William (1610–81) (1st Bart.) = Dorothy St. Quintin (d. 1684)

(Sir) William
(c. 1635–1708) (2nd Bart.)

Cornelius
(of York)
(1644–1734)
= Ann Tyrwhitt
(d. 1734)

Cornelius
(of Hull)
(1692–1779)
= Elizabeth Smelt
(1695–1750)

John
(1730–95)
consul general
of St. Petersburg
= Sarah Cozens
(1732–1803)

(Sir) George F.R.S.
(1773–1857) (6th Bart.)
aeronautical pioneer

Henry
(1768–1850)
merchant
= Maria Antonia Doughty
(1794–1875)

Sophia
(1816–89)
unmarried

William Henry
(1818–19)

Charles Bagot
(1823–83)
unmarried

Henrietta Caroline
(1828–86)
unmarried

Arthur Cayley = Susan Moline
(1821–95)   (c. 1831–1923)

Henry = Kate Dixon
(1870–1949) (1875–1967)
(no–issue)

Mary
(1872–1950)
unmarried

PART 1

Growing Up, 1821–1843

*At a very early age Arthur gave the usual indication by which
mathematical ability is wont first to show itself, namely, great liking
and aptitude for arithmetical calculations.*

—Salmon, 1883, p. 481

Arthur Cayley's dedication to a life of mathematics began in childhood with a fascination for numbers, perhaps encouraged by his father's ability in this sphere. For the first seven years of his life, the young boy lived in St. Petersburg, where his father was a "Russia merchant"—in modern terms a commercial entrepreneur who had daily dealings with money transactions.

The merchants in St. Petersburg lived in a cocoon, insulated from the indigenous population. The Anglican British Chapel to which they belonged, and which formed the linchpin of their lives, reinforced their national identity. They were there to make money and young Cayley's life might easily be regarded as a middle-class English childhood.

When the family returned to England in 1828, the boy attended a private school on the edge of Blackheath village, a district in the course of becoming a London suburb. An upward intellectual spiral followed, in which he attended the three-year course at King's College Senior School in London two years earlier than normal, with this being followed by a glittering undergraduate career at Cambridge, and all of which culminated in his being the champion student—the famed "Senior Wrangler." By this time, his fascination with the basic properties of numbers had grown into a familiarity with the works of leading Continental mathematicians.

An undergraduate paper on the subject of "determinants" from this rather solitary young man signaled a brilliant career ahead. His character of a mathematical explorer was forged and he was among scientific friends at Cambridge, but unlike those who linked mathematics with science, he was addicted to mathematics itself. A Trinity College fellowship followed undergraduate success and he celebrated the prospect of a life devoted to research with a Grand Tour of Switzerland and Italy with his friend Edmund Venables.

CHAPTER 1

Early Years

O n Thursday, 16 August 1821, Arthur Cayley was born at Richmond in Surrey. Within weeks of his birth, the family was on the way back to St. Petersburg following their summer visit to England: his father Henry Cayley, a Russia
merchant, his mother Maria Antonia, the infant, and his four-year-old sister
Sophia. With a fair wind, they would see the island of Kronstadt lying off the
estuary of the Neva in little over a fortnight.

The Cayleys of Yorkshire

In the eleventh century, Osborne de Cailly from Normandy accompanied
William the Conqueror on his expedition to England and was rewarded for his
fidelity to the King. The Cayley family (also known by the variants Cailly,
Caley, Caly, and Cayly) took possession of lands around Massingham in Norfolk. In the fourteenth century they acquired land at Thormanby in North
Yorkshire and settled there. In the seventeenth century, members of the family
served Charles I in the English civil war. In one branch of the family, by then
quite numerous and established in different pockets of Yorkshire, all four
brothers were killed on the Royalist side. In another branch living at Brompton, a small village eight miles southwest of Scarborough, Thomas Cayley was
killed while his elder brother William survived. At the beginning of the war,
William had been knighted by Charles I at his palace at Theobalds in Hertfordshire; some twenty years later in April 1661 he became the first Cayley
baronet in acknowledgment of his services in the reestablishment of the
Monarchy.[1]

William Cayley, Bart., established a seat at Brompton and married
Dorothy St. Quintin. Their family of two daughters and four sons ensured the
line would continue. The sixth baronet, a direct descendent of William, was
the inventor and pioneer aviator Sir George Cayley. This Regency figure was
an ingenious man with a wide capability. In 1825 he invented the caterpillar
tractor and patented it under the title of a "Universal Railway." His other
inventions included an instrument for testing the purity of water and a device
for harnessing electric power to machinery. He was also active in social

reform and carried out a cottage allotment system on his lands in Yorkshire. Sir George was an influential figure in the city of York. He was chairman of the Whig Club and a supporter of the establishment of the British Association for the Advancement of Science when it was founded there in 1831. As a member of Parliament, he spent time in London and in 1838 was chairman and guiding light in the formation of the Polytechnic Institution, Regent Street. Throughout his life he experimented with aircraft design, or "aerial navigation" as he deemed it, and in 1853 he constructed a fixed wing glider that transported his reluctant coachman 500 yards across the Cayley estate at Brompton.[2]

Sir George is now a well-known scientific personality, but his fourth cousin, the subject of this biography, is barely known beyond the world of professional mathematics. Indeed, the two men have been confused with each other in the biographical literature. Both were Fellows of the Royal Society and, for a few years, were members at the same time, though they did not move in the same social circles. Arthur Cayley was only distantly related to the northern Whig family and actually expressed little interest in his antecedents. Yet, such a background gave its members solidity in Victorian society. It gave them a sense of belonging and allowed them to move easily in a society in which social rank was such a useful appendage.

Over the years, members of the wider family (motto, *Callide sed honeste,* with skill but with honor) occupied prominent social positions.[3] There were members of Parliament, members of the judiciary, and some held high office in the Anglican Church. They cannot be claimed entirely by the English establishment, however. Sir George was brought up by a Unitarian mother and educated by nonconformist tutors. Arthur Cayley's great uncle Cornelius Cayley was himself an eminent eighteenth-century nonconformist preacher who for some time served as Clerk in the Treasury of the Dowager Princess of Wales. After a tour on the Continent, he published his *Tour through Holland, Flanders, and parts of France* in book form in 1777 and the following year published *The Riches of God's Free Grace, displayed in the life and conversion of Cornelius Cayley,* a book which was reprinted on four occasions during the nineteenth century.[4]

When the House of Commons met for the first time in 1833 following the passing of the Great Reform Act, there was familial involvement with the social upheavals of the time, during which the franchise widened and there was a shift of political power away from the aristocracy and landed interests. Sir George Hayter's vast oil painting depicts the occasion and shows the assembled MP's in debate. The Duke of Wellington and Prime Minister Earl Grey are on the floor of the House, but in the Upper Gallery, Sir George is standing next to kinsman Edward Stillingfleet Cayley, just newly elected for the North Riding of Yorkshire, a seat he would hold for thirty years.[5]

Closer bonds were forged between these two branches of the "Yorkshire Cayleys" when Sir George's daughter Emma married Edward Stillingfleet. No Prime Minister could count on the support of this independently minded MP who let it be known that he was "not a Whig, but a Reformer" (but one, it might be mentioned, who was against the extension of the franchise and opposed "Free Trade" during the 1840s). His farming background enabled him to claim the allegiance of the smallholders, and to represent them, to the annoyance of the leading Northern families. This group sided with Tories or Whigs as it pleased them but no faction could claim their automatic support. Edward's sons were resident at Cambridge University when Arthur Cayley was there in the 1840s. The elder, another Edward Stillingfleet Cayley (who wrote *European Revolutions of 1848* and *The war of 1870 and the peace of 1871*), and the younger George John Cayley, poet and author (who produced *Las alforjas or the bridle roads of Spain* and *The working classes; their interest in administration, financial and electoral reform*) became part of the literary life of London in the third quarter of the century.[6]

Arthur Cayley descended through the Cayley family branch known as the "St. Petersburg Cayleys." This line made its mark as traders in Russia and the Baltic countries during the eighteenth century.[7] It descended from Arthur Cayley's great-grandfather Cornelius, son of the first Cayley baronet. Living in Kingston-upon-Hull in Yorkshire, Cornelius achieved notability as Recorder during the first half of the century. Hull was a primary English port for the main trade with the Baltic countries. Prominent local families, such as the Cayleys, Smiths, Raikes, Thorntons, and Wilberforces, became very wealthy through their trading and banking interests, which were interwoven with the Baltic trade. The familial network, formed by a complex pattern of intermarriage, included antislavery campaigners of the influential eighteenth-century reformers associated with the Clapham Sect, respected City of London financiers, and a governor of the Bank of England.

A Russian Inheritance

John Cayley, son of Cornelius Cayley of Kingston-upon-Hull in East Yorkshire, was the first of the family to go to Russia.[8] The British traders worked through the Muscovy Company, a chartered trading company granted a monopoly to trade in Russia by Elizabeth I toward the end of her reign. Like other trading companies, such as the East India Company, their function was to set rules for the merchants to follow. At the headquarters in London, the company was led by a governor aided by four consuls and a court of twenty-four elected assistants who met each month. Though the Court of Assistants was the ruling body, the merchants traded on their own capital and for their own profit on an individual basis. In St. Petersburg, there was a strong contingent of English merchants operating in a part of the city known locally as the British Factory. Trade with Russia was carried out not only in St. Petersburg but also from trading outposts such as Riga (now in Latvia) and Archangel, a port on the White Sea to the north of where the Cayleys also had family connections.

Around 1747, John Cayley left Hull for St. Petersburg and apprenticed himself to the company, serving his seven years before becoming a "freeman by servitude." In 1756 he married Sarah Cozens and started a family. First there were four daughters, of whom two survived infancy, and then five sons, the last of whom was Arthur's father, Henry Cayley. The family prospered in the cosmopolitan capital city, Russia's window on the West.[9] John Cayley was a prominent figure among the trading fraternity that conducted business in St. Petersburg, and for a time he traded as a partner of the firm of Thornton, Cayley and Company. He was clearly ambitious, one of the "movers and shakers" of the city of expatriates. Freemasonry was soon established and the "Perfect Union" Lodge acknowledged by the Grand Lodge of England. John Cayley was elected an officer (the Senior Warden) and a few years later, in 1777, he became its Master. The creation and introduction of the English Club was less secretive but no less exclusive and among this highly selective membership we also find John Cayley.[10]

The position of British consul general of St. Petersburg was a much coveted prize and, when it became vacant, John Cayley gained the all-important nomination of the Russia Company in support of his candidature. In the *Court Minute Books of the Russia Company,* it was noted that "it is the opinion of this Court of Assistants [in London], from the well known Character, the long Residence in Russia and great experience in Trade of John Cayley Esq., Merchant at St. Petersburg, who engages to quit business, that he is fully qualified to succeed to the office of Consul General and Agent of the Russia Company in Russia." He was appointed in 1787, a position held until his death in 1795. He continued in the capacity as Agent for the Russia Company but his new position gave him entrée to the Russian Court Circles in the last days of Catherine the Great's reign. When asked to fulfill ambassadorial roles at the Winter Palace, he bathed in the popularity of the English presence and was a trusted friend at court. He was a slightly unconventional figure who stood out at formal occasions and contributed to the image of the English held by the indigenous Russians. As a young English visitor wryly observed in writing home to his father, "I find we, I mean our nation, are generally honored by the Russians with the title of Mad Englishmen."[11]

St. Petersburg was a thriving center of trade in the Baltic. Besides the 190,000 Russians living there, the largest group was the German contingent. There were approximately 17,660 Germans, 3,720 Finns, 1,860 Swedes, 2,290 French, and 930 English and other smaller groups.[12] The British presence in the colony, though not the largest numerically, was the utmost of importance in trading terms. Of all the European countries trading with Russia, Britain was their best customer by a clear margin. By 1790 there were twenty-eight British merchant houses each directed by two or more partners. As to the environment of the old imperial capital in the late eighteenth century, there are a few historical accounts written by visitors. One traveler described the renowned architecture of the city: "The ASPECT of the residence is gay and chearful. Straight, broad and generally long streets, frequently intersecting each other in abrupt and sharp corners—spacious open squares—variety in the architecture of the houses—in short, the numerous canals and the beautiful river Neva, with their substantial and elegant embankations, render the general view brilliant and inchanting."[13]

Much the same picture of a busy commercial center in a hardy climate off-set by a graceful living style is painted by other English travelers.[14] The focal point of St. Petersburg was the river Neva with its long quay and frontage of palaces and private homes on the embankment. In this fashionable part of the city, along the "English Line," the *Angliyskaya Nab Erezhnaya*, and within a short distance of the Hermitage and Winter Palace, lived the wealthiest of the English merchants. To the west were the places of their trade: the hemp and tal-low warehouses, the wharves and the workshops. The Neva was as wide as the Thames at London Bridge and St. Petersburg itself was likened to a trans-planted City of London in miniature. Visitors were so impressed that the civic buildings and palaces were even thought more magnificent than those found in London.

Life in St. Petersburg for the English contingent was very much the expatri-ate life of the British overseas except that it was also cosmopolitan. It was gen-erally reckoned as an unglamorous corner of the British trading empire, per-haps on account of the severity of Russian winters and the robustness required of its inhabitants. Yet the lively cultural life and an element of graceful living compensated. The daily existence of the expatriates was eased by the availabil-ity of accoutrements of life back home. Most important, there was a plentiful supply of household servants.

St. Petersburg's northern locale rendered a short summer season, but in some years, merchants would return home for the English summer. Holiday visits were restricted from May to October and visits entailed either a three-week sea voyage or a long overland journey through Poland and Germany. After the Napoleonic Wars, travel in Europe became less hazardous and visits by the British to the European mainland became more frequent. On the occa-sions when the Cayley family remained in St. Petersburg for the summer, time passed leisurely and the feeling of freedom was enjoyed. Said one visitor: "Among the peculiar charms of the summer here are to be reckoned the bright and generally warm nights. . . . walking parties are met every where, frequently attended by music: on the smooth surface of the Neva, and on all the canals, boats are gliding, from which resounds the simple melody of the popular ballads, as sung by the watermen."[15] In common with colonial life elsewhere, the wealthy Europeans were separated from the indigenous Russian population. The British in St. Petersburg were a close-knit group who jealously guarded their national identity and preserved their social customs despite being officially Russian subjects. The merchants of the British Factory were at the center of active social life organized around the English Club.

In the week John Cayley accepted his commission as consul general he also received a houseguest from England. James Brogden, a 22-year-old ex-Etonian, was on his first trip abroad and eager to learn of Russia in general and the Russia Company in particular. His father, who had had a long association with the Company, provided an introduction to the Cayleys. James Brogden's letters home provide a sidelight on the way of life in the Cayley household. In terms of age, Brogden was a contemporary of John Cayley's son William and only a few years older than Henry. On arrival in the city, he wrote: "When I came within 7 Versts [about five miles] of Petersburg we met Mr. Cayley, who informed me that Stephen Thornton, who had just heard of my arrival, was coming to meet me, which he did in his Chariot & 4 drove a L'Anglois, into which he insisted upon my entering & drove me to the English Inn."[16] Stephen Thornton, from a wealthy Hull family and one of the Cayley-Smith-Raikes-Thornton network, was the son of John Cayley's business partner Godfrey Thornton.

Brogden observed that the expatriates lived in style, "5 servants at table." He was impressed by John Cayley's welcome, and he duly reported to his father that "Nothing can equal the kindness of Mr. Cayley." He was also impressed by the relaxed way of life in the Cayley family home: "They always expect me to breakfast, but at dinner, which is always ready at 3 oclock, if I am there at the time they will be glad to see me; if not, they suppose I am engaged. Supper is brought on at 9 oclock upon the same conditions . . . I conform without scruple to this easy manner." When the harsh winter duly arrived, the family was well protected: "Mr Cayleys servants have been very busy in putting in double windows & making other preparations for the Winter Campaign," wrote Brogden. "I don't know how I shall like the cold weather which must be expected, for I cannot say that I think the Stoves are such good preservatives against its effects as I have been taught to expect. Fire places are become much more common than they were by your account when you were in this part of the World. I have hardly been in a house yet that has not an English fire Place where they burn new Castle [sic] and Scotch coal, both of which being brought as ballast is cheaper here than in London."[17]

Aided by several of his sons who had followed him into the business, John Cayley had achieved the eminence of an eighteenth-century grandee, though not one who adopted an overtly opulent lifestyle. The family's way of life contrasted with the ornate lifestyle of the Russian aristocrats who lived in the city. Brogden noted: "The houses of the English & the other civilised Europeans are fitted up with every convenience, even at the expence of appearance, by which as in all other things we are so peculiarly distinguished from the Russians, whose prevailing passion seems to be the desire of shew. The English are not afraid to use their legs if they think it will contribute to their health, & boldly make use of them." In contrast, the leisured upper-class Russians went everywhere by carriage. Brogden, though, quickly tired of the high life he saw in Russia: "Petersburg is a very agreeable residence for a few months, but is much too gay for me. Balls, Plays, dancing, Eating & drinking seem to be the most important concerns of life to the majority of its inhabitants."[18] The life of John Cayley and his sons was no doubt liberally sprinkled with social activity but this should not disguise the central fact regarding the British in St. Petersburg. The expatriate life was purposeful: they were there to make money.

The traders were solidly middle class, and the British Chapel, funded by the Russia Company, formed a focus for their closely connected social group. It was an integral part of expatriate life and, though it came under the jurisdiction of the Bishop of London, its chaplain was elected by the Russia Company. In their midst was William Tooke, the historian of Russia, who was elected as chaplain in the 1770s and served until 1792. He was a member of the Imperial Academy of Sciences in St. Petersburg and allowed access to the library. During his time in Russia he wrote about his travels and the people he met—on one occasion he was introduced to Immanuel Kant. His eldest son, Thomas Tooke, who became a noted economist, was a contemporary of John Cayley's sons and became a partner of Thornton's firms in London and Russia.

French was almost universally read and spoken in the expatriate community, and even the numerous German group were obliged to converse in it. France had increased its influence in Russia during the eighteenth century and, after the Revolution, there was an abundant supply of impoverished French aristocrats looking for a new home. This presence bolstered French as the universal language of diplomacy and trade in St. Petersburg. The Russian language was usually too difficult for members of the expatriate community. Brogden recalled his attempts to learn it, adding: "I find even Mr [John] Cayley's Sons, who were born in the Country, by no means master of it. All the Russians who are above the condition of Peasants speak either German or French, & the intercourse with the com[mon] people is so small that very little advantage is to be derived from it."[19] The expatriate families had no real need to learn the indigenous language in depth. There is little evidence that the young Arthur Cayley spoke Russian. He published no mathematical papers in that language, but he wrote many of his mathematical papers in French, which he wrote effortlessly. Significantly, it was the language he experienced during his childhood in St. Petersburg.

Henry Cayley, Russia Merchant

Henry Cayley, the ninth and youngest child of John and Sarah Cayley, was born in 1768. Sons of merchants usually followed their fathers into the trading firm at a young age, and a boy entering the family business at the age of fifteen was not uncommon. Henry was no exception; he learned the conduct of business during the 1790s from his father and brothers. When John Cayley was ready to hand over the reins of his business affairs, his sons were well prepared. Resigning as consul general, he left Russia for the last time and settled in Richmond, near London, the home of his recently widowed daughter Elizabeth. He died there in 1795, a gentleman "universally regretted [*sic*] for his amiable manner; and excellent qualifications."[20]

With two of his elder brothers, John and William, Henry now managed the family trading firm. The unstable political situation in the early years of the century ensured that Henry operated in interesting times. In the summer of 1807, Napoleon advanced east toward Russia and administered a resounding defeat on the Russian army at the battle of Friedland. With his borders threatened, Czar Alexander entered into an alliance with Napoleon, a move which made Alexander unpopular in the salons of St. Petersburg. For the English, this pact with Napoleon threatened their trade in raw materials. England faced the prospect of making peace with the emperor on his terms or risk losing trade with Europe through a Continental blockade.

The vital currency for buying raw materials or wielding influence in the imperial court was silver or gold. Henry was deputed by the British government to act as intermediary for a million ounces of silver bullion to be deposited in the Imperial Bank of St. Petersburg after delivery by His Majesty's warship, the sloop *Wanderer*. Henry was to organize the best way for it to be made available to the British ambassador through "bills of exchange, ducats, dollars or any such convenient coin" as he could advise on. Perhaps the money was to be used as a bargaining chip with the young czar or simply as a way of obtaining hard currency at a time of high inflation. Ultimately, the bullion was sent back to England, but Henry played his part as a money broker.[21]

Henry Cayley was an important member of the British community and for some years served as treasurer to the British Chapel. While he and his brothers became traders, there was no obvious path for his two sisters to follow, other than marriage. Although it was not unknown for the English to marry outside their community, it was unusual. Going against the trend, Henry's eldest sister Elizabeth married Count William Henry Poggenpohl in St. Petersburg in 1782 and they eventually returned to England, where Poggenpohl became an attaché at the Russian Embassy in London. It was more usual for the traders to marry within the English community and in 1785 his other sister, Sarah, married the trader Edward Moberly, who had come to St. Petersburg from Knutsford in

Cheshire.[22] With Sarah there is a mathematical connection of passing interest. The mathematician Georg Cantor was born in St. Petersburg in 1845, spent his first eleven years there, and later recalled the friendship between his father Georg Waldemar Cantor, a prosperous merchant in St. Petersburg, and Edward Moberly and his sons.[23]

On 18 February 1814, Henry married Maria Antonia Doughty in the British Chapel in St. Petersburg. During Arthur Cayley's life some mystery had attended his mother's background. Some speculated her family was of Russian origin, a view which Arthur Cayley did little to dispel. Maria Antonia was born in St. Petersburg in 1794, the daughter of William Doughty of Greenwich, England, and Wilhelmina Johanna Doughty. Wilhelmina was the daughter of James Smith, an early settler, and Antoinetta Juliana—indicating Dutch ancestry on Arthur Cayley's distaff side. The first child of Henry and Maria, a daughter Sophia, was born in September 1816. A son, William Henry, was born in June 1818 but survived only one year.

Arthur Cayley was born on 16 August 1821 at Richmond during one of the family summer visits to England. Richmond-upon-Thames was a genteel town outside London inhabited by a wealthy upper middle class and was home to Henry's eldest sister Elizabeth. In the 1820s the town expanded, and travel between London and Richmond became possible by steamboat, but it was still remote and peaceful. George IV had become king the year before after a nine-year period of standing in the wings as the prince regent, and, in the month preceding the child's birth, he was crowned in Westminster Abbey.

After the summer visit the Cayley family returned to St. Petersburg. Richmond itself played no further role in Arthur's story but visits of his family to their English base remind us of the background of his early life and the nature of the family circumstances. A second son, Charles Bagot, was born in St. Petersburg in 1823, almost certainly named after the British ambassador to Russia at the time of his birth. The merchants moved in court circles but the fine detail of Sophia's, Arthur's, and Charles's infant years in the colony is speculative. It is likely they lived a life within the "family" that extended to the whole group of British expatriates in St. Petersburg. Of greater certainty is that for the first seven years of Arthur's life in Russia, the Cayley children received their early education at home. By the 1820s British governesses and male counterparts occupied teaching positions in affluent Russian and expatriate families.[24] Private tutors and nannies were familiar figures in St. Petersburg and if they were not French, they invariably spoke French.

After sixty years abroad, Henry Cayley returned to England permanently in 1828. He was then in his sixties with a wife in her mid-thirties. Maria Antonia's father had died two years prior and no doubt the death of her mother in February 1828 precipitated their decision to set up home in England with their young family. Her parents had lived in St. Petersburg their entire lives and their

deaths marked a turning point in the family circumstances. Of Henry's brothers who had become merchants with him, William had died many years before
and John had already returned to England. It was time for Henry to think about
his young sons' education. One option would be to stay in Russia. When an
English head of family worked abroad, sons were traditionally sent home to
board at expensive public schools for their secondary education. While Arthur
received his early education in St. Petersburg, the idea of separation of parent
and child may not have appealed to Henry and Maria Cayley.

On arrival in England the family took up temporary residence at Eliot Place,
Blackheath, a respectable byway at the top of Blackheath village. The youngest
child of the family, Henrietta-Caroline, was born there in August 1828. The village was on the point of growing into London's first suburb and it quickly
gained popularity among the middle classes as travel to London became easier
with the growth of road traffic and the metropolitan rail network. Positioned
on a plateau overlooking London, the air was fresh in Blackheath and the
wealthy could escape the pollution of the city. And Henry Cayley was one of the
wealthy, one of the rising mercantile class who would displace the aristocratic
oligarchy as industrialization gained a foothold.

Many of this class lived in magnificent houses surrounding the large area of
open heathland of Blackheath, but given Henry Cayley's still active business
interests and the lack of railway travel in the 1820s, it was not an ideal place to
set up permanent residence. Almost at once, he decided to live in London and
he bought a fashionable newly built house in York Terrace on the south side of
Regent's Park. In the 1830s the park lay on the boundary between city and country, and the elegant houses set back from the Marylebone Road were convenient
both for entry to the City of London and for escaping from it. The mansions of
Mayfair enjoyed the social prestige of the capital but the professionals who
inhabited the colonnaded Nash terraces in Regent's Park included the wealthiest of the merchant class.

The elderly Henry was an energetic man and even at the age of sixty was not
ready for an existence of pure leisure. In his new life in England he pursued a
range of financial interests: the Baltic Trade, insurance, and investment banking. He was active in proposing loans for the building of Britain's embryonic
railway network. Having lived abroad, he had a wide experience to draw on in
his mercantile dealings. The trade between Russia and London was largely concerned with raw materials. Tallow, for candlemaking, was the main import;
Petersburg yellow candle, or P.Y.C., was a superior tallow compared with the
product from other countries. Any product to make money: in winter, ice was
also harvested from the frozen river, cut and sawn, and transported all over
Europe. Merchants also traded in such commodities as iron, oils, flax, and
hemp. Large speculations in the prices of these materials necessitated merchants making shrewd deals and controlling their flows. Conversely, London-

based merchants negotiated manufactured articles and luxury items for which there was a growing demand in St. Petersburg.

Merchants returning from Russia easily fell in with the commercial life of the City of London. Thomas Tooke, Henry's contemporary, had a similar background in St. Petersburg with trading firm partnerships and became governor of the Royal Exchange Assurance Corporation when he returned to London. Henry found various niches where his experience was useful and was able to adapt to changing economic circumstances. By the end of the eighteenth century, the importance of the long-established Russia Company as the sole agent trading with Russia had declined; by the early 1820s, in an effort to control speculation and regulate the Baltic trade, the most influential traders combined together and formulated rules for membership of the Baltic Coffee House. It was the forerunner of the modern Baltic Exchange, a jewel in the financial life of the City of London. On his return to England, Henry joined this group of merchant traders and became an influential member.

Henry Cayley's mainstay was as member of the Court of Directors of the London Assurance Corporation (now part of the Sun Alliance insurance group), which was established in 1720 by royal charter. The corporation was a leading City of London company that specialized in fire, life, and marine insurance. He was elected in July 1828 and remained on the board of directors until his death in 1850, regularly attending meetings of the Court.[25] His friend Henry Blanshard and the insurance pioneer George Henry Gibbs served on the same board. A large number of bankers held shares in the corporation on behalf of their clients while the company's solidity at this time is perhaps best indicated by the presence of the celebrated financier Nathan Meyer Rothschild as a major shareholder. Henry Cayley's position as director coincided with a period of extensive expansion in the corporation's business interests.

The London Assurance Corporation conformed to the Dickensian mold of a sober London counting house that conducted its business with an unswerving eye to the balance sheet and paternal supervision of employee conduct. In Henry Cayley's company, clerks had to sign an oath of fidelity to the corporation while those with long and faithful service were treated generously on their retirement. This was the bargain with the patrician court of directors but clerks and other employees were kept up to the mark by the necessity of having to regularly petition the court to be allowed to continue their employment at their previous level, a practice which lasted until 1847. In its turn, the court was a pillar of the establishment and paid careful attention to the framing of messages to the monarch, of congratulations or condolences at auspicious moments of her reign.[26]

Henry showed a great interest in the Great Western Railway project, a venture which was launched in August 1833. When its construction was first discussed it was thought to be a high-risk investment. It was an ambitious plan and completely novel as an engineering challenge but he quickly realized its potential as a highly profitable investment. The project of laying a railway of over a hundred miles between Bristol and London will always be linked with its illustrious engineer Isambard Kingdom Brunel; its completion was one of the great feats of Victorian railway engineering and proved highly successful. Henry was an original director of the London committee that formed to consider the project, and he attended its initial meetings. As well, he remained active in the Baltic Trade. In 1836 he occupied offices on Broad Street in the City and shortly afterward moved his trading business to Great Winchester Street, where he maintained premises in the company of other Russia merchants.[27] In 1837 he was chairman of a meeting that decided to substantially revise the original rules of the Baltic Coffee House. The coffee house dealing system in London was restructured to take advantage of technological developments: the telegraph, steamships, and the advent of the railways.[28] A life such as the elderly Henry was leading in the 1830s does not square with retirement from business as has been suggested.[29] One of his eventful weeks occurred in January 1838. On the morning of the 11th, he stepped into the charred remains of Royal Exchange, London's Stock Exchange, which had been completely destroyed by fire during the previous night. A week later, on the 17th, at a director's meeting of the London Assurance Corporation, he proposed that the Great Western Railway be provided with a loan of £50,000.[30]

In today's terms, Henry Cayley was a venture capitalist, a way of life whereby one could make and lose vast sums by speculation. He appears to have been a man who was prepared to make a carefully judged gamble. The Russia trade, centered in Threadneedle Street in the heart of the City of London, continued to be unpredictable and the Baltic Coffee House gained considerable notoriety in the 1840s when commodities trading was dominated by high-risk ventures. Large speculations in tallow, for example, were rife and fortunes were both made and lost.[31] The principal political issue of the day, however, was bread. The updated Corn Law of 1815 protected the land-owning home producers by levying a tax on imports of grain. The protectionists were set against the free-traders, who desired a repeal of the Corn Laws. In this conflagration, Henry Cayley showed himself to be on the side of the protectionists by siding with a conservative argument: Why enter into the unknown and jeopardize the economic growth made in a period when the Corn Laws were in place?[32] The merchants and traders backed the protectionist Lord Stanley against Prime Minister Sir Robert Peel, thus adding to the turmoil in the Tory Party. It is evident from this that the 70-year-old Henry kept his business interests very much alive during the 1840s. Both he and his wife enjoyed good health and apparently suffered no harm from the Russian winters they had endured. A man whose mental and physical capability was synchronized with the intensely competitive business environment, his obvious capacity for hard work is at one with that characteristically associated with the age. That he passed this quality to his son will become clear.

Arthur seems to have inherited his artistic temperament from his mother. She was a competent sketch artist and watercolorist, and both activities were taken up by her son.[33] One aspect of young Arthur's life though is less speculative. It would have been natural for Henry's son to have been exposed to arithmetic at a young age. His father had always dealt with the money side of the trading business and his son would have had firsthand experience with the culture of the "counting house." Indeed, throughout his life, Arthur Cayley took the utmost care with money matters and exercised his capacity to "watch every penny." In light of his upbringing, the booklet he produced on *The Principles of Book-keeping by Double Entry* in his last years is not so much a surprising appendage to his collection of mathematical papers but an expression of the practical character woven seamlessly into the fabric of his life, one which started at a young age.

An English Schooling

Returning from Russia, Henry Cayley, a staunch Anglican, would have met the Reverend George Brown Francis Potticary in the daily round. A clergyman involved with a range of parochial duties around Blackheath, he also ran a private educational establishment. At the end of Tranquil Vale, which wound its way up from the village center, the school was housed in a large Georgian house facing the heath; it was this school that the 10-year-old Arthur and his younger brother Charles attended.[34]

This preparatory school ("Prep School") was a family concern, for before him, George Potticary's uncle, John Potticary, a nonconformist minister, had owned and operated a private academy for the education of young gentleman in Eliot Place for twenty years. For a brief period, the young Benjamin Disraeli, its most famous pupil, attended there. When the uncle died, George succeeded him and allied himself to the Anglican communion. After graduation from Oxford in 1823 following a fairly undistinguished undergraduate career, this energetic young man with a genuine interest in education created a school with a lively atmosphere and made it a going concern. There were two terms in the school from autumn to Christmas and from the new year to mid-summer in common with David Copperfield's Salem House, "down by Blackheath," though there is no reason to suspect that Cayley's experiences in an early Victorian school mirrored those of Dickens's young hero.[35]

Photographs of the school buildings at the end of the century show a large four-storey building, easily imaginable as one of Dickens's forbidding scholastic academies. Potticary's school housed a warren of classrooms, common rooms, and living quarters connected by a network of unplanned corridors going off in all directions, from the basement to the rooms in the loft. Little is known of school life there but a boy attending around the same time as Cayley did shows the pupils' reliance on close family support by the provision of food, clothing and pocket money: J. P. Ashburnham wrote to his sister, "I should be very much obliged to you if you will send me a basket of grapes and some other things of that sort next week, . . . if you send me a parcel, tell the coachman it is to be sent from the Roebuck, and send me the rest of my pocket money for the rest of the half year, because I want to get something. And send me a pair of single horse reins. And tell mama that she is to make my pigeons tame enough to feed out of my hand. I like French very well but Latin I detest."[36]

Arthur's start at Eliot Place in 1831 coincided with the school's expansion. Potticary purchased the lease of Number 9 Eliot Place, "Pagoda Villa," for a sum not less than £2,400, a large amount in those days. The school catered to ages from eight to fifteen years, and ten years later had fifty-two pupils on the books. As with other private schools in the area, it aimed to provide a suitable education for young gentlemen.[37] On a smaller scale and within a stone's throw of "Potticary's Green," one Edward Wilshere offered an education for day pupils to which any middle-class parent would give assent, an education for those "whose general improvement in the various branches of really useful and liberal Education, will be efficiently promoted, by means, which, during more than twenty years, have obtained the approbation and secured the confidence of discriminating Parents, anxious for the moral welfare of their Children."[38] Providing an education in the 1830s was not cheap. Judging from its advertising, a school more comparable to Potticary's run by the Reverend J. O. Squier in nearby Deptford posted annual boarding fees of £30 for 10-year-olds and above, with reduced rates on a sliding scale for those under ten, for day boarders, and for day scholars. Pupils at this school would receive instruction in reading, writing, arithmetic, geography, mathematics, Latin and Greek (French and drawing optional extras).[39] In an up-and-coming area, Blackheath had another attraction for middle-class parents: when cholera appeared in London in the early 1830s, a parent may well have thought his offspring safer in one of the many boarding schools and academies which ringed the open heathland.

From an early age, Arthur showed a pronounced mathematical ability. This might be inferred from later success though concrete evidence of it existing in the boy is difficult to obtain (no records of the school have been found). He had a liking for arithmetical calculations and had developed a great aptitude for them, but more revealing of his character is his teacher's remark recalling him habitually asking "for sums in Long Division to do while the other little boys were at play."[40] It suggests that he found comfort in the company of "numbers" and the security of a timeless world separate from the real world in which he was immersed. It may be that the voluntary separation from his classmates and his self-sufficiency was learned as a small boy growing up and being educated by private tutors in the British community in Russia. In Blackheath, he was in surroundings which allowed his talent to grow and when he went home to Regent's Park, he went to a family home with a well-stocked library, from which he and his brother benefited.

When later in life, Francis Galton, the scientist of eugenics, asked Cayley about the origin of his interest in mathematics, he responded that he "had an early taste for arithmetic" and a partiality "for long-division sums," thus confirming his teacher's independent evidence.[41] This is pertinent, for long division is a point where arithmetic gives way to real mathematics and such a playful activity could have easily fuelled the embryonic mathematician. Although ability to perform long calculations does not make a mathematician, in Cayley's case we can imagine the technical skill was combined with an acute sense of mystery and beauty, a source of wonder that remained with him all his life. Coupled with his natural gifts, this aesthetic sense would have impressed itself on the young boy. Undoubtedly his liking for this kind of arithmetical exercise was a factor in determining his future career.

At the age of fourteen, Arthur left Potticary's school and in August 1835 enrolled as a day pupil in the senior department of King's College in London. Because he was proposed by a proprietor of the College, his parents qualified for the lower annual fee of £18-17s-0d, but it was still a sizeable sum.[42] From its beginning, it was a private school but less expensive than the great English public schools such as Eton or Harrow and it offered a broader curriculum. Eton and Harrow catered to the sons of the landed aristocracy while schools such as King's provided an education for the middle classes. The college provided for the "middling rich," and while Arthur had distant aristocratic connections, it was his middle-class credentials coupled with a mercantile parentage which identifies him with the typical King's entrant. A few boarding students were admitted, but King's was primarily a day school. It offered a curriculum in the Classics, of the type which might have been found in any public school of the period, but considerable time was spent on mathematics, science, modern languages, history, and practical subjects connected with the professions.[43]

King's College had been proposed as a new institution at a meeting chaired by the prime minister, the Duke of Wellington on 21 June 1828, and gained the royal patronage of the king. At this inaugural meeting, the passing of the first resolution set a stamp on the underlying religious philosophy and purpose of this distinctly Anglican school founded in the heart of London. The resolution proposed that King's should be a school "in which the various branches of literature and science are made the subjects of instruction and it shall be an essential part of the system to imbue the minds of youth with a knowledge of the doctrines of Christianity as inculcated by the United Church of England and Ireland."[44]

King's received its royal charter in 1829 and opened its doors on 8 October 1831. A brand-new building in the classical style had been constructed off The

Strand, which in those days was at the center of a smart residential district. It occupied an unused east side of the large Somerset House site on the north bank of the Thames. At its center, the ornate College Chapel reminded its pupils and staff of the ultimate purpose of education. Twice a week for the entire three years of the regular course of study, the principal and the college chaplain delivered a series of theological lectures at which attendance for all boys was compulsory. The *College Calendar* advised parents that the day began with "Prayers in the Chapel at Ten o'clock precisely, when the presence of the College Students, is required."[45] High Anglican ideals were strictly carried out; the school's intention was to provide a place where culture was as important as a training in science but in the strong Protestant atmosphere, usefulness was stressed "beyond utility."

The educational fare offered at King's would be described in today's terms as "highly academic," its sense of order and purpose maintained within a strictly disciplined environment. The course of study in the senior department was strenuous for students and staff alike, with lectures every morning and afternoon, including Saturday. The grouping of subjects comprising the department headed General Literature and Science in the 1830s was the forerunner of the faculties of Arts and Natural Science in King's College, which became part of the federal structure of the University of London toward the end of the nineteenth century.[46]

King's College was founded by the Anglican establishment as a counterweight to University College on Gower Street, which had adopted an educational philosophy based on utilitarian principles. Thomas Arnold called it "that godless institution in Gower Street" while others referred to it as "that Cockney College" or even "Stinkomalee" on account of being sited on a former rubbish tip. It was nonetheless highly successful in gaining pupils and, much to the irritation of its critics, in gaining a secure reputation for high academic standards.[47] It might be speculated that had Henry Cayley been other than a loyal Anglican he might have sent his sons to Gower Street. It was only a short walk from the family home in Regent's Park. There the young boy would then have been taught by Augustus De Morgan who, apart from a short break, spent his entire professional life there and was desperate for contact with students with mathematical talent.

Alarmed by the secular spirit of University College, the founders of King's were determined to redress the balance in favor of the dominant Anglican establishment. That King's was part of the state's apparatus is perhaps best illustrated by the fact that it was literally the *King's* College. William IV, the newly crowned monarch, was patron, the archbishop of Canterbury its visitor while the "great and the good" of English life filled out the ranks of the perpetual governors, the life governors and ordinary governors. Even with this resplendent superstructure, King's had also to survive financially and it began life as a joint stock company and in its early days suffered financial difficulties through lack of support. While the college looked to the sons of the prosperous London middle classes, it was not restricted to that segment of society. If the fees could be paid, there were no class barriers. Charles Kingsley, an exact contemporary of Cayley's at King's, in his *Alton Locke,* observed one boy at "King's College, preparing for Cambridge and the Church—that being now-a-days the approved method of converting a tradesman's son into a gentleman."[48]

King's was organized in two sections, a junior department, which offered normal schooling of an elementary kind, and a distinctive senior department. Students entering the senior department were usually of three types. The first were those who entered for classes in general literature and science (the Classical students), and they formed an annual entry of about one hundred and twenty each year. Cayley was one of these. The second were the medical students and thirdly there were the occasional students, who could attend part of the normal three-year course. For most, the King's College course in the senior department led to a career in business, a career in one of the professions or progression to Oxford or Cambridge.

John Ruskin was an occasional student at King's, and in 1836, at the same time as Cayley, he attended the English literature class before moving on to Oxford in January of 1837. Ruskin described King's and remarked on its academic ambience:

It is a very sudden change from the unlearned, mercantile moneymaking, abominable bustle, which prevails in the Strand, to the sublimity of silence which reigns in that spacious court: long vistas of Corinthian columns, lofty arches and tall porticoes echo your tread and when you enter the tall door way, mount the noble staircase, and leaning against the base of a massive column, whose richly carved capital extends above your head the graceful leaves of its acanthi, look down long passages on either side, arch beyond arch retiring in infinite perspective, the light falling with variable force through the upper windows, and striking on the figures of the gownsmen flitting up and down among the shadows of the corridors in silent and abstruse meditation . . .[49]

Ruskin's peaceful atmosphere of learning, at odds with the London world of business and the law just outside its doors, was maintained for more than a century. The senior department appears more like an Oxford or Cambridge college than a secondary school. Normally a student of Cayley's age would enter the junior department and only progress to the senior department at the age of sixteen years. On this point the regulations for entry to the senior department were quite strict: "The Principal will inquire into their attainments and former conduct; and, except in cases of remarkable proficiency, of which he will be the judge, none will be admitted under the age of Sixteen years."[50] The authorities clearly recognized the boy's unusual academic prowess and, with his "remarkable proficiency," he was granted direct entry to the senior department at the age of only fourteen.

The three-year regular course in the senior department comprised religious instruction, Greek and Latin Classics, pure and mixed mathematics, history and English literature, plus a modern language if desired. It was tightly prescribed in the first year but considerable choice was possible afterward. In the second year, students could attend additional lectures if they wished, but in the third year they were expected to select "broadening" lecture courses. There was a wide range available. In science alone, there were courses in natural philosophy and astronomy, experimental philosophy, geology, chemistry, botany, and zoology.

An aspect of the senior department was the provision of "professorial" teaching, even though the teaching was based on standard texts and did not go much beyond them. The professors of the college were erudite men who carried out research and gave public lectures, some of which were very popular. In the early years, the college was able to attract excellent staff, what the college authorities proudly proclaimed as "men of parts." They were not just specialists, but had something of the Victorian "professor of everything" about them. The school's emphasis on practical studies for the professions is indicated by the various professorial chairs in existence. From the first there was a chair in law and in 1833 a department of oriental languages to cater to those destined for the Indian Civil Service. In the 1830s and 1840s, a professorial appointment at King's was not always full-time and some professors combined their teaching duties with other employment. The mathematician Matthew O'Brien, for example, was at one and the same time, professor of natural philosophy and astronomy at King's and a lecturer at the Royal Military Academy at Woolwich.[51] Professors received a basic salary but their system of remuneration was based on a bonus system, whereby they were paid a proportion of student fees. Poet Laureate Robert Southey was invited to fill a professorial chair as an adornment to the college and promised a basic salary of £200 together with £4 out of the £6 fee for every pupil above an enrollment of fifty. In this way of inducement the college would receive a handsome profit and a draw-card in the bargain. King's was a joint stock company after all.[52]

Great care was taken by the school authorities to appoint to its staff safe and orthodox Anglicans. It was mandatory that professors be members of the Church of England and the only exception were the professors of modern languages and those who specialized in oriental studies. If prospective candidates met this qualification, a next step was dependent on approval by the Bishop of London. Surprisingly, the religious restriction did not diminish the academic quality of staff. Charles Lyell was professor of geology for two years when the college was first founded. J. F. Daniell (of the Daniell battery cell) became the professor of chemistry, a post he combined with a lectureship at the East India Company Military Seminary and, for good measure, he combined both with being foreign secretary of the Royal Society.

Among the King's offerings, chemistry was a subject that attracted Arthur's attention and one which he linked to his mathematics many years later. In Daniell's course the chemistry syllabus included organic chemistry, its organization and the details of its products as well as the processes of organic analysis.[53] This was the beginning of his lively interest in the subject. Lessons learned in youth are not easily dispelled and it is significant that chemistry, the progressive science seen as continuously and rapidly evolving, played a part in suggesting and enriching some important mathematical problem areas in his own research of the 1870s and 1880s. Daniell is also interesting for the methodology he employed in teaching. Chemistry for him was exploratory science and he made a special point of teaching using the "Inductive Method" of reasoning from the chemical phenomena observed and not from preordained theories.

Several other professors were fellows of the Royal Society. Henry Moseley, who graduated from Cambridge in 1826, was elected to the Royal Society in 1839. He became the professor of natural and experimental philosophy and astronomy at King's and was on the staff from 1831 through 1844, conducting research on the stability of ships and on glaciers. Charles Wheatstone took over the experimental side of Moseley's work and became professor of experimental philosophy in 1834. He was a man of wide scientific interests; though he wrote one elementary paper on mathematics, he showed little more than a passing interest in the subject.[54] Wheatstone became part of the wider circle of scientists Cayley socialized with in the 1850s. Richard Jones, the professor of political economy at King's, advocated the setting up of a Statistical Society in his inaugural King's address given in February 1833. Though the college found itself in dire financial straits in the early years, it survived and these men established a highly regarded scientific ethos that paved the way for future scientific careers for themselves and several of their pupils.

On the side of the arts, Gabriele Rossetti was the professor of Italian, and thus exempt from the rule which restricted college appointments to Anglicans. He was the father of Dante, William, and Christina Rossetti, and the Cayleys forged lifelong links with this artistic family. It was Charles, Arthur's younger brother and another outstanding student at King's College, who gained a consuming interest in Italian literature from this remarkable professor.[55] Robert William Browne was the professor of Classical literature. A prolific author who managed to convert life experiences into print, Browne was also the assistant preacher for the nearby Lincoln's Inn Chapel, where he gave sermons that he also had published. In his lectures at King's he introduced his students to the breadth of Greek thought—to Homer and Aristotle—and years later Kingsley thanked him for introducing him to "the study of the works of Plato, . . . which had a great influence on his mind and habits of thought.[56] An authority on the history of Greek and Roman literature, Browne wrote a two-volume work on each subject, in which he wrote about the philosophy of Thales, Anaximenes, Heraclitus, and Pythagoras and described the theories of rhetoric expounded by Gorgias, Socrates, and Plato.

With the Classics, there was an implication for science. Browne emphasized Plato's object in science, "which is not only to make observations, and to collect facts, but to generalize and classify—to discover a law." The lesson learned by the impressionable minds who sat before him at King's was that to pursue science according to Plato's teaching was to pursue "the true, the eternal, the immutable."[57] At King's, the young Arthur would be well versed in the "Glory that was Greece," of the kind imparted by the archdeacon, particularly Plato's views on science, which became a distinctive facet of his own mathematical thought.

Mathematical Fare

When King's opened in 1831 there was, from the outset, a chair in mathematics. Its first occupant, Thomas Grainger Hall, occupied the position for almost forty years. Hall had graduated at Cambridge in 1824 as the fifth wrangler, and for the first few years, passed his time in Devon working as a tutor. A true King's "man of parts," in his early years he doubled as the professor of history in the young institution. Primarily Hall was a writer of mathematical textbooks, but he also wrote articles. He turned out pieces for the *Encyclopaedia Metropolitania* on the calculus of variations and the calculus of differences. He appears to have been one of those valuable teachers of mathematics who are always thorough but not overly pedantic. He endured a heavy teaching load; he taught mathematics to each of the three years of the King's course and was occupied each morning of the week during each of the three terms of the year. Meticulous and accurate, his courses did not appear to change much from year to year. Unlike some of the other professors, he did not publish an extensive syllabus for his courses. The year Cayley entered the school, he purported to embrace merely "all the branches of the mathematics usually taught in the Universities, and [such] will be delivered in the order which may be found most expedient."[58] Cayley thus received a comprehensive course, and one which was already at university level. In the first year course, Hall taught Euclid's books 1–4, 6, 11 (on lines, circles, areas, simple regular figures, and solid geometry), principles of algebra, plane trigonometry, the use of logarithmic tables, and descriptive geometry. In the second year, he added conic sections, application of algebra to geometry, spherical trigonometry and, probably, the first three sections of Newton's *Principia*. The third year course was reserved for such topics as differential equations and the analytical parts of hydrostatics, optics, and astronomy.[59]

Of Hall's textbooks, one which deserves special mention is his *Treatise on the Differential and Integral Calculus;* it was published in 1834 and came out in a second enlarged edition three years later, thus the preparation, writing, and enlarging took place during the period Cayley attended the college. With further editions appearing later, it must have stirred young pupils who could master its contents. Obviously drawing on the material taught in his lectures, Hall's presence in the college would have been a stimulus to mathematical thinking. Hall himself had benefited from the reforms of Cambridge mathematics instigated by the Analytical Society and it is quite natural to find that his book promulgated their views on the basis of the calculus. After a good deal of preparation, in later chapters he dealt with Lagrange's Theorem, the cornerstone of the calculus of operations and which governed the algebraic approach to the calculus that was to thrive in England for the whole of the century. Cayley would have been attracted to this subject when he started his mathematical career, and he would follow its methods and principles throughout his later research.[60]

It is significant that Cayley would have obtained his introduction to advanced topics at a young age. Moreover Hall was one of the few persons in England who would consider teaching these topics at school level. Students could even go on to more advanced subjects under his supervision. When they completed his book, he recommended them to study Lacroix's smaller treatise *Differential and Integral Calculus* (1816) and George Peacock's *Collection of Examples* (1820). Cayley's schooldays coincided with Hall's greatest effectiveness in mathematics. When he entered King's, Hall was in his mid-thirties and still active as a teacher before the time he was appointed as prebendary of St. Paul's Cathedral in 1845 when the activities of a church career replaced mathematical enthusiasms. In the 1830s, Hall's young assistant and college tutor in mathematics was the ascetic and well-meaning John Allen, the college chaplain. While Hall was the professor, Allen would have provided a mathematical spur. He entered Trinity College, Cambridge, in 1828 and graduated in 1832 and was a mathematics lecturer at King's for the period of 1832 to 1839, whereupon he became one of the first inspectors of schools. Allen was a close friend of the authors Edward Fitzgerald and William Makepeace Thackeray at Cambridge, and he knew Tennyson and other distinguished contemporaries.[61]

From Thomas Hall, and possibly from John Allen, Arthur would have gained more than an elementary appreciation of Euclid. As Hall pointed out, the topics he taught at King's were usually the ones taught at Oxford and Cambridge. The fare was clearly in advance of anything generally available in the great English public schools. At Eton, a leading public school, mathematics was not compulsory until 1837 while before that time the teaching of mathematics was practically nonexistent. The public schools adhered to the terms of their foundation and emphasized Classics. From 1837 at Harrow, another of England's leading public schools, all boys had to study Euclid to some extent, but in the two hours per week allotted to mathematics, the teaching hardly went beyond arithmetic, a little algebra, and the modicum of Euclid, despite the presence of young J. W. Colenso on the staff from 1838 through 1842, then the writer of elementary mathematics books. Even as late as 1867, only three hours a week were devoted to the study of mathematics in these schools, and mathematics was in an inferior position relative to Classics.[62] It is a measure of the awareness of King's that it should take the mathematical preparation of its pupils seriously, it being *the* essential subject when a degree course at Cambridge was considered. King's College and the "Godless" University College on the other side of the City, with the marvelous Augustus De Morgan at the mathematical helm, appeared to offer the most favorable environments for the study of mathematics in the whole country. With De Morgan, Cayley might have studied more pure mathematics, but the King's course involved a substantial proportion of "mixed mathematics," the application of mathematics to physical problems.[63]

No doubt Arthur would have taken it upon himself to become familiar with those parts of Euclid which did not appear on the King's syllabus. In the third year of the King's course, considerable freedom was offered to pupils in the courses they studied. It was something like the old Grammar School "third year sixth form" in which adolescents remained at school in order to prepare for entrance examinations to Oxford or Cambridge. It is probable that he would have consulted Euclid's *Book V* on the theory of proportion of Eudoxus. This would have been too difficult for most, but Cayley was an unusual student, and the Greek mathematical tradition would have shaped his mathematical thinking and perhaps his choice of mathematics as a vocation. In later years, his love of Euclid's *Elements* continued to manifest itself and he put the case for its study at school level with no hint of equivocation. With unbridled enthusiasm he proclaimed (in 1883): "there is hardly anything in mathematics more beautiful than his wondrous fifth book; and he has also in the seventh, eighth, ninth and tenth books fully and ably developed the first principles of the Theory of Numbers, including the theory of incommensurables."[64]

A Flying Start

Apart from the distinct aim of educating young men for direct entry into the commercial life of the City of London, the life which lay immediately outside the college doors of King's, it also prepared them for the next step on the intellectual ladder—a university place at Oxford or Cambridge. It was a natural starting point since entry to the ancient universities was generally restricted to Anglicans and the wealthier section of the population. Durham University and the University of London were "new" universities, but Oxbridge remained the natural choice for students from King's. It was technically feasible to take a

London degree, for the University of London was granted a charter in 1836 for granting degrees and King's students were eligible to take its examinations. This route was too untried for the King's authorities and, in the initial stages, it shunned the new examining authority. They took the conservative route, and the automatic step for students with degree aspirations was to proceed to one of the ancient universities.

Through their early principals, King's College developed strong ties with Cambridge. William Otter, the first principal was a Cambridge graduate. A man of liberal views, in his youth, he graduated fourth wrangler in 1790. The year Arthur entered King's he was to leave to become the Bishop of Chichester but he would have been aware of the young first-year boy who carried off all the academic prizes at his last annual prize-giving ceremony. A more stimulating younger man succeeded him. Hugh Rose was a leading figure in the Tractarian movement and, some said, "the best preacher in England."[65] He had an established reputation for Greek scholarship and, for a short period, was editor of the *Encyclopaedia Metropolitana*. Rose was educated at Trinity College, had studied mathematics successfully, and was a close friend of William Whewell. It would have been natural for him to think in terms of Cambridge for his young protégé, and Trinity College in particular. Rose's period at King's coincided with a period of expansion of the school; part of his plan was to encourage all his able students to go to university. The opening promise was cut short by his untimely death; he was struck down by influenza and in 1838 died in Florence at the age of forty-three years, after only two years at King's.[66]

Arthur's academic record at King's is one of solid academic achievement. He carried off all the academic prizes in which the college and parents set such stock. At the end of his first year, at the age of fourteen, he was a prizewinner in mathematics, English literature, French, and Classics. In the following year, when most of his competitors were eighteen, he was a prizewinner in mathematics, English literature, and history.[67] No wonder that William Michael Rossetti, just starting at King's himself, saw him as that "incessant prizewinner."[68]

Cayley would have been adequately prepared to enter Cambridge after only two years, but at such a young age this was unheard of. In his third year, the historical record shows that he won no prizes in the traditional subjects. A surprising fact at first, but, after all, there was nothing left to prove as far as academic mettle was concerned. He was almost certainly left to his own devices in his final year at the College. What must have been a period of treading water, he attended the chemistry lectures, and with a gift for quickly grasping the essentials of a subject, he achieved the top place in the examination in the winter term of his third year. At the annual prize-giving in May 1838 he received the Silver Medal for Chemistry from the archbishop of Canterbury, the prize usually won exclusively by the specialist science or medical students.[69] On the eve of the prize-giving he was entered for Trinity College at the age of sixteen.

This step was not taken without hesitation. Henry Cayley wanted his eldest son to succeed him in the family calling, just as he had followed his own father. Henry's background had been so different. He had been brought into the Russia Company by patrimony, since his own father was a freeman of the company. It would have been a very natural step to maintain the tradition of son following father: the young Arthur would occupy a stool in his counting house and from this practical apprenticed position his commercial career would commence. Henry Cayley had expectations of his eldest son; he was a commercial man with expertise in framing regulation and making hardheaded financial decisions. He had an eye to making a conspicuous mark in life and, though a man who owned a fine library of books, he would have had little appreciation of the aspirations of a young man who wanted to engage himself in further study. He might have vaguely been aware of a contemporary of his own age, the son of another member of the Cayley clan, a merchant trader in Archangel, Russia, who had graduated fourth wrangler in the order of merit of the Mathematical Tripos at Cambridge in 1796.[70] Even this would leave little impression on a man who lived and acted out his life according to a practical scale of values.

Arthur entered Cambridge in 1838, when his father was seventy. In an indirect way, William Otter, the erstwhile principal of King's and now Bishop of Chichester had a hand in persuading Henry that his son's future lay in a university career. At least three years enjoying a Cambridge "liberal education" could do no harm and it would provide the "finish" that was the goal of many who proceeded to Cambridge. Otter's advice was supported by another of the church staff at Chichester, Canon C. E. Hutchinson. This man married Lucy Cayley, Arthur's first cousin, who had lived in St. Petersburg in her youth. Through Hutchinson's and Otter's intervention, the elderly father heard that his son's future lay in an academic career and not in the one he had planned or hoped for.[71]

Henry Cayley may have been a disappointed man. But he put his own views aside and was persuaded that his son's abilities would be fostered at Cambridge. Neither Arthur nor his younger brother Charles had any abiding interest in business and commerce, which he had made his life. He had sent his sons to King's College, London, with a mercantile career in mind, but they were rapidly developing academic interests and growing apart from his world.

CHAPTER 2

A Cambridge Prodigy

In October 1838, Arthur Cayley took up residence at Trinity College, Cambridge. The new entrant was about to embark on as glittering an undergraduate academic career as any student at Cambridge during the nineteenth century. Among the dons, the promise of the young man from King's College, London, was probably known in advance, for the ties between the two Anglican institutions were strong. Otherwise, he entered Trinity as did most undergraduates, as a "pensioner," the mode of entry whereby a student pays normal fees—historically those who paid a "pension" or rent for their rooms.

For the "frugal and steady" pensioner, expenses were reckoned at approximately £150 per annum but parents of intending students were warned that with indulgence in "boating, breakfast parties, and other amusements and extravagances, £500 will very soon disappear."[1] Cayley would have been on the lower end of this spectrum but one Cambridge fellow believed that a pensioner at Trinity could usually expect to spend £250, so normal entry to the College was restricted to the wealthier section of the population.[2] Of the College entrants that October, there were more than one hundred pensioners but only a handful of sizars, those less well-off students who entered on reduced fees, some "entrance scholars" who had already won scholarships, and the exalted fellow-commoners, distinguished from the multitude by their academic gowns of blue and silver, who paid higher fees in exchange for dining privileges.[3]

The College Freshman

In his first year at the university, Cayley's college tutor was George Peacock. Appointed as a mathematics lecturer at Trinity in 1815 and as a college tutor in 1823, Peacock provided a link between the metropolitan King's College and Trinity. In the 1830s he was an assessor of mathematics for the "satellite" grammar schools associated with King's and reported on their standards. Peacock enjoyed a high reputation in Anglican circles and was one of eight senior fellows, the "Seniority," who, in concert with the master, Christopher Wordsworth, governed Trinity.[4]

The teaching at Trinity took place in the three "sides" in which the tutors, Peacock, Thomas Thorp and William Whewell, were in charge. Lecturers taught mathematics in classes of varying sizes, some containing as many as forty students from each of these autonomous sides. College society being rather parochial, it was unusual for a student from one side to be taught by a lecturer from a different one.[5] Tutors acted *in loco parentis*, but in addition to dispensing general moral guidance and watching over the financial affairs of their students, they were also responsible for supervising tuition. It was an arduous duty and Cayley's cohort of freshmen could be a serious demand on a diligent tutor's time and energy. It was not expected they would have much freedom for their own academic work during university terms, but with multifarious duties to occupy him, Peacock noticed his young charge from London and guided him.[6]

It was widely known that Peacock took his duties as a tutor seriously, it being rather quaintly said that "his inspection of his pupils was not minute, far less vexatious, but it was always effectual."[7] For the freshman, the college day was leisurely if routine: lectures took place for a few hours in the morning, after which the reading man would visit his private tutor. Then lunch and a long walk, extending to the "Grantchester grind" if one was ambitious on the day. At 3 P.M. the Trinity communal dinner took place; afterward, there was socializing in college rooms until evening chapel. For a reading-man, this was invariably topped off with an evening spent in study.

Trinity College saw itself primarily as a teaching institution with responsibilities toward young students, though to call them pupils would be more accurate since the watchful eye of the tutor and his team was ever ready to circumscribe behavior. When pupils came to Trinity they entered the "family" of the college, and like the junior members of a well regulated upper-class family, they quickly became aware of their own lowly position in its structure and the rules to which they were subject. There was a high degree of supervision in college life and a week-by-week record of a student's presence in the college was strictly recorded in large leather-bound ledgers. The strict regimen of religious observance begun at King's College, London, was continued at Trinity, where undergraduates were required to attend daily chapels at seven o'clock each morning and twice on Sundays. For nonattendance, various levels of punishment were available to the junior dean, with the ultimate one for serial nonattendees being dismissal from the college. The compulsion could not endure for too long in the new age and students were beginning to assert their independence. Straws were in the wind in the year Cayley went to university when there were signs of rebellion against these strictures. In February 1838, there was a lively undergraduate complaint over compulsory attendance at chapel. "The Society for the Prevention of Cruelty to Undergraduates" was formed, but while several tutors were lampooned by the protesters, Peacock's reputation emerged unscathed and he remained "our sweet bird."[8]

Though the Cambridge dons had a eye for talent and encouraged academic ability, the fact that Trinity was run more like a school with advanced subjects than a modern university in which the students' independence is taken for granted meant that reliance on their teachers was the norm. From their lowly place in the hierarchy, undergraduates were taught to strive for excellence and to conform to an ideal of modesty and "lack of show." As a tutor, Thomas Thorp dispensed advice to the freshman gathered before him and reminded them of their obligations as gentlemen in training. There was practical advice on how to adapt to the Cambridge environment, its community, and the adherence to the Anglican religion. Thorp demanded that they should acquire "[a] cap and gown in the first place, and a surplice (which should be sent to the Laundress immediately to have the stiffness taken out, that your Freshmanship may not be unnecessarily conspicuous), must be procured at once, in order to put you in your only proper costume, and enable you to get your dinner in Hall, (the only place where you ever ought to eat it), and to attend the service of Chapel, from which you ought on no day of your life to be absent."[9]

Apart from the huge Victorian communal meals, the college refectory also provided spiritual sustenance. If the long-established Statutes of the College were to be followed to the letter, students would eat in silence, hearing only the appointed Bible clerks reading out selected passages. The religious ambience amply provided, the message uppermost in Thorp's mind was on how students should fulfill their general duty. He cautioned against rebelliousness and warned that "[a]t your time of life, coming here to be trained, agreeably to certain settled rules, to the formation of a sound judgement and wholesome habits of thought, it never can be for your profit to have your mind continually distracted by a captious spirit of doubt and cavil and crude questioning of the laws framed for your direction by the concentrated wisdom of many bygone generations." He urged obeisance to solid Cambridge tradition, which, delivered in his pompous way received a blank response from the student body. They did not want to be told how to dress and took exception to what amounted to an instruction to "[n]ever appear in any of those modern tailor-inventions of P.jackets and Taglionis [dress modeled after ballet dancers], and often uncouth clothing, that savour rather of the stable or the boat-house than the seat of knowledge or the House of Prayer."[10] Fortunately some petty restrictions could be bypassed. Some of the Elizabethan Statutes, destined to be swept away in reforms of the 1850s, were impossible to enforce and appeared rather odd in a country embarking on an industrial age. The one statute, which required members of the college to speak Latin, Greek, or Hebrew during term-time (the only exception being when speaking to outsiders), was an easy target and highlighted the need for repeal.

The existence of such rules shows us that the college was above all a community and when students were admitted, they were bound by the rules of the society within. It was, however, a society not impervious to life outside and Cayley would be exposed to the Victorian world of rank and class. King's College, London, had been a largely middle-class institution, a different milieu from the Cambridge of the 1840s. While Peacock had no sizars in his allocation of the 1838 cohort, his Trinity freshmen included some of the wealthiest in the land. Among Cayley's fellow students were Meyer Amschel de Rothschild, the fourth son of the business tycoon Nathan Meyer Rothschild, and George John Manners, the third son of the Duke of Rutland, two others who came under Peacock's jurisdiction. As a Jew, Rothschild was ineligible for admittance to a degree, while Manners, the son of a peer, was entitled to an honorary M.A. degree after only two years residence without sitting a single examination. How studious these two individuals were does not concern us, but neither could be as serious about an education in mathematics as was the young man from King's.

Mathematical Studies

Ahead of Cayley was a course of mathematics that would last ten terms and culminate in six days of examinations in the Senate House to be taken in January 1842. One term longer than the normal three-year course for the ordinary degree or "poll" degree, Cambridge mathematical tutors were of the view that a third Long Vacation and a tenth term in residence were necessary preparation for the "hard-reading" men, who implicitly declared themselves entrants in the race for top positions in the Tripos order of merit.

The peculiarity of mathematics in the curriculum at Cambridge, at this time and throughout the century, was that it was not taught with a view to providing a base for future study of the subject. There was no conception of the subject as one that should be involved with training future scientific workers in mathematics, let alone future mathematicians. Mathematics was inculcated as training for the mind. According to this strongly held belief, mathematics infused the mind with strength and discipline and was character-forming. From this baseline, the subject was perceived of as the centerpiece of a liberal education that properly equipped a young man for the future hurly-burly of life. Mathematically based subjects such as logic, arithmetic, geometry, and astronomy had traditionally formed the medieval curriculum; though this combination faded, mathematics continued to be appropriate for a gentleman's education when Cayley went to the university.

Mathematics was the Cambridge subject, to the extent that, since the middle of the eighteenth century, when the modern mathematical degree came into being, it was the only route to an honors degree until the 1820s when the classical studies degree became an option. Remnants of the old system remained and candidates for a classics degree were still required to study mathematics to honors level and achieve Junior Optime status in the mathematical Tripos examination before proceeding to the classical Tripos.[11] Many a fine classicist was stymied by this requirement, notably J. M. Neale, a celebrated hymnologist, whose father had been the Senior Wrangler of his day (1812) but had not passed on the "mathematical gene" to his son, who had to be satisfied with an ordinary degree in 1840. Charles Kingsley, who went to Cambridge with Cayley from King's College, London, was bound by this rule and duly suffered. In his youth, he had reason to complain about the narrowness and dryness of Cambridge studies, but in later years, when he was outside the clutches of the system, he could afford to appreciate "her mathematics."[12]

Success or failure to appreciate mathematics, then as now, depends to a large extent on the teacher. Sitting before Whewell would be a very different experience than with Peacock leading the way. In his youth, George Peacock had been a member of the influential Analytical Society at Cambridge and had recognized the influence a college lecturer could wield. In the cause of enhancing the love of science and mathematics, he vowed to use this power. As an ardent reformer in his early years, he showed his ability as an independent mathematical thinker. His work in algebra led to the wider concept of the subject than the one confined to the representation of "quantity" by symbols. In his *Treatise on Algebra* (1830), students were guided in the "principles of symbolic calculation." In the year Cayley went to Cambridge, Peacock was at the point of further codifying these ideas. While occupied with this, his academic career and church career were both advancing. He had been appointed the Lowndean Professor in 1837 and Dean of Ely in 1839, but in the intervening year, Cayley's freshman year, he had an input into the young man's mathematical studies.

During the period of Christopher Wordsworth's mastership, Trinity's reputation in mathematics was not particularly high. The champion student at Cambridge was the Senior Wrangler, the student placed first in the mathematical Tripos order of merit, but Trinity of late had enjoyed only sporadic success of having one in its ranks. Previous ones were Archibald Smith in 1836, and Robert Leslie Ellis in 1840, while other top places in the Tripos list were hardly in proportion to Trinity's importance and size. Its primacy among the colleges of Cambridge during the early nineteenth century was due to its strong classics reputation, which Wordsworth had encouraged and promoted.[13] This catered to their public school clientele, whose early education was informed by the literary studies of Greece and Rome rather than mathematics. In one respect, the superiority of Trinity in classics appears strange. Of the three mathematical professors in the university, two were attached to this college. James Challis and Peacock were both recently appointed professors, Challis as the Plumian Professor of Astronomy and Experimental Philosophy in 1836, and Peacock as Lowndean Professor of Astronomy and Geometry. In reality there were few teaching duties attached to professorial chairs apart from their lectures and they had virtually no influence on the undergraduate teaching curriculum.

Professors were sidelined from the educative process. The senior mathematics professor of the university, the occupant of the Lucasian Chair, which could count Isaac Newton as a previous holder, was even more remote. In his ten years as Lucasian Professor, Charles Babbage had not given one lecture at Cambridge. His record intact, he resigned in 1838 to be succeeded by Joshua King, a doctor of laws and president of Queen's College. King had been Senior Wrangler in 1819 and had served as president of the Cambridge Philosophical Society, but as a mathematician he was inactive and contributed only one short note to the society (on the parallelogram of forces) during his ten years of office. Challis and Peacock had more to offer.

James Challis succeeded George Biddell Airy, who became Astronomer Royal. In his lecture syllabus of 1838, he advertised a course in hydrodynamics, optics and pneumatics, and the mathematical theories of light and sound, the "mixed mathematics" of a Cambridge education. His approach to astronomy resulted in his invention of practical astronomical instruments but his professional life was overshadowed by his misjudgment of the ability of John Couch Adams—when national pride was dented in 1846 by the discovery of the planet Neptune as a result of Leverrier's calculations rather than Adams's predictions. Peacock was elected to the chair of "astronomy and geometry" and though he dutifully gave his first lectures on astronomy, pure mathematics was his strong suit. Taking this into account, he reached an agreement with Challis that he would restrict himself to the "geometry" of his title. Thus, in October 1838, when Cayley arrived as a freshman, it is probable that Peacock's lectures were on algebra and geometry and, moreover, that these prepared the groundwork which led to the production of his magnum opus. The contents of Peacock's two-volume *Treatise on Algebra,* published in 1842 and 1845, contain the maturation of his thoughts on the foundation of algebra.[14] He made frequent reference to the works of Continental mathematicians, a continuation of his mission of bringing Continental works to Cambridge and he emphasized the importance of algebra applied to geometry.

As attendance at professorial lectures was not required for success in the Tripos, we can only conjecture whether Cayley was among the large number who attended Peacock's lectures. That he did is most likely, for a youth with his ambition would surely make a point of listening to the lectures of a leader in his favorite subject in his first year at the university. He would not have this chance again. Peacock's career was diverted to Ely in the following year on appointment as dean, where duties included the everyday administration of the large medieval cathedral and the maintenance of its fabric. The numbers attending his professorial lectures at Cambridge gradually fell away, and after an active start he came to treat his professorship as a sinecure. He continued thinking about mathematics and his quietly composed reflections on algebra were written in his leisure periods away from "the higher duties which . . . I owe to my station in the Church."[15]

Peacock handed over his college tutorship to John Moore Heath, who had been acting as his assistant tutor for the previous six years. It is unlikely that Heath would have been the same mathematical influence on Cayley. He was twenty-seventh in the order of merit of 1830, a time when even a modest acquirement of mathematical knowledge would have gained high places in the wrangler list—which his younger brother achieved by being the Senior Wrangler two years later. Apart from this, J. M. Heath was "far too deaf to hear any question addressed to him."[16] Destined for a parish-living in Enfield, North Middlesex, he was a tutor in the traditional mold. He shepherded his students, placing equal importance on their diligence in study as their obedience to college discipline while all-in-all encouraging them to be "gentlemen of moral habits and strict principle."[17] The transition from Peacock to Heath would have been of little importance to Cayley's mathematical progress, as by then he was safely in the hands of his coach William Hopkins.

That we are entitled to call Cayley a prodigy is confirmed by the range of mathematical reading of this newly arrived undergraduate. The youth now had the riches of the Wren Library at his disposal. But what to read? The first point of guidance was Peacock, who sanctioned his early borrowings. On his first visit we might have expected him to borrow a book that would support his course in mathematics, but in fact he took out the three-volume *Principles of Geology* written by Charles Lyell. This was an enormously popular book, richly illustrated with plates, maps, and diagrams. Its full title, *Principles of geology: being an attempt to explain the former changes of the earth's surface by reference to causes now in operation,* indicates the revolutionary character of its thesis and accounts for its controversial reception in a country where questions such as the age of the Earth were taken very seriously. A student of the geologist William Buckland at Oxford, Lyell had been a professor at King's College, London, for a brief period. That Cayley should have borrowed the book is indicative of an active interest in the intellectual issues of the day. As interesting as it was, three days later he was in the library again.

This time he was there to take out works in challenging mathematics. He had clearly left the fare of King's College behind and the few textbooks that did exist for Cambridge undergraduates would have been of a limited horizon. The 17-year-old now saw works on the library shelves by the French mathematicians of the preceding generation. This was an illustrious band, one which had made its influence felt at Cambridge through the Analytical Society. Cayley first selected *Géométrie descriptive* by Gaspard Monge, and, during his first term borrowed the *Elements de géométrie* by Adrien Marie Legendre, and *Méchanique Analytique* by Joseph-Louis Lagrange. These informed his published papers several years later, but it is notable that the process of absorption began during his first days at the university. After the Christmas break, the French influence continued with *Théorie analytiques des probabilitiés* by

Pierre-Simon Laplace, the *Traité du calcul différentielle et du calcul intégral* by Sylvestre Francois Lacroix, presumably the treatise and not the cut-down textbook. For good measure, a selection from the works of Rabelais in both the English and French versions were borrowed as the spring of 1839 arrived. During his undergraduate career, Cayley made limited use of the college library but what he did borrow shows a balanced diet. In the ensuing years he absorbed other classic works, such as the *Leçons sur le calcul des fonctions* by Lagrange, Augustin-Louis Cauchy's *Exercices de mathématiques*, the *Traité de méchaniques* céleste by Laplace, and the *Analyse des équations indéterminées* by Joseph Fourier. These sources of mathematical enrichment surely suggest a probing intellectual inquisitiveness. How much guidance he received is difficult to judge, though the influence of the dons was felt, no doubt, more so in the first year than later.

Cayley settled into Trinity. After a first term living outside the college precinct, possibly spent in a college hostel as was the custom for freshmen, his second-term accommodation was in King's Court, now called New Court. The occupants of Staircase K changed about during his undergraduate days but his immediate neighbor, Frederick Gell, was there for the whole duration. A product of Dr. Thomas Arnold's Rugby, he was the youngest of three brothers to go to Trinity in the 1830s. "Very short-sighted" and "the reverse of agile," he won college scholarships and obtained a First Class Classics degree, suggesting that Cayley's neighbor was equally studious. Francis Galton, an anthropologist already, described him as having "a grand nose, the finest I have ever seen, and a glory to the College."[18] A Hebrew lecturer at Cambridge, he later partnered Cayley as an examiner before their paths finally diverged. Gell passed through the ranks of the domestic church before appointment as Bishop of Madras, where he remained for forty years.

The Competitive Edge

From the day Cayley's Trinity cohort gathered to sit their matriculation examination in 1838, the subgroup with honors degree ambitions were in competition with each other. This first written examination, which allowed entry to the university, was, for most, a mere formality, but it was effectively the starting-pistol for the honors students. After it, if not before, the "reading men" knew who they were. Whereas the "poll-men"—the "hoi polloi"—enjoyed the social life of the university and the town, the virtue of hard work was brought home to honors students by the necessity of "preparatory grind" and the "getting-up" of a subject, phrases which slipped unnoticed into their vocabulary, as they progressed from examination to examination. The central point of the mathematical Tripos was competition and one of the arguments advanced in favor of the suitability of mathematics as the medium for competition was that it alone dealt with indisputable material. It was thus ideal as providing the arena in which students competed. There was no room for ambiguity regarding mathematics for, it was thought, the conclusions were either right or wrong. For this reason, it was believed that mathematics offered "fair" competition. In theory at least, all students started from the same position at matriculation, though in practice the mathematical backgrounds of Cambridge entrants differed considerably. The students who had studied the syllabus designed at King's College, London, by Thomas Hall were undoubtedly well prepared, for they had virtually covered the first-year Cambridge syllabus before setting foot in the town.

Mindful of the kudos Tripos success conveyed on a college, Trinity authorities nurtured its talent. After the first-year college examinations in mathematics, Latin, and Greek, the top twenty students were placed in a special class, which would shape their readiness for the Tripos contest three years later. Approaching his final year of study, Robert Leslie Ellis was contemptuous of this grooming. He listened carefully to Hopkins's promptings that he could be at least second wrangler with a modicum of effort but "from the bottom of my heart," he wrote in his diary, "I detest the system here—the crushing down of mind and body for a worthless end—*Mais que faire?* How is one to break through the threads by which one is tied to it." In contrast, there was no sign that Cayley felt oppressed; indeed, he coped well and even took to the system of speed learning and regurgitation. In the early summer of 1839, Cayley and Ellis found themselves in the same coach bound for the metropolis (where Ellis's father kept a house at Hampstead). "Off at ten," Ellis wrote in his diary, and of his companion a brief recognition, "Cayley in the coach—the great man of the freshmen. He has my pity—yet probably needs none."[19]

At the end of the Easter term in 1840, there was the one-day Previous Examination, in which all students were tested in the classics and Paley's *Evidences of Christianity*. This examination—the "Little-Go"—was a high hurdle for some, but it would not trouble a "reading man" like Cayley. A more rigorous challenge came a few weeks later when the 19-year-old competed with seventy or eighty others in the examinations for a Trinity scholarship. The written tests included Latin and Greek translations, knowledge of classical civilization, and mathematics. It was a matter of personal prestige for a student to be singled out as a scholar. Joseph Romilly, a member of the "Seniority," recorded the meeting with Christopher Wordsworth when the answer papers were looked at and the scholars chosen.[20] His brief diary entry for 29 April read: "—tea with the Master to arrange the 23 new Scholars—the $>^{st}$ n^o [greatest number] we ever had: only one Westmr [student]. We agreed without much disputing: we rejected two for bad conduct who would otherwise have been elected, viz. [T. I.] Barstow & a Sizar Ottiwell Robinson. We elected 3 King's Coll. [London] students viz. Cayley, Fenn & Shaw:—Cayley is to be Senr W^r in 1842."[21]

What prescience! Romilly's forecast that Cayley would be the champion student of his year was well in advance of the local talent-spotters who made it their business to foresee the results of future Tripos examinations. As a scholar, Cayley was now on the lowest step of the college hierarchy, one step removed from the ordinary undergraduate. In the great ledgers of the foundation he would henceforth be designated a scholar and be placed on the payroll to receive the sum of three shillings and four pence per quarter and an annual gratuity (liberatum) of thirteen shillings and four pence (respectively, one-sixth and two-thirds of a pound sterling). For being responsible to one of the tutors and helping him with teaching tasks, he received an additional allowance toward the rent of his rooms.

In 1841, William Whewell became master of Trinity, the beginning of his 25-year reign. During his stewardship, Trinity enhanced its reputation and "the life of undergraduates and dons was never more vigorous or more varied."[22] He had risen from the ranks as a result of an imposing ability combined with a forceful character. When he was admitted to the college in 1812 as a subsizar, he recognized the twin-track career of Science and University as a means of gaining preferment. By the time Cayley arrived, Whewell was growing tired of his tutorship and the rigor of academic writing had lost its appeal. The university and college administrative careers took on greater prominence.[23] Whewell was an able university politician whose views on matters of curriculum reform by this time had settled into a conservative stance. He had firm ideas on proper subjects for undergraduates and expressed the view that classics, Greek and Latin, were indispensable elements of a liberal education but it was mathematics as a supplement that offered training in logical reasoning. To Whewell, the worthwhile degree subjects were those timeless disciplines. Once these had been mastered, students could study the "progressive" subjects like chemistry. He set great store in mathematics as a permanent and unalterable stock of knowledge. To him, mathematics had the paramount quality of being concerned with indisputable truths, whereas subjects such as chemistry were unable to make this claim. Thus Cayley was studying a prized subject though his attraction to chemistry as a youth would not have found favor in the light of Whewell's stringency.

It is obvious that Cayley enjoyed a seamless success at Trinity. He distinguished himself in the Freshman year, the Junior Sophister year, and the Senior Sophister year. Each year he was placed in the First Class and each year he won a College Prize. It was as expected as it was monotonous for all the others. In 1841, his Senior Sophister year, he scored so many marks over his rivals that the examiner underscored Cayley's name to demarcate his performance from the rest. Those in the First Class were: Cayley, Bryan, Cubitt, Fenn J F, Shaw, Smith B F.[24]

In this list, a strong connection between Trinity, King's College, London, and the academic hothouse schools of Blackheath is again evident. Cayley, Fenn, Shaw, and Smith were all from King's, while Fenn and Shaw had been groomed previously in its feeder school, Blackheath Proprietary. All of Cambridge was aware of the Fenn brothers in the 1840s. The Rev. Joseph Fenn, an evangelical minister of the Blackheath Park Chapel, and leading light in setting up Blackheath Proprietary in 1831, sent eight of his sons to Trinity for their education. Benjamin Smith, who appeared in the list, had scientific interests himself (he joined the Cambridge Philosophical Society) and was an ally of Cayley's whose friendship continued after Cambridge.

University life was not all work and Cayley enjoyed these carefree undergraduate days. In the summer following the college examination he went on holiday to Keswick in England's Lake District, a way of spending a vacation that was to continue—a period he reserved in later years for uninterrupted mathematics. Walking in the hills did not preclude research but encouraged it. While he was not thinking about mathematics—even Cayley needed diversions—he sampled from a broad platform of reading material on offer at Trinity. He showed an affinity for biography, reading J. H. Monk's *Life* of the quarrelsome Richard Bentley, a renowned classical scholar and former master of the College. The Roman historian Suetonius was read more than once and in his liking for scandalous biography, he found an escape into the Rabelaisian medieval world of the law and university intrigue. He seems to have had little attraction to organized sports, as was true of many of the reading men, who were generally content to walk the "Grantchester grind" for their exercise. Yet Cayley's walking exploits became legendary among fellow undergraduates. Though it was not unusual in Victorian times to be a "great pedestrian," Cayley seems to be an extreme example—even in the Captain Barclay league. He certainly would have been capable of joining the band of undergraduates who thought nothing of walking the fifty-two miles from Cambridge to London in a day. As a youth of less than twenty years, he had on occasion "walked twenty miles before breakfast."[25]

While a Cambridge education meant a College-centered education, much more than is nowadays the case, Cayley's mathematical talent was being discovered outside the walls of Trinity. Among students of other colleges, he became something of a Cambridge celebrity even before the ordeal of the Tripos examination. Next in size to Trinity is St. John's and its students were keen to compete for the highest places in the wrangler lists. Robert B. Mayor was a strong candidate but his friend Charles Simpson was the best chance the "Johnians" would have in 1842. Of him, Mayor wrote home: "He [Simpson] is getting on as usual, that is to say he is distancing all his competitors, and though Trinity talks high of one man of the same year, all who have any thing to do with Simpson say that if his health, wh[ich] is improving since he came up here, continues, he will not lose the Senior Wranglership without a struggle. The respective colleges give out, 1st that Cayley, (the Trinity man) is the best man they have ever had since Airy and the Johnians say that Simpson is second to none but Sir John Herschel, that have ever rendered St John's illustrious."[26]

William Thomson, who was sent by his father, a mathematics professor, to Peterhouse college in October 1841, met Cayley at Challis's Christmas party and was impressed by the quiet young man about to sit the Tripos examination. Thomson's own entrée to the university was accompanied by a published paper in the *Cambridge Mathematical Journal* at the age of sixteen and he was quickly admitted into the Cambridge inner circle of "reading men." He and Cayley had a burning ambition to contribute to science and mathematics and the impressionable first-year undergraduate wrote to his father in Glasgow of meeting Cayley "who is to be Senior Wrangler this year."[27]

For the honors men, the examination was the single focus of their study. Students in the university may have won their individual college scholarships and received their bound editions of *Plato* from the master of their college for their academic distinction but they had yet to face the glare of the university-wide examination with the famed crown of the "Senior Wranglership" as the top prize—or, if not that, a respectable position in the order of merit. The order of merit was a national institution and thoroughly in keeping with the ideals of Victorian society, which stressed individualism.

The Senior Wrangler position was the linchpin of the whole system. When the structure of the mathematical Tripos underwent radical change at various times during the course of the century, the existence of the Senior Wranglership was never seriously challenged. Important within the university, the middle and upper classes of England took an interest in "the degrees," with it being an event placed on a par with horse racing during Ascot week. To be Senior Wrangler meant more than success in "mere mathematics." It indicated the caliber of a man—women not being allowed to compete until the 1880s but even then not admitted to a degree. A man's position in the order of merit indelibly marked him for life, and once it had been secured in the once-and-for-all examination it could not be taken away. For a man's future career, a high position in this barometer of academic worth was paramount while the cachet of the very top position was of supreme importance whether he followed a career in the church, the law, or in the City of London. A Senior Wrangler also left something of himself at Cambridge, for a college took immense pride in having him among its number and remembered his success, and therefore their own, in the years to follow.[28]

The first requirement for any prospective Tripos candidate was the ability to get down to the grinding memory work and the continual practice of Tripos questions taken from past examination papers. Memorizing bookwork and practicing the answers to "riders" (those tricky problems tacked onto questions to catch out all but the most able) would be a workable strategy. To make success possible, the private mathematical coaches came into their own and supplied extra tuition for the ambitious "reading man" of the university: "[t]he savage pressure at which the Cambridge coaches of the nineteenth century drove their pupils became legendary," wrote E. H. Neville, "not a day, not an hour was wasted; the perfect candidate should be able to write the bookwork automatically while his thoughts were busy with the rider, and the fingers could be trained even when the brain was weary; above all, curiosity about unscheduled mathematics was depravity."[29] Charles Kingsley at Magdalene College, taking his degree in the same year, referred to examination ordeal as "the violent exertion of the Mathematical Tripos" and the degree which "hangs over my thoughts like a vast incubus keeping me down."[30] The whole process was so arduous that it drove many students to despair, an ordeal happily accepted by those who administered it since the imagined outcome in terms of character formation was generally regarded as beneficial.

The coaches kept an eagle eye on the Tripos questions as they appeared from year to year and passed on the very best in examination technique to the pupils who paid them. They knew the proportions of time to spend on each topic, but above all, they continually tested their pupils until they could reproduce Tripos answers on demand. Kingsley has left us with a side-glance of Cambridge during this period, through passages in *Alton Locke*, a social critique clothed in a boy's adventure story, which evolved against the backcloth of Chartism, Christian Socialism, and Cambridge. Kingsley's undergraduates were of the opinion that the formal provision of the university was of little use, for the dons charged high fees for lectures not worth attending and "any man who wanted to get on, was forced to have a private tutor, besides his college one."[31] It was extra expense but no ambitious student could be without the additional classes geared to the examination. It might be thought that private coaches were narrowly based nineteenth-century crammers worthy of Dickens but this was not the case. They were usually men of wide erudition but, more crucially, men who had succeeded in the Cambridge system themselves. There could be no better choice for Cayley than the popular William Hopkins.

William Hopkins

Once a gentleman farmer from Suffolk, William Hopkins went on to study for a Cambridge degree at the age of thirty after the death of his first wife. He realized he had little taste for farming but he did have an interest in academic study. From this new beginning he graduated seventh wrangler in 1827 (a strong mathematical year that included Augustus De Morgan), married again, and settled down in the town as a private tutor. A successful coach such as Hopkins could become wealthy, for the income derived from tutoring was substantial and far in excess of the stipend paid to the occupants of the prestigious Cambridge chairs. Hopkins became widely known as a "Senior Wrangler maker" and could count many notable Victorian scientists among his clientele. In a 22-year period, he coached the top student on no fewer than seventeen occasions, a feat exceeded only by his own pupil E. J. Routh, who dominated the coaching scene in the second half of the century. When not engaged with coaching, he studied science generally but owed his scientific reputation to research in geology. In the mathematical arena, his forte was in applying mathematics to physical science.[32]

A man of refined temperament, Hopkins took a careful interest in his students, assessing their ability and future place in the Senate House examination and counseling them on future prospects. They in turn were attracted by his reputation, a factor which ensured his teaching was highly sought after. The coaches' reputations suffered if their pupils were not placed in a high position in the order of merit, and on one occasion, even the great Hopkins suffered a setback. "Master an't placed this year!," his servant was heard to mutter, but he rebounded the following year with a string of successes, including Routh and James Clerk Maxwell in the two top positions.[33] The relationship between master and pupil was an easy one for Hopkins, and his house in Cambridge became a focus for social events among the new scientific breed, dons and students alike—certainly they were lively. Eligible bachelors attending "at homes" were judiciously paired with the daughters of the local gentry by his second wife. The social mix of these events provided a backdrop to an energetic Victorian evening complete with orchestra and waltzing couples.

In his role as a coach, Hopkins taught differential calculus, the calculus of variations, astronomical instruments, hydrostatics and dynamics, lunar and planetary theory, mechanics, optics, sound and light. He prepared his students for the Tripos with his impeccable lectures and challenging problem sheets and all facets of his teaching delivered in a strictly methodical style. It was an activity he never wearied of, despite the inevitable repetition over the years. Each year he found new challenges and claimed to be less tired of giving his lectures "than I should have been of the daily toil of any other engrossing avocation." A self-proclaimed teacher of mathematics, Hopkins lamented that the art of teaching was generally disregarded at Cambridge. This was not an individual failing since he realized only too well that "the art of teaching is not given to every one to whom ability and knowledge are given."[34] In his pupil rooms in Peterhouse he employed the time-honored Socratic method of drawing out the thoughts of his pupils from their own experience. He was also quite modern in encouraging his students to give voice to their ideas, and his sensitivity and awareness that they unconsciously taught each other was a learning scheme he used to good effect.[35] One of his obvious strengths was that he could relate mathematics to a wider experience of the physical world and to philosophy. He gave his pupils a rounded view of the subject though he was wary of "spoonfeeding" them, which would blunt their natural curiosity and circumvent their necessary struggle with complication. For Cayley and all the budding mathematicians and scientists who attended his course of study, it was a rigorous experience but one which had its lighter moments. In his classroom, Galton reported that he "tells funny stories connected with different problems and is no way Donnish; he rattles us on at a splendid pace and makes mathematics anything but a dry subject by entering thoroughly into its metaphysics. I never enjoyed anything so much before."[36]

When moves were afoot to reduce the rigors of mathematics in the Tripos, Hopkins counseled against watering down the syllabus. He opposed the move to omit physical astronomy, physical optics, partial differential equations, advanced three-dimensional geometry, and most of hydrodynamics. He adamantly disagreed with the proposed reduction, thereby making it simply a basis for the award of a qualification—of service to the generality but neglecting the gifted high-flyers among the honors degree students who came to him for coaching. By maintaining a broad basis at a high level of study, Hopkins believed he was preserving the reputation of the university through its education of the ablest students. A man of conservative leanings, Hopkins took the view that reform should not be entered into lightly, for the university was an institution impossible to think of "without a deep reverence for the hallowed impress which time has stamped upon it."[37] To Hopkins, the Tripos was an integral part of that "impress," with a firmly established mathematical tradition that had been generally beneficial over the years.

Hopkins went further and mathematics was much more than a component in a liberal education. Mathematics to him was much more than "mental gymnastics" that enabled students to jump through the hoops of the Tripos. He believed in mathematics, not only as the supreme instrument for studying science but as the *only* instrument. When it came to formulating and understanding abstract physical concepts, mathematics came into its own. Being a communicator, he knew how to convey this to his junior audiences. As he was fond of saying, mathematicians would as soon fall into the trap of confusing such concepts as force and momentum as measuring "heights in acres, or arable land in cubic miles."[38] We can easily imagine Hopkins imparting this homily in his classroom as his students worked toward the Tripos examinations, but, as he continued to argue, mathematics was important in its own terms. It was, in fact, the key to a deeper purpose:

> It is only when the student approaches the great theories, as Physical Astronomy and Physical Optics, that he can fully appreciate the real importance and value of pure mathematical science, as the only instrument of investigation by which man could possibly have attained to a knowledge of so much of what is perfect and beautiful in the structure of the material universe, and in the laws which govern it. It is then that he can form an adequate conception of the genius which has been developed in the framing of those theories, and can feel himself under those salutary influences which must ever be exercised on the mind of youth by the contemplation of the workings of lofty genius, in whatever department of science or literature it may have been called into action.[39]

As a by-product, Hopkins's students gained a research attitude in their work and this engendered "spirit of enquiry" was perhaps his greatest gift to his students. There is a strong sense in which Cayley received a first impetus in "mixed mathematics" from Hopkins, but while other of his students took to the path of mixed mathematics as their initial research interest (Stokes to hydrodynamics, for example), Cayley found the challenge in mathematics per se.

During the period he taught Cayley, Hopkins published work on physical geology and was preparing papers on the "Physical Investigations on the Motion of Glaciers." His dedication to geology did not prevent a close connection developing between master and pupil, for both shared the common characteristic of being theoretical, preferring the conceptual development of a subject to empirical questioning. This philosophy suggests the kind of contributions Cayley would make to physical theories, but he could also throw the direction of routine scientific method into reverse and investigate how physical problems might suggest problems in pure mathematics. Hopkins exposed his students to up-to-date theories but few of his students could have been absorbed in J. L. Lagrange's *Berlin Mémoires* and the *Mécanique Analytique*. From Cayley's reading of these works, he showed his coach a new theorem he had deduced on multiple integrals, and this highly technical work resulted in two papers published in the *Cambridge Mathematical Journal*.[40] Hopkins had students of this caliber in mind when he thought of the future generations of dons who would preserve and enhance the character of the university.

The Tripos Test

As usual, there was something of a carnival atmosphere at the degree examinations in January 1842. The whole university took part, from the dons who examined the students to the vice-chancellor who admitted them to degrees, and not forgetting the spectators who liked a flutter on the hot favorite. It was the time of year when the college porters could offer the best tips from the intelligence they assiduously gathered over the year. The favorite was Cayley, suggesting that Simpson's setback had become common knowledge. A man who in one week was reputed to have put in twenty hours a day, with two months to go, appeared to be buckling under the strain. Writing home, Mayor commented on the breakdown of health in his friend and apparent loss of confidence. Simpson "did not in the least expect to recover. I am afraid that he has come up, in a state quite unfit for study—For five weeks he had done nothing, and even now he is only able to read for an hour or two a day so that we now put him down as Second."[41]

There was still the prospect of a close finish. The small market town, cut off from the rest of the world—the railway had yet to arrive—as ever, was cheered by the prospect of a new young hero at the beginning of 1842. If a stranger to the town doubted the importance of the event, they would find a corrective in the contents of the *Cambridge Advertiser and Free Press* of the following weeks.[42] The Extraordinary Edition would carry the order of merit, which one might expect, but also the full text of the examination papers. It was not every week of the year that the denizens of Cambridge, The Isle of Ely, Marshland and the Bedford Level would be apprised in their local press of the technical details of the higher reaches of pure mathematics, "mixed mathematics," and astronomy.

Since its emergence in the middle of the eighteenth century, the Tripos examination for the bachelor's degree had evolved into a well-oiled system. Oral examinations, which amounted to little more than superficial debating contests conducted in Latin, had gradually given way to a system of written examination with printed question papers. Though it would lengthen in later years, in 1842 there were six days of continuous examination comprising thirty-three hours of written papers. Each day, including Saturday, they sat examinations for two and one-half hours in the morning and for three hours in the afternoon.[43] The intermediate Sunday was thankfully a day of rest. For the undergraduates who congregated in the Senate House there was often no way of telling whether the furious quillwork was the result of cramming or cultivation. Undergraduates competed both for themselves and for the prestige of their college; this year there were 120 candidates, of whom 20 were from Trinity. By far, the majority of students in the university took the much less demanding examinations for the B.A. degree without honors. Since 1828 this degree had a separate set of question papers based on a very limited syllabus, though the illusion of the poll men taking the honors degree examinations was maintained.

The mathematical Tripos has endured a good deal of modern criticism but the content of the examination papers for 1842 is undoubtedly wide-ranging. The questions indicate depth and rich diversity; to answer them successfully, both factual knowledge and a degree of skill in exposition was needed, and, in addition, some facility with mathematical technique. To be sure, speed was also required and success in many instances revolved around the acquisition of satisfactory training. Given skillful coaching, it would have been possible to answer questions quickly in those parts of the papers where the questions were predictable and dependent on prelearned bookwork. With all its weaknesses, the examination of 1842 represented a formidable intellectual challenge. Several years later, the scope of the Tripos was reduced but the questions Cayley was required to answer were still selected from a very broad syllabus.

Each year the university appointed a quartet of Cambridge dons to administer the Senate House examination, two moderators to oversee the process and set questions plus two examiners to assist. The senior moderator was Thomas Gaskin, supported by D. F. Gregory, the editor of the *Cambridge Mathematical Journal* and, according to some, the finest mathematician in the university.

Gregory's interests coincided with Cayley's, and we shall hear more about him; Gaskin was a fellow of the Royal Society, elected in 1839, at a time when to be interested in science was almost a sufficient condition for admittance. The examiners who assisted in 1842 were Alexander Thurtell and Richard Potter and, though the style and form of the questions had a certain inevitability about them, these two would have their own ideas of how to test undergraduates. Thurtell was the younger and more of the Cambridge type. Graduating fourth wrangler in 1829, he was a Cambridge don in Holy Orders who would eventually retire from Cambridge to a 30-year stint as a Norfolk rector.

The senior of the two examiners, Richard Potter, was anything but ordinary, physically and intellectually: a bulky man "with a face like a woman's and a piping voice."[44] A fellow of Queen's College, his loyalty was divided between Cambridge and his newly gained professorship of natural philosophy and astronomy at University College, London. Potter's former life was that of a Manchester corn merchant who also had conducted himself as an amateur physical scientist, an activity fostered by the newly founded British Association for the Advancement of Science. With encouragement, he entered Cambridge and graduated sixth wrangler in 1838, not far short of his fortieth birthday. The year before he would have seen George Green, another distinguished mature student, graduate fourth wrangler. If Potter eventually became a pitiable worn-out figure at University College, London, driven to supplement his income for his old age who daily muddled his mathematical arguments in front of his class, in 1842 he was a "self-made" professor with a proven track record in optics and mechanics. The candidates in the Senate House could expect to be tested by his expertise in "mixed mathematics."

In the January examination of 1842 (beginning on Wednesday, 5 January, at nine o'clock in the morning and lasting until the following Tuesday), the undergraduates sat on hard benches in academic dress bunched along trestle tables supervised by invigilators dressed in their gowns and mortarboards. By holding the examinations in January—a custom abandoned only in the 1880s, candidates had to sit the examinations in bitterly cold weather. But this could be coped with as collegiate rooms such as libraries were left unheated as a matter of course. The sharpness of this particular winter was aggravated by the fear of shortages in the coal supply. No matter, this was unremarkable for the students who emerged to the Cambridge courts from their Spartan conditions. They were prepared by the exhortations of their tutors that a Cambridge man was a "healthy mind in a healthy body": *Mens sana in corpore sano*.

They were all young men. Cayley was barely several months past the age of 20, not even at the age of majority, when he sat ready in the Senate House in the shivering cold. Sitting near him were William Castlehow and Samuel Carter from Emmanuel College, two men close in ability hoping to shine for their college and themselves. Edward Cook and James Clubbe from Trinity's rival

St. John's would hardly be a challenge, and where Cayley would complete his task in a measured way, they would pick up marks where they could. The same was true of Benjamin Cobb nearby. Poor Cobb ended up sixty places below Cayley in the order of merit and perhaps his preparation suffered through his habit of migrating from college to college during his undergraduate career. He redeemed himself by taking a Third Class Classics degree after he had qualified by appearing on the mathematical order of merit. All these took Holy Orders in the Church, but Cobb chose the Law, where he continued his migratory habits by moving around the Inns of Court. But as they were sitting cheek by jowl in the Senate House in January of 1842, this was all before them, as they waited for the beginning of the Tripos contest.

The first examination paper was introductory and carried the rubric: "The Differential Calculus was not to be used." In the printed paper, itself a recent development, candidates were invited to answer as many questions as they could, hence the advantage to the speed-writer. Bookwork formed the bulk of the Tripos examination generally but problem solving was also tested and an important reason for the wide respect in which the Cambridge mathematics degree was held. One contemporary remarked: "In my opinion it is this continuance of solving problems, this general course of not only acquiring principles but applying them, that at last makes the senior wrangler, who perhaps at the time is one of the most expert mathematicians in existence."[45] Quickness and readiness were both skills Cayley could draw on. He could also summon up a peculiarly mathematical skill, nowadays regarded as rather old-fashioned and best left to machines. "Don't be afraid to slog out very hard problems" would be a piece of advice handed out by an experienced coach to his pupil when faced with long mathematical expressions that appeared formidable. In Cayley's case, it was suggested his ability in dealing with them was not an acquired skill for he "had an instinct for the management of the most complicated processes" to an extraordinary degree.[46]

The examination questions set before Cayley reflected subjects which would be topics of his future research.[47] They were evenly balanced between "pure" mathematics (on such topics as the differential and integral calculus, analytical geometry) and "mixed" mathematics (which included questions on optics, hydrostatics, astronomy, and Newton's *Principia*, an obligatory study in which students were required to know in part, and quote giving chapter and verse). If we imagine sitting with Cayley in the Senate House, we might gauge the examination task ahead and glimpse some of his future research:

Enunciate and prove Newton's Lemma X. [from the *Principia*]

Investigate Sturm's Theorem for separating the roots of an equation. [published by Sturm in 1835]

Explain how the attraction of the Moon on the Earth, each being supposed station-
ary, would raise the waters of the ocean on the side of the Earth nearest the Moon,
and also on the side farthest from it. Assuming the expression for the height of the
tide at a given time and place, find the time of high tide.

The study of astronomy was an important part of the curriculum and was
examined thoroughly. "Astronomy," wrote Whewell, "is not only the queen of
the sciences, but in a stricter sense of the term, the only perfect science," so that
it was an admirable subject for examination from this viewpoint.[48] There were
some general questions of the type: "Investigate the alteration of the major axis
of the disturbed orbit of a planet." Some questions were more specialized: "Find
the horary motion of the Moon's nodes in a circular orbit; and shew that the
mean horary motion of the nodes is half the horary motion when the Moon is
in syzygy."

With a range of questions devoted to the mathematics of astronomy, it is
hardly surprising that Cambridge graduates who took up a serious study of
astronomy continued this tradition when they left the university and became
influential members of the Royal Astronomical Society. Cayley was one and
made many contributions related to the motion of the moon and geometrical
problems related to astronomy.

A development in the Tripos with immense importance in Cayley's mathematical education was the rise of analytical geometry. The "Analytical Revolution," which had received encouragement from the short-lived Analytical Society, had gradually displaced descriptive geometry. In the first thirty years of the
century, this algebraic approach to geometry was firmly established in England.
Several textbooks, either Continental themselves or modeled on Continental
texts, appeared. These (notably Henry Parr Hamilton's *Principles of Analytical
Geometry* [1826], John Hymers's *Analytical Geometry of Three Dimensions* [1830]
and his *Treatise on Conic Sections* [1837]) treated the material from an advanced
standpoint and helped to firmly establish analytical geometry to a prominent
place in the Tripos syllabus. In the 1842 examination, Cayley had the opportunity
of answering questions such as, "Find the equations to a tangent line at any point
of a curve of double curvature [a space curve]: find also the equation to the
locus of all tangent lines which can be drawn to a surface at one point."

The mathematical course at Cambridge also required students to study
"mixed mathematics." They were tested on bookwork questions such as:

Explain the object and advantages of rifling the barrel of a gun.

State and prove the principle of Virtual Velocities; apply it to find the condition of
equilibrium when three forces act on a point.

Explain fully the cause of distortion of images formed by lenses and mirrors, and shew in what case the extremities of an image formed by a lens are more magnified than the central parts, and in what case they are less magnified.

Some questions were topical in a decade that saw the chaotic growth of the British railway system. For instance: "A carriage moves on a railroad with a given velocity round a curve of given radius: find the amount by which the outer rail must be elevated above the inner one, in order that the carriage should not be overturned towards the outside."

The twelve examinations sat by candidates were each composed of questions like those above, and on most papers there were as many as twenty-four of them. To use De Morgan's phrase, it was indeed the "great writing race." But the end was in sight. On Tuesday, 11 January, the last day of the Senate House examination, the final question on the examination paper related to the figure of the Earth. Assuming Cayley answered all nine questions on this paper (a fairly safe assumption), he had about twenty minutes to answer the "bookwork" question:

Assuming that, in the expression for the radius of a nearly spherical stratum of the earth, considered as a revolving heterogeneous mass originally fluid, all Laplace's coefficients [now termed Legendre polynomials] vanish except that of the second order; prove that the strata of equal density are all concentric spheroids of revolution, whose axes coincide with that about which the earth revolves.[49]

State the other results of the hypothesis of the Earth's original fluidity, the subsidiary assumption made in obtaining them, and the evidence for the truth of the theory arising from the comparison of its results with observation.

This was a piece of "mixed mathematics" that tested the undergraduates of the 1840s but which would be jettisoned shortly in an effort to slim down the scope of the examination. It did not lie down completely, and in the emaciated examination following the reforms, "Laplace's coefficients" and the "figure of the earth" became a rallying point for future reformers.

Cayley was the perfect candidate for the Tripos. His prowess was beyond question and, from Hopkins, he had acquired the essential examination technique. Unlike many who competed for mathematical honors, he could afford to take the prospect in his stride. It was rumored that Simpson kept himself afloat during the examination with frequent doses of ether and other stimulants. Albert Pell, an undergraduate in the same year, was a witness to Cayley's performance. Pell, who afterward became a member of Parliament, remembered the occasion of the Tripos examination of 1842:

In the year I took my degree the usual rivalry between Trinity and St. John's for the Senior Wrangler's place was as keen as ever. Our man was Cayley. I forget the name of the Johnian [Charles Simpson]. While the examination was going on in the Senate House a small crowd was frequently in attendance outside by the door, discussing the merits of the examined and waiting to get the latest intelligence of their work. I, among them, went for this purpose, and on a Johnian in our group saying, "I wonder how our man is getting on," [T. I] Barstow [an undergraduate examinee] in a loud, contemptuous voice said he did not know, nor did he care, but he could tell him that "Cayley had finished his papers a quarter of an hour ago, and was now licking his lips for more."[50]

After the examination, Cayley put down his quill pen and walked out of the Senate House. The Tripos grind was over, but how should he celebrate the new-found freedom on the final day? Most of his fellow students would be traditional about it, free at last to celebrate with relief, unleashed exuberance, and student pranks.

Instead, Cayley went to the college library. Five days of continuous examination had not blunted his mathematical appetite and he took out *Géométrie de position* by Lazare Carnot. Aristotle's *Politics* was another borrowing, but the one taken for relaxation was *Lettres de Madame de Sévigné, de sa Famille, et de ses Amis,* by the Marquise de Sévigné, with her racy tales of the French nobility. The whole fuss of the examinations seemed to have been lightly put aside. Later on, Benjamin Smith informed him that Simpson had completed all the questions well within the allotted time. In itself, this was surprising since candidates were not expected to finish any of the papers, but the news failed to cause Cayley any consternation. "Oh," he is reputed to have remarked, "well, I cleaned up that paper in forty-five minutes."[51] When Thomas Gaskin exercised his right to hold a viva voce examination, he told Cayley he was acquainted with his work and need ask him no question.[52]

With the Tripos results imminent, Cayley made plans to go home to London. Sitting on top of the night coach, he is reputed to have had the results handed up to him but "he quietly put it into his pocket, resigning himself very contentedly to the necessity of waiting till the morning light for a knowledge of its contents."[53] The order of merit comprises three classes, Wranglers, Senior Optimes, and Junior Optimes, with all students appearing in a numbered list. The wranglers were the students of the First Class, the name perhaps due to the "wrangles" or oral disputations that were the method of examining in the eighteenth century. In 1842, the first dozen wranglers of the thirty-eight in the wrangler class (in the order of merit, which contained 114 students as a whole) were:

1. CAYLEY   Trinity
2. Simpson   St. John's

3. Mayor R. B.	St. John's
4. Fuller	St. Peter's
5. Bird	St. John's
6. Jarvis	Corpus Christie
7. Shortland	Pembroke
8. Austin	St. Peter's
9. Fenwick	Corpus Christie
10. Jones	Clare
11. Frost	St. John's
12. Parnell	St. John's

John Couch Adams, two years older than Cayley, who had arrived in Cambridge off a Cornish farm and was admitted as a sizar in St. John's wrote in his diary: "Went to the Senate House this morning and found that Cayley was the Senior Wrangler."[54] But Cayley was a rather isolated hero and Adams would have celebrated the fact that approaching half the wrangler class were from his own college In this contest, only a handful were from Trinity. Adams was a year below and would repeat Cayley's triumph in "his year." In 1842 he admired the ability of the 20-year-old youth who conducted himself with such assurance.

After his virtuoso performance, the result confirmed the predictions of the Cambridge academic community. The *Times* duly reported: "Mr Cayley, the Senior Wrangler, is a mathematician of extraordinary powers, nor are they limited to the particular branch of study in which his present honors have been won, but extend equally to other objects of academical pursuit."[55]

With Simpson as second wrangler, the competition for the Tripos in 1842 was a two-horse race. It was also a close contest and Simpson was said by Galton to have surpassed Cayley on the bookwork questions but left 200 marks short when the problems were taken into account. Broadly in line with this, Robert Mayor wrote home: "Simpson's place is due entirely to one paper in wh[ich] he was far from well, and was 200 marks below Cayley—In every other he was either equal or above him—Cayley was [finally] 150 above Simpson,—Simpson was 800 above myself. I am satisfied that the difference was not greater. Every one I know seems pleased with the result. Only one man has been plucked [failed]."[56]

Cayley and Simpson were far in advance of the rest of the field, and all that was needed to be in the wrangler list that year was to score 500 marks in total. As runner-up, Simpson gained a fellowship of St. John's College the following year and immediately entered Lincoln's Inn to study for the bar.[57] Next down the list, Frederick Fuller became a tutor at Peterhouse. Soon after the Tripos, he was assigned to tutor William Thomson and, later, James Clerk Maxwell before he became professor of mathematics at King's College, Aberdeen. He confessed to having no aptitude for original research but saw himself as a teacher rather than a research scientist. He became a force in education and, in recognition of his teaching influence, his students at Aberdeen were known as "Fuller's Men."

Of other appearances in Cayley's list, William Austin, a grandson of the political economist David Ricardo, became a prominent engineer and champion of London's Metropolitan Railway. In the crop of 1842 were men who went out to run the Empire—a future prime minister of New Zealand and a future bishop of Sierra Leone. Most were callow youth endowed with a gift for mathematics but without much experience of life outside the cloistered colleges of Cambridge. An exception was Peter Shortland, a mature student who graduated seventh wrangler at the age of twenty-seven. He had served as mate on HMS *Rattlesnake* in Australian waters during the years 1836–38—the same vessel in which T. H. Huxley was to set sail as ship's surgeon ten years later—and who went on to a naval career that included making a complete survey of Novia Scotia and achieving the rank of admiral.

A Cambridge honors degree in mathematics was an entry point to a secure position in life, and even the man at the foot of the order of merit (distinguished by the award of the "wooden-spoon" on degree day) was a member of an elite corps. The year's "wooden-spooner" was Thomas Irwin Barstow. This was the youth who had not endeared himself to his tutors at Trinity. In the Senate House examination, he observed Cayley finishing early while he, himself, struggled on: he just scraped into the Junior Optime class, thereby qualifying for the classical Tripos, in which he did well. He went on become a magistrate in London's Clerkenwell Police Courts.

In January 1842 it was Cayley who was the focus of attention when he took his place in the Senate House as the celebrity of the day. The academics donned their ceremonial gowns and mortarboards, the bunting was hung liberally between doorways and buildings, and the bells rang out. John Couch Adams "went to the Sen[ate]. House to see the men take their degrees. Shouted for the Senior Wrang^r, for Simpson &c."[58] William Thomson went down there with all the others, and experiencing the Cambridge ritual for the first time, wrote home to his father: "We had a great deal of cheering the Senior Wrangler, the Wooden Spoon, the [newly born] Prince of Wales, Dr Pusey (who however got far more hisses & groans than cheers)."[59]

Cayley was presented to the vice-chancellor with all the pomp Cambridge could muster: "The senior wrangler is led up alone amid a diapason of cheers that shake the building. He has obtained the highest honour that the University has to bestow—a unique intellectual distinction, as all the world acknowledges."[60] Charles Bristed, recorded the occasion of Cayley's triumph: "Our Trinity Senior Wrangler (we have one so seldom that he is prone to be an object of curiosity and a pet) was a crooked little man, in no respects a beauty, and not in the least a beau. On the day of his triumph, when he was to receive his hard-earned honors in the Senate House, some of his friends combined their energies to dress him, and put him to rights properly, so that his appearance might not be altogether unworthy of his exploits and his College." Bristed was impressed by the mathematical accomplishments, though Cayley himself regarded the "dressing-up" story as apocryphal. Bristed, a graduate from Yale College in the United States, and while at Cambridge the winner of a university prize for essay writing, drew attention to another other side of the Senior Wrangler's character not generally known around Trinity. He recalled that Cayley had gained "the reputation of being a mere Mathematician, which did him great injustice, for he was really a man of much varied information, and that on some subjects the very opposite of scientific—for instance he was well up in all the current novels, an uncommon thing at Cambridge, where novel reading is not one of the popular weaknesses."[61] Thus Cayley clearly went some way toward personifying Whewell's desire to have at Trinity—rounded men who also knew their Greek and understood the classics. With the enthusiasm of a new master, Whewell wrote to a friend in 1842: "I hope we shall never become a College of mere scholars and mere mathematicians."[62]

Smith's Prizeman

From success in the Tripos, Cayley went on to win the First Smith's Prize, a monetary prize awarded after another weeklong schedule of competitive examination. The difference between the Smith's Prize examination and the Tripos examination was the balance in favor of questions in natural philosophy compared with questions in mathematics.[63] In Hopkins's view, the Smith Prize tested the "philosophic character" of a man's mind. In 1742 Dr. Robert Smith, the master of Trinity, had made a bequest to the university to fund two prizes to be awarded annually to the "two best proficients in Mathematics and Natural Philosophy."[64] Under the conditions of the bequest, Smith stipulated that, other things being equal, candidates of Trinity College were to be preferred, so Cayley had a head start. The monetary award (of two prizes of £25 each) was quite unreliable as they were dependent on the performance of Old South Sea stock, but as their financial value declined, the kudos to be gained in winning the prize increased. As with the Tripos, the question papers were recorded in the pages of the *Cambridge Advertiser and Free Press*, presumably, for the benefit of the university readership.[65]

The Smith's Prize examination was known as the "professors' examination." In 1842, the examiners were Samuel Earnshaw, an accomplished mathematician, Henry Philpott, standing in for the ineffectual Lucasian professor, Challis, as well as Peacock, and Whewell. The examination was under the direction of the professors, each of whom set wide-ranging question papers, these sat by the handful of top students who had only just emerged from the Tripos ordeal; in total there were 113 questions offered in five examination papers, one per day for a week. As with the Tripos, each paper was balanced between pure analysis and mixed mathematics and there was no restriction on the number of questions attempted. In contrast to Tripos questions, which were based on a definite syllabus, the Smith's Prize allowed the examiners to show off their erudition. As a result, the papers were more challenging and in some measure unpredictable.

Even Whewell, the driving force for placing the applications of mathematics at the center of the Cambridge curriculum, was bound by the system to examine pure mathematics, and Peacock had to draw on his mixed mathematics as a source of questions material. Looking at the papers now, the standard demanded was high, but whether the examiners received "quality answers" is more difficult to determine. The quality probably varied from year to year. One year (1854) Routh and Maxwell were in the ascendant and judged joint winners of the prize only for standards to plummet in the next. Whewell noted with disgust that "the candidates were the worst by far that he had ever examined,— that they knew nothing of the History of Mathematics, had not read Newton, & knew nothing but Cram Mss."[66] This kind of charge could not be leveled at an aspirant such as Cayley in 1842.

The questions set for Cayley's year again suggests links with his future interests. From astronomy, he was asked (by Earnshaw) to "[p]oint out the principal difference in the analytical treatment of the lunar and planetary theories. State the distinction between the secular and periodic variation of the elements of a planet's orbit; and describe the analytical process by which the former are separated from the latter, and their value and period obtained." The idea of secular motion as opposed to periodic motion of the Moon was part of the educational training at Cambridge. There were some geometrical problems drawn from optics: "If a spider's web be placed in sun-light, it will exhibit prismatic colours. Explain the phenomena on the principles of the undulatory theory [wave theory of light]."

Whewell asked questions on the tracing of curves and probed the foundations of the calculus of operations: "Are the consequences of the separation of symbols of operation from those of quantity, [are] universally true? or within what limits?" His geometrical questions had a practical twist: "Find the equation to the surface of which the edge of rebroussement [edge of regression, or cuspidal edge] is given. Apply this to the case when the given curve is the thread of a common screw. Trace the section of this surface made by a plane perpendicular to the axis of the screw; and also, by a plane parallel to that axis." Whewell sent his paper to De Morgan for an opinion and received a strong response. De Morgan remarked on one question, which required candidates to prove that the sun shining through an aperture in the shape of a quadrilateral throws a circular image, but it was the length of the paper that caused him to bristle. "It is a terrible paper for three hours," he wrote in reply, and it is hard to disagree with his judgment.[67] In three hours, the candidates were invited to answer twenty-four questions of this kind and answer as many as possible.

Peacock required answers on the mechanics of the steam engine, while at the other end of the spectrum of his erudition, he asked what the candidates understood by the resolution of the polynomial equation of order five: "All equations whose coefficients are rational, are resolvable into simple or quadratic factors. In what sense is it said that no equation above the fourth degree admits of resolution?" Among pure mathematicians, the solution of this polynomial equation was one of the leading questions of the day. It had attracted the attention of the ablest mathematicians and its solution established the mathematical reputation of Niels Abel and Évariste Galois. Peacock had drawn attention to the problem in his 1833 British Association *Report on Analysis* and had furnished a resumé of the progress made of the resolution of equations including this, the "quintic equation."[68] Peacock also asked a question on his own special interest on the distinction between arithmetical and symbolical algebra.

Cayley took the examination in his stride and even had time for serious mathematics during the Smith's Prize week. On the penultimate day of the examination he took out volume three of Legendre's *Traité des fonctions elliptiques* from the college library, perhaps preparing for that hiatus between the Tripos/Smith's Prize series of examinations in January and the fellowship examinations to be held in autumn. Simpson, runner-up in the Tripos contest, became the Second Smith's Prizeman. On cue, the title of "First Smith's Prizeman" became an added adornment to Cayley's title of Senior Wrangler.

An illustrious group of mathematicians and scientists graduated at the top of the lists at Cambridge in the early 1840s. Robert Leslie Ellis was the Senior Wrangler in 1840, while the following year George Gabriel Stokes had been the top student. Adams stepped into Cayley's shoes in 1843. The lives of these men revolved around Cambridge, and they became close colleagues and leaders of Victorian science.[69] All of them sat the Tripos when it was compulsory to be broad. Moreover, they were also the first cohorts who were subject to the full rigors of the Tripos when the examination in the Senate House became recognizable as something akin to the modern examination system. In the early 1840s, oral disputations were dispensed with, printed papers were the norm, and examiners actually marked the papers, giving up their reliance on an "impression" as they had done in the previous decade.

The reforms of 1848 would increase the number of days of examination to eight but would dilute the syllabus; the mathematical theories of magnetism, electricity, heat, the heterogeneous figure of the earth, Laplace's coefficients would be discarded.[70] With the reforms that would take place at Cambridge, mathematics would lose its imperial position. It had been the center of experience for *all* students when Cayley sat the Tripos, but this was effectively ended at the beginning of the 1850s. The mathematical Tripos remained the degree with the highest prestige for many years but when the reforms came into play, students were no longer required to be on the order of merit as a "permission" for entry to the classics degree examinations. In 1852 the natural science Tripos and the moral science Tripos came into existence and these offered other avenues for students to bypass the travails of the mathematical Tripos.

Juvenilia

Cayley's first contribution to the literature of mathematics was made as an undergraduate in the final year of his degree studies. The paper, published in May of 1841, was made anonymously "from a Correspondent," but the signature "C" indicated its authorship to the *cognoscenti* at Cambridge—to give the author's full name was regarded as bordering on self-advertisement and generally frowned on. The geometrical problem concerns the placement of points in space, or in the plane, and the consequential relationships which exist between their mutual distances.[71] The vertical bar notation for determinants Cayley used to express these relationships altered the course of analytical geometry and is still with us today. A determinant is a technical name given to a formula used to extract a single number from a spatial array. They have many applications in geometry and algebra, in which they "determine" geometrical relationships or the solution of equations. In modern mathematics, determinants play only a subsidiary role but they are too useful to be abandoned completely. The study of them in the nineteenth century represented a vast area of mathematics; their history, in which Cayley occupied a leading position, is summarized in five large volumes written by Sir Thomas Muir.

Cayley could hardly have wished for a better start in his mathematical career. It was, moreover, indicative of his way of considering geometrical problems: "On a Theorem in the Geometry of Position" is a paper relating algebra to geometry.[72] It is significant that at such an early age he should have recognized a problem that offered so much potential—it was, indeed, as has been suggested by Julian Coolidge, a "singularly fruitful bough." It turned out to be a novel treatment of an old problem.[73]

If five points are placed arbitrarily in Euclidean three-dimensional space, or on the surface of a sphere, how are the mutual distances related? If four points are placed on a circle how are these related? The charm of the question lies in the apparent paradox that if points are placed *arbitrarily* then how could they be related at all? Cayley's elegant solution was written (in the case where four points, denoted 1, 2, 3, 4, are placed arbitrarily on the circumference of a circle, and their mutual distances denoted $\overline{12}, \overline{13}, \dots$ etc.) as the relationship expressed by the vanishing of a determinant:[74]

$$\begin{vmatrix} 0 & \overline{12}^2 & \overline{13}^2 & \overline{14}^2 \\ \overline{21}^2 & 0 & \overline{23}^2 & \overline{24}^2 \\ \overline{31}^2 & \overline{32}^2 & 0 & \overline{34}^2 \\ \overline{41}^2 & \overline{42}^2 & \overline{43}^2 & 0 \end{vmatrix} = 0$$

The problem had been treated by J. L. Lagrange, Lazare Carnot, and Cauchy's close friend, Jacques Binet, but Cayley's paper marked the beginning of the custom among English mathematicians of expressing geometrical relationships using determinants. He emphasized the notation of writing the elements of a determinant in a square, and while admitting it was not concise, it was clearer and conveyed more information than any abridged notation. In his mathematical work generally, Cayley favored extensive notations and was generally averse to abbreviations. He worked with coordinates and avoided notation that needed to be "unpacked" in order to be interpreted, an attitude learned in the 1840s which became ingrained.

Though this result was a known one, possibly taking the lead set by Cauchy of arranging elements of a determinant in a square array, Cayley was the mathematician to place vertical lines on each side of the array in order to denote a determinant.[75] In a field of algebra in which every conceivable notation was invented, his choice was described a hundred years later "as the greatest single contribution towards the stabilization of notation."[76] His thinking of determinants in terms of an array was a useful step and shows his enduring fascination with the correspondence between geometry and symbols, in which spatial pattern plays a part. This concern for "spatial notation," which runs throughout Cayley's work, is very characteristic of his mathematical style.[77]

Determinants attracted some attention in England as a subject for investigation in the 1840s. In the previous number of the *Cambridge Mathematical Journal*, published in February 1843, one James Joseph Sylvester, a mathematician unknown to Cayley, published a paper on algebraic elimination that made a novel use of determinants, but he made no attempt to assign a spatial notation to them.[78] Sylvester was professor of natural philosophy at University College, and in the summer of the same year, he resigned his chair for a post in the United States. Surely this paper would have been noticed by the eager young undergraduate who was so avidly reading all that was being done in pure mathematics. It would be a matter of discussion when they did meet in the mid 1840s and formed their enduring liaison.

Sylvester referred to the geometric problem of "five points in space" as presented in Cayley's "juvenile paper," when he examined it a decade later.[79] With the problem of the "five points," we have the opportunity to illustrate Cayley's habit of periodically returning to previous investigations. In 1853, he perceived that there was a "dual" expression for the resulting determinant.[80] In 1860 he gave an expansion of it and consequently obtained a formula for the area of a triangle and the volume of a tetrahedron.[81] Two years later he used the result to obtain a geometric formula (connected with inscribed and circumscribed circles) Descartes had mentioned in a letter to Princess Elizabeth.[82] As a diversion from more serious work, he employed the result to prove an inequality set by a writer to the *Educational Times*.[83] Finally, in the late 1880s, he weaved the problem into two educational notes to the *Messenger of Mathematics* while he was reviewing early work for the compilation of his *Collected Mathematical Papers.*[84]

The problem of "five points in space" is a problem of consequence. It occurs at other stages in the mathematical literature and reflects Cayley's ability in choosing interesting and rewarding problems.[85] Perhaps the most curious reoccurrence of the problem is in *modern* invariant theory.[86] His juvenile paper arose before he had discovered the basic ideas of invariant theory and the geometrical result he discovered as a young undergraduate played no part in his development of this vast subject. But he would not be surprised at its presence, for he saw mathematics as a harmonious unity.

CHAPTER 3

Coming of Age

C lear of the mathematical Tripos and with a fellowship in prospect, Cayley had time to devote to real research. The "paying work" aimed at gaining marks in the examination had been required but he could now apply himself exclusively to "adult" mathematics. Finally liberated from the routine drill of working through examination bookwork and problems, it had not all been futile. Of immediate benefit was the tutorial work with Hopkins, which had run deeper than any Tripos question. The papers Cayley soon wrote on potential theory (which deals with the attraction between objects and finds applications in electricity and magnetism, gravitational theory, and other branches of physics) were inspired by Lagrange's *Mécanique Analytique* and a problem from one of Hopkins's examination papers. Cayley's work was technical and involved multiple integrals treated over n-dimensional volumes; from this first glimpse, we see potential theory was to be a continuing theme in the mathematical world he was to explore. These new results, worked out while an undergraduate, were quickly written and published in the *Cambridge Mathematical Journal* to follow his juvenile offering on determinants.[1]

Another step on the journey to mathematical adulthood was to join a serious scientific society, and on Monday, 11 April 1842, he was elected to the Cambridge Philosophical Society. With its founder Adam Sedgwick in the chair, Cayley heard two papers read: one by James Challis on the differential equations of fluid motion, and the other, out of his subject area, but one which would interest him judging from his previous reading of Lyell's *Principles of Geology*. It was given by Richard Owen, the well-known naturalist, then a 38-year-old, who spoke "on the fossil remains of a new genus of Saurians [lizard] called Rhynchosaurus discovered in the Red Sandstone of Warwickshire."[2]

The Assistant Tutor

In the Trinity hierarchy, Cayley was in limbo, betwixt and between the undergraduate "scholar" and a full member of the foundation. This was the period when "bachelor-scholars" sought to become fellows, and, for three years, until their bachelor's degree could be converted into a Master of Arts, they would be allowed three attempts at the fellowship examination held annually in September. After the Tripos examination in January, there was no immediate pressure and Cayley took steps to refresh his general education. During the interlude he read widely, and if there was any doubt as to philosophical leanings, he took down nine volumes of Plato from the college library shelves. For lighter reading, he took out the works of Laurence Sterne, but he seems to have taken the sources of Sterne's work more seriously than any casual reader. The debt Sterne owed to John Locke, acknowledged in *Tristram Shandy,* evidently set a diversion in Cayley's reading; on his visit to the Wren Library the following week, he took out a work by the English philosopher.[3]

Though Cayley became an assistant tutor, the three-year appointment required few formal duties and he was only nominally attached to one of the sides of Trinity. While he took his turn with undergraduate tutorial duties, he did not attempt to take a large number of pupils but concentrated his attention on study and research.[4] A memory of him at the time conveys an unusual character. Here was a young man occupied with his own thoughts and almost unaware of others. Quietly confident in mathematics, he was diffident and totally lacking in what the Victorians would call "show." The indulgence of drawing attention to himself was absent. From his appearance and unassuming demeanor, it was difficult for outsiders to judge his abilities, but Galton brought insight: "Never was a man whose outer physique so belied his powers as that of Cayley," he wrote. "There was something eerie and uncanny in his ways, that inclined strangers to pronounce him neither to be wholly sane nor gifted with much intelligence, which was the very reverse of the truth."[5]

Cayley evidently enjoyed the ambience of the summer reading party, an adjunct to the normal course of study at Cambridge, which took place in the Long Vacation. Reading parties composed of a few students in company with paid tutors went off to such places as the Norfolk Broads, the English Lakes, or the remoter parts of Wales or Scotland for study and recreation. Away from the cloistered existence of college, where formalities were ritually observed, a more relaxed regimen could be followed. The teaching-ratio was highly favorable inasmuch as six students was thought a large number for two tutors. Reading parties were not the sole preserve of the reading-men but the poll-men, desperate for tuition as the examinations loomed, grasped at straws, and their plight meant additional pocket money for the tutors. In the 1840s, reading parties were very popular, while today they remind us of the leisurely lifestyle of Victorian undergraduates.

In June 1842, with his friend Edmund Venables and their group of students, Cayley embarked at London for a sea voyage to Scotland for a visit that was to last until September. After a 50-hour journey in an overcrowded vessel exposed to rolling seas, the party arrived in Dundee and went overland to Perth and onward to Aberfeldy. Cayley's traveling companion shared a middle-class upbringing. His father was a prominent man of business in the City and an MP for London during the crucial years of 1831-32 for politics. Edmund was educated at the Merchant Taylor's School, where he had been captain of the school. At Cambridge, an interest in ecclesiology was serious enough for him to become a foundation member of the Camden Society, a group which promoted the study of church architecture and decoration, and which Cayley would also join. Their friendship cemented at Cambridge, the studious Venables became a lifelong comrade, present at Cayley's elbow at the turning points of his life and providing an antidote to intensive mathematics. Graduating a lowly wrangler in the 1842 Tripos contest, he was amused that his friend could take mathematics so seriously. But he was at ease with Cayley lost in mathematical thought, and he grew accustomed to a companion he acknowledged as "naturally shy and silent."[6] His own creativity found expression in the writing and editing of religious tracts and handbooks to the English shires, composed during a church career in which he rose to become canon of Lincoln.[7]

The six "cantabs" on board ship were well connected socially but were not destined for top places in the Tripos order of merit.[8] Such teaching of elementary mathematics did not suit Cayley's temperament but he gained the students' confidence. The tutorial duties consisted of teaching them subjects on a par with the theory of logarithms. Francis Galton, one of the party of students, recalled that he proved a popular tutor and there was little about him that conformed to the popular image of the dusty and remote Cambridge don: "Cayley is unanimously voted a *brick* and most gentlemanly-minded man."[9] Venables was more prosaic when writing of the experience years later: "Cayley gained a hold on the reverence and affections of his pupils which he never lost."[10] If only the students could match-up, but at the time Venables wrote about them being "tolerably industrious" but in no "danger of killing themselves by work."[11]

The break gave Cayley time to pursue his own research. Sitting amidst some of the finest scenery of Scotland, Venables wrote to George Gabriel Stokes from the village of Weem on the opposite bank from Aberfeldy:

> We are surrounded by some of the highest mountains in Scotland, among which the most beautiful is certainly Schiehallion famous in the history of gravitation; I have [been] trying in vain to induce Cayley to repeat Dr Maskelyne's experiments on its summit, and to acquire to himself a never dying fame; but alas he has no desire for notoriety, and has a rooted aversion to experiments & calculations of all kinds: he is now sitting by my side carrying on those dreadful investigations commenced in the [*Cambridge*] *Mathematical Journal*, in which having exhausted all the letters of the Greek & English alphabets, he is fain to turn his $\Delta \varepsilon \lambda \tau \alpha \varsigma$ [Deltas] topsy turvy, & have recourse to the old English.[12]

Nevil Maskelyne, who became Astronomer Royal in 1764, was himself a fellow of Trinity College, and ten years later he made the journey to Scotland in the company of the Scottish mathematician John Playfair. His plumb line experiments on Schiehallion, the "hill of the Caledonians," gained him esteem in the scientific world. From his observations, and courtesy of a long calculatory effort by the mathematician Charles Hutton, Maskelyne was able to obtain a value for the gravitational constant G, and from it the density of the Earth. Cayley was generally disinterested in such questions especially since the matter had been settled already. He was a dedicated research mathematician wanting new results, but when teased about posthumous fame by Venables, he appeared disinterested and only interrupted his calculations to send greetings to Stokes and refer him to some of Augustin-Louis Cauchy's work on exact differentials, which would be of use in the Irishman's hydrodynamics.

The daily routine of the reading party involved study in the morning coupled with extensive walking in the afternoon and organized social events: on one evening they gave a ball at the Breadalbane Arms in Aberfeldy and invited the local gentry. Cayley made friends locally, felt welcome, and enjoyed the elaborate celebration of a highland wedding from "3 in the afternoon till 4 the next morning" with only a short break in the celebrations.[13] He and Venables explored the natural world in the way of "botanizing," searching for "new" flowers in the surrounding countryside, and to supplement this gentle activity, he undertook more arduous exercise. In recalling a frail appearance, Galton gave substance to Cayley's physical prowess: "One morning he coached us as usual and dined early with us at our usual hour. The next morning he did the same, all just as before, but it afterward transpired that he had not been to bed at all in the meantime, but had tramped all night through over the moors to and about Loch Rannoch."[14] "That country" around Loch Rannoch was as Robert Louis Stevenson described it, "lying waste as the sea; . . . much of it was red with heather; much of the rest broken with bogs and hags and peaty pools."[15] One thing was sure, it was a treacherous stretch of land, and crossing it was only to be attempted when the weather was right. But then, Cayley would not wait for ideal conditions and was prepared to undertake extensive hikes on the moors, in the hills, and in the mountains in all weathers. Galton recognized a kindred spirit, for like himself, his reading-party tutor was an explorer of the wider intellectual terrain, a man who pressed ahead without the inclination to dwell or even pause. Galton chose different subjects for his long journey in science but, like Cayley, he was always an adventurer.

Cayley was an obvious choice for a fellowship by virtue of his success in the Tripos and the award of the First Smith's Prize, but such an appointment was not automatic. Trinity College was unique among Cambridge colleges in requiring aspirants to acquit themselves in further academic tests.[16] These were the competitive examinations for which there were usually twenty to thirty

candidates actively seeking selection each September. Candidates had to sufficiently impress the Seniority to allow them to join their elite corps and it was
most unusual to succeed at the first sitting. Admitting new entrants to be "one
of them" was a duty taken very seriously by the governing body of the college;
for a bachelor-scholar the preparation and expectancy could be an unsettling
experience. If a bachelor-scholar was admitted a "Fellow of Trinity College"
within his three allotted attempts, it was the next step of an academic career
and the ultimate reward for all those who gained scholarships and attained
high places in the Tripos order of merit.

Some preparation for the fellowship examinations was required yet Cayley
was not overanxious. He visited the library on the Friday before examination
week and took out a slew of books signed out with the permission of the
master himself. The classical texts by Aristotle, Homer, and Cicero would
help him with the generalist papers but those by Euler, Lagrange, Newton,
Laplace, Monge, and Archimedes, were beyond any questions which might
appear on the mathematics papers. These papers were castigated among
the new breed of Cambridge mathematicians for their low standard. An
acknowledgment of the triviality of mathematical questions was evidently
Cayley's borrowing of John Leslie's *Elements of Geometry & Plane Trigonometry* (1820), no doubt chosen to put him into a mode of "examination thinking" of a kind he had left behind years before. Thus the prospect of a series
of examinations on which so much depended did not appear to demand his
whole attention.

Beginning on Wednesday, 24 September 1842, the fellowship examinations
were a test of proficiency in classics as well as mathematics and they included
papers on metaphysics and historical subjects as well.[17] There were twenty-two
candidates for seven fellowships, favorable odds which would not be matched
when a dearth of vacancies became the norm in the next few years. The exercise
itself was a grown-up version of the Trinity scholarship examinations taken by
undergraduates, and each day for four days, from 9:00 until 1:00 and from 3:00
to "¼ p. 6," the candidates would be writing in the same hard and determined
way. This would not trouble the reading men, since as of the day they first set
foot in Trinity they were on an examination treadmill. For the fellowship they
would be translating Greek prose and poetry into English, from English into
Greek iambics, and translating passages to and from Latin. The mathematics
papers were so simple as to be tricky, but if the attention did not wander, they
would be straightforward (for example, to find the moon's variation, or a question requiring the candidate to merely trace out the form of an algebraic
curve). The more discursive papers demanded the history of Greece and Rome.
The philosophy paper set by Whewell involved questions on Plato, Aristotle,
Bacon, Locke, and Butler. In this, Cayley was called on to answer a selection of
a type rooted in Trinity's classics culture:

We pass by induction from special facts to more general truths: do we thus obtain *universal* truths? Of those who answer affirmatively, give the different explanations of the grounds of this universality.

Give an account of Plato's two dialogues, the "Euthyphro" and the "Theætetus." Who are the interlocutors in each; what is the subject; what is the course of the argument, and what the conclusion arrived at? What are the classes into which Plato's dialogues have been divided, and to what classes do these two belong?

One seems to have been tipped by Whewell:

Compare the "Polity" and "Laws" of Plato with the "Ethics" and "Politics" of Aristotle, as their object and doctrines. Is Cicero's Treatise "De Republicâ" an imitation of any of these works? What is the object and plan of Cicero's "De Legibus"?

Based on the answer scripts, the Seniority of the College voted on those they wanted to join them. Scoring well in the examinations though was not an absolute guarantee of success and voting depended on subject lines as well, though the Senior Wrangler would have a distinct advantage in succeeding. Obtaining a fellowship could postpone a career decision, though the intention of the college was to elect men who would fulfill its teaching aims and to provide priests for the parishes it administered.[18] The immediate advantage of a fellowship to Cayley would be a regular income, a life without oppressive obligation, and the fact that he would be living at the center of mathematics in the country.

Cayley's admittance to a Trinity fellowship was announced on 1 October 1842. He was twenty-one years of age and elected at the first sitting, itself a rare event. Not only this, but he was youngest man to be admitted at Trinity during the nineteenth century.[19] He was among six others, all two or three years older than himself who had graduated in previous years.[20] Crossing the divide between undergraduate and graduate was significant, but to be elected a Junior Fellow was a conspicuous step on the academic ladder. Though he gained membership to the Trinity academic hierarchy, he was still a junior (or minor) fellow, a man who dined with the scholars and could not, as of right, dine at high table. A new pathway of subtle distinctions opened up—of minor fellows, major fellows, and the ruling Seniority headed by the Head of the community.

At the apex of the college governance, the master set the tone of the college and, newly elected to this permanent position, Whewell quickly made his mark. His position was further enhanced by his election to the vice-chancellorship of the university in 1842. With his authoritarian manner, junior fellows tended to be treated as schoolboys and he dictated petty restrictions, such as denying them a private key to the college and forbidding them to smoke in the Combination Room. Set against this, they enjoyed the freedom to pursue studies on any subject they chose, wherever they chose (with permission of the master), and, most crucially, backed by the financial security of the College Chest.

Initiation

For Cayley, there was a matter of initiation into the wider mathematical community. In May 1842, the editor of the *Cambridge Mathematical Journal* received a submission from Gunthwaite Hall, Penistone, near the town of Barnsley in Yorkshire, on the subject of elliptic functions (which can be regarded as a generalization of the theory associated with the common trigonometric functions). Problems in elliptic functions arose through the classic pendulum problem, the motion of a fixed body around a fixed point and the measurement of arc length of an ellipse and other curves. The subject had been in existence long enough to be regarded as a central concern of mathematicians by the 1840s, and during the nineteenth century it became of colossal proportions. The paper on elliptic functions from Penistone was on a subject on which the author seemed competent. The Rev. Brice Bronwin had been writing papers on mathematics since 1828 and was seasoned in the art. He had a range of mathematical interests but elliptic functions was a specialty. In his submitted paper, he purported to show that some formulae established by the German mathematician Carl Gustav Jacob Jacobi failed in certain cases and it was to be hoped his results would be published in the Cambridge Journal.[21]

On his return from Scotland and thinking about Bronwin's assertions, Cayley penned a sharp "Note" refuting them. Confident, as befitting a twenty-one-year old Senior Wrangler living at the center of mathematics in England and home of the *Cambridge Mathematical Journal*, he noted: "Jacobi's formulae . . . are perfectly correct. His [Bronwin's] own do certainly fail in that case, and the reason is obvious enough."[22] As with his first undergraduate paper, Cayley signed it "C." Bronwin was not entirely happy with the treatment he received at Cambridge, and, in his next article on the same subject, which appeared in the *Philosophical Magazine* in the new year, he declared: "On this Note [bearing the signature C.] I must now make a few observations."[23]

Approaching his late fifties, Bronwin's scientific interests spanned the wide terrain from elliptic functions to the theory of the tides and astronomy. Within the previous five years he had overseen the building of a new church in Denby Dale in Yorkshire when the original had been declared miserable, filthy, and in a ruinous state by the visiting church hierarchy. Yet, in his pastoral role he was unfulfilled "not having care of souls"—he was a curate—without a church living of his own. Perhaps he was aware of Cayley's rising reputation and was attracted by the possibility of putting the Senior Wrangler firmly in his place. Both men were commenting on Jacobi's *Fundamenta Nova* and their disagreement centered on the evaluation of a constant in a transformation of elliptic functions. In his new paper, Bronwin advanced his argument and concluded: "I feel compelled to say that this Note [of Cayley's] is perfectly absurd at every step of it; and if the author had proved what he aimed at, and which is really true, it would have been nothing to the purpose."[24] Cayley replied in the next number of the *Philosophical Magazine:* "notwithstanding the utter contempt Mr Bronwin expresses for it, I must confess myself to be the author." Explaining the contents of his note again, Cayley added: "This is in substance the note in question, only I was guilty of an oversight not affecting the argument, which Mr. Bronwin has very correctly pointed out."[25] The error was all a consequence of the rush-rush of his early productions, but Cayley was young, impetuous, and wanted to make his mark.

The matter did not end there and the debate between Cayley and Bronwin rumbled on for another two years.[26] In this little excursion in the world of English mathematics, Cayley could not expect special treatment even if he was a comparatively young man. Whereas he received plaudits for his undergraduate paper at Cambridge, he had now moved into the adult mathematical world and out into the country, to the Manse and Rectory where resided some accomplished practitioners. He found that mathematical work could be vulnerable to criticism and that it could not always be guaranteed gentlemanly treatment. Cayley did have the advantage of being attached to a "network of expertise." He was a favorite son at Cambridge and in contact with other mathematicians through the *Cambridge Mathematical Journal* and elsewhere. He also had the activity at the "Cambridge Philosophical" on his doorstep.

Bronwin was the outsider who was to write to a colleague: "I live entirely out of the Mathematical world, and know but little of what is going forward in it."[27] Moreover his work was treated with skepticism at Cambridge and barely tolerated by the editors of the Cambridge journal. "Was not Bronwin expressly invented for the purpose of vexing the editor of the mathematical journal?" wrote Robert Leslie Ellis, then acting as the standby editor.[28] There is nothing to suggest Cayley treated anyone unfairly, and certainly not Bronwin whose expertise he came to value. In the transformation of elliptic functions, Cayley expressed his thanks to Bronwin for detecting an omission in his own work that required further explanation.[29]

Although little more than a storm in a teacup, the episode with Bronwin is indicative of Cayley's *modus operandi*. He was up-to-date with the latest results, contributed his own findings immediately, and did not wait to polish his work. The prospect of being in error did not seem to worry him unduly—after all, a mistake could easily be rectified in a subsequent corrective "Note." He was constantly looking for the next question to settle. Not dwelling on the existing state of knowledge, he prepared the way for the next advance. Though he maintained an interest in the historical threads of his mathematical ideas, it was always from the practical view of wanting to build on the past.[30]

First Offerings

Installed as a Trinity fellow with a reputation to enhance, Cayley was about to embark on his life's work. He had before him the world of mathematics. A mathematician of today might expect a narrow specialism to be selected but the choice he made was to not restrict himself. In the 1840s, the age of specialization had to wait another forty years to arrive. Hard choices were unnecessary, Cayley was under no restraint, and if "blue-skies research" means anything, this was it. He followed a range of leads thrown up by previous work and he exercised free rein. For the first few years, for instance, his work included elliptic functions, curves and surfaces, the theory of integration, and the study of determinants. Compared to later generations, he was a generalist but a specialist when compared to his immediate forefathers. At least he confined himself to mathematics. The previous generation, of which Whewell is the most extreme example of a "generalist," were men who did not limit their palette in any way. Widely known for his "omniscience," Whewell could point to thought-out views on architecture, science, philosophy, and mathematics.

Analytical geometry was a focus for Cayley's initial work, and in the second half of 1842, he worked on the theory of plane curves. In this area, Julius Plücker wrote works of great power and thoroughness. His two volume *Analytisch-Geometrische Entwicklungen* (1829, 1831) set the scheme for the major statement he made in two works on the theory of curves, *System der analytischen Geometrie* (1835) and the *Theorie der algebraischen Kurven* (1839), in which he describes in detail the case of curves of the third order. A work we know did influence Cayley directly was *Aperçu historique sur l'origine et le développement des méthodes en géométrie* by Michel Chasles. Published in 1837, this work established Chasles's reputation as a pure geometer and historian of mathematics. Though Cayley was primarily an analytical geometer—one who studies geometry using algebra—he was also influenced by pure geometers and he read the *Aperçu*. In particular, he noted the theorem on cubic curves: "If a curve of the third order pass through eight of the points of intersection of two curves of the third order, it passes through the ninth point of intersection."[31]

This was a delight to the 21-year-old and it became one of his favorite results, its proof, he thought, one of "extreme simplicity."[32] A fascination was the way it could be used to prove the apparently unconnected "Pascal's theorem" on hexagons. Prior to Chasles, a more encapsulating theorem involving curves of order n had been given by Plücker and Jacobi.[33]

In his paper, "On the Intersection of Curves," published in the *Cambridge Mathematical Journal* in February 1843, Cayley proposed an extension[34]: "A curve of order n, passing through all but $\frac{1}{2}(l + m - n - 1)(l + m - n - 2)$ of points common to two curves of order l, m will necessarily pass through the remainder."[35] Here is a "theorem" which is not true in all cases but contains enough truth to make it interesting. In those cases where there is *one* remaining point to be passed through (as for the case of cubic curves given by Chasles), the theorem is true, but there are many cases where the theorem fails.[36]

In late 1842, Cayley worked "On the Motion of Rotation of a Solid Body," an expository paper based on Olinde Rodrigues's formulae for a rotation in space.[37] The material given was not important, the problem having been effectively solved by Rodrigues, but in the 1840s Cayley described the rotation in several ways and he generalized the whole problem to advantage. "Rotations in space" would link to the new algebra of quaternions when discovered by W. R. Hamilton in the following year and to skew determinants. At the Monday meeting of the Cambridge Philosophical Society on 20 February 1843, Cayley read his first paper to the society "On the Theory of Determinants." George Peacock was in the chair and Matthew O'Brien read a paper on the absorption of light by transparent media.[38]

The subject chosen reveals the essential part determinants played in Cayley's mathematical thought. They were connected with so much of his thinking: rotations in space, the theory of linear equations, analytical geometry, the theory of numbers and, as he later acknowledged, in every part of mathematics. Determinants were intimately linked with the theory of elimination, the theory that was to lead him into his great development of "invariant theory." The reading of the "Theory of Determinants" marks out the wide range of reading Cayley had surveyed between this and the publication of his "juvenile" undergraduate paper "On a Theorem in the Geometry of Position" in May 1841. In the earlier paper, there is little evidence of influences, but one and one-half years later, the range is impressive. For one so young, his authority and mastery is astonishing. He was familiar with the eighteenth-century work of Étienne Bézout, Gabriel Cramer, Laplace, Lagrange, and Alexandre Théophile Vandermonde. He knew the early nineteenth-century work of Jacques Binet and Cauchy, he referred to Johann Carl Friedrich Gauss's *Disquisitiones Arithmeticae* published in 1801, and he cited the contemporary work of Victor Lebesgue and Jacobi.

In the "Theory of Determinants," Cayley developed two distinct subjects. The first part was devoted to the subject of bilinear forms, those forms which involve two sets of variables. The second offered a generalization of ordinary determinants—by their nature linked with two-dimensional arrays—to a notion of n-dimensional determinants. Cayley showed these n-dimensional determinants obeyed a product rule as did ordinary determinants. In the next year, he was to remark to George Boole, a young mathematician living in Lincoln, about these curious extended determinants: "I attempted some time ago in the *Cambridge Philosophical Transactions*, to investigate the properties of some such functions [n-dimensional determinants], formed by a permulatory rule ... thinking they might be applicable to the general theory of elimination, but they do not seem to possess much importance."[39] Contrary to this expectation, he found them useful in invariant theory. First found, these new "hyperdeterminants" were an isolated curiosity, but he was later able to locate them as an essential element in invariant theory. It was a case of the prepared mind being able to see the relevance of what he had already studied. One of Cayley's mathematical trademarks was passing from a theory involved with one set of variables to a theory in which *multisets* of variables occur; from a linear theory to a multilinear theory.[40]

Cayley's regard for determinants as one of the most important areas of mathematics is borne out by a view he expressed to Felix Klein in later life: had he to give fifteen lectures on the whole of mathematics he would devote one of them to determinants.[41] This suggests their value to his way of thinking and the importance in which they were held during his lifetime, besides the range and breadth of mathematical knowledge he was to acquire. To give *one* whole lecture to determinants is indeed indicative of their central place in English mathematics. Strangely, the "Theory of Determinants" was the only paper he published in the Cambridge Philosophical Society journal before becoming professor at Cambridge twenty years later, perhaps because this society at Cambridge was rather generalist whereas the nature of his work required specialist mathematical journals.

Cayley's initial success in mathematics was to link algebra with geometry. This link exemplified his forte, for he was fascinated by the connection between the two disciplines. Consequently, he became a herald of analytical geometry. Early work in analytical geometry is intimately linked to the geometry of position, a subject which exercised a particular hold on English and Irish mathematicians. The "five points in space" problem had been the first example, but another was the famous theorem known as Pascal's Hexagon Theorem: if six points A, B, C, C', B', A' are placed on a conic, such as an ellipse, certain constructed points can be shown to be always collinear. Cayley's proof in his "Demonstration of Pascal's Theorem," published in November 1843, made use of determinants.[42] With delight he had referred to the proof of this theorem using the properties of intersecting curves in the previous volume of the *Cambridge Mathematical Journal*, and now he had an alternative proof based on determinants. Further refinements to the "Pascal configuration" reveal additional relations explored by Cayley in the 1840s. In a series of papers on the geometry of position, Cayley showed acquaintance with the contents of *Gergonne's Journal*, in which "arithmetization of geometry" had gathered momentum in the late 1820s.

Cayley's early abilities were so developed that he learned best by reading the works of the established "mathematical giants" of the past. As we have noticed, he had been reading Lagrange's mathematical papers and the *Mécanique Analytique* as an undergraduate but it was C. G. J. Jacobi of his own time who exerted the greatest attraction. Jacobi wrote on subjects which appealed to Cayley: he introduced functional determinants ("Jacobians") in 1829, and in 1841 wrote a long memoir on determinants. With Jacobi, who was then thirty-nine, Cayley shared a liking for calculation and compiling mathematical tables. Like his "illustrious" Jacobi, his interests touched on all branches of pure mathematics, applied mathematics, mechanics, and astronomy.

"Coming before the World"

Cayley had made an impressive start, but we should notice that he was not yet fêted as England's premier mathematician. He was a young man of twenty-two years with talent. But other Senior Wranglers and Tripos successes could measure up using this yardstick, and yet the achievement of being included in the wrangler fold after the Tripos would be the high-water mark of their mathematical accomplishment. Clearly not all sought a life with mathematics as a daily companion, but the question remains: why was Cayley so successful? One explanation is his absolute dedication, but he was also living in the "right place at the right time." William Wallace referred to Cambridge as the "Holy City of Mathematics" and though ironically meant, it was the main focus for mathematical activity in the country.[43]

As a young graduate, Cayley was in the company of men just ahead of him who were determined to develop mathematics and science. An important figure in this respect was Duncan Farquharson Gregory, great-great-grandson of the mathematician James Gregory. Born in Edinburgh, Gregory entered Trinity College in 1833, where he graduated fifth wrangler in 1837 and became a fellow in 1840. Gregory was a serious scientist not confined to mathematics. Initially he acted as an assistant to the professor of chemistry, James Cumming. He was one of the founders of the Chemical Society in Cambridge, and he put his chemistry to use by making photographic paper, the technology of the moment. His knowledge of chemistry was another link with Cayley, who had also excelled in the subject at school. His pursuits included botany, and he was widely acknowledged as the best mathematician at Trinity College despite his not being higher than fifth position in the Tripos.[44] When William Thomson went to the university he reckoned him as "undoubtedly the best and most original math[n] in Cam[bridge]."[45] At the end of 1836 Archibald Smith wrote to Gregory proposing that a periodical be founded to publish short papers in mathematics. He readily agreed and the *Cambridge Mathematical Journal* appeared in November 1837.[46] The formation of this journal acted as a catalyst and while many of its papers have none of the grandeur of lasting memoirs, its appearance enabled the young tyros to gain experience in the art of publishing work. In particular it gave Cayley his "first opportunity of coming before the world."[47]

In Cayley's mathematical Tripos of 1842, Gregory was one of the moderators, and for a six-month period as of October 1842, both were together as tutors at

Trinity. As a tutor, Gregory showed an interest in those with an aptitude for science and mathematics. He was quite informal in his dealings with them and was popular among students and dons alike. William Thomson, then a second-year student, wrote in his diary entry in March 1843, of dropping into Gregory's rooms and finding him reading *Piers Plowman* and spending time discussing old words with him. When the conversation turned to electricity they enjoyed a light discussion "in which Faraday and [John Frederic] Daniell got (abused)[n]." Another brief entry noted that he "wined with A. Cayley" in his rooms.[48]

Just when Gregory's reputation as a bright young star was being transformed to one of academic weight, his time at Cambridge was drawing to a close. He had been ill for three months and left Cambridge in the spring of 1843; one year later, he died in Scotland aged just thirty years.[49] In mathematics he was perhaps best known for his work on the calculus of differential operations and the foundations of mathematics. His paper "On the Real Nature of Symbolical Algebra," in which he drew on Peacock's outlook, goes further than the Lowndean professor. It gives a view of symbolic algebra, which, though not formally acknowledged by Cayley, had a profound influence on his thinking. Gregory declared it the subject in which symbols were not defined in themselves but "by the laws of combination to which they are subject."[50] In symbolical algebra, the key step identified by Gregory was the "separation of symbols." Typically, symbols were operators θ which operated on a quantity x to give $\theta(x)$. In Gregory's paper, an essential element was "in following out the principle of the separation of symbols [θ] of operation from those of quantity [x]."[51] To a mathematician today this seems unremarkable, but in the 1840s, the notion of a function had not reached its present state of refinement. Writing to his mathematician father, William Thomson reported on Gregory's work in finding the values of definite integrals by "a very curious way by the separation of symbols."[52] Throughout Cayley's career, the phrase reoccurs, and, as we shall soon see, he immediately found that algebraic invariants "may easily be expressed explicitly, by means of the known method of the separation of symbols."[53] The symbolic algebraists, which included both Gregory and Cayley, saw that algebra as a "science of symbols" reached into other compartments of mathematics.[54]

Gregory's influence was felt by Cayley not only in algebra, but also in analytic geometry. In 1842 Gregory had begun writing a textbook on geometry but never lived to see it published. When it appeared as a collaborative effort, edited by another of the young men, William Walton, Cayley remarked on the brilliant section on differential geometry written by Thomson and two friends Hugh Blackburn and Frederick Fischer. In its writing, Cayley's advice was also sought as it dealt with his own specialty, the application of algebraic equations and expressions to the geometry of three dimensions.[55] An objective of the book was to classify quadric surfaces using the Cartesian equations for the surfaces, but wherever possible to strive for "symmetry" of expression in the equations.[56]

Cayley too strove for symmetry of expression in his early papers, but this took a different form. His symmetry was that obtained by the use of homogeneous coordinates in preference to ordinary Cartesian coordinates. It was an idea advanced by Julius Plücker, whereby a conic drawn in the plane, for example, was written: $ax^2 + by^2 + cz^2 + 2hxy + 2gxz + 2fyz = 0$. Cayley appeared to be the only mathematician to use homogeneous coordinates in the early volumes of the *Cambridge Mathematical Journal.*

Another of the "young Turks" promoting the cause of scientific research at Cambridge was Robert Leslie Ellis. Entering Trinity College four years earlier than Cayley, he graduated as the Senior Wrangler of 1840 despite interrupted years due to illness. Like Cayley, he was tutored by Peacock and coached by Hopkins and was regarded as a "natural" for the highest Tripos position despite his contempt for the competition. When Gregory became ill, Ellis took over the editorship of the *Cambridge Mathematical Journal,* though it was not a role he particularly enjoyed. He kept up a steady stream of erudite work until chronic illness overtook him some years later. He was a perfectionist in all he did and a man of wide literary and scientific interests. An aesthete by inclination, his intellectual air no doubt contributed to his local reputation of being rather a cold fish and difficult to know.[57] Judging from letters to friends this judgment appears harsh; his communications reveal a warm individual with a fine turn of wit. In his mathematical and scientific writings are papers on Roman aqueducts, the formation of a Chinese dictionary, the structure of bee's cells, and the theory of vegetable spirals. Ellis was deeply religious and shared his views freely with his friend—in one letter to Cayley, he ended: "[I have] written four times as much as I intended: just as I used to stay four times as long in your rooms as I meant to do."[58]

In their mathematical style, Cayley and Ellis were quite different. Ellis is more identifiable as a widely read scholar than a rugged mathematical researcher in the Cayley mold. His extensive report "On the Recent Progress of Analysis" for the British Association meeting of 1846, written in an easy-flowing style, concentrates on exposition. It gives a panoramic picture of the state of the art of elliptic functions in the 1840s, alluding both to Abel's Theorem and to Cayley's work. He was grateful to Cayley "to whose kindness I have been, while engaged on the present report, greatly indebted."[59] During these two years, elliptic functions and integrals formed a central interest for Cayley. Despite Bronwin's intervention, Cayley's work on this subject was regarded as one of the most important English contributions to the theory of doubly periodic products and their connection with elliptic functions.[60] Cayley cut his teeth on "elliptic function theory" and it would be a subject only second to his first love: invariant theory.

For young researchers in the 1840s Cambridge was an invigorating place to be; a spirit of exploration among the resident group of young academics was in the air. While Cayley was capable of leading a solitary existence, he was by no means a recluse. He enjoyed the fellowship of college life and made friends among fellows and undergraduates alike. In mathematics, the group was supported by the all-important *Cambridge Mathematical Journal* as distinct from the *Transactions of the Cambridge Philosophical Society*, a more senior journal catering mainly to the physical sciences. For mathematicians, a journal represents the lifeblood of their subject and the *Cambridge Mathematical Journal* was a vibrant success in its early years, both financially and mathematically. It shaped and defined a community of active mathematicians where before its emergence there had been a less participative society. It provided an outlet for many previously unpublished authors and offered a forum for many of Cayley's early papers. Augustus De Morgan, professor of mathematics at University College, London, praised it as "full of very original communications" while adding with some understanding, "[i]t is, as is natural in the doings of young mathematicians, very full of symbols."[61]

The *Cambridge Mathematical Journal* was "full of symbols" for good reason. In Britain, the 1840s was a fruitful period in algebra, and this generated vigorous expression in the journals. Symbolical algebra, to use the nineteenth-century term, was introduced and formalized by George Peacock. In distinction to arithmetical algebra (ordinary algebra), in which symbols stood for numerical quantity, in symbolical algebra, the symbols were assumed to be general and unspecific. The advantage of this "science" was that symbolical algebra could be applicable in diverse areas of mathematics—from differential equations, geometrical problems, to the emergent "new algebras." George Boole applied it to logic, in which symbols represented logical statements. The broader conception of symbolical algebra was a recognized field in the 1840s, and its adherents were known in the mathematical community as the "symbolic algebraists."

Symbolical algebra had to take account of the "calculus of operations," an algebraic approach to the calculus. This had its roots in France at the beginning of the century, championed by among others, L. F. A. Arbogast, the author of *Calcul des dérivations* (1800). It was avidly taken up by the Cambridge Analytical Society when it was formed in 1813 and the calculus of operations became the centerpiece of their heirloom to the succeeding generation.[62] Lagrange's theorem $\Delta^n u_x = (e^{hD} - 1)^n u_x$ that links the difference operator Δ with the differential operator D (the case $n = 1$ it is just a restatement of Taylor's theorem) typified the "limitless" approach to the calculus that gained credence in Britain in the first half of the century. Cayley would have touched on this theorem at school, where Lagrange's theorem was presented in texts written by Thomas Hall, in charge of mathematics at King's College, London. In one of his own first papers, published in May 1843, Cayley wrote "On Lagrange's Theorem," in which he gave various alternative expressions and extensions of the result.[63]

In "coming before the world." Cayley's debt to his tutor George Peacock is apparent. Symbolical algebra apart, Peacock's erudition and sure grasp of the antecedents of mathematics displayed in his very extensive "Report on the Recent Progress and Present State of certain Branches of Analysis" (1833) pinpointed some of the areas that attracted Cayley. From elliptic functions to the section on the resolution of algebraic equations, Peacock's graceful account took note of the work of Ehrenfried Walter von Tschirnhausen, Gian Malfatti, Paolo Ruffini, Gauss, Lagrange, and he summarized Abel's and Cauchy's contributions. In particular, Peacock greatly admired Lagrange for his analysis on possible approaches to the solution of the quartic and higher order equations.[64]

In an attempt to establish foundations for algebra, it was Peacock who divided algebra into arithmetical algebra and symbolical algebra, basing the permitted procedures in the latter with those commonly accepted in the former. In crossing over from arithmetical algebra to the speculative symbolical algebra he appealed to a "principle of equivalent forms." This principle is largely forgotten but it was designed to legitimize statements in symbolical algebra. While Cayley's vision of algebra was wider than Peacock's, taking into account the "new algebras," "matrices," and "group theory" (but yet to be articulated), he believed there were instances when Peacock's principle could be employed rightfully as a means of justification.[65]

If Cayley had relied on arithmetical algebra as a suggestive mechanism for symbolical algebra, as had Peacock, his present-day reputation would probably be on a par with that of his tutor. But he saw no need for accepting this limitation. His view differed from Peacock's in other ways. While he was schooled in Peacock's formalism, he was far from being a true formalist, namely, one who deals only with the rules of algebra and the formal manipulations according to these rules. Neither was he ultimately a foundationalist, meaning one who is content with foundations alone. When philosophical squabbles loomed and mathematics was in danger of being pushed to the periphery, Cayley lost interest in the minutiae of metaphysics.

While Cayley was an audacious mathematician who selected a range of research directions and invariably made a quick contribution, he was not a firebrand of the kind who disregarded what had gone before. In the language of the time, symbolical algebra and modern geometry were progressive subjects, whereas the *Elements* represented established knowledge. Progressive subjects were in a state of flux and invited debate and courted controversy. There was also a conservative side to Cayley's character. His being in the vanguard of mathematical "expansion" in symbolical algebra and modern geometry can be contrasted with an attachment to Euclid. In its maturity, Euclid's *Elements* represented a settled state of affairs, and, for the Victorian mathematicians it was the revered "ancient geometry." Cayley believed in the magnificence of the *Elements* and would agree with his tutor that it was the "only one elementary work which is entitled to be considered as having made a very near approach to perfection."[66] As we have seen, it was *Book V,* containing the theory of proportion of Eudoxus that most excited Cayley's admiration. Seeing him as an extreme respecter of the "sacred book" might leave the impression that he most naturally sided with the status quo, but, there was also Cayley the whiz-kid. Without qualm he accepted "new algebras" with rules quite unlike those of ordinary algebra, and he would come to see that Euclid's *Elements* was only a part of a wider vista of geometry.

A New Age

While Cambridge was gradually reorienting itself to a technology-driven age, and Elizabethan vestiges were falling away, movement was slow and not always obvious: in time-honored fashion, traditions were clung to until wrenched away. Whewell assumed the mastership of Trinity College in 1841 and in the new regime, the college underwent the changes institutions do when the old order retires. His style of leadership was different from that of the previous incumbent, Christopher Wordsworth, a high Tory "who lived like a hermit" among the college fellows, who were generally liberal in their politics.[67] Whatever his petty dictatorial customs, Whewell invested energy in reforming the College Statutes, which had remained in place since the sixteenth century. He also had to deal with mathematics. The 1840s represented a high water mark for the position of mathematics in Cambridge studies. All students, "poll" and honors students alike, had to study the subject. The dominance of mathematics caused its role and position in the curriculum to be questioned. A proposal for setting up of a Board of Mathematical Studies to oversee the curriculum of the mathematical Tripos was made in 1843, but it appeared there was no urgency and five years were to elapse for the first meeting to take place.[68]

In the early 1840s, the ambitious Cambridge undergraduates had been encouraged to study the latest mathematical work in treatises and journals. When Galton went to the university, his coach had him reading recent French works in his first term and this regimen was not uncommon for the Cambridge "reading man." This caused Whewell consternation, for he believed undergraduate education should be circumscribed by the "great works" of such established men as Newton, Euler, Lagrange, Laplace, Monge—works blessed as being classics. It was his view that students were not to spend time "hunting out the last [latest] novelty."[69] In taking this more limited approach, Whewell was arguing against those who wanted undergraduates acquainted with the *whole* of mathematics and perhaps he was aware, more than most, what this objective would entail.

Cayley would not have been constrained by the limitations favored by Whewell; he took the "unlimited" approach to his study. Encouraged by his coach William Hopkins, and the spirit fostered by the *Cambridge Mathematical Journal*, Cayley rose above a system of education geared toward success in the final examination of the Tripos. It tended to produce absorbers of information and not students with the skills that would be needed forty years later when the university became more research minded. Cayley was of a different cast of mind, one is tempted to say "ahead of his time." The scholarly fine-tuning of received information was not for him; rather, he was a seeker after "the last novelty" and the impetus to derive new results. He hardly could have been accused of carrying out research with "one foot on the fender."

How different was the 22-year-old from Mr. Potts, another member of the Cambridge mathematical community and a new neighbor on the floor above in Nevile's Court, the inner part of the college which connects the Wren Library to Trinity's Great Court.[70] Robert Potts was almost forty years of age, having graduated the twenty-sixth wrangler in 1832, a meritorious position but not high enough to make him an obvious choice for a career as a don. He remained in Cambridge without a fellowship, a proactive campaigner for university reform while eking out a living as a private coach and author of religious works and student guides. With Cayley setting about establishing his research career in the floor below, Potts was preparing a large edition of *Euclid*, a work which made his name remembered by generations of Victorian schoolchildren. It achieved a large circulation in Britain and also in America, where recognition brought him an honorary doctorate. When Charles Dodgson, a mathematical don at Oxford, (otherwise known as the author of *Alice in Wonderland*) contemplated teaching geometry to his sister, he reckoned it to be "the only edition worth getting."[71] Potts was doing valuable work but decidedly not the sort of contribution Cayley wanted to make.

In his rooms in Nevile's Court, Cayley was part of the research community, but in another sense, he was alone. Of the group of young scientists who emerged from the mathematical Tripos in the early 1840s—Ellis, Stokes, Cayley, Adams, Thomson—only Cayley remained within the boundaries of pure mathematics. He was addicted to abstruse mathematics in a way the others were not. Thus a degree of insularity was automatically enforced, and, perhaps it was compounded by his own solitary character. In these circumstances, he developed an independence of mind, yet beneath the accentuated modesty he had a surely based self-confidence bolstered by early success; an authority was in the making. On the surface, his was a life with few pressures and constraints. There was no necessity to carry out research save the inner drive, though this was overwhelming in his case. With no formal lecturing duties required of him, Cayley could fully indulge his intellectual interests.

His first papers already published in the *Cambridge Mathematical Journal*, as a next step Cayley turned to the Continent to communicate his results. His early contributions to the Cambridge journal would reach an audience, but it was one not fully attuned to the sort of mathematics he was doing. Moreover, he now had no mentor at Cambridge who could steer him at the next stage of his career. Hopkins had completed his duty at the Tripos level while the mathematical seniors at Trinity College were men who had already completed the best of their mathematical production and were at the stage of their lives in which many turned to a position in the Church or to College administration. Whewell and Peacock were such men and modern Continental ideas in mathematics did not come into their reckoning. Early on, the Analytical Society with Peacock, Herschel, and Babbage in their youth, had been instrumental in receiving Continental mathematics but had not thought of transmitting it the other way—for the good reason that they had a belligerent Napoleonic France within recent memory. It was Charles Babbage's judgment that England was cut off from science in the rest of Europe and he painted a rather dismal picture of England's position, as he saw it in the 1820s: "[it] is vain to conceal the melancholy truth. We are fast dropping behind. In mathematics we have long since drawn the rein, and given over a hopeless race."[72]

The young Cayley stood at the threshold of a different age. The threat of continued war with France had receded, and Britain was adapting to a world defined by new technology. Soon the railway, which came to Cambridge in 1845, would open the way to express travel so that the fifty miles to London could be traveled by train in two hours. In the days of the early 1840s, before the railways came to the university town, the journey to London was long and arduous. A traveler to London from Cambridge could board the *Rocket,* which traveled the Epping Road, and take five and one-half hours before arriving at the George and Blue Boar Inn, Holborn, near to the City of London.

Cayley was one of the new men when placed alongside men such as De Morgan, Peacock, and Whewell. They were more inclined to remember journeys to and from Cambridge by stagecoach. They belonged to a less mobile era—to the turnpike days rather than the modern railway era that was just beginning.[73] By the end of 1843, when "Railway Mania" was approaching its height, there was a glimpse of the future: "Towards what new continent are we wending?" asked William Makepeace Thackeray of the transition from the age of the horse-drawn stagecoach to the age of steam-driven locomotives.[74] Cayley, for his part, was a young mathematician who saw new investigations and new designs.

Continental Rambles

With few commitments to engage him, there was time for a Continental tour. This was looked forward to with keen anticipation, for on his own admission Cayley was a "good walker and fond of mountaineering."[75] Journeying across Europe was easier, whereas ten years earlier, travel had been both slow and expensive. This was the year Thomas Cook started his railway tours and the era of the tourist beckoned. Cayley made his preparation by reading *Specimens of Ancient Sculpture* [Dilettanti Society], G. L. Taylor and E. Cresy's *Architectural Antiquities of Rome* (1821–22) and T. L. Donaldson's *Pompeii Illustrated with Picturesque Views* (1827) illustrated with engravings by W. B. Cooke. In the second half of 1843, again in the company of his friend Edmund Venables, he embarked on an extensive tour of Switzerland and Italy. At the beginning of their tour in August, he and Venables were climbing in the Swiss Alps.

Climbing the Swiss peaks was a favorite pastime among Oxford and Cambridge men in the 1840s and the Alps were still uncluttered and free from the presence of nonclimbing tourists. Mountaineers congregated at Dr. Lauber's little inn in Zermatt for climbs in the region of the Matterhorn and surrounding peaks. In *The Traveller's Book* at the inn, Cayley recorded: "Macugnaga to Sion by the Mt. Cervin [The Matterhorn] and Bre[ithorn]. The view from the Schwarzsee is splendid."[76] Macugnaga, on a tributary of the Val d'Ossola, was an isolated place but breathtaking. John Ruskin visited there shortly afterward and described the "mountains of all sizes within the stretching of an arm" and "the roar of the cascades without number all around me."[77]

Like many middle-class young Victorians, Cayley shared Ruskin's accentuated attachment to Switzerland and to northern Italy, and it is unfortunate that he was thwarted in carrying out his most ambitious climbing plans. In 1843 he noted in the *Traveller's Book:* "Jean Baptiste Brantschen, who passed over the Weissethorn eight years ago, makes no difficulty whatever about accompanying one to the top of it and returning to Zermatt. I was prevented myself from accomplishing it by the Rifelhorn which is exceedingly well worth visiting."[78] Cayley records what is a matter of fact and it would be unreasonable to expect anything else in a hotel visitor's book. But if we were in possession of his pocketbook, we would probably find the same lines. He had not the prose of Ruskin ("seven thousand feet above me soared the needles of Mount Blanc, splintered and crashed and shivered, . . .") but unlike this man who was content to wander the Alpine meadows, Cayley was a serious climber intent on conquering peaks in the High Alps.[79]

After the late summer in Switzerland, Cayley and Venables continued their tour from September to December 1843. They followed the classical Grand Tour through Italy. Starting in the north of the country, they traveled through the Italian Riviera to Florence and from there they visited Rome and Naples.[80] The author of his favorite book on architecture, George Edmund Street, wrote that "an architectural tour in Italy seems to afford about as much prospect of pleasure and information combined as any which it is possible for an English student to take."[81] As the architect for the Law Courts in the Strand and dozens of churches across the land, Street's Gothic credentials were impeccable and he combined this expertise with a talent for pen-sketches of his observations of architecture in Northern Italy. During his life, Cayley gained a thorough knowledge of Italian architecture and, like many Victorians, treated the amateur study of monuments in a serious way. In painting he gained a love for Ruskin's favorites, the early Italian masters, such as Giovanni Bellini. In Masaccio and Pietro Perugino, Florentine masters pre Raphael, he chose artists who displayed a mastery over perspective and the use of space in their paintings. In his intellectual tastes, Cayley emerges as a well-rounded individual, seeking a cultural existence outside the confines of his own subject. He was no "mere mathematician," of the breed Whewell so sincerely hoped Trinity would avoid.

PART 2

New Vistas, 1844–1849

I found myself, getting back always to my favourite subjects—linear
transformations and analytical geometry.
—Arthur Cayley to George Boole

S ocial and political unrest in the 1840s coincided with a pathbreaking period in British mathematics. The rather limited
"old algebra" was widened to a "symbolic algebra" in which new
avenues opened up with connections to other parts of mathematics, such as differential equations, logic, and *n*-dimensional geometry. In a distinct though connected development, the algebraic
system of quaternions (a kind of "super number" with four components) was discovered by Ireland's premier scientist and mathematician Sir William Rowan Hamilton.

In his early twenties, Cayley contributed to each of these
mathematical "hot spots" simultaneously. He had the audacity to
publish a paper on the quaternions shortly after Hamilton's discovery and at the same time collaborated with George Boole
on the "algebra of forms." This latter endeavor, known later as
invariant theory, mushroomed into a leading subject of the nineteenth century. It became Cayley's dominant interest and provided a theme that united many of his mathematical activities. He
did much to encourage mathematical research through the
medium of the new *Cambridge Mathematical Journal.* When this
was expanded to become the *Cambridge and Dublin Mathematical*

Journal in 1846, he supervised many of the contributions in pure mathematics.

The prospect of a life either as a churchman or a Cambridge don did not appeal and he entered Lincoln's Inn to train for the bar. He traveled widely, to Scandinavia, France, Germany, and Ireland, and was active in an emerging international network of mathematicians. In England, he provided leadership and helped to remove the isolation of British mathematics, a situation which had existed during the eighteenth century.

Based in London as a law pupil, Cayley met the 32-year-old James Joseph Sylvester studying at the Inner Temple, thus beginning one of the most fruitful liaisons in the history of mathematics.

CHAPTER 4

A Mathematical Medley

By the beginning of 1844, Cayley had returned to Cambridge from his six months on a grand tour of Switzerland and Italy. In accordance with wise Cambridge advice to scientists with promise, he took few pupils and focused on research. There was potential theory, the geometry of position, determinants, and the calculus of operations to consider. Two months prior, Sir William Rowan Hamilton had given quaternions to the world, a point of intense interest to a mathematician of Cayley's leanings. This subject was a new branch of algebra and it became a magnet for nineteenth-century mathematicians, particularly those from the British Isles. With all these subjects to occupy Cayley, he found an embarrassment of mathematical riches on his own doorstep. There would also be George Boole's papers on linear transformations to consider, but first he restarted his theory on the intersection of planar curves and sketched out some ideas on extending analytical geometry to *n*-dimensions.

Curves and Surfaces

In spring of 1844, Cayley studied planar curves in a continuation of his paper "On the Intersection of Curves," which had been published the previous year. He came to regard this work as rather specialized and of little significance, but it contained his first acknowledgment of Julius Plücker's "General Theorems on Equations with Several Variables, of any Degree between any Number of Unknowns," published in 1837. Plücker's generality in treating geometry and his intention to apply algebra to it was appealing.[1]

Plücker's work became a benchmark for Cayley and when the German mathematician turned away from mathematics in 1846 to take up experimental work in magnetism, Cayley continued the research program in analytical geometry. Plücker had introduced homogeneous coordinates (a superior way of labeling geometrical objects) and studied the principle of reciprocity (in which points correspond with lines and, in reverse, lines with points), two leading ideas Cayley adopted. His program in *analytical* geometry counterbalanced one in pure geometry, which dwelt on the study of figures without, its disciples suggested, the artificiality of the algebraic devices found in analytical geometry. Cayley was unmoved by such attitudes and, taking Plücker's scheme further, was able to improve it by using determinants to express geometrical relationships. His plan gained even greater coherence in the way he was able to apply the results of his study of what he called "hyperdeterminants" to geometry, but this application was yet to come.

In May, Cayley published "Chapters in the Analytical Geometry of (n) Dimensions," a paper which clearly illustrated the importance he placed on homogeneous coordinates as a tool for dealing with the geometry of n-dimensions. The paper in analytical geometry is in line with Plücker's program. He began, "I take for granted all the ordinary formulae relating to determinants."[2] The paper concentrates on the solution of linear equations using determinants and has little to do with the *philosophical* implications of n-dimensional geometry. Organized in four "chapters," chapters 1 and 2 treat systems of linear equations and arrays, while chapters 3 and 4, describe a reciprocal set of equations, bilinear forms, and show how the results apply to geometry. He referred to no other writer on this subject, though the title of Plücker's 1837 "General Theorems" paper suggests an influence.

The title of Cayley's paper is faithful to its content, but its title misled Sir W. R. Hamilton. When the Irishman was informed of it, he thought it might contain some results he had obtained via quaternions, but there was in fact no tangible connection. Perhaps Cayley had second thoughts about raising the subject of "n-dimensions," for he later suggested a more correct title for the paper in which of "n-dimensions" was avoided. In his view, a better description would have been "A Generalisation of the Analysis which Occurs in Ordinary Analytical Geometry."[3]

At the end of the paper, Cayley specialized his generally stated results to three-dimensional geometry and showed how algebraically stated theorems were equivalent to familiar results found in Euclid's *Elements*. In this way he emphasized his view of "higher" objects in n-dimensions as generalizations of ordinary geometric objects such as straight-lines, planes, and spheres, which reside in three dimensions—or in *ordinary* space, as he would say. The *mathematical space* "of n-dimensions" represented an extension of three dimensions to an ideal realm, a step which entailed no physical reality for values of n greater than three. Cayley developed wide-ranging algebraic theories for n-dimensions (determinants of order n, transformation of coordinates in n-dimensions, n-dimensional determinants), but to make physical sense of geometrical results, he invariably specialized n to the value three to make concrete conclusions about real spheres, ordinary cones, and real planes.[4]

Cayley's tendency of developing results for n variables was not particularly novel. George Green had treated potential theory in terms on n variables in his paper on the attraction of ellipsoids published in 1835.[5] Augustin-Louis Cauchy began a systematic study of n-dimensional geometry in the 1840s and, independently, Hermann Grassmann, a Gymnasium schoolmaster, was working on an algebraic description of n-dimensions. His *Ausdehnungslehre* (1844) was generally unknown, but Cayley knew of Cauchy's work.[6] Prior to these efforts, August Ferdinand Möbius introduced the idea of four dimensions in his *Der Barycentrische Calcul,* which was published in 1827. As we have seen, Cayley himself treated multiple integrals over n-dimensional volumes in papers published in the month following his Tripos examinations. In the context of n-dimensions, he was contributing to the mathematics of the moment by accretion, and he would not have claimed any priority of discovery.

Cayley was aware of the barren metaphysics that could arise insofar as n-dimensions was concerned. The crux of his acceptance of higher dimensional geometry as sound mathematics was the extension from three dimensions to four. On reflection about making this step, perhaps more accurately described as a jump, he remarked: "one may in effect, *without recourse to any metaphysical notion in regard to the possibility of space of four dimensions* [his italics] to reason analytically."[7] This is perhaps in response to remarks by Plücker, who rejected four-dimensional space of points as too metaphysical, and to Möbius, who thought that four dimensions could not be imagined.[8] For Cayley the algebraic expressions mattered most and not philosophical speculations on the nature or existence of n-dimensional entities.[9]

In Cambridge the perceived abstruse character of these researches in n-dimensions and potential theory (and his editor's penchant for the Gothic alphabet) became a rich source of amusement to his friends. Frederick Fuller wrote to William Thomson in Glasgow of "scientific news" in the market town: "Cayley has extended his geometry of n dimensions to $(n + \frac{1}{2})$ dimensions, and Dirichlet's Theorem from (German n) $\mathfrak{n}$ independent variables to (Chinese n) 魔 variables. This latter triumph over the difficulties of notation has obtained the prize of the Chinese Board of Longitude, and his celestial majesty has ordered the Charge d' Affaires Hi-ho-hum to present Cayley with a japaned tea-tray bearing portraits of the principal members of the celestial family."[10] Within the humor, is the belief of his friends that Cayley was in the vanguard of new mathematical discoveries.

Cayley's ambition required that he publish his work further afield and to a more expert audience than the one afforded by the *Cambridge Mathematical Journal.* He was thinking of Continental journals but by June 1844 had not taken the step of publishing in one. There was certainly a need to open new avenues for British mathematics and thus encourage research. The *Philosophical Magazine,* a general scientific journal with roots in the eighteenth century, put a definite limit on the number of papers in pure mathematics it was prepared to accept, its editor fearing a decline in readership if there was too great a supply of "lofty mathematical subjects."[11] This reluctance was a problem for mathematicians and a similar attitude to the publication of abstract mathematics prevailed in journals published by the Royal Society. Cayley needed more space for his rapidly produced thoughts. There was little precedence for publishing abroad. In the 1840s English mathematicians rarely published in the *Journal für die reine und angewandte Mathematik* founded by A. L. Crelle (1826) or in Liouville's *Journal de Mathématiques Pures et Appliquées* (1836), the two most important journals in mathematics. These two, nicknamed *Crelle's Journal* and *Liouville's Journal*, were to become the two most important Continental journals for his published work, but the time was not yet ripe.

There were ambitions to extend the coverage of the *Cambridge Mathematical Journal* and put it on a par with the prestigious Continental productions. William Thomson, a third-year undergraduate and in-waiting to succeed Robert Leslie Ellis as editor, proposed that the scope of the journal be expanded. He wrote to his father: "If the plan be carried out at all, the great object of course would be to make the journal as general as possible in this country, and to get it made known on the Continent. I have been speaking to Cayley since, and he quite enters into the plan. One great assistance he thinks would be, that there is at present no journal of the kind in this country, and that the want is very much felt by math[ematica]l men."[12]

Thomson's plans were realized with the *Cambridge and Dublin Mathematical Journal*, which first appeared in 1846. Before this happened, Cayley took his own initiative and sent his paper on cubic curves to *Liouville's Journal*.[13] Published in August 1844, Cayley had rediscovered many results originally attributed to Colin Maclaurin and also a description of the surface described by the algebraic equation:[14]

$$\frac{\alpha}{x} + \frac{\beta}{y} + \frac{\gamma}{z} + \frac{\delta}{w} = 0.$$

The surfaces studied by Cayley were invariably determined by equations. If the degree of the equations was two, the surfaces were quadric surfaces (such as cones, ellipsoids); of degree three, cubic surfaces (as the one above, which may be rewritten as a equation of degree three); degree four, quartic surfaces; and degree five, quintic surfaces.

An "Addition" to Cayley's paper contained a description of the curve that eventually became known as the Cayleyan, a planar curve connected with the theory of reciprocity (the Cayleyan curve reciprocal to the classic "folium of Descartes" is a rectangular hyperbola plus an isolated point). At the point of discovery, Cayley called the reciprocal curve the "Pippian," but seeing its importance at a later date, Luigi Cremona christened it the "Cayleyan." The theory of reciprocity was yet another branch of geometry in which Cayley demonstrated his virtuosity.

Cayley's publications to date, encompassing the analysis of curves and surfaces, the theory of integration, and symbolic algebra, are indicative of his lifetime's performance, one in which he hopped from subject to subject.

Research on cubic surfaces and the Cayleyan curve linked algebra to geometry, and this was a "Cayley strength," but, as yet, his research lacked a focus—a subject to absorb him and at the same time be worthy of his talent.

With all his extensive reading of Continental journals, he found a subject near to home in George Boole's work on linear transformations, which was published in the *Cambridge Mathematical Journal* in 1841. Boole, a self-taught mathematician living in Lincoln, had made contact with the Cambridge mathematicians, and, encouraged by D. F. Gregory, contributed to their journal. The idea considered by Boole that captivated Cayley's imagination was the peculiar property of invariance, whereby some formulae are preserved when a "parent" algebraic form is itself transformed.

George Boole in Lincoln

Boole was a decisive influence that elicited young Cayley's talent. At twenty-six, Boole earned his living as a schoolmaster in charge of his own school, a small educational establishment lying within the city walls of Lincoln in the shadow of its imposing cathedral. The city had been important in the history of Britain but in the nineteenth century; it was rather an isolated place on the broad Lincolnshire plain and an uncomfortable reminder to Boole of his own mathematical isolation.

Although Boole's parents lacked money, they maintained a strong appreciation for learning and passed this on to their son. He showed exceptional ability and established a local reputation for educational accomplishment, but he was an outsider in the world of science dominated by the progeny of the ancient universities. He felt his lack of formal training keenly. Fortunately, he had a supporter in Sir Edward Thomas ffrench Bromhead, Baronet of Thurlby Hall near Lincoln, a man who took his duties of patronage seriously. Bromhead was a Cambridge graduate who was on intimate terms with Charles Babbage and John Herschel and mixed in scientific circles through his membership of the Royal Society. His own education in mathematics made a deep impression on him and Boole benefited from his patronage. Bromhead loaned him books from his collection and took an active interest in his mathematical writings.

At one stage, Boole was himself attracted to the idea of reading for a mathematics degree at Cambridge but the high financial demand and the duty he felt toward his dependent parents probably caused him to abandon the idea. Happily, Boole was able to learn independently. By the time Gregory recognized Boole's talent, he had already displayed a thorough knowledge of the classic mathematical works of Laplace and Lagrange as well as recent Continental work. He was passionate about mathematics and described his intellectual experiences to a boyhood friend: "The processes of the higher analysis are also in themselves essentially beautiful and the study of those who pursue it sufficiently far assumes a very high and increasing degree of fascination."[15] In 1840, he published a paper on differential equations and, during the period between 1840 and 1842, this work culminated in "On a General Method of Analysis," published in the *Philosophical Transactions of the Royal Society*. For this he was awarded a Royal Society Gold medal in November 1844.[16]

Living and working in his small school, Boole had cause to regret the scarcity of special books and worked under circumstances he regarded as unfavorable for mathematical investigation. During this period, he was thinking about the theory of linear transformations and published introductory papers on the subject; he saw that linear transformations and the idea of invariance was applicable to the solution of polynomial equations and that it could also illuminate analytical geometry. Though Boole perceived that the embryonic idea had potential, he chose not to follow it. The idea interested him, but it did not captivate him and cause him to devote what little time he could muster exclusively on linear transformations. On 21 October 1841 he closed the second of his introductory papers by indicating his wish to abandon it: "It is not my intention to enter into the subject in this place, nor have I leisure either to pursue the inquiry, or to elucidate my present views in a separate paper. To those who may be disposed to engage in the investigation, it will, I believe, present an ample field of research and discovery."[17]

Cayley saw this passage and he avidly took up the invitation. He had written nothing on the subject prior to his grand tour but had demonstrated a deep knowledge of the closely related subject of determinants—a subject he surveyed in his "Theory of Determinants" published the year before. Moreover, he could see that invariance was intimately connected with analytical geometry, his growing passion in mathematics. On his return to England from his Continental tour, he was primed and receptive to Boole's ideas.

The Apprentice

The 22-year-old Cayley did not delay in communicating with Boole. In June 1844, the earnest young mathematician wrote an introductory letter: "Will you allow me to make an excuse of the pleasure afforded me, by a paper of yours published some time ago in the [Cambridge] Mathematical Journal, 'On the theory of linear transformations' and of the interest I take in the subject, for sending you a few formulae relative to it, which were suggested to me by your very interesting paper; I should be delighted if they were to prevail upon you to resume the subject, which really appears inexhaustible."[18]

The idea of invariance can be detected in the work of other mathematicians before the 1840s, by Gauss and Lagrange, for example. As a method for treating geometrical questions, it was also appreciated by the Trinity mathematicians in the 1830s.[19] The young Cayley saw an opportunity and, in his hands, invariant theory, as it came to be known, was amplified, formalized, and became the most important idea of his mathematical life.

Invariance is central to science and mathematics. In the most profound theories, the question of "what abides" has been a natural one to ask. In the conservation laws of physics, invariance plays a central role: physical configurations may change in time and space but the conservation of mass and conservation of energy are basic. In the theory of relativity, to use a more modern example, the equations of motion remain unchanged under Lorentz transformations.[20] In mathematics, quite simple properties of geometric figures can be described in terms of invariance. For example, the *area* of a triangle on the base *AB* is unaltered if a vertex *P* is moved on a line parallel to its base: the triangle *APB* changes shape but its area remains unaltered (see figure below).

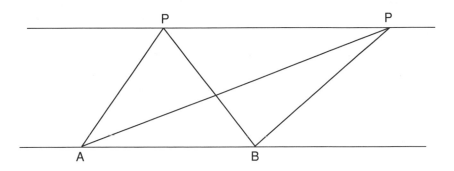

This idea of "permanence amidst change" has been such a recurring idea in religious, philosophic, and scientific thought that its explicit appearance in mathematics would have acted as a magnet to the young mathematician on the lookout for new subjects. Had he not assimilated the Platonic vision of scientific research as taught at King's College, London? Classics, supervised by R. W. Browne at King's, taught that Plato's interpretation of the object of science as the search for "the true, the eternal, the immutable" was the correct vision and that to succeed in this cause "is to know intellectually the essence of things absolutely."[21] Cayley took on this Platonic vision: the study of permanence amidst change.

There was an added attraction which caused him to take up its study with such alacrity: the transparent property of a mathematical invariant offered a scheme for uniting disparate areas of mathematics and a key to progress in several branches of analysis and geometry. An invariant carries information that is independent of that belonging to the variables.[22] In geometry, this information is independent of the coordinate system employed and therefore of especial interest to the geometer. The utility of "invariance" was a feature of the theory that Cayley advertised from the beginning. In the hands of Gauss, Lagrange, and the young German mathematician Gotthold Eisenstein, the property of invariance was also seen as remarkable, but while it was duly noted, it did not deflect them from their own research agendas. C. S. Peirce observed, that as they "strolled over the ground, ... they did so without recognising upon what goldmine they trod."[23]

The question of *algebraic* invariance, the subject of Boole's papers of 1841, revolves around algebraic forms and the linear transformation of their variables.[24] The simplest algebraic form is the quadratic expression of school mathematics $ax^2 + 2bx + c$.[25] He would most naturally consider *this* in its homogeneous state (a homogeneous polynomial): $ax^2 + 2bxy + cy^2$. Corresponding to the area of the triangle unchanging (see fig. 1), the algebraic invariant in this example is $ac - b^2$. This expression is essentially conserved when the binary form is transformed and therefore deserves its name *invariant*. The invariant $ac - b^2$ is denoted by Δ_2 to indicate its alignment to the quadratic binary form, which is of order two. When higher order binary forms are considered, the corresponding formulae can be extremely lengthy. The value of Δ_3, corresponding to the case of the binary cubic, $ax^3 + 3bx^2y + 3cxy^2 + dy^3$, contains five terms and is $a^2d^2 - 3b^2c^2 - 6abcd + 4b^3d + 4ac^3$, but Δ_6, corresponding to the binary form of order six (the binary sextic form), contains as many as 246 terms.

These formulae were well known to the small group of algebraists who worked on the theory of equations in the 1840s. In a short paper submitted to *Crelle's Journal* on New Year's Day of 1844, Ferdinand Eisenstein gave them as by-products to his explicit solutions of polynomials of the first four orders.[26] In another paper he noticed Δ_3 had the property of invariance and other properties. In considering the same elementary equations, Boole came to similar conclusions about the remarkable properties their researches uncovered.[27] Cayley concurred and was in good company.

Cayley's opening letter to George Boole stirred a common interest between the two mathematicians. The value of Δ_4, for instance, was subsequently calculated by each scholar using different methods.[28] The calculation of Δ_n (called the discriminant) was of both theoretical and practical interest and its calculation, for various values of n, recurs throughout the history of invariant theory; the discriminants Δ_5 and Δ_6 were yet to be investigated. Cayley needed the explicit expression for Δ_5 several years later in connection with geometrical researches on developable surfaces.[29] The expression for Δ_6, with all its 246 terms, was also calculated, but discriminants of higher order forms are so extensive that it is unlikely they were ever calculated—even by the Victorians, who were not deterred by the prospect of massive feats of calculation and often relished them.[30] For all the modern criticism of the Victorians for their penchant for calculation, the binary quintic form and the binary sextic—polynomials of order five and six in modern language—are of small degree and therefore quite basic. To know all their invariants explicitly would be a natural desire, and this task formed the first immediate challenge for Cayley.

That the binary quadratic is so elementary but the binary quintic so intricate is in itself intriguing. It is a remarkable fact that Cayley's life's work in invariant theory was dominated by the study of the binary quintic and even foundered on it, as will be seen. It was only toward the end of his life that he could count his investigation of this form as approaching completion. Yet even then the calculations presented difficulties, tested his mettle, and caused him lingering doubts.

Cayley's attachment to the problem of analyzing the binary quintic form stemmed from the classical problem of solving the polynomial of order five: $ax^5 + bx^4 + cx^3 + dx^2 + ex + f = 0$. The question of whether the general quintic polynomial equation was solvable using only basic arithmetical operations is one of the great problems of mathematics. The fact that the cubic and quartic equation can be "solved" (by finding a formula for their solution in terms of $+$, $-$, $\times$, $\div$, the qth root $\sqrt[q]{\ }$) was established by such mathematicians as Girolamo Cardano in the *Ars magna* (1545) and Rafael Bombelli in the *Algebra* (1572). Since that time it spurred many to investigate the problem of the "quintic." In the nineteenth century, the quintic was a live issue and attracted the efforts of the finest mathematicians of the period.

Cayley was well aware of the general result—that it was *not* possible to solve *all* polynomials of order five, a result obtained by the Norwegian Niels Abel in the 1820s—but thought that invariant theory offered another window on this great problem, *the* problem of nineteenth-century algebra.[31] Through Peacock's erudition and his report on the progress of analysis made to the British Association (1833), the result itself was common knowledge at Cambridge, if not among British mathematicians generally. The Irishman Robert Murphy, who studied at Cambridge in the 1830s but died before fulfilling his promise, investigated the roots of the quintic through symmetric functions without much success but realized its inherent difficulties.[32] Well into the 1850s and 1860s, it continued to attract the attention of British amateurs bent on showing that all quintic polynomial equations were soluble.[33]

Cayley quickly discovered that the solution of the quintic problem could be tackled using the new invariant theory. The stimulating correspondence with Boole charts the path of his first steps in this theory, where the prompting and suggestions from Boole were answered by his constant industry and application. Cayley benefited immeasurably from guidance but while Boole was fascinated, he was driven, and while the elder mathematician contemplated the subject, the younger rushed on. His letters vividly illustrate the mathematician at work—before the printed work appears and the scaffolding dispensed with. An important aspect of his early steps was the way Cayley anticipated future results. He conjectured freely and was unmoved by the occasional wrong turning. The modern concern for "proof" played little part in his initial ideas, but the interplay of calculation, conjecture, and analogy was all-important.

Having taken the initiative of opening a correspondence, Cayley was anxious about the reply he might receive. Though Boole had been to no university, he was known in the scientific world as a mathematician of substance, being the holder of a Royal Society Gold medal. Cayley was understandingly cautious: "Hoping you will excuse the liberty I have taken in writing to you," he wrote with deference.[34] The attitude of the apprentice did not persist, he soon gained assuredness in the new subject, and the two entered into an equal discussion on a wide range of mathematical topics. They discussed analysis and optics, but Cayley was most interested in the theory of linear transformations. Occasionally Boole tried to tempt him with other questions, but "linear transformations" remained his consuming passion and he was not easily diverted. For his part, Boole was sufficiently stimulated to return to the theory he had previously abandoned. After Cayley's first letter, Boole calculated the discriminant Δ_4 hoping that the result might have some application in the theory of algebraic equations but also remarking on the tedious process of calculation, which appeared necessary.[35]

Cayley proposed a substantial extension of invariant theory. Whereas Boole considered only homogeneous polynomials in *single* variables, Cayley went one stage further and considered algebraic forms in *sets* of variables.[36] Thus, Cayley would replace terms such as x^3 by multinomial product terms like $x_1 y_1 z_1$. He was led to this generalization by the invariance property he found in Boole's papers: "[i]n attempting to demonstrate this very beautiful property it occurred to me that it might be generalised by considering for the function U, not a homogeneous function of the nth order between m variables, but one of the same order, containing n sets of m variables, and the variables of each set entering linearly."[37] In the context of Cayley's discovery of a multilinear function θU (to use Boole's notation) with the property of invariance, he had written out a simple example for Boole's benefit:[38]

A simple example of the function is the following: [For]

$$U = ax_1 y_1 z_1 + bx_1 y_1 z_2 + cx_1 y_2 z_1 + dx_1 y_2 z_2$$
$$+ ex_2 y_1 z_1 + fx_2 y_1 z_2 + gx_2 y_2 z_1 + hx_2 y_2 z_2,$$

and [is] the [multilinear hyperdeterminant]

$$\theta U = a^2 h^2 + b^2 g^2 + c^2 f^2 + d^2 e^2 - 2ahbg - 2ahcf - 2ahde$$
$$- 2bgcf - 2bgde - 2cfde + 4adfg + 4bech,$$

which satisfies for instance

$$\left(a\frac{d}{de} + b\frac{d}{df} + c\frac{d}{dg} + d\frac{d}{dh} \right) \theta U = 0$$

and five other analogous equations.[39]

In this letter, Cayley outlined his immediate line of research. He sought some algebraic device, like the determinant, for writing down what he called a hyperdeterminant (a "higher" determinant) without recourse to differential operators or long-winded elimination algorithms: "I am not in possession at present of the law of the coefficients of θU," he wrote, "it is very desirable that this should be ascertained, and thus θU, [its] terms and coefficients, developed independently of any elimination."[40]

While Boole offered help, he was not the sole influence on the young mathematician. Cayley was an avid reader of Continental journals and took careful note of the previous work of Eisenstein and Otto Hesse, the former student of C. G. J. Jacobi. This combined to impress on him a wider view, a scenario for invariant theory that unified Boole's ideas with those of the German mathematicians. Practical results demanded different treatment. Although Cayley had created a theory for multilinear forms, this was not his primary goal. It was in Boole's specialized theory of homogeneous polynomials that he saw the prospect of immediate progress of a practical sort and the completion of the task of listing invariants.

The significant advance offered by Cayley's multilinear theory, or as he called it, the "tantipartite theory," was that it shed light on Boole's "unipartite theory." A metatechnique frequently employed by Cayley was his movement from "single variable theory" to "multivariable theory." His significant discovery at this stage was that multilinear invariants reduce to ordinary invariants as a special case.[41] More important, he discovered that the multilinear theory gave rise to new invariants not obtainable from within Boole's specialized theory. For the binary quartic form, he discovered the perennial $v = ae - 4bd + 3c^2$, an invariant which Boole could not obtain from his more limited phrasing of the theory.[42]

From the outset, Cayley was intent on calculation. It was not enough to merely suggest the existence of invariants. From his viewpoint, they had to be calculated and displayed. There was thus an implied need for the theory to be based on algorithms, which could be used to generate the invariants. In practice, the calculations were often more tedious than difficult, but they required a certain frame of mind. Though Boole balked at the thought of calculating extensive invariants and hoped someone with enough leisure would undertake their calculation, he did not question the usefulness of such calculations.

By the end of August, Cayley wrote to Boole again: "[w]ill you excuse my troubling you with another letter." Ostensibly, it was to correct a "mistake" in his first letter but he was keen to report the progress he had made. The magnitude of the research, both in calculatory terms and its difficulty, was all too evident: "For [a quartic multilinear function with] four sets of two variables there is one value of [the multilinear invariants] F, of the second order, another perfectly independent one of the sixth order, the completely developed form of which consists of 232 terms, which I have succeeded with a good deal of difficulty in working out."[43]

The work did not progress smoothly and through his desire to calculate, he was quickly led into difficult combinatorial problems. That Cayley found them puzzling, to some extent indicates their absolute difficulty. For despite his protests, he really was a mathematician of extreme competence where combinatorial problems were concerned: "I have not been able to work out any thing further on the subject of linear transformations, indeed I almost felt myself come to a standstill for the present and have hardly attempted it. The question now, gives me the idea of requiring some rather complicated combinatorial analysis and I doubt whether I have enough of that, to bestow upon it."[44] This impasse was a direct consequence of him wanting to calculate invariants explicitly and he had to overcome these combinatorial problems if the subject was not to be stillborn.

Mathematical Diversions

Cayley's work on invariant theory occupied the summer of 1844 but by September had come to a standstill. In that month he was diverted to other rich pastures, problems associated with the newly discovered system of quaternions and further problems in analytical geometry. First, he recognized the importance of quaternions, a new kind of "number" consisting of four components written in the form $a + bi + cj + dk$. These constituted a system of algebra which did not fit with the notion of algebra as it was commonly understood.[45] When quaternions are multiplied together, the order in which this is done is vitally important, thus breaking the rules of ordinary arithmetic, for which order is immaterial.

The discovery of the quaternions was made by Sir William Rowan Hamilton, Ireland's foremost mathematician and scientist. With a pronounced precocity in classics, mathematics, and a facility with languages, he had shown early signs of academic brilliance. The oriental languages to which he was attracted might be of service in a future career as a clerk in the East India Company—so his father thought. Parental plans were subsequently derailed, for scientific thoughts were too appealing. While still an undergraduate, he was elected to a professorship in astronomy at Trinity College, Dublin, and the accompanying title of Royal Astronomer of Ireland with responsibility for the Dunsink Observatory. At the British Association meeting in Dublin in 1835, he was knighted by the Lord Lieutenant, the King's representative in Ireland. While his early work was in optics, he had a passion for mathematics itself, and in particular for the quaternions.

Hamilton's discovery of the quaternions came in a "flash of inspiration," a flash underpinned by a 15-year period in which he sought a system that would extend two-dimensional complex numbers to higher dimensions—the complex numbers comprise a system of numbers with two components $a + bi$, which obey the rules of ordinary arithmetic.[46] Hamilton first tried "triplets" of the form $a + bi + cj$, but no satisfactory system resulted. When he came down for breakfast each morning, his sons used to ask him, "Well, Papa, can you *multiply* triplets?," and he was bound to answer that he could only add and subtract them.[47] Success came rather unexpectedly: triplets could not be multiplied satisfactorily but "quartets" could and the word "quaternions" was imported into mathematics (hitherto the word "quaternion" had designated a unit of stationery).

On 16 October 1843, at the age of thirty-eight, Hamilton wrote in his note-book: "I, this morning, was led to what seems to me a theory of *quaternions*, which may have interesting developments."[48] With the date scored into his mind, the subject became an obsession. It was the great discovery of his life and it became a dominant line of research in British mathematics. Hamilton would never forget that day when quaternions "started into life, or light, full grown."[49]

Hamilton immediately presented his results to the Royal Irish Academy, but on the English side of the Irish Sea, readers of scientific journals were made aware of the discovery in the following July. It was the first announcement to appear in print and came with an opulence unusual for the *Philosophical Magazine*:

On Quaternions; or on a new System of Imaginaries in Algebra. By Sir WILLIAM ROWAN HAMILTON, LL.D., P.R.I.A., F.R.A.S., Hon. M. R. Soc. Ed. and Dub., Hon. or Corr. M. of the Royal or Imperial Academies of St. Petersburg, Berlin, Turin, and Paris, Member of the American Academy of Arts and Sciences, and of other Scientific Societies at Home and Abroad, Andrews' Prof. of Astronomy in the University of Dublin, and Royal Astronomer of Ireland.

1. LET an expression of the form

$$Q = w + ix + jy + kz$$

be called a quaternion when w, x, y, z which we shall call the four constituents of the quaternion Q denote any real quantities, positive or negative or null, but i, j, k are symbols of three imaginary quantities, which we shall call imaginary units, and sup-pose to be unconnected by any linear relation with each other.[50]

Hamilton presented the rules for addition and multiplication of quater-nions and remarked on their unusual properties: "though it must seem strange and almost unallowable, to define that the product of two imaginary factors in one order differs (in sign) from the product of the same factors in the opposite order $ji = -ij$. It will, however, it is hoped, be allowed, that in entering on the discussion of a new system of imaginaries, it may be found necessary or con-venient to surrender *some* of the expectations suggested by the previous study of products of real quantities, or even of expressions of the form $a + bi$, in which $i^2 = -1$." Hamilton's statement obviously was designed to make an impression.[51]

Cayley saw Hamilton's announcement and was impressed. He wrote to Boole in early September:

I was very much interested lately, by a short paper of Sir William Hamiltons in the philosophical Magasine, On a new system of imaginary quantities, He considers what he term[ed] quaternions, expressions of the form $x + yi + zj + wk$, i,j,k being imaginary symbols satisfying

$$i^2 = -1, \quad j^2 = -1, \quad k^2 = -1$$
$$ij = k, \quad jk = i, \quad ki = j,$$
$$ji = -k, \quad kj = -i, \quad ik = -j.$$

The remarkable part of which is evidently that the factors of a product are not convertible [commutable], but as he observes, why should they be. The results that the supposition leads to are certainly quite consistent with each other and some of them very remarkable.[52]

Cayley immediately accepted the notion of quaternions as belonging to symbolical algebra. He accepted their anticommutivity (that $ji = -ij$) without hesitation. The idea that order mattered when writing down the composition of functions was a familiar one to those who had studied the work of Lagrange and his associates. This experience found resonance in Hamilton's quaternions and Cayley immediately grasped its significance.

At this early stage, Hamilton also observed that the quaternion could be used to express a rotation in three dimensions, so they were not merely algebraic contrivances but could be interpreted geometrically. This fact added importance to their study since they could be seen as a natural extension of complex numbers, which could be interpreted as planar rotations. With Hamilton's enthusiasm, quaternions rapidly became the topic of the moment. Sensing that something newsworthy had been discovered in Ireland, a puzzled member of the Anglo fraternity was heard to mutter, "what the deuce are these quaternions anyway." At the British Association meeting at York at the end of September, Hamilton gave an exposition of their properties and followed his "algebraic" announcement in the *Philosophical Magazine* of July with a "geometrical" announcement in the October issue.[53]

While Cayley drew inspiration from Hamilton's announcement of the quaternions, so had De Morgan. In a paper on triple algebra read to the Cambridge Philosophical Society in October of 1844, he wrote: "The systems which I shall examine differ entirely from that of Sir William Hamilton, both as being triple instead of quadruple, and as preserving, in their laws of operation, a greater resemblance to those of ordinary Algebra."[54] De Morgan was intent on studying the algebra with three base units, the triplets which Hamilton had considered but discarded. In adopting a conservative viewpoint, De Morgan wanted to preserve a connection with ordinary algebra: "Sir William Hamilton seems to have passed over triple Algebra altogether on the supposition that the modulus, if any, of $a\xi + b\eta + c\zeta$ must be $\sqrt{a^2 + b^2 + c^2}$. It is certain that there cannot be a system of triple Algebra with such a modulus; but it is by no means requisite that the modulus should be a symmetrical function of a, b, and c. I should also notice that in Sir W. Hamilton's quadruple Algebra there is a complete departure from the ordinary symbolical rules: AB and BA have different meanings."[55] De Morgan produced systems of triplets which obeyed the commutative law but not the associative law. With quaternions, a modulus defined by

$$|w| = \sqrt{a^2 + b^2 + c^2 + d^2}$$

satisfies the law of moduli $|w_1||w_2| = |w_1 w_2|$, but the modulus of De Morgan's triplets failed this test. The triplets generally lacked the harmonious symmetry of the quaternions. While De Morgan looked for a geometric application of triplets, nothing was found to compare with the elegance of the quaternions. Sitting in the audience of the Society in Cambridge, Cayley listened to De Morgan's abstract being read.

Cayley lost no time. At the beginning of December 1844, he submitted his paper "On Certain Results Relating to Quaternions" to the *Philosophical Magazine*.[56] His own realization that quaternions could be considered as a rotation was perhaps suggested by Hamilton's "geometric announcement," but to him, the geometric result seemed "rather a curious one." He found it especially interesting for he originally perceived the achievement of Hamilton's discovery as within symbolical algebra and not necessarily connected with geometrical ideas. Cayley's brilliant observation readily showed that if a pure quaternion $xi + yj + zk$ represents a point with coordinates (x,y,z), the product $q^{-1}(xi + yj + zk)q$ yields the new coordinates of a rotated point.[57] This succinct expression corresponds to the rotation given by a long-winded formula attributed to Olinde Rodrigues. Cayley was well prepared for his own discovery since only the year before, he had written a paper on Rodrigues's formula *without* using quaternions.[58] Seeing the correspondence between rotations and quaternions and the interconnections with Rodrigues's formula heightened his interest. As a first step, he applied these ideas to questions in mechanics.[59]

Hamilton remarked that the application of quaternions to geometry "had indeed occurred to himself previously; but he was happy to see it handled by one so well versed as Mr. Cayley is in the theory of such rotation, and possessing such entire command of the resources of algebra and geometry."[60] Cayley's quick contribution to the theory of the moment was a splendid piece of opportunism and is indicative of his fecundity imposing itself on so many new investigations. Hamilton admired Cayley's work, but he may have recognized that any claim over quaternion territory should not go unchecked. In a Supplement to the December issue of the *Philosophical Magazine*, Hamilton published his account of the discovery.[61]

Whereas Hamilton advertised quaternions as a universal tool of use in mathematics and physics, Cayley was not so extravagant. He saw the quaternions as a very beautiful discovery, but they could never supplant the traditional methods of analytical geometry. Indeed, he regarded the quaternions as an inferior medium for analyzing geometrical problems. While he believed that quaternions were by nature geometrical, since they could be used to represent three-dimensional rotations, they could not replace coordinates as a means for studying analytical geometry. To Cayley's mind, coordinates and algebraic equations represented the most tangible and easily understood representation for curves and surfaces, something that could never be achieved by the compact quaternionic notation.

Cayley took the opportunity of raising algebraic queries. What would happen if standard mathematical notions framed in terms of "ordinary quantity" are reconsidered in relation to quaternions? There followed myriad new questions. Determinants, for example, might be allowed to have quaternionic entries. About this possibility, he wrote to Boole: "The properties of determinants for instance are modified most curiously. But like, I forget what Jewish writing it was said of, the idea would require Camel loads of commentaries for its developement; every [word] would require to be rewritten, and the new version would be ten times as long as the original, if all the formulae of analysis had to be adapted to the cases of the symbols it contained, denoting quaternions."[62] Using the normal definition of the determinant, Cayley wondered what properties were transferred. There were some obvious peculiarities, which pointed toward a rich seam. How did he deal with these questions? The short answer is that he adopted a quasi-empirical approach. His experimentation with determinants of small order was typical and his habit of declaring results true in general without formal proof was learned at an early age.[63] He not only suggested the extension of the notion of determinants, he also put forward the idea of functions of a quaternionic variable, noting that a first step in this direction would be the establishment of quaternion exponentials. The somewhat sketchy outline of these important ideas indicates a pursuance of novelty, so much a feature of his work. For Cayley, the subject of quaternions constituted an interesting theory and provoked more questions than it could answer. He returned to the subject many times, but as a youth, it was one theory among the many that entranced him.

Also in September, Cayley was casting about for other research subjects and a promising line was suggested by Green's potential theory. George Green, the son of a Nottingham miller, had entered Gonville and Caius College, Cambridge, as a

40-year-old mature student in 1833 and graduated fourth wrangler. His serious mathematical studies had begun on joining the Nottingham Subscription Library in 1823 and, like Boole, had benefited from Sir Edward Bromhead's patronage. He read widely and in 1828 published An Essay on the Application of *Mathematical Analysis to the Theories of Electricity and Magnetism* for private circulation. Influenced by Siméon-Denis Poisson in particular, Green set out a theory of "potential functions," in which he included the now famous "Green's integral theorem."[64] Subsequently he published important papers in the *Transactions of the Cambridge Philosophical Society*. Boole had sent one of Green's memoirs on "Attractions" to Cayley but, on receipt, Cayley claimed he had not understood it.[65] A week later he wrote to Boole again: "I have not courage to undertake any thing in the theory of attractions; it is too difficult to be taken up merely as a secondary subject,—and being unconnected with any thing at which I have been working for some time past. I have hardly used the [integral] sign $\int$ for a long time."[66] Cayley had published work on multiple integrals following graduation but afterward had veered off in other directions. As we have seen, one of these paths led him into the theory of algebraic curves in the two-dimensional plane.

Cayley now began to think of curves in three-dimensional space, then known as curves of double curvature (but also as space curves, skew curves, twisted curves, tortuous curves), a subject originally studied by Alexis Clairault in the first half of the eighteenth century. Toward the end of September, he asked Boole: "Have you ever done anything with curves of double curvature, I should be particularly grateful for any new lights upon that subject."[67]

In October he sketched out some ideas on elliptic functions and doubly infinite products and sent them to the *Philosophical Magazine*.[68] He also reconsidered Green's memoir on attractions, but the main work was invariant theory. He prepared a final draft of his paper "On the Theory of Linear Transformations" for the *Cambridge Mathematical Journal*. He wrote to Boole with a note about the binary quartic form and its discriminant Δ_4 (here written $\theta(q)$): "I have been quite delighted with the results you have just sent me. The $\theta(q) = v^3 - 27V^2$ is a particularly interesting one."[69] Here he met the problem of *dependence* and *independence* of invariants when he discovered that Δ_4 could be expressed in terms of the invariant, $v = ae - 4bd + 3c^2$, which he discovered, and Boole's invariant: $V = ace + 2bcd - ad^2 - b^2e - c^3$.

As he advanced in invariant theory, the calculations inevitably became more tedious, though he was not deterred. Indeed, his unruffled stance in the face of calculatory work is seen from his letter to Boole: "I have a plan just now of effecting to a certain extent, the elimination of the variables between three quadratic equations: the result as far as I mean to expand it will contain about 7 or 800 terms, each [*sic*] of them the product of four determinants of the third order and linear in the coefficients of each of the equations: I can do it without any excessive trouble by a method given by Hesse ... and I think the result will be a valuable one."[70]

Two months after finding combinatorial problems barring his way, he had finished his first paper. "I have just finished a paper on linear transformations for the next N° of the journal," he wrote to Boole, "which I believe is to be printed soon. I shall be very anxious to hear your opinion of it."[71]

Incubation

On completing the first invariant theory paper, Cayley immediately began to work on its sequel. He had been working on "Hyperdeterminants" for only a short while (intermittently over a period of five months) yet he had acquired a superb intuition for the subject, as is so admirably indicated by his numerous speculations. Analogy was an important tool, though sometimes dangerous. For example, in the case of a binary form of order six, he had found invariants of order two and of order four and speculated on the existence of others: "I should rather think there was one of the *third* order," he wrote to Boole in November, "which I know nothing about however; it is only on the analogy of the theory of functions of the fourth order that I imagine it to exist."[72]

While Cayley strove for generality and system, even at this stage, it is significant he approached invariant theory through specific examples and informed "guess-work." In trying to gain headway, he proceeded on the basis of intuition of what might be true. Mathematical proof, the canon of twentieth-century mathematics, was invisible. From a letter he wrote to Boole toward the end of the year, we glimpse Cayley at work. At this stage he knew isolated invariants, for example, the relatively short invariant G for the binary quintic had already been computed:[73]

$$G = a^2f^2 - 10abef + 4acdf + 16ace^2 - 12ad^2e + 16b^2df$$
$$+ 9b^2e^2 - 12bc^2f - 76bcde + 48bd^3 + 48c^3e - 32c^2d^2.$$

He used such established results for binary forms of order two, three, and four to speculate on invariants for binary forms of orders five, six, and seven. In his letter to Boole he drew up a table, showing the state of the theory:

The theory of even functions [binary forms of even order] appears the simplest and I conjecture that the following would be the number and degrees of the transforming functions [hyperdeterminants]

Eqns of the order	Transforming functions [hyperdeterminants] of the order [degree]
2	2
4	2, 3
6	2, 3, 4

Then might that for odd functions be

3	4
5	4, 6
7	4, 6, 8.
.	.
.	.

since clearly the orders must be even [The order of an invariant of a binary form of odd degree must be even]. This is only founded on the instances $n = 2, 3, 4$ and it contradicts your results for $n = 5$, still it seems so natural that the number of the functions [hyperdeterminants] should depend very simply upon the value of n. . . . The question then comes is there any practicable mode of finding the functions for the fifth and sixth orders. If my supposition about the degrees [of invariants] should be correct, it ought not to be so very difficult but with a function of the 12th order, one is afraid to think of it.[74]

After summarizing known results, Cayley then moved to the case of the binary forms of the ensuing orders to make conjectures. Inductive thinking from known instances was his strategy. To a large extent, he was in the position of ignorance. For the binary quintic, for example, he knew the existence of the invariant G of degree four (above) but was unaware of the invariants of degree eight, twelve, and eighteen at this stage. In 1844 the binary quintic form was *terra incognita* and his knowledge of it quite rudimentary.[75]

Cayley's character as an exploratory mathematician is very clear while his boldness in basing part of his argument on experience is typical. He was at the beginning of his great enterprise and the art of making conjectures was indispensable. A young tyro, he showed himself to be a master of speculation. He did not progress slowly and surely with inevitable logic but took the route of conjecturing and testing speculative statements. Hunches of what might be the case were published alongside definite results; some would not be touched by mathematicians for more than fifty years and even then, many were left unanswered and remain. If he failed to make immediate progress with a question, he moved swiftly to another line of attack.

The practicalities of invariant theory continued to resurface, and he wrote to Boole in December 1844: "Do you see any way of calculating, in rough, the degree of labor that would be necessary for forming tables of Elimination; Sturms functions, our transforming functions [invariants] &c. . . . If one could get to any practical results about it, and they were not very alarming, it would be worth while I think presenting them to the British Association: but I am afraid the limit of possibility comes very soon: suppose one ascertained a result would take a century to calculate, it would be rather a hopeless affair."[76] At Cambridge, one of the problems he endured was relative isolation. The *Cambridge Mathematical Journal* was a success, but it was difficult to find serious mathematicians with common interests. Boole was a boon, but were there others? He was aware of one and alerted Boole to his existence: "a propos of equations, did you ever see Sylvester's forms for Sturms functions?"[77]

Photographic portrait of Arthur Cayley (1821–95) around age 30

Lithograph of King's College, London, in the 1830s

Engraving of St. Petersburg in the 1820s

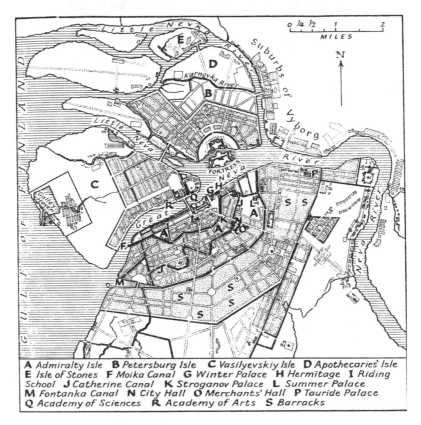

A *Admiralty Isle* B *Petersburg Isle* C *Vasilyevskiy Isle* D *Apothecaries' Isle*
E *Isle of Stones* F *Moika Canal* G *Winter Palace* H *Hermitage* I *Riding*
School J *Catherine Canal* K *Stroganov Palace* L *Summer Palace*
M *Fontanka Canal* N *City Hall* O *Merchants' Hall* P *Tauride Palace*
Q *Academy of Sciences* R *Academy of Arts* S *Barracks*

Map of eighteenth-century St. Petersburg

Annual Presentation of Prizewinners, King's College, London, in the 1830s

Cayley family home in York Terrace, London

Sir George Cayley, aeronautical
pioneer and fourth cousin to
Arthur Cayley

George Peacock, Cayley's tutor
at Cambridge

William Hopkins, Cayley's coach
at Cambridge

Senate House,
Cambridge,
setting for the
Tripos examinations

Newspaper announcement of Tripos results
and order of merit for 1842

MATHEMATICAL TRIPOS,

1842.

MODERATORS.

Rev. THOMAS GASKIN, M.A., *Jesus college.*
D. F. GREGORY, M.A., *Trinity college.*

EXAMINERS.

Rev. ALEXANDER THURTELL, M.A., *Gonville and Caius college.*
RICHARD POTTER, M.A., *Queens' college.*

WRANGLERS.

1	Dr. CAYLEY	Trin	21	Dr. Westmorland	Jesus
2	Simpson	Joh.	22	Dumergue	Corpus
3	Major, R. B.	Joh.	23	Bryan	Trin.
4	Fuller	Pet.	24	Shears	Joh.
5	Bird	Joh.	25	Greenwell	Joh.
6	Jarvis	Corpus	26	Suffield	Caiu.
7	Shortland	Pemb	27	Middlemist	Christ's
8	Austin	Pet.	28	Davies, H.	Queens'
9	Fenwick	Corpus	29	Cook	Joh.
10	Jones	Clare	30	Penny	Inn
11	Frost	Joh.	31	Davies	Joh.
12	Parnell	Joh.	32	Eastwood	Caius
13	Johnstone	Joh.	33	Venables	Pemb.
14	Castlehow	Emman	34	Baily	Christ's
15	Carter	Emman	35	Light	Joh.
16	Wilson	Joh.	36	Walker	Sidney
17	Smith, B. F.	Trin.	37	Tandy	Joh
18	Finn	Trin.	38	Kinder	Trin.
19	{ Anger } Æq.	Joh.			
	{ Goode }	Pemb.			

SENIOR OPTIMES.

1	Dr. Vidal, J. H.	Joh.	27	Dr. Little	Christ's
2	Fitz Gerald	Christ's	28	Green	Caius
3	Hey	Joh	29	Hughes	Queens'
4	Parkinson	Queens'	30	{ Cobb }	Corpus
5	Ottley	Caius		{ Shaw } Æq.	Trin.
6	Allen	Trin.	32	Powell	Christ's
7	Metcalfe	Sidney	33	Parr	Cath.
8	Vidal, O. E.	Joh.	34	Sharples	Joh.
9	Inchbald	Cath.	35	*Hopwood	Pet.
10	Penrose	Magd	36	Buckham	Joh.
11	Riley	Trin.	37	Thrupp	Trin.
12	Brooks	Joh	38	Atkinson	Clare
13	Gillett	Emman.	39	Kingsley	Magd.
14	Walpole	Caius	40	Lloyd	Jesus
15	Rowton	Joh.	41	Pattie	Corpus
16	Wolfe	Joh.	42	Woodford	Pemb.
17	Morse	Joh.	43	Thornall	Sidney
18	Clubbe	Joh.	44	Blake	Jesus
19	Ommaney	Trin.	45	Montague	Caius
20	Ridley	Jesus	46	Balderstone	Joh.
21	Douglass	Trin.	47	Boyce	Sidney
22	Hogg	Emman.	48	Rothery	Joh.
23	Marix	Queens'	49	Salkeld	Pet.
24	Tabor	Trin.	50	Gordon	Pet.
25	Swann	Christ's	51	Stansfeld	Joh.
26	Haslehurst	Trin.	52	Munro	Trin.

JUNIOR OPTIMES

1	Dr. Twisaday	Joh.	12	Drs. { Pratt } Æq.	Joh.
2	Maul	Joh.		{ Ramsay }	Trin.
3	Vaughan	Christ's		{ Hutchins } Æq.	Trin.
4	Wilkinson	Joh.	14	{ Slade }	Joh.
5	Teague	Emman.	16	Conybeare	Pet.
6	Kerry	Joh.	17	Yeoman	Trin.
7	Sheringham	Joh.	18	Hough	Caius
8	Nugee	Trin	19	Sheepshanks	Trin.
9	Peter	Jesus	20	Shackleton	Cath.
10	Webster	Emman	21	Finnia	Queens
11	Smythies	Emman.	22	Worfledge	Clare
			23	Barstow	Trin.

ÆGROTAT.

Onslow, Emman.

Bishop	Corpus	McNiven	Trin
Fowler	Clare	Parminter	Trin
Hamilton	Trin	Parry	Christ's
Knipe	Pemb	Wyer	Joh.

* *Died since the Examination.*

Oil painting of the presentation of the Senior Wrangler, Arthur Cayley,
in 1842

Watercolor graduation portrait of Arthur Cayley as the Senior Wrangler, 1842

Pen and ink sketch of Stone Buildings, Lincoln's Inn, London

Photograph of Cayley's barrister's chambers at 2, Stone Buildings,
Lincoln's Inn

Photograph of C. G. J. Jacobi,
Cayley's "illustrious Jacobi"

Sketch of
Ferdinand Gotthold Eisenstein,
mathematical contemporary
of the young Cayley (Sketch)

Pencil sketch (1847) of George Boole,
who set the scene for Cayley's
escapades with invariant theory

Oil painting of James Joseph Sylvester
at age 26, Cayley's friend and confidant

Photograph of William Thomson,
editor of the *Cambridge and Dublin
Mathematical Journal*

Photograph of George Salmon,
Cayley's colleague in Ireland

Photograph of Arthur Cayley
at about 35 years of age

Arthur Cayley's certificate for
membership of the Royal Society
from June of 1852

Victorian Royal
Society in London

Cayley's "original" group table
for the symmetry group
of a triangle

	1	α	β	γ	δ	ε
1	1	α	β	γ	δ	ε
α	α	1	γ	β	ε	δ
β	β	ε	δ	α	1	γ
γ	γ	δ	ε	1	α	β
δ	δ	γ	1	ε	β	α
ε	ε	β	α	δ	γ	1

Dear Sylvester
        Every Invariant
Satisfies the partial diff.
equations
$(a\frac{d}{db} + 2b\frac{d}{dc} + 3c\frac{d}{dd} \cdots + nj\frac{d}{dk}) U = 0$
$(b\frac{d}{db} + 2c\frac{d}{dc} + 3d\frac{d}{dd} \cdots + nk\frac{d}{dk}) U = \frac{1}{2} ns U.$

(s the degree of the Invariant) &
of course the two equations
formed by taking the Coeff.s
in a reverse order. This will
Constitute the foundation of
a new theory of Invariants.
        Believe me yours very
                Sincerely A. Cayley
5th Dec. 1851.

Cayley's historic letter of 5
December 1851 to J. J. Sylvester
announcing new discovery in
theory of invariants

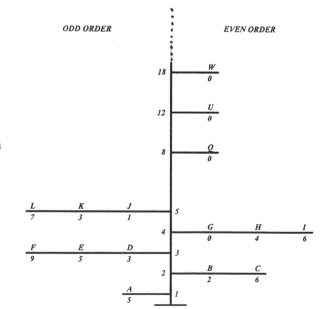

Mathematics on the front of an envelope (1856)—the nature of partnership with Sylvester

Degree

ODD ORDER   EVEN ORDER

Invariants and
covariants of the
binary quintic as
known to Cayley, in
part, as of 1854

CHAPTER 5

From a Fenland Base

B y the beginning of 1845, Cayley had made progress on several topics, but the coming year, his second as a fellow of Trinity, would be special. Invariant theory was important, but it was one interest among many. In the ferment of activity surrounding Hamilton's discovery of the quaternions, Cayley took a leading part; in analytical geometry and the theory of elliptic functions, he also made contributions. With the precocious energy of a young and ambitious mathematician, he flitted from one subject to another, as each competed for his attention. He was in Cambridge, but his main lifeline continued to be Boole in Lincoln. They had not met in person and Cayley wrote on 17 January: "I wish I could manage a visit to Lincoln, I should so much enjoy talking over some things with you—not to mention the temptation of your Cathedral. I think I must contrive it some time in the next six months,—in spite of there being no railroad, which one begins to consider oneself entitled to in these days."[1]

The New Year

Only the week before, Cayley had framed a response to Brice Bronwin in York-shire about their ongoing controversy on the subject of elliptic functions. As an afterthought, on 16 January 1845, he added a postscript about a completely differ-ent topic, on the algebraic system known as the octaves. He had just returned to Cambridge after time away but the matter was on his mind. It was the briefest of postscripts attached to his paper "On Jacobi's Elliptic Functions, in Reply to the Rev. B. Bronwin; and on Quaternions" and it turned out to be a gem. His obser-vations on elliptic functions proved fallacious, but the postscript has endured and is widely recognized as one of his finest discoveries. He began with an assertion followed by a question: "It is possible to form an analogous theory with seven imaginary roots of (-1) (? with $v = 2^n - 1$ roots when v is a prime number)."[2]

Once the complex numbers $a + bi$ with two components and quaternions $a + bi + cj + dk$, with four, are seen in a doubling-up sequence, the generaliza-tion to eight units was a natural one to take. An octave or octonion is an expres-sion of the form $w = a + bi + cj + dk + ep + fq + gr + hs$. They have since been studied in depth and found to have unusual properties.[3] Cayley's reward for always adopting a policy of prompt publication was that the system of octaves have become widely known as Cayley numbers. Quick action was necessary if priority was to be confirmed and not snatched away, as the case of the octaves amply demonstrates.

This generalization was made also by John Graves one year earlier, shortly after Hamilton had made his discovery of the quaternions. He was a classmate of Hamilton's, his valued friend, and well placed to make a discovery of this sort since it was to him that Hamilton immediately wrote detailing his momentous revelation.[4] Today, if Graves is remembered at all, it is for the magnificent "Graves Mathematical Library" he bequeathed to University College, London.[5] Coming to England from Dublin, he qualified for the Master of Arts degree at Oxford and was called to the bar. He established many contacts in the scientific world, and, elected to the Royal Society in 1839, he served on its council. His avocation for mathematics was accommodated by his duties as chair of jurisprudence at University College from 1839 to 1843. De Morgan was professor of mathematics, and the two were guiding spirits of the Society for the Diffusion of Useful Knowledge. Like so many Victorians, Graves carried on two careers simultaneously: on the verge of discovering the octaves himself, he was preparing and delivering a series of lectures at University College on the Law of Nations.[6]

Beginning on Boxing Day 1843, Graves wrote Hamilton a series of letters in which he enclosed a multiplication table for the octaves; the names "octaves" and "octads" were coined by Graves. In the middle of January of 1844, he wrote again: "I should be much gratified by your mention of my theorem as to the product of two sums of eight squares (as a theorem suggested by your theorem), if you should send a paper on the subject to the *Phil[losophical]. Mag[azine]*."[7] The octaves could not compete with quaternions in importance for their multiplication table exhibited an odd feature: when three octaves were multiplied together, the answer depended on which two were multiplied first—they were not associative. Graves thought this too great a divergence from ordinary algebra, a view which Hamilton shared. While the discoverer of the quaternions positively welcomed anticommutivity ($ij = -ji$), and indeed celebrated it, he was not prepared to abandon too much and did not desire too much novelty. He reflected on the nature of the octaves and came to the conclusion that, in algebra, the preservation of the associative law should be maintained: "To this *associative* principle, or property of multiplication, I attach much importance,... The *absence* of the associative principle appears to me to be an *inconvenience* in the *octaves* or octonomials of Messrs. J. T. Graves and Arthur Cayley."[8] Hamilton touched on the octaves out of curiosity but quaternions were his absorbing interest. Indeed, he was so besotted with his quaternions that he failed to take much notice of the "minor" discovery of his long-standing friend. When Graves entrusted Hamilton with setting the historical record straight, the matter escaped his attention until it was too late.[9]

To not publish immediately was rather a wrong move in an atmosphere of high activity among mathematicians focused on the construction of new algebras. An added element in the race to discovery was the "obvious" pathway to the octaves once the quaternions were seen as a twofold extension of the complex numbers. Exactly one year after Graves made his discovery, Cayley submitted his postscript. He did not highlight the lack of commutivity and associativity of the new system but merely warned the reader that the octave expressions "requires some care in writing down, and not only with respect to the combinations of letters, but also to their order."[10] Cayley was quite willing to jettison the commutative *and* associative laws, a position others could not take. He was the youth among older men and not weighed down by years of schooling in the traditional thinking of ordinary algebra. He accepted the validity of symbolical algebra without the semblance of geometrical interpretation (though it is significant that he looked for it) or whether it conformed to the usual rules of ordinary algebra. His observation that the octaves satisfy the law of moduli ($|w_1\|w_2| = |w_1 w_2|$) implicitly solved the historically important "n-squares problem" (to decide whether the product of two sums of n squares can be expressed as a sum of n squares) in the case $n = 8$.

On seeing Cayley's postscript, John Graves became alarmed and wrote to Hamilton: "I find that Cayley is near my octaves. I must publish my extension of Euler and Lagrange's theorems of products of squares."[11] The n-squares problem was one of high status in the mathematical world. Not only was it unsolved generally but the particular problem with $n = 4$ had been solved by the esteemed Leonhard Euler, so there was an opportunity to follow in the steps of the master. The omission of an explicit statement of an eight-square theorem in Cayley's postscript gave Graves hope. Accordingly, there was a short and pungent *post*-postscript from Graves to his paper on "normal couples" published in the *Philosophical Magazine* for April, 1845:

> P. S. "The product of two sums of eight squares is a sum of eight squares." To Euler is due a corresponding theorem relating to sums of four squares, which was extended by Lagrange, who added coefficients to the squares. As Euler's theorem is connected with Hamilton's quaternions, so my theorem concerning sums of eight squares may be made the basis of a system of octads or sets of eight, and was actually so applied by me about Christmas 1843. I mention this in consequence of the idea suggested by Mr. Cayley in the last Number of this Magazine. But the full statement and proof of the theorem concerning sums of eight squares, and of several other new theorems connected with the doctrine of numbers, must be reserved for another time.[12]

The discovery of the octaves was such a magnificent discovery that Graves must have been bitterly disappointed at seeing the honor slipping away. Nor could he claim to be first with the eight-squares theorem, for he later found it had been established by the Danish mathematician Carl Ferdinand Degen twenty-five years before.[13] After all this frustration, Graves harbored no ill-feeling and maintained a high opinion of Cayley as an accurate and profound mathematician. In a further "Note on a System of Imaginaries" published in 1847, Cayley pointed out that certain base units of the "octuple system of imaginary quantities" i, j, k, p, q, r, s satisfy an "antiassociative" law (for example, $i(rs) = -(ir)s$), which corresponded to the anticommutative law for quaternions.[14] He concluded that in the system of octaves "neither the commutative nor the distributive [associative] law holds, which is a still wider departure from the laws of ordinary algebra than that which is presented by Sir W. Hamilton's quaternions."[15]

Fresh from his work on the octaves, Cayley picked up De Morgan's triple algebra based on the three base units in a letter to Boole of February 1845:

> I have not seen De Morgan's paper on triple Algebra; It is I suppose a good deal connected with Sir W. Hamilton's quaternions, which is a most interesting theory. I think I mentioned it to you in a former letter; you can easily work out the fundamental results if you are so inclined, from the mere idea—which is that of considering the general symbol $\alpha + \beta i + \delta j + \gamma k$, i, j, k being imaginary roots of -1, which combine according to the laws $i = jk$, $j = ki$, $k = ij$, $-i = kj$, $-j = ik$, $-k = ji$. What are de Morgan's analogous assumptions! I heard an abstract of the paper read:—which was just sufficient not to tell me what I wanted to know about it.[16]

Boole took a passing interest in De Morgan's triple algebra, had written to him about it, and evidently to Cayley also.[17] But Boole was lukewarm on the mathematics of the new algebras generally and hoped the attention shown to them by John Graves and Hamilton would be a passing phase. "They will soon, I should think, get through their quaternions and triplets," he wrote to Thomson, "and, interesting as the subject is, I must confess that I should be glad to see them turning their attention to the Int[egral] Calc[ulus] and to physical science."[18]

In the spirit of De Morgan, Cayley investigated "algebraic couples." In April, he read Graves's Post-Postscript on the eight-square theorem attached to a paper on a theory of "normal" couples, (symbolic expressions of the form $ai + bj$). Graves assumed the ordinary algebraic rules of commutivity and associativity to apply to normal couples and excluded "anomalous" couples but staked out that subject for future work. Cayley's reaction was to submit his own "On Algebraical Couples" before the end of the month. He made no assumption about imaginary symbols obeying the laws of ordinary algebra, only that the system be closed: $(ia_1 + jb_1)(ia_2 + jb_2) = (iX + jY)$. Different definitions of moduli for algebraic couples were considered, but no attempt was made to examine the properties of the individual systems.[19] Cayley's intention to classify systems is evident. He began but did not dwell on the study of these algebras; there were other attractions and he could not resist their allure. The foray into the new algebras had been short-lived but incisive. Barely four months had elapsed since he first learned about the quaternions, yet he had rapidly written a paper on the subject and capped it with his brilliant discovery of the octaves. Now it was time to move on.

At the beginning of 1845, Cayley wrote to Boole about his papers on curves in space. Having expressed a desire to visit Lincoln, he added a note on immediate investigations to be undertaken regarding these curves:

> I am hoping just now to investigate some of the properties of curves of double curvature & developable surfaces—the Analytical theory seems very difficult for there is no means of expressing the equation of a curve of a given order. I rather think it would be better to begin with developable surfaces, for since every curve is the edge of regression of a developable surface, the theory of the curve depends upon that of a single equation $\psi = 0$ which denotes the dev[elopable] surface.[20]

After giving the relevant equations he continued,

> By the by there is a very beautiful form for that equation given by Hesse. If ψ be made a homogeneous function of 4 variables, the determinant formed with the second diff[erentia]l coeffs of ψ, $= 0$. But to develope all this is quite out of my power. The particular theorems I want about curves I think I can manage by some demi geometrical methods (*methode de la nouvelle Geometrie,* I suppose it would be in French), which are mere extensions of the reasonings in Plückers theory of Algebraical curves.[21]

Cayley's difficulty was rudimentary: a space curve generally cannot be represented by a single algebraic equation or even as the intersection of two surfaces.[22] He ruminated on this difficulty and continued his study of planar curves. By the beginning of February, he had completed another paper on cubic curves.[23]

Despite Cayley's claimed neglect of the "integral sign ∫" made five months earlier, Boole's encouragement to study Green's memoir on "Attraction" bore fruit and he completed a paper on the evaluation of multiple integrals. In substance, it was a way of approaching integrals found in Green's work and that of J. P. G. Lejeune-Dirichlet, both of which had been published in *Crelle's Journal.*[24] As he mentioned to Boole:

> I have been looking, a propos of your paper, once more at Green's Memoir on attractions, and have most satisfactorily ascertained that I cannot understand it at all... The equation may of course be integrated in the way (and a very beautiful one it is) in

which Green has done his—but then I do not see what to do with the result when one has it,—for how are the constants to be determined. —It is curious that Green's integrable case is different from Dirichlet's, (his density being $(1-x^2/n^2-\ldots)^{(n-2)/2}$ $Rat[iona]l.func(x, y, \ldots)$ and Dirichlet's simply $Rat[iona]l.func(x, y, \ldots)$ —the results too are in an utterly different form, since Green has that curious function H_0 in the denominators of his integrals: I wish I could understand the paper; one is so willing to believe that the true way of resolving the question introduces some such functions, analogous to Laplaces functions for the sphere, —but one is sorry to see them as denominators especially as they seem to become evanescent [reduce to zero] for particular values of the variables. —I shall be much obliged to you, if you can help me about the aforesaid equation—not that I much expect to go on with the subject at all.[25]

The work on this done, he reported to Boole, "I have given up Green's memoir."[26] Cayley's study of potential theory proved a recurrent theme during the 1840s, and eighteen months later he wrote to William Thomson that he understood it "perfectly."[27] In the 1870s, he would attempt to unite Green's and Dirichlet's contributions within a single theory.

A member of the inner scientific circle at Cambridge, it was widely predicted that Thomson would be the Senior Wrangler in 1845. The story may be apocryphal, but when he sent a servant to the Senate House to find out who was the second wrangler, he brought back the rather unexpected news that "You are, Sir." Thomson was mortified, for himself, and his father who was professor of mathematics in Glasgow. Cayley wrote to Boole that he was "very much disappointed in Thomson's being second [wrangler] only. A Johnian [Stephen] Parkinson is first: we can only boast of a fifth wrangler."[28] Cayley's own college allegiance aside, it was very rare for a pupil of William Hopkins to fail to gain the top spot. Master of Trinity William Whewell wrote to Airy that Hopkins's "expected senior wrangler Thomson, is beaten by one of those 'black horses' which the Johnians so often have in their stables."[29] Around Cambridge, there was no doubting who was the superior mathematician, notwithstanding the results of the writing race in the Senate House, in which the well-drilled candidate could triumph over the real mathematician. After the Tripos events, Thomson and his friend Hugh Blackburn traveled to Paris to meet the French *savants*.

The day before leaving Cambridge, Thomson secured a copy of Green's "Essay on Electricity and Magnetism" from Hopkins and read it on the coach journey. His reaction to Green's work was quite different from Cayley's. He saw it as mathematical analysis combined with physical insight, a brilliant piece of mixed mathematics, but to Cayley, as we have seen, it offered promise in pure mathematics. Cayley's paper on the subject, and one on cubic curves, were destined for *Liouville's Journal* with Thomson acting as courier. Cayley was already acquainted with the Continental mathematicians and introduced Thomson to them. From Paris, Thomson wrote to his father:

I called on Liouville, shortly after I arrived here, with a paper which Cayley had given me to carry to him. We very soon became acquainted, and he began directly to work with pen and paper at various subjects of conversation, and when I went away he invited me to return again "pour causer de toutes ces choses." [to chat about all these things] I called again, bringing with me another paper wh[ich] I had received from Cayley. I found Chasles there, to whom Liouville presented me, and we had a long conversation on mathematical subjects.[30]

Cayley's letters to Thomson informed him of life back in Cambridge. There was some Cambridge gossip, such as the news of the Cambridge Camden Society, a highly influential society devoted to ecclesiastical architecture and antiquities. Cayley was a member for two years and had joined in 1843, encouraged by his friend Venables, who was a foundation member.

In this episode of the Camden Society, especially, "Cayley" is not merely the name of a mathematical theorem or concept that has survived in the modern mathematical texts. He is a young man living in a period of bitter religious controversy, where those at the university were compelled to conform to religious orthodoxy. Adherence to the Thirty-nine Articles of the Church of England had to be sworn before undergraduates received their Cambridge degrees. However, there is little to suggest Cayley was other than an orthodox Anglican. His friendship with Venables would suggest this, and that with John Pilkington Norris, another of his walking companions. Norris was a classics scholar at Cambridge and another churchman who became a school inspector. He married into the Lushington family, known for its part in founding the Society for the Diffusion of Useful Knowledge and the anti-Slavery movement. Later he achieved notice for helping to save the fabric of Bristol Cathedral.

The Camden Society published the *Ecclesiologist,* held meetings, and organized excursions to churches around the country. Its object was to further the "science" of church-building and the society was sympathetic to the controversial object of reintroducing high Roman Catholic ritual into the Anglican church. Founded by two Trinity College undergraduates, their tutor, Thomas Thorp became its president in 1839.[31] There was a furor over the society's restoration of the Norman church of St. Sepulchre (the "Round Church") in Cambridge and the Camden Society proposal to reinstate the church altar was too close to Romanism for many—it was proposed to close the Society.[32] About the closure, Cayley wrote to Thomson: "My only piece of Cambridge news for you, is that the Cambden [sic] Society is to be discontinued,—as Thorpe [sic] announced, with a good deal of conscious humility at the last meeting—how that it was setting a most commendable example of obedience and soon that it should be so &c. —and so it is to go on till the anniversary meeting when its dissolution will be formally arranged."[33]

Cayley was a social being and not the reclusive mathematician who hid from the world. Friends visited his college rooms and his membership in the Camden Society and the Chess Club indicates a balanced personality, that of one taking an active part in college life, if at times he took himself off on solo expeditions to remote parts. He was interested in those around him and made an especial effort to make contact with mathematicians. He was particularly interested in Thomson's experiences in Paris and the people he met there. As he had read the work of Michel Chasles, he was naturally curious about him:

> What sort of a person is Chasles? it must have been interesting meeting him—I should have found it particularly so, at any rate. Pray do not forget the [Chasles's] *Aperçu Historique*, if you do meet with a copy, which I am afraid there is no chance of. Have you looked at it at all. How do you like Paris, it is an unfortunate time to be there, it is a place one can so well enjoy walking thro[ugh]' on a fine day in Summer— the Louvre you have been studying of course; I was a little disappointed with it myself after some of the Italian Collections, but that depen[ds] so much upon what ones favorite style is. What part of Paris is your abode in—North or South of the Seine.... With kind regards to Blackburn—whose loss our Chess Society regrets most deeply.[34]

When he did meet Chasles in London two years later he remarked to Thomson that he "liked him very much, all that I saw of him."[35] Chasles was to become a key figure in Anglo-French mathematical relations and an inspiration for English mathematicians.[36] He not only explored the history of geometry but involved himself with the history of mathematics generally. Robert Leslie Ellis reminded a friend of his qualities as a mathematical thinker: "you know he [Chasles] is a great man about the history of Arithmetic & thinks that he has proved that our system is virtually contained in that of the Greek Abacus."[37] As we shall see, Cayley heartily approved the award of the Royal Society Copley Medal to Chasles in 1865.

Cayley's mathematical papers delivered to the Paris mathematicians by Thomson were published in *Liouville's Journal* in March and April. In the following months, he wrote an addition to his paper on multiple integrals and, making progress with his projected theory of space curves, contributed a paper that extended Plücker's equations, which express connections between the singularities of plane curves to those of curves in space.[38]

The Arthurian Quest

In February 1845, the "Note sur deux Formules données par M. M. Eisenstein et Hesse" was published in *Crelle's Journal*. It was the first of Cayley's papers published in the German journal and, for several years, the first by an English author. In it he announced his intention of writing out a theory of hyperdeterminants, or, as we have referred to this theory anachronistically, by invariant theory, the name by which it subsequently became known: "The property of the function u that I have just set out, is related to a quite general theory, of a new kind of algebraic functions which I am currently looking at and which, because of their analogy with determinants, I believe one may call 'Hyperdeterminants.' I propose to set down the first fundamentals of this theory in a memoir which will appear in the next Number of the *Cambridge Mathematical Journal*."[39]

In all, Cayley produced four papers in invariant theory during 1845 and 1846.[40] His announcement in *Crelle's Journal* was followed by his pioneering "On the Theory of Linear Transformations," which appeared in the same month in the *Cambridge Mathematical Journal*. He had overcome his earlier combinatorial difficulties on the material, about which he had been "very anxious to hear [Boole's] opinion," and the paper was published.[41] Straightaway he started work on its sequel, "On Linear Transformations," which he was to submit to the Cambridge journal in March 1846 (the fourth paper was a translation into French of these two and appeared in *Crelle's Journal*). Taken in the round, the papers of these years inaugurated the two great methods of invariant theory in the nineteenth century and are pivotal in the history of algebra. If he had done little else, this work would have afforded Cayley a high mathematical reputation. As with much of his early work, there is an impatience to set down his discoveries and a sense of urgency to progress to other investigations.

For the sequel to his first paper, Cayley found a completely new method for calculating hyperdeterminants, which he called the method of "hyperdeterminant derivation." In March 1845 he wrote to Boole about this new discovery: "I have just found a property of hyperdeterminants, which like most of the others gives *another* method of determining them (one would be glad not to have so many) and which seems to me perhaps the most curious of all."[42] In his paper he formally announced this new method and set out the intended scope and realistic possibilities for the investigation: "[i]n continuing my researches on the present subject [linear transformations], I have been led to a new manner of considering the question, which, at the same time that it is much more general, has the advantage of applying directly to the only case which one can possibly hope to develope with any degree of completeness, that of functions of two variables [binary forms]."[43] The method involved the hyperdeterminant operator Ω, of which a special case is:

$$\overline{12} = \begin{vmatrix} \dfrac{\partial}{\partial x_1} & \dfrac{\partial}{\partial x_2} \\ \dfrac{\partial}{\partial y_1} & \dfrac{\partial}{\partial y_2} \end{vmatrix}.$$

To obtain the invariant for the binary quadratic $\Delta_2 = ac - b^2$, for instance, Cayley's "Ω-process" requires the calculation $\overline{12}^2 u_1 u_2$, where u_1, u_2 are copies of the binary quadratic form. The process is mainly of theoretical significance and not particularly suited to calculation. Using it to obtain invariants is a tedious procedure, even for simple examples, and he was never overly fond of it as a calculating mechanism.[44]

With ideas crowding into his conscious thinking, Cayley was making rapid progress in the infant theory. It was not the absence of ideas and results that perplexed him but their organization: "I am quite at a loss how to combine into order, all the results which have been found, and one is continually meeting with new ones," he wrote to Boole.[45] He thought about the overall goal for the subject, a perspective quite remarkable for a 23-year-old. While many embryonic mathematicians are precocious and can solve problems delivered at their door, Cayley's work on invariant theory indicates both a broad vision and an enterprising attitude.

It is a measure of Cayley's mathematical maturity that his aims were conceived of so boldly. From his brief and direct statement, it is evident that he was striving for complete generality: "To find all the derivatives [invariants] of any number of functions [algebraic forms], which have the property of preserving their form unaltered after any linear transformation of the variables."[46]

When applied to binary forms of low order, this question is relatively straightforward and can be disposed of quickly. The case of the binary form of order two, the binary quadratic, there is only one invariant, namely, the discriminant Δ_2. For the binary cubic, there is the discriminant Δ_3, and for the binary quartic, he and Boole had already found three invariants, namely, the discriminant Δ_4, and the forms v and V. For higher order forms, the answers are not readily forthcoming, and indeed the quest to discover them would concentrate Cayley's energies for more than fifty years. While the statement of aims, concerned as it is with cataloguing invariants, is of a familiar type (find *all* the curves, find *all* the surfaces,...) and is clearly ambitious, he already had a shrewd idea of what was actually possible. Indeed, he seems to have realized that he could do little more than scratch the surface of the general problem.

Cayley's quest of mapping the "derivatives" of algebraic forms was in character, the mathematical equivalent of the human genome project. While the mapping of genes is now "complete," the same cannot be said for invariant theory. Even today, with the advantage of abstract algebraic techniques in harness with powerful computing machinery, the quest Cayley outlined is still in its infancy, even for binary forms. The statement of its aims allows for single algebraic forms of any number of variables (not just binary forms) and combinations of these forms. He was looking for the evolution of a mathematical theory but was equally concerned with compiling a "work of reference" wherein one could look up the expressions for invariants. Their listing was a primary mission.

Compared with the task of *finding* invariants, Cayley's interest in providing "proofs" was slight. Of greater interest were the algorithms—the methods by which invariants could be discovered. His objective for the theory was couched in startlingly general terms, and he was wise to introduce a caveat into the overall goal: there was only ever real hope for success in the case of binary forms. This proved a deep insight, as, according to a leading twentieth-century practitioner, H. W. Turnbull, the second part of the nineteenth century constituted the "binary era" of invariant theory.[47]

A recurring theme was Cayley's fascination with the binary quintic form, quite lowly in the general scheme, but a particular case that challenged him consistently. As to the establishment of concrete results during the nineteenth century, binary forms of order up to and including eight were examined with a degree of success by its end, but even in this case the results were incomplete. It is a measure of the problem outlined by Cayley as a youth in 1845 that steady progress in the twentieth century has led only to results for binary forms up to order twelve but after this point the results are uncertain. Where the calculated results have been established as trustworthy, apart from a few cases, it is where Cayley left them.[48]

Though the finding of invariants was the principal objective in 1845, Cayley added: "But there remains a question to be resolved, which appears to present very great difficulties, that of determining the *independent* derivatives [his emphasis], and the relation between these and the remaining ones. I have only succeeded in treating a very particular case of this question, which shows however in what way the general problem is to be attacked."[49]

Invariants could be found, but a complicating factor arose concerning the relationships which could exist between them. As we have seen, in the case of the binary quartic, the invariants Δ_4, v, and V, Cayley had encountered were not independent. They were, as Boole had pointed out, connected by the expression $\Delta_4 = v^3 - 27V^2$. Thus Δ_4 is dependent on the fundamental independent invariants v and V. and these constituted a minimal set that could be used to generate *all* the invariants for the binary form of order four. In the 1840s, Cayley set out to find the independent invariants. He soon found that intricacies began with the invariants of the binary quintic form. The full extent of this complication was unknown to Cayley in 1845 but a brief consideration of a future discovery indicates the type of problem he would very soon face in the "Arthurian Quest." There are four basic invariants, G, Q, U, W, of the quintic form. It is not possible to write one of these invariants as an algebraic function of the others, so that in Boole's sense, they are independent. For example, it is not possible to write W as an algebraic function of the others. A rather surprising discovery of the 1850s was that the composite invariant W^2 could so be written, $W^2 = \frac{1}{16}(GQ^4 + 8Q^3U - 2G^2Q^2U - 72GQU^2 - 432U^3 + G^3U^2)$. Such a relationship was called a syzygy (the term borrowed from astronomy when planets are in line). The specter thus raised by this interconnectivity phenomenon, the "problem of the syzygies," was a complication that bedeviled the whole history of invariant theory.

There were also other algebraic forms with the property characteristic of invariants, which involved both coefficients *and* variables. Otto Hesse worked with these in the 1840s but they came into prominence in Cayley's work in the 1850s. These forms were then named covariants and these had also to be determined. By this time, the goal of invariant theory was sharper than it was in 1845: to find the minimal generating set of invariants and covariants and to determine the connecting syzygies.

In the 1840s, Cayley's objective of listing invariants was not pursued in a systematic way and he took results where he found them. Harvesting the low-hanging fruit, he computed the most accessible invariants. The results were sporadic while the "highest" invariants calculated were for a binary form of order nine. In this case, he found two invariants of degree four that were independent, and it was the binary form of lowest order in which this phenomenon occurred. "It is not pleasant having to go to the ninth order for one's simplest example," he remarked to Boole.[50] The discovery of invariants "along the scale" of the order of the binary forms eventually became a workable strategy, but when he progressed from the binary quartic form to the binary quintic and higher-order forms, he experienced a vast increase in complexity. It was perhaps anticipated since the solvability of the quintic equation was intricate compared with the relatively simple cases of the cubic and quartic equations. Cayley met these problems head-on but whenever he examined minutiae he inevitably became involved with tedious calculations covering many pages. Mathematicians of the period were justly regarded as "leviathans in symbols" by the scientists of Victorian society, who had recently witnessed the launch of Isambard Kingdom Brunel's iron-hulled *Great Britain* in July 1843.

In early May of 1845, Cayley was coming to the end of his initial incursions into the embryonic theory. He allowed himself a reflection on the history of the subject and expressed his debt to Boole as its originator: "None of the papers of Hesse & Eisenstein that I know of, refer professedly to linear transformations; you have I think the completest claim to originating the subject. Eisenstein's papers are principally on the theory of numbers & Hesse's on Analytical Geometry."[51] Boole himself had migrated to other interests and could no longer supply a stimulus as he had done previously. He now expressed an ambivalent attitude toward pure mathematics, not regarding himself "a genuine mathematician,…for I am not one myself," he wrote to Thomson, a remark which indicates that the Boolean fervor of a few years before had to some extent been lost.[52] His lack of appetite for the fashionable higher dimensional algebras and his diminishing interest in invariant theory no doubt contributed to a disaffection with pure mathematics generally. He still shared a commonality with Cayley in the techniques of multiple integration and the application of calculus to physical science, and, in return, Cayley appreciated his *Logic* and his theory of probability.

By the end of May 1845, Cayley had completed the work for his two foundation papers in invariant theory. While he was anxious to circulate his work widely, there was little reaction to it among English readers. The Cambridge mathematicians with their interest in analytical geometry may have taken a passing interest, but no one published papers in the subject in the early issues of the *Cambridge Mathematical Journal.* As with many of his ideas and projects, his was a lone voice initially. One deterrent to a budding reader is easy to determine; for any but the hardiest, the papers presented a formidable challenge. Cayley's habit of peeling away motivation from abstract subjects would have put off most, but the abysmal typography of the local journals combined with an extensive use of the emboldened Gothic alphabet would have clinched disaffection. This had a predictable consequence. Without the lifeblood of communication, his attention to invariant theory waned from the intensity he felt in 1844 and 1845.

To gain Continental interest, a translation of his two papers was done, and this appeared as the opening article for the 1846 volume of *Crelle's Journal.* He kept a watchful eye on the subject and noted the utility of invariants where this occurred, but he did not give it the same concentrated attention until the end of the decade. The theory even lacked a generally accepted name. His foundation papers were on "linear transformations" and the "theory of hyperdeterminants." Awaiting rediscovery in the 1850s, the subject would then make an impact, and, indeed, would become the basis for the mid-Victorian's modern algebra. By then James Joseph Sylvester would have fully arrived on the mathematical scene, active and bursting with energy; seizing on the immutability of hyperdeterminants, he would christen the venture the "Theory of Invariants."[53] Only then would it come into its own and Cayley's contribution of the 1840s be recognized as a colossal imprint.

As Cayley's subsequent memoirs unfolded, the Herculean nature of the whole enterprise became increasingly apparent. The ambition to "find" all the derivatives of "any number of functions" was a quest that required the assembling of vast algebraic expressions and, *prima facie,* there were an infinite number of them. Yet, for the moment, this imperial adventure lay in the future.

With his first steps in invariant theory drawing to a close, there was the prospect of seeing the British "men of science" gather at his university. In June 1845, the British Association for the Advancement of Science was to hold its annual meeting in Cambridge and he made plans to attend.

A Cambridge Pilgrimage

The British Association annual meeting, the gathering of the British scientific establishment, was a regular fixture in the scientific calendar and had been since 1831 when the association was founded in York. The annual pilgrimage to a chosen city in Britain, and later in the century to one in the Empire, was also a social occasion.

The Cambridge meeting promised to equal any of its previous meetings in pomp and illustrious visitors, with many members taking the opportunity of returning to their alma mater. They looked forward to being cosseted in their colleges where dinner and supper parties would be numerous and the "delicacies of the season" readily available. Cambridge had not quite entered the modern age, but in the next month, the railway would remove some of the fenland isolation. Living in the town, Cayley attended the meeting, a young man of twenty-three, and very likely the first opportunity he had to attend the annual jamboree.[54] The year was proving to be a vintage one, one in which he published sixteen papers, including extensive ones on elliptic functions, hyperdeterminants, geometry, and quaternions.

Mathematics was high on the association's agenda and many practitioners were present. Boole had made the journey from Lincoln. With his correspondence with Cayley exactly one year old, they would have the opportunity to meet and discuss hyperdeterminants, if Cayley had his way. Sylvester took the opportunity of visiting Cambridge again after years spent in London and America. While he was at the York meeting of the previous year and had read a paper, he did not read one at Cambridge. Though Cayley had taken note of his method of elimination, it is most likely they had not yet met. Victorian formalities had to be observed and Cayley was almost from a different generation, something akin to a postgraduate student, while Sylvester was seven years his senior.

Among the throng, two members of the Analytical Society were conspicuous. Peacock was the retiring president of the association and Sir J. W. F. Herschel, foremost scientist and able mathematician, about to become his replacement. Many European specialists had been invited, but Peacock was disappointed that Gauss, "the great patriarch of magnetic science" could not be with them. A young John Couch Adams, newly admitted fellow of St. John's College, wrote home that he "had the pleasure of seeing for the first time some of our greatest scientific men Herschel, Airy, Hamilton, Brewster. . . . [t]here were many Men of Science from the Continent who honoured us with their presence, and on the whole I understand that this has been a most satisfactory meeting."[55] Boole was equally impressed and penned a poetic vision on the occasion:

> 'Twas something on the banks of Cam, to see
> Men known to Science, known to History;
> Airy, who trod the subtlest paths of light,
> And Herschel, worn by many a watchful night,
> Whewell and Brewster, Challis, large of brow,
> And Hamilton, the first of those to know.[56]

From the presidential chair in the Senate House, on the evening of 19 June, Herschel painted an optimistic picture for the future. He noted the success of the *Cambridge Mathematical Journal* and the new developments in algebra and he encouraged further exploration. He referred to Gregory's work on the calculus of operations, De Morgan's papers on the foundations of algebra, and to John and Charles Graves's contributions to the new algebras. He noted Peacock and Warren's pioneering thoughts and alluded to Hamilton's quaternions.

These were the seniors, but Herschel also referred to Cayley and his work on the new algebraic subjects—quite a compliment to a young man from a man of Herschel's position: "Conceptions of a novel and refined kind have thus been introduced into analysis—new forms of imaginary expression rendered familiar— and a vein opened which I cannot but believe will terminate in some first-rate discovery in abstract science." As well as Hamilton's research, no doubt Herschel had in mind Cayley's papers on quaternions, octaves, and algebraic couples then recently published. It was surely vindication of Herschel's belief in the excellence of the Cambridge mathematical training, and, perhaps thinking of Cayley: "[a]nother instance of the efficacy of the course of study in this University, in producing not merely expert algebraists, but sound and original mathematical thinkers (and, perhaps, a more striking one, from the generality of its contributors being men of comparatively junior standing), is to be found in the publication of the *Cambridge Mathematical Journal*."[57]

The specialty of the meeting was magnetism, where George Biddell Airy was a leading light. Airy, regarded by his colleagues as a "distinguished mathematician," delivered a lecture on progress in terrestrial magnetism in "his clear distinct manner"[58] Boole read a paper on a Laplacean differential equation solved without using the calculus of operations. Hamilton, the great protagonist of quaternions, was once more signaling their universality. Quaternions were Hamilton's great discovery and he made the most of them, as year after year he came to the annual British Association meeting to promote them. In keeping with the theme of the meeting, he argued for their applicability in electricity by pointing out the analogy between the opposing rotational effects of opposite electrical currents and his fundamental quaternion equations $ij = k, ji = -k$.[59] In his lecture to the mathematics and physics section "On the Elementary Laws of Statical Electricity," young William Thomson advanced a mathematical theory of electricity to explain experimental results obtained by Michael Faraday. Charles Graves, brother of John, produced a paper on algebraic triplets.[60] Cayley was silent but the British Association would become a forum when he had ascended the rungs of the scientific establishment and been formally elected as a "gentleman of science" F.R.S.

A few weeks after the close of the British Association meeting, Cayley was admitted to the Master of Arts degree and, consequently, to a major fellowship of Trinity—no longer a minor fellow but a full-fledged member of the foundation. It was a natural progression and with it went full dining rights at the Trinity high table. It also meant he had seven years' grace to remain a fellow before the momentous decision to either leave the College or take Holy Orders—the seven-year regulation, which applied to major fellows, was originally intended as a period for preparation for Holy Orders and, during this period, the aspirant was provided with income from the College Chest. On elevation to his major fellowship, the college stipend was raised to thirteen shillings and four pence per quarter and a liberatum of £1-6-8d. These were fairly paltry sums, even then, but what made a fellowship financially attractive was the annual dividend. This payment was dependent on the Trinity rental incomes for the year and seniority in the college hierarchy. For Cayley, this meant a substantial annual income.[61]

Meanwhile, the new *Cambridge and Dublin Mathematical Journal* was about to be launched with Thomson at its helm and taking over from Robert Leslie Ellis, who was anxious to drop the editorship of its forerunner, the *Cambridge Mathematical Journal*. Thomson was keen to make it a success and appealed to his friends for support. He could always rely on Cayley for a ready supply of papers. He wrote to Boole to send him something "as it would be very pleasant to have some good things in the first number," adding, "I have got several papers from Cayley for the first and second N^{os} and I hope to get something from Ellis to put at the beginning."[62]

Breaking Away

In August Cayley went off on another Continental tour, this time without his regular traveling companion Edmund Venables. Instead of heading south toward the Swiss Alps as he had done two years previously, he traveled to Scandinavia. Hiking around the Norwegian fjords, he stayed at Gjeudin, a lake in the Johunheim Mountains. Thomson could not resist a postscript to Boole updating him on their mutual friend's progress through mountains and mathematics: "I received a letter a few days ago from Cayley, from Gjeudin (lat 61°. 20′, long 26° from Ferro) which he says is a single hut, being a 'Saetter' or dairy, and at the time he wrote, was solely inhabited by himself. After giving me some details of his movements, he commenced 'mathematical notes.'"[63] The isolation of a remote Saetter was evidently conducive to mathematics and Cayley sent short papers back to Thomson for the Cambridge journal. One related to homography and the effects of a homographic distortion on a surface, conclusions owing to John Graves's work originally. The other contained results on quaternions.

Cayley continued his tour to Berlin, which he reached at the beginning of September. J. Steiner and Johann P. G. Lejeune-Dirichlet were based there, and C. J. G. Jacobi, for whom he had the highest regard. Among the younger mathematicians, ones of Cayley's age, were Eisenstein, Ferdinand Joachimstal, and

Carl W. Borchardt. A student of Steiner and Dirichlet, Joachimstal taught analytic geometry at the university; another subject he shared with Cayley was the application of determinants to geometry. While in Berlin, he met Cayley and showed him a theorem on the theory of elimination as applied to geometry.[64] Cayley did not meet Steiner, but it is conjectural who he did meet, though he would have had subjects in common with many of them. Borchardt, who was to later to become the editor of *Crelle's Journal*, worked on the same subjects as Cayley and would have had much to discuss. Never at rest, while in Berlin during the summer, Cayley submitted a paper to *Crelle's Journal* on the geometry of position and another on analytical geometry.[65]

Returning to Cambridge by October, Cayley's treatment of Abel's doubly-infinite products would appear in *Liouville's Journal*.[66] As a young mathematician, he developed an affiliation to Abel and much regretted his premature death. Much later he included his work in his own lecture courses at Cambridge but he made his first mathematical tribute to the tragic hero by showing that Abel's "double factorial expressions," published posthumously in 1841, are elliptic functions in the way Jacobi defined them. The work was the culmination of several earlier sketches on elliptic functions made in 1844 and 1845.[67] In contrast to the more speculative invariant theory, elliptic function theory was mainstream, and his guiding spirits were Abel and Jacobi, who had extended Legendre's treatment of elliptic functions with real variables to those with complex variables. It is in connection with this research that Cayley made a study of Cauchy's theory of complex integration, the subject generally made known in England by D. F. Gregory and A. De Morgan.[68]

On 6 November, Cayley was elected to the Council of the Cambridge Philosophical Society, an appointment which demonstrates his willingness to take his share of scientific administration. His scientific field of inquiry was rapidly expanding. During the years following graduation, he had avoided narrow specialization and took all of mathematics as his field of endeavor. He saw the beauty of the idea of invariance and its wide applicability. The new algebras were opening the field of symbolical algebra and he had no difficulty in abandoning the strictures of ordinary algebra. About this time, he was beginning work on *déterminants gauches* (skew-determinants) and setting in train another important sequence of papers that would inform later work in the theory of matrix algebra.[69]

Despite having a secure position at Trinity, Cayley was seriously thinking of leaving and taking up the law. The teaching duties in college were light but it was dispiriting work. He taught both ordinary and honors degree students, writing up model solutions to Tripos-type questions for the earnest young competitors. He was not a gifted expositor and whether students saw this or they regarded this tutoring as inessential, his lectures were not well attended. On one occasion when two students composed the entire audience, he sat down with them and worked through questions on mixed mathematics where the solution was concerned with a question on the stability of ships.[70] One is bound to question whether he wanted to perform this type of work for the years to follow. There was another reason for leaving, for life in the college was not as congenial as it might have been. The authoritarian manner adopted by the new master of Trinity showed no sign of ameliorating and the atmosphere generated was hardly conducive to a long stay. Safely insulated at Peterhouse, Thomson wrote to his father of the "despotic Whewell" and spelled out the disharmony felt by the Trinity men who "are most seriously incommoded by Whewell at present, who meddles with everything in a most mischievous manner."[71] The alternative was London. In town, there was an active group of mathematicians successfully carrying on professional and scientific careers in tandem.

If he had thought of relinquishing his fellowship, it would have been good news for one aspiring bachelor of arts, for fellowships were in short supply.[72] There were only three vacancies in 1846 and in the following year, and difficult choices were imposed on the Seniority. On hearing of Cayley's plans, Ellis wrote: "The report I heard about your plans & which I was impertinent enough to mention to you, is I believe a source of much consolation to the bachelors who are nearly in despair about vacancies."[73] Cayley probably had no intention of taking the drastic step of a full resignation, for why should he? Having a fellowship was the ideal and the commonly accepted way to finance training at the bar. The annual share-out from the College Chest would be maintained, he could continue to benefit, and by maintaining links with academe, he would be in a strong position if a suitable academic position became available. Trinity would also gain by not cutting the umbilical chord; on its books, it would have a barrister *and* a mathematician of growing repute.

The most important consideration for Cayley was mathematics and publicizing his work. He was keen to make his mark outside Cambridge and his sights were set firmly on writing for an international readership. In 1844 he had become the first English mathematician to publish regularly in the *Journal de Mathématiques Pures et Appliquées*, founded and edited by Joseph Liouville in Paris. In this journal the only British mathematicians to publish before him was the Scot James Ivory (in 1837) and the Irish mathematicians James MacCullagh (in 1842) and William Roberts (in 1843). Liouville saw Cayley as a talented mathematician and a welcome contributor to his journal, even if his style had yet to mature in terms of precision and presentation.

The elderly editor and founder of the *Journal für die reine und angewandte Mathematik*, August Leopold Crelle, regularly received submissions from the Cambridge tyro. Beginning in 1845, Cayley was the first British author to regularly publish in this prestigious journal and the sole British representative until 1850, when he was joined by William Thomson. This editor evidently experienced similar difficulties with Cayley's papers as did Liouville. Frederick Fuller, who had competed against Cayley in the wrangler race of 1842, was in Berlin in 1850 and reassured Thomson that he had seen his paper safely in print "in the midst of an ocean of Cayley's symbols, which I am happy to say overwhelm Crelle as well as most sensitive Englishmen."[74] The only other British contribution to this journal had been by W. H. Miller, the professor of mineralogy at Cambridge, and Charles Brooke, surgeon and scientist, who jointly published three brief mathematical notes in 1835. These last two authors had tasted mathematics at Cambridge, published a little, but then given up the subject.[75] Cayley was a man of a different ilk; in his unassuming way he placed a high premium on publication in European journals. In so doing, he was the man who broke the isolation of English pure mathematics from the Continent, and his "oceans of symbols" in *Liouville* and *Crelle* were the beginnings of a relentless publishing record in Continental journals.

Mathematical links between Britain and the Continent were gradually strengthened, but it would prove a slow process. An original intention of the *Cambridge Mathematical Journal*, founded in 1837, was to abstract the works of the leading Continental authors. In the course of the following years, readers were exposed to a limited range of Continental mathematics (for example, from Joseph Liouville and Joseph A. Serret). In reverse, few British mathematicians were in a position to regularly publish in the leading Continental research journals. Cayley was the exception. He was in the vanguard, but few could follow in his wake.

CHAPTER 6

The Pupil Barrister

In April 1846 Cayley took the decisive step and entered Lincoln's Inn as a pupil barrister. Fellows who remained in College and took the ultimate step of entering the Church could remain there for the rest of their natural days. The path of the Cambridge bachelor don, which conformed to the ideal laid down by statute, had a certain inevitability about it. From the ranks of the undergraduates the "reading man" was seen to progress serenely along the rails: "After three years, he comes out as a high classic, or a wrangler; takes pupils, obtains a fellowship, and dies ultimately at an advanced age in the possession of a college living, virtuous, ignorant, happy and beloved."[1] It is difficult to imagine a young man of Cayley's intellectual restlessness falling in with this scheme.

A Legal Career

Of the 500 students who went to Cambridge in Cayley's year, more than half went into the Church, but the Law was the next most popular choice.[2] The question for fellows of the colleges was whether to stay on. It was no secret that D. F. Gregory had intended to leave, while R. L. Ellis did leave but was spared the life of a working lawyer by inheriting a private income. Taking Holy Orders did not seem an option Cayley would seriously consider, and his Anglican orthodoxy not strong enough to incline him toward the priesthood. His confidant Venables, who did follow a Church career, said quite simply that Cayley "felt no vocation for holy orders."[3]

One model of the "don turned lawyer" Cayley had before him was Arthur Shelly Eddis. When Cayley graduated and moved into his new rooms in Nevile's Court, Eddis was a resident tutor living on the same staircase. A few years older, he was a brilliant classicist and the winner of scholarships and medals. A year later, in 1843, he left for Lincoln's Inn to train for the bar. Taking the pathway to a legal career allowed successful barristers leisure and financial independence to pursue a range of interests outside the professional life of a lawyer. Frequently the diversions were scientific and mathematical. In this respect, Archibald Smith, the Senior Wrangler in 1836, was another example for the young Cayley to think about. The instigator of the *Cambridge Mathematical Journal,* Smith was offered the life of a don on several occasions but declined it. The availability of a Scottish professorship might have tipped the scales toward an academic life but, ultimately, he too entered Lincoln's Inn to read for the bar. He became an Equity draftsman of some standing but maintained a vigorous program of scientific research in the application of the theory of magnetism to sea navigation.

A college teaching position represented a choice of sorts but occupants inevitably became part of an education system in which cramming and low-level teaching was the norm. The majority of students at Cambridge were candidates for the ordinary degree. Branded as poll-men, they came to Cambridge for the social life and had little interest in academic study and among them academic standards were certainly low. For their teachers, there was no natural career path other than progression through the college hierarchy. It was theoretically possible to be elected to the sought-after Cambridge chairs, but these prestigious professorships were few in number and their availability usually depended on the expiry of the incumbent.

The professors were insulated from the daily grind of teaching elementary mathematics in the colleges and were given a stipend. They had the freedom to marry and live where they pleased with minimal duties attached to their chairs. College fellows who remained in college had less freedom, lived a closeted existence, and were under an obligation to conform to a communal way of life. If they decided to marry, they were forced to resign and retire to a church living, usually one in the gift of the college—in Trinity's case there were seventy available. If they took this step, and happened to be interested in their subject, they would have to cultivate this academic work cut-off from the support of a community of the scholarly minded. For those with academic leanings but without substantial inherited wealth—as was the case with Cayley—the law offered the most obvious escape route from a college cul-de-sac.

The law presented an attractive route, for a man would be able to conduct research in parallel with the occasional legal business over which he would have total control. By remaining on the Trinity Foundation, he could train with the financial support provided by the annual dividend and this would be available when most needed. In effect, he was being paid by the college to train for the bar with provision for the postqualification period when a young barrister established himself and times could be lean.

For over four hundred years, the Inns of Court in London, the function of which was to administer the law, had offered the only alternative to a university education. In the nineteenth century, the education function of the Inns had declined, and latterly it was a way of prolonging student life and "finishing" an education obtained elsewhere, most usually at the two ancient universities. For a student progressing from one cloistered institution to the other, it was a seamless change from one privileged existence to another. The four Inns that survived to the nineteenth century (Lincoln's Inn, Inner Temple, Middle Temple, and Gray's Inn) had the ambience of a Cambridge or Oxford college. Instead of dons and undergraduates, the Inns housed benchers (members of the Inn's ruling Council), barristers, and students. The Inns each had a library, a chapel, and deans who acted as administrators as they did at Oxbridge. Members of the Inns wore gowns, and the same learning methods were transferred, to the extent that students' study was supervised by working barristers rather than college tutors or private coaches. Students fulfilled attendance regulations and were then called to the bar, at which point they would become barristers of the lowest rank.

Lincoln's Inn

As the young man of twenty-four years was admitted to Lincoln's Inn, his father Henry, then in his late seventies, was to take a less active role in the London commercial world. He had sold his magnificent house in the Nash Terrace in Regent's Park and would live in an eighteenth-century mansion—Cambridge House—close to "The Point" in Blackheath with its spectacular views over London.[4] He was a wealthy man and the financial backing required for his son to train for the bar would be readily forthcoming. There was family precedence in the law as his paternal great-grandfather, Cornelius Cayley, had been a prominent member of the bar in the eighteenth century and had risen to become the Recorder of Kingston-upon-Hull in Yorkshire.

The young Cayley became a pupil of the leading conveyancing counsel and barrister J. H. Christie. A barrister working for Christie's firm would be involved with the Chancery Division of the High Court concerned with land and wills. To become a "conveyancer" or real property lawyer (*real,* as in real estate or realtor) was a recommended route for trainee barristers and many who planned to treat the law as a part-time occupation took this route. The transfer of land was staple fare for barristers and was the bar's exclusive right until 1881, when the law was changed to allow solicitors to take part. In the 1840s, barristers like Christie, unlike his solicitor cousins who occupied an inferior position in the legal profession, took the work and prospered. A pupil could learn much from Mr. Christie.

Jonathan Henry Christie put a notorious youth behind him. He gained a bachelor's degree at Oxford in 1813 and in February 1821 acquired the distinction of being one of the last men to fight a duel by pistols on English soil. It was fought in defense of the honor of his close friend John Gibson Lockhart, editor of the Tory *Blackwood's* literary magazine, who had left London leaving his unfortunate friend to see the matter through. The moonlit duel involving the London *literati* took place at Chalk Farm at the north end of Regent's Park.

Christie's opponent, John Scott, editor of the *London Magazine,* appeared with his second, Peter George Patmore, in attendance. Unhappily, Scott was mortally wounded in the engagement, and Christie was indicted for murder but subsequently acquitted at the Old Bailey trial.[5]

Christie left behind his eventful youth and settled down to his legal career. He became a member of the "Institute," a self-selecting dining club of barristers—in limiting itself to forty members, it became known as the "Forty Thieves" among the legal fraternity.[6] As a man of the world, Christie can hardly have been overly impressed by the shy young man who threaded his way through the maze of London streets in the spring of 1846 to arrive for his first day at Lincoln's Inn:

> Mr. Cayley arrived at Stone Buildings, sent in his card, was admitted, and asked to be taken as a pupil. Christie inquired whether he had any introduction; the reply was, No. Had he been at a University? Yes. Christie, who seldom had a vacant chair in his pupil room, used to describe himself as not having been very favourably inclined towards this monosyllabic applicant. However, he had been at the University and might be worth while to inquire further. Christie did so, and by successive and separate questions elicited the information that Cayley's University was Cambridge; his college Trinity; that he had taken a degree; in honours; in mathematical honours; that he had been a Wrangler; that he had been Senior Wrangler.[7]

For Christie this was evidently sufficient. The halo of the Senior Wrangler was a prized asset in the barristers' chambers which ringed Lincoln's Inn Fields. Traditionally, Cambridge honors graduates were thought ideally suited to the law. Those who appeared on the order of merit were reckoned to have the life skills the Tripos was supposed to inculcate: clarity of thought coupled with its accurate and brief expression. By virtue of his position as top of the Cambridge honor list, Victorian society decreed that the Senior Wrangler had these skills to a superior degree, and Christie may well have thought that in Cayley he was interviewing a future legal star. That the young man was wedded to mathematics would not become obvious to Christie until later, but Cayley gave good service and he became a very favorite pupil and one of Christie's elite.

Cayley too had chosen well, for Christie was a widely respected legal counsel. He was a congenial man who made friends easily and counted a long list of the most eminent Victorian lawyers among his pupils. One recalled his attachment to literature and remembered that he "loved works of fiction, and many a half-hour have I spent with him in discussing Balzac, when his confidential clerk was under the impression that we were finalising the draft of some marriage settlement."[8] Christie wrote for *The Quarterly Review* and in his youth had passed up the opportunities of a writing career in favor of the law. Well known at the Athenaeum, the most prestigious scientific and literary gentleman's

London club, he was a close friend of the classical scholar Benjamin Jowett of Oxford. With his own liking for novel reading, Cayley would have found a happy working environment with such a mentor.

In entering Christie's pupil-room, Cayley became part of the practical legal culture and was among men who valued qualities not always found in the academic world. The regulated working day for sharp legal minds was occupied largely with the perusal of legal documents and the drafting of new ones. They prided themselves on being rapid, concentrating their attention on the matter at hand, and above all ensuring that a piece of work was gone through once and once only. This practice meant that a legal draft could then be "settled" and the next document dealt with, a lesson absorbed by the callow youth who sat in the pupil-rooms of their mentors. They learned to act quickly in dealing with long legal documents, for which an eye for detail and a long memory were essential assets. The trainee barristers did not need to be flamboyant or good speechmakers.

In Christie's pupil-room, Cayley remained in the background. Remarking on his retiring modesty, a fellow pupil established in the chambers for six years recollected: "he had one of the most unsophisticated minds I have ever known; jokes, and the badinage of the pupil-room, seemed to be delightful novelties to him, and his face beamed with amusement as he listened to them without taking much part in the conversation, being content to devote his time assiduously to work which I suspect was not altogether congenial to his taste."[9] He would far rather spend his time with mathematics, but Christie's chambers, though run by a thoroughly reasonable man, was still a place where repetitive duties represented a large slice of the day. Another of Christie's pupils referred to "the gloom of a London atmosphere without, the whitewashed misery of the pupil's room within—both rendered more emphatic by what appeared to us to be the hopeless dinginess of the occupations of the inhabitants. There stood all our dismal text-books in rows—the endless Acts of Parliament, the cases and the authorities, the piles of forms and of precedents—calculated to extinguish all desire of knowledge, even in the most thirsty soul."[10]

In these new circumstances, the immature face Cayley presented to the legal world might well contrast with his assuredness in dealings with the mathematical community. This unimposing legal novice sitting in pupil-rooms off Chancery Lane, with his "piles of forms and precedents" drew inspiration directly from such eighteenth-century masters as Laplace, Cramer, and Vandermonde, and he confidently compared his own published work with such established senior mathematicians of the day as Cauchy in France, Hesse in Germany, and Hamilton in Ireland.

Training for the bar could be flexible and the time spent learning basic legal skills varied widely. The pupil-rooms were open from ten in the morning until ten at night but attendance was self-selected and training could make minimal demands. Many treated their pupilage as a pastime and a few hours of legal work was a diversion from a busy social calendar. Wilkie Collins, author of *The Moonstone,* who entered Lincoln's Inn a month after Cayley, claimed to have spent only six weeks studying law before his call to the bar more than five years later.[11] Others spent their whole waking hours on legal tasks. As a barrister's apprentice, Cayley learned the basic rudiments but judging from his continuing mathematical activity, he spared his legal books. There were no compulsory lectures to attend or examinations to pass. Students were merely required to read exercises at the Lincoln's Inn bar table and gain the approval of the supervising barrister. Cayley's legal training was very informal. The custom of giving lectures to trainee barristers commenced only in 1852 when a Council for Legal Education was established, whereupon lectures on a range of subjects were delivered, such as the series on *Equity* given at Lincoln's Inn by the Reader in the Law of Real Property. Even after this reform, there was a degree of latitude, there being no written examinations until 1872 when trainee barristers were required to submit to a set of undemanding tests.

In Cayley's year at Cambridge, George Denman also took the university route to Lincoln's Inn. As "Captain of the Poll," the top student of the "Polloi" who sat for the nonhonors degree contest, he qualified for the classical Tripos by virtue of being the son of a peer (although he was opposed to this loophole in the Cambridge system in principle). A brilliant classicist, Denman became a fellow of Trinity in the year following Cayley and went on to become a celebrated judge. Summing up the requirements for the bar in the 1840s, he wrote: "[b]ut though nothing in the shape of a legal education was required, there was a distinction made in the time required for qualifying between graduates of the universities and other mortals. And I think in the case of the latter there was a sort of examination in Latin. But it was also usual for the law-student to enter the chambers of some learned pundit or pundits and work or not as he pleased, and as much or little as he pleased in those chambers during the years of his studentship."[12]

Cayley thus entered the society of Lincoln's Inn by the privileged route as did most Oxford and Cambridge M.A.'s. "Other mortals," as Denman put it, were subject to a five-year qualifying period and had to pay a deposit of £100 to set against communal expenses. The training provided by Christie, on payment of fees of something like hundred guineas a year, was practical and concentrated on the technicalities of the law—it was not a theoretical study of the law as a branch of knowledge. The educative element in a man's background could be supplied elsewhere and it was widely held that the ideal qualification for entry into the profession was a liberal education assimilated at Cambridge.[13] Cayley had merely to wait three years, eat "Twelve Terms Commons" (three meals per term for twelve terms) in Lincoln's Inn, pay fees, and perform a time-honored rigmarole at the ceremony when he was "called" to the bar.

The Lawyer's Life

When Cayley began his pupilage in April 1846, his output of mathematical papers momentarily dropped. Certainly he could not match the rate of production of the previous year, but his pace soon recovered and increased steadily. Not knowing the system (or lack of it) for aspiring scientists in Britain, Joseph Liouville was startled by the thought that Cayley was proposing to give up a promising research career by entering the law. Since 1844, Liouville had promoted Cayley's work in his *Journal des Mathématiques* on a variety of topics, including elliptic functions, algebraic curves and surfaces, and multiple integrals. In some anxiety, he wrote to William Thomson, perhaps worried he was in danger of losing one of his most dependable contributors: "I have been told that he was giving up mathematics and that he wanted to become a lawyer. This would be a real misfortune for science. Nature has done everything for Mr Cayley who must help it by work and patience. By endeavouring to put a little more order and above all a little more clarity in his writings, he would soon be placed among the most distinguished analysts of the times. England owes it to itself and Mr Cayley owes it to his country and to all those who love geometry not to allow such a clear obvious vocation [in mathematics] as that to be lost."[14]

Academic positions did arise (though not enjoying a high prestige, as they did in France), but Cayley does not seem to have been attracted. His congenial position as a trainee barrister had to be taken into account, and academic opportunities in Glasgow, Cork, and even the part-time Gresham professorship in London were allowed to slip by. His eyes were fixed on a legal career and he quickly adjusted to this way of life, suggesting a delicate balancing act between the two lives he was living. But he was young and energetic and managed the two activities brilliantly. He had developed prodigious powers of concentration and could work intensely at long stretches. About one two-week period at this juncture he noted: "At the age of twenty-six during fourteen days, [I] was only three hours per night in bed, and on two of the nights was up all night preparing for [mathematical work]."[15]

Invariant theory continued to hold Cayley's imagination and he found several areas of application for his "hyperdeterminants." These reinforced the view that invariant theory was central—not only to specific mathematical topics but to mathematics generally. Cayley saw invariant theory as a *universal* method that could be applied to problems in both algebra and geometry, and this made it important enough for him to devote so much of his energy to its progress. In its application to the solution of polynomial equations, it provided a method in stark contrast to the theory of Évariste Galois, each theory answering different kinds of questions. Though different in character, both theories yielded striking results.

Around the beginning of 1846, Cayley found the calculation of invariants necessary for his study of elliptic integrals. Invariants were also present when he considered geometric duality, or as it was better known then, geometric reciprocity. This topic was one for which he had especial high regard, remarking on the "endless variety of propositions and theories" of Plücker's idea of duality in geometry. He discussed this in the context of his study of the Tetrahedroid (a surface of order four in which the intersection with the faces of a tetrahedron is composed of two conics), and he touched on a special case of it, Fresnel's wave surface, in connection with a problem suggested by Thomson. In the July issue of *Crelle's Journal*, he proved the intriguing result that the reciprocal surface of a Tetrahedroid was a surface of the same kind.[16] In this problem, he was involved with invariants, but he found other contexts in which they also appeared. No opportunity was missed to note their utility, and again, in the same year, he reiterated their importance:

> I have only put forward these theorems [in geometry] (without seeking to prove them), so as to show their link with the theory of elimination and with that of hyperdeterminants; the latter in particular that one needs to use, I believe, to demonstrate the formula given above . . . and in order to find the hyperdeterminant by means of which one determines the points of contact of the double tangents. I would be very pleased if this research were able in some way to facilitate the solution of the problem of reciprocals of *surfaces:* which still remains in complete obscurity.[17]

Though these were important applications, his progress in invariant theory, was intermittent. He continued to grapple with difficulties caused by the independence question—the "problem of the syzygies."[18]

The summer of 1846 found Cayley on a visit to the North Yorkshire moors, which he combined with tutoring a Cambridge reading party. At Kettleness, just around the headland from Whitby, the country and coastline in this isolated part of England is exceptionally rugged. He enjoyed the sea-fogs and the freedom of the Martian landscapes in that part of the world. Somehow he coped with unreceptive pupils but he had limited success with one William Eccles. The son of a Blackburn cotton merchant in his first year at Cambridge, this student would become a candidate for the poll degree. "I can make nothing of Eccles," reported Cayley, "e.g. he has to find $\log(2 - \sqrt{2})$, calculates $\sqrt{2}$ by logarithms, and cannot for the life of him discover that the next step is to subtract it from 2."[19] It is pleasant to record that Eccles was more than satisfied with his learning experience and afterward wrote to Thomson: "I had a very pleasant 'long' [vacation] with Cayley and owe you my thanks for having recommended me to him."[20]

The remoteness of Kettleness was ideal for thinking about mathematics and Cayley used the opportunity to attempt some problems stemming from Jacobi's papers. His plans were ambitious and he appealed to Boole to solve a problem would lead to "the general formula for multiplication of elliptic functions."[21] Boole too had thoughts of a summer break since the demanding teaching schedule at his school in Lincoln caused his health to suffer. By coincidence, he planned a holiday in Yorkshire but at the seaside resort of Hornsea farther south than Cayley. He too enjoyed the summer prospect of putting some distance "from the resorts of fashion and the noise of the great world."[22]

Thomson spent the summer of 1846 organizing his application for the Glasgow Chair of Natural Philosophy, garnering support from the widely dispersed mathematical community. If he had not met potential backers through his duties as editor of the *Cambridge and Dublin Mathematical Journal*, there was a fair likelihood he had corresponded with them on scientific matters, such was his expertise in networking. De Morgan gave him a testimonial on this basis of his acquaintance through the editorship, but Cayley was able to write a more penetrating reference from his firsthand knowledge of Thomson's ability:

> From having been for some years past, continually in the habit of working and discussing mathematical questions with Mr Thomson, and so having had considerable opportunities of perceiving, as well his enthusiasm in such pursuits, as his complete familiarity with the modern theories of Mathematical physics, & the fullness & fertility of his mind upon these subjects; judging also from his already published researches, some of them mere hints and outlines of investigations with which he is occupied; I am convinced, that in addition to his perfect capability, of discharging the ordinary duties of a Professorship of Natural Philosophy, he is most likely to contribute largely to the future progress of Science & to add to the reputation of any institution with which he may be thus connected.[23]

During his ramblings in Yorkshire, he kept Thomson informed of his movements and asked for news of the Glasgow chair. He was so confident that Thomson would succeed that he could afford to tease him about future European prospects:

> Where are you, and what have you been doing—your Canvassing & testimonializing etc. has not been occupying you all this time I hope; do you know anything at all more certainly about it yet—what a comfort it will be to Crelle to be able to direct to you "Professeur de Mathématiques à l'Université de Glasgow"—do announce it to him. Have you seen any Crelle's or Liouvilles, or Comptes rendus—I am out of the world here of course—mathematical or otherwise—Deighton [a Cambridge bookseller and publisher] did send me some [of] Cauchy's [publications] as I particularly directed him *not* to do; I am beginning to think with you that Cauchy does write too much by half; and I generally care very little for the other half. I came here intending to do little mathematics,—which I have so far kept to, that I have been very unsuccessful, and done very little good in them, but I have spent time enough about it.[24]

Through their extensive links, the Deighton brothers kept Cambridge academics supplied with works from Continental publishing houses. No work was too specialist for them, so that "Cauchy in Yorkshire" was all part of their efficient service. It is notable that Cayley, of all people, should complain about Cauchy's overproduction, however lightly he intended the remark that Cauchy wrote "too much by half."[25] He too suffered from Cauchy's affliction and could not stop himself, and his intention to abstain from mathematics while holidaying in Yorkshire was impossible to meet. He was a driven man. Continuing the letter to Thomson, he reported: "I succeeded in proving a formula of Jacobi's for the multiplication of Elliptic functions—and have been trying to integrate his partial dif[ferential] equation, but it wont do—there are very odd breaks in the law of the terms, and I can make nothing of it."

Putting aside their own research, the other two of the trio, Boole and Thomson, were more intent on obtaining academic positions. A week later, Boole wrote to Thomson: "A letter which I yesterday received from Mr Cayley contains, after a due measure of *linear transformations* and *elliptic functions*, an extract from a letter of yours in which you are so good as to suggest that I might possibly succeed in obtaining one of the new Irish professorships."[26]

From the North Yorkshire moors, Cayley journeyed south to Boole's home in Lincoln. He was settled in training for the bar at Lincoln's Inn, but Boole needed an academic appointment. The tip from Thomson that there might be a chance in Ireland encouraged him to apply for the new position at Cork, one of Sir Robert Peel's projected Queen's Colleges (the others being at Galway and Belfast). In their setting up, sectarian controversy was not far below the surface and the mix of religion and educational matters highly contentious. The ideal of the colleges being open to all (and thus branded as "Godless Colleges"—this time the charge came from the Roman Catholic hierarchy) was an attraction for Boole. He enlisted Cayley's help and on the day of the request, a glowing testimonial was sent by return:

I am persuaded, from all I know of Mr. G. BOOLE's Mathematical acquirements, (as well from his published memoirs in the *Philosophical Transactions* and the *Cambridge Mathematical Journal* as from my own correspondence and personal intercourse with him,) that they are such as cannot fail eventually to insure him a high place among those who have devoted themselves to these studies—and that his appointment to a chair of Mathematics, at the same time that it placed him in a more favorable position for the pursuits in which he is so likely to distinguish himself, would be a sure means of adding to the efficiency and reputation of the institution in which he held. The principal subjects to which Mr. BOOLE has applied himself have been multiple definite Integrals, and ordinary and partial differential Equations, subjects (it is almost needless to mention) of the highest interest and importance, for their own sakes and in reference to physical science (with a view to which in a great measure Mr. BOOLE has been studying them) and from the amount of attention which modern Mathematicians have bestowed upon them; but I may also notice the general theory of Symbolical operations, to which many of Mr. BOOLE's researches (in the memoir which obtained the gold medal of the Royal Society, and elsewhere) relate, and his very original and ingenious investigations on Linear Transformations.[27]

Other testimonial writers (including Thomson, A. De Morgan, Robert Leslie Ellis, Phillip Kelland, Charles Graves) gave supportive reports of Boole's prowess, but Cayley, who was most familiar with his work, could supply first-hand information to the Board of Electors. It is characteristic of him to assume the electors would be interested in the detail and he fondly imagined that they would be at the forefront of Boole's specialisms as well. Of particular interest is his assessment of Boole's "original and ingenious investigations on Linear Transformations."

Cayley was hopeful Boole would succeed at Cork, both for the well-being of mathematics and for Boole himself. The *official* testimonial written, Cayley appended a covering note for Boole's attention: "I wish it may be of any use; believe me, I should be most sincerely rejoiced in your success, if you obtain an appointment, for which I believe you to be so well fitted, & which will give you such opportunities of going on with mathematics & be I should think so pleasant for you: I wish I was more practised in the art of writing them, I am afraid I have not made half enough of it."[28] Cayley's understanding of Boole's situation went beyond his appraisal of Boole's technical abilities in mathematics. He wrote to Thomson of the possibilities at Cork: "it would give him so much more leisure & opportunities for going on with his studies; he will be badly off I am afraid as far as the mere number of testimonials goes but it is to be hoped whoever appoints will have sense enough not to regard that—however he [Boole] seemed very well satisfied with his present employment—gets on well with his boys he says, which after all must be what would make the pleasantness or unpleasantness of it."[29] Of the three friends, Thomson, the youngest at only twenty-two years, was the first to land a position in the academic world, and the good news of his election to the Glasgow chair came through days later. Boole had to wait a few years longer for Cork but Cayley had seventeen years before his return to academe.

Working from different perspectives, both Thomson and Boole sparked off Cayley's genius and suggested new lines of exploration. Boole regarded Cayley's work highly, not only in invariant theory but also in the theory of integration. When he wrote to the new professor at Glasgow, he commented on the inter-leaving of his and Cayley's work: "Cayley has been working a good deal at the subject [of integration of functions] since I showed him my formula and I enclose you two letters which I have received from that most indefatigable and I may add most favoured wooer of the Nymph Mathesis. Have the goodness to return them. Cayley has gone over nearly the same ground that I have done diverging a little to the right where I have gone to the left and vice-versa. My theory of fractional differentiation is, I think, a little more general than his but I have not completed it."[30]

Cayley's very earliest papers (his second and third) were concerned with the integration of functions and he made subsequent contributions spurred on by Boole's encouragement. In a continuation now, he further explored the link between Thomson's and Boole's work on multiple integrals and their evaluation in terms of the gamma function.[31] Boole tempted him to study optics but his response shows he was reluctant to wander too far from pure mathematics: "Physical Optics seems to be a fatally seducing subject, it has attracted so many to itself just now: my only acquaintance with the sort of analysis that occurs in it, is derived from the theory of heat: & some memoirs of Cauchy's that I have looked thro[ugh]' sufficiently to get an idea of them, without properly studying them. . . . I have remained tolerably constant to linear transformations."[32]

Despite his protestations, Cayley was encouraged and within the next few months he published a paper on caustic curves caused by reflection in a circle. With the occasional "Note" placed in the *Philosophical Magazine,* his output appeared in *Liouville's Journal* and *Crelle's Journal* and also in the revamped *Cambridge and Dublin Mathematical Journal* under Thomson's editorship. Cayley provided expertise in judging submitted papers, and he provided reports that were invaluable to Thomson. He could also send him his own papers in case the editor was short of copy—in one postscript he added for good measure: "I have 14 papers for you—which I will not trouble with just now, as you seem to have stock enough."[33]

With the new year, Cayley read Jacobi's "very elaborate" memoir in *Crelle's Journal* "Theoria novi multiplicatoris systemati aequationim differetialium vulgarium applicandi" ["A New Theory of the Multiplier for a System of Ordinary Differential Equations and Applications"] and found a connection with one of his earlier paper on skew-symmetric determinants, known as *déterminant gauches.* His letter to Thomson in February 1847 informed him: "I have learnt some very curious things from Jacobi's paper Theoria novi multip. &c about my 'Determinans gauches' (in this case luckily not interfering in the least with what I had written, I must make a note about it in Crelle to explain the connection), so my perseverance in reading the paper has been rewarded in more ways than one."[34] Cayley showed how the notion of the ordinary determinant could be extended to include the Pfaffian, an algebraic form Jacobi named after the German mathematician Johann Friedrich Pfaff. This extension was different from the earlier *n*-dimensional determinants—Pfaffians can be considered as *half-*determinants. This duly appeared in *Crelle's Journal* and contained his result: that a symmetrical skew-determinant is the square of a Pfaffian, generally regarded as one of Cayley's most elegant theorems at the time but now somewhat arcane.[35]

Jacobi's paper on differential equations prompted another response from Cayley, this time on the application of differential equations to dynamics. Thomson wrote to Boole in February: "I have received recently from Cayley a paper wh[ich] seems exceedingly good on differential equations, (such as those of dynamics) with reference to properties proved by Jacobi, about the possibility of integrating one of a system by a factor. I also received a short paper from him yesterday on a geometrical theorem of Jacobi's." The link between dynamics and differential equations, ostensibly suggesting mixed mathematics, was mostly a piece of camouflaged pure mathematics. When papers on genuine pure mathematics arrived on his editor's desk, particularly those submitted on symbolical algebra, he passed them on to Boole or Cayley for judgment. On one occasion he wrote to Boole: "I have some compunction about the way in which I save myself trouble & put it on you and Cayley; but still in such cases the benefit is not merely a saving of trouble, as I could not come to sufficiently satisfactory opinions on some of these papers not being *au courant* of the subjects."[36]

By May 1847 Cayley had succeeded in giving a direct explanation for a result of Jacobi's on elliptic functions in the *Fundamenta Nova* and during the year continued with the subject. He sent his paper to Thomson with "a few short ones that I want to knock off as soon as I can—not that I suppose you will have room for them—but please take care of the wee things."[37] Cayley was adjusting to life away from Cambridge, but he maintained contact with his alma mater. His membership in the Cambridge Philosophical Society continued and during the whole period of his law pupilage he served on its council. With Trinity College, he acted as their senior examiner for the annual college examinations.

In June, the British Association met at Oxford and the mathematicians were well represented. Many of those who had been at the Cambridge meeting two years previously were also at Oxford. Hamilton came across from Dublin, this time to connect his quaternions with the theory of the moon, and Boole gave a paper on the theory of integration. William Spottiswoode, the young Oxford mathematician, Stokes, and Thomson were also present. Adams was on speaking terms with Leverrier after the "race for Neptune," a *cause célèbre* in the previous year portrayed in the popular press as a French victory. Thomson urged Cayley to attend, not least to meet the French contingent and Carl Borchardt who was visiting the Parisian mathematicians, and planned to attend: "Le Verrier has been here [at Oxford] from the commencement, and has had a good deal of conversation with Adams. He is not working at Neptune at all at present, nor I think has he been since he (Neptune) was seen. [C. W.] Borchardt (who has written on hyperdeterminants, &c) is here at present, & wishes much to see you. I promised him an introduction to you. He has been working at a great many of your subjects. He has a short paper here on Jacobi's last multiplier, but it is very different from your paper in the last number [of the *Cambridge and Dublin Mathematical Journal*]."[38]

Following the Oxford meeting, Thomson planned a Swiss mountaineering expedition. After his own sojourn in Switzerland in the latter half of 1843, Cayley was able to offer useful advice on the worthwhile mountains to conquer, even with the help Thomson might gain from John Murray's "Handbooks for Travellers":

> If you want to immortalise yourself, go over the Weissenthor—or you may try my more modest & unpretending scheme (which turned out a failure however [on his visit of 1843] thanks to a soft morning) of getting up to the top of it on the Zermatt side and returning. But I hope I shall be able to talk *all these* routes over with you, and give you my best advice upon them.—you know a story of an Irish boatman wishing to show his respect to somebody, for the love he bore the family—"I'll take you *nearer the falls*—than any one ever went before"—and so I can promise to recommend you all the *soi disant* [so called] (or murray *disant*) dangerous and difficult passes. Heigh ho. I wish I was going too; I should like to finish off with the Col de Erin, and the Col de geant for instance.[39]

Instead of the Continent, Cayley planned a tour in Scotland, playfully suggesting to Thomson that "it is very unpatriotic of you not to come there too." He returned to Chancery Lane after a five-week trek in the highlands, which covered an extensive figure-of-eight route with its crossover at Inverness and its northernmost point at Ullapool. The tour, which reached into the heart of the eastern and western highlands, began and ended with Loch Lomond, a walking adventure he recounted for Thomson's benefit: "Considering LochGoilhead as my starting point I went by Inverary, Glen Etive &c. to Fort William—then a long cut across the moor of Rannoch & by Loch Ericht & across country to Foyers & then to Beauly, Ullapool, Loch Maree—the isle of Skye (charming weather it was there, a perfect hurricane of rain & wind. . . . From Skye thro[ugh]' Glen Affaric to Inverness, thence to Aviemore, & across to Braemar. I blundered about the route, thanks or no thanks to my map, and climbed Cairn Toul by mistake for Ben Muckdhui [*sic*]—not that there is much difference in the height."[40] These mountains in the Cairngorm range situated in North-East Scotland are both more than 4,000 feet and only 150 feet less than Ben Nevis, the highest peak in Britain.

Through bracing "horizontal rain" Cayley progressed to Aberfeldy, which he had visited with his reading party after the Tripos in 1842. A solitary existence often accompanies a mathematical pathway, and this is in evidence in the hilly country where Cayley spent time around Glen Lyon. So too we can sense Cayley's ability to finish what is at hand, as his letter to Thomson suggests: "There is a good deal of pleasure in meeting with some human beings one knows, I must confess travelling by one's self is occasionally a little dreary—& so it made an agreeable episode, coming to a place one knew & where one was known. I walked thro[ugh]' Glen Lyon to Tyndrum—and then, for I hate spinning out a tour when one gets out of the country one likes, in short not finishing when one has done, I came as fast as I could home."[41]

Such extensive hikes offer opportunities for the body to weary and rest while the mind unconsciously reflects—and mathematicians are very aware of the right idea striking when least expected. The mind of an active mathematician is always receptive to a happy mathematical thought but on this tour, as Cayley continued his letter to Thomson: "I have been guiltless of any Mathematics all the time except a note for Crelle which is subscribed 'Loch Rannoch' & which will puzzle his geography a little I expect, & that was something I had in my head before I started, so I really was not visited with an idea the whole time."[42] Out in the hills, inspired by the landscape, can we really take Cayley's not being "visited with an idea the whole time" at face value? It is inconceivable.

James Joseph Sylvester

In London, only a few minutes walk through the labyrinthine passages of the Inns of Court, another young man had begun his pupilage. In appearance James Joseph Sylvester was of small stature, barely five feet in height with a shock of black hair and sharply defined features. The flimsy pair of wire-framed spectacles worn while reading indicated this quick-witted man had something of the scholar in his makeup. Desiring a law qualification, he entered the Inner Temple as a pupil in July 1846, only a few months after Cayley entered Lincoln's Inn.

In today's terms, Sylvester was a mature student and one far more experienced with the ways of the world than Cayley. It was not the mere fact that he was seven years older but he had far more worldly experience. He had been a professor twice over, was a Fellow of the Royal Society, and, having lived abroad in America, had experienced several "hard knocks" in crossing the threshold into adulthood. In December 1844, he secured a position as an actuary of the Equity and Law Life Assurance Society, becoming its secretary four months later.

Sylvester is one of the most extraordinary personalities in the history of mathematics, and his life is inseparably intertwined with Cayley's. From his early days he showed a remarkable precocity for mathematics, while his great fortune was to obtain a student's place at Cambridge where he could realize this mathematical promise.

Born of Jewish parents into a family of six boys and two girls in the East-End of London in 1814, James Joseph was their youngest son.[43] His parents were financially secure, and he was educated in several private schools for Jewish boys in North London, and at Neumegen's boarding school in Highgate, where his mathematical skills were noted by the school's proprietor, himself a man with mathematical credentials. At the age of eleven, he was sent to be tested by

Dr. Olinthus Gregory, professor of mathematics at the Royal Military College Woolwich and one of the moving spirits in the establishment of University College, London. Gregory was so impressed by the boy's abilities in algebra that he asked to be informed of his subsequent progress.

At the age of fourteen, Sylvester spent five months at University College, where he was one of De Morgan's pupils. Life was far from easy for the boy. His precocity in mathematics and his Jewish background contributed toward his being the butt of practical jokes by class bullies. On one occasion, he was pushed too far and took a knife to his tormentors. It was a spirited defense so characteristic of his impulsive behavior. The scrape caused his withdrawal from the college and he went to Liverpool, where he lived with aunts and attended the Anglican school of the Royal Institution. He was then fifteen and at Liverpool his mathematical abilities ripened. Like Cayley at the same age, he won a range of the school prizes.

Sylvester remained in Liverpool for less than two years, yet another completed passage in an unsettled existence. The youthful experiences of being moved between schools in different parts of the country encouraged a restlessness and a roving existence. In adult life he became a cosmopolitan figure and traveled at every opportunity—he could leave at a moment's notice so natural was it for him to be on the move. After Liverpool, he returned to London, where he studied with Richard Wilson, a fellow of St. John's College, Cambridge, a move which proved fortuitous. Wilson was also a leading Hebrew scholar and one with a social conscience. He would have been interested in helping a boy of Sylvester's background and supporting his application to enter his old college. Though Cambridge University was an Anglican stronghold, it never closed its doors on talent.

In 1831 Sylvester was admitted to St. John's College as a sizar, a level of entry reserved for the less well-off section of the population. This amounted to a reduction of fees but by the 1830s did not include the duties of servitude as required in earlier centuries, although it did provide a daily diet passed down from the fellow's table. At St. John's, he struck up a friendship with another sizar, John William Colenso, who became known for his textbooks on elementary mathematics but later achieved greater notoriety in Victorian society as the heretical Bishop of Natal. It is a mark of Sylvester's sense of loyalty that he actively supported the Bishop when he came to England to defend himself in the 1860s.

Like Cayley, Sylvester was no "mere mathematician" and at Cambridge he attended lectures in classics and science. In chemistry, he was taught by the professor of chemistry, James Cumming, an eminent man who specialized in thermo-electricity. In mathematics he was taught by John Hymers, the writer of several texts on geometry, and also by Philip Kelland, then a Cambridge coach before his election to the mathematical professorship in Edinburgh.

Kelland forecast that his rather apt pupil would take "an eminent position among the mathematicians of Europe."[44] But Sylvester's path was never smooth and his Cambridge studies were interrupted, possibly owing to illness. Eventually, in January 1837, he took the Tripos examination and was placed second wrangler. When the results were posted, Joseph Romilly recorded in his diary: "s^t Johns has the first 3, viz. [William] Griffin [the Senior Wrangler], Sylvester (a Jew!!!) & Brummell. Sixty years ago Trin^y had the 1st 3 . . . [George] Green of Caius (son of a miller) who was expected to be S.W. was only 4th. [Wrangler]."[45] Had Romilly continued he would also have written "Gregory of Trin^y the fifth Wrangler."

Following the mathematical Tripos examination, but without being admitted to the bachelor's degree, nor being allowed to compete for a Smith's Prize, since he was a Jew and could not subscribe to the Thirty-nine Articles, Sylvester quickly showed his ability in research by publishing original work. On leaving the university in 1837, he applied for the Professorship of Natural Philosophy at University College, London, which had become vacant. De Morgan remembered him at the school ten years earlier and informed the appointing committee that "he never, before or since, saw any mathematical talent so strongly marked in a boy of that age."[46] On appointment his first research papers were written on suitable subjects for a professor of Natural Philosophy: "Analytical Development of Fresnel's Optical Theory of Crystals," "On the Motion and Rest of Fluids" and "On the Motion and Rest of Rigid Bodies."[47] It was partially on account of this work in physics, or as close to mathematics as physics would allow, that Sylvester was elected to the Royal Society in April 1839 at the young age of twenty-four, the "author of several papers on Physical & Mathematical subjects."[48]

While Sylvester shone in research, he had difficulty coping with the lecturer's craft. He was a man of ideas, inspirational above all, while simple practical matters such as drawing diagrams on a blackboard utterly defeated him. De Morgan was not so impressed with his lecturing performance in the classroom at University College: "[w]hen he was with us he was an entire failure: whether in lecture room or in private exposition, he could not keep his team of ideas in hand."[49] That was the way with Sylvester but there was no denying he was inspirational. In 1841, his restive spirit came to the fore once again. He had battled with the college authorities over the curriculum and the altercation may have prompted him to think of his London professorship as no more than temporary. He was a young man and the fact that an elder brother had settled in New York City may have suggested America as a possible haven.

The chance came when Sylvester was offered a one-year trial appointment as the professor of mathematics at the University of Virginia in Charlottesville. His arrival in November 1841 augured well; welcoming bonfires were lit by the students and they anticipated much from the young English professor. Academic standards were high in Virginia and, for his part, Sylvester was excited about the prospect of spreading the mathematical word so far from home. His transplantation to American soil was at one with a role in which he was totally at ease and at his most effective—of being a mathematical missionary.

Another side of the adventure was not so pleasant. When Sylvester accepted the position, he was going to a university where serious student rebellions had occurred and in which murder and the destruction of faculty homes had been recorded. There was opposition to his appointment expressed in the *Watchman of the South* on account of him being both a Jew and an Englishman, and therefore, it was argued, unable to instill the "vastly improved discipline" required in the student body. Into this highly charged Virginian culture stepped Sylvester from his classroom in London. In England the learned and learners were gentlemen, and participants knew their proper place: he was the professor and they were the students who sat at his feet. He was ever conscious of "position" and expected his pupils to be respectful, obedient, and deferential. In Virginia one William H. Ballard, a student hailing from New Orleans did not fit this mold and was to be Sylvester's nemesis. The professor complained about his behavior and each time Ballard responded in kind. Perhaps fearing further trouble from the student body the faculty failed to give Sylvester the support he desperately needed. Whether he was asked to resign or whether he voluntarily decided to leave is uncertain, but at the end of March 1842, he left Virginia and traveled to New York to stay with his brother. In the spring of the following year, he unsuccessfully attempted to obtain a post at Columbia College.

Sylvester returned to England in the summer of 1843. His stay in America had been disastrous to his mathematical creativity—he experienced a desultory period and published little. In the upheavals of his domestic life, some of his best years had been lost. By returning to England, a limited mathematical existence was possible, but it would be in his own time without the financial means and leisure provided by an academic post. In attempting to retrieve his creative mood, he attended the British Association meetings at York in 1844, and, as we have seen, at Cambridge in 1845. He spoke at the York meeting on number theory but the way back to mathematics was difficult. There was a false dawn in late 1845 after the Cambridge meeting when his spirits were briefly revived. The expansion of the *Cambridge Mathematical Journal* to include Dublin in its title caught his eye and he wrote to Thomson: "I hail with much pleasure the auspicious conjunction of Cambridge and Dublin Mathematics under your able guidance—as likely to lend in a most material degree to introduce a Catholic Spirit among your readers and exhibit an agreeable and instructive spectacle of the union of the Sister Sciences of Analysis & Geometry: the one [Analysis] the creation of a Metaphysical & Utilitarian Age—the other [Geometry], the revival under a maturer form and animated by a bolder genius, of the beautiful speculations of a more objective and imaginative period in the world's history."[50]

Sylvester started as he went on, to signal first and deliver second. After the meeting he mentioned "some old papers lying by me, which I think it would be a pity should be consigned to the moths & worms, as they relate to a department of thought [combinatorial mathematics] which must some day take its place among the Mathematical Sciences. . . . The ideas are entirely new and will I believe be found at some future day, to admit of important theoretical & also practical applications. I lay claim to the honor of bringing up, the Third & Unbethought of Grace (still a mere child) who is wanting to complete the Mathematic Braid—Will you kindly take her by the hand & introduce her to the world?"[51] Mathematical problems, such as these, involving games of chance and combinations of numbers are very often formidable and require insight and clear thinking. Sylvester excelled in this combinatorial aspect of mathematics, an ability he shared with Cayley, and it was an essential one they both applied widely. Regrettably, on this occasion, his new mathematical work was left incomplete.

Though there was a hiatus in Sylvester's published work, he never gave up hope of returning to the creative scientific world of London and Paris. He placed importance on personal contact and made it his business to meet the leaders of his field. The array of testimonial writers he could assemble attests to this: even on the way to Virginia, he had stopped off in Dublin to meet Sir William Rowan Hamilton at Dunsink. By the beginning of 1847, he had met Michel Chasles, who in 1846 had been appointed to the new chair of modern geometry at the Sorbonne. As a sign of friendship, Chasles gave him a treasured copy of Descartes' *Géometrie* from his collection. It was "a recompense for a great service," possibly the smoothing of Chasles's plans for a trip to England which he would make in the following November, or perhaps for an offer to translate his geometrical work into English.[52] Sylvester valued his scientific contacts in France and worked hard to sustain them, but he had also to make a living. He busied himself with his pupilage at the Inner Temple and the life of an insurance man in the newly established position with the Equity and Law Life Company at 26 Lincoln's Inn Fields.

Sylvester was exceptionally busy. As well as his work for the Equity and Law Life Company, he took private pupils and is said to have taught mathematics to Florence Nightingale. Through his erstwhile tutor Richard Wilson, a guiding force in the establishment of the College of Preceptors, he was appointed as one of the first examiners. The college was set up as an examining authority to make awards and oversee the education of schoolmasters and thus safeguard standards in the education of children. Wilson was its dean for eleven years and editor of its journal, the *Educational Times,* for a decade after its founding in October 1847.[53] Through the college, Sylvester maintained contact with the world of education. In the business sphere, he was an active member of a committee instrumental in setting up the Institute of Actuaries in 1848—as a "scientific and practical association" for managers and actuaries of the country's Assurance societies. He was one of its first vice-presidents and also acted as an examiner of this fledgling institution.[54] Sylvester was a man to seek new enterprises, new ideas, and in the first breaths of excitement about any proposed venture, this seeker after novelty flourished most naturally.

Common Bonds

Apart from them being mathematicians, there was no obvious reason why Cayley should have been aware of J. J. Sylvester around London's Inns of Court or that their paths would cross. Of late, the law had become a popular career choice, and the Inns of Court were swarming with new entrants. The two had also chosen different placements, Cayley at Lincoln's Inn and Sylvester at the Inner Temple. While both had been to Cambridge and had there distinguished themselves, they had been there at different times and had resided in different colleges; they existed outside each other's networks of intimate links and, consequently, there would be few shared memories. Then there was the difference in pedigree. With the assurance of birth, Cayley was planted squarely in the middle-class stratum of English society while Sylvester only yearned to be accepted into this fold of privilege—his Jewishness prohibiting free entry to a group in which Anglicanism was the norm. What they did have in common was an unqualified attachment to mathematics and, seeing each other's papers in the journals, it was perhaps inevitable that they would one day meet. As they trained for the bar, they formed a liaison that was to last for fifty years.

Of the hundreds of written communications that passed between Cayley and Sylvester, most are concerned with mathematics. Sylvester allowed the tribulations of life to intrude, but Cayley hardly ever did so; Cayley valued the partnership but he felt no need to wander too far from mathematical interests in swiftly written notes to his confrère in the adjacent inn. In his letters to William Thomson, we observe Cayley's wittiness and sense of fun, but they were Cambridge friends with broad-church interests in the promotion of "science" but nonoverlapping in the specifics. Thomson would have no need for an in-depth account of Cayley's technical mathematics and would not want it since he was generally antagonistic to purely abstract mathematics. Sylvester, on the contrary, was intensely alive to these issues and mathematics would naturally be at the top of the agenda.

While Cayley moved easily in the Lincoln's Inn environment and his career was on an upward path, Sylvester had to adjust to starting a new life once more. He had thrown away one academic position at University College, and when he returned to England from America, he would have seen others with a claim on the kind of position he had relinquished. Opportunities to enter the academic world were few, even for Anglicans, and they were practically nonexistent for

Jews. The prospect open to him, perhaps obvious in hindsight, was to launch himself as a man of business in which he could use his mathematical skills. To be a barrister-at-law would give him an entrée into society.

Cayley had followed Sylvester's work in the mathematical journals and no doubt the openness of his personality would have placed Sylvester at ease. They had their mathematics in common and the convenience of being daily in the same part of London; for relaxation, they might engage in a game of chess for both were avid players. Sylvester's name cropped up again in connection with some French work on geometry when Cayley wrote to Thomson:

> I have been making acquaintance with Chasles since I saw you; he has been in London for a few days & to Cambridge & Oxford and started on his return [to Paris] this morning I believe. I liked him very much, all that I saw of him. He is going to publish the lectures he is now delivering [his current lectures on geometry]; Sylvester is going to translate them so that at last one will have an English book on Modern Geometry: I hope it will get introduced into the Cambridge Course. —He seems to have been mostly working about the geodesic lines on the Ellipsoid & mentioned some very curious theorems about them.[55]

At this stage, Cayley did not know Sylvester well and evidently mistook his style in conducting mathematical research—which was to make his own investigations and not afford scholarly translations of existing work. In any event, Sylvester did not complete the task of providing a text on Modern Geometry.

On Wednesday 24 November 1847, Sylvester sat down to write one of his first letters to Cayley. It began with the formal "My dear Sir," the two yet to be on surname terms, which indicated the Victorian seal of friendship and acceptance. His letter contained a host of mathematical problems aligned to the theory of numbers. Sylvester invited Cayley to consider one particular question, indicating: "[i]t occurs to me that this investigation is likely to prove very congenial to your present line of thought and I make you a Cadeau of the subject not doubting that it will turn to good account in your able hands."[56] This was in the morning, and, in a scenario Cayley would grow used to, he would receive later communications on the same day containing further thoughts.

In the afternoon, Sylvester appears to have received his first impulse of algebraic invariance, and he rushed off his thoughts to his newfound friend: "Ought there not to be some name given to Functions of one or more variables having the property of as it were casting their skins—i.e. of throwing off a factor exactly similar to themselves on substituting certain functions of the variables for the variables themselves? The class of such functions must be very extensive."[57] Though Cayley had noted Sylvester's work previously, he now referred to it in print—in papers on Elimination and Sturm's functions (connected with the solution of polynomial equations).[58] These were topics for which Sylvester had made his reputation and coupled with his work on number theory, he was briefly fired with research ideas, though they languished. His "Cadeau" to Cayley proved an isolated gift and for most of his law pupilage he was mathematically silent.

In the pupil-room of J. H. Christie, Cayley fulfilled his duties, pursued mathematics, and generally kept his own counsel. Sylvester was a man-about-town involved with the world of science obliquely, but establishing his credentials as a man of business. Always conscious of position, he found social pathways difficult to negotiate. Life was a battle for this young man of obvious mathematical ability but he was one unable to attune himself to the sympathies of the Victorian world of class, status, and manners. This was most obvious in his dealings with authority but being sensitive to social position he was also keen to put servants and retainers in theirs. Still, did he have to make an enemy of the assistant secretary of the Royal Society? To make an enemy of someone in the back-office indicates a certain lack of judgment: Mr. Walter White reported that he had "[c]alled on Mr. Sylvester, F.R.S., for his subscription due to the Society—he [Sylvester] took offence at what he called my 'peremptory prescribing of a time' in which he should pay. An atrocious mistake, to call it by no worse name, on his part. He insisted also that I should address him as 'sir' with every sentence, or query, or rejoinder. What a paltry vanity."[59] Making enemies among mathematicians through blatant assiduity in such matters as priority was not so damaging since they grew accustomed to his outbursts and made appropriate allowances. Sylvester had yet to learn the social ease of a gentleman.

In Cayley's measured existence, the common bonds of mathematics were maintained with an immediate circle by correspondence. There was a flurry of letters with Boole on receipt of a copy of his recently published *Mathematical Analysis of Logic,* a publication which was a great fillip to the symbolical algebraists who claimed that algebra was the key to a wider mathematical vista. The Macmillan brothers had set up as booksellers in Trinity Street, Cambridge, but expanded their business and Boole's book was one of their first ventures in book publishing. Thomson was also sent a copy and while he could glance at it, and put it aside, it being on a subject outside his main interests, he recognized the achievement: "The advocates of 'symbolical algebra' must be delighted with such an unlooked for extension of the class of subjects, for laws of symbolical operations."[60] In the forefront of the "symbolical algebra" movement himself, Cayley was an ardent reader. "I hope now you have set to work to examine my principles you will not stop short but prove them to the bottom," Boole wrote to him, "I do not fear the result. I had rather have one such reader as you than a thousand who take everything for granted."[61]

Cayley's and Boole's discussion spilled over to geometry, in which subject, Cayley had well-formed ideas of his own, in particular on the wrong way to use *imaginary* in geometry. Matters raised by Boole in the *Logic* led him into a discussion of this controversial subject: "I wonder we should never have stumbled in our previous correspondence on the subject of my *utter disbelief* of the received 'English' theory of the geometrical interpretation of $\sqrt{-1}$. I would much more easily admit witchcraft on the philosopher's stone."[62] In the "English" theory, originally suggested by John Playfair, as it happens a Scot, we are asked to envisage a plane containing an "imaginary axis" at right angles to the original plane. It was an interpretation of $\sqrt{-1}$ advocated by Cayley's associates at Cambridge, D. F. Gregory and William Walton.[63] While Cayley had a deep-rooted notion of a unique three-dimensional physical space that was Euclidean, he quite freely admitted the notion of mathematical space and mathematical constructs in this space (imaginary points of intersection, points at infinity, n-dimensions, real and imaginary distances) and was adept at employing them. Cayley castigated Playfair's interpretation of the imaginary linked to a plane at right angles to the plane of reference. Disregarding it, he appealed to his oft repeated Greek "ουτως οντα" (really real), a phrase which he used in distinction to plain *real* which was reserved for use with complex numbers. "My own theory as far as I can express it is that a distance *x, whether real or imaginary* is an 'ουτως οντα' capable of being *measured* (that word won't do I admit, but capable of existing) in any direction real or imaginary: somewhat as if beings whose space was of two dimensions only (which I think is inconceivable) had by their science of geometry arrived at the notion of a third dimension."[64]

Cayley's position is almost that of a modern algebraic geometer. "In fact I should admit no distinction between real and imaginary," he continued, "it is only when you draw the figures that the difficulty arises. This is rather idiosyncratical I am afraid, and I do not expect or wish to convert you."[65] Continually at pains to reject this "English" interpretation, ten years later he disparaged it as "the wholly different point of view from which I look at imaginary quantities in geometry."[66] Though Cayley definitely excluded the "English theory," he was prepared to allow a variety of interpretations for "imaginaries" in geometry. They could be explained by the principle of continuity (two disjoint circles "intersect" in imaginary points, for example) but they were also linked to quaternions. Six months after writing to Boole, he returned to the quaternion "imaginaries" and showed how three-dimensional rotations, which he had earlier expressed in quaternion form, could be combined. Boole rejoined with some "philosophical" observations on "Mr Cayley's ingenious researches" which "have recalled to my mind some speculations of my own on the same subject" but he did not attempt to criticize Cayley's mathematics.[67]

During his pupilage, Cayley made every effort to introduce himself to mathematicians with common interests. On the threshold of his mathematical career, he had his Cambridge contacts, his links in Paris and Berlin, Boole in Lincoln, and Sylvester was at hand in the Inns of Court. Now he opened a correspondence with the solitary Rev. Thomas Penyngton Kirkman, living in the north of England. This Anglican minister was forty-two years old and from another social milieu—the son of a cotton and waste dealer from Bolton in Lancashire. Kirkman had been successful at Bolton Grammar School and after nine years working for his father became a student at Trinity College, Dublin, where he obtained a degree in 1833 and opted for a church career. After six years of curacies in Lancashire and Cheshire, he was made rector at the new country parish of Croft-with-Southworth in Lancashire. For the length of his fifty-year posting in the village of Croft he eked out a living, raised a large family, wrote religious polemics, and pursued mathematics. Near Warrington, Croft was too insignificant for the Victorian Ordnance Survey map, while today, standing at the junction of two major national motorways, the world rushes by.

Cayley became directly aware of Kirkman's existence when he was asked to referee a paper "On a Problem in Combinations" submitted to the *Cambridge and Dublin Mathematical Journal* in December 1846. He was a discriminating critic, especially in his young days, and in the same report rejected a submission on geometry which treated the subject in "that way without any reference to general geometrical theories or without any attempt to make a 'Zusammengesetzung' [a composition] of the whole mass of theorems one obtains, it is very uninteresting work." With an appreciative eye, Cayley judged that "Kirkman's paper is decidedly interesting and his main result a very elegant one when it is separated into its distinct cases."[68] The strong recommendation was acted on and the paper published.

Kirkman's object was to investigate systems of triples that can be formed with x symbols so that no two triples contain the same pair. Kirkman showed that x must be of the form $6n+1$ or $6n+3$ and that for such values these systems do exist. This landmark paper became a beacon for "combinatorialists" and has been the inspiration for countless similar problems, some of which are unsolved to this day.[69] A formulation of the "systems of triples" investigation beloved by Kirkman, who developed a flair for posing problems in a quirky literary form, was the related "schoolgirls problem" through which he is best remembered: fifteen young ladies of a school walk out three abreast for seven days in succession; it is required to arrange them daily so that no two shall walk abreast more than once. Kirkman gave one solution and Cayley subsequently solved it himself and linked his solution to the algebra of octaves and to the existence of higher dimensional algebras.

Cayley wrote to Kirkman about the new algebras, which were then being enthusiastically explored and in which Kirkman took a serious interest. Branching out from purely combinatorial problems, Kirkman wrote about the algebra of *pluquaternions,* which "is the fruit of my meditations on Professor Sir W. R. Hamilton's elegant theory of quaternions, and on a pregnant hint kindly communicated to me, without proof, by Arthur Cayley, Esq., Fellow of Trinity College, Cambridge."[70] He rediscovered the octaves for himself, and opened up questions about pluquaternions having more than seven imaginary units and referred to the "*n*-squares problem."[71] Further thoughts found their way into print when Cayley was solving the "schoolgirls problem" after Kirkman had challenged the readers of the *Lady's and Gentleman's Diary* with it. Cayley discussed it with Sylvester and noted "to my knowledge, [it has] excited some attention."[72] Indeed it had, and it was even set to poetry:

> *A governess of great renown*
> *Young ladies had fifteen,*
> *Who promenaded near the town,*
> *Along the meadows green.*
>
> *But as they walked*
> *They tattled and talked,*
> *In chosen ranks of three,*
> *So fast and so loud,*
> *That the governess vowed*
> *It should no longer be.*
>
> *So she changed them about,*
> *For a week throughout,*
> *In threes, in such a way*
> *That never a pair*
> *Should take the air*
> *Abreast on a second day;*
> *And how did the governess manage it, pray?*

Cayley's own solution differed from those of Kirkman and other solvers of the problem; it has been discovered subsequently that there are seven fundamentally different solutions.[73] Cayley's solution did something more than the one published by Mr. Bills from Newark and Mr. Levy from Hungerford, for he linked it with algebra conveyed to Kirkman in the "pregnant hint." He identified a property of the system of seven triples with the existence of the seven imaginaries of the octaves and suggested that a permutation property associated with the "fifteen schoolgirls problem" proved the nonexistence of an algebra with fifteen imaginary units. This type of extension of the algebra of octaves became a problem to which Cayley periodically returned.

Kirkman shared many interests with Cayley, including combinatorics, geometry (in particular, "Pascal's Hexagram theorem"), group theory, and the theory of polyhedra. He tackled significant problems tenaciously, and, although he was an isolated figure who remained outside the metropolitan fold, Cayley respected him and responded to his research. He published his work in curious places, as if to seal priority but not caring if the work would be noticed by the mathematical community. In the pages of the Tory newspaper, the *Manchester Courier,* he listed twenty-five theorems on the Hexagram and related topics. Following this tour de force, Cayley showed the existence of twenty lines which passed through the sixty "Kirkman points" three-by-three in a symmetric manner.[74] An appreciation both shared was the power and pervasiveness of algebra when applied to geometrical problems of this sort.

The Irish Connection

In June 1848, an uncertain time for traveling, Cayley went to Dublin. The European "year of revolutions" was in full sway, and the Paris uprising in February had served to put the British establishment in a state of readiness. England's great Chartist demonstration took place on 10 April 1848 and the authorities in London were nervous. Special Constables were sworn in to keep the peace and Lincoln's Inn provided a cohort. Of the 85,000 special constables guarding public buildings in the capital, some eight hundred from Lincoln's Inn volunteered their services. Armed with staves, furnished with white ribbons as their uniforms, organized in groups of ten to twenty "specials," they went out to guard London against the anticipated mayhem. It all came to nothing and according to Hugh Blackburn (writing to Thomson), "We have been all protecting the metropolis here. I marched up and down the Strand in the rain for some time in a body of Specials, What a hoax it all was."[75]

The crisis averted, the British government, which failed to come to terms with the Irish potato famine, had a further problem on its hands. When Cayley arrived in Dublin, the civic authorities were faced with an imminent Irish rebellion against British rule. Cecil Woodham-Smith painted a picture of a city on high alert: "More and more troops were poured into Dublin; the 71st Highland Light Infantry from Scotland, the 48th Regiment of Foot from Belfast, the 31st from Manchester . . . and under the direction of the Duke of Wellington, Trinity College, the old Parliament House, the Post Office and the Custom House became strong points."[76]

In what can only be described as siege conditions, Cayley attended Hamilton's lectures on quaternions at Trinity College. The high walls that divided Hamilton's lecture rooms from the streets of Dublin were more than enough to keep the trouble out and aficionados of the quaternions safely in. At the beginning of the first lecture, Hamilton publicly praised Cayley's work and acknowledged him as the first mathematician to publish on the quaternions— after himself. Hamilton was keen to differentiate his position from any precocious youngster snapping at his heels. But Cayley did not go unrewarded. As Hamilton recalled a few years later: "Arthur Cayley, Esq., Fellow of Trinity College, Cambridge, who first (except myself) has publicly used the quaternions:

for he published in the *London, Edinburgh, and Dublin Philosophical Magazine* a Paper entitled *Results respecting Quaternions,* of which I have had the pleasure to acknowledge publicly the importance. My second lecture on the same subject was delivered on Friday, June 23, 1848, after meeting Mr. Cayley at breakfast at the rooms of the Messrs. Roberts, in College, where also Mr. Jellett and Mr Townsend breakfasted."[77]

Hamilton adopted an elder patrician's stance in praise of the young man. He wrote to Thomson that he was "delighted that Mr Cayley is writing so vigorously for the [Cambridge] journal."[78] The two schools of mathematics, Dublin and Cambridge, were drawn together by the recently created *Cambridge and Dublin Mathematical Journal* and Cayley's visit can be seen as an effort to cement the union. The difference in mathematical style was immaterial and both Trinity Colleges stood to gain from cooperation. Cayley was an ideal emissary. The young man made a firm impression and Salmon later remembered "how we in Dublin were struck by his proficiency in pure geometry, a subject then much cultivated with us, but which we had been accustomed to look on as too little esteemed at Cambridge."[79] In pure mathematics, the active Dublin school had cultivated pure geometry in distinction to the analytical geometry favored by Cayley. It was not just the quaternions on which Hamilton was so impressed by the young pretender so obviously skilful in "Analytics." Richard Townsend, an authority in pure geometry and principal go-between in Dublin, reported on Hamilton's opinions to Thomson: "—you say the present Number [of the *Cambridge and Dublin Mathematical Journal*] is being *Analytical,* perhaps I might not like it as well as either geometry or physics but there are others who will like it far better in consequence—for instance, *I* do not feel as much interest in Mr Caley's [*sic*] papers as in your's but there is no writer of the present day that one of our first Analysts, if not our very first, Sir William Hamilton, praises like Mr Caley [*sic*] nor one whose papers he reads with greater pleasure—So I dare say he will enjoy a great treat in the coming Number."[80]

The other Irish pure mathematicians, including the twin brothers Michael and William Roberts and John Jellett, were devotees of geometry while Michael Roberts acquired a research reputation in invariant theory and proved some fundamental results. In Dublin, Cayley found an opportunity to comment on their work and add to it. He generalized one of Jellett's results on integration and added to William Robert's work on the transformation of curves.[81]

George Salmon

Most important, on his visit to Ireland, Cayley met George Salmon. He was two years older than Cayley and in this year appointed Donnegal lecturer in mathematics. Salmon was born in 1819 of a middle-class Cork family and attended the college from the age of fourteen. Four years later he became the First Senior Moderator in Mathematics, the equivalent of the Senior Wrangler at Cambridge. Influenced by James MacCullagh, who held the mathematics chair at Trinity, he was elected to a fellowship at the age of twenty-two and three years later was ordained in the Anglican ministry.[82]

Salmon settled down to a life of teaching mathematics and divinity. Unfortunately, he found himself in a position of teaching elementary mathematics to the ordinary degree students, the teaching to honors aspirants being reserved for a lecturer with more seniority. Generations of capable mathematics students at Dublin were thus deprived of Salmon's keen intellect—but they occasionally experienced his originality in the "interesting" questions he set for college examinations.[83]

By his Dublin colleagues, Salmon was seen to have exceptional potential as a geometer. Townsend had promised Thomson a paper on "Reciprocal Polars"—geometric duality—but thought it better to introduce Salmon to the editor: "that a far abler hand is at present occupied with that subject and has thrown out some new and very original views respecting it,—Mr Salmon—fellow of this College. He intends preparing a paper containing these views for the Journal which he will send you—but in any case my smaller and more ordinary paper will appear as a digression from the Main Subject—I hope to be able to prevail on him to do so quickly—but he is very unwilling to set to work in earnest at it, tho[ugh] I have rarely met one gifted with higher abilities."[84]

Salmon set to work and when the paper "On the Degree of a Surface Reciprocal to a Given One" was produced, Cayley as referee for the *Cambridge and Dublin Mathematical Journal* was struck by Salmon's ability. In dealing with one of his own favorite topics in geometry—the theory of reciprocity—it provoked his unstinting admiration: "I have been very much interested with it & have learnt a good deal from it."[85] Salmon's introduction to the paper indicates his breadth of view and the source of his interest: "[o]f all the additions which modern investigation has made to the ancient geometry, none seems more important than the method of reciprocal polars, by which our knowledge of extension is at once doubled, and we are enabled from any known property of curves or surfaces at once to deduce another correlative one." The paper touched on the difficult problem that had taxed Cayley, namely, of assigning equations to space curves, a problem he "thought a good deal about without arriving at much."[86]

Salmon was at the commencement of his mathematical career. He was in need of contacts outside Dublin and in the wider mathematical community and he saw the advantage of links with England. Cayley forged an immediate bond with him, for both wrote their first mathematical papers on the geometry of position. In this subject, the theorem referred to as "Pascal's Hexagram theorem" was one example that was an enduring interest for both; Cayley had discovered several approaches to it soon after graduating and Salmon published a paper on its extension to three dimensions as his first contribution. In the structure of the hexagon's "points and lines," they independently found the system of twenty special lines linking together the recently discovered "Kirkman points."

Cayley and Salmon's first joint work, conducted by correspondence, was in connection with the straight lines that lie on a cubic surface, a surface specified by an algebraic equation of order 3. What is surprising, and perhaps counterintuitive, is that a "curved" surface should contain any straight lines at all. It does, and Salmon showed that there are precisely twenty-seven on a nonsingular cubic surface. Together they showed how the lines might be represented as equations. Their joint research on cubic surfaces was ahead of Jacob Steiner's published in 1857. Cayley's contribution "On the [45] triple tangent planes of surfaces of the third order" came out in May 1849. A triple tangent plane is one which intersects the surface and contains three of the lines lying in the cubic. Cayley gave the equations of all the forty-five tangent planes, a remarkable feat of calculation and one which illustrates his patience and commitment. Expressed in homogeneous coordinates (x, y, z, w), each tangent plane is potentially of the form $Ax + By + Cz + Dw = 0$, but in his listing of them there is no symmetry or pattern and he performed his calculations in the absence of a good notation. Here he showed his pragmatic side. It was the result that mattered; let elegance follow later. It is noteworthy that Cayley had difficulty envisaging the complete figure formed by the configuration of twenty-seven lines and forty-five tangent planes. The construction of a model, as well a suitable notation would have helped, and he had to admit "[t]here is great difficulty in conceiving the complete figure formed by the twenty-seven lines, indeed this can hardly I think be accomplished until a more perfect notation is discovered." As imperfectly as he understood this intricate geometrical system without an adequate notation, he published the results, for "in the mean time it is easy to find theorems which partially exhibit the properties of the system."[87]

Cayley openly made conjectures and suggested future investigations. He did not attempt to keep ideas to himself but would publish papers finished or unfinished. On seeing his paper on the forty-five triple tangent planes, Andrew Hart, Salmon's Dublin colleague, produced a better notation for the lines and confirmed several of the results. Hart's polished notation reaped Cayley's praise but we should not forget that Hart's refinement was predicated on having the rough-hewn results in the first place. The cubic surface continued to fascinate Cayley with its configurations lying within the surface: quadrilaterals, pentagons, hexagons formed from the various combinations of the twenty-seven lines. Here was a fund of theorems which "might be multiplied indefinitely, and the number of different combinations of lines or planes to which each theorem applies is also very considerable." This was prophetic. The configurations were a constant source of exploration in the nineteenth century for mathematicians generally and gave rise to a substantial literature to which Cayley contributed definitive work in the 1860s.[88]

Cayley's skill in pure geometry, alluded to by Salmon, demonstrated his insight into the configurations which could exist in physical space, a facility which contrasts with his algorithmic side so clearly in evidence in the machine-like formulation and calculations of invariant theory. While the algebra of invariant theory required an element of routine, this was not the case with projective geometry, which needed spatial intuition of an uncommon type. Cornelius Lanczos reminds us of "the elegance of the methods [of projective geometry] should not blind us to the fact that the results were obtained by sheer invention and not by the application of systematic analytic procedures." The essence of the subject was "ingenuity" rather than "turning the crank."[89]

Salmon became a prodigious writer of textbooks. In 1847 he opened his account with A Treatise on Conic Sections. Of all the mathematical texts that were to flow from his pen, this was the most famous and enduring. Its innovative style lay in the way it dovetailed intuition in pure geometry with analytical methods based on coordinates and equations—how coordinates could be used effectively in the study of projective geometry—the basic tenet of Cayley's geometrical work. It was said that had Salmon done nothing else, he would have gained a lasting reputation for this single book; it went through numerous editions and was translated into the major European languages. When most textbooks lasted only a decade, "Salmon" was the "one outrageous incredible glorious exception" and lasted a century.[90]

When Salmon perused Cayley's founding papers on invariant theory, he recognized the theory as geometry. In later editions of his Conic Sections, he demonstrated how invariant theory could be applied to geometry and promulgated Cayley's work to generations of students. While Cayley came from a strong Cambridge tradition in analytical geometry and was unlikely to change his general stance, Salmon in Dublin might well have been swayed by the zealousness of Hamilton's mission of applying quaternions to geometry and thereby creating a "symbolical geometry." Though he came to appreciate the "admirable consistency and harmony of the whole scheme" of the quaternion calculus, he remained steadfastly a student of analytical geometry based on coordinates. Salmon lived in Dublin but his affinities were those of a Cambridge pure mathematician.

The Irishman became a close personal friend of Cayley's, a friendship which lasted throughout their lives. Apart from mathematics, they shared an appreciation for remote Continental journeys and chess. Salmon's strong Protestant leanings, with a suspicion of the rising Tractarian movement, might have been a potential difference with a member of the Cambridge Camden Society, but it was one that was never realized. Furthermore, in mathematics, too, their relationship was calm. With Sylvester, there was always the competitive element, but with Salmon there was wholehearted cooperation. It did not manifest itself in jointly written mathematical papers, such things were unheard of at the time, but Cayley contributed whole chapters to Salmon's textbooks and, of all the British mathematicians, Salmon was closest to his point of view on geometry.

Called to the Bar

Cayley had been at Lincoln's Inn for three years but the law hardly absorbed him. He was a member of the group of mathematician/lawyers who seemed to work harder at their diversions than at the real business of the law. They were all aware of each other: Benjamin Gray, Richard Mate, Hugh Blackburn, Archibald Smith, Henry Wilbraham, George Hemming (a Senior Wrangler and rowing friend of Thomson's) were all apprenticed at Lincoln's Inn, while Sylvester, James Cockle and Robert Moon were members of inns nearby. The personal relationships established in the Cambridge colleges were transplanted to the collegiate setting of the Inns of Court in London. Richard Mate, a friend of Robert Leslie Ellis and William Walton, for example, was elected to fellowship of Trinity at the same time as Cayley and had joined in the emigration from Cambridge. On one occasion, Cayley wrote to Thomson: "I made an engagement with Mate, whom much to my surprise I met in the Hall at Lincolns Inn (really every body seems to have intended or begun or gone on with Law. 'Qui que tu sois voilà ton maître. Il l'est, il le fut ou il le doit être' [No matter who you are, there is your master. It is, it was, or it must be your master]—which was not said of Law however) and made an engagement with him to come up for a chess meeting."[91] The law was never Cayley's master or that of the gentleman scholars who partially worked for "Legal London."

If the right opportunity arose, academe would no doubt be preferable to some of the scientific lawyers, but there were few newly created positions, leaving death of an incumbent as the most likely source of opportunity. On 12 January 1849, Thomson's father, James, the professor of mathematics at Glasgow, died from cholera at the age of sixty-two. Hugh Blackburn saw an opportunity to abandon the law and leave the coterie of mathematicians at the Inns of Court. He was in communication with William Thomson on mathematics and in March reminded him of his famous pendulum:

> Do you remember my pendulum vibrating with diff[erent] periods in planes at r[igh]t angles? Cayley told me the other day (not knowing of my private interest in the subject) that he had been to see a new sort of engine turning which traces all these curves. I often thought of getting such a machine myself. Since then we have been busy (when we ought to have been drawing acts of parliament and such sublunary matters) in constructing the developable surface generated by the tangent to the

curve of intersection of the st[raight] cylinder and sphere of double its radius, but though we have a tolerable notion of it, it is rather difficult to make it of paper as it runs through itself in 6 curves.[92]

Thomson and Blackburn had gone to Cambridge together in October 1841, Thomson via Glasgow and Blackburn via Eton and Edinburgh. They were close undergraduate friends—in the middle of winter, they went outdoor swimming and spent evenings reading Faust and swinging on Blackburn's pendulum. Now Thomson hoped Stokes would fill the vacancy with Blackburn a second choice. Thomson had a high opinion of Blackburn's abilities and when his father was alive had petitioned him to look out for a "scotch professorship" for him.

Whether or not Cayley entertained the thought of applying himself, he ultimately supported Blackburn's appointment by supplying a short testimonial—written economically and to the point, but bearing no comparison to ones he wrote for Thomson or Boole. In distinction to all the others Blackburn brought together, it mentioned the mathematics to which Blackburn could justly lay claim: "From what opportunities I have had of judging of Mr. BLACKBURN's Mathematical attainments, (his high place in the Tripos list was gained in a more than ordinarily good year [1845], and he is the discoverer of some singular theorems communicated to his friends at Cambridge upon Developable Surfaces,) I believe him to be well qualified for discharging, with credit, the duties of a mathematical Professorship."[93]

To prove his competency to occupy the Glasgow professorship, Blackburn was required to write an essay on mathematics in Latin, and then to subscribe to the religious tests of the Presbyterian Church. The second requirement would have debarred Cayley for the post since he had already subscribed to the Thirty-nine Articles of the Anglican Church; the Latin essay would have seemed juvenile. Blackburn was appointed and took up duties in April 1849. Though he was a man of all-round ability, he served Glasgow as an administrator with an accentuated penchant for organizational detail. His interests, which spanned the study of university regalia and academic dress with only a general interest in mathematics, added little to the scientific reputation of Glasgow.

Cayley maintained contact with Cambridge by taking on the role of examiner for the regular Trinity College internal examinations. His seven years up, Robert Leslie Ellis was on the point of leaving, but Cayley asked him one last favor on the style of examinations he should set. He was well into his stride as a pure mathematician but he could turn his hand to most things mathematical, and his training in mixed mathematics had left an indelible imprint. His lack of assertiveness, marked by his recurrent claim of not knowing a subject remained: "Will you have the kindness to look over as you did last year my examination papers. I have besides those I send you the Mechanics (2nd year higher paper) and Newton 3rd year . . . [t]he Newton is I suppose the first three and the ninth and eleventh sections (all which I have rather forgotten) and there is not very much choice, but with the Mechanics I really might miss all I ought to set and vice-versâ. When is the examination to be? I quite enjoy the prospect of coming up [to Cambridge] for it."[94]

In the summer, after his three years in pupilage, Cayley received the "Call to the Bar." Admission to full membership of the Society of Lincoln's Inn mirrored the passage from undergraduate to admission to a Cambridge degree. The rowdiness of undergraduates in the Senate House and the bow toward the vice-chancellor was surpassed only by the ritual of entry into the barrister's profession. The rites of passage involved attendance before the Benchers of Lincoln's Inn and intoning the fatuous legal line, "I hold that the widow should have the estate." On 3 May, Cayley spoke them, and, with his father at his side as sponsor, and the stamp duty of £50 paid, he received the "Call."

In August, after a long wait, the 33-year-old Boole heard that he had been appointed as professor of mathematics at the Queen's College, Cork. Cayley promptly congratulated him: "I suppose your duties do not begin for some time yet and that the commendable custom of a five or six months vacation in the summer is to be kept up there. Shall you be in town at all, I am where I used to be [58 Chancery Lane] and should be delighted to see you again."[95] Thus Cayley saw his friends about to enjoy their summer. After a busy teaching year in Glasgow, Thomson could "look forward to the six months vacation with great pleasure."[96] Cayley began to make plans for his life as a practicing barrister. In July and August, outstanding pieces of mathematics were cleared up and submitted to *Crelle's Journal,* on elliptic functions, the geometry of position, and one paper on conics.[97] Barristers also enjoyed the summer recess equivalent to the Long Vacation in which chambers closed their doors, and, as Dickens remarked, in summer "the bar of England is scattered over the face of the earth."[98]

Cayley formed his mathematical perspectives in the 1840s and made the first contributions to areas that he would return to over the years. His attachment to analytical geometry, classical analysis, and his first steps in invariant theory had all been commenced. Combined with inexhaustible energy, his precocity spawned a myriad of topics to attract him further: elliptic functions, determinants, combinatorial mathematics, the geometry of position, and the analytical geometry of curves and surfaces. He had, moreover, established a network of mathematical friendships that would continue to expand. Various Cambridge, Irish, and Scottish connections had been made, and when he had visited Paris and Berlin, he made himself known to mathematicians and editors of mathematical journals. He knew Joseph Liouville and had likely met the doyen Augustus L. Crelle on his visit to Berlin. He had met Michel Chasles in London, a mathematician who became well known to English mathematicians. Julius Plücker would have been an ideal correspondent had he not given up the study of geometry for physics—just the previous year he had given a lecture to the British Association at Swansea on the effect of magnetism on the optical properties of crystals.

Cayley's manner was one of an unusually modest young man. A youth versed in the ways of the academic world when he entered Lincoln's Inn, he experienced legal life among a group whose focus in life was not academic. Over the three years of his legal training, their culture was transmitted to the young man, and in the "whitewashed misery of the pupil room" he actually enjoyed the experience. In letters to his closest of friends, he revealed a waggish sense of humor delivered in his neat angular handwriting. The everyday repartee and close working offered new avenues to the undergraduate regarded by all as the Cambridge mathematical genius. Taken up with the new life, he easily could have left mathematics behind, but he remained loyal to his first love. Amidst the humdrum of legal training in Chancery Lane, he remained alert and when a mathematical idea visited him, he could still ask Thomson to hold a page in the *Cambridge and Dublin Mathematical Journal,* as on one occasion for an idea which had "just struck me about Elliptic integrals."[99]

Now qualified as a barrister, he was intent on pursuing his mathematics further. He was already a mature mathematician; in terms of published work he had contributed on equal terms with Eisenstein, Hesse, Cauchy, Jacobi, and Hamilton. The inner circle of his British friends perhaps knew his work best. He had corresponded with them all, and their alliance coalesced around their attachment to mathematics. In the real world, they led separate lives: James Joseph Sylvester was pursuing a business career nearby in Lincoln's Inn Fields, Thomas Kirkman tended his parishioners at Croft in Lancashire, George Boole was establishing himself as the new professor in Cork, and George Salmon was instilling the bare rudiments of mathematics to his young pupils in Dublin. In their "other lives," these gentlemen of mathematics were enthusiastic as creative mathematicians, and they shared a common belief that the subject was moving forward. Cayley was a mentor for this loosely connected group. He had done much, yet this was only the beginning.

~ PART 3 ~

The Rising Star, 1850–1862

Mr Cayley appears to me to be one of those who are powerful in throwing out blocks for the next generation to work up.
—Philip Kelland, 1861

The 1851 Great Exhibition in Hyde Park, London, ushered in the high Victorian period, a decade which coincided with Cayley's establishment in the London scientific world. While working as a barrister with chambers in Lincoln's Inn, he produced some of his most fruitful and innovative mathematical work. The Royal Society was intent on shedding its amateurish image, and he was in the new wave of young scientists to be elected. In his early thirties, his natural abilities combined with increasing maturity and a wide knowledge of mathematics and its most recent developments.

From the start, Cayley was a successful lawyer but he spurned a legal career and used his free time to pursue the path of being a highly active mathematician. In invariant theory, he joined with James Joseph Sylvester in a remarkable burst of creativity at the beginning of the decade. He firmly established his adult mathematical reputation with a series of memoirs on "quantics" and toward the end of the decade made a major discovery that was to underpin the classification of the various types of non-Euclidean geometry. He wrote on group theory, a subject which occupied him from time to time, and contributed papers on matrix algebra. These subjects, just part of a growing "modern algebra," only gained prominence as a mainstream theory in the twentieth century.

As the decade progressed, Cayley hoped for one of the few available positions in the British academic world. He expressed interest in professorial chairs at Aberdeen, the Western University of Great Britain (to be situated in South Wales), the Lowndean chair at Cambridge, and one in Glasgow. His failure to secure an academic post ended when he was elected in 1863 to a newly created Sadleirian chair at Cambridge.

CHAPTER 7

Barrister-at-Law

A rthur Cayley began the 1850s as a newly qualified barrister. He left behind the years of apprenticeship, and, as if to mark his newly found status in the London legal world, migrated from rooms in Chancery Lane to Stone Buildings situated within the inner quadrangle of Lincoln's Inn. This elegant Georgian building accommodated the chambers of his sole conduit for legal work, his former pupil-master Mr. Christie.

2, Stone Buildings

Number 2, Stone Buildings, Lincoln's Inn, would become Cayley's signature in the years to follow.[1] From this address he began to think in terms of writing not only single mathematical papers but whole series of papers united by a common theme. These constituted measured responses in distinction to previous "bulletins from the front." They would consolidate his youthful gains and break new ground, in keeping with a gentleman of science beginning his mature years; he was approaching thirty years of age. His capacity for long stints at the desk working on mathematical problems was undiminished as he was about to begin a career that removed him from the servitude of student life.

In appearance, Cayley was of average height, 5′ 7½″, but even then his spare figure was marked by a slight stoop. The black cutaway frock coat he wore over a waistcoat and pinstriped suit was the outfit of the professional class in town, and his hair was worn long over a high collar swamped by a bow-tied neck cloth. The stovepipe hat donned for the street was the piece of the uniform in which he seemed prepared to stand out. When William Rossetti visited his home, Arthur to him was the famous mathematician and brother of his friend Charles, but "before my mind's eye," he wrote, "is a hat of his that was a sight to see—or *not* to see."[2] And, when the magnificent bushy beards associated with the Victorians became de rigueur after the Crimean campaign a few years later, he remained clean-shaven.

Cayley was in the prime of life, intellectually vigorous and physically energetic and joins other Victorians whose walking exploits were legendary. He once told Francis Galton, "when about thirty-two, [I] walked forty-five miles."

Presumably on a different day—though one cannot be sure—at one of those wonderful Victorian party evenings when the carpet would be rolled back after dinner and the trio would strike up, Galton was informed, "[I] dined and danced till two in the morning without fatigue."[3]

London, at the beginning of the 1850s, was the capital of the most prosperous country in the world measured by national income.[4] The turbulent 1840s had passed and the threat of political revolution had been dispelled. The year of European political revolutions, 1848, seemed remote as the Victorian middle-classes settled in for a twenty-year period of their greatest prosperity in the days of a laissez-faire economy and expanding markets at home and overseas. There were occasional bumps but historians are generally united in their view of the midcentury as a period of general economic success. The "Ten Pound" franchise (a rateable value of that amount which qualified householders for the vote) established under the Great Reform Act of 1832 enfranchised the middle class, and the requirement would stand until the Second Reform Act of 1867. As the 1850s unfolded, the temper of society changed to one of solemnity. Even clothing reflected this. The colorful fashions of the 1840s gave way to a preponderance of gray and black, and life in the age of industrialization became altogether more sober and serious. Ambitious building projects that changed the face of London were based on clear-sightedness, Victorian thrift, and public subscription. The cornerstone of King's College Hospital was laid (and Cayley was a £10 subscriber to the Building Fund) at a ceremony without frills in the summer of 1852, but many of the great Victorian enterprises lay in the future. The mid-Victorians still moved in a city lit by gaslight falling on cobbled streets, and in winter they were blinded by the smogs, as in this year, a great one. There was no circular metropolitan railway and Sir Joseph Bazalgette had yet to unearth the capital and put down his famous sewerage system.

Lincoln's Inn was adjacent to one of the busiest parts of London. It was, and still is, strangely cut off from the commercial life outside the high walls, which serve to protect it. Once inside, there was a "sudden quietude" said Dickens. Barristers occupied their chambers in peace, a place that effectively constituted a home-away-from-home. Once called to the bar, they were free to go their own way and avail themselves of their Inn's facilities. They could keep their names on the membership list for life and dine there as they pleased and, of course, the call to the bar provided a mark of social standing.[5] Those from the landed aristocracy studied law to protect their property, while the ambitious advocate would represent clients in court and set out on a legal career that might lead to the bench or high political office. Many never intended to practice as advocates but chose to draft legal documents, often acting in this capacity as a junior to a leading barrister, as did Cayley.

Barristers with income were free to accept work or decline it as they wished, leaving themselves free to engage in individual pursuits, which were often academic. The Inns of Court in nineteenth-century London provided a convenient shelter for scholars quietly pursuing erudite scientific, literary, and bibliographic research. A man like Trollope's Mr. Wharton Q.C., an habitué of Stone Buildings, could reflect on hours spent as he wanted "in the centre of the metropolis, but in the perfect quiet as far as the outside world was concerned."[6] Cayley's work as a barrister was largely concerned with the drafting of documents relating to trusts, legacies, and property, much of it was based in the Chancery Court, a division of the High Court. Before the building of the Law Courts in the Strand, the Chancery Court sat in Lincoln's Inn Hall out of term time, though the work of a legal draftsman required few appearances in court if any at all. The conveyancing barrister was a "secluded being" and within the collegiality of the barristers' profession, he was an anonymous figure.[7]

Legal drafting required as close attention as any piece of mathematics. It was demanding work, involving the wording of tightly written documents, a process requiring a high degree of precision with legal language, for the English law of property in the middle of the century was archaic and convoluted. While moves were afoot to simplify the transfer of property, these proposed reforms were successfully blocked by the aristocracy, who fought for the preservation of their landed estates. The system of transfer, when combined with a client's specific requirements, required individual drafting, especially when it applied to large estates. Every case was different and invariably required the expertise of specialized legal counsel. Once qualified, the profession of conveyancer could be a lucrative profession for the able barrister. If the foundation law professor at King's College, London, did not have his tongue in his cheek, the practice of conveyancing even had an intimate affinity with mathematics: "there is *nothing* which a conveyancer of consummate skill cannot do, and that with a certainty that it will effect its object. No novelty can appall him, no complication can check him. He has scientific principles for every thing; he has logarithms for every possible ratio. His powers, like those of mathematical science, are, humanly speaking, infinite; they can never be exhausted. . . . So infinite is the variety of English conveyancing."[8] The auguries were overwhelmingly in place for Cayley's future legal career.

As a student and college fellow at Cambridge, and even as a pupil barrister at Lincoln's Inn, Cayley could retire to his study and quietly pursue his mathematical researches without hindrance. Being newly qualified, it was essential to take on work and establish his legal credentials, but with a growing reputation in the scientific world as well, it also became obligatory for him to adopt a more public persona, a role that did not sit easily with a retiring disposition. Even with the competition in the legal profession brought about by an oversupply of young barristers, he was fortunately successful. He was able to set his own professional agenda, and thus the constant stream of mathematical papers continued. He maintained contact with university affairs, served as senior examiner at the annual Trinity College examinations, and took on the responsibility of being senior moderator for the mathematical Tripos in 1851 and senior examiner in 1852.

In the Tripos of 1851, the young Norman Macleod Ferrers scored top marks and just beat off his nearest rival for the top position in the order of merit at Cambridge. He immediately entered Lincoln's Inn as a pupil but, on being called to the bar, decided against a law career and returned to Cambridge as a lecturer. In the following year, Cayley would have been impressed by the top two candidates, William Steele and Peter Guthrie Tait.[9] The twenty-year-old Tait from Edinburgh set the record of being the youngest to win the Senior Wrangler title, four months younger than Cayley at the same stage himself. If Ferrers was "solid" and altogether a future Cambridge don, Tait was a flamboyant bon vivant with an open disregard for convention. His large physique draped in a "sack coat" set off with a soft velour hat suited his combative style and unfortunate tendency to use intimidation as a form of argument. Tait had Victorian industry and he quickly set a formidable publishing record in motion. Two years after graduation, he went off to Queen's College, Belfast, as professor of mathematics, but Cayley would hear from him again.

While the legal profession provided stability, Cayley could not escape a career decision. He had been called to the bar, but would it be law or mathematics? Would he want a regulated life coupled with progression from barrister to judge and beyond? To make the decision more difficult, in his first year as a qualified conveyancing barrister, he found his legal skills in demand and his scale of fees indicating a lucrative private practice ahead. Would he choose a life with the prospect of few financial cares or the uncertain life of an academic? "Men of science" carried out their avocation by sheer dedication and without hope of material reward. T. H. Huxley, beginning his career in science, grumbled to his sister: "Science in England does everything—but pay. You may earn praise but not pudding."[10]

To succeed in mathematics and law, even for a man of Cayley's obvious ability, would be difficult if not impossible. Some lawyer-mathematicians gained a measure of success in one but it would be an unrealistic aspiration to reach the pinnacle in both. There were eminent lawyers who kept their mathematics going, like the well-regarded James Cockle. A contemporary of Cayley at Trinity College (who occupied an adjacent staircase during their student days), he was part of the "new algebras" movement of the 1840s. While Hamilton had quaternions, Kirkman the "pluquaternions," and Cayley his octaves, Cockle had "tessarines." He also devoted much energy to the solution of the equation of order 5, not trusting the proof given by Abel that a formula which yielded the solution for *any* such equation could not exist. Though he wrote over a hundred papers in mathematics and once ventured to publish in *Liouville's Journal*, he was never at the forefront of mathematical research. Instead, he acquired a reputation for solid legal work in Cayley's field of drafting legal instruments, and, when a career choice was imminent, he attracted the attention of Sir William Erle, chief justice of the Court of Common Pleas who "named" him as chief justice for Queensland in Australia. Cockle did not have Cayley's flair for mathematics, but any man responsible for the drafting of more than 150 colonial statutes and bills for the administration of the fledgling state would inevitably have little time for scientific work at the highest level.

Charles James Hargreave was another mathematician well known to Cayley. Only six months his senior, he specialized in land law and took legal briefs from Christie as well. His was a curious career, perhaps symptomatic of the lawyer-cum-scientist during the 1840s. Hargreave had been a pupil of Augustus De Morgan at University College and at twenty-three years of age succeeded John Graves as professor of jurisprudence in 1843. In the following year, he was elected to the Royal Society, and two months later called to the bar. He had published one paper in the *Philosophical Transactions* on the "figure of the earth," but his 1848 essay on differential equations earned him a Royal Society Gold Medal. A year later, he contributed a paper on the distribution of prime numbers.[11] Hargreave's career took a dramatic turn when he was appointed one of the three commissioners to administer the Encumbered Estates Act (1849) in Dublin, legislation designed to ease the sale of land in Ireland following the famine. Cayley was enthusiastic about Hargreave's new position and would have been impressed by the annual salary of £2,000.

Hargreave spent his ensuing career in Ireland, but the law failed to stretch him intellectually. Like Cockle, he was attracted to finding a method for solving the quintic equation. His administration of the Land Law Court left little time for mathematics, but, even with this encumbrance, he never gave up on the problem of the quintic and died thinking he had solved it.[12] Unaware that his quest was fundamentally misconceived, he had doggedly continued his researches oblivious of the fact that the mathematical world had moved on. Mathematics and the law were each evolving at a quickening rate so that to excel in both spheres was becoming impossible. At the onset of the second half of the nineteenth century, the decision for Cayley whether to pursue mathematics "professionally" to the exclusion of all else would have presented a dilemma.

Had Cayley settled for a halfway house as had Cockle or Hargreave, he undoubtedly would have matched their achievements. Like them, however, he would not have succeeded in establishing a lasting reputation in either activity. He evidently decided to be a competent lawyer, to do the minimum legal work required, and to devote himself wholeheartedly to mathematics. He had the security of being a fellow of Trinity College and, for a few years at least (until 1853), he had the financial cushion of the annual dividend. In working solely for Christie, he made no attempt to obtain work from other legal quarters. He did not "take silk" and become a Q.C. like other serious-minded barristers of his own age. To wear the silk gown of a Queen's Counsel in the English law profession is the mark of a senior barrister, but he seemed content with the stuffgown of the junior counsel. Supplied with conveyancing briefs, he was content to merely "devil" for Christie. He struck up a good working relationship with his employer, and in the years that followed, he worked exclusively for him. He took few legal briefs and jealously guarded the time reserved for mathematical pursuits.

Though Christie had a thriving legal practice, he was evidently satisfied with the arrangement of having an additional part-time barrister on his books. He was able to accommodate mathematicians, for, as one of his pupils remarked to a friend: "We are a very intellectual circle at Christie's. Only two Senior Wranglers at present, but, no doubt, more are coming."[13] Christie would have been sympathetic to Cayley's mathematical aspirations and taken pride that his chambers housed a mathematical celebrity. He valued Cayley's legal expertise, though "clients and fees" did not exert the same urgency on Cayley as on the majority of career barristers who moved hurriedly about the inns. Ultimately, perhaps the career choice for Cayley was no choice at all: he was a compulsive mathematician who had enjoyed early success and he would have pursued his craft whatever the alternative.

A selection of Christie's celebrated cases gives us an idea, albeit incomplete, of the legal work required of a real property lawyer and of the ambience of legal work at 2, Stone Buildings. Cayley remembered the cases of the Law Life Assurance Co., the Will of the Marquis of Bute, the Black Sluice Drainage, the Portarlington Estates in Ireland, the Settlements of the Staffordshire Estates of the Earl of Shrewsbury, and the Lancashire Estates of the Earl of Crawford and Balcarres, head of the Lindsay family, whose interests centered on the iron and coal industry around Wigan in Lancashire.[14] The Will of the Marquess of Bute is perhaps the most celebrated case that came into Christie's chambers. John Crichton-Stuart, the second Marquess of Bute, was probably the wealthiest man in Britain. Though Bute is in Scotland, he possessed extensive property, particularly in Cardiff in South Wales. When he died suddenly at the age of fifty-five, the entire property passed to his son, an infant of only six months. With such a fortune at stake, it was a coup for Christie to gain the ensuing legal work for his chambers and be able to call on the services of his pupils and young barristers like Cayley, who spent their energies drawing up the long documents connected with such complicated cases. Christie was a popular master, and when Cayley eventually left the law, he had good reason to remember him: "I shall always remember with affection and gratitude the great and unfailing kindness which I received from Mr. Christie during the time of my pupilage, and afterward in the fourteen years for which I worked for him."[15]

During the 1850s, Cayley was leading two lives. Primarily, he was the mathematician of great virtuosity, but amidst mathematics, he made a reputation among lawyers for his skill in drafting legal documents. It was a reputation made incidentally—for being competent in the work for which he was trained. We are reminded of "Cayley the lawyer" when we review some titles from his private law library. He owned the four-volume set of Sir William Blackstone's *Commentaries on the Laws of England* (21st edition, 1844), the eighteenth-century work which links him with students of the law past and present. To be effective in his legal career, a mastery of the history of both the legal system and knowledge of its present conduct were required. Mandatory reading included the practical manuals found in most conveyancers' offices—such as J. R. McCulloch's *Treatise on the Principles and Practical Influence of Taxation and the Funding System* (1845) and Andrew Bisset's *Practical Treatise on the Law of Partnership* (1847), the latter including the law relating to railway companies and joint-stock companies. Present on his shelves, these books remind us of the law apprenticeship served and the working life in chambers. Essential for daily reference were such volumes as James Traill Christie's *Concise Precedents of Wills* (1857) and Joshua Williams's *Principles of the Law of Real Property* (1852).

Conveyancing was Cayley's main chore and here he had the expertise of Charles Davidson at his elbow. This Cambridge graduate had been eighteenth wrangler in 1832 and a fellow of Christ's College, Cambridge, until 1844, but he had no interest in taking mathematics further. He was able in what was required for the Tripos, but with this hurdle cleared, his dedication ceased. He found a release for his practical talents in the law and eventually acquired a large conveyancing practice and a small fortune. One indicator of his practical skills was his completion of Thomas Martin's unfinished reference text on conveyancing. Both the landmark *Martin's Practice of Conveyancing: With Forms of Assurances* (1844), in five volumes, and *Common Forms of Conveyancing, Including Recitals* (1846) were present in Cayley's legal armory.

The legal instruments dealing with land and assets in the 1850s were cumbrous documents drawing together recent reform with ancient phraseology. In practice, lawyers made use of model legal drafts written in the appropriate style, which could then be amended and altered to suit specific needs. When Davidson joined forces with Jacob Waley and Thomas Key to produce the voluminous *Davidson's Precedents and Forms*, Cayley's name received more than an honorable mention.[16] Of this legal manual, which passed through many editions, it was the third volume on "Settlements" that was regarded by the professionals as exemplary legal work and that provided the greatest benefit to the conveyancing profession. On a question of family settlement, the editors entered a personal footnote alluding to Cayley's ability in the complexity of English land law: "The writer ventures to call attention to the remarkable skill exhibited in this settlement [of Family Settlement by Father and Son], the work of Mr. Arthur Cayley." This legal piece is a document concerned with the retention of family estates: Davidson's Family Settlement XXXV. The opening paragraph of this model legal draft (which ran for fifty-five pages) indicates Cayley's grasp of the law and the conduct of his daily work:

FAMILY SETTLEMENT by FATHER and SON (a) comprising ESTATES and COL-
LIERIES subject to a POWER of JOINT APPOINTMENT by them, and other estates
and collieries belonging to the FATHER absolutely. LIMITATIONS to the intent that
the FATHER may appoint by way of MORTGAGE for securing a specified SUM, and
that if the father shall not exercise the power to its full extent the SON may exercise
it in respect of the deficiency, and subject thereto, to the use that the SON may
receive, during the JOINT LIVES of himself and his father, a RENTCHARGE, to be
augmented in amount if he shall cease to reside with the father, with powers of dis-
tress and entry, and a POWER for the son or his personal representatives within a
year after his death, to LIMIT a TERM to secure the rentcharges. LIMITATIONS to
the FATHER for LIFE, with remainder to the son for LIFE, with remainder to
TRUSTEES for raising portions, with remainder to the FIRST and OTHER SONS of
the son successively in TAIL MALE, with remainders over. POWER for TENANTS
for LIFE to WORK MINES and have POSSESSION of COLLIERY PLANT, subject to
the obligation of keeping up its value. . . .

And it continued with . . . Covenants, . . . inheritances, . . . right to assign, . . . and the document concluded: "for the time being interested in the same premises shall be reasonably required. In WITNESS, &c."[17] Cayley held the phraseol-
ogy of the law in the highest regard, for it could express nuances in the most precise manner. Yet, there was a limit, and time had to be made for mathemat-
ics. Generally, he did not care for the works of Charles Dickens, but he would have felt some sympathy with Dickens's satire on the legal documents which accrued in the Chancery Court: "bills, cross-bills, answers, rejoinders, injunc-
tions, affidavits, issues, references to masters, master's reports, mountains of costly nonsense."[18] Wanting time for mathematics, the working practices of the Chancery Court could be trying.

Cayley was a remarkably calm person and Sylvester remembered only one occasion when he saw him out of sorts with the conflict of the two lives he was living: "Entering his office one morning, intent on some new mathematical thought which he was discussing . . . , he opened the letter-box in his door and found a bundle of papers relating to a law case which he was asked to take up. The interruption was too much. He flung the papers on the table with remarks more forcible than complimentary concerning the person who had distracted his attention at such an inopportune moment."[19] Cayley needed to find a niche in academe, away from the law and without these unwanted legal diversions.

In July 1850, Henry Cayley died. "My mother is still in very low spirits," his eldest son wrote to Thomson from Stone Buildings a few weeks later, "but otherwise well, [and] the rest of us are quite well. I have given up my Scotch plan,

I do not think I shall do anything more than perhaps spend a few days with [Edmund] Venables or B. F. Smith." Scotch plans usually meant walking tours combined with a scientific meeting, and that summer the British Association was meeting in Edinburgh. But he was required in London and he quickly moved to being head of the family and running its affairs. Work was an antidote to the sorrow of his loss, and in the gloom of bereavement, mathematics was soon resumed. "Shall you be in town again," he continued to Thomson, "I send you with this a paper which I have been working at this last week containing the generalisation of your formula on the attraction of Spherical Shells which I think will interest you."[20]

After a lifetime in commerce, Henry Cayley left the next of kin in comfortable circumstances. Having reached his majority, Arthur inherited his own share of his father's bequest absolutely.[21] The family maintained a level of financial independence and, for a few years, his widow Maria, who gave her "occupation" as a "Share holder in Railways" in the census of the following year, kept up a large rambling Georgian mansion in Blackheath.[22] Perhaps she scaled down her outgoings, for their retinue consisted of only two servants in 1851, a Norfolk-born cook and a housemaid.[23] For a few years on, Arthur, his brother Charles, and sisters Sophia and Henrietta-Caroline lived at this imposing residence, curiously named "Cambridge House." The two sisters never married and stayed at home with their mother. The elder, Sophia, was "handsome and striking, ready and amusing in conversation" according to Christina Rossetti, in comparison with Henrietta who apparently inherited the family trait of being "stolid."[24] All things indicate a fairly comfortable existence near Blackheath's "Point" with its open heathland on one side and panoramic views of London on the other.

My Brother Charles

As the new Industrial Age advanced, the men of "brass and work" who proclaimed the modern world coexisted with those who floated through it, living off their capital without the need for paid employment. London had its fair share of gentlemen scholars, men who rose in the morning with the prospect of an uninterrupted day spent reading, writing, or translating. Such a person dwelling in these ample pastures of time was Charles Cayley.

Charles followed his brother from school at King's, London, to Trinity College, Cambridge, where he gained a second-class classics degree in 1845. Afterward, he returned to London, where he became an official at the Patent Office in Chancery Lane. In those years, he regularly visited his former teacher Gabriele Rossetti for private Italian lessons. Charles became part of the Rossetti family and was always in and out of the company of Dante Gabriel, William Michael, and sister Christina; from this association, he became a fringe member of the pre-Raphaelite circle founded in 1848 by Dante Gabriel, John Everett Millais, and William Holman Hunt.

A receptacle for all kinds of knowledge, those who met Charles could not fail to be impressed by his encyclopedic range once it had been prised out of him. He was also a capable linguist and he gained a reputation in the learning of remote languages. One of his accomplishments, and one which contributed to a peculiar artistic notoriety, was his ability to converse in the Iroquois language as one of his literary projects had been a translation of the New Testament into that language. Charles was shy, absentminded, and thoughtful, according to William Rossetti: "A more complete specimen than Mr Charles Cayley of the abstracted scholar in appearance and manner—the scholar who constantly lives an inward and unmaterial life, faintly perceptive of external facts and appearances—could hardly be conceived."[25] Rossetti knew none to match Charles "who lived the intellectual life with so little obstruction from 'the world, the flesh and the devil.'" Siblings can be polar opposites, but in physical appearance at least, they bore a close resemblance. According to William Rossetti again, Charles "was decidedly good looking with a very large cerebral development, dark hair and eyes, ruddy cheeks and fairly regular features."[26] Arthur as the London barrister could not afford to be so bohemian as his brother but the two men from Blackheath cut distinctive figures in London's artistic circles. When Thomas Archer Hirst was introduced to Charles, he noted that he was "even more singular looking than the mathematician himself."[27]

Charles was not particularly inclined toward mathematics and acknowledged that he found standard skills such as the ability to visualize geometrical planes and solid figures "difficult." The skill of picturing lines of figures and holding them in the mental field was, for him, "very imperfect."[28] This sharply contrasts with his mathematician brother, who possessed an exceptional visual sense and had developed a superb geometrical intuition. Charles Cayley's confessed inability in mathematics, however, should not be taken at face value. Just to qualify for the classics degree, he had first to reach the standard of an honors degree in mathematics, and in this qualifying task, he was thirty-eighth in the Senior Optime division (the one below the wrangler class), effectively seventy-sixth in the mathematical order of merit of his year. Charles also put substantial numerical skills into his literary work. In his translation of the *Iliad of Homer* (1877), William Rossetti observed that he "never spared himself; his Iliad, for instance, being not only in hexameters, but in quantitative hexameters."[29]

But how could Charles begin to compete with a brilliant brother? After graduation, he tried his hand at publishing his work, and perhaps at his brother's suggestion, he submitted an article to the *Philosophical Magazine,* a rumbustious journal that took articles on all kinds of scientific subjects. There he appeared as a "Fellow of Trinity," a status he would have most desired but which in real life eluded him. A second-class classics degree was not good enough to survive the fierce competition for fellowships that existed in the 1840s, or even for admission to the examination competition. His article in the *Philosophical Magazine* displayed further evidence of a mathematical turn of mind and an appreciation of the subject that went quite beyond "quantitative hexameters." In effect, it was on mathematical linguistics, in which he suggested a symbolic notation for phonetics and proposed further scientific work on the analysis of vowel sounds.[30] At home or on the new railway to and from their places of work in London, one brother in Lincoln's Inn and the other nearby in Chancery Lane, they would have had the opportunity to discuss the connection between mathematics and language and particularly the application of "algebraic notation" to linguistics.

Charles Cayley's sound reputation for scholarly work has almost been forgotten, but in his day he was widely respected as a translator of merit by writers and critics, though none of his works contributed to his financial well-being. Between 1851 and 1855, he published Dante's *Divine Comedy* in four volumes, translated in the original ternary rhyme.[31] This was his first and best known work, and though he was a prolific author just like his brother, there is not the sense of "rushing on" in his literary works compared with his brother's rapid and diverse mathematical productions.

Translating Dante was a popular pursuit among artistic Victorians and the less well known literati with leisure to spare. Charles Cayley's translation was actively discussed by Alfred Tennyson, who had just become poet laureate. Coventry Patmore admired the work, and John Ruskin regarded this version of Dante as that one most faithful to the original. Just as his brother was overtaken by his addiction for mathematics, Charles had been imbued with the love of classical literature derived from his days as a school pupil at King's College, London. When he sent his translation to John Henry Röhrs, a scientist who knew both brothers and had attended King's College, London, with him, this man who lavished praise sparingly described it as "the best translation of Dante in the language, admirably *done*." Moreover, he declared: "Cayley (C.B.) is a master of English, & as great a scholar as his brother is an analyst; he is a beautiful but rather obscure writer—It is strange he is not better known."[32] Robert Leslie Ellis, who read the translation, was not so impressed (clever but unsatisfactory) but took from it a favorable impression of Charles's character. Not knowing him personally, Ellis was quite "willing to believe he is like his brother."[33]

A member of the gentlemanly Philological Society, Charles regularly communicated papers with titles such as "On Certain Italian Diminutives" (inspired by Gabriele Rossetti), "On the Aspects of the Verb in Russian Grammar" (connected with his own Russian background perhaps), "On Greek Pronunciation and the Distribution of Greek Accents," and one read to the society shortly before his death in 1883, "On the English Name of the Letter Y." He took his turn in the administration of the society and was an esteemed colleague of the architect of the great *Oxford English Dictionary*, J. A. H. Murray, who took time off from the Scriptorium to note his passing.[34]

In Charles, the elder brother's money-sense is missing, as is his man-of-business attitude, but both shared a capacity for abstract thought and a lack of concern for the trivialities of life and its cosmetic details. In love, he was unlucky. He proposed marriage to Christina Rossetti but her religious principles would not allow her to accept. "She inquired as to his creed, and she found that he was not a Christian; either absolutely . . . or so far from religious orthodoxy that she could not accept him."[35] They remained close companions and she acted as his literary executor. In a quiet withdrawn existence, punctuated with long afternoons in the British Museum Reading Room, many did not take Charles Cayley seriously. Though he was often seen as a figure of fun by the London literati, he was respected and valued and especial note was made of his "unswerving loyalty to truth."[36] Arthur took familial pride in his literary productions and in their academic turn of mind; in aptitude for intellectual pursuits, the two brothers were not dissimilar.

A Boon Companion

James Joseph Sylvester was called to the bar in 1850. His address in New Square, Lincoln's Inn, and the formal connection with the law gave him the status he most desired. Though he stipulated that he should be known professionally as a barrister-at-law, he remained as an actuary with the Equity and Law Life Assurance Company and did not practice law. He may have seen barriers to success in that arena, on account of being Jewish. An immediate cause for this handicap lay in the *Test Acts,* which prevented non-Anglicans from entering Parliament, a disability repealed for Jews only in 1858. Benjamin Disraeli was a well-known politician of the day but he was baptized an Anglican and in adulthood was a Jew by birth, but an Anglican by subscription. Sylvester was not in this position. Ineligibility to enter Parliament did not affect him directly but that such an "exclusion" existed would have placed him on the margin in subtle ways. Sylvester brought many difficulties on himself, but living in a country of rigid class divisions, his Jewish faith was a handicap.

To a Victorian, status was immediately obvious and "social station" decided welcome or rejection. Being a barrister, Sylvester would have been recognized as a gentleman; but he did not possess the social ease and assurance of Cayley, whose family background linked together the twin bastions of aristocracy and commercial success. Sylvester lacked the quiet confidence of Cayley, who represented stability and mathematical authority, an enviable position when viewed from Sylvester's world, and one accentuated during periods when he was haunted by depression. Limpet-like Sylvester fastened himself to his alter ego "who habitually discourses pearls and rubies."[37]

While Sylvester's mathematical life floundered in the 1840s, it blossomed at the beginning of the new decade. He returned to the mathematical fold properly and rediscovered his enthusiasm and zest for publication. His superabundant energy was channeled into mathematics without the diversion of having to establish himself in the actuarial field, for he was well known in that world after having taken a leading part in the setting up of the Institute of Actuaries. In his newly found inspiration for mathematics, he could sign off a letter to Cayley: "Yours ever in the love of Analysis."[38] The two entered into their long association manifested by a vital correspondence that was to flow between them. There was no doubting his friend's part in his recovery to mathematical life, for it was Cayley, he declared to the world, "to whom I am indebted for my restoration to the enjoyment of mathematical life."[39] When Sylvester was an old man happily secure in Oxford, he assessed the value of his friend's influence. Admittedly, it was written in the golden afterglow of a turbulent life safely negotiated, but it reflects his reliance on Cayley throughout the years, for "there is no one I can think of with whom I ever have conversed, from my intercourse with whom I have derived more benefit."[40] This was a reflective Sylvester surveying the fruitful years that had passed, whereas now, at the beginning of the 1850s, we have the energetic Sylvester in his mid-thirties, vibrant and ready for the challenge.

Sylvester was a natural research mathematician, quick in mind, impatient with any lack of progress, and always ready to change to a new line of attack. When success came, it was coupled with childlike wonderment. This phase was transitory and he quickly set to work in preparing his manuscript for the printers. Any delay carried the risk that the idea might be discovered by others and this he could not endure. Whether a particular idea was worth relating or conserving was of secondary consideration when priority was at stake. His wide-ranging interest in literature was put to good effect in the way he peppered his papers with literary allusion. The lines of poetry or the brief classical phrase that often laced his mathematical exposition were not mere decoration but expressed his intense feelings about the work at hand. His "poetic" driving force naturally found its way into the rapidly written mathematical papers and to poetry alike. In him we see a reflection of William Blake's "[w]e of Israel taught that the Poetic Genius (the inspired voice of the imagination) was the first principle and all others merely derivative."[41] While his mathematical readers could smile at his couplets, a reaction he still provokes, the reader of Victorian poetry might lack the taste for mathematics in the midst of his literary renderings. Whatever the case, Sylvester's writings demanded attention and between all his lines, whether mathematical or poetical, was his cry for recognition.

By the spring of 1850 Cayley's and Sylvester's collaboration was established. Though awkward on occasions—dealings with the moody Sylvester could be

unpredictable—they had much mathematics to discuss. Cayley could acknowledge their discussions of Kirkman's "Schoolgirl's Problem," in which Sylvester
latched on a generalization, while Cayley reflected that the combinatorial problem had "excited some attention" among the mathematical public and commenced to deliver a solution for it.[42] This was a little diversion, for both were
occupied with geometrical subjects. Cayley would shortly provide a paper on
skew surfaces (surfaces generated by a moving straight line, like a cone), which
would prove pathbreaking and open a rich vein for mathematicians when his
substantial series of memoirs on the subject was launched in the 1860s. On their
geometrical researches, both were working independently, but Cayley made
helpful suggestions and assisted Sylvester in his choice of mathematical
terminology.[43]

Cayley and Sylvester made no conscious joint decision to work together on
one subject to the exclusion of all else. The Victorian cult of "individual accomplishment" prevented wholly open and frank exchanges, but they came closer
to cooperation than most and they protected each other from isolation. At the
beginning of the 1850s, both were free spirits pursuing individual paths but this
passage of time caught them traveling in the same direction. Sylvester analyzed
various geometrical problems using analytical geometry and he was grateful for
the "hints" and "remarks" as well as the "penetration and sagacity" of his
friend.[44] Afterward their paths diverged and, though they often touched on
connected subjects and continued their correspondence, they acted as individuals and never again achieved the same unity of purpose.

From his office at 26, Lincoln's Inn Fields, rapidly written notes from
Sylvester—as many as four in a day—found their way through the arteries of the
Inns of Court seeking Cayley's observations on his most recent idea. While Cayley could remain in his chambers and calmly prosecute his work, Sylvester was a
sociable being and was not happy with seclusion. He learned by personal communication, but Cayley, the lawyer, learned his mathematics from the printed
page. Sylvester needed the human spirit and personal inspiration and, in such
times, he would turn to his friend: "Would you favor me with your company to
tea this evening at 8 for mathematical conversation if so inclined at the time and
if not not."[45] On matters mathematical, he depended on him utterly: "I must
send this to Cayley" or "Cayley has pointed out a difficulty" were expressions
overheard by Galton as Sylvester fretted over some recalcitrant problem.[46]

In solitary moments, Sylvester recorded his experiences for the world to witness. Even during periods of relaxation, his quizzical frame of mind was receptive to new ideas. One had come to him "in a walk before breakfast by the side
of the ornamental water in St. James's Park (a time and place by no means,
according to my experience, unfavourable to the inspirations of the analytic
muse)." There, Sylvester tells us, "I had the satisfaction of falling upon the
rather *piquant* demonstration."[47] Perambulations in the park were sufficient for

him to obtain a few ideas to take back to the Athenaeum followed by a note to Cayley announcing a new discovery. This was not the ruggedness of Cayley engaged in his massive forty-five mile hikes, but it worked for Sylvester.

In moments of optimism, Sylvester was fulsome in his praise of Cayley. In explaining his combinatorial approach to invariant theory, at the start of his mathematical rejuvenation, he wrote: "I have succeeded [using an umbral notation] in reducing to a mechanical method of compound permutation the process for the discovery of those memorable forms invented by Mr Cayley, and named by him hyperdeterminants, which have attracted the notice and just admiration of analysts all over Europe, and which will remain a perpetual memorial, as long as the name of algebra survives, of the penetration and sagacity of their author."[48] Sylvester relied on his friend's practical support, too, especially in regards to the preparation of manuscripts, a skill in which he was completely inept. On one occasion, Cayley received a supplicant letter from Sylvester in Madrid describing his hurry leaving London: "My cab was waiting at the door when I went through the first half [of a manuscript] which swarmed with errors. I beg that you could have the kindness to use your discretion in changing all that appears unintelligible in the Mss and moreover that you will be so good as to strike out the whole of a foolish paragraph (one sentence) in which I say something about 'Analysis looking down with contempt upon Geometry.'" The manuscript Cayley was called on to read was the one in which Sylvester first uses the term *matrix*. Perhaps the effort required to cut through the embellishments and decipher the text made the word to remain with him.[49]

By the beginning of the decade, both Cayley and Sylvester were simultaneously but separately on the point of combining the previous work on invariants into a new subject—invariant theory, the subject of their Modern Algebra. Mathematics is rarely linked to everyday events—"it is far too abstract"—but here the aspirations of these two mathematicians coincided with undercurrents of national life. Preparations were under way for the staging of the Great Exhibition of 1851 and the midcentury was being heralded as a turning point in the fortunes of the country. Prince Albert declared the period as one of most wonderful transitions and the world could witness it when the Great Exhibition was opened by Queen Victoria on 1 May 1851. Sylvester caught this mood in his "Sketch of a Memoir on Elimination, Transformation, and Canonical Forms" published that month. He alluded to the discovery that determinants have the invariant property: "There exist (as is now well known) other functions besides the determinant, called by their discoverer (Mr Cayley) hyperdeterminants, gifted with a similar property of immutability." Moreover, Sylvester continued, "I have discovered a process for finding hyperdeterminants of functions of any degree of any number of letters, by means of a process of Compound Permutation." This he followed by a privately printed essay on "canonical forms," in which he worked out the theory for binary forms of odd order.[50]

Sylvester was on song and, in such a mood, while seeing the possibilities for invariant theory, he heaped praise on Cayley for his pioneering analysis. He saw that the subject would not exist but for Cayley's early forays. When John Couch Adams was offered a knighthood and a pension as a reward for his part in the discovery of the planet Neptune, Sylvester publicly argued that Cayley also warranted recognition: "what has been done in honour of the discoverer of a new and inexhaustible region of exquisite analysis?"[51] Of course nothing had been done, or would be done, for the truth was that a setting out a fresh branch of mathematics does not have the éclat of finding a new planet in the night sky.

While Sylvester was generous in praise, Cayley and Sylvester competed as well as cooperated, and in this climate of rivalry, Sylvester was especially wary of sharing ideas in case proper credit was not fairly apportioned. Any hint of another receiving praise was reason enough for him to pick up his pen in support of his own reputation, and a statement of what he had done in the subject would be duly issued. Letters could be used as evidence of priority, but the ultimate guarantee was the appearance of results in print—and to be the first in print. A modern arrangement such as a joint paper might have cleared the air, but this was not the custom in Britain in the middle of the nineteenth century. Both young men nursed mathematical ambition, but they were individualists and though they discussed problems, they kept close guard on their own ideas. In Cayley's vast production, there is not *one* jointly written paper with Sylvester, nor indeed with any other author, a matter to his later regret.

Priority disputes of the most virulent kind were commonplace among mid-century scientists. The "first past the post" reaped the scientific accolades but the unfortunate coming in two minutes later was reduced to an also-ran and promptly forgotten. Charles Darwin felt a twinge of anxiety on learning that Alfred Russell Wallace was on the point of publishing theories of natural selection similar to his own. Mathematics was not exempt. As with Darwin's theory, there was little prospect of commercial success in this endeavor and, perhaps for this very reason, feelings ran deep. Mathematics is a subject where simplicity may conceal years of thought and where peer recognition is the only tangible reward this side of posthumous fame. Any talk of simultaneous discovery was treated as highly suspicious. In this climate, a man like Sylvester could easily quarrel with his most reliable friend.

In relation to the discovery of a "mechanical method" for generating invariants, the principle of "Compound Permutation," Sylvester warned Cayley not to publish first. This method involved a kind of n-dimensional determinant (commutants) and the fact that Cayley had discovered them eight years earlier made little impression. Sylvester's letter in May 1851 is worth quoting as it indicates the volatile side of his character and the irritating predicament for Cayley: "As you appealed to me on the subject, I must say that I do not think that you are justified in publishing your views on the Method (as applied to Hyperde-terminants) of *Compound Permutation of Umbral elements* founded on or sug-gested by my communications to you on the subject which were meant as con-fidential, until I have first published my own account of the matter. It will then be right for you, to point out whatever part of the idea you may think is included in your former printed papers and to suggest any generalisations." Sylvester continued by outlining the different methods that might be used for generating invariants and his assessment of the correct balance of credit to be accorded to Cayley and himself:

> As regards the principle of *restricted* permutations, I was aware of & even as you will see by my notes had enunciated the general notion thereof. Indeed I believe you acknowledge your inspirations on that point arose out of accidental observations on my part—but I owe to you the first simplified statement of its application in a par-ticular case which however I repeat, it is quite certain from the direction my researches had taken, I must in algorithmizing the Permutant for the Hyphers [invariants] of odd degrees have necessarily arrived at. To put out the method as your own and as only owing certain improvements of nomenclature to my sugges-tion would not I think be quite fair.[52]

The paper "On Permutants" was an important one for Cayley—one in which he was chiefly concerned with the organization of the algebra of forms. He assembled the various types of determinants he had dealt with previously and grouped them together under one umbrella concept, the algebraic form he called a permutant. Perhaps suffering from the illusion that to think of the term was sufficient, Sylvester failed to recognize Cayley's earlier idea. Not content with issuing a private rebuke, he brought the matter into the open with a printed statement. Responding to this, Cayley put out a postscript in an attempt to clear the air: "I wish to explain as accurately as I am able," he wrote, "the extent of my obligations to Mr Sylvester in respect of the subject of the present memoir."[53] Sylvester was still unhappy and Cayley, only wanting to ease his friend's pain, rushed out a correction. Fortunately, the matter rested. In reality, it was no more than a storm in a teacup, but the incident must have made Cayley more cautious in his dealings with his "priority-conscious" friend.[54]

There were other occasions, but Cayley was generous to a fault and did not allow Sylvester's outbursts to cloud the appreciation of his genius. When others experienced Sylvester's behavior firsthand and recounted it to Cayley, he would simply smile and make no comment. Those who did not know Sylvester but received his referee's reports would know they were dealing with a man unable to separate objective remarks from hubris: one recipient drew the conclusion that the great man was of a "somewhat peppery disposition."[55] All this was known to Cayley and discounted. In a biographical sketch of Sylvester, written in 1889, he even slipped in the lines: "there is always a generous and cordial recognition of the merit of others, his fellow-workers in the science."[56] While Sylvester was a natural competitor, Cayley thought more in terms of advancing the science and less on personalities, though he too was anxious to ensure that due recognition should be accorded.

Disagreements with Sylvester were forgotten as quickly as they arose. It was the prospect of being caught in a sudden squall that made people wary, but the storms were invariably short-lived. The current contretemps on "Compound Permutation" was put aside and a few weeks later, Sylvester invited Cayley to join him in fresh investigations. "I wish you could so *study* with me," he wrote, "in seeking the reduction of the Even Degrees [orders of binary forms]—there is a *secret spring* somewhere, if one knew where to lay one's finger upon it—In the meanwhile failing to pick the Lock one must endeavor to make an entry by some side passage."[57] The problem with binary forms of even order $2m$ was that they could not be reduced to the sum of m powers in general, unlike the binary forms of odd order $2m-1$, which could be reduced.[58] It was a fact gradually realized that binary forms of odd order were more straightforward to deal with than binary forms of even order.

The fever pitch of Sylvester's addiction to invariant theory at this time is admirably conveyed by his recollection of the instance when the muse visited him: "I discovered and developed the whole theory of canonical binary forms for odd degrees, and, as far as yet made out, for even degrees too, at one evening sitting, with a decanter of port wine to sustain nature's flagging energies, in a back office in Lincoln's Inn Fields. The work was done, and well done, but at the usual cost of racking thought a brain on fire, and feet feeling or feelingless, as if plunged in an ice-pail. *That night we slept no more.*"[59] In reducing binary forms of even order to a canonical form (in effect, a simpler form), Sylvester made some progress. He did not fully succeed with those of order 4 and 8 but made headway with the binary form of order 6. These details formed part of a general thrust; in his substantive paper, "On the Principles of the Calculus of Forms," published during 1852 and 1854, he vividly surveyed the subject, proposed generalizations, and introduced terminology. He alluded to orthogonal invariants and gave his *Law of Inertia for Quadratic Forms* dealing with orthogonal substitutions. He even suggested studying invariance under nonlinear transformations.[60]

At this point, Sylvester relinquished the old terminology of hyperdeterminants, "not without high sanction" of his friend. "Invariant theory" was ushered in as the name of the new subject when he announced: "I may notice that the Calculus of Forms may now with correctness be termed the Calculus of Invariants."[61] His long paper, which bestraddled several issues of the *Cambridge and Dublin Mathematical Journal*, started with the title "On the Principles of the Calculus of Forms" and ended with the qualifying phrase: "otherwise the theory of invariants."[62] His dictum was that the theory of invariants summarized all that was valuable in mathematics: "As all roads are said to lead to Rome, so I find, in my own case at least, that all algebraical inquiries, sooner or later end at the Capitol of Modern Algebra over whose shining portal is inscribed 'Theory of Invariants.'"[63] Sylvester's conversion was complete.

A New Foundation

In November 1851, Cayley sent a note to *Crelle's Journal* in which he employed Sylvester's new terminology of invariant and covariant for the first time. This work was based on the "hyperdeterminant derivative," and it contained a further development of his youthful work on linear transformations.[64] Two weeks later he made a discovery that appeared to hold more promise, and he announced it in a brief note to Sylvester:

> Every Invariant [U] satisfies the partial diff[erentia]l equations
>
> $$\left(a\frac{d}{db} + 2b\frac{d}{dc} + 3c\frac{d}{dd} + \ldots + nj\frac{d}{dk} \right) U = 0,$$
>
> $$\left(b\frac{d}{db} + 2c\frac{d}{dc} + 3d\frac{d}{dd} + \ldots + nk\frac{d}{dk} \right) U = \tfrac{1}{2}nsU,$$
>
> (s the degree of the Invariant) & of course the two equations formed by taking the coeff[icien]ts in a reverse order. This will constitute the foundation of a new theory of Invariants.[65]

The existence of this landmark letter was commented on by Cayley in his *Collected Mathematical Papers,* and it is clear that he attached some value to it.[66] Sylvester remembered receiving it and regarded it as a vital piece of evidence in establishing Cayley's priority against any outside challenge: "Mr Cayley's communication to mc was made in the early part of December last [1851] and my mcthod (the result of a remark made long before) of obtaining these and the more general equations and of demonstrating their sufficiency imparted a few weeks subsequently—I believe between January and February of the present year [1852]."[67] Many of Cayley's letters to Sylvester are missing, as a consequence of Sylvester's roving existence, but this letter may have been deliberately saved to mark a milestone in invariant theory.[68]

Cayley assembled results using the new method and dispatched the paper to the editor of *Crelle's Journal* on 23 February 1852; they were eventually published as "Nouvelles recherches sur les covariants" in 1854.[69] This discovery dictated his subsequent approach to the subject. Moreover, it was a method that could be adapted easily to deal with "covariants," algebraic forms that are more general than invariants. The method was, said Sylvester, "an engine that might-iest instrument of research ever yet invented by the mind of man—a Partial Differential Equation, to define and generate invariantive forms."[70] Cayley's thinking appears to have been entirely algebraic. In retrospect (1889), he noted: "I believe I actually arrived at the notion [of the differential equations] by the simple remark, say that $a\partial_b + 2b\partial_c$ operating on $ac - b^2$ reduced it to zero."[71] This explanation tends to reinforce a formalist or manipulative approach to algebra, but it has to be treated with some caution since it was a mental recon-struction made some forty years after the event.

That an invariant can be defined by differential equations was an idea Cayley expressed in his pioneering papers on linear transformations of 1845 and 1846; in fact, he was rereading these papers two weeks before writing the "landmark" letter to Sylvester.[72] The upshot of the "new" discovery is that Cayley decided to make it henceforth the foundation of the theory and relinquish the "Ω-process" governed by the notion of the "hyperdeterminant derivative" operator, which had been his favored approach when he was starting out in the 1840s.

In retrospect, it was the wrong road taken since the Ω-process held more theoretical potential and was the basis for the future German symbolic calculus developed in the 1860s.[73] Again from hindsight, we can now see that he should have developed his hyperdeterminant derivative and pioneered an abstract cal-culus, but this did not happen. Hindsight is a fine thing but it was not available to Cayley struggling with the initial concepts of a new subject. By the end of the century, one of his disciples, P. A. MacMahon, concluded that Cayley "had little fancy for the [hyperdeterminant derivative] method, which, dropping from his hands, was carried on with great results by the mathematicians of Germany."[74] From the time Cayley began his memoirs on quantics, he rarely dealt with the Ω-process again; as his method for finding invariants and covariants, it was quietly forgotten, though it did not disappear from his writings completely.[75]

Why did Cayley abandon the hyperdeterminant derivative? He gave no rea-son but I believe he dropped it because it was ultimately a poor instrument for calculation, and calculation for Cayley was a high priority. In the 1840s he had applied the hyperdeterminant derivative in the course of finding invariants of degree less than or equal to four, and, as he found no invariants of higher degree at this stage, he may not have fully realized then that the hyperdetermi-nant method was unsatisfactory in general. The invariants and covariants of higher degree were the ones that posed greater calculatory problems. The new synthesis, discovered in November 1851, seemed to hold greater promise when calculation was concerned. In comparison, using the new method, he remarked "one finds easily, the covariants, by the method of undetermined coefficients."[76]

With Cayley's reformulation of invariant theory, there was also the prospect of another priority dispute in the making. Sylvester was aware of the impor-tance of the new approach and he remarked: "it is clear that all the general properties of invariants must be contained in and be capable of being educed out of such [partial differential] equations."[77] He had the same idea himself but obtained Cayley's two equations from a different standpoint. Actually, Sylvester discovered the equations by a more transparent route, one of continuous or infinitesimal variation.[78] In later years, Cayley recognized that this insight of Sylvester's was "very important."[79] On this occasion, Sylvester resisted the temptation to claim priority and held back to allow his friend to publish first.

The Invariant Trinity

In the 1850s, Cayley and Sylvester were joined in their invariant theory quest by George Salmon from Dublin. The trio were dubbed the "Invariant Trinity" by Charles Hermite and they became known in the mathematical community by this sobriquet.[80] Cayley met Salmon at Hamilton's Dublin lectures in the 1840s, but Sylvester appears to have met him in 1852 on a visit to England, whereupon his enthusiasm for invariant theory drew the Irishman into the net. While Cayley and Sylvester met frequently in London, Salmon would slip across from Dublin when his heavy teaching duties allowed. On one occasion, Cayley conducted him on a tour of Lincoln's Inn Fields, Fleet Street, and the City of London and Salmon provides a marvelous cameo of their society together. The maze of crooked streets and hidden courtyards to the south of High Holborn and east of Kingsway was very much Cayley's home territory. On his way back to Dublin, Salmon penciled a note to Sylvester from his railway carriage: "Give Cayley a good scolding for me. He undertook to walk part of the way home with me last night & we progressed Westward in a satisfactory manner till we got to the church (I forget its name) [St. Clement Danes] just beyond Temple Bar— There we made a complete circuit of the church. He had got on some interesting topic so that I never observed it & he left me with my face turned to the East & I did not discover my mistake till I found myself in St. Paul's Church Yard."[81]

Passing through Temple Bar, Salmon walked the length of Fleet Street, across Ludgate Circus and up Ludgate Hill before finding himself at St. Paul's Cathedral. Was he thinking about Cayley's "interesting topic," almost certainly mathematical, as he walked along oblivious of his new destination? More certain was that their friendship grew over the years and Cayley had frequent occasions to call on Salmon's results in geometry and invariant theory.

Cayley's early papers of the 1840s on invariant theory provided the initial focus for the three mathematicians. While it is doubtful whether Sylvester pored over their contents to any extent, he flamboyantly declared them of major importance, "where first dawned upon the world the clear and full-formed idea of invariants (the most original and important [idea] infused into analysis since the discovery of fluxions)."[82] Thus he attempted to push Boole into the background and, unlike Cayley and Salmon, Sylvester could never accept Boole as the planter of the acorn from which the great theory of invariants grew. When Cayley sent Sylvester off to read Boole's papers in order to resolve a specific query, you can sense Sylvester's irritation. "[a]ll that I find in Boole relating to the matter" was one which gave "so much uneasiness."[83] When Sylvester found an opportunity to set out his views on the history of invariant theory he gave generous credit to Eisenstein, Aronhold, Hesse, and Cayley, but dismissed Boole's papers as being of little value.[84] On the other hand, Salmon's debt to Boole is very apparent since he claimed his work gave him his first clear ideas of the nature and objects of invariant theory. Being more elementary, they have a directness and simplicity Cayley's foundation papers lacked. Salmon read Cayley's foundation papers in the 1850s but found them difficult, as did most mathematicians. He was impressed with the 1846 paper "On Linear Transformations," but admitted, "I am ashamed to say I never studied [it] carefully before."[85]

Treating geometry through invariant theory and "coordinates" separated the Cayley-Sylvester-Salmon alliance from Hamilton, who wanted to tackle geometrical questions using quaternions. The differences between those who favored quaternions as a method of geometric analysis and those who favored coordinates emerged quite early and well ahead of the squabbles that took place toward the end of the century between the camp loyal to coordinates and the one who thought quaternion vectors were superior. On hearing about Cayley's proof of Pascal's theorem, which employed determinants, Hamilton delivered a public statement in January 1852 outlining his own motivation: "But while gladly acknowledging the great mathematical learning and originality exhibited in that and every paper by Mr Cayley, Sir W. Rowan Hamilton thinks it right to state, that he was led to his own results, . . . between *ten points on the surface of the second order,* not by any system of *coordinates,* but by considerations of [quaternion] *vectors,* and by seeking to extend to *ellipsoids* the results respecting *cones,* . . . as derived from the Calculus of Quaternions."[86]

Cayley never wavered in his support of the idea that geometry should be studied using coordinates with invariant theory in attendance. Those who followed Sir William were equally adamant that the short and concise quaternion notation was superior. Maintaining his own position, Cayley always remained on the best terms with Hamilton, whose preoccupation with quaternions left him little time to appreciate Cayley's work in algebraic geometry.[87] Hamilton acknowledged the support he received from the Englishmen on his quaternionic investigations: "Now, Herschel, Cayley, Donkin, Peacock, yourself and others in England, to say nothing of my Dublin friends, have, as it seems to me, stepped out of their own ways to recognize and encourage my exertions."[88] Thus their professional relationship was asymmetrical: Hamilton was relatively ignorant of Cayley's researches, but his own work was known in detail by the English lawyer.

For a remarkably short time, a matter of only two years, the mathematical activities of the "invariant theory" trio were at their peak. Sylvester the catalyst was in full flow and discovering new results every day. Cayley was on the point of formalizing the goals and basis for invariant theory more comprehensively than had been done, and Salmon joined them as a student of both, using Cayley's new foundation to underpin his extraordinary feats of calculation.

The Royal Society

In April 1852, Cayley submitted his first paper to the Royal Society: "Analytical Researches Connected with Steiner's Extension of Malfatti's Problem." He had been interested in this problem previously and knew its history. The Italian geometer Gian Francesco Malfatti posed the problem of constructing three circles in a given triangle, each of them touching the two others and two sides of the triangle. His solution was published in 1803 and the problem and its variants gained a following among mathematicians. Jacob Steiner considered a generalization in which he replaced the three straight lines by three sections of a quadric surface and solved it by pure geometry, but in his own paper Cayley used analytical geometry to investigate this generalization.[89]

At the Royal Society Council meeting two weeks after submitting it, Cayley was included in the fifteen candidates proposed by the council for election. In the same list was the scientist John Tyndall and Oxford mathematician Bartholomew Price ("Bat" Price, the tutor of Charles Dodgson). Since Cayley's call to the bar, three years previously, he had achieved much and his proposal for membership was long overdue, so necessary was it for the society to elect active scientists and mathematicians in the first years following reform.[90]

The reforms of the constitution of the Royal Society of 1847 enabled the society to gradually cast off its amateurish image. Instead of being a club for the aristocracy and the well-connected, it now intended to promote the membership of "scientific men." To carry out this object, Sir William Robert Grove and his party of reformers wanted a stricter selection of fellows. This would reduce the proportion of nonscientific fellows and raise the standards of research, it was hoped. Huxley, who had been elected the year before Cayley, made a note in his journal of the tighter election procedure: "I have been elected into the Royal Society at a time when that election is more difficult than it has ever been in the history of the Society."[91] Following the reforms, the British Government began to recognize the important position held by the Royal Society and it provided modest monetary grants for the support of research. The Government Grant Fund was used by mathematicians and Cayley would use it to defray expenses in the construction of geometric models and in making the great calculations of invariant theory.

He read his paper on Malfatti's problem to the Royal Society in May 1852. At their council meeting on the 3 June 1852, he was elected at the first ballot. His candidate's certificate is completed in the unmistakable hand of Sylvester, who was his principal proposer. His qualification for membership was listed as the author of "various papers" in the *Cambridge Philosophical Transactions, Crelle's Journal, Liouville's Journal,* the *Cambridge and Dublin Mathematical Journal,* and the *Philosophical Magazine.* For the answer to the question: *Discoverer of ?,* Sylvester simply entered: "Hyperdeterminants" and to the question *Eminent as a ?,* he added "geometer and Analyst." As might be expected, the list of supporters of Cayley's candidature is a long one, and no fewer than twenty-one signed the certificate.[92]

Those who welcomed him to the society were the people who had played important parts in his life. Included were his close friends from Cambridge, William Thomson and the two Senior Wranglers who flanked each side of his own year, George Gabriel Stokes and J. C. Adams, in 1841 and 1843, respectively. Next came his contacts from University College, London: Charles Hargreave, the mathematician-lawyer, and John Graves, now a Poor-Law Inspector, who had once been anxious that Cayley would pip him in the race for the octaves. Charles Wheatstone had been one of his teachers at King's College, London, and was now a friend. Signatories from Cambridge included his mathematical coach William Hopkins, who had tutored him for the Tripos, and Whewell, the formidable master of Trinity. The naturalist Richard Owen, a friend of Whewell's since childhood, supported Cayley. Another supporter, John Shaw-Lefevre, duplicated Cayley's own pathway but from an earlier generation: Senior Wrangler in 1818, a legal conveyancer before a life of public office took over, including a twenty-year stint as the vice-chancellor of London University. John Gray with a Blackheath connection (a keen promoter of the Mechanical Institution there) and Keeper of Zoology at the British Museum, who had published over a thousand papers and memoirs, also signed Cayley's certificate. All these endorsed Cayley's certificate by reason of Personal Knowledge, while those who signed from General Knowledge included such scientific luminaries as G. B. Airy, with whom he would clash on questions of mathematical education, the scientist of optics David Brewster, and the meteorologist James Glaisher (the elder).

Mathematicians such as George Peacock, who had introduced the Continental methods into Britain in the early years of the century, were handing over the reins to the coming men, who announced the new "Modern Algebra" to British audiences. Peacock, now the dean of Ely but still Lowndean professor at Cambridge, was called on to referee Cayley's offering on Malfatti's problem. At the age of sixty-one, Peacock had done his work in mathematics and was now using his considerable energy in the service of Ely Cathedral and as a royal commissioner looking into the reform of Cambridge University. "I will very frankly confess also that the class of investigations with which this paper is connected are in advance of me & I have not for the last three or four years kept pace with what has been done by Cayley and some of the Dublin and German analysts though I have no doubt, even from the slight examination which I have made of it, as well as from my knowledge of the character of its author, that this paper may be very safely printed." Although Peacock harbored misgivings about the appearance of pure mathematics in the *Philosophical Transactions* and argued for the principle of accepting only mathematical contributions linked to physical science, he was prepared to make an exception for Cayley: "In the case of Cayley's paper [on Malfatti's problem] his high character as an analyst would exempt it from the application of this principle, more especially as it is the first which he has presented to the Society."[93] As a gentleman of science, Cayley had all the right credentials. He was obviously a favorite son.

At the time of his election, Cayley embarked on the most fruitful part of his mathematical life, during which he published virtually everything he produced. His publication in home-based journals increased and he began to bring out long memoirs in the *Philosophical Transactions of the Royal Society*. The amount of his work appearing in Continental journals was maintained but home-based journals carried equal prestige with Continental journals, if not more. As a fellow of the Royal Society, Cayley was virtually assured of publication of his work and he was able to issue whole series of memoirs of fairly abstract subjects of great length in the *Philosophical Transactions*.

At the beginning of the 1850s the avenues for publishing mathematics in Britain were, if anything, diminishing. The editors of the *Philosophical Magazine* were wary of too many pages filled with symbols, fearing a swollen postbag of readers' complaints. Financially, the *Cambridge and Dublin Mathematical Journal* was entering choppy water and *The Mathematician*, a journal based at the Woolwich Military Academy, had recently folded. In bidding farewell to *The Mathematician* after only three volumes, its founding editor T. S. Davies painted a pessimistic picture for the journeyman mathematician: "As a people [the British] we are indeed become so practical and so physical, that there is now no journal open to the discussion of pure geometrical science:—except, indeed, it present itself occasionally under the guise of 'practical utility' or 'physical application'; or else assume a form so transcendental as to bewilder thoughtful readers, and excite the wondering admiration of the 'oracle' of the social party."[94] There should be no suggestion that publication difficulties stood in the way of Cayley's career. He was hardly the journeyman mathematician. With his background, he had the Cambridge journals at his disposal and he now had access to the *Philosophical Transactions*.

In London, his career entered a new phase. In addition to his legal activity, he took part in the organization of science in the capital. There, with the Royal Society, he became involved with the administration of the Royal Astronomical Society. The British Association for the Advancement of Science was another body he supported. With all this, he became something of a scientific man-about-town. Membership in the Royal Society was especially convenient.

In the early 1850s, the society was housed in Somerset House, a five-minute walk from his chambers in Lincoln's Inn. It was only the short distance along the Strand, one of the busiest thoroughfares of Victorian London. He was on familiar territory, dividing his time between the twin centers of his scientific and legal lives; between the two places lay King's College, which he had attended as a school pupil.

In June 1852, a few days after his election to the Royal Society, Cayley was in Sylvester's rooms in Lincoln's Inn Fields. In the outer office, Thomson was writing a letter to Stokes. He did not share Cayley's enthusiasm for pure mathematics and thought it wasteful of his gifts. He was impressed by Cayley's mathematical talent but believed it should be put to service to the world instead of such rarefied domains as pure algebra. For Stokes's amusement, he added that his letter "was written in the dark, with Cayley and Sylvester talking about invariants all the time."[95] The incessant talk in the inner office was concerned with the establishment of their priority in the discovery of the new foundation for invariant theory. Sylvester had heard Eisenstein was to publish similar ideas in *Liouville's Journal*. If Continental competition was afoot, a claim for his and Cayley's discoveries had to be made, a task for which Sylvester was especially equipped. They were among a group of young men who met formally at the Royal Society but scientific business was also transacted on such social occasions as their bachelor evenings of the 1850s and "at homes" later. Following Thomson's eavesdropping, William Spottiswoode, by profession the Queen's printer but also an active gentleman of science, invited Cayley to dine with him at his magnificent home off Buckingham Gate. Thomson and Stokes, themselves elected to the Royal Society were also invited, while Sylvester, an old Royal Society hand, was in attendance. The host of the evening would be elected in the following year. It was an influential group and Stokes would soon be elected a secretary of the Society, a position that put him at the center of metropolitan scientific activity for thirty years. Here was a "mathematical dinner" of the kind that lubricated the scientific network in London in the 1850s.[96]

After only a year's membership of the Royal Society, Cayley was proposed for the award of its prestigious Royal Medal. It was made on the strength of his past work, for following his paper on Malfatti's problem, he had submitted nothing further to the *Philosophical Transactions*. On 2 June 1853, James Booth, an Irishman based in England, who also specialized in analytical geometer, proposed him for the award. This was seconded by William Hopkins, and the minutes recorded that "Arthur Cayley be placed on the list of candidates for a Royal Medal for his Analytical Researches published in the *Philosophical Transactions*, *Cambridge Journal* and *Philosophical Magazine*." At the same meeting, the chemist August Wilhelm von Hofmann and Charles Darwin were also proposed. Stern opposition and more competitors were to follow. Edward Frankland the chemist, John Tyndall the physicist, the botanist John Lindley, and Robert

Bunsen, another chemist, were added to the list at a subsequent council meeting. Surprisingly, Booth later proposed Sylvester. He had been a fellow for fourteen years and the move, perhaps, was calculated to smooth ruffled feathers. In the event, Charles Darwin and John Tyndall were recommended to the Queen, but for Cayley to be put forward for a Royal Medal after one year of membership was a significant recognition by his peers.[97]

CHAPTER 8

A Grand Design

In the summer of 1853, Cayley went on a walking tour around Machynlleth, an ancient market town near Aberystwyth, which lies at the junction of the coach roads in that western part of mid-Wales. Out in the hills, in the "wetness, windiness and storminess of the weather," he ruminated over a problem of probability theory set by George Boole. On returning to the town, he immediately dispatched a paper to the editor of the *Philosophical Magazine*.[1] It was his thirty-second birthday.

A Wide Compass

Cayley's probability problem is known today as "Boole's Challenge Problem," an inverse probability question framed in terms of n causes.[2] The work gave rise to a controversy with Boole over foundational issues and drew in Richard Dedekind and another Lincoln's Inn conveyancer, Henry Wilbraham. However interesting, and it surfaced again ten years later, it was a digression for Cayley. His concern for the foundations of probability was more than matched by his need to contribute to the outstanding algebraic and geometric problems of the day. Only the previous month he had submitted a paper on Poncelet's polygon problem of inscribing a polygon in a given conic so that it circumscribes another.[3] This question attracted him on later occasions though his mathematical thinking frequently lay beyond the solution of individual problems, no matter how compelling these were.

A driving element in his activity was the organization of algebraic subjects. Underlying this was his early schooling in the calculus of operations, in which symbols were allowed to stand for symbolic operators, differential operators, quaternions or matrices, or simply as undefined symbols.[4] The divide which existed between the "old algebra" and the new algebra was expanding. In the old familiar algebra, variables represented "ordinary quantity" but in the new symbolical algrebra they could take on many meanings or none at all. In the case where variables represented operators, the calculus was described by Augustus De Morgan to a nonmathematician as a way of calculating with "symbols which do not represent quantity, but action upon quantity."[5] Potential sources of confusion arose but these are perhaps felt more by modern readers than was the case for the nineteenth-century framers of the new algebra. The early practitioners experienced ambiguities but, curiously, these could be turned to advantage in widening their scope.[6] In particular, they arrived at operator identities that were useful to them but have since fallen into abeyance.[7] Without a clearly defined basis, the calculus of operations now appears something of an art, but its practitioners were able to handle long and complicated expressions with dexterity, even if the foundations were weak.

Groups and Symbols

When Cayley returned to London from his summer's travels, he wrote to William Rowan Hamilton thanking him for sending the compendious *Lectures on Quaternions*. In his letter, Cayley enthusiastically described his work both on the polygon problem and on caustics, the patterns generated by rays of light incident on surfaces. Hamilton was interested in both geometrical investigations. In optics, especially, he was expert. As a young man he had made his mark with his paper "On Caustics" (1824), though it would be the second part of Cayley's letter that would now interest him more. From his researches on caustics, Cayley had found "by accident" a system of symbols 1, α, β, γ, δ, ϵ, satisfying certain relations analogous to Hamilton's quaternion symbols i, j, k.[8]

Cayley set to work to formalize the idea, and he connected the six symbols with group theory. In November he submitted two papers on group theory to the *Philosophical Magazine*. These were published in 1854, followed by a third in 1859.[9] The titles were identical: "On the Theory of Groups as Depending on the Symbolical Equation $\theta^n = 1$," a title which reveals the intimacy of the link between the emerging theory and the calculus of operations. Cayley's name is now well known through his contributions to the theory of groups and these papers have attracted a good deal of modern attention, but in their own time, they went almost unnoticed. Now seen as a pioneer in abstract group theory, Cayley's attention to this subject was slight in comparison with his dedication to analytic geometry and invariant theory, and it amounted to barely 1 percent of his published work.

Cayley acknowledged Évariste Galois's work on groups of substitutions and the importance of it when he wrote in the opening paper: "The idea of a group as applied to substitutions is due to Galois, and the introduction of it may be considered as marking an epoch in the progress of the theory of algebraic equations." Galois had introduced the concept of a group in 1830 but his "Memoire sur les conditions de résolubilité des équations par radicaux," which contained the idea, was only published in *Liouville's Journal* of 1846.[10] Through his close contact with Continental mathematics, Cayley would have been aware of the various references to group theory by Joseph Liouville, Joseph Alfred Serret, and Enrico Betti.[11] Another impulse may have been given by Charles Hermite, who communicated with him on invariant theory in 1852; Hermite may have discussed Galois's theory, for it was a subject on which he had detailed knowledge.[12]

Cayley was well aware of new developments. He may also have read Betti's commentary on Galois's papers on the solution of algebraic equations published in 1852 in Barnabas Tortolini's new *Annali di Scienze mathematische e fisiche.* This was published prior to his own work, and he only noted Betti's contribution at a later date. Reference to Galois's work had been made in Joseph Serret's *Cours d'algèbre supérieure* in the first edition (1849) and was repeated in second edition (1854), but Serret appears to have had only a limited grasp of the details of Galois's work initially. Though Cayley was familiar with this book, he would have learned precious little about Galois theory from its first two editions.[13]

Cayley's first paper on group theory is partly expository and concerned with informing readers of the latest ideas in mathematics. There is a new ingredient, however, for he *generalized* the notion of a permutation group and recast the definition of a group in his own terms. In this, his background in the calculus of operations was central[14]; the much-vaunted "separation of symbols" technique in this theory, by which an operator was detached from its operand, was crucial to Cayley's viewpoint: "Let θ be a symbol of operation, which may, if we please, have for its operand, not a single quantity x, but a system so that $\theta(x, y, \ldots) = (x', y', \ldots)$. Where $x', y' \ldots$ are any functions whatever of $x, y, \ldots$, it is not even necessary that $x', y' \ldots$ should be the same in number with $x, y, \ldots$, In particular $x', y' \ldots$ &c. may represent a permutation of $x, y, \ldots$, &c. θ is in this case what is termed a substitution."[15]

He noted that the addition of these *set operators* θ (as he later termed them), a legitimate operation in the calculus of operations, was superfluous in the definition of a group. He made a special point of singling out the identity operator: "the symbol 1 will naturally denote an operation which (either generally or in regard to the particular operand) leaves the operand unaltered. . . .," but there was no need to stipulate an associative law since it automatically holds for operators. In the first paper, he also introduced the useful idea of describing a group by a "multiplication" table, the Cayley group table, showing how pairs of elements are combined.[16]

There are enough similarities between Cayley's definition of a group and the modern abstract axiomatic definition to claim him as the grandfather of abstract group theory in some vague sense, but he was not doing abstract mathematics in the modern axiomatic way. This emerged much later, fifty years after Cayley published his group theory papers. In 1853 he is more accurately the generalizer of the previous notion based on permutations. Cayley did not pursue abstraction as did the little known polymath A. J. Ellis, who graduated sixth wrangler at Cambridge in 1837 in the same year as D. F. Gregory, James Sylvester, and George Green. Duncan Gregory was a friend; from him Ellis learned the importance of operators in algebra and attempted to put the calculus of operations on a proper foundation "without any metaphysical or *a priori* reasoning."[17]

In illustrating the utility of the group idea, Cayley emphasized its wide applicability. He found examples of groups occurring in the theory of equations, the theory of quadratic forms, the theory of elliptic functions and even an occurrence in the theory of matrices, as such existed in 1853. The presence of the group of permutations on three letters was noted in a problem in geometrical optics he was working on in parallel. In the 1850s, he resumed his interest in caustic curves (those curves generated as the envelope of rays of light); on the same day he delivered his initial papers on group theory, he also submitted a paper on caustics generated by refraction in a circle.[18]

Group theory, as Cayley perceived it in the 1850s, was part of the symbolical algebra, the view of algebra promulgated by Gregory and George Peacock. Its scope was ably set out by Boole in his *Mathematical Analysis of Logic,* published in 1847, a book Cayley had studied intensively when it was first produced. The position of group theory within symbolical algebra fitted the scheme described by Boole in his *Logic:* "the same process [in symbolical algebra] may, under one scheme of interpretation, represent the solution of a question on the properties of numbers, under another, that of a geometrical problem, and under a third, that of a problem of dynamics or optics. The principle [of the validity of algebra being independent of interpretation of symbols] is indeed of fundamental importance."[19] Cayley's grasp on the potential for algebra was acute. He was able to move easily between fields of knowledge and expose connections: indeed, it is this feature of his work that illustrates his power as a mathematician and contributes to his reputation as a pioneer. His encyclopedic knowledge of mathematics, even in the 1850s when he was a young man, enabled him to connect groups with Sir William Rowan Hamilton's study of quaternions, which was just beginning to take root in the consciousness of the mathematical community in Britain. He was no doubt encouraged to make connections by other current research. Around the time he submitted his papers on group theory, he established a relationship between rotations in four-dimensional space and quaternionic multiplication in which a point x is rotated to a new point pxq where p, q are quaternions. This was an extension of his earlier result expressing a rotation in three-dimensional space expressed as $q^{-1}xq$. This supplement was far-reaching and Felix Klein was later to refer to it as both "remarkable" and "beautiful."[20] Fifty years later these transformations were rediscovered by Albert Einstein as the Lorentz transformations of the Special Theory of Relativity. This is, of course, in the future and beyond Cayley's world of the 1850s, a period to which we now return.

Being active in so many different areas, Cayley was in a unique position to make links between subjects, encouraged by his continued attention to the subject of quaternions, which was coterminous with his group theory. In his first group theory paper, he discarded the idea of algebraic expressions formed from group elements: "It is, I think worth noticing, that if, instead of considering α, β, & c. as symbols of operation, we consider them as quantities (or, to use a more abstract term, *cogitables* such as quaternion imaginaries; the equations expressing the existence of the group are, in fact, the equations defining the meaning of the product of two complex quantities of the form $w + a\alpha + b\beta + \ldots$"[21] Here Cayley generalizes the notion of a linear combination of quaternion imaginaries, i,j,k, to a (finite) linear combination of his group elements, which are to him the symbols of symbolical algebra. Thus the new "complex quantities" are analogous to the quaternion $a + bi + cj + dk$. Cayley did little with them, though by setting down $w + a\alpha + b\beta + \ldots$ he was in a good position to deal with group algebras. Instead, he observed connections. To him, algebra was an organic entity and a unity, but one in need of organization.

The Brotherhood Gathers

At the annual British Association meeting held at Hull in September of 1853, Sylvester acknowledged that his intention to prepare a report on determinants had been rendered redundant by William Spottiswoode's publication on the subject two years prior. Nonetheless brimming with enthusiasm, he promised the association a report on invariant theory: "The much vaster subject of Invariants, which includes the theory of Determinants as its simplest case, has at present no chronicler or editor."[22] Such ambitions could act as a spur to Cayley to commence his own memoirs immediately before Sylvester became that editor. As Sylvester suggested, there was a need for a clear authoritative outline of the basis for the theory and a review of its achievements to date. Cayley did not delay in getting the formal presentation underway. He was utterly dedicated to invariant theory and, following his discovery of the new foundation toward the end of 1851, it became his principal research area. He paid constant attention to it, and though Sylvester was an incessant campaigner in 1852, his attention wavered and his dedication became intermittent.

In the 1850s invariant theory attracted mathematicians from all the leading European countries, and like the wares displayed in the Great Exhibition, Cayley's and Sylvester's mathematics attracted the interests of those on the Continent. Invariant theory would not be displayed in *The Great Exhibition of the Works of Industry of All Nations* held in Hyde Park, London, in 1851, but it would be a branch of mathematics that would actually have an effect on the future developments of mathematics on a worldwide scale.

As the century progressed, some of the most prominent mathematicians of the age joined the quest, and in the 1850s, the invariant trinity lay at the center of a growing circle of followers. The subject was already beginning to acquire a substantial body of knowledge and also a history.[23] It could be applied to both geometry and algebra, as Cayley had clearly understood as a young man, and to many disciples, it appeared to provide an underlying theory to both areas. The outstanding problem in algebra to which it could be directed was the solution of the polynomial equation of order 5—the solution of the binary quintic. This problem became a beacon that often motivated the new theory, a wholly distinct one from the group theoretic approach applied to the same problem, as scribbled down by Galois on the eve of his death.

Siegfried Aronhold, from Königsberg in Prussia, a student of Otto Hesse (and hence a "grandson-student" of Carl Jacobi), studied invariant theory from an algebraic viewpoint as distinct from the geometric one of his mentor. Aronhold read the work of the English algebraists and, in the course of his work, adopted some of their terminology. He introduced a new succinct notation in comparison to Cayley's amplified notation based on coordinates. Cayley regarded Aronhold's 1850 paper as a very important step in the theory and could count him among his correspondents. Another who became a long-standing correspondent was the Swiss mathematician Ludwig Schläfli. Initially, mathematics was a spare-time activity for this man (along with Sanscrit and the ancient language of the *Rigveda*), who was a schoolteacher at the start of his career before he became professor of mathematics in Bern in 1853. Among the papers he read was Cayley's 1845 paper on elliptic functions as doubly-infinite products and he thought highly of it. In invariant theory, he wrote a paper on resultants, a way of combining algebraic forms. Cayley opened a correspondence with him and helped him publish papers in English journals.[24]

Hermite began a close association with Cayley in the early 1850s. He contributed to invariant theory for a decade and adopted the English terminology and notation. In the subject he made two fundamental discoveries. One was a law of reciprocity, which established a one-to-one correspondence between the covariants of order m of a binary form of order n, and the covariants of order n of a binary form of order m. This is the sort of theorem that appeals to the mathematician. It may be useful in making calculations, but it is the simplicity and unexpectedness that is appreciated. For Cayley it was not merely the law of reciprocity (which, incidentally, he attributed to Sylvester as well) but the *beautiful* law of reciprocity.

The other of Hermite's landmarks was his startling discovery of a new invariant of the binary form of order 5. Before Hermite's conspicuous revelation, Cayley had thought this form possessed only three fundamental or irreducible invariants.[25] He had found expressions for them, and labeled them G (found in the 1840s), Q, U, the invariants of degrees 4, 8, and 12. Hermite's example, to be labeled W by Cayley, was a skew invariant of degree 18.[26] Its appearance surprised Cayley and ran counter to his previously held belief that the degrees of invariants in the case of the quintic were multiples of four. On its appearance, he remarked that its existence removed the "erroneous impression which I had been under from the commencement of the subject." Following Hermite, George Salmon, who became renowned for his feats of calculation, produced an algebraic expression for W, a formula which occupied five quarto pages.[27]

Hermite also put his finger on the "problem of the syzygies," which was to intrude on Cayley's further researches. As noted earlier (chap. 5), the invariant W could not be expressed as an *algebraic* function of invariants G, Q, U, yet there was a syzygy $16W^2 = GQ^4 + 8Q^3U - 2G^2Q^2U - 72GQU^2 - 432U^3 + G^3U^2$. Hermite wrote:

> We see by this that an essential character of forms of order greater than four is revealed, which is that the invariants may not in general be expressed as a rational function of a certain number of supposed algebraically independent ones. M. Cayley, M. Sylvester and myself had for long time thought that in general the invariants of forms of the mth order should be expressible from among $m-2$ whole functions and this had prevented M. Sylvester to seek a demonstration of the law of reciprocity of which he had also presumed the existence, for a necessary contradiction would be clear between this law and that of the number of fundamental invariants.[28]

Binary forms of the first four orders have relatively simple systems of invariants and covariants, but the binary form of order 5 presented a formidable challenge.

At Hermite's urging, mathematicians were encouraged to study polynomial equations from an invariant theory perspective, a point of view he claimed would offer a fertile field of research. Invariant theory offered a rival method for dealing with the algebraic solution of polynomial equations quite separate from the method suggested by the emerging Galois theory. It produced a marked contrast and suggested new insights, and it purported to answer different questions. For instance, under what circumstances are the roots of a polynomial real? The answers in terms of invariants were less elegant to be sure, but amidst the elongated formulae worthy of the Steam Age, it could also offer arresting answers. In the case of the quintic equation, for example, the vanishing of Hermite's skew invariant ($W = 0$) is a condition for it to be soluble.[29]

In Italy, Cayley's line of research found favor when Francesco Brioschi energetically took up the subject and readily pursued study of binary forms. Brioschi contributed to the determinants and the theory of invariants in a series of papers beginning in 1854, which were published in the *Annali di Scienze matematiche e fisiche* and later in the *Annali di Matematica pura ed applicata,* newly created in 1858.[30] He too adopted Cayley's methods, for both the basis of invariant theory and its notation. His textbook on determinants, *La Teorica dei determinanti,* appeared in 1854 and was translated into French and German. For Brioschi, invariant theory became a lifelong pursuit dispersed within broad interests of both pure and applied mathematics leavened with a contribution to political life in the emergent Italian state. In turn, he influenced Enrico Betti, who produced a paper on invariant theory dealing with binary forms.[31] Betti's mathematical interests in algebra coincided with Cayley's but invariant theory proved a brief flirtation. Another Italian mathematician, Francesco Faà di Bruno, was attracted to invariant theory and published work in Britain. In 1859 he published *Théorie générale de l'élimination,* which again adopted Cayley's viewpoint.[32]

In England, Cayley's most prominent disciple, Sylvester apart, was the Oxford-educated William Spottiswoode. At university, Spottiswoode had been taught by Bartholomew Price, who recalled that from the first his student "showed an extraordinary liking for, and great skill in, what I may call the morphology of mathematics."[33] But the man who showed him the "magnificence of mathematics" was W. F. Donkin, the Savilian professor of mathematics. Spottiswoode's first publication *Meditationes analyticae* (1847) demonstrated a wide range of interests and a facility with mathematical ideas. He focused on pure mathematics and published a slim but influential *Elementary Theorems Relating to Determinants* in 1851. A second edition, largely extended and rewritten, was published in *Crelle's Journal* in 1856.[34] Following this lead, Richard Baltzer published *Theorie und Anwendung der Determinanten* in 1857 and a French edition appeared in 1861.

In the 1850s Spottiswoode brought his considerable organizational skills to bear in rescuing the family printing firm from the financial ruin following his father's bankruptcy. Even then, the "Queen's Printer" reserved time for science. This respectable patrician also made it his business to conduct scientific lectures for the benefit of his employees, while for his social circle, "at homes" in his palatial London residence were well attended, as we have seen. He was a founding member of the X Club in 1864, a group of scientists who promoted science and their own careers in tandem. He was constantly active in the cause of science and an inveterate member of societies. He joined the Ethnological Society, the Royal Asiatic Society, the Society of Antiquaries, the Royal Astronomical Society, and acted (with Galton) as a secretary of the Royal Geographical Society during a difficult time in its history. He was elected to the Royal Society in 1853, the year following Cayley's election, and later became its president.

Spottiswoode's generalist scientific career throws into relief Cayley's single-minded application to mathematics. Being at heart a polymath, Spottiswoode pursued diverse scientific and historical interests while Cayley was a comparative rarity on the British scientific scene at midcentury. While Spottiswoode published research in invariant theory in 1854 and coupled it to his interest in geometry, his contributions to the subject were sporadic. In the 1870s he turned his attention to physics and the polarization of light, but in 1883, shortly before his death, in coming full circle, he made a final contribution to his youthful endeavor.

The link Cayley forged with the Irish mathematicians during his visit to Dublin in 1848 encouraged invariant theory in that country. Of the Roberts twins, Michael took a leading part in pursuing the theory. His range in pure mathematics coincided with Cayley's and he published papers in invariant theory during the period 1855–70. Salmon was the principal disciple, but while he was a mathematician of considerable talent, Sir William Rowan Hamilton reigned supreme in Dublin. Salmon tried to recruit him to the invariant theory cause, but, since that momentous day in October 1843 when he carved the quaternions on Brougham Bridge, Hamilton increasingly saw mathematics through a quaternionic haze. Indeed, he wished to convert the young Salmon to quaternions, writing to a colleague, "[he] is getting on so awfully fast in Quaternions that if I don't take care we shall get into some contest of priority. It is a genuine pleasure to me to believe that in Salmon I shall have a worthy successor, and may he much excel, even in quaternions, myself!"[35] For Hamilton, even invariant theory was reducible to his own great algebraic theory. A more mature Salmon hoped to reverse this lopsided standpoint, though Hamilton would undoubtedly claim in private that the invariant theorists were themselves myopic. Salmon's efforts to sway his famous Dublin neighbor were strenuous: "I hope you will not be lazy about the Covariants and Contravariants," he reminded Hamilton. "I have no doubt that great fruits will come from the marriage of the Quaternions with the Calculus of Forms [invariant theory], and no one but you ought to give away the bride."[36] Hamilton did not respond and invariant theory was left at the altar.

Farther south of Dublin, George Boole, established as professor of mathematics at Queen's College, Cork, was a significant departure from the invariant theory fold. While he had initially inspired Cayley by the papers he published in 1841, he made his last contribution ten years later. In this, he calculated an invariant of the binary quintic and indicated an intention of taking a "connected view of the methods and the results already obtained."[37] Boole had actually grown tired of invariant theory. His early enthusiasm had fallen away and he now regarded the subject as a "peculiar and rather isolated branch of analysis."

Boole was drawn into propounding his new theory on probability and he really wanted to get on with this and the study of logic. He wrote to a friend in 1851: "The Logic has been at a stand still for some time as I have been writing papers [on invariant theory] for the Cambridge Journal the first part of which has just appeared. As professor [of mathematics] I feel bound to keep up some correspondence with the purely mathematical public but my heart is with the Logic."[38] This was part of Boole's more general disaffection with mathematics itself. He now felt an intense devotion to a technical branch of mathematics was limiting and cut one off from other cultural endeavors. Clearing the decks of this kind of mathematics, his work on "Reciprocal Polars in Differential Calculus," which he discussed at length with Cayley, was approved of by his friend and sent for publication.[39] Cayley was ever supportive of Boole in his endeavors, whether it was invariant theory or other subjects. When *Investigation of the Laws of Thought* did appear in 1854, Cayley wrote to him, "I have to thank you for the copy of your long expected book. I have been particularly delighted with the logical part of it and the generality you have given to the theory as to the probabilities the whole subject seems so paradoxical that I hardly know what to make of it."[40] While probability theory was an enduring interest and his controversy with Boole a desideratum, invariant theory loomed much larger in his scale of priorities.

The Treasure Hunt

As Victorian naturalists went out and gathered their scientific specimens, Cayley sought invariants and covariants. The prime duty of a pure mathematician is to prove theorems, but to impose this modern view on Cayley would lead to an unbalanced picture of his mathematical activity. In thinking of him embedded in his own times, he more accurately fits the description of a nineteenth-century botanist "who collects specimens to swell his herbariums, gives them barbarous names, and tries to arrange them in a system which some other botanist will then dispute. The unit of science is the species."[41] While Cayley actually indulged in the Victorian pastime of going out to hilly pastures and collecting new species of plants, he duplicated this leisure activity in his mathematical endeavors.

The initial development of invariant theory has much in common with the establishment of botany, zoology, and geology, what the Victorians termed the *great sciences of observation*. These were overtly classificatory. Carl Von Linné (Linnaeus) in botany and Friedrich Mohs in mineralogy were scientists who made classification in terms of genera and species the object of their science during the eighteenth century. In the nineteenth century, these sciences were concerned with the accumulation and classification of vast and very detailed information.

The great organizational maps of fields of inquiry and their evolution loomed large for the Victorian scientist. Near Cayley's intellectual home, J. B. Henslow, professor of botany at Cambridge and mentor of the young Charles Darwin, presented two cabinets of his entire collection of British insects and shells to the Cambridge Philosophical Society; T. H. Huxley revised his course of lectures for 1854 in order to begin with a new classification of the five sub-kingdoms of animal life into protozoa, coelenterata (endoderm and ectoderm), vertebrate, and annulosa. And we must not forget John Gray, Cayley's supporter for membership of the Royal Society. Gray published zoological catalogs such as *Synopsis of British Mollusks* (1852) and many others, including charts on monkeys, lemurs, and fruit-eating bats and classifying divisions of mammals into those which are carnivorous, ruminant, pachydermatous (thick-skinned), and edentate (incisorless).

Such classifications were carried out in the most minute detail. Scientists typically saw themselves separated from an objective reality of the world and it was their task to investigate this reality. This separation of mind and matter was fundamental. Cayley's position was not unlike that of his scientific colleagues, "on one side lay the field of inquiry, the material world, or nature, standing off from the inquirer and preserving an essential and unalterable *reality* independent of the thoughts and wishes, hopes and fears, goodness or badness of the minds that contemplated it. On the other side were those minds, whose scientific function was solely to discover, first by observation and experiment and then by rational deduction, what the material world contained and how its course was ordered."[42]

The search for irreducible invariants and covariants closely parallels the activity of chemists in filling out the table of chemical elements. The chemists were putting the finishing touches to their endeavor in the 1850s and 1860s; Robert Bunsen discovered the new elements rubidium and caesium and William Crookes discovered thallium. In an analogy between mathematics and chemistry, *composite* algebraic forms would correspond to chemical *compounds*, and, being less basic would be of lesser interest. If one substitutes "irreducible invariants and covariants" for "chemical elements," one comes close to Cayley's point of view of invariant theory. These algebraic forms represented reality, and it was the task of Victorian mathematicians to describe their properties. Like the empirical scientists with their representative specimens, they had first to find them.

Naturalists would first find their specimens; second, order and classify them; third, name them; and, finally, place them in their "herbariums." It is in this sense that Cayley is a naturalist mathematician—explorer, collector, and classifier. A comment by A. N. Whitehead writing about *Natural Science* is applicable to this activity: "Classification is a half-way house between the immediate concreteness of the individual thing and the complete abstraction of mathematical notions. The species take account of the specific character, and the genera of the generic character."[43] With his patient cataloguing, Cayley bears comparison to the great Victorian classifier of botany, Sir Joseph Hooker, who over the twenty-five year period from 1862– to 1883 (with George Bentham), compiled the monumental three-volume work *Genera Plantarum* based on the collection of botanical specimens at Kew Gardens, the famous plant repository situated on the fringes of London. The taxonomic work, written entirely in Latin, containing thousands of species, aimed to set a rigorous standard in the classification of plants.[44] Cayley's work is as detailed as that contained in *Genera Plantarum*, but, as well, the quest for the invariant theory of binary forms had the potential of dwarfing any number of volumes. Botanists may have believed their species were finite in number but Cayley believed he had an infinite task before him.

This methodological link between mathematicians and other scientists is revealing of Cayley's motive. Both groups looked on scientific advance as a *process of gradual accumulation.* The Victorian ideal was simply: "Knowledge once acquired remained for all time; fresh knowledge was added to it, but without changing it in any way," to quote Herbert Dingle's analysis.[45] This is particularly relevant to the problem of putting Cayley's work into a temporal context with respect to the way he remorselessly collected invariants and covariants. Without it, the collection of labeled tables of invariants and covariants is too easily dismissed as lacking point and even of being "wrong-headed"; it is quite different from the modern ideal of crisp deductions made from clearly stated axioms. An added fillip to the collection of invariants and covariants lay in the attitude of the Victorians to certainty in mathematics (mathematics, unlike many other subjects, was regarded as absolutely certain and permanent). Once a result was found, it was there for all time. Results might require laborious work to uncover, and many of the calculations of invariant theory were indeed of massive proportions, but once calculated the hard-won results would never have to be recalculated. The invariants and covariants of the binary form of order 5 occupied Cayley for a half-century, but once achieved, the long formulae could be placed inside the storehouse of knowledge and regarded as permanent information. He had no reason to believe that succeeding generations would be ungrateful.

The style of Cayley's papers served this purpose of collecting and classifying. They are typically discursive, contain little formal proof, and many of them simply assemble the specimens. Perhaps there is no other alternative in the early days of a new theory, even a mathematical one. Ponderous formulae can only be safely jettisoned when the underlying concepts are understood and the theory refined and initial clumsiness forgotten.

Sylvester was quite graphic in the way invariants could be found, examined, and dissected. About the ever-present invariant $I = ae - 4bd + 3c^2$ of the binary quartic form, which Cayley and Eisenstein had brought to light, he wrote: "In the accidental observation by Eisenstein, some score or more years ago, [1844] of a single invariant (the Quadrinvariant of a Binary Quartic) which he met with in the course of certain researches just as accidentally and unexpectedly as M. Du Chaillu might meet a Gorilla in the country of the Fantees, or any one of us in London a White Polar Bear escaped from the Zoological Gardens. Fortunately he pounced down upon his prey and preserved it for the contemplation and study of future mathematicians."[46]

Paul Du Chaillu's accounts of his African exploration excited great interest in Victorian society in the 1860s. His great success was finding the gorilla in equatorial West Africa, a discovery at first met with disbelief. A blatant self-publicist, he startled his audiences at the Royal Geographical Society with his traveler's tales of places no European had yet ventured—apart from himself. Whether a more appropriate stage would have been a circus sideshow need not concern us, but he had enough scientific credibility to sell his collection of specimens to the British Museum. Yet, in his wildest dreams, Sylvester could not have expected to match the efforts of Du Chaillu, who once boasted: "I shot, stuffed and brought home over 2000 birds . . . more than 60 new species . . . and I killed upwards of 1000 quadrupeds of which 200 were stuffed . . . not less than 20 of these are species hitherto unknown."[47]

Sylvester was keen to tell his audience that mathematicians were having comparable success in more northern climes as they conducted their exploration of the invariant theory kingdom, wishing to stress the unity of approach of mathematician and scientist. Cayley too, with less beating of the drum, took the idea of classification and naming of *genera* and *species* as a natural way of proceeding. With considerable advantage, invariant theorists began to speak of binary cubics, ternary cubics, and quaternary cubics as different types of algebraic forms with two, three, and four variables. The use of two-part nomenclature also had been an important step in the *genera-species* nomenclature of the empirical sciences.[48] Both groups of workers made frequent allusion to species and genera in their work, and the discovery of a new invariant or covariant specimen could be of comparable significance as the appearance of a new animal or plant species. Cayley was concerned with "convenient rules for selection," his guiding principle being to catalog specimens by selecting the *simplest* invariants and covariants.

At the Baptismal Font

The activity of naming and choosing mathematical notation was an important part of the classification procedure. Naming was also a necessity actively pursued by scientists in other domains, such as Alexander Williamson working at University College, London, in the 1850s, as he involved himself with the new language of organic chemistry. Cayley's outlook regarding questions of nomenclature is conveyed by a seemingly casual remark he made about invariant theory. Noting that the "modes of generation of a covariant are infinite in number," he saw the need to "single out and to define and give names to new covariants."[49]

Faced with a vast array of these new forms, classification by *name* was a natural way to organize the infant theory. It accounts largely for the introduction of a remarkable vocabulary. Cayley was hardly abstemious but Sylvester took the lead: "Progress in these researches," he wrote at the beginning of the 1850s, "is impossible without the aid of clear expression; and the first condition of a good nomenclature is that things shall be called by different names."[50] He took his own advice to heart and dubbed himself the "Mathematical Adam," advocating the naming process which "[t]urns them to shapes, and gives to airy nothing/ A local habitation and a name."[51] He laid claim to creating *all* the technical terms of invariant theory, barring "quantic," which was indubitably Cayley's, and "eliminant," which neither of them liked. Of course, he did not act without Cayley's approval and new terminology was not introduced "without high sanction," for on matters of notation Cayley chose carefully and might even be described as a stickler.

Names of invariants and covariants enabled these semiprecious stones to be collected, recorded, and carefully placed in the "mahogany cabinet" with all the other specimens. Sylvester was hard at work in Lincoln's Inn Fields distributing the honors, and on the point of immortalizing Cayley, happy to announce his intention to the powerless recipient: "I find that there are objections on principle to call the Hessian of one function bordered with the 1st derivations (I mean differential coefficients) of another function, the Hessian of the 1st relative to the 2^{d} as I had proposed—and as I must have a specific *name* and you appear to have been the first to turn this kind of form to practical account I shall henceforth call such form the Caleyan [*sic*] of the 1*st* in respect to the second."[52] "Since then," Sylvester noted many years later, "we have lived to see Jacobians and Caleyans and Pippians and Quippians—and many other more or less meaning or unmeaning names take rank alongside of the Hessian."[53] Though the "Caleyan" is not so well known, the "Hessian" is now firmly anchored in mainstream mathematical terminology. Cayley himself was circumspect about the introduction of new terminology and his own choices were often sober, but he supported Sylvester in his endowment of spectacular terms. In common with other sciences, both derived their terminology from Latin and Greek in the same way Linnaeus had favored the classical languages for botany. The chemists of England during the 1850s and 1860s made their terminology by fusing Greek and Latin prefixes to Latin terminations.[54] In similar vein, Cayley and Sylvester did the same, hoping to achieve a "state of fixity" for mathematics as was striven for in the natural sciences.

The initial effect of this efflorescence of mathematical language was a communication problem. Felt most keenly abroad, the general reaction of Continental mathematicians to the new notations and terminology emanating from the Inns of Court in London was utter bemusement. While visiting the Continent, the English mathematician Thomas Archer Hirst recorded Liouville's views: "He acknowledged their ability but he protests against their wilful obscurity. He considers Cayley and Sylvester to be in some measure the disciples of Cauchy in this respect. In order to attain a broader view of the subject, they lose precision. Ordinary phraseology hampers them, and without hesitation they coin a language of their own, useful to them no doubt, but for others decidedly inferior to the ordinary language. To be precise and clear is equivalent in their eyes to being tedious." Thinking of Cayley's and Sylvester's imperial procession, Liouville added: "Rather than march over their difficulties and through their conquered territory with a firm, steady step, they leap and turn somersaults. It is possible that by so doing *they* are able to take a rapid and sufficient view of their subject, but others decidedly see better with their heads upwards."[55]

There were catalecticants, concomitants, and combinants. From Sylvester's pen flowed other marvels such as: bezoutoid, co-bezoutiant, rhizoristic, syrrhizoristic, meio-catalecticis, all of which have long since disappeared from mathematical usage. Other of his creations remain and can be found in current dictionaries of mathematics: cogredient, contragredient, covariant, invariant, discriminant, Hessian, Jacobian. Names of mathematicians (such as Hesse and Jacobi) were the source of some, but more interesting are the terms borrowed from botany, where neologisms were plentiful. Sylvester's rhizoristic with its Greek prefix is a term associated with a plant having a perennial root structure and perishing stem, while both rhizocephalus and syziganthus designate genera described in the *Genera Plantarum*. Of them all, it was "syzygy," borrowed from astronomy, which was the most used and most memorable. When in one of his evangelizing moods, Sylvester declared: "Think the meaning of the word 'Syzygy' and the logic of algebra has become part of your being."[56]

Some choices could even confuse members of the inner circle. Salmon on one occasion appealed to Sylvester for clarification, or, was he turning the lighter side of his devastating wit on his friend? "[I] am stopped at the outset by not knowing your terminology," he wrote on one occasion, "[a]n evectant is $x^n(d/da) + \ldots$ but what is a d*evectant*?" There was little hope for those outside the brethren's coterie. To their credit, Sylvester published a glossary of "new or unusual terms" and Cayley too gave one on "recent terminology in mathematics" for the *English Cyclopaedia*.[57]

A Matter of Calculation

From his treatment of the infant invariant theory in the 1840s, Cayley concerned himself with calculations. These could be onerous, but he gained an impressive fluency. In dealing with one problem, he noted, "it is almost needless to remark, [the calculations would] be very laborious, but the forms of the results are easily foreseen, and the results can be verified by means of one or two coefficients only."[58] From a modern perspective, the extensive calculations seem futile and barren, but from his own, they were perfectly natural.

Given his Platonist perspective, Cayley was in the process of finding the "objects of mathematics," and the building blocks of the subject itself. Boole, who did not entertain too much long-winded calculation, was also of the view that such results were important: "I have no doubt that the time will arrive when the tabulated forms of analysis [i.e., invariant theory] will be as important for reference to the mathematician engaged in original researches, as the tabulated values of different analytical functions are to the calculator now. Mr Cayley's present memoir must be considered in connexion with his previous ones, and with reference to this end."[59] Boole was commenting on one of Cayley's exercises in the theory of symmetric functions, a related field in which the calculations could also be extremely lengthy, but also routine. It was good fortune Cayley had George Salmon alongside. Salmon entered into such long calculations with enthusiasm. In addition to his calculation of Hermite's five-page skew invariant W, he also processed the expression for the discriminant Δ_5 wanted in connection with Cayley's research on surfaces generated by a variable plane. Invariant theory produced an abundance of this work and, next on the agenda, Cayley needed the expression for the discriminant Δ_6 for his work on caustics. Though not adverse to calculation himself, he gladly relinquished the repetitive work to others and in particular welcomed the assistance of Robert Harley.[60]

Robert Harley, a Congregationalist minister, was one of the heroic self-taught "amateurs" of mathematics. As a young teacher, he was encouraged by James Cockle through their mutual interest in solving problems in the *Lady's and Gentleman's Diary*. When he became a part-time professor of mathematics and logic at Bradford College in Yorkshire, Boole was a frequent visitor, and from him he acquired an absorbing interest in logic. He became a disciple of Boole's, but he would be known in the academic world for his publications in the theory of equations; his twin passions were suitably summarized as temperance and "quintics." On coming to live in London, he was welcomed into the scientific community by Cayley and enjoyed his support for election to the Royal Society in 1863. Harley enjoyed a wide range of friendships, in the Church, in politics, and in science. His varied life included educational reform in Leicester, a teaching position at the Mill Hill School in North London, and for a spell he was a pastor in Sydney.[61]

While Harley and Salmon could be harnessed for the occasional calculations of invariant theory, they could not calculate all that was needed. For the systematic calculations required for the compilation of extensive tables, "human computers" were employed, servants who occupy an honorable place in the history of mathematics. In astronomy they were frequently called on to solve large sets of linear equations arising from the method of least squares. There was an active squad of fourteen computers at Greenwich working under George Biddell Airy's close supervision. Cayley relegated much of the humdrum of invariant theory and lunar calculations to hired helpers, who were employed through grants awarded by the British Association and the Royal Society. As he advanced to the binary form of order 5 and onward to binary forms of higher order, the services of these gentlemen, who sat on their stools and repetitiously turned the crank, became indispensable. Increasingly, they were needed in other areas of his work, particularly for astronomical calculations, where Cayley's two favorites, Messrs. Charles Creedy and William Davis, were engaged in preparing the very extensive tables in the theory of elliptic motion.[62]

William Barrett Davis, in particular, was a calculator of some ability. Five years Cayley's senior, he was a tailor by trade who lived in Marylebone, an area of London near Regent's Park. Cayley valued him highly for Davis was equally at home with invariant theory and astronomy. He was also adept with calculating tables for the use in number theory and produced a list of large prime numbers, which, in Victorian times, was an astonishing calculatory feat. This gentleman was not a calculating genius of the traditional type, fluent with reckoning but otherwise illiterate. He possessed a historical turn of mind and gave a paper on Joseph-Louis Lagrange's theorem to a British Association meeting.[63] Creedy, too, had cut his mathematical teeth and had written an article for the new *Quarterly Journal of Pure and Applied Mathematics* on "Kepler's Problem." Undoubtedly he is the "Creedy" who worked on the difference engine and analytical engine projects alongside Charles G. Jarvis in Charles Babbage's drawing office. Earning 2s 6d an hour, Jarvis was "a first-rate draftsman" and "made all the beautiful drawings for the analytical engine," reported Babbage's son Major-General H. P. Babbage. Unfortunately, Creedy did not measure up in this nineteenth-century military mind: "Creedy knew high mathematics and was an enthusiastic analyst, but he was unfortunately very irregular in his habits."[64]

While Cayley was intensely interested in calculation for invariant theory, he showed a distinct lack of interest in what is now referred to as numerical analysis. In his thinking, there existed a sharp division between solving an equation numerically and the problem of finding a theoretical *algebraic* solution. Cayley wanted to know the invariants and covariants as they were of theoretical importance, but the methods which were used for solving equations numerically held little attraction. On one occasion, Sylvester was involved with a numerical quadrature problem, but he could not interest Cayley and issued a mild reproach: "Why should you *despise* this subject?" he asked, "or regard it only as a mere matter of Numerical Approximation? Your question 'et puis?' would have choked many a grand theory in the bud."[65]

With all the intensity of discovery in invariant theory by an expanding circle of researchers, what the subject lacked during the early years of the decade was a well-defined core of papers that would summarize progress, establish a bridgehead, and suggest further work. Cayley was now in his early thirties and about to answer this challenge, and he set out a framework for his most ambitious project yet, the famous and definitive series of "memoirs on quantics." His decision to publish the work in the *Philosophical Transactions* was probably dictated by the anticipated length and nature of the memoirs. It was not a specialized mathematical journal and few Royal Society fellows would gain much benefit, since even the selected referees found the material difficult and unfamiliar. In its favor, the *Philosophical Transactions* could accommodate bulk while the editor of *Crelle's Journal,* in whose journal he placed shorter works, would surely have demurred.

The Memoirs on Quantics

Cayley's "Introductory Memoir on Quantics," received by the Royal Society on 20 April 1854, dealt with the language of invariant theory and with definitions. In this first paper read before the Royal Society since his election two years before, the whole basis for invariant theory was remodeled. Invariant theory was removed from its historical roots and acquired a degree of abstraction. The opening paragraph set the scene for the projected series. His first creation was the word "quantic" itself: "The term Quantics is used to denote the entire subject of rational and integral functions, and of the equations and loci to which these give rise; the word 'quantic' is an adjective, meaning *of such a degree,* but may be used substantively, the noun understood being (unless the contrary appears by the context) function. so used the word admits of the plural 'quantics.'"[66]

This leading paragraph underlines the purpose of stating afresh the leading ideas of invariant theory and to unify it under the new principle of characterizing invariants and covariants as algebraic forms annihilated by differential operators. They had now become those algebraic forms that are reduced to zero by the application of a differential operator. All reference to linear transformations was suppressed and the earlier term used for homogeneous polynomial (rational and integral function) was replaced by the simpler "quantic." This introduction was criticized by the referees, but Cayley felt a need for replacing the traditional term. His formal style is, in part, due to his legal training.

By defining the extent of the word "quantics," his opening paragraph parallels the preamble of Cayley's legal draft on "Family Settlement." Both placed limits on the projected field and explained terminology and both are analytical in tone. He saw the need to place limits and make a start on its reorganization; the generality of the "Introductory Memoir" set out the territory while the ensuing detail supported this grand scheme and indicated its richness and immense proportions. For Cayley, "quantics" were as fundamental in invariant theory as "quantities" were in ordinary algebra.

If anything can be guaranteed to raise hackles among specialists, it is the introduction of new terminology in which the once familiar takes on an unfamiliar air. Objections were raised by Boole who thought "quantic" a poor choice on etymological grounds since it was a Greek termination to a Latin adjective "which expresses nothing." John Graves, another of the referees, noted the paper's abstract character and believed the introduction of new terminology unnecessary. If Cayley's formality is evident in the "Introductory Memoir," so too are glimpses of his intuition. He took due note of Hermite's law of reciprocity, but he explained it in his own way, suspecting, as mathematicians do, that Hermite's proof was identical with his own, "I am inclined to think."[67]

Outside Lincoln's Inn and outside of mathematics, the country was gripped by international instability, and by the end of March 1854, Britain and France had declared war on Russia. Income tax was doubled to meet the War Budget and the 26th of April was set aside for a national day of fast, humiliation, and prayer, in order to awaken the population to the seriousness of the situation. April and May were busy months for Cayley, too. Just before the day of fast, he submitted his "Introductory Memoir" and on Thursday, the 4th of May, he read it before the Royal Society. Colonel Sabine, treasurer and vice-president, was in the chair and Cayley's paper followed William Thomson's account of his researches on thermoelectricity. He put the binary form of order 5 under the spotlight and struggled with the search for its irreducible invariants and covariants. He finished off a number of researches and gave progress reports on others. A month after he submitted his "Introductory Memoir on Quantics," he submitted a block of seven papers to *Crelle's Journal:*

> "Réponse à une Question proposée par M. Steiner";
> "Sur un Théorème de M. Schläfli";
> "Remarques sur la Notation des Fonctions Algébriques";
> "Note sur les Covariants d'une Fonction Quadratique, Cubique, ou Biquadratique à deux Indéterminées";
> "Sur la Transformation d'une Fonction Quadratique en elle-même par des Substitutions linéaires";
> "Recherches Ultérieures sur les Déterminants gauches"; and
> "Recherches sur les Matrices dont les termes sont des fonctions linéaires d'une seule Indéterminée."

These took up the bulk of the fiftieth volume of *Crelle*, published in 1855, the year following the publication of the groundbreaking work on elliptic functions by Karl Weierstrass in the same journal. Cayley's papers were on conics, algebraic resultants, the matrix notation, elementary invariant theory, quadratic forms, a related one on skew determinants, and finally one on analytical geometry. In the submitted manuscript for the paper on the matrix notation, he intentionally used the notation

$$\begin{pmatrix} a, & b, & c \\ a', & b', & c' \\ a'', & b'', & c'' \end{pmatrix}$$

as a natural *extension* of the notation for a quantic $(a,b,c,\ldots\!\!\!(x,y,z\ldots)^n$. It was not a clean break from the past and the use of the unusually shaped "verticals" was not a limitation imposed by the printer as he used the same notation in handwritten letters.[68] Cayley was particular about the notation he used and he showed how the matrix notation, outlined in "Remarques sur la Notation des Fonctions Algébriques," facilitated the writing of linear equations and bilinear forms.

From scraps of evidence, one can piece together Cayley's mathematical activity in the last week of May. On reconsidering the binary form of order 5, he was halted by the question of existence of covariants of degree 5, a sticking point as his strategy was to investigate this binary form by listing covariants by successive degrees. Corresponding questions for the cubic and quartic binary form are straightforward and their irreducible systems easily written down. For example, the binary cubic has a covariant (the Hessian) of degree 2 and order 2:

$$H = (ac - b^2)x^2 + (ad - bc)xy + (bd - c^2)y^2.$$

The key to a covariant is its leading term $(ac - b^2)x^2$ and from this the full expression for H can be recovered. In connection with the binary of order 5, Cayley wrote a note to Sylvester about an irreducible covariant of the fifth degree in the coefficients and order 15: "There is no covariant of an order higher than 15, and if there is a covariant of the order 15 there is only one and the leading term is $(a^3cd - a^2b^{[2]}d - 3a^2bc^2 + 5ab^3c - 2b^5)x^{15}$. It is of course directly ascertainable whether this is or is not but I have little doubt it will turn out to be a covariant."[69]

The seven papers sent to *Crelle's Journal* on the Wednesday of the week, he soon realized his intuition about the putative covariant was wrong and he wrote back to Sylvester on the following Saturday.

The leading term $(a^3cd - a^2b^{[2]}d - 3a^2bc^2 + 5ab^3c - 2b^5))x^{15}$ breaks up it is obvious into $(ac - b^2)x^6$. $(a^2d - 3abc + 2b^3)x^9$ i.e. there is no [irreducible] covariant [with leading coefficient] x^{15}, by continuing the calcul[ati]on I find that there is no covariant x^{13}, x^{11} or x^9 or so that the covariants are necessarily x^7, x^5 or x^3 & x^1 [respectively Cayley's L, but an irreducible covariant of *order* 5 does not exist], K, J; As regards x^1, we know that there is one case and one case only viz. Hermite's linear covariant [Cayley's J]; all that remains to be found is therefore whether there are any covariants x^7, x^5 and x^3. I shall accomplish this without much trouble.[70]

With the publication of the "Introductory Memoir," the invariant theory stage was set. The ground had been cleared as to definitions, but now Cayley needed theorems to provide the superstructure and expose the interplay of the definitions and the richness of the tapestry.[71] In particular, he asked two questions of *binary* quantics:

- Is there a general *formula* for invariants and covariants?
- What is the *number* of linearly independent (asyzygetic) invariants and covariants of a specified degree and order?

Eureka!

To both these problems, Cayley found solutions. To the first, he gave a formula (Cayley's Formula) and by October 1854 had found the combinatorial formula for the number of linearly independent invariants and covariants (Cayley's Law). It was a triumph since, he admitted, they were problems that "had always resisted my efforts." Given the magnitude of his success, his long letter to Sylvester was a restrained response.[72]

There was but one word of celebration—"Eureka"—and the mathematics followed. The two-part theorem, comprising both the formula and the law, encapsulates Cayley's approach to invariant theory and it provided the practical means for the construction of the long expressions for invariants and covariants. He was able to reveal the presence of invariants and covariants using the law and, knowing them to exist, calculate them using his formula. It was an algorithmic procedure of a very detailed kind.[73] At the point of discovery, Cayley brushed aside one difficulty in his confident manner: "there is no reason for doubting that these equations $[XA = 0]$ are independent."[74] No doubt he was bolstered by the success of his calculations, which required the solution of these equations. He was delighted to find a proof of the law, appeared satisfied about the truth of the "gap" in the argument, and was content to rest his belief on intuition.

The "Second Memoir on Quantics," the central memoir of the whole series, was submitted to the Royal Society on 14 April 1855, exactly one year after the "Introductory Memoir" had been submitted. Delay in publication may have been a result of illness suffered in 1854, hinted at in a letter to William Thomson.[75] He visited the Lake District, presumably for recuperation, but his publishing record indicates that he submitted nothing for publication in the latter half of 1854.

Following the generalities of the "Introductory Memoir," the "Second Memoir" combines theoretical work and painstaking calculation. In his referee's report, James Booth wrote: "I had to devote many a long day's labour to getting up on the subject generally before I felt myself in a position at all competent to take a broad view of that wide untrodden field opened up to us by the researches of Messrs. Cayley, Sylvester, Spottiswoode and Boole."[76] Sylvester read the "Second Memoir" but found Cayley's proof of the formula and the law difficult and judged it "exoscopic" but wrote later: "I now perfectly understand y[our]r exoscopic proof about $XI = 0$ involving $YI = 0$ and wonder it ever occasioned me any difficulty.[77] The theorem was a breakthrough for Cayley and as a theorem in mathematics it has proved remarkably durable. Cayley also put a third question, a different and deeper question about irreducible invariants of a binary quantic. Are they finite in number, or are they, like the prime numbers, of infinite extent? In a letter to Sylvester (unusually labeled R.S.V.P.), he wrote:

> Is there any reason a priori, why the number of *irreducible* invariants of a quantic $(\otimes \lbrack x, y)^m$ should be *finite*. It is so we know for $m = 2, 3, 4, 5$ & 6—I know nothing about $m = 7$ but the formulae I have obtained for $m = 8$ seem to show that there are an infinite number of irreducible invariants. The question is merely this—Can there not be an *infinite* number [of] quantities I, rational functions of $m + 1$ elements and such that any I is an irrational function of any $(m - 2)$ I's say of $I_1, I_2, \ldots I_{m-2}$ but so that there is no finite number of I's of which any other I *whatever* is a rational function.[78]

Here Cayley is examining binary quantics of succeeding orders for invariants in the first instance, but he was also thinking of the wider question for invariants and covariants. He continued: "My results for $m = 8$ are I must confess of a very paradoxical form, I find that there is one irreducible invariant of each of the orders 2, 3, 4, 5, 6, 7, 8, 9, 10 [—] one syzygetic equation of each of the orders 16, 17, 18, 19 [—] one irreducible invariant of each of the orders 25, 26, 27, 28, & 29 which is as far as I have carried the developement." This was the point where Cayley started to go astray and the method of enumeration had been leading him into the realm of mathematical fiction. As he admitted, his results were "of a very paradoxical form," so that his incipient error was not merely a slip of the pen. He reached the conclusion that there were an infinite number of irreducible invariants for a binary form of order 8. For the binary form of order 5, and above, he also concluded that the number of irreducible covariants was infinite. In effect, he concluded that only binary forms of order less than 5 had finite systems of invariants and covariants.

While Cayley had inaugurated his memoirs on quantics, he appeared to have little support from others in the invariant theory circle of researchers. The inner quintet of Cayley, Sylvester, Boole, and Spottiswoode and Salmon, which had formed the nucleus of British invariant theorists, had broken up. Only Cayley and Salmon remained committed to its study while the others had drifted away by the time the "Second Memoir" was published. Sylvester, who had taken up invariant theory at the beginning of the decade with such brio, just as quickly abandoned it around 1854. The famous skew invariant W, discovered by Hermite, had been within his grasp but he failed to recognize it before Hermite published the result, and, in dejection, he abandoned invariant theory for ten years. He reminisced on this lost opportunity: "I had but to think the words 'Resultant of Quintic and its Canonisant,' and the octodecadic skew invariant [W] would have fallen spontaneously into my lap."[79] His having missed the discovery of W might have been a contributing factor to his general malaise but it was not the sole cause.

The London world of business increasingly sapped Sylvester's creative energies and he began to seek an academic position. In 1854, he was passed over for the professorship of mathematics at the Woolwich Military Academy in favor of Mathew O'Brien, a body blow aggravated by his low regard for O'Brien as a mathematician. He was in no mood to delve into invariant theory and what attention he could muster was concentrated on the subsidiary combinatorial questions the theory suggested. Sylvester lost contact with Cayley for a brief period and, writing in August 1855, voiced his despair: "I truly regret that I have been able to avail myself of late so little of your society which is to me of inestimable value. But it has been, as I have said, the necessary result of my present unsettled state of mind & the destitution of my habits of 10 years standing—I am beginning to bend my mind again some little to mathematical study—but my own work is at an absolute standstill—Happy you who never flag and never weary!"[80]

When Sylvester recovered from his depressive state, he became besotted with this subsidiary combinatorial study and, for a time, his view of invariant theory became somewhat restricted. In contrast, Spottiswoode broadened his scientific interests so that invariant theory and determinants became just one scientific activity among many. In his scientific work, he included work on the calculus of operations and mechanics but spared time for the history of mathematics, geology, and astronomy. Boole had long forsaken invariant theory, but Cayley worked on. There was still the problem of the binary form of order 5 to fully explore.

The Quintic Campaign

At the time Cayley published his "Second Memoir," he had found a number of irreducible invariants and covariants of the binary quintic form and speculated on others. His progress with this form in the 1850s can be judged from the "quintic system" (see "Invariants and covariants of the binary quintic as known to Cayley, in part, as of 1854" on p. 12g of the gallery). The system is visual evidence of the quintic's rich structure and the intricacy suggests a plausible reason for its resistance to the best endeavors of mathematicians over the course of three centuries. Binary forms of low orders have trivial structures but with the binary form of order 5, the mathematics becomes a challenge.

Of particular note is Cayley's *historical* path in the exploration. He simply "climbed the quintic system" from branch to branch calculating invariants and covariants by successive degrees (the more vivid term "quintic tree" would be more appropriate here, but Cayley himself used the term "tree" in a different context). By the end of the 1850s, he had calculated the invariants and covariants as far as degree 5. It was clear, but a fact which would leave him undaunted, that as the degrees of invariants and covariants increased, so did their length. With degree 5, he was on the nursery slopes of the endeavor, but what faced him for higher degrees was, at this stage, unknown territory.

The whole process was in keeping with his role as a collector and classifier. In the quintic campaign, we see Cayley at his speculative best in the way he freely conjectured on the existence and nonexistence of covariants. In effect, he was in his scientific laboratory sifting the evidence and making judgments. If this seems primitive, it was; he had none of the sophisticated streamlining of a century later when the full power of modern abstract algebra was available.

At this stage in the story of the binary form of order 5, Cayley was studying invariants and covariants of quite low degree. He was attempting to list them, and though he believed the list to be infinite, he was undeterred. He meticulously continued calculating, collecting, and cataloguing by successive degrees, an activity which might be compared with number theorists who examine specific prime numbers knowing them to be of infinite extent. Miscellaneous results were scooped up and placed in the herbarium, which soon became well stocked.[81] In the "Second Memoir," Cayley displayed twenty-four specimens for binary forms neatly labeled, including fifteen invariants and covariants for the binary form of order 5, and they were designated by the letters of alphabet *A* to *L*, *U*, and *W*.

Problems with Partitions

The head of steam generated by this ambitious invariant theory program inevitably suggested other research; one question settled generates ten more to be answered. Consequently, Cayley became involved with various problems in partitions. A partition of a number *n* is a way of composing it as a sum. The number 11, for example, can be written as $3 + 4 + 4$, or as $1 + 2 + 3 + 5$, but there are many other ways too. Euler had tackled the problem of counting the exact number of partitions of *n* by using generating functions and their expansion, but it was a highly laborious method. This was the approach Cayley adopted with such vigor. Indeed, the spring and early summer of 1855 has been described as a highpoint in the theory of partitions with both Cayley and Sylvester each offering different viewpoints on the theory. Problems of this kind were attractive for their own sake, but invariant theory provided the motivation.[82]

Cayley's work depended on finding equivalent forms for generating function that could be easily manipulated. While his method "followed in the footsteps of Euler," Sylvester's method depended on Bernoulli numbers and used different methods, and he became so obsessed with the theory of partitions that he regarded it as his own preserve.[83] When Cayley entered into partition theory himself, it almost amounted to trespass and caused another ripple in their relationship. In the week between submitting his "Second Memoir on Quantics" to the Royal Society and it being read, Cayley received an abrupt letter from Sylvester, who was about to prove a "remarkable theorem" in partition theory. It appealed to him *not* to help: "[A]t all events I am so close upon the complete solution of this great question of partitions and am so well aware of your *terrible* sagacity in seeing to the bottom of a matter when once you have adjusted it to your focus of vision that I think it better to request you, (in the way of warning) not to communicate to me any discoveries you may make in the subject for the present as such anticipation would place me in an awkward position in drawing up my account of the result of my own researches." Cayley, of course, complied. Sylvester was in an excitable state and continued: "I think even the Venerable Majesty of Euler would have vouchsafed a benignant smile," he concluded, "upon being made cognizant of the altered shape of the question."[84] He need not have worried about competition, for Cayley was too bound up with invariant theory to pursue this theory in any detail.

Three years later, Sylvester's interests had diverged significantly from Cayley's. He branched out from his study of single partitions to problems in double partitions and onward to problems of multiple partitions, effectively seeking the number of integer solutions to a system of linear equations. Again it was natural for him to write to Cayley: "The more I ponder over my Theory of multiple partition, the more astonished I feel at the difficulties of the analysis admitting of being so easily overcome." In a postscript, he was satisfied the problem was solved: "I have sent a short account of my discovery to the *Phil[osophical]. Mag[azine].* under the title of 'The Virgins Unveiled' (in allusion to Euler's description of the problem as being that 'referred by the Arithmeticians to the Regula Virginium')." Two weeks later, he felt the need to issue a caveat, though he started off excitedly: "I have discovered a proof of my theorem—it is almost *ridiculously simple*—involving no work of calculation but merely algebraical logic (!) of a sort—too you & myself are quite familiar with viz. depending on the involution principle," he wrote, but did allow, "there may be some little corners in my proof which still require more closely looking into."[85]

His paper "On the Problem of the Virgins, and the General Theory of Compound Partitions" was published and he revealed to his readers that the virgins "who appeared to Euler, but with their forms muffled and their faces veiled, have not disdained to reveal themselves to me under their natural aspect." He ignored the broad questions of invariant theory and devoted himself to partitions for the remainder of the decade.[86]

Marking Time

It was becoming increasingly clear to the academic community that Cayley's lack of an academic appointment was an anomaly. He belonged in an academic environment and should have been spending time with the finer points of mathematics rather that those of the law, a profession which was diverting him from his true vocation. His academic credentials were impeccable and his scientific reputation assured, both at home and in the wider European scientific community. His election to the Royal Society in 1852 was followed by membership of its committees. In the British Association for the Advancement of Science, hardly a year went by when he did not contribute a paper to its annual meeting.

Cayley was living the life of a barrister, but it was not the life he wanted. A further complication was the financial setback suffered by his younger brother Charles, who unwisely invested his capital in a railway advertising venture that went bankrupt. The family rallied round, but this added responsibility would have inevitably placed pressure on his brother as its head. In the years to come, Charles found it difficult to scrape out a living, and several years later his elder brother attempted to secure a position for him at the Royal Society as an assistant secretary.[87] Seeing Charles, provided for by his late father, but now plunged into a moneyless existence would have sharpened his care in dealing with his own money matters. Financial considerations may have gently suggested he supplement his barrister's income by taking on other work. It was quite common for "men of science" to hold more than one appointment. Stokes seems to have been particularly able in this regard during the 1850s, as he was the holder of the Lucasian professorship at Cambridge *and* the professorship of physics at the Government School of Mines in London and various other academic appointments of a minor nature.

Cayley applied to the Board of Commissioners for the Affairs of India for membership on the board of examiners, which would administer the newly announced competitive examinations. Instead of direct nominations from Haileybury College, appointments to the Civil Service of the East India Company were to be made by open competition in written examinations. For the first examinations, there were about 140 candidates. A new breed of professional civil servants (the *competition-wallah*) was due to go out to India. Young men between eighteen and twenty-three years old were to sit a battery of examinations that included mathematics (pure and mixed), natural science (chemistry, electricity and magnetism, natural history, geology and mineralogy), moral sciences, Sanscrit, English literature, and history.[88]

The first meeting of the examiners was held at the India Board, Cannon Row, on 2 July 1855. Members were drawn from an elevated section of the educational world. Sir James Stephen and Frederick Temple were selected for English literature, and for science, G. D. Liveing from Cambridge and the Irish botanist G. J. Allman were appointed. Evidently, Boole applied for mathematics, but Cayley and Stokes were the only ones appointed and they prepared the examination papers together.[89]

In the summer, Cayley went on a tour of Scotland to visit of old haunts, preceding his attendance at the annual British Association meeting at Glasgow. The island of Arran just off the northwest coast was within striking distance of Glasgow, and it was an opportunity for some serious walking. He wrote to Thomson:

> I like this place very much; but I have not been walking a great deal, I had a good day up Goatfell [the highest point of Arran]—and yesterday I was up a hill at the head of Glen Rosa—opposite Goatfell,—where I remained in a pelting shower waiting for it to clear which it did at last, and I had as good a view as I could wish for.—I then went on over the ridge at the head of Glen Rosa, and ought to have had a very good view, into, I believe, Glen Sannox, but it was thick mist again and I saw nothing, and made the best of my way, keeping on the top of the hills, to my present abode, the farest up house in Glen Sherrick. I ought to have taken advantage of the fineness of today to have repeated the walk but it is too late now, and I must trust to having another fine day.[90]

He returned to Glasgow for the association meeting and contributed a paper on an enumerative problem in geometry connected with the Poncelet polygon problem.

News came from Sylvester in London that "poor O'Brien is dead!. . . It is scarcely a twelvemonth since he was appointed at [the Royal Military Academy] Woolwich," he wrote, "I am *again* an applicant which keeps me in town."[91] Behind the scenes, he again enlisted the support of Lord Brougham, who wrote to Secretary of War Lord Panmure in support. Patronage was the way and this interception by Brougham had been successful when only the month before Panmure had appointed him to be an examiner at Woolwich. "My Dear P.," Brougham wrote in a private letter to Panmure: "My learned excellent friend and brother mathematician Mr. Sylvester is again a candidate for the professorship at Woolwich on the death of Mr. O'Brian [sic] who carried it against him last year. I entreat once more your favourable consideration of this eminent man who has already to thank you for your great kindness."[92] This time Sylvester was successful and took up his new position at Woolwich on 15 September. He was now part of the military machine, complete with a residence in K-quarters (one of the last residential professors at the academy) in officer style and a salary of £550 with rights of pasturage on Woolwich common.[93] The employment, which was to last fifteen years, and which was mostly turbulent for all parties, may well have caused the military authorities to question their own judgment in appointing him.

Sylvester's appointment to the "Shop," as the academy was known, brought him near to Cayley's family home on the fringes of London, convenient enough for the two to meet and discuss their latest mathematical news. The impressive residence at Blackheath was also a short walk down a steep hill to the bustling port of Greenwich, where the Molines lived. Whether or not Cayley had met Susan Moline at this stage, the prospect of marriage was unrealistic for a barrister who concentrated his attention on mathematics. With the responsibility of a mother and family to consider, no middle-class Victorian male would consider marriage without a regular income. After the death of his father and the changed family circumstances, they moved from the large rambling Georgian mansion with its grand staircase decorated with hand-painted murals. They left a house with unparalleled views of London from the "Point" on the western end of the heath, an extensive parkland on a windswept plateau, to move to a smaller and more modest home in Montpelier Row on the eastern side nearer the village and, as it happened, within a stone's throw of the Potticary's School he had attended as a boy.

From the middle of the century, Blackheath was connected to London by rail but it was sufficiently separate for the residents to maintain their lives independent of the metropolis. It was an intellectual community with a lively interest in education, the arts, and science and during the nineteenth century, it was inhabited by the wealthiest sections of society. Those residents who entered Cayley's story included James Glaisher (the elder), J. W. L. Glaisher, his son, John Stuart Mill, and the Astronomer Royal George Biddell Airy, nearby at the Royal Greenwich Observatory.

Through the interlocking network devolved from Cambridge and the Royal Society, Cayley was drawn into fruitful discussions with scientists and astronomers. Stokes appealed to him in geometrical questions and discussed curves and surfaces determined by algebraic equations. While Cayley had specialized in pure mathematics, Stokes admitted to him that he was an "amphibious animal with an ambiguous character, at present, a sort of Jack-of-all-trades and master of none."[94] Cayley's dedication to pure mathematics left him little scope to make the same claim of "ambiguity" and he was not "amphibious" by inclination. Occasionally there was the opportunity to claim a dual nature, usually brought about as a reaction to an inquiry from a friend. He brought his mathematics to bear on a popular topic of the time—on vision and the eye—and he discussed a linear mathematical model for color-blindness with James Clerk Maxwell and Stokes. Cayley stayed strictly within his mathematical brief to the extent of describing one investigation as being "a very pretty problem for a physicist," but not one for him.[95] Stokes would tease him as a matter of course: "As physics are physic to you, here is the mathematical problem, or at least one to which it is readily reducible, containing only so much physics as facilitates the enunciation."[96]

Given closer inspection, several of Cayley's papers with optics or electricity in their titles turn out to contain pure mathematics as their main result. While this is also the case for astronomy, Cayley has a fair claim to be a mathematical astronomer of the first rank. He began to take a professional interest in astronomy around 1855 when he read the work of Peter Andreas Hansen, a Danish mathematical astronomer widely respected in England. Cayley's paper, "On Hansen's Lunar Theory," was his first on astronomy.[97] He found the work challenging but admitted to the Oxford-based W. F. Donkin: "I find Hansen a very troublesome book [*Fundamenta Nova* (1838, in Latin)] to read principally from the difficulty of grasping the whole system of the formulae which are in use at the same time, but I am making my way through it, and I hope to explain the method of integration in a continuation of my paper."[98] In writing his paper, Cayley received encouragement from Britain's premier astronomer J. C. Adams and support for the method he proposed to use: "I am very glad to find both your letter and your paper on Hansen's Method, for which I have to thank you, that you are now turning your attention to the Lunar Theory. I am not aware that this theory has ever been systematically treated by the Method of Variation of Elements. The method is theoretically a very beautiful one, especially when the disturbing function is so simple as in the case of the Moon."[99]

Cayley was concentrating his energies on invariant theory but continuing to make himself available as a consultant in the application of mathematics. That invariant theory would have practical ramifications would have been surprising. It was generally regarded as an esoteric and abstract slice of mathematical theory. Of all people, a request for help came from Thomson, an able mathematician himself but with a tendency to criticize pure mathematicians "who don't take their part in the advancement of the world." At the end of 1855, Thomson discussed a problem concerning three quadratic forms calling for the joint invariants to be computed. This work was outside the mainstream of Cayley's research on binary forms, but there was no denying it was invariant theory. On 14 December, Cayley advised Thomson, "I do not think your theorem is correct," stating one invariant of a system but not allowing that others could exist. Three days later, Thomson replied: "I think I am right this time. I worked out the actual transformation and found the theorem verified." The physicist's calculations carried out in his minute hand are worthy of the "clatter of the co-ordinate mill" but they proved the point. A fortnight later, from his chambers in Stone Buildings, Cayley agreed: "I must apologise for objecting to your invariants, my last letter in fact proves that you were quite right—for it is clear that $x+y+z$ and $x'+y'+z'$ are each of them invariants."[100]

Cayley was not cut off from the world of applied science. On this occasion help was limited to mere support, but coming to the aid of Thomson was typical of Cayley. He was always ready to help scientists, though he was a specialist mathematician and generally reluctant to stray outside its borders in conducting his own research. Increasingly, he reached out and advised his colleagues from a specialist's perspective and continued working away at group theory, matrix algebra, invariant theory, and much else within his "grand design" for mathematics itself.

CHAPTER 9

Without Portfolio

The ideal position for a man of Cayley's leanings and abilities would be a professorship with minimal duties and leisure to carry out research. Evidently, he resolved to find one and if "without portfolio" in terms of having a role in academe, what better way to enhance his credentials than to establish a scientific reputation beyond question. His rate of publishing accelerated quickly: around 1856 there was a threefold increase in the number of papers published.[1] Block submissions to journals at home and abroad were not unusual. In the rare event a paper was not accepted by one journal, he placed it in another, as happened in a piece he wrote on surfaces, in which he touched on "Talbot's curve," a curve of order 6 named after the pioneer of photography, W. H. Fox-Talbot.[2]

The Job Search

While laying the foundations of invariant theory, Cayley was engaged in myriad mathematical activities. At the beginning of January 1856, he completed a paper on the enumeration of "tree" diagrams, a subject stimulated by Sylvester's ruminations on an obscure sidelight in the differential calculus.[3] He discovered that a particular Eulerian generating function could be used for enumerating geometric "graphical notation," or the "rooted trees" which arose in the composition of differential operators.[4] The work was decidedly pure mathematics and he was not to know that he would return to its study to solve problems in organic chemistry twenty years later. Such is the long gestation of mathematical ideas that he would never know that his "rooted trees" would be fundamental in present-day computing theory concerned with the analysis of data structures.

He was very aware that work in one area suggested problems in another. Partition problems are linked to the task of listing symmetric functions, providing yet another connection with invariant theory. He complemented the calculations of symmetric functions started by Meier Hirsch, whose very successful book appeared in a new edition, and, with others, he discovered the law of conjugate partitions.[5] The forty-five papers published during the following breathtaking year included such topics as symmetric functions, polyhedra, geometrical caustics, lunar theory, general theoretical dynamics, the three-body problem, elliptic functions, number theory, and the theory of partitions.

Cayley was active as a barrister, but his desire for a position in the academic world evidently became more pressing, and the concentration on mathematics must have resulted in a reduction of legal income. The curtailment of a share in the Trinity College dividend, which had lasted until the end of 1853, would have added to his need—the loss of a guaranteed income of £240 per annum, which supplemented his "refreshments" at the bar, would be uncomfortable. Working at the legal bar did not stem his flow of mathematics, but he may have regretted the time spent on drafting legal documents. Of his contemporaries at Cambridge in the 1840s who had made their mark in science, he was practically the only one still outside of academe.

Yet there were grounds for optimism. The universities were changing and modernizing, and the "Cambridge Act" passed in July 1856 sought to improve the governance of the university, which would signal expansion. The act of Parliament repealed the Elizabethan Statutes (of 1570), which had set the legal framework of the university since that time, and candidates for most degrees were no longer required to swear an oath of allegiance to the Anglican Church. While acts of Parliament indicated change, albeit slow change, the reality for academics in England was that there were still few suitable posts available. Undoubtedly a man of Cayley's attainments, had he been a citizen of France or Germany, would have been safely tucked away in a university. The accented phrase for Cayley was a "suitable appointment."

Though James Joseph Sylvester had been appointed at the Royal Military Academy in 1855, and was grateful for it, the position was not really "suitable" to any of the parties. It was no comfort to the military management that his intellectual life lay outside the walls of Woolwich and that he answered the call to research before he thought of transmitting mathematics to the gentleman cadets in his classroom. Bending to the demands of "authority" was always difficult for Sylvester and on a day-to-day basis he found the teaching at Woolwich irksome. The management must have realized too late that they had a prima donna on the payroll, and one who could not conform to their stereotypical view of a mathematician and buckle down to teaching simple sums. The resulting conflicts no doubt contributed to the bouts of depression that too often acted as a brake to his mathematical creativity and deflected him from his true calling.

Sylvester took exception to view that he was an expert on all things arithmetical and that his Woolwich colleagues had little idea of higher mathematics—or his place in it. He expected more understanding and respect from his colleagues at "the Shop"; he painfully recalled being asked "to answer some question as to the number of cubic inches in a pipe" and of the colleague who "expressed his surprise that I was not prepared with an immediate answer, and said he had supposed that I had all the tables of weights and measures at my fingers' ends."[6] Sylvester had actually been appointed both professor *and* lecturer but happily the professor of practical astronomy came to the rescue and relieved him of some lecturing duties. In August 1856, without the extra classroom duties, he wrote to Cayley of being restored to mathematics:

> I suppose you are still at the lakes but trust to this being forwarded to you from your chambers. You will find me restored to Mathematics on your return & I shall hope to see a great deal more of you than I have been able to do in the last twelvemonth as I am relieved from that portion of my work which pressed so intolerably upon my spirits and hindered me from entering into any mathematical speculation.... Kirkman has sent me a letter [on his theory of polygons] asking for information which you can give but which quite transcends my mathematical lore: I shall hope again to engage some long mathematical evenings with you as a free man, my mind being rid of its nightmare and feeling no longer any call to labor at the vain task of transforming myself into a Practical Man.[7]

With Sylvester at the Royal Military Academy, their discussions on mathematical topics, initiated around the Inns of Court, were disrupted but carried on elsewhere. Cayley would journey southward from his chambers in Lincoln's Inn and Sylvester from "K-quarters" at Woolwich to meet in Greenwich or Lewisham on the south side of the Thames. Alternatively, if Cayley was at home, the walk would be shorter: "[i]f you are inclined to walk Blackheath ways viâ the Shooters Hill road the whole way i.e. from your side of the [Woolwich] common, on Sunday, I will walk that way to meet you," Cayley wrote as a postscript to a letter one Friday.[8] There were several such excursions in the winter of 1856 and a renewed Sylvester responded to such invitations: "I shall be very glad to see you on Saturday as you propose and to walk over with you afterwards to dinner at Blackheath. I have been very much baffled and disappointed lately in my algebraical matters. I shall be glad to talk to you about the differential equations to Isonomial Plexures ..."[9]

As the decade progressed, Cayley was recruited to serve on various Royal Society committees and since the beginning of the year had been member of its council. The young "professionals" of scientific London were displacing the old men who sat there by virtue of aristocratic connections. In all, Cayley served five periods of office on the council.[10] While taking his part in administration, he was no Royal Society politician of the type who was active behind the scenes organizing the affairs of the society through membership of its dining clubs and ginger groups. This was not his way, but he was influential and actively sponsored able mathematicians for membership if he thought them worthy enough. If they were able, he encouraged them irrespective of social background or nationality.

Cayley championed other causes and, contrary to the view of him as a "purer than pure" mathematician, he maintained an interest in technology. A case for his support arose in the mid-1850s. A matter of great topical interest was the Scheutz (second) difference engine, which was being shown in London and Paris. From the beginning, the project augured well. In the Swedish workshop where it was made, it was observed that "[t]he very first occasion it was assembled, it worked so well that no adjustment was necessary."[11] It had come to London and was exhibited at the Royal Society toward the end of 1854, and it was promoted by Charles Babbage and the civil engineer William Gravatt. At the beginning of the 1855, a committee headed by George Gabriel Stokes and consisting of Michael Faraday, Charles Wheatstone and Cambridge professors William Hallowes Miller and Robert Willis, concluded that it could be used for the cheap calculation and recalculation of mathematical tables. It was demonstrated to Prince Albert and appeared on the front page of the *Illustrated London News*. In the summer of 1855, the machine went to the Paris Universal Exposition and the constructors were awarded a gold medal. Further honors came, and in February 1856 Georg Scheutz was elected to the Swedish Academy of Sciences.[12]

In his first year of council membership, Cayley proposed Scheutz for a Copley Medal.[13] He might even have concurred with Charles Babbage, who had the temerity to address King Oscar I on Sweden's great achievement: "The science of mathematics is becoming too vast in its details to be completely mastered by human intellect and the time is approaching when the whole of its *executive* department will be transferred to the unerring power of mechanism."[14] With Cayley's interest in mathematical machines that could calculate and those that could draw curves, it was quite natural for him to propose Scheutz for the society's highest honor. That he made the proposal indicates his preparedness to argue the corner for mathematics on the council. In the following year, he proposed Julius Plücker for the same medal but neither this nor his proposal for Scheutz resulted in success in the ensuing competition.

Cayley began to take interest in the workings of the British Association for the Advancement of Science and joined several of its committees. One at Glasgow concerned itself with the question of constructing a scientific bibliography. Prompted by Joseph Henry, the director of the Smithsonian Institution in Washington, its remit was to investigate the advisability of publishing a serial index of all scientific papers (physical and mathematical) published since the year 1800. The initial committee comprised Cayley, Stokes, and the astronomer Robert Grant, but General Sabine sought the cooperation of the Royal Society. The list of scientific works originally were intended to be produced only in manuscript form, but it grew into a serious undertaking with the government providing the initial funding for printing and publication. A larger committee was formed (including Augustus De Morgan as a member) and on its recommendation, the idea of a *Royal Society Catalogue* was born. Not only did the committee advise that it should go ahead but they set out how it might be done. Indeed, it was to be one of the great Victorian publishing projects, which could be placed alongside the *Dictionary of National Biography* and the *Oxford English Dictionary* in terms of its ambition and comprehensive proportions. The first volume of this massive work eventually was published in 1867.

Cayley was moving in influential circles with increasing responsibilities for the organization of scientific activity in the capital and the country. How much longer could he go on doubling as a lawyer? At Cambridge the relevant professorial chairs equal to "his station" were the Lucasian Professor of Mathematics, occupied by Stokes since 1849, the Plumian Professor of Astronomy by James Challis, and the Lowndean Professor of Astronomy and Geometry by George Peacock. Once obtained, these professorships were chairs for life. If Cambridge was Cayley's unstated ambition, there was simply no opening.

Scotland's Highlands

Cayley became interested in the chair of natural philosophy at Marischal College in Aberdeen, which became vacant on the death of David Gray on 10 February 1856. There were drawbacks to the post, for he was not inclined to the idea of professorial lectures, which, in the Scottish university system, would be attended by large classes. His friend Hugh Blackburn, appointed as professor of mathematics at Glasgow in 1849, had ninety students when he started but more than three hundred by the end of his thirty-year tenure. A feature of the post, with which he would have no quarrel, would be the income considerably bolstered by fees paid directly by students attending his lectures. Another welcome compensation in the teaching system north of the border would be freedom from the dominance of anything like the Cambridge examination system.

Cayley wrote immediately to Thomson enquiring about the duties "in particular the number and subjects of the lectures or courses of lectures, and whether the emoluments would be not less than what is stated to be about the value, £300 a year and whether this does or does not include a house or rooms. The great inducement to me would be your seven months vacation."[15] The income would indeed be greater than £300 and though there was no house attached, the duties would be slight. The professor was required to lecture for two hours daily from the beginning of November to the beginning of April on such subjects as statics, dynamics, hydrostatics, optics and astronomy, but there was complete freedom for the holder to choose the curriculum.

Sylvester registered surprise that Cayley would be willing to quit London:

> We (I mean your friends) shall all be sorry to lose you by your going to such a
> distance as Aberdeen but if you think such a change would conduce to your advantage

or happiness I can only wish that you may succeed in obtaining your wishes—I thought that you used to express an objection to the Scotch system of lecturing and to taking anything but a pure Math[emati]cal Professorship—I suppose the complete appostcy [*sic*] which they boast of & the prospect of being some hundred leagues & odd nearer to the Highlands are among the inducements which have operated to make you change your mind.[16]

Among the applicants for this post was the 24-year-old James Clerk Maxwell, then a fellow at Trinity College, Cambridge. James Forbes, professor at Edinburgh and a personal friend, urged him to apply while admitting that he had no influence in the appointment since it was in the gift of the crown through the Lord Advocate of Scotland and English Home Secretary. "[I]t would be a pity were it not filled by a Scotchman," and, furthermore, Forbes volunteered, "you are the person who occurs to me as best fitted for it."[17] Maxwell's father encouraged him with practical advice, "I believe there is some salary, but fees and pupils, I think, cannot be very plenty. But if the *postie* be gotten, and prove not good, it can be given up; at any rate it occupies but half the year."[18] Maxwell was interested to see if Cayley would be a competitor. In a letter to Stokes requesting a testimonial for the Marischal appointment, he asked: "Is it true that Cayley is going in?"[19] He was surprised Cayley wanted a professorship knowing he was successful in the law and thinking him content. He asked Thomson: "How came Cayley to be wanting a professorship. I thought he was a great lawyer."[20]

The reality of the appointments system worked against Cayley. Leaving aside the obvious fact that he was not a "Scotchman," his expertise hardly suited him for a chair of natural philosophy. Any testimonial writer would have to take the nature of the post into account and in terms of physical science. Sylvester wanted to help Cayley find a post in academe: "[i]f any letter from me would be likely to be useful you of course have only to say the word & I will write my opinion which may carry more weight from the official position [at Woolwich] I fill than it would from the private merits of the writer."[21] A testimonial from Sylvester could hardly camouflage the obvious drawbacks to Cayley's candidature of being a pure mathematician and not being "physical enough." There was also Maxwell's suitability in this regard. William Hopkins, who taught them both, was unequivocal: "During the last thirty years I have been intimately acquainted with every man of mathematical distinction who has proceeded from the University, and I do not hesitate to assert that I have known no one who, at Mr Maxwell's age, has possessed the same amount of accurate knowledge in the higher departments of physical science as himself."[22]

At the end of April Maxwell learned he had been appointed and he left for Scotland, where he was formally inducted as a regent of Marischal College in August.[23] Perhaps Cayley was fortunate after all. Life in Aberdeen would be pleasant enough, but it was an intellectual backwater and cut off from London.

Moreover, Cayley would not have the lecturing skills necessary in dealing with the large lectures, which were a feature of the Scottish higher educational system. It would also be a short-lived appointment. Maxwell was not to know that rationalization was afoot and only four years later, the University of Aberdeen fused their two constituent colleges, Marischal and King's. Maxwell's post was eliminated and the college made him redundant. He came south again and was appointed to the chair of physics and astronomy at King's College, London.

Cayley worked on. His "memoirs on quantics" was progressing smoothly. The "Third Memoir," completed in March, was concerned with geometrical questions—an interest he shared with Ludwig Schläfli: "I hope soon to have the pleasure of sending you a second memoir upon Quantics—I have also written and presented to the Royal Society a third memoir upon the same subject but it will be some time before that is published, it contains the system of the Covariants and Contravariants of a ternary cubic, and gives I think the materials for a complete discussion of the theory of curves of the third order."[24]

As the "Third Memoir" was important for the theory of curves, being a study of algebraic forms of three homogeneous variables, a further object was to extend the theory to surfaces. These are defined by algebraic forms with four variables. In this regard, Cayley cemented his connection with Schläfli, who had written to him about cubic surfaces.[25] The Swiss mathematician probed the structure of the twenty-seven lines that lay in this surface and discovered the famous "double-six" configurations which existed among them. Laying at the heart of the cubic surface, the double-sixer comprises twelve paired lines with intricate intersection and nonintersection properties. Their discovery and Schläfli's notation for the "Doppelsechser" was regarded as epoch-making by geometers fascinated by the cubic surface.[26] Cayley translated Schläfli's paper, which also dealt with classification of cubic surfaces, and it appeared in the *Quarterly Journal of Pure and Applied Mathematics*. It was a preliminary consideration of the cubic surface that would presage his own work in the 1860s.[27]

Cayley spent the summer of 1856 at Grasmere and Keswick in the Lake District. Kirkman's summer activity in Croft, in Lancashire not far from him in the Lakes, included a new paper "On the k-partitions of the r-gon and r-ace." The challenge before him was to provide a formulae for the number of nonintersecting diagonals of a polygon that can be drawn from vertex to vertex. By November he had sent his solution of this stubborn problem to Cayley, and, in the absence of a reply, wrote to Sylvester, "I hope you will find it a more agreeable companion than I have found it. Of its truth I am certain, & my proof in Cayley's hands will convince the reader."[28] Cayley's assessment of Kirkman's paper was encouraging: "I have considered Mr Kirkmans paper on the k-partitions of a polygon and polyace. The paper appears to contain a complete solution of the problem which is one the difficulty of which could hardly have been appreciated without the knowledge of Mr Kirkmans researches and I think that the paper well deserved to be published in the Transactions of the Society."[29] A formula for a polygon of n sides and m diagonals was given, and Cayley attempted to provide a proof of it. Neither was able to give a cast-iron proof of the formula and none was forthcoming for more than a century.[30]

Kirkman's paper would have been further evidence to Cayley that the rather isolated clergyman living in Croft deserved a place at the high table of science. Always willing to promote talent, he immediately started the process of proposing Kirkman for the Royal Society, and, in the same year, rectifying an anomaly, he also proposed George Boole for membership. He wrote out Kirkman's certificate and canvassed signatures in support with the time-honored phrase for "one who is attached to science and anxious to promote its progress." For the benefit of nonmathematical fellows, he introduced Kirkman as the author of "various memoirs in the *Cambridge and Dublin Mathematical Journal*, the *Manchester Philosophical Transactions*, the *Phil. Magazine* and the *Philosophical Transactions* relating to Permutations, the Partition of Numbers, Quaternions and Pluquaternions and the enumeration and classification of Polyhedra."[31] Sylvester was fully supportive and sent back the certificate "which I have great pleasure in signing."[32] Thomson was also enthusiastic, and, with such support, Kirkman was elected on 11 June 1857. The remoteness of his parish in Lancashire made it virtually impossible for him to take advantage of his elevation but the outward sign of acceptance into the ranks of men of science would have ameliorated a sense of isolation. But, "buried in Croft," he was hurt by imagined slights, and little more than six months later, he wrote to Sylvester: "Cayley does not write to me, a single line. Did I offend him by my mode of referring to your discoveries & his in my paper of the 7-partition of x? I am afraid he has cut me."[33]

The Royal Astronomical Society

On 10 July 1857, Cayley became a fellow of the Royal Astronomical Society, a metropolitan society founded in 1820. Augustus De Morgan was one of his main sponsors; Robert Grant also signed his certificate of application. They had all been members of the Royal Society committee, which reported in June, to advise on the feasibility of publishing a catalogue that would list and organize the scientific memoirs being produced.[34] Membership of the Royal Astronomical Society turned into a lifelong commitment. The society valued such men with administrative energy willing to take their turn with committee work as well as carry out erudite research. Cayley quickly threw himself into its activity and in March of the following year attended his first meeting as a council member. During the year, he took a leading part in the revision of the society's constitution and the neatly made annotations to the draft bylaws bear the authority and judgment of the experienced barrister.[35] His membership of the council continued until 1893, a continuous service of thirty-five years.

Cayley's contributions to the society's *Memoirs* after his election consisted of several long and very detailed papers on the disturbing function associated with elliptic motion and the mathematical results applicable to the lunar and planetary theories.[36] His expertise was respected within the astronomy community, and within two years of R.A.S. membership, he found himself a referee on work submitted by the Astronomer Royal: "On the Movement of the Solar System in Space."[37] He was widening his horizons and gaining influence in the world of astronomy and a qualification for professorial chairs that included "astronomy" in their titles. His lasting reputation is in pure mathematics, but it is sometimes forgotten that he carried out highly valued research in astronomy, albeit at the very pure end of mathematical astronomy.

Closely connected with astronomy were Cayley's contributions to "dynamics," a subject which could include problems involving the motion of the moon or the figure of the earth. In October, he submitted two papers on the three-body problem, another famous problem which falls into this category. He began to compile and annotate a bibliography of the principal mathematical works written on dynamics.[38] For the British Association, he duly produced a 42-page "Report on Theoretical Dynamics," which itemized progress from Lagrange in 1788 to the then-current work of J. L. F. Bertrand. It included reports on the work of S. D. Poisson, Sir William Rowan Hamilton and Carl Jacobi.[39] This work is often singled out as one of Cayley's highest achievements, and it was well received at the time. William Whewell praised the compilation for distilling the complexity of the original papers "for enabling us compendiously and easily to understand what has been done and how it has been done."[40] Thomson singled it out as one of the most useful technical reports to have been published by the association, and it must have given Cayley additional pleasure to hear him praising it for its "practical utility."[41] De Morgan recommended it to Hamilton: "I dare say Cayley will send you—or has sent you—his Report on dynamics— You will enjoy it."[42]

The report was followed by an even more extensive sequel (1862) devoted to a selection of *special* problems of dynamics. In this, Cayley reviewed progress in the intervening period and corrected his omission in the first report of Mikhail Ostrogradsky's work, which had been published in St. Petersburg in 1850. The sequel contained a summary of work on the transformation of the equations of motion from Lagrangian to the Hamiltonian form and of systems which arise from arbitrary problems in the calculus of variations, and the three-body-problem.[43] In one of his last articles, Maxwell evaluated the reports jointly as "intended rather as a guide in reading the original authors than as a self-interpreting document, though, of course, besides the criticism and the methodological arrangement, there is much original light thrown on the mass of memoirs discussed in it. It will be many years before the value of this report will be superseded by treatises."[44] Fifty years after it was written, Felix Klein was to recommend the 1862 Report to his students for its treatment of quaternions applied to dynamical problems.[45]

Overall, Cayley's papers in dynamics were concerned with what he called *direct* problems—the study of motion under certain circumstances. Planetary motion, and in particular the lunar theory, fitted this category. These might be considered great problems, in that they were of interest to astronomers generally; for Cayley, these involved the solution of differential equations. He also made remarks about lesser-known problem areas, like the motion of a falling chain. The paper "On a Class of Dynamical Problems" has since been seen as contributing to rocket motion and continuous impact problems.[46] In this paper, Cayley examined the motion of a variable mass and drew attention to the class of problems that "have not been considered in a general manner."[47] His contribution to *inverse* problems, those involved with the calculus of variations, such as finding the curve of swiftest descent ("brachistochrone problems") were few, though he reported on Jacobi's work in this subject.[48] "Dynamics" was one area where Sylvester would not follow him to any real extent, but newly installed professor at the Royal Military College in Woolwich, he was keen to show off his applied credentials to Cayley by explaining why a rotating unboiled egg did not rise. He had some boxwood eggs made for inspiration and let slip to Cayley: "I shall carry my wooden egg in my pocket with me to the next select committee."[49]

South Wales

Not succeeding in securing the professorship at Marischal College in Scotland, Cayley turned his attention nearer to home—to Neath in South Wales. In September 1857, in an attempt to find a position more congenial to his mathematical interests, his name became linked with the ill-fated "Western University of Great Britain." The planned institution was to be modeled on the privately organized École Centrale des Arts et Manufactures founded in Paris in 1829 for about two hundred pupils. The "university" in Wales was to be adapted to the "Wants of the Age."[50]

The aim of the projected university was to complete the industrial education of young men. To afford the fees, these youth—to be taught the "accurate knowledge of needful things"—were required to have wealthy connections. Typically, they would be sons of landed proprietors, owners of mines, or sons of men drawn from the professions. The crude link between making money and education should have made Cayley treat the scheme with some caution. The university prospectus was signed by the Resident Council of the College, a body whose members included the entrepreneur William Bullock Webster, the agent for the Gnoll estate in Neath and a man with a ready eye for speculation, Lewis Hertslet who had been the Clerk to the Metropolitan Commission of Sewers but retired on grounds of ill health, and a London-based map publisher, Trelawny Saunders.

Cayley was to be one of seven resident professors. The position had its attractions, not least a salary of £500 compared with a similar sum for a well-endowed professorship at Oxford or Cambridge. Professors in the "Western University" would also have the assistance of special lecturers, which, in mathematics, probably meant that Cayley would be exempt from routine lower-level teaching. The entrepreneurial spirit has to be admired, for Webster and his committee had spotted a niche in the educational market. The only college in the principality was the tiny St. David's College at Lampeter, which was founded in 1822 and given a Royal Charter in 1852, which allowed it to award bachelor's degrees in divinity. Gnoll College was conceived on a grander scale than this.

An advertisement for the newly proposed institution appeared in the *Times* on 10 September 1857:

> GNOLL COLLEGE will OPEN in October, to complete the education of young gentlemen in the practical applications of science to the management of land manufactures and commerce, to the public service, the professions and other pursuits. The staff includes professors of mathematics, mechanics, physics, chemystry, natural history, human history and design; also examiners, tutors and special lecturers. The household arrangements are adapted to the habits of the best society. The College will be Incorporated as the Western University of Great Britain and will grant diplomas. Fees 200 guineas per annum; no extras. Apply to the Resident Council of Gnoll College, Vale of Neath, Glamorganshire.[51]

In the *Times* of the following day, the names of the professors were given. Cayley was the most eminent, being the only fellow of the Royal Society, but Thomas Spencer Cobbold as professor of natural history was a successful parasitologist and was to achieve eminence as a biologist and election to the Royal Society in 1864. But generally, compared with the eminencies drafted in to oversee the competitive examinations for India, the candidates for Gnoll College were drawn from a different milieu. The selected professors of physics, human history, and mechanics came from school backgrounds and were all Anglican ministers: Alexander Bath Power, the principal of Norwich diocesan normal school; Andrew Wilson, the master of Leamington College; and Charles B. Wollaston, an Oxford graduate, was a Dioscesan School Inspector. J. E. D. Rodgers, the projected professor of chemistry, had been employed as a chemist for seventeen years by St. George's Hospital in London. Edward Wehnert, the professor of design, was a respected watercolorist and member of the New Society of Painters.

The advertisement for Gnoll College, which appeared in the local Welsh newspaper, capitalized on Cayley's reputation: "Arthur Cayley, F.R.S., Barrister at Law; late Fellow, Assistant Tutor, and Mathematical Lecturer of Trinity College, Cambridge; Senior Wrangler and Smith's Prizeman, 1842, and since Moderator and Examiner in the University; Author of several papers in the Philosophical Transactions, etc. etc."[52]

In a flamboyant marketing exercise, it was argued that Neath was at the hub of the United Kingdom with sea and railway connections to Ireland and other centers of population. It was truer then than now, as Neath was on the direct route from Milford Haven, the gateway from Ireland to the seat of government in London. The home of the proposed university was described with some accuracy, as a practical place with canals, ports, mines, naval dockyards, and an arsenal in close proximity.[53] The "Western University" group who leased Gnoll Castle issued a plan that was nothing short of grandiose: "In order to distinguish Gnoll College from numerous schools which assume the collegiate title, and to place it on a footing equal to its organization, as well as to the character of its professors and the pursuits of its students, it is intended that the College shall exercise the function of an independent University, so as not to be involved in the arrangements of any other body. The originators of Gnoll College require freedoms of action in undertaking to combine the highest instruction with careful regard for religious, moral, and social habits, apart from unnecessary exclusiveness."[54] In keeping with the philosophy of the proposed venture, the practical aspects of mathematics were to be stressed by such intermediate courses as descriptive geometry and higher calculus and final courses in such topics as astronomical observation and trigonometrical surveying. Mathematics was seen as an essential tool for understanding science and the arts as it was at the École Centrale. Mathematics was a service subject, to support final-year practical subjects such as hydrostatics, railways, and steam engines.

Unhappily, the plans never went beyond the drawing board. Accepting failure for the establishment of university in South Wales, the local newspaper lamented: "Gnoll College has apparently gone to that abyss where all failures of abortive schemes lie in dull oblivion. One knell more and all shall be forgotten."[55] Bullock Webster vanished from South Wales but not before accusations of fraud and theft were leveled against him. He reappeared in the Middle East, but creditors last heard from him in supplicant letters written from one of Queen Victoria's prisons. Six years later, he appeared again, and undefeated, he floated a new money-making scheme in Malta.

In retrospect, Cayley would have done well to heed the advice offered by Sylvester, who was more experienced in the twists and turns of academic chicanery. He was anxious to offer immediate help: "I have only received your note about an hour since. I am sorry to see the unprivileged use the Gnoll people have made of your name. I never thought well of the scheme—I like it now much less than ever and most earnestly trust & hope that you will not associate yourself with it. I shall be in town tomorrow purposely to see you and we will then talk the matter over at length. I am sure that were Gnoll College all that it professes to aim at being you could do much better for yourself by waiting a little time longer."[56]

The Remarkable Matrix Algebra

A month later, Cayley was to make a discovery that would prove to be the greatest highlight to his reputation. Of all subjects on which Cayley wrote, matrix algebra is the one most closely linked to his name. The subject was not his single-handed creation, far from it, but his two memoirs, written with great prescience, remained a beacon for well over a century. At their heart is a theorem that is no less than a mathematical jewel.

On Thursday 19 November 1857, he wrote Sylvester a letter which opened: "I have just obtained a theorem which appears to me *very remarkable*."[57] The theorem, that a matrix is a solution of its own characteristic equation, was a result that became the centerpiece of his famous "Memoir on the Theory of Matrices" submitted three weeks later and read to the Royal Society in January 1858.[58] It has been a surprise for generations of mathematicians, for its succinctness, but most of all for its unexpectedness. Hardly one to be publicly too excited about many mathematical results, by declaring this one to be "very remarkable," Cayley signaled it as a find of the highest order.

The idea of a matrix as an array had already been known to Cayley. As noted previously, one of the papers in his "tour de force" of seven papers in the fiftieth volume of *Crelle* published two years earlier with the title "Remarques sur la Notation des Fonctions Algébriques" introduces a matrix as a succinct way of writing down many equations as one. The idea of a matrix occupied his thinking in the years from 1854 to 1858 and, simultaneous with the presentation of his famous "Memoir," in another paper which deals with a geometrical problem, he noticed that the result "may be represented in the notation of matrices by a single equation."[59] To be sure, the matrix notation had been absorbed within the wider mathematical community at this time. The underlying theory of matrices had been investigated under the guise of quadratic forms, in advance of their semblance as rectangular arrays. In the modern form of an array, they were known to Gotthold Eisenstein in the 1840s.[60]

Cayley was in prime position to appreciate the array notation as he had used them in his juvenile paper of 1841. The time was now ripe for such a development in the case of linear substitutions. In the same year as he was preparing his "Memoir," Cayley had reproduced and commentated on Eisenstein's work in the theory of numbers and lamented that so much was left undone after his early death of tuberculosis in 1852: "Eisenstein is now, alas! dead; too soon for the complete development of his various and profound researches in elliptic functions and the theory of numbers."[61] In his theory of numbers, Eisenstein had made use of the matrix notation explicitly.

It is probable that Hamilton had an understanding of matrices though he produced no papers on the subject explicitly.[62] Another point of contact would have been Charles Hermite. He corresponded with Cayley on a range of problems in the 1850s and, in turn, Hermite had also met and corresponded with Eisenstein. Hermite had used the matrix notation for linear transformations used to transform Abelian functions two years previously:[63]

$$\left\{ \begin{matrix} a_0 & a_1 & a_2 & a_3 \\ b_0 & b_1 & b_2 & b_3 \\ c_0 & c_1 & c_2 & c_3 \\ d_0 & d_1 & d_2 & d_3 \end{matrix} \right\}.$$

In this small band of mathematicians, all in contact with each other in various ways, the emergent ideas of matrix algebra crystallized in the 1850s.

A paper written by P. G. Lejeune-Dirichlet, which related matrix algebra to the algebra of forms, was submitted for publication in August 1857. It used the binary matrix notation to denote binary forms and linear substitutions.[64] Cayley followed Dirichlet's work; very early in his career, Cayley had read the German mathematician's work on multiple integrals and in the 1850s had made and published observations on his work in number theory.[65] Dirichlet was sixteen years his senior and, being eminent—he succeeded Gauss at Göttingen in 1855—Cayley had corresponded with him in seeking a testimonial in connection with his quest for an academic post.[66]

The term *matrix* appears to come from Sylvester, who had used it as early as 1850 to designate any rectangular array of coefficients, but his focus then was naturally on the more polished subject of determinants.[67] He alluded to Cayley's undergraduate paper on determinants in a letter: "A consideration of the principle of your paper in the Cam Journal 1841," he wrote in 1852, "has led me to the following agreeable extension of the common rule for the multiplication of Determinants which extends also to the Combination of Rectangular Matrices."[68] In his paper on group theory of 1853, Cayley hurriedly referred to inversion and transposition in a theory of matrices but remarked merely: "I do not stop to explain the terms as the example may be passed over." If only he had stopped, though his explanation may be deduced: a matrix was a useful piece of notation for expressing linear equations and bilinear forms.

In his "Memoir on Matrices," Cayley's intention was to map out mathematical territory: "The term matrix might be used in a more general sense, but in the present memoir I consider only square and rectangular matrices."[69] The paper is concerned mainly with square matrices as distinct from broad and deep matrices, which, he suggested, were less important than square matrices. As was the case with his other algebraic enterprises, the influence of the calculus of operations is evident. Acting as a referee, Boole saw the memoir as "an application of what has recently been termed the Calculus of Operations" and a matrix as "a complex symbol denoting the operation by which from any set of quantities x,y,z, we form linear functions of these quantities."[70] Cayley indicated an *active* operator M defined by $M(x, y, z)=(X, Y, Z)$, followed by an appeal to the "separation of symbols" leaving the matrix M.[71] It was the same process he had adopted when dealing with the set operators θ in the theory of groups when he had earlier discussed that theory.

Cayley now saw matrices as part of a mathematical system distinct from a lifeless notational device for denoting algebraic forms and he was aware of their importance from the outset. He stressed the distinction between matrices and determinants and he emphasized the "[matrix] theory which might have preceded that of a determinant; the matrix, is, so to speak, the matter of a determinant."[72] As with group theory, he saw connections between matrices and Hamilton's quaternions. In so doing, he practically fulfilled Hamilton's aim "to impregnate all existing algebra (including the Differential and Integral Calculus, Calculus of Variations &c.) and all pure and applied geometry with a NEW ELEMENT of thought and calculation."[73] It is a measure of Cayley's breadth that he saw the connections between the two distinct traditions of matrices and quaternions, which were in their infancy. But to Hamilton's view of quaternions as the key to all things, he was resolutely opposed.

The Cayley-Hamilton theorem was the spark that motivated Cayley to write his "Memoir." Part of its lasting charm is its succinct expression: A square matrix A satisfies its own characteristic polynomial. At the point of discovery, Cayley did not have an adequate notation for expressing it this way. He used a notational-device as a way of distinguishing A thought of as matrix array from it thought of as a scalar, in which case it was denoted by $\tilde{A}$.[74] The "Cayley-Hamilton theorem" was expressed by Cayley as $Det. (\tilde{I}.A - \tilde{A}.I) = 0$. That he did not prove his theorem in the general $n \times n$ case is well known. He verified it in the 2×2 case, merely stating that he had verified it in the case of 3×3 matrices and had "not thought it necessary to undertake the labor of a formal proof." Nevertheless, he asserted the theorem true for $n \times n$ matrices. Boole was critical of this hand-waving, for "if generally true [the theorem] ought to admit of a symbolical proof not involving much complexity but resulting from the first principles of symbolical algebra—this being the kind of proof, which according to analogy and from the intrinsic character of the theorem ought to be sought for."[75]

The absence of a rigorous proof causes the present-day reader more consternation than it would Cayley. In attempting to prove the so-called fundamental theorem of algebra (that every algebraic equation has a root), for example, he merely showed it was equivalent to a geometrical statement, the truth of which

"I conceive, [is] a matter of intuition."[76] The relatively modern "can you prove it?" would certainly have less resonance than the question: "is it true?" Once he had convinced himself of the veracity of a mathematical statement or formulae, verification in elementary cases would follow, and, this would decide the question of whether it were true or false in general. The general statement asserted, he moved quickly on. He had learned this mode of progressing from particulars to generalities at a young age, his thinking perhaps typified by a casual aside made about linear transformations as a young student: "[i]t is obvious that every step of the preceding process is equally applicable whatever be the number of variables."[77]

The real mystery surrounding the "Cayley-Hamilton theorem" concerns Cayley's *discovery* of the theorem. Why should he take the illegitimate step of substituting a matrix in place of a scalar? I believe this permission stems from his familiarity with the calculus of operations, where the substitution of operators for scalars was stock-in-trade. Another instance of this is in the passage from $(a+b)^n$ to $(D+b)^n$, in which the scalar a is replaced by the differential operator D. For Cayley, if not for us, it was quite natural to substitute an operator for a scalar and a familiar process in passing from those versed in arithmetical algebra to symbolical algebra. The now standard form of the theorem was not as general as the one he discovered in November 1857, a few weeks before submitting his "Memoir." He wrote to Sylvester alluding to a more general and "symmetric" theorem in terms of the characteristic "binary quantic," which arose from contact problems in conics.[78]

Hamilton's name is linked to the theorem through a version of it written in terms of quaternions and published in his *Lectures* (1853). Cayley had reviewed this book and, presumably, had been aware of Hamilton's treatment of the theorem, though Hamilton's exposition is completely different and submerged within pages of calculation. Hamilton pointed out his prior discovery but there was no dispute between the two. He admitted to Salmon his feelings about Cayley's work more generally: "[I] can assure you that Cayley has written heaps of things, which I could not without deep study read."[79] When the challenge came, it was many years later, and emanated from P. G. Tait, who had taken on the role of Hamilton's posthumous champion and executor of his intellectual estate. In 1894, the querulous Tait attempted to suggest that Cayley had obtained his matrix ideas from Hamilton's quaternions but Cayley emphatically refuted this:[80]

I certainly did not get the notion of a matrix in any way through quaternions: it was either directly from that of a determinant; or as a convenient mode of expression of the equations:

$$x' = ax + by$$
$$y' = cx + dy.$$

Cayley's "Memoir," which could have been a useful starting point for further developments, went largely ignored.[81] Spottiswoode and Henry Smith took note but neither the memoir nor its companion on the transformation of bilinear forms made much impact.[82] Cayley himself even relapsed into using matrices as a mere notational convenience, even blurring the distinction between matrices and determinants on occasions. At the point of discovery of the Cayley-Hamilton theorem, he saw the potential of matrices and had the presence of mind to codify the main ideas. In the companion memoir published at the same time, he solved a problem known as the Cayley-Hermite matrix problem.[83] His habit of instant publication and not waiting for maturation had the effect of making the idea available even it was effectively shelved. With his unswerving eye for what is important in mathematics, he characteristically had something to say about matrices.

High Points and Lows

The year 1858 began well and hard on the heels of Cayley reading his matrix memoir to the Royal Society; in February, he submitted the fourth and fifth memoirs on quantics for publication in the *Transactions*.[84] Work continued apace in potential theory, and in March he submitted a short paper to the *Philosophical Magazine* on a theorem that became known as Cayley's theorem on elliptic functions.[85] Although illness had laid him low in 1854 and delayed production, minor inconveniences did not interfere too much. In the early summer, he wrote to the Royal Society about his and George Salmon's joint researches: "I have been laid up the last three weeks with a boil in the neck & am not fit for much but if you will send me Mr Salmon's paper [on curves of the third order] I will make an abstract of it."[86] Soon better, he worked on his long papers on disturbed elliptic motion as it applied to lunar theory. But, this summer, London was not a pleasant place to be: the "Great Stink" emanating from the Thames was in full sway, in which the combination of a low rainfall, hot weather, and lack of a sewerage system had the obvious effect. It was a good time to escape to the Continent for fresh air and recuperation.

During the summer, Cayley was mountain climbing in Switzerland. It was a popular pursuit among the young middle-class men and closely fitted the Victorian ideal of being "character forming." The new Alpine Club had been established in London in the previous year. The Club's most prominent scientific member was John Tyndall, who had been elected to the Royal Society on the same day as Cayley and who made the first ascent of the Weisshorn, rivaling Edward Whymper, author of the classic *Scrambles in the Alps* and the first to scale the Matterhorn in 1865. Of mathematicians, Spottiswoode was already a member when Cayley joined several years later.[87] The club, which helped to achieve popularity for climbing in the 1850s and 1860s, was very different from the general run of Mayfair dining clubs. Cayley was hardly a man to fritter away time, and membership in this club with its clarity of purpose did not involve "entering into society" and engaging in niceties. It was a favorite haunt for those engaged in the law or who had close associations with the universities of Oxford and Cambridge. It was not open to all as qualification for membership required tangible mountaineering experience. Cayley could offer his exploits in the 1840s and his climbs of 1858, when he narrowly missed being the first to scale the Dom.

Le Dom is a mountain in the Mischabelhörner range and is the highest peak which lies entirely within Switzerland's borders. Its appeal to mountaineers was its boast of being barely three hundred meters shy of Mont Blanc, the highest peak in Western Europe. It was a prime target for climbers from Britain, who set store by being the first to scale the Swiss mountains. This peak was notched up by John Llewelyn Davies on 11 September 1858. Like many of his background, he was a keen mountaineer and a founding member of the Alpine Club; in addition to his conquest of Le Dom, he was the first to scale the neighboring Täshhorn. The brother of Emily Davies, suffragist and founder of Girton College, Cambridge, he was an associate of F. D. Maurice, the Christian Socialist. Educated at Trinity College, he was elected a fellow and private tutor prior to accepting a Crown appointment in 1856 as rector of Christ Church Marylebone, a position he occupied for next thirty-three years. In this parish with its endemic poverty, he drew adherents by his brilliant preaching, in which he promulgated Maurice's ideals of social reform.

In the account of Llewelyn Davies's Dom adventures, he paid a generous tribute to Cayley's attempt the month before: "In the summer of 1858, Mr Cayley attempted the ascent from the latter village [of Saas], and would no doubt have succeeded, had not a mist come on, which stopped him not far from the summit."[88] Climbing was a serious business and as the records of the Alpine Club show, climbing in the snow and ice of the High Alps could be perilous. Gentlemen took little in the way of equipment as this lack formed part of the ethos of Victorian climbing—to be worth anything, a climb had to be accomplished by fair means. Cayley climbed alone, thus putting himself in line for the highest accolade, for the highest prestige was reserved for climbers who scaled the peaks without companions.[89]

Cayley put his mountaineering to good effect in the service of mathematics. It was expected among the scientific fraternity that climbs would also be put to scientific use. Observations about glaciers and other scientific objectives, such as magnetic soundings, from the scientific men ensured that climbing was not pure pleasure but also had a serious purpose. Mathematician Cayley drew inspiration from the undulations of mountains and valleys, for it was his clear belief that the common objects of life suggested geometrical thinking: here it was the sugar-loaf shape of Le Dom, likened to a conically molded mass of sugar, which suggested ideas. It is refreshing to find Cayley's mountaineering experience hidden away in one of his little-known papers. In the year following the exploits in Switzerland, he wrote, "On Contour and Slope Lines" for the *Philosophical Magazine.* The style is discursive and his intention was to outline the meaning of new mathematical terminology for use in the study of topography.[90] In considering such concepts as contour lines (lines connecting points of equal height) and slope lines (paths of steepest descent) he unintentionally invites his readers to share in his walking and climbing and to add to the picture by giving the French equivalents *lignes de niveaux* (contour lines) and *lignes de la plus grande pente* (slope lines).

The final section is sufficient to convey the link between mathematical adventures and time spent on the mountains during a day's climbing. He explained that a series of *ridge* lines, which pass from summit to summit, define a watershed—the line of separation between two adjacent river systems: "In the case of an isolated conical or dome-shaped mountain, and in general when the contour lines are all of them closed curves, there is no definable watershed; but in the case of a chain of mountain summits, the watershed runs from summit to summit through the heads of the passes over the connecting cols, i.e. it is made up of a series of ridge lines each extending from a summit to a summit through an intervening knot [saddle-point]. And the course lines are, as nearly as may be, the beds of the streams which flow from the heads of the passes down the lateral valleys."[91]

Cayley's return to England after Switzerland presaged the announcement of one of his most outstanding mathematical ideas. In late September a small audience assembled at the mathematics and physics section of the British Association at Leeds and listened to an introduction from Whewell—who complained about the smallness of the room but suggested that presentations in this section were of such sublimity, speakers "thin the room very speedily." Whether or not there was an audience at all, Cayley gave a short abstract of a recent discovery "On the Notion of Distance in Analytical Geometry," in which distance is defined in relation to a conic, which he called the absolute:

> [T]he principles of modern geometry show that any metrical proposition whatever is really based upon a purely descriptive [projective] proposition, and that these principles contain in fact a theory of distance; but that such theory has not been disengaged from its applications and stated in a distinct and explicit form. The paper contains an account of the theory in question, viz. it is shown that in any system of geometry of two dimensions, the notion of distance can be arrived at from descriptive principles by means of a conic termed the Absolute, and which in ordinary plane geometry degenerates into a pair of points.[92]

This became incorporated in Cayley's famous "Sixth Memoir on Quantics," which was completed and submitted to the Royal Society on 18 November and published in the following year.

Toward the end of 1858, the Lowndean chair of astronomy and geometry at Cambridge became vacant following the death of George Peacock. Cayley thought he had a chance and wrote to Sir John Herschel for the obligatory testimonial: "I take the liberty of acquainting you that I am desirous of offering myself as a candidate for the Lowndes' Professorship at Cambridge and of requesting that if you should think fit so to do, would have the kindness to give me a testimonial in support of my application."[93] Cayley had all the right credentials. He was a Cambridge man and the authority to profess his subject. His lack of experience in observational astronomy was offset by the papers in the mathematical branch, which he was beginning to publish regularly. De Morgan, a member of the Royal Astronomical Society and a fellow member of Royal Society Catalogue committee, commended him to Herschel without reservation: "Cayley is a capital man for it."[94]

Events were moving quickly and if Cayley held high hopes of the Lowndean chair, they were about to be curtly dismissed. The competition for this chair was strong and it came from an unexpected quarter. J. C. Adams, the Senior Wrangler after Cayley but two years older, may have seemed a disinterested party, for he had recently taken a chair in Scotland. There was no doubting his academic strength. He had been president of the Royal Astronomical Society and already had been awarded the highest medal of the Royal Society, the Copley Medal, when most have to wait a lifetime, if it comes at all. In the course of the discovery of Neptune, he had become a national figure and declined a knighthood. In the previous March, he had been offered the chair at St. Andrews and had accepted the appointment from the Home Secretary.

Cayley soon learned of Adams's intention of returning to Cambridge and seeing him better qualified evidently switched his attention to the chair Adams intended to vacate. Though Cayley knew about the conditions attached to chairs north of the border through his abortive application at Aberdeen, he received unpleasant news from Adams on his assessment of the duties that would await a new professor in St. Andrews: "The work here is rather hard compared with that of a Cambridge Professorship, as the teaching of the men entirely depends on the Professor. I have to give 13 lectures every week, & the teaching is very elementary indeed, all the students in the same year being taught in the same class. At present the three classes belonging to the different years have about 20 students in each. I shall be happy to give you any further information on the subject of the Professorship in case you should wish for it." This may have caused Cayley to think twice about Scotland, but the final decision on the Lowndean chair was made elsewhere. Shortly afterward Adams informed him of the inevitable: "I have received a letter from Lord Derby announcing that the Electors have agreed to appoint me to the Lowndean Professorship & I write a line, in accordance with your wishes, to inform you of the fact."[95] It was abrupt but there appeared to be no ill will; in the next month, they were corresponding with each other on lunar theory.

When Cayley heard that he had failed to secure the appointment at Cambridge, Sylvester was supportive. He had employment in the academic world but could readily appreciate his friend's position after his experiences in the United States, his initial failure at Woolwich, and the unsuccessful attempts to be appointed for the Gresham professorship of geometry in the City of London. At the time, he was fully occupied with preparing examinations for his cadets at Woolwich and battling with the military authorities on the methods for awarding marks in examinations. Indeed, he could see another row looming, which might lose him his academic position, but, as in his support for Cayley in the "Gnoll affair" in South Wales, his response was immediate. He commiserated with Cayley on his failure: "I had not heard before your note of your disappointment—it is most vexatious and it has annoyed me as much as a personal disappointment could do. I shall try & call upon you very soon. You know how much pleasure and profit it always is to me so to do."[96]

In his attempt to gain employment in the teaching sphere, and earn money, Cayley looked to the possibility of obtaining private pupils. He had some experience at this lower level of mathematical instruction, though it appeared to be mostly limited to undergraduate supervision at Trinity College, examining duties at Cambridge, and the Indian Civil Service competitive examinations. Sylvester was helpful again: "I shall be only too glad to make known your wishes about the pupils where the knowledge can be made available to affect the end."[97] But would Cayley really desire this? Would he want to endure a succession of pupils like poor Eccles in the landscape of the Yorkshire moors struggling with elementary calculations?[98] By not having an academic post, he was stranded. It was a desperate situation for England's leading mathematician.

A Cayley letter regarding a discovery
in matrix algebra

Photograph of Charles Hermite, an
editor of the *Quarterly Journal of
Pure and Applied Mathematics*

Early home of Cayley family, during the 1840s, at Blackheath

Montpelier Row, 1850s, home to the Cayley family after Henry Cayley's death

Photograph of 5 Montpelier Row, Blackheath, Cayley family home (1855–75)

Photograph of Charles Bagot Cayley, Arthur's brother

Photograph of Thomas Archer Hirst, geometer and close friend to Cayley

Photograph of Paul Gordan, the "king of the invariants"

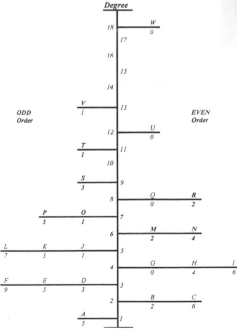

Invariants and covariants of the binary quintic as known more completely to Cayley in 1871

Lithograph of George Biddell Airy,
Astronomer Royal

Photograph of P. G. Tait, Cayley's
correspondent in Scotland

Photograph of W. K. Clifford, Cayley's
first student

Photograph of
J. W. L. Glaisher, Cayley's
second protégé

After Lowes Cato Dickinson's
painting of Arthur Cayley as a
student at his desk

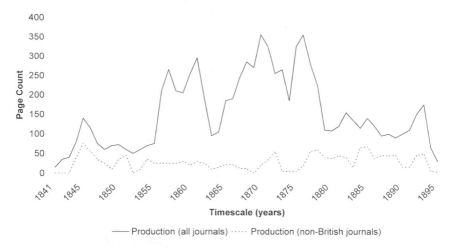

Chart of Cayley's publication record in journals

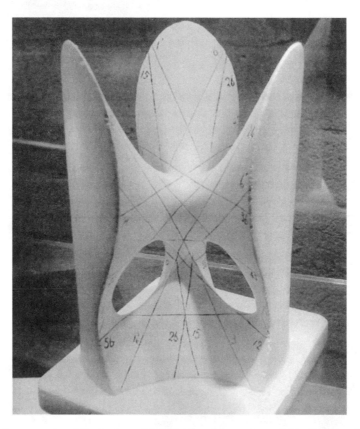

Photograph of Cayley's and Salmon's cubic surface and its "27 lines"

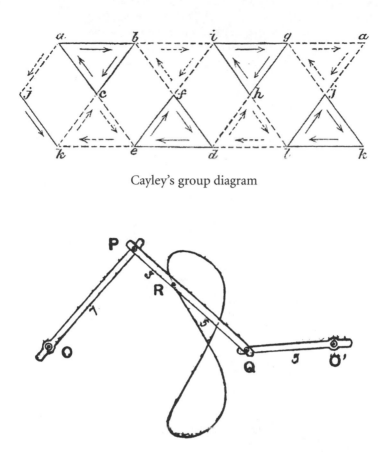

Cayley's group diagram

Three-bar motion and curve of order six, with the form of the curve shown

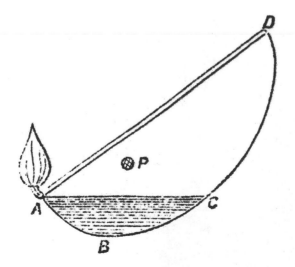

Isaac Milner's lamp and the shape of its bowl

P. A. MacMahon's discovery in invariant theory

P. A. MacMahon, Cayley's
collaborator in invariant theory
in the 1880s

Newnham student, during Cayley's tenure (1880–90) as president of
Newnham College

Newnham Hall, 1880s

Photograph of A. R. Forsyth,
Cayley's successor to the
Sadleirian Chair at Cambridge

Cayley family portrait at Garden House, Cambridge, in the 1890s (*from left:* Mrs. Cayley, servant, Arthur seated, servant, Mary Cayley seated in front; servant in upstairs window)

Garden House Hotel (modern site)

Photograph of
Henry Cayley,
Arthur Cayley's son

Photograph by
A. G. Dew-Smith
of the elderly
Arthur Cayley (c. 1890)

Last notes written to Lord Kelvin on polyhedra (dated 7 December 1894)

Arthur Cayley's headstone as it existed in 1980; no longer in existence

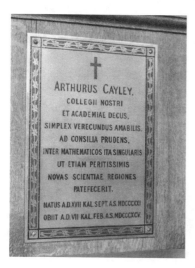

Plaque dedicated to Arthur Cayley in Trinity College Chapel

CHAPTER 10

The Road to Academe

By the beginning of 1859, Cayley's need for an academic position was plain, to himself and to all his friends. They were mostly in suitable posts; William Thomson, George Gabriel Stokes, and George Boole had had professorships for a decade and with an academic position, they were not diverted by the need to earn an income elsewhere. George Salmon was without a chair, but, being an active research-minded lecturer at Trinity College, Dublin, he was on a path which might lead to one. James Joseph Sylvester had to make do with inconveniences but at least he was in a quasi-academic environment at the Royal Military College at Woolwich. Cayley's position was less certain. He had forgone an upward career in the law for mathematical work, but this was hardly valued in the utilitarian Victorian society. For how could it value such a mathematician as Cayley?—"a solitary student in his closet, the result of which will not so much as cheapen one yard of calico."[1]

Primus Inter Pares

Cayley read his "Sixth Memoir on Quantics" to the Royal Society on the 6 January 1859. The "Sixth Memoir" is perhaps the best known of the memoirs on quantics, and apart from the "Introductory Memoir," it is the only one that does not involve substantial calculation. As he noted at the beginning, "[t]he chief object of the present memoir is the establishment, upon purely descriptive principles, of the notion of distance." It is here that he introduced his famous absolute. Though he largely confined his attention to the geometry of one and two dimensions, central to the "Sixth Memoir" was the meaning he attached to "space." In a category apart from physical space was his understanding of a mathematical space, which he regarded as an *ideal* one: "I consider there is an ideal space of any number of dimensions, but of course, in the ordinary acceptation of the word, space is of three dimensions; however, the plane (the space of ordinary plane geometry) is a space of two dimensions, and we may consider the line as a space of one dimension."[2]

The "Sixth Memoir" contains Cayley's definition of distance in projective space. With the ordinary Euclidean distance obtained as a special case of this "projective metric," a different light was cast on the relationship of Euclidean geometry and projective geometry. Projective geometry had usually been thought part of Euclidean geometry and all its geometrical constructions based on Euclidean definitions of length and angle.

Cayley turned this thinking around when he suggested that Euclidean geometry was a special case within the broader notion of projective geometry, or, as it was then called, descriptive geometry. The underlying geometrical idea is deep and a layman's description perhaps lends it an air of mystery. C. S. Peirce, the American philosopher, logician, and mathematician, as well as a great admirer of Cayley, attempted an explanation in nontechnical terms: "there is in space a certain individual Something [Cayley's absolute], circular in shape, which though it lies in a plane not only at an infinite distance but also in that unseen, inward region of space of which our universe is but the rind, is yet in intimate relation with everything we see, and cuts at two points even the smallest circle which can be drawn around any point." That "something," the absolute, Pierce declared, is a "short cut to the solutions of innumerable practical problems."[3]

Cayley's discovery of the absolute and the underlying principle was extremely significant in the theory of geometry. It is a recognition that the properties of a figure depend on the space in which it is immersed. This was the principle recognized by Cayley at this early date: "the metrical properties of a figure are not the properties of the figure considered *per se* apart from everything else, but its properties when considered in connexion with another figure, viz. the conic termed the Absolute." Thus the proper way to develop geometry, according to this principle outlined by Cayley was first to consider the properties of geometrical figures irrespective of its metrical properties (for example, properties of quadrilaterals irrespective of whether they are squares or rectangles). In this way, projective geometry logically precedes geometry in which a measure of distance is required. This discovery of Cayley's was the source of his best-known quotation, which reflects this idea: "Metrical geometry is thus part of descriptive geometry, and descriptive geometry is *all* geometry."[4]

He had made a rich find. He utilized results he had proved in 1843, and results on spherical geometry made a few years later, in 1846, but the idea may have crystallized in his mind as he climbed in the Alps shortly before its announcement at the Leeds Meeting of the British Association.[5] In the "Sixth Memoir" there was the potential for a systematic investigation of non-Euclidean geometry, but he did not realize this fully. He was also somewhat reticent in proclaiming the discovery in too loud a voice, and emphasized that it was made "*in my own point of view*" [his italics].[6] The year 1859 was a "creative year" for Cayley and likewise for others more generally. At the end of the year, Charles Darwin published *On the Origin of Species by Means of Natural Selection,* but while Darwin was met with either intense enthusiasm or hostility, Cayley's work was met with less than a flutter of recognition.[7]

In the same month of his presentation of the "Sixth Memoir" to the Royal Society, Cayley wrote a significant paper on polyhedra. Four new "star" polyhedra had been added already to the five classical Platonic solids of the Greeks. Johannes Kepler found two stellated dodecahedrons, with star-shaped faces, and in 1809 Louis Poinsot added two others with regularly shaped faces but with star-shaped vertices. Cayley called Kepler's specimens the small "stellated" dodecahedron and the great "stellated" dodecahedron and Poinsot's the great dodecahedron (with twelve pentagonal faces) and the great icosahedron (with twenty triangular faces).

Euler's famous formula $V + F = E + 2$ (relating the number of vertices (V), faces (F), and edges (E)) was satisfied by two Kepler-Poinsot polyhedra, but it failed for the other two. Cayley generalized Euler's formula to one of the form $mV + nF = E + 2D$, which applies to all five Platonic solids and four Kepler-Poinsot polyhedra without exception. Two weeks after the publication of his paper, he read Augustin-Louis Cauchy on polyhedra (of 1810) and, with admiration, commented on it. More on the subject of polyhedra was to follow, and while Cayley's attention to "topological" geometry in connection with mountain climbing and hill walking has been noted, he studied "topology" from a more abstract point of view in his "partitions of a close" two years later.[8]

The work on polyhedra reflected Cayley's concern for a theory of a pure character but his down-to-earth practical side runs alongside and reacts with it. His fascination for machines manifested itself from time to time, like the time he left legal matters aside to go and see "a new sort of engine" that traced out curves.[9] A curious incident occurred in April when he drew attention to the construction of a mechanical device that might be used for copying engineering drawings. With Sylvester as intermediary, he sent plans for the mechanism to Elim Henry D'Avigdor, an 18-year-old who had just begun training with John Hawkshaw's firm of civil engineers, a large firm involved with the construction of the railways. D'Avigdor was a member of a prominent Jewish family and Sylvester probably had met him at Woolwich, where the young man had entertained the hope to join the Royal Engineers.[10] He appreciated the idea put to him by Cayley (possibly acting for a third party) and was flattered to be consulted by such an eminent mathematician, but he expressed doubts to the method being applied in practice. Though the scheme was "ingenious," he believed it would not cope with the detailed drawings of the type usually found in engineering offices. Whatever its absolute merits, D'Avigdor's assessment appears cursory, but at the age of eighteen, he may be forgiven for making an immature judgment—it seems the matter did not progress.[11]

Cayley's interest in mechanism and machinery was not out of character, and it shows him to be a man of his time living in a great age of transition that clamored for technological progress and invention. Among the weird and the wonderful, the "Patent Daylight Reflector" (basically a mirror) was a Victorian contraption at work in his chambers at 2 Stone Buildings, and other ingenious mechanical devices were constantly being offered to the public, even those with relevance to mathematics.[12] His going "to see a new sort of engine," which drew curves, and his championing of the Scheutz calculating engine had shown a keen level of appreciation, and his latest foray, intervening on the side of the practical drawing mechanism makes this side of his character even more intriguing.[13] If Cayley were alive today, he would embrace the current technological age and advocate the place of computer software for discovery in the world of mathematics. In his own lifetime, Cayley employed designs of geometric-mechanical devices for sketching curves. When James Thomson (brother of William), professor of civil engineering in Belfast, designed a mechanical device for calculating integrals, Cayley commented on it.[14]

The next academic opportunity for Cayley arose when the chair of astronomy at Glasgow and the directorship of the observatory fell vacant in the autumn of 1859 due to the death of John Pringle Nichol. The salary was £300, but there was an official house on the grounds that was "very good, & delightfully situated," noted William Thomson. The duties involved experimental work and physical observation, requirements which immediately placed Cayley in a weak position, for the appointee would be obliged to conduct and direct physical and optical laboratory experiments and manage astronomical observations. He could not step into Nichol's shoes either, for the former director was one of astronomy's most enthusiastic popularizers. It was only too obvious to the people involved that Cayley was exclusively a mathematician and uninterested in physical observation. Thomson urged Stokes to apply: "Cayley is thinking on being a candidate, but it will probably be considered, & I believe justly, that he is not physical enough. I say this to you in confidence, knowing that you will not misunderstand what I say of so good a friend of us both. I think Cayley ought to be provided for by the country—mathematician laureate would be his right post—but there is no doubt but that popular or physical lines of science are not in his way, & that in a situation where either may be required, he might not be well placed."[15]

It was to be an elaborate game of professorial chairs. Thomson wanted Stokes out of London and Cambridge, "those great Juggernauts under which so much potential energy for original investigation is crushed." Thomson hoped Stokes might resign his professorship at Cambridge, thus clearing the way for Cayley to slip into the Lucasian chair. "I think this last would be a much better thing for Cayley," wrote Thomson, "who is so much more devoted to mathematics than to any kind of physical observation."[16] Stokes had no intention falling in with this plan and Thomson's precious calculation failed. The astronomer Robert Grant, a fellow member of the Royal Society Catalogue Committee, was appointed and took up the post at the end of the year.

Cayley had to be satisfied with an appointment to conduct evening classes at the place of his alma mater in the Strand. The evening classes at King's College,

London, were one of the first ventures in adult education in England and they were highly successful. More than three hundred students enrolled in the first two days of registration for the winter courses starting in October 1859, and more were expected. The course in "beginning Greek" was so oversubscribed that the authorities had to divide the class into two sections. In mathematics, the *Times* was "happy" that a King's old boy had come home and "now most kindly offered to give to the students of the evening classes, on two evenings in the week, the benefit of his great mathematical powers."[17] It was not a substitute for professorship in Scotland but it gave Cayley some the vital teaching experience, which he generally lacked.

Robert Grant's appointment had further repercussions for Cayley, and when Grant resigned his post as editor of the *Monthly Notices of the Royal Astronomical Society,* Cayley succeeded him in December.[18] His own term as editor lasted until November 1881, with the exception of the years 1872-74, when he was the society's president. The transition also meant a change of emphasis. Grant was a physical astronomer (he wrote a *History of Astronomy*) while Cayley was the mathematical astronomer par excellence. As editor, he was in prime position to comment on the leading issues of the day. Under his guidance, the *Monthly Notices* became a separate journal specializing in the latest news on astronomical researches.

On taking on the editorship, Cayley inherited an incipient controversy. A French Army officer, Philippe Gustave Doulcet, comte de Pontécoulant, who had retired at the age of fifty-four, had devoted himself to astronomy. In publishing an article in the *Monthly Notices* the previous May, he challenged Adams's astronomical calculations, which had been published in 1853. They were, according to him, nothing but a "truly analytical hoax." Adams had claimed that a result in Laplace's important calculations in the lunar theory should effectively be halved. The calculation of the secular acceleration of the moon, the nonperiodic increase in the moon's mean motion (the average angular velocity between perigee and apogee), was a historically important astronomical problem and it was about to involve bitter controversy between the English and French astronomers.[19] Laplace found that the mean motion increased by approximately eleven seconds of arc per century, a tiny amount but significant in astronomical terms. Whereas he had discovered this secular phenomenon by consideration of the eccentricity of the earth's orbit, Adams took on the additional consideration of the solar eccentricity: he came to the conclusion that 5.70 seconds of arc was closer to the true value. Without taking the solar eccentricity into account, Pontécoulant produced a wholly different result. Added confusion resulted when the physical astronomers entered the fray.

The newly appointed editor of the *Monthly Notices* became closely involved. Cayley knew of Adams's value before the storm broke, for Adams had sent the

1853 paper to him with an accompanying letter stating that "he had shown how all previous theorists who have attempted to find more than the first term of the expression for the Moon's secular acceleration [as had Laplace], have gone wrong." Cayley wrote to Adams in June of 1860 from Stone Buildings: "I am trying to work out the lunar acceleration by a method which appears to me a good one & which has given me without much trouble all your new periodic terms, but I have not yet obtained from it the disputed term in m^4 of the acceleration. You may perhaps be interested in seeing [it] & I send herewith the fundamental equations."[20]

By an independent method, using a series solution of the differential equations, Cayley corroborated Adams's result and, with an editor's prerogative, and perhaps for Pontécoulant's ultimate benefit, gave the extensive calculations involved in the *Monthly Notices*.[21] They took up sixty columns of the journal and he later reflected, perhaps ruefully, that "everything in the Lunar theory is laborious."[22] When the dust had settled, and the vociferous controversy was forgotten, and Sir John Lubbock, W. F. Donkin, Giovanni Plana, and Charles Delaunay (by two distinct methods), and Cayley himself had proved Adams correct, the summing up from this imperturbable man might be guessed—he merely remarked that the validity of the correction "was a good deal discussed."[23]

London Life

The mathematicians of London still had no meeting place exclusively their own. The Mathematical Society of Spitalfields, which had existed in the manufacturing district of the East End since 1717, dissolved in 1845 owing to declining membership and financial difficulties. The "old clay and pewter days" alluded to by Augustus De Morgan had failed to survive into the new industrial age, and the East End Society, with its pockets of erudition, was absorbed by the Royal Astronomical Society.[24] Although the Spitalfields Society had hardly been a research-oriented society but, rather, an organization where working men might meet and solve puzzles and conundrums in convivial surroundings, its failure left London without any society for mathematicians for twenty years until the London Mathematical Society was founded in 1865. During the interlude, better-placed mathematicians met at the Royal Society then established at Burlington House off Piccadilly.

Mathematicians and scientists knew each other through other connections. There was the Oxford and Cambridge network and the links between the legal societies of the Inns of Court. Informal meetings took place, such as the one which occurred in June and July 1859, when Sylvester gave a series of seven lectures at King's College, London, on the partition of numbers, in which he presented his first serious attempt at framing a multidimensional partition theory.

Since his disengagement from invariant theory around 1854, partitions loomed large in Sylvester's life. The little contretemps with Cayley, generated by Sylvester about priority in the theory, had evaporated as soon as it had begun. Regarding his forthcoming lectures, he wrote to Cayley: "I shall be proud to have you as an auditor at my lectures. . . . & if I obtain an audience of 6, I shall be quite content." In fact, he obtained many more. Henry Smith and William Spottiswoode and an audience of forty to fifty were present. During the course, he wrote again: "I owe you many apologies for failing in my appointment for yesterday. Excessive fatigue the rain and the necessity of consulting some books at the R[oyal] S[ociety] must plead my excuse. Today and tomorrow I have to get ready my lectures for Monday as the [Royal] Society keeps a day's holiday on the 1st July. The lecture on Monday will be on Simple Partition. The *closing* one on the Geometry of Disposition."[25]

This lends credence to his admission, made many years later, that the lectures were prepared "hand-to-mouth," as were many things by Sylvester, the rap on the door being his greatest spur to creativity.[26] That aside, the lectures were regarded as exceedingly important and "full of inspiration."[27] Thomas Archer Hirst was impressed by Sylvester's obsession and noted that he "talks incessantly about his partition of numbers."[28]

Salmon in his own charming way wrote to Sylvester that he was willing to forgo the details of some mathematical creations written out in the "very heat of discovery" but the work on partitions was too valuable to lose. "I shall be content to let you off some of these," he wrote, as a man used to the solemnity of the pulpit,

> if you will do justice to what you have done on the subject of partitions. I wish you would seriously consider whether it is not a duty everyone owes to Society, when one brings a child into the world, to look to the decent rearing of it. I must say that you have to a reprehensible degree, a cuckoo-like fashion of dropping eggs and not seeming to care what becomes of them. Your procreative instincts ought to be more evenly balanced by such instincts as would inspire greater care of your offspring and more attention to providing for them in life, and producing them to the world in a presentable form. Hoping you will mediate on this homily and be the better for it.[29]

Cayley likewise recognized the importance of the "Geometry of Disposition," calling it no less than the "theory of permutation of space."[30] He urged Sylvester to complete the work, and a year after the lectures had finished, gently chided, "Have you begun working on the Partitions yet?"[31] Perhaps he received a negative reply, for he took the matter into his own hands and published several of Sylvester's results in the *Philosophical Magazine*, himself.[32]

That there should be a substantial demand for public lectures in advanced mathematics is indicative of a growing following for research in mathematics in Britain in the 1850s. The Spitalfields Society had existed for relaxation and amusement after a long day's work but mathematics was now more serious and less parochial. Cayley and Sylvester continued to cultivate contacts abroad by publishing in Continental journals, engaging in correspondence with French, Italian, Swiss and German mathematicians, and making frequent visits to the continent. Both had begun this in the 1840s; when Cayley visited Berlin on one occasion, he found himself the "centre of scientific attention," with the leading savants lined up to meet him—much to his consternation.[33] The Continental links were continued and strengthened in the 1850s. In the 1860s Sylvester, ever the rolling stone, traveled to Pisa where he met Enrico Betti and to Heidelberg where he listened to lectures by Otto Hesse, reporting back and carrying messages to Cayley on both occasions. That both attended Joseph Liouville's lectures on potential theory at the Collège de France was symptomatic of their good standing in the French circle in Paris.[34]

The English mathematicians were now claiming serious attention as an identifiable group of scientific workers. The setting up of the *Quarterly Journal of Pure and Applied Mathematics* for communication of current research work was indicative of this though after it succeeded the Cambridge journals it had to overcome a stuttering start before recovery—before going on to serve British mathematics for a further seventy years. Cayley and Sylvester were the two London based promoters of this journal, Sylvester as an editor (with N. M. Ferrers) and Cayley on the "board" as an adviser and assistant. Meetings between writers and readers in and around the capital were informal and the talk of mathematics. Hirst provides a valuable record of one which took place under Sylvester's roof in the grounds of the Royal Military College at Woolwich in the early 1860s:

> On Monday evening I went to Woolwich to dine with Sylvester. Cayley was there to meet me and we had a very pleasant and very simple dinner. Sylvester's researches on Commutants and the Integration of Equations of finite Differences formed the principal subject of conversation. Sylvester was full of brilliant ideas. Cayley pulled him in incessantly to obtain greater precision for Sylvester lacks the power of placing himself in his hearers position and appreciating what it is necessary to explain in order to bring them to his point of view. I understood the matter but imperfectly. On Elementary mathematics both threw out ideas which I must find time to examine, one by Sylvester was on the development of $\cos n\theta$ and $\sin n\theta$ in terms of powers of $\sin \theta$ and $\cos \theta$.[35]

Hirst's remembrance of the evening party gives an insight into Cayley's mathematical thinking. Although he was a generalizer of mathematical concepts and did not emphasize proof to any great extent, he contemplated a logical deductive approach to the new ideas of modern algebra and geometry. Hirst recalled for his diary: "Cayley's suggestion [for future work] had reference to the theory of determinants which he would define as a linear function of the elements in a row (or column) which changes sign when any two rows are interchanged. From this definition all properties might be easily deduced." This is indeed a hint of Cayley's awareness that an "axiomatic" approach to the theory of determinants was possible and even desirable. He did not pursue it and an antidote to the plethora of mathematical papers on special determinants was lost in the talk of the dinner table. Except for Hirst's brief diary entry the idea was promptly forgotten.

The London Clubs gave another avenue of communication, but Cayley did not belong to this essential part of middle- and upper-class London society. He was not "clubbable," as was Sylvester, who gained entrance to the Athenaeum, the London Club favored by men active in the arts and the world of ideas generally. Influential Victorians met there and set the agenda for the arts and science in the capital. Sylvester had been elected to this prestigious club in 1856 under "Rule 2," whereby the committee were entrusted to annually elect up to nine men eminent in science, literature, and the arts. There was intense competition for admittance, and hopefuls were kept waiting before they faced the critical ballot. While Sylvester was busy organizing an honorary membership for Jean Duhamel, a professor of mathematics at the École Polytechnique under a special rule for the admittance of foreigners, he thought of Cayley. "I wish," he wrote, "you were one, I mean a 'foreigner' *ad hoc,* so that you might come among us; you would thus meet all the people you would be most likely to wish to meet and know. Pity! that you are obstinately bent on declining a privilege that many rich men would give £1,000 or £2,000 to obtain or even more. You have only to speak the word and the Committee would bring you in among the 9 Muses over the heads of about 1500 expectant Candidates."[36]

Nor did Cayley join the popular dining clubs of the type attached to the Royal Society. In one, The Royal Society Club was set up in the eighteenth century for the "scientific members" when the generality of the members' fold included many who owed their place to influence rather than scientific achievement. Sylvester was a member as was Adams. The Philosophical Club, another inside group which counted Sylvester a member, did not attract Cayley either.[37] In one sense, their roles were strangely reversed. We might expect that Cayley, through his middle-class family background, would be part of the social hub and Sylvester, with his Jewish antecedents, existing on the fringe. The contrary was true: Cayley was the self-imposed outsider and Sylvester, a man who drove hard to be accepted as an insider.

Most of Cayley's friendships appear to have been forged through his connections with mathematics: notably Sylvester, Salmon, and then Hirst. This geometer had acquired an impressive mathematical pedigree during his meandering on the Continent, studying under Carl Friedrich Gauss and Wilhelm Weber in Göttingen, J. P. G. Lejeune-Dirichlet and Jacob Steiner in Berlin, Michel Chasles and Gabriel Lamé in Paris, and Luigi Cremona in Italy. He stopped in Germany long enough to gain a doctorate, many years before it became fashionable for the English to take their acquisition seriously. With such a background, Cayley looked forward to meeting him. One can see why they became friends. In the openness of Cayley's associations, he could match Hirst, who sought friendship among those who were devoted to science "pure and free untrammelled by religious dogmas."[38] As Cayley's communication was usually confined to mathematics, there were few opportunities for controversy, but apart from mathematics, he and Hirst shared an interest in climbing. When recently widowed, Hirst had climbed two-thirds of Mont Blanc in the company of John Tyndall and Thomas Henry Huxley in August 1857, the same month Cayley was on an alpine tour and the year before he made his attempt to scale the Dom.

When Hirst wrote to Cayley from Italy, the event was significant enough to warrant a brief diary entry: "Yesterday I wrote to Cayley, 'England's great Mathematical luminary' as Sylvester styles him, for the first time." He corresponded with Cayley and looked forward to their meeting, so that on his return to England in July 1859, a visit became a priority. In the hot summer of that year, Hirst traveled across town to the chambers in Stone Buildings: "I went alone to visit 'the great Cayley' [and] I found a thin weak-looking individual with a large head and face marked with small-pox; he speaks with difficulty and stutters slightly, rarely looks you in the face and under no circumstances impresses you with any adequate notion of his own ability. He is extremely modest and at the same time generous in his praises of others. We had a long and very pleasant talk and I left him both pleased and instructed."[39]

It was Cayley's work in geometry that cemented their friendship. Hirst held Cayley as worthy of his highest admiration and was somewhat in awe of him. They met again in December and Hirst provided a cameo of Cayley in restive spirit at a moment of reflection:

> I explained what I was doing in which he expressed some interest. I was a little amused and encouraged too by his asking me for a definition of the rectifying plane. The great geometer had forgotten it for the moment. What a wonderful head he has, not merely round but spheroidal with largest diameter parallel to his eyes, or rather to the line joining his ears. . . . He never sits upright on his chair but with his posterior on the very edge he leans one elbow on the seat of the chair and throws the other arm over the back. Yet he is a keen sighted and extraordinary man, gentle I think by nature and at once timid, modest and reticent. Often when he speaks he shuts his eyes and talks as if he were reading from an unseen book, and talks well too that one has to sharpen one's own wits to follow him.[40]

Hirst's tangential remark on the shape of Cayley's head reflected the general interest in phrenology and the belief that cranial measurement could be used to confirm personal qualities and mark out human "types." That Cayley had a "wonderful head" went with a powerful intellect in phrenological terms. Even if Hirst did not follow the popular pursuit of "weighing heads" to its conclusion, his vivid pen-portrait captured the way Cayley could concentrate the whole of his attention on a mathematical problem to the extent of forgetting the definition of the rectifying plane. For the record, the circumference of Cayley's head was a modest twenty-three inches, quite normal, but he undoubtedly cut a distinctive figure. His ability to detach himself from ordinary goings-on was a valuable asset. His brother Charles had the same capacity and was known among the London *literati* for his wool-gathering, shabby dress, politeness, and utter devotion to the pursuit of learning; William Rossetti observed that Arthur appeared to "be almost as regardless of externals" as his younger brother.[41]

Hirst's views on Salmon touched on the way Cayley was perceived by other mathematicians. As a geometer, Hirst had also corresponded with Salmon, and when he returned to England, Salmon had invited him to visit Ireland. On their meeting, Hirst found to his surprise that "he was not a great reader—that Cayley to him is just as difficult as to the rest of us and that it is only on those subjects upon which he has himself worked that he can even read Cayley." Hirst regarded Salmon as "a great calculator, fond of calculating for its own sake," but he did not place him in Cayley's class. "I do not class him amongst the high mathematicians however. The mere ready-reckoning element is too prominent in him. I had often noticed that his books although excellent as a collection of theorems gave no compact rounded view of the subject and this defect was at once explained when I learnt that he writes his books in a fragmentary manner beginning to print before he had concluded what shall be the precise nature of the book. . . . He is just beginning a book on Surfaces which he is writing in his usual manner."[42]

Hirst's interest in pure geometry coincided with that of Olaus Henrici and Henry Smith as its principal proponents in England. He became an important figure in British mathematics and supported reforms of geometrical teaching at the school level. Cayley, who specialized in analytical geometry, respected Hirst's abilities in dealing with geometrical problems but teased him about selecting the most difficult geometrical problems to solve.[43] Hirst was elected to the Royal Society in 1861, a year in which Cayley was a council member, and, ever scrupulous in his dealings, Cayley evidently chose not to sign the membership certificate himself. His support came through a welcoming note informing him of his inclusion in the council's fifteen recommendations. Hirst rejoiced in his election and particularly so since he would be in contact with such an inspiration as Cayley: "One of the greatest advantages of this position is that I meet Cayley nearly each week and always this remarkable man has something to communicate to me. The more I know him the more I am amazed at his great force in mathematics. No matter what part of mathematics one is speaking to him of always he has brilliant and instructive ideas on it."[44]

With Cayley at Blackheath and Sylvester at Woolwich, Hirst took his turn to entertain the bachelors about town: "Yesterday, Saturday, Cayley, Sylvester and Harley dined with me. Tyndall was not present. It was without question to me the most interesting dinner party I ever gave and I believe one of the most successful, at least all appeared to enjoy themselves. I contrived to give my three guests opportunities of communicating their latest results. Cayley explained his late controversy with Boole on a question of Probabilities. . . . At 9.30 P.M. we all adjourned (in a cab) to Sabine's Soirée at [the Royal Society] Burlington House."[45]

Modern Theories

If Cayley impressed Hirst, he was also gaining recognition at the top of the British scientific establishment. Since a first proposal that he be awarded a Royal Society Royal Medal in 1853, the occasion when Charles Darwin was honored, he had been proposed in two further years, in 1854 and four years later.[46] In June 1859 his name was mentioned once again and this time with success, the recognition primarily owing to his work in invariant theory. At the Royal Society meeting in November, President Sir Benjamin Brodie (the elder) delivered the oration. An eminent surgeon and chemist, his own observations on Cayley's mathematics would ordinarily be of little value, but his son may have briefed him. Benjamin Brodie (the younger) gave up the law at Lincoln's Inn to study chemistry, having studied mathematics at Oxford. For twenty years he endeavored to construct a "Mathematical Calculus of Chemical Operations" based on the mathematical calculus of operations.[47] Brodie senior was able to deliver a substantial tribute in praise of Cayley: "Mr Cayley is among the foremost of those who are successfully developing what may be called the *organic* part of algebra into a new branch of science, as much above ordinary algebra in generality as ordinary algebra is itself above arithmetic. The effect is a vast augmentation of our power over the comparison and transformation of algebraical forms, and greatly increased facility of geometrical interpretation."[48] Thus Brodie was applauding Cayley's work in invariant theory, the "living" part of algebra, which was being organized into a new science of symbols. Through the memoirs on quantics, the theory with its application to geometry became an organized entity, systematically set out within a single framework.

In the same year, invariant theory received another impulse, this time for potential students. If Cayley's work was to be passed on to the next generation, it was vital it be transmitted at a suitable level in an understandable and coherent form. This was a task for which Cayley was ill-equipped. His lack of contact with students was a drawback, added to which he never really appreciated the difficulties others experienced in learning mathematics.

Salmon stepped into the breach and played the vital part in evangelizing. All the years teaching the ordinary degree students at Dublin were put to good use in the way he gathered together the ideas, concepts, and results of invariant theory. With his effective publishing technique of seeing the printed word of the earlier chapters emerge before the later ones had been written, the *Lessons Introductory to Modern Higher Algebra* was rapidly published. To the finished volume, he prefixed the dedication: "To A. Cayley, Esq., and J. J. Sylvester, Esq., I beg to inscribe this attempt to render some of their discoveries better known, in acknowledgement of the obligations I am now under, not only to their published writings, but also to their instructive correspondence."[49] Without Salmon's text it is doubtful if modern algebra, as it was understood by the invariant trinity would have been so widely appreciated. The members of the trinity complemented each other, for in Cayley it had a mathematician who could relate the theory to many parts of mathematics and provide leadership, with Sylvester a proselytizer, but with Salmon, it had a man who could write with clarity and be an effective conduit to generations of students.

Brodie's allusion to organic algebra implied a reference to other parts of symbolical algebra, to the new algebras, and the emerging group theory. Cayley and Thomas Kirkman were the principal contributors to group theory in England. Both attended the famous 1860 Oxford meeting of the British Association, which opened at the end of June. An overflowing audience attended Section D when Bishop of Oxford Samuel Wilberforce and T. H. Huxley "debated" Charles Darwin's *On the Origin of Species by Means of Natural Selection* published the previous November.

In Section A of the British Association meeting, Cayley gave a paper on curves of the fourth order and Kirkman outlined a theory entitled "On the Roots of Substitutions" and identified his groups within a list provided by Cayley.[50] In June of the previous year, Cayley had submitted the last of his group theory trilogy to the *Philosophical Magazine,* in which he included a listing of the groups of order 8 and provided yet another link with the new algebra of quaternions.[51] "There is nothing else on the subject in English," Kirkman wrote to a publisher, "except a masterly little paper by Cayley in the Phil. Mag. and by this reprint [submitted], readers will have a better chance of understanding both Cayley there, & me in a short paper in the Brit. Assoc. Report for 1860. It has been hitherto quite a French question."[52]

After the British Association meeting, Cayley was stimulated to rethink his own generalization of permutation groups. By mid-August, he went on his annual retreat in the Lake District and took Cauchy's work on substitutions with him. Cauchy had written a series of articles on permutations between 1844 and 1846, in which he referred to his own long interest in these substitutions. It was the chapter published in the 1844 *Nouveaux Exercices d'Analyse et de Physique Mathématique* that we find Cayley reading in Patterdale.[53] He greatly admired Cauchy's work and declared the section on conjugate permutations *"exceedingly beautiful."* He made some observations on notation and suggested an improvement on Cauchy's representations.[54]

Cayley approached group theory from a different angle. He reiterated his conception of a group of operators from the more limited notion of "corporate group" or group of permutations: "I consider a substitution, when applied to an arrangement, as *corporified,* and using the word group as primarily applicable to substitutions (or to my more general set operators as I propose to call them) I use the expression *corporate* or *corporal* group to denote a group of permutations. But the word group may be understood as denoting or including corporate group."[55] He also pointed out to Sylvester the important result [Cauchy's theorem on groups]: "Let *M* be the order of a system of conjugate permutations on *n* letters and *p* a prime less than or equal to *n*. If *p* divides *M* then among the conjugate permutations there is at least one regular permutation of order *p*."[56]

Cayley was thinking of his own set operators. He noted that "the whole course of the investigation [by Cauchy] is peculiar to substitutions, and I see no way of extending it to set operators in general—otherwise it would be the very theorem which I mentioned to you that I was in want of—& which I wish you would consider—viz. *that in any group whatever of set operators, every prime factor of the order of the group presents itself as the index of at least one operator."*[57] He did not prove this conjecture but it is notable that he made it. After his 1850s trilogy on finite groups and his afterthoughts in 1860, he applied himself to group theory intermittently.

Out in the country, Kirkman was also active and thinking about group theory. He had competed for the French Academy medal with a submission on the new subject in 1860 without success. He felt hard done by: "I think the Academy have shown an intolerable disdain of the Foreigner, in suppressing *all my results,* even while they have ungraciously confessed . . . that they are both new & important in both *the matter and the method* of the enquiry. I am told that much discontent is felt at the behavior of the Academy in the matter of their prizes, to foreigners, especially their medical prizes, which are their greatest: and it is as well that facts should be before those who are able to judge."[58]

Kirkman and Cayley shared many interests. Group theory is intimately linked with the study of polyhedra, another of Cayley's favorite subjects. Perceiving links between the two subjects, Cayley noted that rotations of polyhedra formed a group and he investigated subgroups associated with the Platonic solids. In the wake of his failure for the Paris prize in group theory, Kirkman considered with care whether he should send a submission in competition for the 1861 French medal on the subject of polyhedra, or as he called the study, Polyedra. He drafted a paper in French, in which he generalized former results achieved for polygons to more difficult questions concerning polyhedra. In any event, he decided against getting his fingers burned in Paris and published in England instead.

Kirkman promised the publisher of the Royal Society Proceedings an abstract "of my *complete theory of the Polyedra*. The prize question for this year at Paris is an "*quelque point important*" [some important point] of this theory. I hope you will have in a week or two in your hands the whole bird of which they have a feather at Paris."[59] The "whole bird" consisting of his outline of the theory, arrived a few days later, but the truly plumped-up bird was a manuscript of twenty-one huge sections, and this was carried through the portals of the Royal Society during the festive season of the New Year 1862.[60] It was intended for the *Philosophical Transactions,* a journal that could take bulk, but not the extreme bulk that was dispatched from the out-of-the-way rectory in Lancashire.

Kirkman's treatment of the esoteric theory was elaborate and perhaps complete, but it was written with a terminology so novel that he was practically the only person capable of reading it. As an appointed referee, Cayley recommended that the paper should not be published *in extenso,* a judgment Kirkman thought was a case of haughty treatment. He vented his fury in an outburst sent to Sir John Herschel: "[Cayley] has done all he can to prevent me getting the French Medal, for reasons that I can only conjecture. At first he addressed the R. S. not to print it. Afterward he consented to the printing of the first two of 21 sections and has shelved the rest."[61]

Despite such uncharitable thoughts, Kirkman maintained contact with Cayley and even invited his cooperation on group theory. Surely Cayley would have approached this proposal with trepidation. He would have been especially wary of working so closely with such a volatile character, for there was no telling where it might lead. Kirkman was an aggrieved man, as his letter to Cayley makes clear: "I have done 10 times as much as the Academy asked for in their Group question—& I should have done it sooner, had they not so *unfairly* refused to let me fight it out. As it is, I have done it, in less time than they usually allow their Prizewinner. I will print instantly at Manchester and I will demand *before the world* in the bold, to style that medal of 1860 as my lawful spoils. I told you long ago that a pretty quarrel was brewing [about the award of the 1860 Medal]. I will teach them manners."[62] Kirkman's failure with the Académie des Sciences and their Grand Prix de Mathématiques was deeply felt and he sounded his exasperation to Cayley: "I hope somebody will yet do me justice, after the insults I received from those Frenchmen." In 1863 Kirkman learned that the competition for the unawarded 1861 polyhedron medal was to be reopened with a submission date of the New Year's Day 1864. He had some hope on this occasion, as he expressed it to Cayley: "[t]hey may yet be driven by their necessity to give it to me, thro[ugh]' no competitors," yet it was a false hope and there was no prize for the clergyman from Lancashire.[63] Cayley's election as a corresponding member of the Académie des Sciences in the following month only highlights their respective positions in the scientific world; a blow to Kirkman perhaps ameliorated by a reading of the *Times* that Cayley was elected to the department of astronomy.[64] Ten years prior to this honor, Cayley had taken up astronomy as a second string but now had an established reputation in the subject; in his election to the French Institute he was preferred over the Russian Otto Struve well known in astronomy for his observations of double stars. Cayley's standing in astronomy meant he could now apply for an academic position in this field with some hope of success, yet there was nothing on the horizon for this specialty. He could thus afford some time for his favorite subject, pure mathematics.

While group theory and matrix algebra were avant-garde in the 1850s and 1860s, there were the problems associated with the algebraic solution of polynomial equations for Cayley to consider. There were two traditional approaches to these problems and both involved invariant theory. A direct approach to the solution of polynomial equations was to transform the equation so that the second coefficient vanished. This method originated with the German-born potter-cum-mathematician Ehrenfried Walter von Tschirnhausen, a contemporary of Newton, who developed various transformations in attempting to solve the general quintic equation. Cayley transformed it to one of the form $y^5 + Cy^3 + Dy^2 + Ey + F = 0$ (with the "missing" term in y^4) and where the coefficients are covariants. The calculations were onerous and use was made of a Government Grant Fund to employ the two in-house computers Messrs. Davis and Otter.[65]

From another vantage point, Cayley employed the indirect approach attributed to Joseph-Louis Lagrange, in which use is made of an auxiliary equation (the "resolvent") the roots of which could be expressed as rational functions of roots of the original equation. In March 1860, Cayley applied this method to analyze the solution of polynomial equations of low degree, the cubic and quartic equations, referring to the previous eighteenth-century work of both Lagrange and Edward Waring. The following year, he considered the solution of the quintic equation, the real problem at issue since the cubic and quartic equations are relatively trivial.[66] This extended the investigations carried out by James Cockle and Robert Harley, whose work he greatly valued.[67] In this, an auxiliary equation of order 6, $z^6 + Gz^4 + Hz^2 + Jz + K = 0$, in which the coefficients are semi-invariants. Whereas Cockle had dealt with the special form $x^5 + ax + b = 0$ (the so-called Jerrardian form), Cayley dealt with his general form. In November of the following year he discovered this auxiliary equation had been published by Carl Jacobi twenty-five years previously.[68] Despite this duplication, the paper was regarded as one of Cayley's most important papers.[69]

The work on polynomial equations around this time was fuelled in part by the English mathematician G. B. Jerrard. He was devoted to the solution of polynomial equations and in his twenties had devised a method for removing terms from such equations, as in the previously mentioned "Jerrardian form," work which resulted in his gaining high standing both in his native country and on the Continent. In 1858 he published *An Essay on the Resolution of Equations,* in which he claimed that *all* quintic polynomial equations were solvable. It was a position he appeared reluctant to relinquish, despite the efforts to dissuade him of it as published by Cockle and Cayley in the pages of the *Philosophical Magazine.*

Laid low by the illness, Jerrard attempted to refute all objections point by point. The two lawyers pointed out that his error was tantamount to an assumption made about the auxiliary equation. Cockle made his argument through prolix articles, while ten years at the bar had the opposite effect on Cayley's prose, which was unpolished but directly aimed.[70] In his final response to Jerrard, who had accused him of being irrelevant, he wrote in September 1862 in a manner such that one could tell he was still a lawyer: "But if my objection be (curiously or otherwise) *irrelevant,* then the proposition I contend for might be admitted without prejudice to Mr. Jerrard's results . . . Mr Jerrard, in his reply, contends for the contradictory proposition . . . and he thus in effect treats the objection as a *relevant* one. It appears to me that the objection is not only a relevant one, but that the proposition therein contended for is completely proved; at any rate the issue is so narrow a one that it seems useless to argue it further, and it is not my intention to do so."[71] There the matter dropped. Cockle's legal skill in drafting the *Juristiction in Homicides Act* (1862) attracted the attention of Sir William Erle and he was appointed chief justice of Queensland and took up his post in Australia in February of the following year. Jerrard, who remained unconvinced by the weight of established theory, died in November.

In the early 1860s, Cayley submitted his "Seventh Memoir on Quantics" for publication, a paper relating invariant theory to the study of cubic curves in which he compared his work with Siegfried Aronhold's.[72] That he duplicated the German's results did not worry him unduly, and, as was his custom, he drew up a table showing the correspondence between the two notations.[73] He absorbed Aronhold's work and used it as a counterpoint to his own. For Cayley, mathematics was not a matter of pure competition where the supreme prize was wrested by being first. He strove to assign provenance of ideas but ultimately others were free to use his work and, by the same token, he felt as free to utilize their investigations. In his civilized mode of conducting himself, they were all brothers united in the cause of mathematics. The "Seventh Memoir" marked a halt in the steady train of memoirs on quantics that appeared between 1854 and 1861.

The series drew to a close, a premature one as it turned out. They averaged more than one per year, and if one could disregard all the other papers interleaved, it would still be a performance that sets him apart, and all this while he doubled as a Lincoln's Inn barrister with legal duties to perform and money to be earned.

The whole series of ten memoirs published between 1854 and 1878 (the last three trailing in 1867, 1871, and 1878) was Cayley's longest series of memoirs on a single subject. When contemporaries thought of his work, they naturally thought of the memoirs, a body of work which bridged a quarter of a century and established his reputation. While Boole's support was important in the 1840s, the memoirs were his own accomplishment. Others may have been alongside, but he dictated their content and style independently. It was a carefully measured response and a tour de force in its wide ranging-compass and design.[74]

Invariant theory was the subject of the moment. Henry Smith, from Oxford, gave a good idea of the need for more work as it related to the theory of numbers:

> Our knowledge of the algebra of homogeneous forms (notwithstanding the accessions which it has received in recent times) is far too incomplete to enable us even to attempt a solution of them [problems in number theory depending on algebraic forms] co-extensive with their general expression. And even if our algebra were so far advanced as to supply us with that knowledge of the invariants and other concomitants [such as covariants] of homogeneous forms which is an essential preliminary to an investigation of their arithmetical properties, it is probable that this arithmetical investigation itself would present equal difficulties.[75]

In number theory, the binary quadratic form had been thoroughly researched, the binary cubic form a little, and Cayley was aware that higher order binary forms had scarcely been touched.[76] Invariant theory was turning out to be universal and it found applications in other areas of his work, with Cayley actively seeking them. His solution of Poncelet's problem of the inscribed and circumscribed polygon solution drew on his expertise in invariant theory and elliptic functions.[77] Spottiswoode, a man for the wider view of science and mathematics, saw invariant theory as uniting seemingly disjoint branches of mathematics. As well as the theory of numbers, the theory of equations and algebraic geometry he saw the omnivorous theory reaching out to the calculus of variations, molecular physics, and mechanics. At the beginning of the 1850s, invariant theory consisted of a disparate collection of papers written by such mathematicians as Cayley, Sylvester, Hesse, and Aronhold, but by the end and at the beginning of the 1860s, the situation was quite different. Others had entered the field, and by that time Cayley's memoirs had, according to Spottiswoode, become classic.[78] They provided a nucleus on which the theory could develop and grow.

In the context of evolving "modern theories," Cayley's geometrical work is central. In June of 1860, he returned to a problem he had encountered as a young man in his discussions with Boole—that of representing a space curve by algebraic equations. It was well known that such curves in three dimensions were not simply the intersection of two surfaces. He submitted a paper that suggested a solution, in which he proposed to consider a space curve as the intersection of a surface with cones having variable vertices. In the course of this argument, he showed how a straight line in space may be described by six coordinates.[79] This problem of space curves was of great interest to him. He opened a correspondence with the Italian geometer Luigi Cremona in 1861, sending him some work on plane curves, but adding, "I am very interested in the other theory which you are occupied with, curves of the third order in space; and I still hope to study this with closer attention."[80] Cayley was now rethinking the case of cubic space curves, a species of the curves of double curvature he had briefly studied in the 1840s.

Cremona had recently been appointed as professor of higher geometry at the University of Bologna, a time in which he altered his focus from analytical geometry to one of pure (or synthetic) geometry. Results for cubic space curves that Cremona had obtained by analytical methods he now reviewed as results in pure geometry, whereupon he proclaimed he had "opened his eyes to the sun."[81] While this shift marked a departure from the analytical treatment of geometrical questions Cayley preferred, Cremona was not a diehard pure geometer like, for instance, Jacob Steiner. He believed that the different standpoints mutually supported each other, and this flexible attitude meant that the letter-writing between Cayley and Cremona in the 1860s was to their mutual benefit: Cayley discovered a curious cubic surface, which suggested future directions for Cremona, but Cayley for his part could draw inspiration from pure geometers.

Motivated by a geometrical result owing to Michel Chasles, Cayley referred to the work of his compatriot Thomas Weddle. Ten years earlier, this promising young mathematician had drawn attention to a remarkable quartic surface that was generated by quadric cones and now known to geometers as the Weddle's Surface.[82] As a young man living in one of the more remote spots of Northumberland, he came to the notice of the Royal Society for a paper outlining a numerical technique for solving equations. After a succession of appointments, he became professor of mathematics at the Royal Military College, Sandhurst, where he died of tuberculosis at the age of thirty-six.[83] The author of twenty-nine papers in pure mathematics and acknowledged as a fine mathematician by his own generation, he is forgotten today. From among his works, Cayley noted Weddle's ingenuity with regard to one of his own favorite problems, that of constructing the ninth point of intersection of two cubic curves which pass through eight points.[84] Seeing the solution offered by Weddle, he simply reproduced it in his own notation.

In February 1863, Cayley produced a study of surfaces which proved epoch-making.[85] Building on a paper produced ten years earlier on skew surfaces—like a hyperboloid of one sheet—he submitted a memoir for the Royal Society *Transactions* on these surfaces, which he now called scrolls. Scrolls became a central plank of Cayley's contributions to geometry, and, as in his other investigations in algebraic geometry, the subject was developed in correspondence with George Salmon. Yet, just when Salmon's mathematical career faltered, having failed to secure the Dublin mathematical professorship, Cayley's was about to rise.

Cambridge in Prospect

In 1860, Cayley was approaching forty years of age. He had acquired some experience in teaching but this was limited. He had been examiner for the Tripos, but even at Cambridge, he did not have everything his way. E. J. Routh, on his way to becoming the famous coach in succession to Hopkins, was appointed examiner for the mathematical Tripos in 1860 and, judging from Sylvester's reaction, it seems in preference to either he or Cayley: "You of course have heard that Routh has been judged *dignior* [worthy] than either of us—I regret your disappointment *at least* as much as my own."[86]

To supplement his income and gain experience, Cayley again attempted to take private pupils, a wish remembered by Sylvester: "Some time ago [at the end of 1858] you said you wished to obtain pupils in Mathematics to read with you—I need not say that I think you fitted for much higher and more remunerative work but do not feel myself at liberty to judge for you in this matter. Please therefore to let me know if I am at liberty to mention your name to a party who has applied to me to recommend some one with whom a young friend can read 'the higher Mathematics' to qualify him for a competitive examination."[87] There were hopeful stirrings at Cambridge. The Royal Commission inquiring into the state of the university at the midcentury drew attention, among other things, to shortcomings in the arrangement for teaching mathematics. Since the eighteenth century, college lectureships for the teaching of algebra had been endowed through the Will of Lady Sadleir, whose husband had been a member of Emmanuel College.[88] The Sadleirian lecturers, one for each college (except at Emmanuel where there were two) had become ineffective. Hopkins doubted "whether a single undergraduate would estimate the advantage which he has derived from them at the value of the smallest coin in the realm."[89] With the income from Lady Sadleir's will, Hopkins advocated that the "utterly useless" college lectureships be discontinued and replaced by university public lecturers. In the wake of the Royal Commission, the eye of Parliament was on them and the university was expected to modernize. It did act, but not in the way suggested by Hopkins.

The establishment of the Sadleirian Chair of Pure Mathematics was set in motion following Parliament passing the modernizing Cambridge University Act (1856).[90] A proposal was made to the newly constituted Council of the Senate of the university that a "new direction should be given to the [Sadleirian] Endowment" and this was approved on 26 November 1857; the restriction to algebra of the original lectureships was removed to take into account modern developments in mathematics. Not all were in favor of a new professorship; Stokes felt that existing professorships should be properly funded before creating others.[91] Notwithstanding objections, Queen Victoria formally approved the establishment of the chair and, after a decent period had elapsed, a formal advertisement appeared in the *Times* in May of 1863:

> An election of a person to fill the office of Sadlerian Professor of Pure Mathematics will take place at Clare College lodge on Wednesday, the 10th of June, at 10 o'clock in the morning. All candidates for election to the said professorship are requested to communicate with the Vice-Chancellor [Edward Atkinson, Master of Clare College] on or before Saturday, the 6th of June. The electors are the Vice-Chancellor, the Master of Trinity College [William Whewell], the Master of St. Peter's College [Henry Cookson], the Master of St. John's College [William Bateson], the Lucasian [George Gabriel Stokes], the Plumian [James Challis], and the Lowndean [John Couch Adams] Professors.[92]

Leaving nothing to chance, the very same day Cayley made his application. It was the opportunity for which he had hoped: "I beg respectfully to offer myself as a candidate for the Sadlerian Professorship. I venture to hope that my connection with the university and the special direction of my studies to the advancement of Pure Mathematics may be considered as circumstances in my favor." There was no room for slippage and he quickly organized the printing of his list of publications. The curriculum vitae sent to the vice-chancellor was in keeping with one framed by a highly motivated and ambitious modern-day academic. It was clearly designed to impress, and how could it fail but to do just that? The quickly printed booklet listed no fewer than 318 of his publications.[93]

Seven other candidates declared themselves. They included Norman Macleod Ferrers, Edward John Routh, and John Clough Williams-Ellis, young men without a realistic hope of success but putting down markers for later preferment. Routh, a private tutor, had been Senior Wrangler in 1854, and Ferrers had been so in 1851 and was now a lecturer and joint editor (with Sylvester) of the *Quarterly Journal of Pure and Applied Mathematics*. Both Routh and Ferrers showed "promise," particularly Routh, who was ambitious and a man who sought promotion. At the age of thirty, Williams-Ellis was the youngest candidate in the field. The third wrangler in 1856, he was a Sadleirian lecturer but had little chance in the professorial contest.[94]

In a former age, the resident Cambridge dons Thomas Gaskin, James G. Mould, Percival Frost, and Isaac Todhunter would have been shortlisted, each with a real chance of being appointed. Gaskin had been the moderator of the Tripos when Cayley had graduated, but he published little in mathematics and had spent the latter years as a private coach. Frost, second wrangler in 1839, was a well-regarded coach who wrote books on the geometry of curves and would achieve the distinction of teaching W. K. Clifford several years later. Todhunter at the age of twenty-eight was the Senior Wrangler in 1848, having first graduated from the University of London, and, like Routh, had been a student of De Morgan. Todhunter was an ambitious man who had succeeded in the academic world through hard work. He wrote numerous school-level texts, and when he applied for the Sadleirian chair, he had just completed a history of the calculus of variations. In Todhunter's favor was the fact that he had recently been elected to the Royal Society and "distinguished for his acquaintance with the science of Mathematics and its history—eminent as a mathematician and for varied erudition," with his certificate signed by Cayley.[95]

Todhunter was not expecting success, but he had reservations about Cayley's suitability. He acknowledged he was the favorite but encouraged Boole to apply: "You may perhaps have seen in *The Times* a notice that on Saturday June 6th, or before, candidates are invited to send in their names to our Vice-Chancellor at Clare College, for the Sadlerian Professorship of Pure Mathematics which is now to be instituted. It is supposed that Mr Cayley will come forward and be elected. The electors are Professor Stokes, Professor Adams and Professor Challis, and three heads of colleges, namely those of Trinity, St. John's and St. Peter's, and the Vice-Chancellor." Furthermore, Todhunter suggested he would rather have Sylvester in a post which demanded the explaining and teaching of mathematics, and he continued his letter to Boole: "I have of course no objection to Mr. Cayley; but it is obvious that he cannot teach or explain any thing, and I do not myself estimate his work I think so highly as you do. Had I a vote, I should give it to you, and I have told our Master I consider that you are the proper man for the post . . . I should like above all to see you in the post, and next to you I should like Sylvester, and next to Sylvester I am willing to have Mr Cayley."[96]

Boole was not persuaded. Perhaps he was too aware of Cayley's desire to obtain a permanent academic post, though he must have known his application would not have succeeded. Boole was a Christian, but not an orthodox Anglican, and this would have placed an automatic barrier against Cambridge aspirations. Surely Sylvester would not have applied for the post in competition with his mentor, and Todhunter should have known that Sylvester, being a Jew, would not be eligible.

At a time when the university was beginning to turn its attention to research, none could match Cayley's proven achievement. He had all the right credentials, an impressive research record to say the least, and a pure Cambridge pedigree.

The electors gave no votes to any of the other candidates and his election proved a formality. A small notice appeared in the *Times* on June 11: "Mr Arthur Cayley M.A. of Trinity College, has been this day elected Sadlerian professor of pure mathematics."

In the week following, Cayley went to Clare Lodge, signed the vice-chancellor's book, "made the declaration of office, and was on his knees admitted Professor by the Vice Chancellor."[97] Mr. Cayley, Barrister-at-Law, became Professor Cayley and the long road to academe had reached an end. No longer would he have to work undercover in the warrens of Lincoln's Inn. Legal briefs with tight deadlines would become a thing of the past. One of his former legal colleagues from his former chambers wished him well, mindful that "one of the first of living mathematicians, whose rare capacity and attainments have now happily been transferred to a more congenial employment."[98]

PART 4

The High Plateau, 1863–1882

*... it shall be the duty of the Professor to explain and teach the
principles of Pure Mathematics and to apply himself to the
advancement of that Science.*
—Terms of the Sadleirian chair

Cayley was elected to the Sadleirian chair of mathematics at
Cambridge in the summer of 1863. He married Susan
Moline, and with a more settled existence in prospect, immersed
himself in some of his most extensive studies. He embarked on his
great geometrical period, ceaselessly writing hundred-page memoirs on difficult topics in the theory of curves and surfaces. At
Cambridge, he became involved in controversy with the Astronomer
Royal when he was summoned to defend pure mathematics as a
serious object of study for undergraduates.

In 1868, German mathematician Paul Gordan proved a theorem
that became a watershed in invariant theory and which confirmed
that one of Cayley's earlier conclusions was erroneous. Cayley frankly
acknowledged the power of Gordan's method and, uniting his own
work with that of the German school, brought his long series of
memoirs on quantics to a close in 1878. His book *Elliptic Functions*,
brought out in 1876, enjoyed success and has been influential.

Cayley's mathematical range, always wide, increased in his
"mature" years. He touched on many subjects, including curves
and surfaces, invariant theory, group theory, link-work, graph

theory, decision theory, and he supplied some initial thoughts on "fractals." Chemistry and astronomy were two areas where he applied his mathematics. During the 1870s, he served as president of several important British scientific societies. Beyond question, Cayley was *nonpareil* the pure mathematician of the Victorian era.

A keen supporter of the movement toward the higher education of women, he quietly but effectively acted behind the scenes to bring this about; his sympathy for the cause combined with a practical ability where his skills in administration acquired during his days in the law were brought into play. At James Joseph Sylvester's instigation, he visited the United States at Johns Hopkins University in 1882, where he lectured on the theory of Abelian functions and graph theory.

CHAPTER 11

The Mathematician Laureate

A rthur Cayley was elected as a professor at Cambridge at the age of forty-one on 10 June 1863.[1] At first sight, the change of career might have seemed unwise: the annual stipend of around £500 for the Sadleirian chair was subject even to the vagaries of agriculture on which its financial provision depended. But, in being elected to a Cambridge chair, he enjoyed a high social standing, this being determined by birth, education, and professional status, as well as salary.[2] Professors at Cambridge were comfortably off and could afford large houses and servants. The social class of those whose incomes fell in the range £200–£1,000 included well-to-do clergymen and the lesser gentry, while the upper middle class with incomes in excess of £1,000 included the higher clergy and successful merchants. It would be unrealistic to expect Cayley to accumulate as much money as he might have done in his former occupation, for the financial reward for a professor did not bear comparison with the sum a topflight barrister might command.[3]

Cayley was satisfied, and for its side of the bargain, the university gained an eminent mathematician. "Perhaps no professorship was ever founded more opportunely" was the retrospective judgment from a close colleague.[4] The personal advantage lay in a newfound leisure to pursue his own academic interests and the relative security which the university position offered.

The Lawyer Turned Don

The most important consequence of the appointment was that Cayley could settle down. With his longstanding friend Edmund Venables officiating, he married Susan Moline at the newly built parish church of All Saints, Blackheath, on 8 September. As was the custom in Victorian society, middle-class men waited until they had the prospect of a secure income before matrimony was considered. It was only then they could establish themselves by acquiring their large house and complement of servants.

Susan Moline was the youngest daughter of Robert and his Somerset-born wife Mary Pritchard. Robert Moline was a banker and woolstapler whose early career began in the wool town of Godalming near Guildford in Surrey. By the early 1850s, the family had moved to Nelson Street, Greenwich, when he became manager of a branch of the London and County Bank. They lived "above the shop" in a commercially busy Georgian Row next to the river frontage, but by the 1860s they had progressed to Westcombe Park in Blackheath. Of their six boys and six girls, the two youngest were Susan and Lewis, the baby of the family.[5]

Like the reality of many Victorian women who dutifully lived out their lives in the shadow of their husband's career, biographers often have little option but to leave them there. This is the case with Susan. Venables described her as "a help meet" for her husband but offers nothing more. She ran a home with domestic servants, brought up a family, and fretted about her husband over-working with mathematics. Later, when he was involved in the cause of higher education for women, she gave support to the young students making their way in an all-male world and was on the committee that set up a women's residence in Cambridge. She was popular in Cambridge and assumed her place as a professor's wife in the academic society. When he took the opportunity of spending six months in America in the early 1880s, she accompanied her husband and the Baltimore ladies "were quite taken with her, 'love at first sight,'" according to Sylvester. A. R. Forsyth, the author of Cayley's principal obituary draws a veil over their home life, for he was a Victorian, too, and in his terms, life within the family home was a private affair. He did speak of "its singular happiness, based upon the affection felt by its members for one another."[6]

For Cayley, the move to Cambridge was almost a continuation of a former life. He knew the town and the ways of a university, which had hardly changed since the 1840s when he was there as a student. As professor, the formal demands made by the university were minimal. More would be expected than of his eighteenth-century counterpart, who might go to work "tracing a fluxion, peeping through a Telescope or sauntering in the Garden," but residues of this leisurely existence still existed. In the early nineteenth century, many professors had adopted a leisurely approach to their university chairs; why should they do otherwise? They were under no obligation to teach, influence education, or even live in the vicinity of Cambridge. Charles Babbage, who was elected to the Lucasian chair in 1828 famously never gave a lecture there. Attitudes in the 1860s were beginning to shift, albeit at a snail's pace. Cayley was a "new professor," to use Sheldon Rothblatt's term, men who could direct serious research and distance themselves from the amateurish image that had been a feature of British science for too long. The idea of university chairs as a reward for services rendered was on the way out; instead, they would be filled by those with "promise" of future feats of scholarship that would bring glory to the university. The perceived success of the German scheme of higher education encouraged reform and some encouragement was given to "Germanize" the University. The dons of the 1860s who had previously concentrated on teaching duties alone could not now count on a professorship.[7]

That Cayley was a different species of academic is plain when one compares him with contemporaries. Of the leading competitors for the professorship, Isaac Todhunter's biographer noted: "When Arthur Cayley was appointed to the newly-created Sadlerian chair, Todhunter did not barely acquiesce in his own defeat (for he thought it right to be a candidate); he exulted in it, and would have resented any other decision."[8] E. J. Routh was unsuccessful in his application for the foundation chair of mechanism and applied mathematics at Cambridge in 1875. An erudite man, he became the most successful coach in the history of the Tripos, and in his long career, he produced twenty-eight Senior Wranglers and wrote widely respected texts in dynamics and statics. Percival Frost was a highly regarded teacher of mathematics who graduated as the top Tripos student in 1839. He wrote books on geometry, including a popular book on elementary curve tracing and late in life was elected to the Royal Society. N. M. Ferrers continued his long stint as editor of the *Quarterly Journal of Pure and Applied Mathematics* and became a well-known Cambridge figure. He was master of his college and later elected to a term as vice-chancellor of the university.

Much was expected of the new professor, the returning son who brought his prestige back to his alma mater. William Whewell, a discriminating man with a high regard for mathematical talent, rated him highly. Newton apart, the master of the college bracketed him with George Biddell Airy as the two finest mathematicians Trinity had ever produced. Shortly after the election, he wrote to his old friend Augustus De Morgan: "I am glad that we have a chance of seeing Cayley among us. He will help us to some of the last [i.e., latest] mathematical novelties."[9] As professor, Cayley would be required to both carry out research and teach. The Mathematical Board set up by Whewell in the 1840s endeavored "to encourage attendance at the lectures of the Mathematical Professors, and to secure a correspondence between those lectures and the mathematical examinations of the University" and the scheme for the professorship was aligned to this ideal.[10] The first duty asked of the Sadleirian professor was "to explain and teach the principles of Pure Mathematics." In his new position, Cayley was enthusiastic, and he busied himself with the preparation of his first lecture course, which he chose to be on his favorite subject, analytical geometry, a subject he would choose for the first three years of his tenure.

The Inaugural Lecture

Cayley rarely spoke about elementary mathematics but he did so in his first lecture at Cambridge on 3 November 1863. Before him lay the prospect of "explaining," a task which posed a challenge. He was adept with research pursued in the solitary confines of his barrister's chambers but he had little experience in conducting lectures.[11] The opening one on analytical geometry is an overview of his first Cambridge course. His first lecture began with the preamble:

In reference to the subject of my present course, it gave me some pleasure to remark that the (as appears by the will of Dame Mary Sadleir dated 25th September 1701) the Sadlerian lectures were instituted "for the full and clear explication and teaching that part of mathematical knowledge commonly called Algebrare or the method and rule of contemplating quantities generall with particular application and use of it in Arithmetic and Geometry either according to the method of D[es]Cartes or any of those who have best improved it since." The Professorship to which I have been appointed replaces to some extent these lectures and the duties of the Professorship regulated not by Lady Sadleir's will [but] by the Statutes for the Mathematical Professorships sanctioned by an Order in Council of the 7th March 1860—it is thereby provided that it shall be the duty of the Professor to explain and teach the principles of Pure Mathematics and to apply himself to the advancement of that Science.

The lecture was pitched at an introductory level, but Cayley used the occasion to discuss the tenets of analytical geometry and his interpretation of them. At the commencement he indicated his preference for analytical geometry as opposed to synthetic, or pure, geometry, which made no appeal to algebra:

I have selected Analytical geometry—the creation of Descartes—for the subject of my course—and I propose to consider it, indeed also theoretically, but in a great measure historically, for the present attending chiefly to the period beginning with the *Géométrie* of Descartes, and in order to fix a termination ending with Cramer's *Traité des lignes Courbes,* the included period being 1638 to 1750. I have said *Analytical* Geometry—not that I intend to exclude from consideration any of the methods and theories which belong to Modern Geometry—but partly because I do intend to dwell chiefly on the analytical view of the subject—partly because I am not able to satisfy myself that the Modern Geometry does not rest essentially on an analytical basis.

René Descartes' *La géométrie* and Gabriel Cramer's *Introduction à l'analyse des lignes courbes algébriques* were high points in the growth of analytical geometry. Modern geometry was a term used in distinction to the geometry of Euclid and that of other Greek geometers of the ancient classical period. Of the two faces of modern geometry, analytical geometry and synthetic geometry, Cayley did not merely prefer analytical geometry, he championed it. In articles on geometry that required representation of both sides, he wrote almost exclusively on this side of the subject.

As Cayley outlined the leading concepts of geometry, it is noticeable that he was impatient with prolonged discussion. Though he reflected on the philosophical underpinnings of his subject (as will be shown by his attention to it in his 1883 Address to the British Association), he also indicated a reluctance to be burdened by them. The philosophy of mathematics was perhaps the only subject connected with mathematics that did not claim his concentrated attention.[12] With debate on the foundation of "number" still current and argument over such controversial ideas as the use of ideal elements in geometry and the principle of continuity, Cayley adopted the stratagem of the working mathematician and concentrated on the technicalities of the subject.

For Cayley, analytical geometry was concerned with figures in a mathematically n- dimensional space in which coordinates assume complex number values. For this reason, complex numbers themselves were placed at the center of his geometrical thinking. When he referred a curve to real projective space, he would make such observations as "three points of inflection are real," though these were secondary remarks:

> The imaginary space of Modern Geometry exists for the purposes of the Science—but does not exist, as an "ουτως οντα" [really real]—I use the Greek expression merely because the word real as above remarked is appropriated as the contradictory of imaginary. In support of the opinion that it does so exist it may be argued, according to the theory we adopt of ordinary space, that on the one hand that if space is only a form of the perceiving mind, then, since in the development by geometrical science of the notion of space imaginary space has presented itself to the mind, it has as good a right as ordinary space to be considered as a form of the perceiving mind—on the other hand if ordinary space be an existence exterior to the mind, then there [is] no reason why imaginary space should also exist exterior to the mind, incognisable to the senses, and only arrived at as a necessary substratum for the truths of geometry. But the question is [one] which hardly admits of any profitable discussion.

The echo of his "ουτως οντα" or "this is how it is" made to George Boole in 1847, indicates his unaltered views on the distinction he made between physical space and mathematical space. For Cayley, the "really real" space or physical space was "ordinary" three-dimensional Euclidean space, while in a separate category, mathematical space was typically the n-dimensional space of modern geometry with its ideal elements, such as "imaginary points," and "points and lines at infinity." Cayley believed that mathematical space had an equal claim to be considered as existing though it was not necessarily "really real." To go much further than this would take him down the track of philosophical debate and to such questions as whether imaginary points were real or the actual existence of points at infinity, and so on. He regarded endless speculation on these things as a dismal prospect.

In his lecture, Cayley discussed the importance of projective geometry, the subject which provided the focus of his geometrical studies:

> Among the theories of Modern Geometry may be mentioned one which presents itself under various forms; the theory of related figures. The most simple case but a

very important one, is the relation of Projection; . . . But what I wish particularly to remark as to the principle of Projection is that it may be used in two opposite ways;

(1°.) to simplify a figure, so as for instance, to obtain the circle instead of the ellipse— or parallel lines instead of lines meeting in a point—or what includes this, to project some line of the original figure into the line infinity; we have then in the simplified figure properties which are frequently self evident or nearly so—and we thence arrive at the existence of corresponding properties in the more complicated original figure. This is in fact the way in which the principle has been most used—in particular in Poncelet's Classic work the "*Traité des Propriété's projectives*" (1817) [*sic*]; and I may in passing refer to the very beautiful theory of the in-and circumscribed polygon. But the principle of projections may be used in the opposite way—

(2°.) we really understand the simplified figure better—and we escape from a very great number of special forms, which would otherwise need to be examined—by considering the simplified figure merely as a case of the original more general one. Thus for instance the notion of a conic touched by a line is an easier one than that of the parabola considered as a conic touched by the line infinity.

He explained to his audience the central idea of his "Sixth Memoir on Quantics": that ordinary Euclidean geometry could be considered a special case of projective [descriptive] geometry:

It would at first sight appear that a broad separation exists between metrical and descriptive geometry; whereby metrical geometry is to be understood all that relates to magnitude—including therein equality and perpendicularity: by descriptive geometry all that is wholly independent of magnitude. As an instance of a purely descriptive theorem (take Pascal's theorem for two lines). But in fact this is not so; it has gradually been becoming evident that descriptive geometry includes metrical geometry, as for instance, the circle instead of being considered as the curve all the points whereof are equidistant from a centre, is considered as a conic passing through two fixed points, the circular points at infinity; lines at right angles to each other—are in fact lines which are harmonics in regard to the lines drawn through the two circular points at infinity etc. And the ultimate conclusion appears to be that descriptive geometry is in fact all geometry, the difference being that in descriptive geometry the figure is considered apart by itself, in metrical geometry in connexion with a certain conic wherein the notions of linear and angular distance have respectively their origin. But so far as the distinction is recognised at all I shall be chiefly concerned with descriptive geometry.

To underline his claim as an "analytic geometer," his final words were concerned with the power of the analytic method:

With regard to the Analytical Method generally, it is to be noticed, that given a geometrical theorem of any kind, or say a supposed theorem only, you may always—potentially at least—by a direct application of the method of coordinates demonstrate the truth or falsehood of such [a] theorem: this is of course a perfection in the method, but it is the worst way of using it, so to apply it to the demonstration of an isolated theorem, considered apart from the geometrical theory to which such theorem belongs and the various other theorems with which it is connected. And the analysis required for the demonstration of an isolated theorem would in many cases be far more difficult and complicated than would be necessary in order to [give] the demonstration of some more general theorem of which the other is a mere Corollary.

Finally, he referred to the seemingly unconnected consequences that can be derived from quite general theorems. Here we glimpse what he found so interesting in mathematics: the spectacular connections which may exist between seemingly unconnected statements. An instance is the theorem on cubic curves, a theorem now known as Cramer's paradox.

In ordinary Cartesian coordinates, a cubic curve given by the equation

$$ax^3 + bx^2y + cxy^2 + dy^3 + ex^2 + fxy + gy^2 + hx + jy + k = 0$$

has nine independent coefficients (after factorization) and was thus thought uniquely determined by nine points. That this is not the case was so paradoxical that it made a very lasting impression on Cayley in 1843 as he was just embarking on his mathematical career. He explained this problem in his lecture, and two of the possible results which flowed from its resolution:

"Every cubic curve whatever which passes through eight of the nine points of intersection of two cubic curves passes also through the ninth point of intersection."

Now this includes in itself the two very different theorems.

1. *Pascal's Theorem.* If a hexagon be inscribed in a conic the three intersections of pairs of opposite sides lie in a line.

2. *Maclaurin's Theorem.* The line joining two inflexions of a cubic curve, again meets it in a third inflexion.

This is perhaps somewhat of a puzzle wherewith to conclude the present lecture, but to some of you it would be really a good exercise to think out for yourselves the mode in which the last-mentioned two theorems are really included in the theorem relating to the nine points of intersection of the two cubic curves.

And so the lecture ended. He had answered his own final question in 1843 when beginning his mathematical career and it was one he had revisited prior to his return to Cambridge as professor.[13] Now he put it forward for another generation to solve.

One wonders what impression Cayley's lecture made on his audience.[14] He was not at his best when lecturing. Uncomfortable, prosaic, stiff, and lecturing over the heads of his audience, his nervousness accentuated by those half-furtive glances at his pocket watch. But Todhunter's aside that he couldn't explain anything is too severe and probably inaccurate. To be sure, in front of an audience Cayley gave mathematical results matter-of-factly, without relief of light remarks and probably little humor, the hour not helped by his belief that a lecturing performance did not ultimately count for much. The text of his lecture tells a different story and richly outlines the concerns of an analytical geometer of the 1860s.

Evidently, he encouraged the 18-year-old Clifford, the newly arrived Foundation scholar of Trinity College. His burgeoning talent was publicly visible. Within three weeks of the inaugural lecture, the youngster had written a paper on analogues of Pascal's theorem in geometry, and he turned challenger by setting questions for the newly established *Mathematical Questions with their Solutions from the Educational Times*. In a reversal of roles, two were solved by the professor. Both questions drew on the material in Cayley's course of lectures. The first question was a problem in pure geometry (on conics) and solved by an application of the theory of pole and polar. The second was a problem on the projective cross-ratio, which Cayley treated as a problem in analytical geometry. Clifford built on this meeting of interests and, during the following August, wrote "Analytical Metrics," a topic inspired by Cayley's "Sixth Memoir on Quantics."[15] For young men who elected to attend Cayley's lecture course, the concepts of modern geometry would have appeared very different to their previous diet of Euclid. They would have seen that geometry practiced by a working mathematician did not consist of learning Euclid's propositions by heart and "proofs" in the rigidly enforced prescribed order of the Greek text.

"Paying Work"

At Cambridge, Cayley was required to give one course of lectures for one term per year until 1886 (and thereafter in two terms each year). Thus, he had twenty-three years ahead in which the year from January until October required no teaching duties at all. The lectures themselves could be on a topic of his own choosing. Even when the titles were announced, he did not feel bound by them. The content of lectures more accurately corresponded to his current research and he would often take the opportunity to give an account of one of his latest discoveries before embarking on the advertised topic. With John Couch Adams, the professor of astronomy, the fare was more straightforward, and his audience knew what to expect. "Father Neptune," as he known to generations of students, invariably lectured on lunar theory, each year polishing his notes and perfecting his specialist knowledge.

Unhappily, both mathematical professors were marginalized by the Cambridge system, in which examinations dominated. This was not a new phenomenon, for Cayley had witnessed it himself as a student, but despite Whewell's best efforts, the situation had not improved. So it continued with both Cayley and Adams lecturing to small classes. Their lectures would have little relevance to students whose only object was to score well in the mathematical Tripos examination. Cayley was concerned with mathematics per se, but the authorities wanted competence in a narrowly focused and circumscribed syllabus in which any deviation was considered a waste of time and effort by students. Who could blame them? With examinations at every turn, few would want a professor's enthusiasms on subjects that lay off the examination track, and those who ventured off it were undoubtedly penalized for diluting their Tripos preparation. A novel way to bolster attendance was for professors to lecture to each other, and Cayley increased his audience when he invited Adams to join him for his lectures on the geometrical problem of determining a planet's orbit: "I include in my course a few lectures on the orbit problem; if it would interest you to hear them, I should be very glad to have you at the lectures." The title of these lectures was officially "Dynamics," and as it happened on this occasion, the lectures were a preamble to some of his extensive researches.[16]

James Wilson, who attended Cambridge a few years earlier, noted that one of the functions of the private coach was to deflect students away from professors' lectures. They would no doubt be interesting, but they did not "pay" in the Tripos. Wilson was fascinated by trilinear coordinates and the topical "abridged notation" for use in geometry, but his coach advised that a better use of time would be spent on Tripos practice. This nonattendance was a missed opportunity for many students. Compared with the depth and modernity Cayley could offer, the requirements for a Tripos student were really quite basic. High on Lord Rayleigh's student booklist of the 1860s, for example, were fairly standard and locally produced texts: Todhunter's textbooks on *Differential Calculus, Integral Calculus, Algebra, Conic Sections, Trigonometry,* and *Geometrical Conics.* Also indicated were N. M. Ferrers's *Trilinear Coordinates,* W. H. Drew's *Analytical Statics, Dynamics of a Particle,* Sir J. F. W. Herschel's *Outlines of Astronomy,* George Salmon's *Conic Sections,* and Routh's *Rigid Dynamics.*[17]

Attendance at the lectures of other Cambridge professors was kept artificially buoyant by the university regulation applied to poll degree students, the ones not enrolled for an honors degree. They were required to attend at least one term's lectures from a professor selected from a "starred" list—a list that did *not* include the mathematical professors. During the 1860s, this compulsion did much to swell attendance in the courses of the nonmathematical professoriate. Men like Henry Fawcett, the professor of political economy, had lecture rooms brimful with students whose main concern was to have his signature on their certificates of attendance. By contrast, for honors degree students of mathematics in 1863, the informal position was that the professor's lectures were "of service" but not required. In practice, the recommended study regime for success in the Tripos was to attend college lectures with the necessary element of supervision from private tutors on alternate days. The private tutors drilled their students in "Tripos technique"; and the effective tutors knew the correct balance to place on each well-defined topic. In their pupil-rooms, students received a strictly administered diet of learning, practicing, and endless testing so that by the end of this meticulous preparation they could produce examination answers on demand.[18]

The "paying work" for the students who paraded in Cambridge in their tall hats and frock coats was the work which counted in the examination. This was what they wanted and what their parents expected them to receive. Mathematics was the mark of a Cambridge education and, for the "reading men," success in the Tripos would admirably suit a later professional career. The belief in the utility of some mathematical training—but not too much—was held in high places. While Lord Palmerston, cabinet minister and prime minister, preferred patronage to examinations as a basis for sound appointments, he did acknowledge the value of mathematics and especially Euclid's *Elements,* which he thought excellent training for a diplomat.[19] William Everett, an American student who spent three years at Cambridge and graduated in 1863, noted the characteristics of its mathematical education:

> ... more than any other pursuit does mathematics require faith, implicit faith, and English mathematics most of all. Englishmen hate going back to first principles, and mathematics allows them to accept a few axiomatic statements laid down by their two gods, Euclid and Newton, and then go on and on, very seldom reverting to them. This system of mathematics developed in England, is exceedingly different from that either of the Germans or the French, and though at different times it has borrowed much from both these countries, it has redistilled it through its own alembic, till it is all English of the English. This was the study in which for two hundred years, all, and now more than half, of the Cambridge candidates for honors exercise themselves.[20]

The system of teaching mathematics at Cambridge was so deeply imbedded that it proved relatively immune to the occasional sideswipes from critics. It was one of De Morgan's favorite targets and he did not spare his tongue: "The Cambridge examination is nothing but a hard trial of what we must call problems—since they call them so—of the Senior Wrangler that is to be of this present January [1865], and the Senior Wranglers of some three or four years ago. The whole object seems to be to produce problems, or, as I should prefer to call them, hard ten-minute conundrums."[21] A more stinging attack came from Leslie Stephen—not withstanding his disenchantment with mathematics as a suitable instrument for education. De Morgan had been a student at Cambridge in the 1820s but Stephen had more recent experience. He was twentieth wrangler in 1854 and had been a mathematics tutor at Cambridge for seven years. In 1865 he published the popularly received *Sketches from Cambridge by a Don* by way of a parting shot when he resigned his fellowship in 1862. Stephen purloined the well-known image of the Tripos as a horse race when he described the Senior Wrangler as the winner of the Derby and the stables in which the horses are trained are analogous to the pupil-rooms of the different coaches.

As the newly appointed professor, Cayley ought to have been the most influential voice in any reform but no doubt weighing up how time should be spent, he evidently chose research as his mainstay. He may have taken the view that the system was so ingrained that any attempt to mount a reform campaign would have been futile. Moreover, he was temperamentally unsuited to such a task. He had a talent for administration, and on this count his advice was sought on various university committees, but he was no university politician of the likes of Whewell. Cayley's inclination was to stand aside from the daily hurly-burly of university politics. In matters of syllabus reform, his argument was invariably subject-led with the inclusion or rejection of topics dependent on their importance in the world of mathematics. Using this sole criterion as a guide, he believed the best educational benefit could be derived.

Living in Cambridge was no barrier to an active scientific life in the capital. In April 1866, the Great Northern Railway introduced an express service between King's Cross, London, and Cambridge, which reduced the travel time to one and one-half hours. Cayley and J. W. L. Glaisher regularly took the 4:30 P.M. Great Northern to London to attend scientific meetings. While Cayley was a council member of the Royal Astronomical Society and the London Mathematical Society, it was surmised that the timing of the latter's meetings was changed to suit Cayley's convenience. Indeed, in the 1860s, Cayley found his scientific life in London rather than Cambridge, for during this decade, there was little inducement to research at Cambridge. The teaching dons failed to engage in the quest for new mathematics, and it was frequently argued that it was none of their business for it would deflect them from their teaching function.[22]

With freedom from the daily demands of the law, Cayley could choose research subjects which might result in extensive memoirs. Many were extensions of topics he had already sampled, and he immersed himself in algebra, geometry, lunar theory, elliptic functions, and combinatorial problems. He coupled this extraordinary mathematical production with an increasing statesmanlike role in the scientific life of the country. His position as professor at Cambridge obliged him to become a public figure but he did so with a degree of reluctance, preferring life away from the spotlight with private time to carry out mathematical investigation.

After a year in Cambridge, he moved to Garden House, an imposing residence within a short walk of the Cambridge colleges.[23] In former times, the town miller lived in this quiet backwater, but now it was a "professor's house," set in three acres and built of old Cambridge yellow brick, with a river frontage to the Granta, a tributary of the Cam. The sedate progress of horse-drawn barges hauling grain from the fens was a familiar sight, and beyond the water meadows, the village of Newnham could be seen from the drawing room. In the house, Cayley arranged his library and his workplace, and from this stability he settled down to his substantial researches.

He showed no regret in forsaking a legal career and he effortlessly entered the tranquil existence at Cambridge. Hirst remembered a meeting with him at Garden House: "[a]fter smoking my pipe in the Conservatory by Mrs Cayley's kind permission—mathematical talk with Cayley! I was much pleased with the simple life they lead and with their kindness towards me."[24] The bachelor dinner parties in London would now be a thing of the past, and he would see less of Thomas Hirst, William Spottiswoode, and mathematicians passing through the metropolis. Most of all, he would see less of J. J. Sylvester. At the age of forty-two, he settled to this new serene existence, the one set out in the Greek agenda for a virtuous life outlined by Socrates in the *Theaetetus* of Plato, a work he read and reread, alongside other Greek classics. The words of Socrates to Theodorus on the use of leisure fitted Cayley with the snugness of a kid-glove and throws into relief his new life as against his former existence in Lincoln's Inn. "In the leisure of which you were speaking, and which a freeman can always command," Socrates reminded Theodorus, "he has his talk out in peace, and, like ourselves, wanders at will from one subject to another, and from a second to a third, if his fancy prefers a new one, caring not whether his words are many or few; his only aim is to attain the truth. But the lawyer is always in a hurry."[25]

The years fell into a routine pattern and the "commendable custom of five or six months vacation in the summer," which he had congratulated Boole on achieving at Cork, seemed modest by comparison. The leisure of a Cambridge professor to work at his research can be contrasted with his mathematical neighbor, the hard-driven Cambridge coach. Throughout the entire year, including the Long Vacation, these "merchants of drill" and Tripos success worked themselves into the ground. It was not uncommon for Edward Routh to work a fourteen-hour day without much respite. For his stipend, Cayley performed his light lecturing duties, and the expanse of the nineteenth-century "Long" could be given over to visits to favorite stomping-grounds in Wales, the English Lake District, or a Continental journey. Gentlemen of science had a different agenda to their coaching confrères, and their "Long" was a time for outdoors. Cayley's friend John Tyndall, who had produced *Glaciers of the Alps* (1860) and successfully brought together a climbing and a scientific reputation, reserved the summer for the Swiss mountains. On one occasion, he invited Cayley to join him on a mountaineering expedition: "Give Mrs Cayley my kindest regards and tell her that I am not only going to the Alps this year, but I am perfectly prepared to take charge of you and return you to her safe and sound, if you come with me."[26] But after arriving at Cambridge, the married man gave up membership of the Alpine Club and former Swiss adventures gave way to domestic holidays. A favorite retreat was Keswick, where lay the prospect of a Skiddaw or Biencarthan.

Occasionally Cayley might remember Thomas Kirkman, the hard-up cleric with his unappreciative flock in Croft in Lancashire, just south of the Lake District. Recognizing Kirkman's plight, Augustus De Morgan vainly intervened on his behalf, in a letter sent to William Whewell, hoping for some old-fashioned patronage from the master of Trinity: "Now there is a man of science very poorly provided for, and the mere mention of him can do no harm, and sometimes things come in the way and recommendations might be given if it were known who was in the way. I allude to Mr Kirkman, incumbent of Croft near Warrington. His mathematical papers on polyhedrons, and other things, are very deep and quite at the head (*altus* deep, high) of their department. Cayley can tell you all about him." Putting the facts before Whewell, De Morgan continued: "He is buried at Croft and very much desires a better field of action, in which he may be able to see a little more of intellectual life, over and above his clerical doings. He has barely £180 a year, and is an active man, moderate, and of orthodox repute. His Bishop he says is warmly interested in him, but has no patronage, and would recommend him. If on inquiry you should find him in all other respects worthy, his science certainly deserves a lift. He has worked for many years at subjects, which will not bring him before the general eye, and is a staunch enthusiast and *con amore* mathematician."[27] Whewell confused De Morgan's letter as a plea for College patronage, instead of the few chosen words in the right ecclesiastical ear, which De Morgan intended. The master of Trinity replied in courteous but officious tones: "I should be very glad to help a good mathematician in any way, and especially to do so on your recommendation," but, "I am afraid I am unlikely to have any power to do so in the instance which you mention. Our College preferment is given according to rules over which I have no control, and none but very small livings are ever given to persons who are not 'on the [Trinity] foundation.'"[28]

Unhappily, Kirkman's life did not change, desires of a better income or dreams of international fame were left unfulfilled, and he retired hurt to his country parish. In the years to follow, Cayley would have ample occasions to remember the Rev. Thomas Penyngton Kirkman from the parish of Croft-with-Southworth. Kirkman was not a man to let matters lie and old controversies were stirred in the pages of the *Mathematical Questions with their Solutions from the Educational Times*. Ten years later he was still bitter about the fate of his paper on polyhedra: "They consigned it to the Archives; and they served me right. If a country clergyman, down in the crowd of the Church's 'passing rich,' chooses to read lectures to Imperial Institutes, he must even take what comes from Royal Societies. Later, however, they made me, for an Englishman, and a Divine afflicted with science, quite proud and happy, by printing in the *Philosophical Transactions* the two first of my twenty-one sections."[29]

Around Cambridge, Cayley took an interest in the work of young mathematicians. Clifford is perhaps the most outstanding example. Another was James Stuart, who graduated third wrangler in 1866 and stayed on to become a colleague. Known to his students as "Potential Jimmy" because he taught electricity using the potential function in contrast to the way it was taught by Routh, Stuart was a man with radical ideas. He was active in the University Extension movement, in which university level lectures were given outside the university, and enjoyed considerable success lecturing to large numbers at various venues in the north of England. A successful academic career beckoned, and he became the first Professor of Mechanism and Applied Mathematics at Cambridge, a position he occupied for fourteen years from 1875. But Stuart was too revolutionary for Cambridge ways and his reforming zeal too sweeping for a permanent life in the groves of academe; he eventually left it and became a member of the House of Commons as a Liberal.

While Cayley lived within the collegial world of the university, his friend J. J. Sylvester continued to "fight the world," so much that it was a wonder he completed any piece of research work. But he did, and notably at this time he produced an extensive paper on the quintic equation. Perhaps he was celebrating the paper's completion when he aimed another salvo at the Government's War Department. Stopping one day at the offices of the publisher of his paper, he wrote "I am engaged in a struggle *pro arâ et focis* [for hearth and home], with my implacable foes at the Horse Guards, the so called Council of Military Education."[30] Existing on the fringes of the curriculum, Cayley did not experience the rough-and-tumble of Sylvester's teaching life at Woolwich. Teaching advanced mathematics to a handful of interested students was a far cry from instructing large numbers in the basics of the subject. "Gentlemen cadets" did not always measure up to their status when it came to learning "higher mathematics" taught by Sylvester in his academy on the south side of the Thames. Young men who had an eye to military action hardly thought much of "swot," as they called their enforced study of mathematics. Reluctant as they were, they could not escape it, as mathematics was reckoned only second to artillery in the Woolwich scale of academic worth.

Unlike Cayley with his occasional Clifford or Stuart to inspire, Sylvester had to deal with men pressed into mathematics. When there was frequent pandemonium in his classroom, he could not always rely on the Duty Corporal to return the cadets to order. They invented new tricks for his benefit, some with "success," like filling his ink-bottle with chalk to clog his pens, and generally to make his life a misery: "One plan which was occasionally tried [by the cadets], was for a large number of them to drop down behind their desks. Sylvester would suddenly awake from the solution of some abstruse problem and see the class-room half empty. This made him rush up and down, a movement which was prepared for by sprinkling the floor round his table with wax matches, which went off in succession as he stamped round, driving him quite wild." Like many a recalcitrant, the erstwhile cadet who told these stories was brought to appreciate his teacher years later. Sylvester was then "a splendid mathematician" and "with all his little ways, could teach well if he was allowed his own method, and personally I owe a good deal to him."[31] For his part, Cayley influenced only one soldier, P. A. MacMahon, an outstanding mathematician who benefited from his patronage on return to civilian life.

"Nonpaying Work"

Cayley's work on invariant theory had come to a standstill. After the publication of the "Seventh Memoir" in 1861 and his researches on the solution of the quintic polynomial equation in the same year, he turned his attention to geometrical questions. That Cayley and Sylvester worked as individuals can be seen by Sylvester's reawakening to mathematics. After a ten-year break, Sylvester's intensity toward invariant theory was briefly renewed in 1864 when he published his paper on the quintic equation. An extensive paper of over a hundred pages was presented under the lengthy title: "Algebraical Researches, containing a disquisition of Newton's Rule for the discovery of imaginary roots, and an allied rule applicable to a particular class of equations, together with a complete invariantive determination of the character of the roots of the general equation of the fifth order, &c." In it, Sylvester gave a criterion, based on invariants, for the reality of its roots by applying a method originated by a mentor of longstanding, J. C. F. Sturm.[32]

Cayley praised Sylvester's paper, a work which even merited a short piece in the *Times*. "The rule is a very remarkable one," he wrote. "I think that the paper is of the greatest interest and value, and in a high degree worthy of publication in the Transactions of the [Royal] Society."[33] As a matter of course, Cayley turned out memoir after memoir for the *Philosophical Transactions* but Sylvester published only four papers in this journal. Such productions needed planning and steady work, a medium suited to Cayley's probing of ideas and calm prosecution, while Sylvester claimed instant discovery and relied on immediate transmission to the printed page. After a decade of inactivity in invariant theory, Sylvester was back on the track again and, during these few years, strived to place invariant theory on surer foundations. He was to write to Cayley: "The time may come when it will be needful to turn a deaf ear to the voice of the syren [*sic*] with her endless chant of algebraical developments but the present strain is not of so unreasonable length as to call for the practice of such austerity."[34] But "austerity" did follow and this paper of 1864 was an island in Sylvester's work in invariant theory—between his feverish activity at the beginning of the 1850s and the calculatory phase which ensued after his arrival in America in 1876.

Unlike many British mathematicians before him and many of his contemporaries, Cayley hardly wavered from this dedication to the whole of mathematics; within his encyclopedic work, there were few omissions.[35] But his concentration on pure mathematics did not go without criticism. In the year following his appointment as professor, William Thomson voiced his frustration to Hermann von Helmholtz that such a talent should not apply itself to the theory of electricity: "Oh! that the CAYLEYS would devote what skill they have to such things [Kirchoff's work on electrical conducting plates] instead of to pieces of algebra which possibly interest four people in the world, certainly not more, and possibly also only the one person who works. It is really too bad that they don't take their part in the advancement of the world, and leave the labour of mathematical solutions for people who would spend their time so much more usefully in experimenting."[36] Gustav Robert Kirchoff's theory of electricity, when applied to conducting plates of arbitrary shape, certainly required attention by competent mathematicians, and though Cayley disappointed them on this topic, he did not stand aloof from practical needs generally.

Through his regular contact with John Couch Adams, George Gabriel Stokes, James Clerk Maxwell, and Francis Galton, Cayley was able to help with mathematical problems, and it might be remembered that he corresponded with Thomson himself when the physicist needed assistance with invariant theory.[37] When Isaac Todhunter was studying elasticity with a view to compiling its history, he appealed to Cayley for help in dealing with elliptic functions and their foundations.[38] He helped Lord Rayleigh and his knowledge of classical analysis resulted in a very accurate calculation with Bessel functions.[39] Electricity and electromagnetism was Maxwell's domain and Cayley does not appear to have dwelt on it. He did set out with a paper on the distribution of electricity on a spherical surface, but was seduced by the pure mathematics of it and the paper ultimately produced several results on Bernoulli numbers.[40]

With increasing specialization, mathematicians were being split off from the main body of science, even from their natural allies in mathematical physics. Walter Cannon made the point "that mathematicians themselves began around mid-century that process of segregating themselves and repudiating demands that they be useful [to science] . . . The third quarter of the century saw heroes of the independence of mathematics, such as Sylvester and Cayley arise."[41] Adams, Stokes, Thomson, and Maxwell were all first-class mathematicians who received their training in mathematics at Cambridge but they had emigrated from the mathematical territory. Cayley remained to explore it and was unable to give more than perfunctory consideration to the physical meaning of practical problems. He was broadly based, and although he was undeniably a pure mathematician, he was not a specialist in the modern sense, nor did he cut himself off from scientists as might be inferred from Cannon's statement. He tackled practically based problems, but it was always the mathematics which held his interest, not the application.

Geometry dominated Cayley's early life as a professor. In his middle years, from the time he became professor at the age of forty-one until 1882 when he embarked on a visit to Baltimore at the age of sixty, he enjoyed his great geometrical period and wrote some of his lengthiest memoirs.[42] In February of 1864, appropriately for a newly appointed Cambridge professor of mathematics, he submitted a commentary on Newton's work (1704) on planar cubic curves. Newton's classification, in which planar cubic curves were organized according to the nature of their infinite branches had recently been discussed in a new edition of the work.[43] Newton divided these curves into seven classes, one of the classes, for instance, being the important "divergent parabolas," with equations of the form $y^2 = ax^3 + bx^2 + cx + d$. The division of seven classes consisting of fourteen genera comprising seventy-two species of cubic curves (with later additions making this seventy-eight) was of especial interest for the taxonomic problems it posed.

Cayley noted that Newton's individual classes did not separate singular curves from nonsingular curves. He described a different classification scheme and compared Newton's with Julius Plücker's based on the form of the cubic, which, under Plücker's scheme, resulted in sixty-one groups with 219 species.[44] Cayley's preferred scheme, following August Ferdinand Möbius, was based on singularities in which curves fell into five classes falling under three heads, the nonsingular cubic curves (*simplex* and *complex* in Cayley's terminology), the nodal curves (*acnodal* and *crunodal*), and the *cuspidal* class. Classification was the motivation for much of Cayley's geometrical work and cubic curves offered a rich insight into the complexity of such schemes. Moreover, the investigation of cubic curves was manageable, unlike the neighboring enterprise, the classification of quartic curves.

This family of quartic curves presented a far greater challenge to nineteenth-century geometers.[45] For these, each of ten genera possessed a complex network of further subdivisions. For each genus, there were different species corresponding to types of singularities. For the uninodal genera, for example, there were three further subdivisions into curves in which the single nodes are of different types. Beneath these subdivisions were further subdivisions and even subdivisions of these. The potential wealth of detail for quartic curves caused Salmon to exclaim that "the number of species is so great, and the labour of discussing their figures so enormous, that it seems useless to undertake the task of an enumeration."[46] Coming from a man with such powers of calculation, this is indeed a warning, but he did not dispute the desirability of such an undertaking. Cayley was also wary of an attempt to classify quartic curves; even the quartic curves without singularities. He summarized theorems on the number of circuits for this simplest case but cautioned: "the forms of the non-singular quartics are very numerous."[47]

With his work on the singularities of curves and the latent interest in the geometry of position, Cayley was once again led into Pascal-type theorems he had investigated in the 1840s. That a straight line which passes through two inflectional points on a cubic curve passes through the third inflectional point is Maclaurin's theorem, and Cayley discussed this at his inaugural lecture in 1863. These special points on a cubic curve might be termed ordinary inflectional points. Several years previously he had established that there were twenty-seven "higher inflectional points" on a cubic curve, and that these "sextactic points" gave rise to an intricate geometry and allied theorems. First, he attacked the problem of finding the actual number of the sextactic points on a curve of order m. Initially he was unsuccessful, but at the end of 1864, he established that there were $m(12m - 27)$ such points. In connection with this work, we may note Cayley's rugged algebraic style, for which he was reputed, and his rapidly written prose, as is suggested by its lack of punctuation. Pace is not a quality we might not associate with a mathematical paper but the handling of his "symbols in battalions," observed by Galton, is evident in the style of his delivery with this exploration. With large complicated formless expressions, he proceeded by "substitution," "rejecting" irrelevant expressions, and "throwing out factors" until he arrived at the sextactic points on the curve.[48]

With customary sangfroid, Cayley went about his geometrical studies. The problems of curves and surfaces constituted his principal area of interest, but on 21 January 1865 he published a short note in the *Philosophical Magazine* on Nikolai Ivanovich Lobachevsky's non-Euclidean geometry.[49] Though it was a short note, it has been identified as the first paper in a mathematical journal to recognize Lobachevsky's achievement.[50] Cayley was in "extracting mode," performing a service of abstracting a technical paper; its publication effectively brought the possibility of a non-Euclidean geometry to English readers. In the note, Cayley referred to Lobachevsky's paper, originally published in *Crelle's Journal* in 1837, as a "curious paper" and declared Lobachevsky's interpretation of certain trigonometric equations as "hard to be understood." He reworked the equations but concluded: "I do not understand this; but it would be very interesting to find a *real* [his italics] geometrical interpretation of the system of equations" that defined Lobachevsky's geometry, in which the sum of angles of a triangle is strictly less than π. Cayley wanted a geometrical model of this geometry as part of the *really real* physical space. At this point, he did not connect his own insights into projective geometry and the importance of the absolute (as set out in his "Sixth Memoir on Quantics") with Lobachevsky's geometry. Felix Klein lectured on the Cayley metric in March 1871 and it became central to his systematization of non-Euclidean geometries in relation to different choices of the absolute conic giving different geometries but this organizing insight was not apparent to Cayley when he wrote his "Sixth Memoir."[51] While the excursion into non-Euclidean geometry in his note of 1865 was an isolated event, his work in analytical geometry in the 1860s was continuous. It consisted of extensive geometrical forays resulting in long memoirs on the theory of curves and surfaces, a body of work indicative of the intensity of his geometrical work during this decade.

Geometry did not claim Cayley's undivided attention, and he expressed thoughts on the organization of mathematics is his short paper "On the Notion and Boundaries of Algebra," published in 1864, to which he attached "some importance." Though Cayley did not acquire a consuming interest in semantic questions, the existence of this short essay on the metaphysics of algebra illustrates his occasional concern for foundational issues. Algebra might have arisen, as he once suggested to a nonmathematical audience, from easy puzzles that could be reduced to the solution of equations or from the twelfth-century Indian mathematician Bhaskara with his "picturesque forms of the *Bija-Ganita* [seed mathematics] with its maiden with the beautiful locks, and the swarms of bees amid the fragrant blossoms, and the queen-bee left humming around the lotus flower," but it had changed considerably.[52] It was still evolving and in need of organization. Cayley's paper questions what algebra is, and, in the 1860s how it had widened its domain and how it differed from the days of "ordinary" algebra, in which symbols represented numbers.

Cayley's discussion was concerned with the boundaries of algebra *within* mathematics. Of note are his views on the basis of number, this being in direct opposition to Hamilton's views, in which number was based on "time." "I do not admit the assertion, that the idea of number is derived from that of time," Cayley wrote; "it appears to me that it is derived from that of succession in time and space indifferently. But I would rather say that the idea of cardinal number is derived and abstracted from that of ordinal number."[53] In probing further, he examined the division between *finite* and *transcendental* analysis, which depended on whether algebraic statements could or could not be verified arithmetically.[54] The formal algebra of Euler's generating functions, a subject on which he was particularly expert, was *finite* analysis. In contrast, algebra as transcendental analysis consisted of statements that could *not* be verified in a finite number of steps, for example, in the infinite series expansion of π.

Cayley briefly commented on algebra as an art and a science, a dichotomy which was frequently revisited by various writers during the nineteenth century. Logic had been subjected to the debate in the early part of the century by Richard Whately in his *Elements of Logic* (1826) and by John Stuart Mill in his *System of Logic* (1843). Whewell equated art and science with practice and theory: "Art has ever been the mother of Science; the comely and busy mother of a daughter of a far loftier and serener beauty." In the 1830s, Peacock, Hamilton, and De Morgan had all considered the dichotomy as applied to analysis in mathematics. In some quarters, science was distinguishable from art by its greater degree of rational analysis.[55]

Cayley claimed algebra as an art and a science in that it combined both calculation and theory. As an art, it was concerned with operations that were combinatorial (tactic) or arithmetical (logistical) while as a science it predicted the result of such operations (see diagram below). The separation of art and science as applied to arithmetic goes back to Plato, and it is a measure of Cayley's admiration for Greek mathematics that he should apply these ideas to algebra.[56]

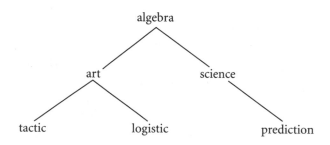

In delineating the boundaries of algebra, a new subject of combinatorics was formally suggested:

> Although it may not be possible absolutely to separate the tactical and logistical operations; for in (at all events) a series of logistical operations, there is always something that is tactical, and in many tactical operations (e.g. in the Partition of Numbers) there is something which is logistical, yet the two great divisions of Algebra are Tactic and Logistic. Or if, as might be done, we separate Tactic off altogether from Algebra, making it a distinct branch of Mathematical Science, then (assuming in Algebra a knowledge of all the Tactic which is required) Algebra will be nothing else than Logistic.[57]

In England, Cayley, Kirkman, and Sylvester were leading exponents and many of their problems in "Tactic" found their way into the journals of the day.

Cayley was concerned with the organization of the subject but he was never the ultimate foundationalist mathematician of the kind who concern themselves, and often content themselves, with answering "what is" questions: "what is a curve?" "what is a number?" The final outcome of such an "essentialist philosophy" is often a list of definitions, a quest that may seen rather barren to a mathematician driven by traditional problems in geometry and algebra. Cayley position is in line with Occam's razor, from a modern perspective, in his giving few definitions and often relying on intuition as a sufficient basis for his work. While he read Cauchy's algebra, for example, he did not adopt the rigors of

Cauchy's analysis or attempt to come to terms with the analytical definition of a limit. The presence of expressions such as 0/0 in his work, in which a theory of limits was needed, is evidence of this, but then, Cayley was an algebraist and geometer, and his dealings with analytical subjects such as elliptic functions were purely algebraic.

Society Life

Outward signs of Cayley's high profile in the academic community were the award of academic honors. In 1863 he was proposed for membership in the Paris Academy of Science for the astronomy section, though ultimately he was chosen for the geometry section.[58] With membership, the *Comptes Rendus,* in which he had already begun to place articles, became a regular avenue for his productions. In June of 1864 he was at Oxford to receive a D.C.L., the degree of Doctor of Common Law (and in the company of Henry Smith breakfasted with Charles Dodgson).

Following the summer sojourn at Grasmere in Westmoreland, Cayley took on his first major responsibility with the British Association for the Advancement of Science. Toward the end of each summer, the annual pilgrimage was made to a chosen venue and in September 1864 it was in Bath in the west of England. Charles Lyell was the president for the year and chose the occasion to present the geological evidence for establishing the age of man on the Earth—and his conclusions departed from the Church of England's position in supporting the literal truth of Biblical teaching on the origin of mankind. At the same meeting, David Livingstone gave an account of his travels in Africa but the real drama of the occasion was provided by the news of a shooting accident that killed another African explorer, John Hanning Speke, on his nearby estate at Box.

Cayley was president of the Mathematical and Physical Sciences Section and, in Hirst's view, made "a very good President" of section A.[59] It was the custom to make a speech, but he chose not to give one, not even one as brief as Stokes had given two years previously. He merely moved on to the business of the day and did not "propose to detain the membership." With little appetite for popular talks, he opted for the technicalities of the subject instead. He gave a paper on his continuing investigation of the inscribed and circumscribed triangles and compared his work with Chasles on the contact problems of conics in geometry.

The event which filled the seats of the Mathematics and Physics Section was James Glaisher's lecture on balloon ascents. Two years previously, with the acclaimed balloonist, Henry Coxwell, Glaisher had made a sensational ascent of 37,000 feet (about six miles) above Wolverhampton. He was saved from paralysis through lack of oxygen only by Coxwell's dramatic descent, which was achieved by tugging the balloon's valve-rope with his teeth. No branch of "normal science" could compete with such "derring-do." It was the perfect subject for an association lecture and, in thanking Glaisher for his able and entertaining address, Cayley expressed the wish that "he might go on making the series of ascents which were necessary to complete his unexampled and most valuable experiments and investigations, with the same success and *pleasure to himself* [my italics] as had attended his previous experiments [loud cheers]."[60] It was a Victorian pleasantry perhaps; scientific meetings were formal occasions to be formally conducted but the phrase "pleasure to himself" reminds us of the reward Cayley found in his own mathematical research, balloon ascents notwithstanding. His reserve and reticence to express his feelings about mathematics obscures the emotion he felt for the subject. Though the bare framework of his papers is often a successful camouflage, the mask does slip occasionally. In admiration of Hirst's rendition of "geometrical inversion," for example, Cayley said it was a piece of work which clarified a result of Steiner's that had often "excited his own wonder." Hirst basked in the praise, for, as he reminds us, "the wonder of the distinguished President of Section A, as every one knows, is not easily excited in geometrical matters."[61]

In November 1864 several of the young men-about-town formed a clique, styling themselves the X Club.[62] Hirst and Spottiswoode were the representative mathematicians in a group composed of Joseph Hooker, John Tyndall, Thomas Huxley, George Busk, Edward Frankland, John Lubbock, and Herbert Spencer, a band with wide-ranging scientific interests. All but the philosopher Spencer were fellows of the Royal Society, but none were educated at Oxford or Cambridge. Through widespread connections, they promoted the cause of serious research in pure science and became highly influential as they gradually infiltrated important positions in the scientific establishment. Hirst may have floated the idea of Cayley becoming a member, but then, Cayley was not a joiner of groups.

Cayley's name would carry weight, but he was wary of supporting political or religious causes of any kind. The organizers of the "Scientist's Declaration," which hoped to establish a harmonious relationship between Physical Science and Revealed Religion, could not hope to gain his signature for their public banner, but then few F.R.S.'s gave it their support.[63] In this ideological battle, the mathematician Robert Harley signed, but he was a minister of religion. Yet Cayley would not mount a fierce attack on the whole idea of such a public display, unlike De Morgan, who expressed vehement opposition. Cayley counted De Morgan, Harley, and Hirst, as allies, but all yoked together in the cause of mathematics. He never courted controversy as did the X Club group, which met at St. George's Hotel on the first Thursday of each month. He was not a recluse but preferred to remain in the background. It was from this vantage point that he became a pivotal figure in the network of mathematicians who moved in European and London scientific circles and, indeed, of those who worked independently and alone in the provinces.

In the same month that the X Club first met, the first informal meeting of the embryonic London Mathematical Society took place at University College, a latecomer to the specialist scientific societies that already existed in the capital. Its first official meeting took place on 16 January 1865 with De Morgan its first president.[64] His remarks at its inauguration are worth quoting:

> Our great aim is the cultivation of pure Mathematics, and their most immediate applications. If we look at what takes place around us, we shall find that we have no Mathematical Society to look to as our guide. The Royal Society, it is true, *receives* mathematical papers, but it cannot be called a Mathematical Society. The Cambridge Philosophical Society seems to fulfil more nearly the functions of a Mathematical Society, but it is in an exceptional position. It is the Society of the place which may be regarded as the centre of the Mathematical world; it is a Society in which almost all the members are able to relish its highest discussions. But in London we have no Mathematical Society at all.[65]

The society quickly found a following. Hirst was involved from the beginning and in June 1865 proposed Cayley as a member. He replied to Hirst's invitation with his customary formality and courtesy: "I shall really be very glad to join the London Mathematical Society; it has always appeared to me that something of the kind was a desideratum; and tho[ugh]' I cannot do it so much as if I had been still in London, I will certainly try to take part in the proceedings; I shall therefore be much obliged if you will propose me as a member." Cayley was formally proposed at the meeting held on 19 June 1865. The society grew rapidly in the first year of its existence and at the end of the year had a membership of sixty-nine and a bank balance of £36.[66] Clustered around such leading mathematicians as De Morgan, Cayley, Sylvester, and H. J. S. Smith were an enthusiastic band of amateurs.

Cayley found himself in the position of guiding the society and alerting members to the latest research, and whenever discussion followed a paper, Cayley's name was invariably reported as having taken part. By their regular attendance at meetings, he and H. J. S. Smith gave the society its research orientation in the early days—he could lead by example and refer to the work of significance in the mathematical world at large. The society met regularly each month and when there was nothing for the forthcoming meeting Cayley could be relied on to supply a paper at short notice. A move to attract papers in applied mathematics was unsuccessful, and the society gradually became a bastion of pure mathematics.

Sylvester's synopsis of his long paper on "Newton's Rule" was the first to be published in the society's *Proceedings,* but "On the Transformation of Plane Curves," which Cayley read to the society on 16 October 1865, was the second. In describing Cayley reading a paper, J. W. L. Glaisher noted his forthright style:

"Nevertheless, he stated the main features of his paper clearly and at a suitable length, but he confined himself strictly to the contents of the paper, so that it conveyed little information to those not already acquainted with the subject. It was a bare statement of methods and results." Cayley regarded the reading of a paper as a formality, a step to take before its printing. With an air of reticence, he found the public display a "disagreeable formality" that had to be suffered. Members of his audience, even mature mathematicians, had difficulty following his train of thought, and it was noted that "unless one knew the subject beforehand, his method of running over the leading formulae was not instructive."[67] He focused all his resources on the production of printed papers and no doubt he gained something of this attitude through his former membership of the legal profession in which the importance of the printed word is absolutely paramount. The value placed on the print medium in science and mathematics was accentuated by the ever-present priority issue which dogged its practitioners.

Cayley continued a close association with the Royal Astronomical Society through his editorship of the *Monthly Notices*. The 1860s has been described as a crucial period for the development of astronomy, a period when the physical side of astronomy grew in stature. There were two main groups in the membership of the society, the observational astronomers who studied the physical properties of planets and galaxies and the strong contingent of Cambridge mathematicians, of which Cayley was a leading member. His papers in astronomy were often concerned with the solution of a geometrical problem wherein the astronomical content is subsumed under pure mathematics and calculation. One paper, ostensibly on planetary orbits, was coupled with a problem on ellipsoids. Another required the determination of the position of a body moving around the sun given only three observations, which he openly admitted was a pure geometrical problem and one with limited utility. A problem in the dynamics of planetary motion was reducible to a problem in the theory of differential equations. Yet Cayley did not see divisions in mathematics generally and, to him, the boundary between the "mixed mathematics" of mechanics and astronomy and the problems of pure mathematics was often blurred. As he remarked when commenting about one problem in statics and dynamics: "We seem to be thus passing out of pure mathematics into physical science; but it is difficult to draw the line of separation."[68]

The London Mathematical Society and the Royal Astronomical Society were important venues for Cayley, but his principal scientific society in London took place at the Royal Society. He continued to sit on its council after his arrival in Cambridge and pressed for Royal Society recognition of British and European mathematicians.[69] Two geometers from an earlier generation who stand out in relation to his geometrical work were the mathematicians who had influenced him in the 1840s, Julius Plücker and Michel Chasles. When the Royal Society Council was deciding on the Copley Medallist for 1865, the shortlist included both. Cayley had proposed Plücker on an earlier occasion in 1857 but had not been successful. Cayley's reply to Hirst's solicitous inquiry was equivocal: "But of course as to the absolute merits both of Plücker and Chasles there can be no shadow of a doubt, and I should rejoice very much at the award of the Copley Medal to either of them—and I should be really sorry to have to decide between the two."[70] Hirst championed Chasles, while Stokes, appreciating both his mathematics and physics, sponsored Plücker.[71] During the final rounds of voting, the council was evenly split, but finally Chasles "won by a neck." The medal was awarded to Plücker the following year for his research in analytical geometry, magnetism, and spectral analysis.

The European Dimension

Of the two, Chasles and Plücker, Cayley had most in common with the German mathematician—a belief in the primacy of projective geometry and a commitment to the analytical method. Around 1866 Plücker turned from his research in magnetism to the theme of his earlier geometrical work. Cayley was then considering higher singularities of curves in the Plückerian terms— of analyzing curves by the simple singularities of node, cusp, double tangents, and inflectional tangents and giving formulae which connected them.[72] The classificatory scheme based on the Plückerian singularities gradually became outmoded, and in the 1860s several new theories applicable to the study of algebraic curves emanated from Germany, Italy, and France. G. B. Riemann investigated the rational transformation of a plane curve in which Abelian integrals and complex function analysis played the leading role, Cremona investigated the rational transformations of the plane itself, and Chasles put forward a theory of correspondence between points on curves.

Cayley worked extensively on plane curves and obtained a generalization of Michel Chasles's theorem on the principle of correspondence. This had been published by Chasles in March 1866 for unicursal curves such as ellipses, the curves with deficiency equal to zero.[73] For example, a one-to-one correspondence is set up between the points of an ellipse by a system of straight lines which cut the ellipse in two "corresponding" points. The case of tangents to the curve gives self-corresponding points, and Chasles's theorem calculates the number of these self-corresponding points for unicursal curves. Cayley regarded Chasles's principle as a most "powerful instrument of investigation which may be used in place of analysis for the determination of the number of solutions of almost every geometrical problem."[74] Writing to Ludwig Schläfli in March of 1866, he noted that the theorem could be applied to more general curves:

> The last thing I have been working at is a theory of the correspondence of points on a plane curve. Since in a unicursal curve or curve [of order m] with the maximum

number $(m - 1)(m - 2)/2$ of double points, the coordinates are expressible rationally in terms of a parameter Q,—it is clear that Chasles' theorem for the correspondence of points on a line applies to any unicursal curve whatever—and we have thence a theorem applying to any curve, viz. if the correspondence of the two points is such that to any point of the first system there correspond a' points of the second system, and to any point of the second system a points of the first system, then the number of self-corresponding points $a + a' + 2kD$, if D be the *deficiency* of the given curve. I have obtained a theorem which enables me to find, if not always, at least in most cases, the value of the coefficient k.[75]

The theorem meant that Chasles's theory could be applied to a wide range of curves, for example, to curves of the Newtonian "divergent-parabola" type known today as "elliptic curves," so-called since their points can be expressed in terms of elliptic functions. Today, these curves play a central part in the very advanced theory of numbers and are linked with Fermat's Last Theorem. The geometrical theorem surmised by Cayley was not proved rigorously by him; but its correctness is a measure of his keen geometrical intuition. The rigor was supplied by Alexander von Brill in 1883. In this subject, Hirst was a natural ally but he was too weighed down by teaching to join in. That the work gave Cayley some pleasure can be gauged from Hirst's *Journal* entry; "Dined with Cayley at 4 Westcombe Park, Greenwich, the house of his father-in-law, Mr Moline. Before dinner, Cayley and I had a walk in Westcombe Park and he explained to me his late very beautiful researches on the correspondence of points on any curve."[76]

In November, Cayley brought his research on caustics to life once more. His next major project after this was stimulated by the geometrical work of Jean-Philippe Fauque de Jonquières, a disciple of Chasles and Jean Victor Poncelet.[77] De Jonquières had just been promoted ship's captain in the French Navy, another step in a career bounded in time by his entrance to the Naval School in Brest at the age of fifteen until he reached vice-admiral in his sixtieth year. Spliced into this naval life, in part pursued in the Far East supervising national interests in Cochinchine (now Vietnam), he produced over 150 papers, mainly in geometry, work regarded as of being of the highest quality by his contemporaries. "Have you seen in the *Comptes Rendus* the very interesting papers by De Jonquières?" Cayley wrote to Schläfli in December.[78] The problem in these papers was to enumerate the number of curves of order r which satisfy specified conditions of contact with a given curve of order m; exactly one conic is determined by five points, while two conics may satisfy five conditions in which four points are given and the conics are required to touch a given line. These "contact problems," in which geometers endeavored to enumerate such conics, became a popular pursuit.[79] For them, Chasles created a theory of *characteristics* and, using this technique, Cayley took up the quest with alacrity.

Cayley transmitted his admiration of Chasles's work to the young Clifford, who delighted in the application of the theory of characteristics to particular problems. Another follower was the Danish mathematician Hieronymous Georg Zeuthen. Both Cayley and Zeuthen worked on the geometry of reciprocal surfaces and are responsible for the Cayley-Zeuthen equations that link together the singularities of these surfaces. When Cayley wrote out a citation in support of Zeuthen's honorary membership of the Cambridge Philosophical Society, he wrote of his completion "in an essential point the general theory (based upon Chasles' theory of characteristics) of the determination of the number of conics satisfying five conditions of contact with a given curve or curves, . . . , and he is the author of various valuable memoirs in relation to the theory of correspondence, the singularities of surfaces and curves, the classification of plane curves of the fourth order, and other geometrical questions."[80] In excruciating detail, Cayley calculated the mutual relationships which existed between the twenty-eight bitangents that could be drawn to a quartic curve, thus taking Zeuthen's work one stage further.[81] These two mathematicians held a common belief: both regarded Euclid's *Elements* and Descartes's *La géométrie* as fundamental to the study of geometry.

Drawing on Riemann's work, the important event in the theory of curves which heralded the modern view of the subject, was the book written by Alfred Clebsch and Paul Gordan entitled *Theorie der Abelschen Functionen* (1866). Gordan had gained his doctorate at the University of Giessen in 1862, the year when Clebsch joined the staff of the university. At that time, he was a specialist in the theory of Abelian functions and was able to guide Clebsch in Riemann's work linking Abelian functions to the theory of curves. It was this subject which gave rise to their joint publication. In its Preface, they referred to the "penetrating studies of Mr Cayley."[82] This was a subject which proved irresistible to Cayley and the theme of the "Clebsch-Gordan theory" of Abelian functions was one which occupied him in the ensuing years. A bonus of the Giessen partnership of Clebsch and Gordan was Gordan's introduction to the study of algebraic forms, a subject in which he made his reputation. With his deep studies in this subject, Gordan would soon spring a surprise on the mathematical world.

Salmon's Bequest

Of Cayley's co-workers in geometry, Salmon was the colleague of longest standing. When he was elected to the Royal Society on 4 June 1863, it was natural for Cayley to propose him. Salmon offered himself as the "discoverer of several new species of curves and surfaces" and "the inventor or improver of several algebraical and geometrical methods."[83] In Cayley's own work on cubic surfaces and "scrolls" (skew surfaces) of the third order, Salmon had been a collaborator and Cayley was beginning to investigate scrolls of the fourth order.[84] For Salmon, this new work was probably out of reach, as around this time, he began to see his future in terms of divinity rather than mathematics. Divinity offered the prospect of promotion as he had narrowly missed the Dublin professorship in mathematics in succession to Charles Graves, who had resigned the chair in 1862. Though admirably qualified for the mathematical chair, his political calculations went astray and the chance evaporated. Michael Roberts, a fine mathematician, and another of James MacCullagh's students, was appointed.

In 1866, at the age of forty-seven, Salmon was appointed Regius Professor of Divinity, a post he occupied at Trinity College, Dublin, for twenty-two years before becoming its provost. His switch to divinity was a loss for mathematics. His *Lessons Introductory to Modern Higher Algebra* was the text that carried Cayley's view of algebra to a wider world, and a second edition was being prepared for publication just at the time of his elevation to the chair of divinity. It was to be a considerable extension to the beautifully succinct slim volume published in 1859. In its preparation, Salmon reported that he had partly calculated the value of the very long expression of the invariant *E* for the binary form of order 6 and promised: "if I find leisure to complete the calculation I will give the result in an appendix." He did, and it occupied thirteen pages.[85] Cayley drew Schläfli's attention to this feat of calculation: "The second Edition has appeared of Dr. Salmon's lessons on the *Modern Higher Algebra* expanded to a thickish volume of 300 pages—there are two great pieces of calculation in it, that of the discriminant $[\Delta_6]$ of the general sextic function, . . . and of the skew invariant of the 15th degree [Salmon's *E*] of the same sextic function: and in connexion with this an interesting speculation as to the form of the invariantive criteria for the reality of the roots of a sextic equation."[86] Salmon's prodigious calculatory work was greatly valued and admired by Cayley. It was perhaps on his initiative that Salmon gained the Royal Medal of the Royal Society in 1868. A man who published over forty research papers and four classic textbooks, he was awarded the medal for analytical geometry especially for elaborating the theory of correspondence for surfaces, which Chasles had established for curves.

Whatever the demerits of Salmon writing his books piecemeal, he moved with the times and the latest work could be incorporated into further editions. In this respect, his unpolished work resembles the way Cayley wrote and published his mathematical papers. A quick response, for example, was his introduction to the latest symbolic notation of invariant theory at the end of the second edition of *Modern Higher Algebra*. Cayley's work reached a wider audience through Salmon's fourth and final book, *A Treatise on the Analytic Geometry* of three dimensions, first published in 1862. Known among mathematicians as simply "Surfaces," the book contained an introduction to Cayley's work and disseminated his view of the subject.

Salmon's books were highly prized in Germany, where he gained a reputation for being a fine geometer. Max Noether noted that he was cast in the English mold of being openly inductive in his ways "and he appears as a pure Empiricist almost more so than Cayley."[87] In a letter to Leopold Kronecker, Ernst Kummer wrote: "Salmon in Dublin has published a *Geometry of three dimensions*, which I have bought and which I study with diligence and pleasure. I have found in it many straightforward explanations on questions which have especially interested me, amongst others on the theory of algebraic curves of double curvature [space curves] and related questions." Kummer cast Cayley as a formalist and continued: "Salmon's manner of treatment of algebraic space-curves is not exhaustive, but it is mostly as we have discussed, very neat and simple development. Get this book immediately and perhaps others by the same author on conic sections, and higher plane curves, which I own, and lessons on higher algebra, which I have ordered for myself, in which he frequently refers to the Geometry and which does not deal merely with the formal business of Cayley and Sylvester." Kummer ended: "Another work from the same author: *Sermons preached in Trinity College Chapel* [1861] will be of lesser interest to us."[88]

The Royal Medal was won by Salmon and recognition afforded as he packed his mathematical bags. He continued with mathematics in a desultory way, but his time was absorbed with the disestablishment of the Irish Church. When he came to Cambridge on future visits, his first duty was increasingly not for mathematical conversation with Cayley but to preach at the university church.

CHAPTER 12

Years of Challenge

Cayley's work on geometry displaced invariant theory as a dominant inter-est in the first half of the 1860s. His favorite subject had run out of steam but, while relegated to the background, it was not forgotten. At its high point in the 1850s, he had set in train the memoirs on quantics, but the burst of activity from J. J. Sylvester in 1864 on "Newton's Rule," announced in the *Times* as a "mathematical discovery," reminded him of the subject once again.[1] The 1860s would bring a challenge to his former research and the beginning of a new approach to invariant theory from the German mathematicians, a thrust which offered an alternative to his own methods. While he struggled with this new possibility and with emerging geometrical theories, there was also an oblique challenge to his views on mathematical education.

A Quiet Interlude

Equal to Cayley's duty to "research," was the requirement of the Sadleirian professor "to explain and teach the principles of Pure Mathematics." With his appointment it was a fresh experience to teach and influence students, albeit few in number. In January 1867 he had the pleasure of seeing his protégé William K. Clifford give a good performance in the Tripos. After the examina-tions, he wrote to Hirst: "I saw Clifford just after the examination, his own impression as to his place [in the order of merit] is 6th or 7th, from which I infer that it will be at any rate 5th."[2]

In the 1860s, Trinity College dominated the top position in the Tripos but a break was threatened in Clifford's year. Robert Kalley Miller was already a Glasgow Master of Arts when he was tipped as the Senior Wrangler in the Peter-house colors. On the day itself, he was unable to compete, and when the results came out, Cayley informed Hirst: "It is a good deal better than I expected—Clifford is second. Miller, who was the favorite, took an aegrotat" (an unclassi-fied degree which may be awarded in British universities to candidates unable to sit the examinations through illness).[3] Next came the examinations for the Smith's Prizes, and Miller offset this nondescript degree reserved for worthy nonrunners by winning the First Prize. Though Clifford took the Second Smith's Prize, he need not have worried about being runner-up. He was an established local star and his fame soon spread to the London salons as a polished communicator.[4] He became a fellow of Trinity College and for a few years his mathematical work focused on projective geometry and invariant theory, two of Cayley's subjects.

The *Royal Society Catalogue* project, which had begun twelve years earlier, and which had involved Cayley from the outset, produced its first volume in 1867. Initially he was skeptical of its worth, informing a colleague "as, of course, one *always* knew what had been done in one's own department of science."[5] His personal knowledge of the mathematical literature was vast and he could instantly assess whether a result was really new. Sylvester, who readily admitted his own mathematics did not come from books, acknowledged the erudition of his friend, remarking during the year: "I shall however consult Crelle [*Crelle's Journal*] and thank you very much for the reference. You are a living library to your friends."[6] Cayley's reputation as a mathematical repository became legendary, but in the 1880s he was eventually forced to alter his view that one could routinely know the relevant literature, and he then recognized the accelerating growth of mathematics. Then it was being generated by a growing army of professionals which made it impossible to know "one's own department."

The *Catalogue* provided an outlet for Cayley's administrative skill. In its immensity and ambition, it was a task worthy of the energies of the Victorians even in organizational terms. It required teamwork, at least in the form of masters and dutiful servants, and, unlike individual projects, it did not raise scientific rivalries and was not controversial. Initially funded by the Royal Society, first financial estimates bore little relation to the actual costs, but fortunately, William Gladstone, then chancellor of the exchequer, was persuaded that it should be funded through the public purse. The Government Minute, which launched the first volume, ended with a Victorian flourish:

> Having regard to the importance of the work, with reference to the promotion of scientific knowledge generally, to the high authority of the source from which it comes, and to the labour gratuitously given by Members of the Royal Society to its production, my Lords consider themselves justified in having the work printed at the cost of the public, with the understanding that, reserving such a number of copies for presentation as my Lords, in communication with the President of the Royal Society, may hereafter determine, the work shall be sold at such a price as it may be calculated will repay the cost of printing.
>
> Their Lordships, however, desire it to be understood that the work shall go forth to the public under the authority of the Royal Society, by the exertions of whose Members this important aid to the study of Science has been produced.[7]

While Cayley's ongoing program of mathematical research was largely a metropolitan activity in the 1860s, life was also lived at a local level. Away from the London bustle, Cambridge mathematicians and scientists had the Cambridge Philosophical Society on their doorstep. This catered for all the sciences, but as Augustus De Morgan had indicated, it was the closest to a mathematical society that existed in England before the founding of the London Mathematical Society in 1865, and indeed he had read many papers to it himself. Cayley read his first paper to the Cambridge Philosophical Society when only twenty-one years old and barely graduated but rather oddly did not communicate another paper to it before his arrival at Cambridge as professor. He would now regularly sit on its council, and counting up, he was a vice-president on five occasions and would be elected president in 1869.[8]

Settled back in the town, one sees him at its Monday meeting in March 1867, where a person attending observed "the careful closing of doors and windows keep out the noise and air of the world."[9] First there was the business part of the meeting, and W. K. Clifford, just graduated, was formally admitted as a member. Cayley himself was an important presence: "[i]t was gaslight. The general audience—among these were Professors Cayley and [W. H.] Miller, and a few ladies who would lend countenance to philosophy—were seated on tiers of benches, and below the general audience sat an audience fewer and more select; a more vital part of the society, I suppose, whether by merit or by office. Among this lesser [i.e. lower and seated at the front] audience are Challis and Adams the chief speakers at the meeting."[10]

It was to be an astronomy meeting conducted by the two Cambridge astronomers, John Couch Adams, the discoverer of Neptune, and James Challis, blamed that his dilatoriness twenty years before had cost Adams his priority. Challis gave a lecture on the difference of longitude between the society's clock and the transit-clock of the Cambridge observatory. Adams spoke mathematically about the great "Leonids" meteor shower in the previous November, the dust trail from the comet Tempel-Tuttel which obits the Sun and which throws off a spectacular show of night-lights as it enters the earth's atmosphere.[11] What was unknown was its orbital period, but, after a vast scale of numerical calculations, of which only mathematical astronomers such as Cayley, Charles Delaunay, and Adams were capable, Adams rejected several possibilities. In conjunction with the work of the American mathematician and astronomer, H. A. Newton, he showed that the comet orbited the sun three times a century.[12]

Of special interest are Adams's views on the nature of space, for it coincided with the way Cayley conceived of the *reality* of what was commonly called ordinary physical space: "I [W. P. Turnbull] observed that Adams treated space as a *reality*:—He spoke definitely of a body's position in *space*. His answer to the question 'Is there such a thing as Infinite Space?' would have been a decided Yes, with no metaphysical limitation. The fixed stars had all their own proper motions, the meteors travelled about hither and thither, and all was not only conceived of, but actually was in space."[13] This reality was the oft-quoted "really real" of ordinary physical space, in which Cayley believed. It was quite different from the *ideal* realm of *n*-dimensional mathematical space, the other aspect of his geometrical studies.

During these middle years, Cayley made an extensive study of curves and surfaces of different types lying in this "ordinary" space. He was particularly interested in the geodesic lines and lines of curvature that can be drawn on the surface of an ellipsoid. He studied ruled surfaces—surfaces generated by the motion of a straight line in three dimensions. Work began in the 1840s was expanded into wide areas of investigation.[14] He continued his study of quadric surfaces (surface of the second order) and set in train a series of memoirs on cubic and quartic surfaces. He was ever fascinated by the cubic surface, with its "complicated and many-sided symmetry" and its twenty-seven lines.

Cayley's intuitive and tactile approach to geometry is heavily camouflaged by formal analysis. He liked making physical models, and when Sylvester wrote asking for advice on whether to apply to the Royal Society for a grant in order that his amphigenous surface should be constructed in pasteboard, Cayley replied with encouragement: "I think it will be very interesting to have a model of the surface—and you may with great propriety apply to the R[oyal]. S[ociety]. for a grant for that purpose . . . I should be very glad if a few more algebraical surfaces could be modelled—the only ones which I know of as having been done are the wave surface and Steiners Surface of the 4th order mentioned in the last number of Crelle [by H. E. Schröter]."[15] Other mathematicians were using models for geometrical studies, and in particular, Julius Plücker, with the physicist's eye for the practical, was a leader. He too adopted an intuitive approach to the whole subject, but Charlotte Scott describes his "hasty intuitiveness" as contrasting with Cayley's measured foresight, and she pinpointed Plücker's failure to take account of the work of others.[16] But then, Plücker was a trendsetter. In 1866 he visited England and described his models of quartic surfaces to the British Association meeting and allowed the models to remain in the country so that they could be copied. In March 1868, Cayley and Sylvester joined forces with James Clerk Maxwell in applying for a Royal Society grant (of £40) for the construction of models of geometrical surfaces. When the Geometrical Modelling Club was set up in Cambridge in the 1870s, Cayley and Maxwell came together again in a common endeavor of wanting to visualize geometrical shapes through the making of physical models.[17]

In his cooperation with Stokes in the 1850s, in which surfaces are described in physical terms with reference to carefully drawn figures and models, Cayley was not merely accommodating a physicist by employing such practical means. He needed models himself in order to sharpen his intuition and further his research. He had written to Sylvester, "I should certainly be much interested in seeing the curve carefully drawn on a scale of sufficient magnitude to perceive the peculiarity of the form of the infinite branches as compared with those of an ordinary hyperbola. An interesting form to trace would be that in which the curve is on the verge of degenerating into the three asymptotes." It is difficult to see what Sylvester would contribute here, for if Cayley regarded a box of drawing instruments as an essential part of the mathematician's equipment, Sylvester never possessed one.[18] An inveterate drawer of curves himself, in which he employed his draftsman's skills to his tinted drawing paper, Cayley firmly believed in testing his theorems against the pictorial evidence he produced. Conveying this work to a meeting of the young London Mathematical Society, he informed the audience that lengths of tinted drawing paper were sold at Messrs. Lechertier-Barbes, Regent Street, at 6d. per yard, or 9 s[hillings] a roll, the paper being of a good quality and taking color very readily—and to this day their name is still attached to art materials. Surfaces in three dimensions were more difficult, but perhaps acting on a suggestion from Maxwell, he annexed representations of cones that could be viewed through stereoscopic glasses.

Not all of Cayley's visual representations were successful, and he rated his drawings of the "Pascal's Hexagram," with its sixty Pascalian lines and forty-five Pascalian points as "almost unintelligible," being composed of points which "are either crowded together or fly off to a great distance."[19] The important point is that he made them at all.

Old Haunts

With the appearance of the "Seventh Memoir on Quantics" in 1861, Cayley reached a temporary halt and the series lost momentum. He had absorbed Siegfried Aronhold's work on cubic curves into his own work and had shown the equivalence of their respective notations, but, six years after its publication, the "Eighth Memoir on Quantics" resumed the theme explored in the second and third memoirs—the binary quintic—and he evaluated Charles Hermite's work on the quintic in comparison with Sylvester's.[20] Submitted to the Royal Society on the 8 January 1867 and read a week later, the memoir contained miscellaneous results, including the study of a surface that could be used to demarcate those quintic equations that had real roots.

In the "Eighth Memoir," Cayley's whole program became enmeshed in detail. At the beginning, he merely remarked that "it was interesting to proceed one step further, viz. to the covariants of degree six." The invariant theory strategy adopted by Cayley was to calculate invariants and covariants by ascending degree, but as he progressed "up through the quintic system" the arithmetic grew increasingly involved. In particular, the problem associated with the linear interconnections between covariants, the "problem of the syzygies," loomed larger. As noted, for the binary quintic there is a single syzygy involving *invariants* (see chap. 8), but he also found single syzygies involving covariants of low degree; in particular, he found one for the three covariants of degree five (L, K, J), but "for the degree six, we obtain two covariants $(M, N,)$ but as many as seven syzygies" (see "Invariants and covariants of the binary quintic as known more completely to Cayley in 1871" on p. 15g of gallery).[21]

The binary quintic form was gradually revealing its elaborate structure, but it tested the mettle of its investigators. In the knowledge that the system of the irreducible invariants and covariants quintic was infinite, Cayley pressed on, realizing that intricacies increased with degree. Unknown to him, there would be even more bizarre results associated with covariants of degree eight and higher. Fortunately, help was at hand from Germany, where mathematicians were also taking a good deal of interest in this research. In an "Addition" to his "Eighth Memoir," submitted in October, Cayley acknowledged a joint paper written by Paul Gordan and Alfred Clebsch on binary algebraic forms for the first issue of the new series of the *Annali di matematica pura ed applicata*.

In the spring of 1867, Cayley submitted two very long connected papers (totaling 100 printed pages) on the subject of curves that satisfy certain conditions.[22] These were the outcome of questions posed by Jean-Philippe Fauque de Jonquières the year before. This stimulated Cayley to adopt a pragmatic attitude: to accept a conjecture as true and work out the consequent "solution." He readily admitted, "There are difficulties which I am unable to overcome; and I have contented myself with the reverse course, viz. knowing in each case the number of proper solutions, I use these results to determine *à posteriori* in each case the expression of the Supplement [an expression introduced to account for special solutions]; the expression so obtained can in some cases be accounted for readily enough, and the knowledge of the whole series of them will be a convenient basis for ulterior investigations."[23] Thus Cayley was content to undertake further investigations even though there was a risk of his premises being false. To press ahead was the order of the day, and whatever else, there was no lack of confidence on his part.

It is clear this work caused Cayley difficulty. He wrote to Luigi Cremona that he was totally occupied and could only find some of the required expressions a posteriori and could not deduce them a priori. A month later he informed Hirst that he was relieved to have finished the papers.[24] Submitted in April 1867, it was evident they were published unfinished and there were lingering doubts. He could hardly call on the assistance of Sylvester, who in other times would have been the first port of call. Sylvester was suffering one of his periodic depressions and unable to enter the mathematical arena. Just when Cayley was struggling with these geometrical problems, he received a pitiful letter: "I was prevented to my regret from being present when your paper was read at the R[oyal]S[ociety]. Since then I have done no mathematics—ever intending and ever putting it off. Thus I have been too much ashamed to call upon or to write to you but have been hoping all along to see or hear from you." This was not the confident Sylvester of a few years before, on top of his subject and the proud recipient of a Royal Medal. The depth of his despair was conveyed by his closing signature: "Ever truly your (bodily well) but mentally or spiritually afflicted."[25]

As an antidote to the view that Cayley found mathematics an effortless activity, we have the evidence from his wife to the contrary. He was still dealing with this problem attributed to de Jonquières when Susan Cayley wrote to Hirst: "For the last three weeks, my husband has been working at something that 'will not come out' and he is in such wretched spirits that I am quite uneasy about him. I believe among all his mathematical friends you are the one with whom he sympathizes most on certain subjects and he says now and then he wished he could talk it over with you or Dr Salmon." A few days later, the outlook seemed brighter and Susan Cayley less distressed: "[a]nd I also want to tell you that my husband has 'made a step' and is now working on cheerfully and hopefully."[26] In his poor state of health, Sylvester was called on to referee Cayley's two memoirs. To his report, he appended an apologetic remark—that he saw that the papers were incomplete but did not wish to stand in the way of his colleague: "In these memoirs as in others I assess on the same class of subjects the laws of strict reasoning or rather of rigorous demonstration are not nor are professed to be in all cases observed. Certain assumptions are made à priori, which are justified with a high degree of probability by the results to which they are found to lead. I do not consider that this fact detracts in any degree from the value of the memoirs presented to the Royal Society."[27] Sylvester expressed the belief that geometry was not just a matter of logical deduction but involved speculation. It did not matter if this was publicly expressed. Cayley was not deterred from putting the work into print even if it employed post-hoc deduction coupled with a heady mixture of experiential evidence and intuition.

With Cayley fully productive, Sylvester continued in the doldrums. His inability to be creative himself only increased his vigilance on questions of priority. Broadsides from Sylvester had abated over the past few years, but in November he was on high alert. Cayley showed him work on geometric transformations and this set off his coiled defenses. First he appealed to natural justice through remarks "which I hope you will not take in illhart [sic] and will weigh in that scale of right which every man carries in his own breast!" He then descended into specifics:

> I put it to you whether it is conceivable that any person having the most superficial acquaintance with the theory of multiplication and transformation could have failed to suspect and to have taken the trouble to ascertain the existence of this connexion [between a geometric transformation and one in elliptic functions]. My unhesitating view on the subject is, that it would have been more consonant with your usual mode of proceeding to have put it to me to ascertain the fact in question—but not having done so and having so far taken the matter out of my hands to whom it properly belonged, your work ought to figure as a note in the memoir which I am preparing

and that all further investigation in that direction ought to be left to myself until that memoir is off my hands and has appeared in *print*. If you do not accord with this view, I should be glad to refer it, to any mutual friend, Hirst, Salmon, Ferrers or any other competent to give an opinion on the subject.[28]

No doubt Cayley absorbed these sharp remarks with equanimity. He understood Sylvester and, moreover, he believed that scientific controversy was of limited value: at best, controversy served to demarcate opposing positions but hardly ever suggested progress. Sylvester's contretemps, of tea cup proportions, was short-lived. A few months later he was dining with the Cayley family circle at Montpelier Row, Blackheath.

Cayley was also working on the connected subject of polyzomal curves—curves connected with geometrical problems such as families of circles touching three given circles, or spheres touching four spheres. In this, George Salmon alerted Cayley to similar work being done by the Irish geometer John Casey. A self-taught mathematician, Casey was the same age as Cayley but from a very different background. After establishing himself in the world of elementary schools around his hometown of Kilkenny, County Cork and Tipperary, he had the temerity to solve a version of Poncelet's polygon problem, his purely geometrical proof bringing him to the notice of geometers at Trinity College, Dublin. They persuaded him to join them as a student and he graduated B.A. at the age of forty-two. A stream of geometrical papers followed and he gained membership of the Royal Irish Academy in May 1866. On hearing from Salmon, Cayley immediately wrote to the up-and-coming academician. In reply, Casey outlined his geometrical results on circle contact problems that had been published by the Royal Irish Academy. The work was connected with Cremona's and it set Cayley on a new tack while carefully acknowledging Casey's priority.[29] When the Irishman's name came up for election to the Royal Society in June 1875, Cayley's praise was unstinting for "what he has already done and of [his] talent as a mathematician."[30] Cayley continued his work on planar curves and by the end of 1867 had completed the massive memoir on polyzomal curves.[31]

Geometry aside, a challenge was brewing in Scotland. Peter Guthrie Tait had left Belfast where he had been professor of mathematics and returned to Edinburgh as professor of natural philosophy. He preferred "mixed mathematics" as his theater of operation and so turned to Cayley for advice and authority in pure mathematics. Was the paper he had written on the "rotation of a rigid body about a fixed point" really new? In August of 1868, Tait asked Cayley whether his result was "merely a shortening of yours." Tait was thinking of the rotation operator $q^{-1}(\ )q$ Cayley introduced in 1844 shortly after Sir William Rowan Hamilton's discovery of the quaternions. The unambiguous reply from Cayley was welcome: "The rotation formulae are deducible by an easy transformation from formulae in my paper [of 1844] . . . But the actual form you have given to the formulae is, so far as I am aware, new; *and a very decided improvement* [his emphasis]."[32] Tait went on to win the Keith prize for his two quaternion papers on the rotation formulae. He was after all "a man who got the Quaternion mind directly from Hamilton," and with Hamilton's death in 1865, he was the man to safeguard Hamilton's legacy.[33] Tait was used to scientific controversies and positively enjoyed them, and for good reason did A. R. Forsyth call his fellow Scot "a 'bonnie fechter,' never reluctant in the use of the controversial tomahawk."[34] Though Tait was now on his best behavior, treating Cayley with civility, his advancement of Hamilton's claims for the quaternions grew more vociferous as the years passed. It was inevitable that, with unswerving dedication to this end, he should make Cayley one of his principal targets.

In November 1868 Cayley completed the comprehensive "Memoir on Cubic Surfaces," in which he compared and extended recent results obtained by Ludwig Schläfli. He had an inside track on this development since he had edited Schläfli's publication in the *Philosophical Transactions*.[35] It represented a shift in view from the one taken by Schläfli in 1858, in which an attempt was made to classify cubic surfaces based on the number of *real* lines from the twenty-seven lying in its surface. In this own memoir, Cayley classified cubic surfaces with respect to their singularities and made reference to geometric duality and "reciprocal surfaces." "I disregard altogether the ultimate division [of Cubic Surfaces based] on the reality of the lines, attending only to the divisions into twenty-three cases depending on the nature of the singularities," he wrote, "[a]nd I attend to the question very much on account of the light to be obtained in reference to the theory of Reciprocal Surfaces."[36]

The pioneering work on reciprocal surfaces had been carried out by Salmon in 1855, but it was precisely this work for which he was awarded the Royal Society Royal Medal as Cayley completed his paper on cubic surfaces.[37] Following Schläfli, Cayley's analysis of the cubic surface included the double-sixer, a configuration of twelve lines intrinsic to the surface. He carried out the laborious work of analyzing each of the equations of the double-six and verifying their intersection properties. By making appropriate selections of twelve lines from the twenty-seven lines available, thirty-six possible double-sixers are possible, and Cayley enumerated all of them in his memoir. Quartic surfaces represented a stiffer challenge but, undeterred, he began a broad survey of them and, again, these surfaces were classified according to their singularities.[38]

Gordan's Surprise

Set against Cayley's intermittent activity in invariant theory, the German mathematicians with their streamlined approach to the subject made greater progress. The school comprised Siegfried Aronhold, Paul Gordan, and the brilliant Alfred Clebsch, newly installed as professor of applied mathematics at

Göttingen. With Carl Neumann, Clebsch was the founder of the *Mathematische Annalen,* a journal that promoted invariant theory. Of this trio—a German Invariant Trinity—it was Gordan in Giessen who became "King of Invariants," a man who devoted himself to the subject for the remainder of his life. Supported by Clebsch, he made *the* breakthrough that altered the whole perspective of the theory.

In the summer of 1868, eighteen months after Cayley's "Eighth Memoir on Quantics" was written, Gordan submitted his paper on binary forms to *Crelle's Journal,* in which he diplomatically referred to Cayley's success with binary forms of the first four orders. Then he announced his theorem: for a binary form of *any* order n, there *is* a *finite* set of irreducible invariants and covariants. This flatly contradicted Cayley's central result in his "Second Memoir" published ten years earlier. The effect on Cayley can only be guessed, but there is no doubting the importance of the theorem. Gordan had established a pivotal point of reference for the whole invariant theory program.

Though Gordan proved his "finiteness theorem," his method did not give a constructive formula for directly listing the irreducible invariants and covariants or even a formula for counting them. Gordan's technique was to generate a complete system of invariants and covariants that was finite (and hence the theorem), from which the reducible ones could be found and discarded. He admitted that "for the general case of the given system is always large but may be reduced in the special case; for forms of the fifth and sixth order I have done the reduction and a possible small system of groundforms established [twenty-three groundforms for the binary form of order 5, and twenty-six for the form of order 6]."[39]

Gordan's irreducible invariants and covariants for the binary quantic were expressed in a symbolic notation referred to by Cayley as the "Clebschian" notation. For instance, the binary quadratic form, written by Cayley as $u = ax^2 + 2bxy + cy^2$ was written in the symbolic notation as $u = (\alpha x + \beta y)^2$. The power of the "symbolic method" lay in the development of a formal calculus built around a transvection operation, a way of combining algebraic forms u, v to produce a third, $(u, v)^k$, called a transvectant.[40] This is closely connected with Cayley's hyperdeterminant operator $\overline{1\,2}^k uv$, which he himself discovered in 1844.

Cayley readily acknowledged the correctness of Gordan's result—he could hardly do otherwise—but the result must have been a shock. C. S. Peirce claimed Cayley had found the source of his error prior to Gordan's announcement and failed to alert the academic community, but this seems unlikely and I have found no evidence to support Peirce's surmise. The result did encourage Cayley to catalog the fundamental system of invariants and covariants for binary forms on his own terms. He wanted them expressed as full homogeneous expressions and was not content with them being expressed in the shortened

Clebschian notation. The German mathematicians lifted the theory onto a more abstract plane, but though Cayley appreciated the power of their symbolic calculus, he did not use it himself nor did he advocate its adoption generally. To him the notation was opaque and at a level of abstraction he was not prepared to accept. Expressing an invariant or covariant in symbolic form did not allow it to be examined in the minute detail he desired. It is perhaps a measure of this attitude that, despite the obvious success of the symbolic approach, it was not accepted in the English-language literature of invariant theory until the 1890s.[41]

While the German mathematicians worked with this partially abstract calculus, they did not eschew calculation, far from it. The preparation of classification schemes was a goal of their research program as it was of Cayley's. While Gordan's approach settled one outstanding problem, Cayley recognized that the method failed to provide an efficient algorithm for determining *irreducible* invariants and covariants. The scale of this weakness was apparent to Cayley, as he estimated that an aggregate of 429 transvection operations would be required for computing the twenty-three irreducible invariants and covariants of the binary quintic. For higher order forms, the number of transvections would be much greater, thus suggesting that the symbolic method was somewhat cumbersome when viewed as a method of calculating. Because Cayley was so concerned with calculation, the efficiency of an algorithm was paramount, thus only adding to his doubt that the symbolic method would form a suitable basis for the theory.[42]

Instead of adopting the new techniques, Cayley adhered to the tried and tested approach in which invariants and covariants were calculated through partial differential equations. This was the method discovered in 1851, which he said would "constitute the foundation of a new theory of Invariants."[43] Gordan's result had been rightly hailed but when the dust had settled, it could be seen that many problems were left standing: the most obvious being connected with binary forms of order greater than 6. The actual calculation of invariants and covariants remained a serious problem. The algorithmic method adopted by Cayley required the solution of difficult combinatorial problems with the use of generating functions, and this was a stumbling block. Gordan's result had settled one question but, as with most mathematical conquests, dozens more arise and these often involve finding more powerful techniques. In an aside, Sylvester wistfully remarked: "But why should we expect to do this [regarding a problem of enumeration in geometry] for all degrees seeing how limited our powers of enumeration extend in the case of Invariants which have been so long the subject of study? ([t]his consoling reflexion has only just occurred to me)."[44]

In November 1868, Cayley was elected the third president of the London Mathematical Society, following De Morgan and Sylvester. One of his first duties was to explain the serious discrepancy between his and Gordan's result, an admission all the more galling since his erroneous result had been broadcast to a wide audience in notes appended to Salmon's expanded *Lessons Introductory to Modern Higher Algebra* of 1866.[45]

At a meeting of the society in February 1869, Cayley drew members' attention to "an important discovery in the theory of covariants."[46] There was no suggestion of regret or the denting of national pride. Later it became a detail and even evidence to his contemporaries that Cayley was human after all—the mathematician so obviously a genius, standing head and shoulders above his British contemporaries, could falter. Sylvester thought it yoked his hero with Newton: "*Apropos* of the mistaken impressions of great men. Did not Newton live and die in the belief of the incurability of chromatic dispersion; Cayley affirm the infinitude of the number of asyzygetic invariants of binary quantics beyond the sixth order, thereby arresting for many years the progress of the triumphal car which he had played a principal part in setting in motion."[47] Cayley took Gordan's result as a new starting point for his own research and applied to the Royal Society for a grant of £10 to employ a human calculator to prepare tables for the four remaining extensive covariants: "According to a recent discovery of Prof. Gordan's the number of covariants (instead of being, as I had supposed, infinite) is = 23. Of these 17 (including the last of all, by far the most laborious to calculate, Hermite's 18-thic, calculated by Salmon) are calculated & published in my Memoirs on Quantics. I have since calculated two more; so that there remain 4 covariants which have yet to be calculated: I should wish to give these to Mr. Davis, to do for me."[48]

W. B. Davis obliged and toward the end of the year, Cayley had verified the calculation of the remaining covariants R, S, T, V at the summit of the quintic system. The twenty-three fundamental covariants had at last been obtained in his own notation.[49] (See "Invariants and covariants of the binary quintic as known more completely to Cayley in 1871" in the gallery on p. 15g.) Moreover, Cayley had found the mistake in his earlier work and explained the "theory being thus in error, by reason that it omits to take into account of the interconnection of the syzygies," and he was led to into making it "by reason that certain linear relations, which I had assumed to be independent, are really not independent, but, on the contrary, are linearly connected together."[50] The error was made for an understandable reason when seen in a historical context. In the 1850s, Cayley had only advanced to covariants of degree 5 (those with *coefficients* of degree 5) for the binary quintic form but for this degree the complicatory factor of relations between relations (secondary syzygies) does not exist. It was only in the higher reaches of the quintic system (covariants of degree greater than or equal to 8) that the new phenomenon occurred. Cayley had no practical experience of these in the 1850s and was unaware of the possibility of "secondaries" appearing. Nevertheless, it is surprising that his error published in the "Second Memoir on Quantics" went undetected for ten years. That this occurred at all is a measure of the intricacies of the binary quintic form and the small number of mathematicians who kept abreast of the subject and had enough detailed knowledge to suspect that Cayley's conclusion could be challenged.

It was not at all obvious to anyone how Cayley's methods could be united with those of Gordan's, for the *exact* connection between the two methods is far from being transparent. In the "Ninth Memoir on Quantics," Cayley presented a synopsis of Gordan's proof, but remarked that the proof was a difficult one to understand: "I cannot but hope that a more simple proof of Professor Gordan's theorem will be obtained—a theorem the importance of which, in reference to the whole theory of forms, it is impossible to estimate too highly."[51] The theorem was not in doubt, but neither he nor Sylvester gave up the hope of finding a "simple" proof independent of the German symbolic method. They saw this method as artificial and their alienation from it was compounded by the fact that they did not understand it sufficiently to make it into a natural way to proceed.[52] They steadfastly believed in their own view of invariant theory, and, partly in an attempt to show their plain algebra held the key, they continually strove for a nonsymbolic proof of Gordan's theorem.

The Astronomer Royal

The debate on whether Cambridge mathematics was an appropriate kernel of a mathematical education continued throughout the decade and beyond. In November and December 1867, Cayley became involved in a heated dispute with Sir George Biddell Airy, the Astronomer Royal and erstwhile Lucasian professor of mathematics. At Greenwich he conducted the affairs of Astronomer Royal with unsparing efficiency, and he earned a reputation as a tenacious bureaucrat. The team of calculators at the observatory knew him as an exacting employer. Airy originally made his mark in mathematics and had been a beneficiary of the reforms brought about by the Analytical Society in Cambridge in the early 1820s. He was the Senior Wrangler of his day and occupied a pivotal position in British science throughout the century, including a spell as president of the Royal Society. He was a man of wide scholarship and recondite intellectual interests, for example, using his knowledge of local tides in southern England he attempted to pinpoint Julius Caesar's beachhead. He was also public-spirited and keen to enter the public domain on various forays; in one incursion he issued a pamphlet suggesting the impossibility of building Paxton's Crystal Palace for the Great Exhibition of 1851.

The clash between the practical astronomer and the mathematical theoretician marked out their positions on the teaching of mathematics. Superficially, their disagreement appeared to hinge on the appropriate balance between pure and applied mathematics to be taught at Cambridge. As the debate unfolded, Cayley's attitudes toward the teaching of mathematics became clearer and at variance with the place of mathematics in a liberal education; it was more than a vehicle which provided "training for the mind." While Leslie Stephen had attacked the abstruseness of Cambridge mathematics and the idea that beauty (as perceived by mathematicians) should be the supreme criterion for judging its worth, Airy's views counted for more among the mathematical fraternity. He was a mathematician and practicing scientist and one with a strong interest in the physical applications of mathematics. He was not satisfied by mathematical sophistication alone and, to this end, directed a vigorous campaign aimed at the introduction of more applied mathematics into the Tripos. Airy was deeply interested in educational matters and waged his long campaign with energy.

Ten years before, Airy had bemoaned sending 22-year-old Cambridge graduates into the world without the slightest knowledge of important physical problems. Among these was the important question relating to the shape of the earth viewed as a rotating body, the ever-present "figure of the Earth" question. This was a problem dear to Airy's heart and years before had been a source of controversy between himself and James Ivory.[53] From the start, his educational campaign was vibrant. He suggested to the vice-chancellor that a university capable of omitting this subject "has already sunk to the position of a second-rate Academy."[54] He also had a personal axe to grind. His two eldest sons, Wilfrid and Hubert, had been in residence at Trinity College and he had become concerned about the mathematics they were *not* studying. Armed with inside information, his letter to the vice-chancellor complained of the lack of practical results and that "the Figure of the Heterogeneous Earth—are absolutely forbidden for examination. Yet the results of this theory are known in the scientific world, not only to the mathematical men, but also to every military or naval officer of moderate acquirements, who has been employed in national surveys or pendulum experiments."[55] This was a rash statement but in keeping with Airy's sense of drama, for there was no disputing the problem's importance. Not only had the "Figure of the Earth" question taxed the greatest minds, it continued to challenge, and even today is far from being solved. Airy seemed to be thinking of the "textbook" solution of a simplified problem, which could undoubtedly be learned for examination purposes. Cayley knew the problem only too well, it having been taught to him as an undergraduate, added to which he had subsequently written several research papers on the subject, a problem he saw as one of "Analysis and Geometry."[56]

A sharper attack was to come from Airy after the Sadleirian professor was installed. In the Senate House where Airy gave the Rede lecture in 1864, he aimed some well-chosen darts at pure mathematics as a useful Tripos subject. By then his elder sons had graduated, but his third son, Osmund, was in residence, having been admitted to Trinity College in the year Cayley was appointed. Of his three sons, Osmund showed the surest sign of mathematical talent (he graduated twenty-seventh wrangler in the order of merit). Now a concerned parent attuned to the teaching offered at Cambridge, he wrote once more to the vice-chancellor, proposing that the Tripos course be extended to mathematical physics. In May 1866, the Board of Mathematical Studies met at Garden House with Cayley presiding. The board (which on that day included Thomson as a guest) made no resolution, but at a following meeting in November, it formed a subcommittee to consider Airy's proposed schemes. The board's report of 8 May of the following year was the outcome.

The board's report set the stage for addressing the dilemma posed by the mathematical Tripos. The problem before the mathematical community at Cambridge was to reconcile the continually changing face of mathematics with the traditional comprehensive nature of the Tripos examination and its long-standing stability. Part of the report proposed reinstatement of applications (heat, electricity and magnetism, elastic solids) lost in the earlier reform of the 1840s, and the introduction of a course on elliptic integrals was also suggested. The board stressed the importance of some of Cayley's subjects and had "already recommended the re-introduction of Laplace's Co-efficients and the Figure of the Earth considered as heterogeneous, in doing which they are partly influenced by the publication of a recent work on these subjects. While curtailed in some directions, the [Mathematical Tripos] course has however extended itself in others, especially in those of analytical geometry and higher algebra, so that it is really as extensive as ever, while yet there are some important subjects which are entirely omitted." It noted that, since "the progress of science is continually enlarging the field of Mathematical knowledge by extending the range of pure mathematics, and reducing fresh branches of physics to mathematical laws," there was a need to continually evaluate the course of mathematics and maintain the comprehensiveness of the examination without making it too onerous.[57] Following the report's publication, the mathematical professors met to consider ways in which the Tripos could be modified and the Smith's Prizes distributed. Ahead of any decisions, Cayley was involved in a short, sharp correspondence with Airy in November and December. Meanwhile, Airy was among the first batch of honorary fellows of Trinity elected in June of 1867.

Cayley, the retiring scholar with the manners of a gentleman from an earlier generation, was matched against Airy, highly organized with a sure gift for pedantry, who had an established reputation for forthright criticism. Of different temperaments, Cayley's astuteness would be required to repulse Airy's outspoken attack; the course might also be unpredictable, for Airy had views on everything and he was fond of giving them. James Stuart, a young tutor at Cambridge described Airy as the "most original person" he had met and, more-over, one who was "colossal minded."[58] In the debate, Airy had already suggested that the large part of "useless algebra" should be reduced and applications of mathematics be introduced. It is part of a familiar theme experienced by generations of mathematicians, who, from time to time, are called on to defend their subject in terms of practicality and Cayley now took his turn to repulse the attack. The appointed authority and standard bearer of pure mathematics was pitted against a man with diametrically opposed views of what constituted useful mathematics. The opening salvo came from Airy in the guise of a discussion on partial differential equations: "I do not know that one branch of Pure Mathematics can be considered higher than another, except in the utility of the power which it gives. Measured thus, the Partial Differential Equations are very useful and therefore stand very high, as far as the Second Order. They apply, to that point, in the most important way, to the great problems of nature concerning *time,* and *infinite division of matter,* and *space:* and are worthy of the most careful study. Beyond that Order they apply to nothing."[59]

The pugnacious Airy continued his assault, deriding the usefulness of subjects such as analytical geometry for the majority of students and raising the value of differential calculus on account of it encouraging a "logical habit." In his reply, Cayley neatly deflected the argument by making direct appeal to the statutory duty of the Sadleirian professor: "to explain and teach the principles of pure mathematics," a useful shield he regularly employed when under attack. He parried the thrust of Airy's remark on partial differential equations: "As to Partial Differential Equations, they are 'high' as being an inverse problem, and perhaps the most difficult inverse problem that has been dealt with. In regard to the limitation of them to the second order, whatever other reasons exist for it, there is also the reason that the theory to this order is as yet so incomplete that there is no inducement to go beyond it; there could hardly be a more valuable step than anything which would give a notion of the form of the general integral of a Partial Differential Equation of the second order." He saw an educational defect in Airy's proposal and seized on its narrow vision: students would not be able to develop their own methods with their limited background.

[T]here is scarcely anything that a student can do for himself:—he finds the integral of the ordinary equation for Sound—if he wishes to go a step further and integrate the nonlinear [differential] equation

$$\left(\frac{dy}{dx}\right)^2 \frac{d^2y}{dt^2} = a^2\frac{d^2y}{dx^2},$$

he is simply unable to do so; and so in other cases there is nothing that he can add to what he finds in his books.[60]

He was on surer ground when he countered Airy's views on the teaching of modern geometry. It was in this subject, Cayley maintained, in which there was room for experimentation and, through such, the opportunity of developing geometrical intuition: "Whereas Geometry (of course to an intelligent student) is a real inductive and deductive science of inexhaustible extent, in which he can experiment for himself—the very tracing of a curve from its equation (and still more the consideration of the cases belonging to different values of the parameters) is the construction of a theory to bind together the facts—and the selection of a curve or surface proper for the verification of any general theorem is the selection of an experiment in proof or disproof of a theory."[61]

It was this blend of induction and deduction of which Cayley was so fond. The Astronomer Royal could hardly lecture Cayley on experimentation, for it was precisely the way he carried out his geometrical investigations. On this point Airy would find it difficult to be convinced, though Cayley bravely asserted that geometry was a developing theory just as much as physics. Moreover, he intimated that geometry developed all the faculties Airy claimed could be achieved only by studying physics. By seeing pure mathematics as subservient to science, Airy argued for the teaching of those parts of mathematics that could help students understand *real science*. He balked at any hint of artificiality, as might be found, he suggested, in a "game of billiards with novel islands on a newly shaped billiard table." His accommodating nature sorely tried, Cayley insisted that pure mathematics was a science in its own right with its own intrinsically interesting problems. These were as real as any in Airy's physical world, he contested and, perhaps unwisely, argued that Airy's exotic extension of ordinary billiards "*if it* were found susceptible of interesting mathematical developments, would be a fit subject of study."[62]

As the debate proceeded, Cayley began to lose sight of the average undergraduate. There could have been few among the annual crop of Tripos contenders with an appreciation of the famous "three-body-problem," and still fewer who seriously contemplated its solution, a problem that had attracted legions of mathematicians, including Newton and Euler. In his essay of 1771, Lagrange had brought elliptic integrals to bear on the problem. In its classical form, the problem required an exact mathematical formula describing the motion of three bodies in space (traditionally the Sun, Moon, and Earth) subject only to their mutually attractive gravitational forces. Cayley fuelled the Airy debate with this example: "But admitting (as I do not) that Pure Mathematics are only to be studied with a view to Natural and Physical Science, the question still arises how are they best to be studied in that view. . . . Now taking for instance the problem of three bodies—unless this is to be gone on with by the mere improvement in detail of the present approximate methods—it is at least conceivable that the future treatment of it will be in the direction of the problem of two fixed centres, by means of elliptic functions, &c.; and that the discovery will be made not by searching for it directly with the mathematical resources now at our command, but by 'prospecting' for it in the field of these functions. Even improvements in the existing methods are more likely to arise from a study of differential equations in general than from a special one of the equations of the particular problem."[63]

It was Cayley's ill-advised sentence: "I do not think everything should be subordinated to the educational element," which caused Airy the greatest consternation. "I cannot conceal my surprise at this sentiment," he wrote, "[a]ssuredly the founders of the Colleges intended them for education (so far as they apply to persons *in statu pupillari*), the statutes of the University and the Colleges are framed for education, and fathers send their sons to the University for education. If I had not your words before me, I should have said that it is impossible to doubt this."[64] When Airy was a student at Cambridge in the 1820s, the University had been primarily a teaching institution, but, with the new professors appointed in the 1860s, the seeds of change were planted, and the "educational element" would not be the sole objective of the university.

Research in science became a watchword but research in mathematics was utterly dormant in the 1860s. William Peveril Turnbull, who gained a Trinity College fellowship in 1865, published *Analytical Geometry* (1867), but the "book was produced at a time when abstract thought was rather at a discount, for physical research was in the ascendant."[65] Twenty years were to elapse before corporate attention was shown to research allied to pure mathematics as an important activity, but, even then, recognition was slow. As the curtain on this Airy-Cayley sideshow was about to descend, Cayley's general views on education crystallized: "my idea of a University is that of a place for the cultivation of all science."[66] This included pure mathematics and it was to be studied for its own sake, his lofty position being that mathematics was a thing of beauty and did not need a practical purpose to motivate and legitimize its study.

The debate over the structure and content of the mathematical Tripos was of national interest and the debate, which took place in the University Senate, in which Cayley argued for pure mathematics, was reported in the *Times*.[67] William Clifford, the young scientific "man about town," came to Cayley's aid when he lectured at the Royal Institution of Great Britain on "Conditions of Mental Development." Putting the case for active involvement in research as the ideal, as opposed to the time-honored, but passive assimilation of established knowledge, Clifford illustrated his argument in the case of abstract mathematics: "I am here putting in a word for those abstruse mathematical researches which are so often abused for having no obvious physical application. The fact is that most useful parts of science have been investigated for the sake of truth, and not for their usefulness."[68] If Airy was not in Albemarle Street to hear the lecture, he would have heard of Clifford's views on the scientific grapevine. His own proposal to judiciously prune Cambridge studies of the somewhat "luxuriant growth of pure algebra [and] analytical geometry" would be an action that would certainly hinder science in Clifford's eyes. Clifford defended the thrust of Cayley's argument: "A new branch of mathematics [abstract analytical geometry], which has sprung up in the last twenty years, was denounced by the Astronomer Royal before the University of Cambridge as doomed to be forgotten, on account of its uselessness. Now it turns out that the reason why we cannot go further in our investigations of molecular action is that we do not know enough of this branch of mathematics."[69]

Many Cambridge dons who taught mathematics would not sympathize with the philosophy that undergraduates should be studying mathematics as a living entity. For them, the subject provided "training of the mind," but it was not a serious subject with its own raison d'être, as Cayley continued to argue. But what of his attitude toward Airy's interventions? Perhaps it is summed up by his oblique reference made to John Couch Adams on one occasion of Airy hearing what he wanted to hear: "I do think his [the Astronomer Royal] wanting advice is very much in the style of the song 'I'll gie you my bonny black hen, gin you'll advise me to marry, The lad I loe dearly Tam Glen'—only there is no bonny black hen to be had for giving it."[70]

Another report was issued, and this time Airy was satisfied and declared it a well considered and valuable document while William Thomson thought the proposed changes "were exactly what was wanted."[71] The practical upshot of the whole debate was that the advanced part of the mathematical Tripos would contain a rigorous mathematical study of such applied topics as heat, electricity and magnetism.[72] At the end of the 1860s these studies were gradually introduced. For the honors degree students, the reforms meant an examination period of nine days, an increase in an already overstuffed examination period. There was no agreement on reform of the Smith's Prizes, but to remove some of the competitive intensity of the system, some reformers wanted the Tripos results list published in alphabetical order instead of the traditional merit order. Here the conservatives had their way, and Cayley for one was firmly of the opinion that the order of merit should be retained. The new regulations for the Tripos were approved by the Senate and came into operation for the graduating class of January 1873.

In a parallel development to the debate in mathematics, the university was considering its position in relationship to *empirical* science, in which the topics of heat, electricity, and magnetism would be treated as branches of physics. This lay outside the competence of the Cambridge professors, who were schooled in mathematics. While suggesting that a little of their subject was essential to science, they swiftly curtailed any possibility that they should be drawn into service on or near any laboratory bench. To a man, the holders of the Cambridge chairs, Cayley, Stokes, and Adams invoked their conditions of service to avoid this possibility but none did it so definitely and tersely as Cayley: "The Sadlerian Professor by the terms of his professorship is required to devote his attention to pure mathematics." Furthermore, the professors "do not think they are able to meet the want of an extensive course of 'Lectures on Physics' treated as such, and in great measure experimentally" and thus they had to propose the appointment of a new professor who could handle experimental physics. Cayley was one of the leading supporters of this idea.[73] Following this, the Cavendish Laboratory was founded and James Clerk Maxwell appointed as its first director. With his tongue-in-cheek brand of humor, Maxwell was thus able to treat the mathematical professors to a lesson on the usefulness of mathematics in the opening lecture of his course: an exposition of the mathematical relation between the Fahrenheit and Centigrade scale.[74] In his own sphere, Cayley went on with his own lecture courses as before and with the ritual of conducting advanced lectures before very small classes.

Euclid in the Dock

At school level, Cayley was also called on to make pronouncements on mathematical education. This duty had little to do with Cayley's experience as a schoolteacher, for he had none, but everything to do with the Cambridge professor speaking *ex-cathedra*. A leading question of the day in the world of mathematical education concerned the position of Euclid's *Elements* in the school curriculum.[75] In the late 1860s there was widespread concern that Euclid's geometry, as taught in English secondary schools, was archaic when compared with the more flexible approach to geometry taught in Continental schools. The French had abandoned Euclid in the Napoleonic era, and teachers in the lycées developed their own courses in geometry. Support for Euclid from mathematicians was mixed; for example, both Enrico Betti and Francesco Brioschi worked hard to establish it in Italian schools, but there was opposition too. In England, teachers were obliged to follow geometrical propositions in Euclid's order without deviation, often leading to the parroting of Euclid without any conception of what it meant to think geometrically.

Pressure for change was increasing and voices were raised against Euclid's *Elements* as it was employed in the schools. The Clarendon and Taunton Royal Commissions of the 1860s looking into secondary education, the Clarendon to the public schools and the Taunton more generally, both pointed to defects in the teaching of geometry.[76] The publication in 1868 of the School's Inquiry Commission Report of the Taunton Commission, in twenty-one volumes, prompted administrative action, and, as a result, in the following year a committee of the British Association was formed to consider the possibility of improving the methods of instruction in elementary geometry.[77]

Cayley was a member and other names were familiar to those in the university world: Clifford (Secretary), Sylvester, Hirst, and H. J. S. Smith. From Ireland, George Salmon and Richard Townsend, and from Scotland, Philip Kelland and Frederick Fuller. The token schoolmasters in this top-heavy committee, R. B. Hayward and James Wilson, were hardly representative of all teachers.[78]

Hayward had been a Cambridge tutor and taught at Harrow from 1859 until 1893. In the year of his appointment to the committee, he enjoyed the distinction of being one of the few schoolmasters to act as an examiner for the mathematical Tripos. A man of combative spirit, who took to controversies in the pages of *Nature* with relish, he was elected to the Royal Society in 1876.[79] "Haycock," as he was known to the Harrovians, suffered from the well-known condition of being an inspiration for the few but a remote figure for the many: he was unable to make contact with the generality of pupils who regarded mathematics as a chore. Euclid was a punishment for most pupils: at Harrow, the lower classes were either required to read the book out loud or to copy it out in writing, but both activities were accompanied with no explanation from the teacher.[80] Wilson was a mathematical master at Rugby, a leading public school, and had been a Cambridge Senior Wrangler. An early member of the London Mathematical Society, he was proposed by Sylvester and Hirst for membership. When we recall Wilson's thirst for new mathematical ideas as a Cambridge undergraduate in the 1850s, his verdict on Euclid as obscure, tedious, and barren can be appreciated easily.

On balance, there was a mood for reform among the university representatives but it was close: Smith, Hirst, and Sylvester were in favor of liberalization while Kelland and Cayley were against change. Both schoolmasters were anxious for change, knowing the dilemma of their thoughtful schoolmaster colleagues forced into a straitjacket of teaching Euclid by heart to those who "though they wrote it all by rote, they did not write it right."[81] Views were strongly held, none more so than by Sylvester. As president of the Mathematics and Physics Section at Exeter, his address indicated the committee debate in store. Glancing around his audience, and perhaps thinking of the range of assembled views, particularly those of his "spiritual progenitor," he reflected: "The early study of Euclid made me a hater of geometry, which I hope may plead my excuse if I have shocked the opinions of any in this room (and I know there are some who rank Euclid as second in sacredness to the Bible alone and as one of the advanced outposts of the British Constitution)."[82] When he sensed there was a case to be made, he summoned the writers of the literary canon to his side. With Euclid, Shakespeare was pressed into service, and quoting Prospero's speech from *The Tempest*, Sylvester proposed to bury Euclid "deeper than did ever plummet sound."[83]

To Cayley, the proper way to teach geometry to schoolchildren was from the sacred book. To him, Euclid's propositions and its logical structure represented perfection independent of any learner difficulties—and this was the problem. The schoolmasters inhabited one community and he another. His thoughts were consumed by geometry of a more advanced character. Only the previous year he had put down his pen on his series of memoirs on surfaces called "scrolls."[84] In the few weeks following Exeter, he laid out a scheme for the study of abstract geometry, both from the point of view of analysis giving rise to geometry and for geometry illuminating the analysis. "I SUBMIT to the [Royal]

Society the present exposition of some of the elementary principles of an Abstract *n*-dimensional Geometry," he wrote:

> The science [of Abstract Geometry] presents itself in two ways,—as a legitimate extension of the ordinary two-and three-dimensional geometries; and as a need in these geometries and in analysis generally. In fact, whenever we are concerned with quantities connected together in any manner, and which are, or are considered as variables or determinable, then the nature of the relations between the quantities is frequently rendered more intelligible by regarding them (if only two or three in number) as the coordinates of a point in a plane or in space: for more than three quantities there is, from the greater complexity of the case, the greater need of such a representation; but this can only be obtained by means of the notion of a space of the proper dimensionality; and to use such representations, we require the geometry of such space.[85]

This was a far cry from the classroom, and at the end of 1869, his "geometrical eye" had a higher focus than the difficulties experienced by schoolchildren. If he thought about Euclid it would be from the perspective of a mathematician: the Eudoxian theory of proportion was sublime, for example, but he could not easily place himself in the way of a child struggling to learn *Book I,* 47 (Pythagoras's theorem) or other propositions they were required to commit to memory.

The debate on Euclid and its position in the school curriculum would be an active debate for the next thirty years with Cayley at its center for the most part. He was the professor at Cambridge and by this fact alone, he was a pivotal figure in mathematics education in the country. Swayed by the perfection of Euclid, he acted on the basis of entrenched views, hardly connecting with the plight of the schoolmaster. He too readily accepted the "absolute monarchy of Euclid" on the basis of its worth as a sacred text.[86]

CHAPTER 13

A Representative Man

The 1870s saw Cayley at the peak of his mathematical production. In the opening year alone, he published a torrent of papers amounting to more than three hundred pages of mathematics. Now in his fiftieth year, he took leading positions in scientific societies and was increasingly recognized as a Representative Man, as Emerson called the rare breed whose exploits defined an era. He was the representative British pure mathematician in the same way that Alfred Tennyson stood for poetry or Charles Dickens or William Makepeace Thackeray for literature. Cayley was allotted his position in the pantheon of science and the arts in an age that demanded greatness of its leaders. If a member of the scientific aristocracy had been asked what it meant to be a mathematician, the reply invariably would have been, "to be like Cayley." The influential philosopher Herbert Spencer proudly claimed the "A1 among mathematicians" as his ally.[1]

At the beginning of the decade, his children Henry and Mary were born, but outside the world of family and academic life, the relative calm of the 1860s led to a more turbulent decade.[2] On the economic front, Britain was increasingly challenged for supremacy by the United States and the emergent German state, which arose after the Franco-Prussian War of 1870–71. At home, after an economic boom year in 1873, a deep agricultural depression began in the next year with a crisis year in 1879. Professors at Cambridge were not immune, as their chairs were partly funded from income derivable from farming rents.

"Drawn Round His Camp"

Cayley was universally admired at Cambridge and it was time that he should be honored with a portrait hung in Trinity College. Lowes Cato Dickinson was the choice of artist, a highly respected portraitist whose subjects included scientists George Gabriel Stokes, William Thomson, and James Clerk Maxwell. From the ranks of the Victorian establishment, those who also sat for him included Charles Kingsley, Richard Cobden, General Gordon, W. E. Gladstone and his 1868 Cabinet, and Queen Victoria herself. Maxwell, who had returned to the university to be director of the new Cavendish Laboratory, addressed the 1874 portrait fund committee in his own way:[3]

O WRETCHED race of men, to space confined!
What honour can ye pay to him, whose mind
To that which lies beyond hath penetrated?
The symbols he hath formed shall sound his praise,
And lead him on through unimagined ways
To conquests new, in worlds not yet created.

First, ye Determinants! in ordered row
And massive column ranged, before him go,
To form a phalanx for his safe protection.
Ye powers of the nth roots of -1!
Around his head in ceaseless cycles run,
As unembodied spirits of direction.

And you, ye undevelopable scrolls!
Above the host wave your emblazoned rolls,
Ruled for the record of his bright inventions.
Ye Cubic surfaces! by threes and nines
Draw round his camp your seven-and-twenty lines—
The seal of Solomon in three dimensions.

March on, symbolic host! with step sublime,
Up to the flaming bounds of Space and Time!
There pause, until by Dickinson depicted,
In two dimensions, we the form may trace
Of him whose soul, too large for vulgar space,
In n dimensions flourished unrestricted.

Maxwell allowed himself to be inspired by mathematical topics such as determinants, the "roots of -1," and the cubic surfaces, so long a subject of Cayley's study. A high point of the geometer's art was "Solomon's Seal," formed from the famous "27 lines" in the cubic surface, a three-dimensional analogue of the six-pointed star of David on King Solomon's signet ring.

Cayley may not have enjoyed the limelight cast on him by this verse—which Maxwell so obviously enjoyed composing—but it shows the affection in which he was held in the academic community. It also suggests a lone hero of mathematics, thinking his own thoughts. It intimates intellectual isolation: from non-scientists naturally, from scientists probably, and even from a greater part of the mathematical literati. The erudition is not in doubt, but who else could escape the "two dimensions" and enter the world where he "flourished unrestricted"? Who in the younger generation could learn from his expertise? There were few, but two who sat at his feet were W. K. Clifford and J. W. L. Glaisher.

Clifford was the first to be taken under his wing. During the period of his Trinity fellowship following the Tripos in 1867, Clifford found his mathematical research topics in Cayley's subjects. The young man who rushed to shock Victorian society with his unbridled atheism simultaneously impressed them with his brilliance in mathematics, science, and philosophy. In his years at Cambridge, he benefited from the Cayley's knowledge and protection. This is seen from the range of Clifford's research but shown best by little instances, as occurred when the young man was growing up mathematically; when he sent off a solution to a problem in William Miller's *Mathematical Questions*, a solution which reaped some criticism, Cayley was quick to defend him. In a letter to the editor, he wrote, "[i]t is rather unfortunate that anything should have been said about Mr Clifford's demonstration as to the roots of an equation— I pointed out to him that it is substantially the same as is given in one of the Notes to Lagrange's *Equations Numériques*—Lagrange uses the theory of roots to show that the degree of the equation resulting from an elimination is $\frac{1}{2}n(n-1)$, and it is of course an improvement to obtain this by a direct process of elimination."[4]

Clifford left Cambridge in 1871 to become professor of applied mathematics at University College in London. In geometry and algebra, Clifford especially valued Cayley's theory of distance and matrix algebra. Cayley's "prepotentials," essentially work on multiple integrals, was thought of by Clifford as "a contribution of great value to this most important branch of mathematical physics."[5] In March 1874 Cayley organized his election to the Royal Society, citing his work on analytical metrics, Miquel's geometrical theorem on systems of circles, biquaternions, polyhedra, and sundry geometrical results in support.

Cayley noted that Clifford "was distinguished for his acquaintance with Metaphysics of geometrical and physical science and as an original investigator in the same."[6] In terms of philosophy, Clifford was a "progressive" and he a "conservative," the young firebrand contrasted with the sober eminence; Clifford rejected Kant's a priori notions while he accepted them. But then, Cayley would make no claim to be a metaphysician and hardly strayed into this territory, which was inevitably controversial and involved debate. It was Clifford's mathematics he valued most. He thought especially highly of Clifford's work on geometry and singled out his work on Riemann surfaces, the different types of contact which could exist between surfaces, and on the flatness of curves in n-dimensional space for special mention. Although invariant theory is implicit in his thinking, Clifford took his most active interest in this at a later stage of his career. In 1878, he suggested a method for computing invariants and covariants for the binary forms of orders 5 and 6 and was captivated by the link between invariant theory and chemistry. It is a measure of both men that when Clifford died at a young age the following year, he was judged to be the "most remarkable [British] mathematician of his generation and promised to be a second Cayley."[7]

The young James Whitbread Lee Glaisher was the elder son of the meteorologist and balloonist James Glaisher. The young man was groomed for science—as a 20-year-old his father had purchased him life-membership in the British Association.[8] Glaisher Jr. graduated second wrangler in 1871 and became a fellow of Trinity College in the same year, the earliest opportunity which regulations permitted. He was elected to the London Mathematical Society (and shortly to its council) and joined the Royal Astronomical Society—he so enjoyed the round of scientific meetings. An undergraduate paper written on the numerical values of integrals was communicated to the Royal Society by Cayley, who wrote in support of his protégé: "it appeared to me that the *Philosophical Transactions* would be the proper place for the not very bulky result of an enormous piece of good honest work, which anywhere else might easily be overlooked, just when the numerical values happened to be wanted."[9] Glaisher was renowned for his accuracy in calculations, his industry in mathematical table-making exceptional, even for the Victorians. He was elected a fellow of the Royal Society before his twenty-seventh birthday, supported by Cayley.[10] At Cambridge, he was Cayley's closest contact in pure mathematics and was to spend his entire career at the university.

Though less distinguished, in many ways Glaisher's mathematical life parallels his mentor. Like Cayley, Glaisher published a mathematical paper while an undergraduate and was to publish almost four hundred during a lifetime of constant industry. His dislike of personal advertisement could have been learned from Cayley, as could the acquisition of a frame of mind which took unremitting industry as the norm. His life was one of continuous service. Apart from the scientific societies he served, he was an energetic editor of the *Messenger of Mathematics* from its inception in 1871 until his death in 1928 (the entire period of its existence), and for the period of 1878 to 1928, simultaneously an editor of the *Quarterly Journal of Pure and Applied Mathematics*. There were differences as well, and the mythical don who spent his life moving serenely along Cambridge rails is more readily identifiable with Glaisher than with his mentor. Glaisher was emotionally attached to college life, and when his friends left the academic community for future careers outside Cambridge, it was a matter of disappointment to him. Cayley had his London years at the Bar and was only briefly a don with rooms in College. He was more focused than Glaisher and would not be diverted by college teaching and the consuming hobbies of live-in Cambridge dons, whether it was a mineral collection, a collection of church music, or the magnificent collection of porcelain Glaisher left to the Fitzwilliam Museum.

Both Clifford and Glaisher graduated in the "old style" mathematical Tripos, before the reform that followed in the wake of Airy's debate with Cayley. The new system encouraged students to specialize and this had a critical effect both on the way the mathematical curriculum was delivered and the staff who were required to teach it. It was no longer possible for the college tutors to be "founts of all knowledge" and the same challenge was posed to the private coaches. What man could single-handedly lead his students through the thicket lined by the new plethora of subjects, let alone be the expert in them? The range encompassed higher algebra, the theory of equations and probability, higher analytical geometry, higher dynamics with precession and nutation, the theory of potential (including its application to electricity, magnetism, attraction and the figure of the Earth), calculus of variations, higher differential equations and finite differences, sound (with the vibrations of strings and rods), dynamical theory of heat, theory of elastic solids, hydrodynamics, elliptic integrals, including their application to the theory of heat. Only rare college lecturers, such as Horace Lamb, famed coaches as E. J. Routh, or university professors like Cayley could effectively lead the students "to conquests new, in worlds not yet created" as suggested by Maxwell in his poetic eulogy.

Coupled with this need for special knowledge, the students faced a bewildering amount of material as the once-and-for-all Tripos examination swelled to nine days and they tended to gain superficial details of each specialism as they scrambled for examination marks. As the new scheme was introduced, it became clear, almost at once, that a reevaluation of the Tripos was a necessity. The education lay with some old hands, William Walton and N. M. Ferrers, and young lecturers Clifford, Glaisher, James Stuart, Richard Pendlebury, and George Greenhill, but they specialized and could not be expected to cover all the subjects as was expected of the traditional teaching don. On a broader canvas, scientific education was in the political spotlight at the beginning of the 1870s, when the seventh Duke of Devonshire, who headed a Royal Commission on Technical and Scientific Instruction, began work in 1870.[11]

The hallowed annual award of the two Smith's Prizes was under scrutiny again. Years before, these served to supplement the Tripos by a written examination of the same type but by testing more advanced material. After the reforms of 1873, they became of little educational value as the Tripos had been extended to cover this material. Moreover, after nine days, which student would be inclined to another week of writing furiously? The prizes could not be abandoned altogether as they were part of the Cambridge fabric: the examining professors had their dining rituals and, though of little monetary value, the prizes maintained their immense prestige. In February and March of 1875 a hotly contested debate ensued between the old protagonists, Cayley and Airy, on the way they should be administered. Airy endeavored to divine the intention of the will of the eighteenth-century Master of Trinity Dr. Robert Smith in setting up the

Smith's Prizes. Once more, he spoke about the "pernicious preponderance—of a class of Pure Mathematics, which is never likely under any circumstances to give the slightest assistance to Physics."[12]

Cayley could not ignore the debate drummed up by Airy. Unlike his opponent, he was a reluctant publicist, but this did not prevent him from entering the lists on this occasion, and the printed "flysheet," in which he employed all the arts of his legal training, is masterful. When it came to the legal niceties of "wills," Cayley was more than a match for the Astronomer Royal. In response to Airy's harangue, he calmly wrote: "Two remarks arise: first, it is rather bold to assume of any class of Pure Mathematics that it is never likely under any circumstances to give the slightest assistance to the really important subject of Physics. Secondly, the testator [Robert Smith] may have considered, and men who have entered into the Science of the world may also consider, that there is some importance in the Science of Pure Mathematics for its own sake, whether it is or is not likely to be of any assistance to Physics." The defense supplied by Cayley was hardly new, and being aware of it from their previous encounter, Airy swiftly dismissed it. While they had both passed through the Tripos system and were outstanding products of it, Airy had become overly concerned with the needs of the moment. His position as Astronomer Royal at his court in Greenwich was concerned with practical results, while Cayley's concern was for mathematics itself and the intellectual contract attached to his own professorial chair. Their relative positions are more evident from the second remark in the continuing paragraph of Cayley's flysheet. Airy would have appreciated that Descartes's contribution was "of some assistance" in physics, but Cayley took the opportunity of reminding him:

> Dr Smith was probably acquainted with, and may have desired to encourage, the study for their own sakes, of such works as the *Geometry* of Descartes, and the *Enumerato linearum tertii ordinis* of Newton:—the former of these has been of some assistance to Physics, and the Cartesian curves in particular may find an application to optical purposes; and I cannot assert that cubic curves will be always useless: it is at least supposable that Dr Smith desired to encourage the study for their own sakes of such developments of Pure Mathematics as should in after times occupy the attention of mathematicians.[13]

The debate quietened but Airy kept the Smith's Prize issue in his sights and sought future occasions to complain of the "purely idle algebra, [composed of] arbitrary combinations of symbols, applicable to no further purpose." One question asked of Smith's Prize students, which was bound to raise Airy's ire, was framed in terms of two circles of "imaginary radii," in which it is required to construct the common chord of intersection. A common chord can be constructed *analytically* whatever the nature of the radii, but an explanation in terms of the principle of continuity, so much favored by Cayley, left Airy unconvinced. His retort was only to be expected: "I am not so deeply plunged in the mists of impossibles as to appreciate fully your explanation."[14]

Cayley battled *sotto voce* to preserve pure mathematics at Cambridge, but though applied mathematics had gained the ascendancy, it was not all one-way traffic. Just the previous January, two young mathematicians had graduated as joint second wranglers. George Chrystal and William Burnside were to make their marks in pure mathematics—Chrystal's *Algebra* is a classic and Burnside's group theory pioneering—but they began their careers more concerned with the application of mathematics, the prevailing theme in the 1870s.

On graduation, Burnside was elected to a Pembroke College fellowship. Later a group theorist with an international reputation in that subject, he was first attracted to applied mathematics and gave a course of lectures on hydrodynamics. This was on the point of becoming a prominent Cambridge subject once more, staging a revival from the 1840s. Cayley's influence on his group theory is discernible. Burnside read his papers and his definition of a group is not radically different from Cayley's.

Chrystal had been educated in Aberdeen and had achieved a first-class degree from that university the year before entering Peterhouse in 1872. He found little to challenge him in the "stodgy" Cambridge course and hardly improved his knowledge base, but, in recompense, he valued being at the center of mathematics and science and making the acquaintance with up-and-coming scientists of his own generation. He seemed content to gain inspiration at a distance, pointing out that "it was no small matter to come even within view of such men as Cayley, Adams, Stokes, and Maxwell."[15] He graduated in 1875 and for two years taught around Cambridge and worked in the Cavendish Laboratory before returning to Scotland.

Scientific Circles

In November 1870, Cayley gave his retiring address to the London Mathematical Society. A society still in its infancy, it was the first meeting in new premises on Albemarle Street in London's West End, a few steps from the august premises of the Royal Institution of Great Britain. The establishment of the custom of presidential addresses at the end of their term was effectively begun by Cayley at this meeting and he chose to give an account of his recent researches on quartic and quintic surfaces.[16]

The choice of its first foreign members reflects the society's adherence to pure mathematics. Michel Chasles, with his long-standing connection with English mathematicians since the 1840s, had been elected in 1867 and, in 1871, Enrico Betti, Alfred Clebsch, Luigi Cremona, Charles Hermite and Otto Hesse joined him. Each one can be identified with Cayley's research. Hermite, in particular, was an adviser for the *Quarterly Journal of Pure and Applied Mathematics* when it was established fifteen years previously. Cayley maintained close links with Continental mathematicians, and when the reviewing journal *Jahrbuch über die Fortschritte der Mathematik* was founded in Berlin in 1868, he provided reviews and became a regular contributor.[17]

There was talk about Cayley serving a second term as president of the London Mathematical Society when J. J. Sylvester proposed him for the office. Public office went hand in hand with political machinations; evidently, the small size of the society was no barrier to maneuvers, and this time it was Sylvester's attempt to block Hirst's election. Given that only one year had elapsed since Cayley's first term had ended, it was a fanciful idea and after a close voting contest in the council, Hirst was elected by one vote.[18] Instead, Cayley became president of the Royal Astronomical Society. These waters were no calmer and in his two years of office, 1872–74, the society was transformed from a chivalrous organization going about its scientific business to one where polemic and intrigue was the order of the day. In the eye of the storm sat Cayley, his peacemaking skills tested to the limit.[19] Matters came to a head at the June council meeting in his first year of office. First, there was Arthur Cowper Ranyard's confrontation with Warren De la Rue over the possession of photographic plates of the solar eclipse of the previous year. The question was important to Ranyard, for in his many-sided career involving the law, publishing, and politics, his reputation as an astronomer rested on his expertise in solar eclipses.[20]

All societies experience minor squalls but other business was more far-reaching and contentious. The issue surrounding the status of physical astronomy was formally raised at a council meeting in May. Lieutenant Colonel Alexander Strange F.R.S. proposed that the president bring the need for a permanent national provision in astrophysics to the notice of the Commission on Scientific Instruction, which was then sitting, and of which he was an influential member. This gentleman had been a leading surveyor in the Indian Army and had assisted with the great trigonometrical survey of India. On his retirement to England, he turned to astronomy, and in his observatory at Lambeth he specialized in the testing of astronomical instruments. He was active in the administration of the Royal Astronomical Society and a zealous campaigner for state support for scientific research. After vociferous arguments for and against the involvement of the Royal Astronomical Society, the council was split between two opposing camps. One was led by Strange and backed by the combative self-taught astronomer and editor of *Nature,* Norman Lockyer. They argued the case for the setting up of an astrophysical observatory in England separate from the Royal Observatory at Greenwich. The opposing group favored a more general approach to astronomy under the control of Greenwich.

Airy, who had acted as peacemaker in the Ranyard–De la Rue confrontation, was drawn into the debate, but Strange's principal opponent was Richard Proctor.

After graduating from Cambridge in 1860, Proctor turned to writing popular books on astronomy after a bank crash. He was highly successful and published fifty-seven books on a wide variety of astronomical topics while at the same time contributing to serious astronomical research. When Cayley assumed the presidency, Proctor temporarily took over the editorship of the *Monthly Notices*, though this accommodation camouflages the fact that he too could also be a thorn in Cayley's side with his pet antagonism toward geometry being taught as a branch of invariant theory. Like Ranyard, his great friend and ally, he was free of institutional ties and spoke his mind.

During the "debate" on the observatory, Cayley had to cope with mass resignations, a "war of circulars" and offensive broadsides published in *Nature* and the *English Mechanic*. The clamor rocked the Royal Astronomical Society and deflected it from disinterested research and scientific activity. This can be judged by the careful statement issued in the annual report which successfully hid the antics of some ostensibly upright members: "After long and careful consideration of this subject [Astrophysics], extending over four meetings, two of which were especially convened for the purpose, and including the discussion of points importantly effecting as well the interests of science as the dignity of this Society, your Council by a large majority passed the following resolutions on 28 June 1872."[21] The society's resolution did not support the setting up of an independent observatory. There was already one at Greenwich, it was argued, and no need for another. The agitation did have some effect, but not before De la Rue, Strange, and Lockyer resigned from the council. The Board of Visitors at Greenwich (of which Cayley was a member) recommended photographic and spectroscopic records of the Sun to be carried out at Greenwich, thus taking the edge off the argument that proposed a new observatory be built for the purpose.

A piece of official business which would have given Cayley greater pleasure was the award of the Royal Astronomical Society's Gold Medal to Simon Newcomb in 1874. In his work at the United States Naval Observatory in Washington, Newcomb was a maker of charts and astronomical tables, and was even in the "Cayley league" as far as the production of scientific papers was concerned, publishing more than five-hundred in his lifetime. Newcomb was being honored for his research on the orbits of Neptune and Uranus, his contributions to mathematical astronomy, and Cayley singled out his "mathematical skill and power" and praised his "good hard work." At the medal ceremony, Cayley suggested that "[t]he formation of the tables of a planet may, I think, be regarded as the culminating achievement of Astronomy."[22] In his *Memoirs* Newcomb recalled how he mistook Cayley for an attendant in the anteroom and was surprised to be introduced to Professor Cayley. Disturbingly, there were also signs of illness, for Cayley's appearance indicated that he was no longer a young man and he looked considerably older than his fifty-three years. The occasion made a strong impression on Newcomb and he later recalled: "His garb set off the seeming haggardness of his keen features so effectively that I thought him either broken down in health or just recovering from some protracted illness" and he had wanted to say, "'Why, Professor Cayley, what has happened to you?'"[23]

In the middle of Cayley's term at the Astronomical Society, other plans were being made on his behalf. Unsuccessful in obtaining a second term for Cayley at the London Mathematical Society, Sylvester had a more prestigious position in mind for his unsuspecting friend: the presidency of the Royal Society—the P.R.S. In November 1872 Airy indicated his intention to resign as president; he was seventy-one years of age and had only held the office for barely one year. There were two leading contenders for the succession and the question reduced itself to whether the president should be a practicing scientist or a man of social rank. Joseph Dalton Hooker, the hardworking botanist, was the choice of the influential X Club members, while the Duke of Devonshire, at that time chairing his Commission on Scientific Instruction, was the favorite of the lobby who felt that the president should be an aristocrat.

Election as P.R.S. could be useful to a career, but it could also sap time from those who wanted little else than to pursue scientific work. Hooker felt this and wrote to Charles Darwin: "but my dear fellow, I don't want to be crowned head of science. I dread it—'Uneasy is the head,' &c—and then my beloved *Gen[era] Plant[arum]* will be grievously impeded. The dream of my later days is to be let alone, where I am and as I am—I want no high position, no dignities nor honors."[24] When Hirst dined with Cayley in the next month, he may have sounded out his feelings about the position and discussed the X Club's intention of asking Hooker to stand. At this point Sylvester stepped in. He took exception to the X Club and its influence, possibly since he thought of himself as a political operator on the London scientific stage and had not been asked to join the coterie. In Sylvester's eyes, Hirst was a member of a cabal, an "inveterate plotter" and definitely not to be trusted.[25]

With these views to the fore, Sylvester wrote to Stokes, the long-serving secretary of the Royal Society, recommending Cayley for the presidency in the strongest terms. "Cayley's claims are immeasurably superior to his [Hooker's] and I confess to having felt a little scandalised at finding a disposition to give [such a one] as Cayley so completely the go-bye—I presume that the present President [Airy] thinks well of Cayley and I still cherish the hope that with his (the President's) yours and Prof ͬ [W. H.] Miller's support he may receive the compliment that is so justly his due—provided, that is to say, that he would be willing to accept the office."[26] If Cayley had these aspirations, Sylvester would hardly have been the agent to choose to advance them. In reality, he stood outside the orbit of influence and would have raised more hackles than votes. In power-broking terms, Sylvester was not taken seriously. "He is a good creature," remarked John Tyndall to Thomas Huxley, who did "not require a hacked edge" but one to be humored is the obvious subtext of his footnote.[27] In the active lobbying which took place, Hooker was persuaded to let his name go forward, and, encouraged by the X Club members, he was duly elected.

The Parliament of Science

If the Royal Society represented the aristocracy of science, the British Association represented the people—the "Parliament of Science." For a few years during the 1860s, Cayley appeared to take a lesser interest in this institution than had been true earlier, but he attended at Exeter in 1869 and was a regular attendee during the 1870s. At Exeter, he discussed a geometrical problem in astronomy and, in the following year at Liverpool, with an admirable sense of continuity, resumed the problem of the in-and-circumscribed polygon. It was a favorite problem, and he picked it up at the same point he left it six years before.[28] At the 1871 meeting at Edinburgh he summarized Paul Gordan's work in invariant theory, preparing the way for his own paper on the subject, the "Ninth Memoir on Quantics."[29]

Alfred Clebsch published a book on the invariant theory of binary forms in the following year and there seemed an opportunity of meeting and discussing mutual interests with the German mathematicians. "Clebsch & Klein talk of being at the Brighton meeting [in 1872]," Cayley wrote to Hirst, "and I suppose I shall come to it, but I am waiting to hear again."[30] It was a chance for the Germans to travel abroad following the Franco-Prussian War, but it is unlikely they made the trip and Cayley made no presentation at Brighton in 1872. Felix Klein was busy putting together his work on the famous Erlangen Program in September and October, and Brighton would have been a distraction. Cayley and Clebsch, two leaders in invariant theory and algebraic geometry would have had much to discuss, but there would not be another opportunity: Clebsch died of diphtheria at the beginning of November.

The Bradford meeting of the association that took place in the following year was of particular relevance to mathematics, a meeting which Oliver Lodge described as a "mathematical orgy." Lodge became a staunch supporter of the British Association and hardly missed a meeting but the impressionable youth was never again "so moved as by that first meeting at Bradford in 1873."[31] There was plenty to interest the emerging scientific class and the assortment of interested amateurs who trailed around the sections hearing of the new wonders being discovered—it was the meeting where Clerk Maxwell gave his famous "Discourse on Molecules."

Henry Smith gave the opening address to the mathematicians of Section A, in which the day's proceedings lasted from "eleven to three" without respite. During this span, Hermite presented his recently constructed proof that the exponential number *e* is transcendental, Robert Stawell Ball talked about the use of dynamometers for measuring force, and Clifford gave a paper on elliptic geometry, a talk which made such a firm impression on Felix Klein, who was present on this occasion. Afterward Ball and Clifford kept Klein up to the early hours discussing geometry and were impressed by his extensive knowledge of the subject. By this time, Klein had extended Cayley's "Sixth Memoir on Quantics" and shown the fruitfulness of the projective metric. While Cayley demonstrated that Euclidean geometry was a special case of projective geometry in the "Sixth Memoir," Klein went further and showed how hyperbolic and elliptic non-Euclidean geometry could also be obtained from Cayley's notion of the absolute.[32] Klein was generous in recognizing Cayley's contribution, writing later: "This whole manner of viewing the subject was given an important turn by the great English geometer A. Cayley in 1859, whereas, up to this time, it had seemed that affine and projective geometry were poorer sections of metric geometry, Cayley made it possible, on the contrary, *to look upon affine geometry as well as metric geometry as special cases of projective geometry: 'projective geometry is all geometry'*" [his emphasis].[33] There were some lighter moments at Bradford and the young Oliver Lodge recalled Cayley experimenting with a Möbius strip: "I remember Cayley folded a gummed strip of paper into a ring in front of the [Mathematics and Physics] section, so as to make a sheet with only one surface. This was done by giving the paper a single twist before it was stuck together into a ring. He then cut it longitudinally right through without dividing it: it remained a whole larger ring. In the discussion it was asked what would happen if the paper was cut again; and then and there the experiment was tried. The result, as I remember it, was two interlocked rings, curiously twisted together."[34]

The committee, which was appointed in 1869 to investigate ways for improving the methods of teaching geometry, gave an interim report at Bradford. The "geometry conservatives" were under such attack from the progressive brigade that the Cambridge don Charles Taylor suggested a "Euclid defense association" be formed, which would include, no doubt, the band of apologists who opposed the abolition of Euclid's "Fifth Book."[35] As a member of the committee, Cayley was utterly resistant to change while Sylvester continued to argue that Euclid be "honourably shelved" in its entirety. There was little inclination to act, though members worked around their disparate views and made an outline for a proposed new treatise on geometry, a short-lived attempt to dictate the content of school geometry in Britain. The interim report came to the unworkable conclusion that the British Association should itself put forward a syllabus. Three years later, the committee presented a full report and advised the retention of Euclid, including the famous "Fifth Book," which set out the much heralded theory of proportion of Eudoxus. In presenting the report, Cayley stood out from the other committee members and staunchly defended the teaching of

Euclid's "Fifth Book" in its original form, unaided by algebra or any other devices that would make it palatable. He was "strongly of [the] opinion that it ought to be retained."[36] Of all the books of Euclid, this one was universally admired by mathematicians, but though the general theory of proportion due to Plato's teacher Eudoxus was revered, it was "unintelligible to modern ears" according to De Morgan. Retreating from total opposition to change, the committee showed some sympathy for the plight of the schoolteacher and were prepared to allow this topic to be taught using algebraic notation in the spirit of Augustus De Morgan.[37]

During the 1870s, Cayley threw himself into the working of the British Association, and perhaps none of its committees on which he sat shows this more clearly than the remarkable *Report on Mathematical Tables* produced for the Bradford meeting. Given a grant of £100, the Mathematical Tables Committee had been formed in 1871 and comprised Stokes, Thomson, Smith, and the young Glaisher.[38] Glaisher was the mainspring of the committee whose purpose was to organize and catalog existing tables, reprint them, and to recalculate them if necessary. Elliptic integrals, for example, had been tabulated by Adrien-Marie Legendre in 1826 but the tabulation of elliptic *functions* had not been treated comprehensively.

Cayley guided the committee and drew up the classification schedule, wherein his erudition is unmistakable. The setting up of the project is further evidence of the Victorian "storehouse principle" and their intention to organize all existing knowledge. This purpose and the high regard in which the Victorian mathematicians held the compilation of numeric and symbolic tables is evident from the aims of the project: "A report on tables differs from a report on any other scientific subject in this—that whereas in a progressive science [for example, Chemistry] the earlier works become superseded by their successors, and are only of historical interest, a table forms a piece of work done, and, if done correctly, is done for all time."[39]

Cayley believed that future generations of mathematicians would find their tables useful. Maxwell concurred and thought the compilation impressive. It was admittedly a dry task but it was valuable work done; it was "worthy," to use a good Victorian word which signaled overall approval. To be done *for all time* elevated the compilations above the status of the transitory memoir in mathematics, and certainly above the memoir in science, which would be made redundant by a later and more successful theory. The belief in the importance of tables partially explains why a mathematician of Cayley's caliber would spend so much of his energy in either performing or directing calculation. During his career, he constructed tables of various sorts, but at the top of the list in order of importance were the invariant theory tables. From Cayley's perspective, these elaborations, so carefully calculated, would be of *permanent* value, since they dealt with the very foundations of modern algebra, the algebraic forms which were the building blocks of the subject. If not the triumph of mathematics, their formation would be a necessary step to take. Accordingly, they would form part of the permanent printed record and be available for later generations to consult.

The Tables Committee continued to meet for many years, but its first report compiled for the Bradford Meeting was its magnum opus. As the 1870s progressed, the possibility of machines widened the scope of the calculatory enterprises. As a member of the British Association committee formed to consider the possibility of constructing Charles Babbage's Analytical Engine seven years after the inventor's death, Cayley was in a position to assess this possibility. When this committee reported at the Dublin meeting in 1878, it reached the dismal conclusion that its construction was "not more than a theoretical possibility."[40] But in a further report, it was almost certainly Cayley who spoke out: though the committee could not support the building of the original Engine, a variant was suggested which might be constructed for the more limited calculation of determinants.[41]

Cayley's affinity for calculation and his attraction to combinatorial questions found an outlet in counting graphical "trees," a study motivated by the necessity of enumerating structures in organic chemistry. He had used the term "tree" in 1857 and had then enumerated root-trees in connection with a problem in pure mathematics.[42] His new research in the 1870s involved the problem of enumerating *unrooted* trees, in which there is no selected node to act as a root.

Cayley's attention to this problem was prompted by Carl Schorlemmer. This organic chemist was born in Darmstadt, studied at Heidelberg, and came to England in 1859, where he worked as an assistant to Henry Roscoe at Owen's College, Manchester. "Little Jollymeier," as he was known to the close-knit group of émigrés, which included Karl Marx and Friedrich Engels, was an active member of the German Social Democratic Party, his closest and oldest friend and drinking pal being Engels. In the laboratory, Schorlemmer was fastidious and took every safeguard in the planning and execution of experiments, but his lectures were less than inspirational, with him reading in a monotone "without emphasis or pause from notes held close to the nose." In his professional life, Schorlemmer gained recognition for his work on the classification and structure of hydrocarbons, and, as a consequence of this work, he was elected a fellow of the Royal Society in 1871. In 1874 he was appointed to a chair of organic chemistry at Owens College, the first of its kind in Britain.[43]

In the study of organic chemistry, Schorlemmer proved experimentally that "dimethyl" and "ethyl-hydride," both carbon compounds with two carbon atoms and four hydrogen atoms, were really identical chemicals which the eminent chemist Edward Frankland had thought were distinct. The two chemicals were unambiguously *ethane,* with formula C_2H_6 and molecular structure:

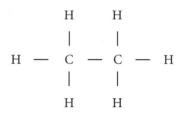

ethane

This diagram with two centers C–C, can be compared with methane, common marsh gas, with formula C_1H_4, or more commonly written CH_4. This has one center in its diagrammatic arrangement of atoms:

$$
\begin{array}{c}
\text{H} \\
| \\
\text{H} - \text{C} - \text{H} \\
| \\
\text{H}
\end{array}
$$

methane

They are both hydrocarbons since they have a common structural formula C_nH_{2n+2} where n is the number of carbon atoms. While the chemicals "dimethyl" and "ethyl-hydride" had the same structural formula, in the 1870s Schorlemmer asked another question: could there be genuinely different compounds with the same number of carbon and hydrogen atoms but differently arranged? This was the problem of *isomerism,* in which compounds with the same number of carbon and hydrogen atoms but with different structures were called isomers.

Schorlemmer's question led to the mathematical problem of counting possible arrangements of carbon and hydrogen atoms. At this point, he failed to interest German mathematicians in the problem and wrote to Cayley, a happy choice since Cayley's affinity for chemistry went back to his youth; he had won the Chemistry Medal at school and had kept his interest alive by attending lectures at the Royal Society given by Edward Frankland—a mountaineering companion of their mutual friend John Tyndall. Cayley answered Schorlemmer within a fortnight and quickly produced a paper on the "mathematical theory of isomers," theoretical work which was well ahead of the knowledge of organic chemists.

The mathematical theory depended on counting the number of centric and bicentric structures for the carbon atoms alone (the hydrogen atoms can be filled in later). For hydrocarbons with five atoms, the possible structures are:

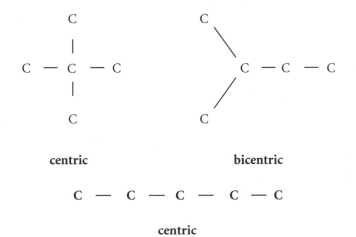

centric   **bicentric**

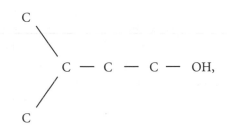

centric

To show the mathematical theory had practical relevance, Cayley applied his mathematical theory to compounds known as alcohols (those with chemical formulae $C_nH_{2n+1}OH$). In connection with these, there were surprising implications. Alcohols are derived from hydrocarbons by replacing a hydrogen H by a hydroxyl OH in their chemical structure. Shorn of its hydrogen atoms, isobutyl carbinol can be represented by the bicentric diagram

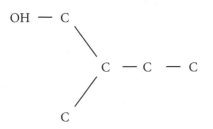

and an "isomeric" one, shown to exist in nature by Louis Pasteur in 1855,

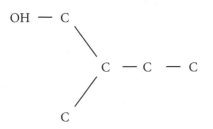

For this case ($n = 5$), the mathematical problem requires the number of distinct trees derivable from the three arrangements of carbon atoms (two centric and the one bicentric arrangement) by adding the hydroxyl OH. In this case, Cayley's theory predicted eight alcohols, many more than were known to exist in nature. He noted "[t]he number of known alcohols is two [discovered by Pasteur], instead of the foregoing theoretic number eight. It is of course, no objection to the theory that the number of theoretic forms should exceed the number of known compounds; the missing ones may be simply unknown."[44] His prescient remark was, in effect, a case of mathematics predicting the future. Now the family of all eight isomers with five carbon atoms (the amyl alcohols) are known to exist chemically and traditionally have been used in the manufacture of perfume. These methods were applied to other compounds.

Cayley's lecture on this work was applauded at the 1875 British Association meeting at Bristol. He had the unusual experience, for him, of addressing the chemistry section of the association though the audience contained a strong mathematical presence since on his admission the "subject he was about to consider was more mathematical than chemical." He published his work in the journal of the German Chemical Society. Counting the number of unrooted trees, although fewer than the rooted variety, poses a far greater challenge for the mathematician. For the structures which appear in the alcohols problem (those of the form C_nH_{2n+2}), his detailed analysis gave numerical results as far as $n = 13$. For this last value, there were 799 topologically distinct unrooted-trees, comprising 419 centric ones and 380 bicentric ones. Further calculations have since given exact values up to $n = 22$, and for this last value there are more than 22 million different structures.[45]

The 1860s and 1870s was ripe for the application of mathematics to chemistry. Sir Benjamin Brodie (the younger) saw this when he attempted to apply the calculus of operations to chemistry.[46] From among the mathematicians, Sylvester and Clifford were convinced that invariant theory was relevant to chemistry. Cayley's work was not so sensational as theirs, but in demonstrating that the theory of trees was really applicable and one which engaged the chemists directly, it was perhaps more useful—it has proved more endurable.

Cayley's service to British Association committees which reported at Bristol was concerned with mathematical notation and printing. Its principle concern was to recommend mathematical notations (such as using the shilled a/b for fractions), which would make the typesetter's job less expensive. On the subject of notation, he had decided views. One was his unfortunate predilection for denoting the elements of matrices and vectors by a single letter notation, believing that mathematics would benefit by avoiding the use of subscripts. "I have a theory that mathematicians in general are too fond of suffixes," he remarked to Klein on sending him a paper, "and it is partly intended to show how by the notation employed, of rows and matrices, it is to a very great extent possible to get rid of them."[47] This now seems a retrograde step, but his matrices were invariably of small dimensions and he hardly needed elaborate notations. He playfully turned the tables on Stokes for not adopting the modern conventions. In editing his *Mathematical and Physical Papers* for publication, Stokes made a feature of employing De Morgan's solidus *a/b* used for printing fractions, whereupon Cayley looked out to see whether he used voguish notation for denoting the exponential function. He responded to Stokes: "I think the *solidus* looks very well indeed and is really a great improvement: it would give you a strong claim to be President of a Society for the prevention of Cruelty to Printers. I ran through the volume to see whether you had adopted *exp.x* instead of e^x: you do not seem to have any exponents complicated enough to make this necessary or even advantageous. I think you do not approve of cosh and sinh."[48]

Geometrical Theories

The interplay between geometry and algebra moved forward apace in the second half of the nineteenth century; the new approaches instigated by Bernhard Riemann, Gordan, and Clebsch and their disciple Max Noether (father of Emmy Noether) would inspire the methods of algebraic geometry and abstract algebra in the twentieth century.

A leading topic in algebraic geometry for nineteenth-century geometers concerned the theory of planar curves. If $\phi = 0$ and $\varphi = 0$ represent two curves in the plane, a linear combination of them, a curve of the form $A\phi + B\varphi = 0$ passes through their points of intersections, but is the converse true? Is it necessary that any curve $\Omega = 0$ passing through the intersections of ϕ and φ be of this form? In his juvenile paper on cubic curves produced in 1843, Cayley merely assumed this without question, and he regarded it now, in the same way as then, as being intuitively true for any curves in which the intersections were simple. Max Noether's first statement of a theorem that gave conditions for a plane curve $\Omega = 0$ to be expressible in this form was published in 1869 with the general result appearing in 1872. The "$A\phi + B\varphi = 0$" theorem marked a transition to the modern view of algebraic geometry and the *Fundamentalsatz* gained Noether a leading reputation among algebraic geometers. The exactness required in its proof lay outside Cayley's knowledge and, one is tempted to say, his interest. At a time when increasing rigor was the order of the day, his assumptions were exposed as unwarranted, belying an uncritical approach to geometry. What was required in 1870 was an awareness that geometry should be placed on a secure basis, a familiar path in the history of mathematics in which foundations acceptable to one generation as being rigorous are exposed as nothing of the sort by a later one. When Ludwig Schläfli wrote to Cayley in August 1870, he replied:

I would have written before, but I was away from Cambridge, among our small scale but very enjoyable mountains in Westmoreland & Cumberland, & you mentioned that you were also away from Bern. I trust that you have had an equally pleasant summer excursion.

As to the equation $\Omega = AP + BQ$, admitting that it exists, the determination of the functions A, B could be effected by indeterminate Coefficients. . . . But I understand your difficulty is that we have no proof of the existence of this equation with integral or fractional values [functions] of A, B.[That they do] appears to me axiomatic.[49]

It is indicative of the differences between generations that Cayley's student Charlotte Angas Scott would write an influential paper supplying a rigorous proof of Noether's theorem thirty years later.[50]

Cayley hoped to extend the same set of ideas relating singularities of plane curves to the singularities of surfaces. In another letter, this time to Max Noether, he suggested this approach:

Please consider yourself quite at liberty to publish any of the results which I have sent or may send you. I am only anxious that they should be useful in the progress of this most interesting question [relating to deficiency of curves and surfaces] to a complete solution, and the best chance of it is in the cooperation of different workers. It would be a great thing to do for surfaces, as for plane curves—find all the simple singularities, such that any higher singularity whatever may be considered as compounded of certain of these. And it is I think only by an examination of particular cases, that we can hope to arrive at this.[51]

In terms of the number of singularities, there was a substantial difference between the theory of curves and surfaces. Cayley knew that the deficiency of a plane curve is always non-negative, whatever the nature of the singularities, but six months before writing to Noether, he found there were surfaces with negative deficiency.[52] This was a result that was not expected and it occurred in the well-known class of ruled surfaces, which comprise scrolls and developables.

The general problem of classifying algebraic surfaces, as conceived of by Cayley, amounts to a research program of immense proportions involving detail of a most intricate type. He had cataloged cubic surfaces in the previous decade, but quartic surfaces presented a challenge of a different order and one that has never been worked out in a way which Cayley would find acceptable.[53] The passage from cubic to quartic for curves was accentuated for surfaces. It was a task that would involve several hundred different types of quartic surface and which occupied those such as Karl Schwarz, Luigi Cremona, Ernst Kummer, and Alfred Clebsch but, writing at the end of the century, Gaston Darboux noted that the study of the general quartic surface still appeared too intractable.[54] Progress was made through specific cases. The quartic surfaces may have as many as sixteen nodes, as Cayley's youthful investigations of the "tetrahedroid" surface had shown, and in long papers written at the beginning of the 1870s, he examined those surfaces with eight nodes.[55] A grasp on the magnitude of this colossal "quartic surfaces" program is gained from observations Cayley made in the 1870s. He was able to identify eight of fourteen "oceanic surfaces" modeled by Julius Plücker. These, also known as "equatorial surfaces," were representatives of a class comprising seventy-eight species of this one type of quartic surface.[56]

Cayley's preferred approach to algebraic curves and surfaces was via Plücker's famous equations connecting basic singularities. Early in his career, Cayley had extended their applicability from plane curves to space curves, and in the 1860s he used them in studying the compound singularities of plane curves. Whereas Plücker's six equations for plane curves involved four types of basic singularity (nodes, cusps and their reciprocals, bitangents, and inflections), the same thinking as applied to surfaces, which Cayley eventually published in 1882, settled on no less than forty-six types of singularities, which were found to be connected by twenty-six equations.[57] This type of work lacked precision and was produced at a time when algebraic geometry was moving toward the study of rational curves and rational transformations of them, a general approach which demanded rigorous analysis.

Following the Bradford Meeting of the British Association, in which geometry had been at the fore, Cayley and a small group at Cambridge broached the subject of setting up a club to make and exhibit geometrical models and invited others to join them.[58] This commitment showed again Cayley's need for a physical realization of geometric entities. In October 1873 the club's list of geometrical models included cubic scrolls, ellipsoids, and ray surfaces. Cayley constructed a model of Jacob Steiner's symmetrical Roman (quartic) surface, made drawings of it, and in December he gave a talk on the subject to the London Mathematical Society.[59] The first meeting of the Cambridge Modelling Club was held on 7 February of the following year, and he was elected its president with W. H. H. Hudson serving as secretary.[60]

The object of the Modelling Club was formally expressed "to promote the making of models, machines, and drawings illustrative of Geometry," and it undertook to display them at public exhibitions. In the classic Victorian way, it was well-regulated and open to all "who have made or assisted in making a model, machine or drawing approved by the officers."[61] Alongside Cayley, Hudson, and Maxwell (the appointed custodian of models) were staff from the Cavendish Laboratory, but admitted into their illustrious company was a sprinkling of keen undergraduates. The range of models could justly claim to rival the "string and sealing wax" of the physical experiments of the Cavendish Laboratory. Members provided elaborate constructions of a wave surface in clay, a twisted surface made from thread, and a plaster model of a surface which illustrated ideas in thermodynamics.

Cayley's own club constructions included a cubic cone with a nodal line, a model of Charles-Nicolas Peaucellier's parallel motion, linkwork made from zinc, models of two of Riemann's triply connected surfaces, and to illustrate how models can be made from paper, a model of a skew surface. In particular, he investigated Ludwig Christian Wiener's celebrated model of the cubic surface.[62] Wiener had constructed a gypsum model of a cubic surface and realized it in stereoscopic drawings showing Salmon and Cayley's twenty-seven lines— the Eikosi-heptagram—to advantage. Among other things, this model showed that in some species of the surface, all twenty-seven lines could enjoy a real existence.

Cayley made a string model of the 27-line configuration to show at a meeting of the Cambridge Philosophical Society. Though it was not completely successful in mathematical terms, it does illustrate the degree to which he was an empirical geometer. Using Wiener's model, he physically measured the coordinates of points in order to write down approximate equations of the lines, which he then adjusted to make them satisfy the theoretical geometrical constraints. His ensuing drawing of the constructed surface was described, with some justification, as being "so quaint in its topsy-turveydom as almost to suggest Mr. W. S. Gilbert as joint author."[63] The twenty-seven lines appeared as a random "bundle of sticks," but his equations, though empirically gained, have since been used to build more sophisticated models of cubic surfaces.[64]

In 1876 the club took an active part in the international exhibition of laboratory and teaching apparatus held at the South Kensington Museum opened by Queen Victoria. There was a section on three-dimensional geometrical models, representations by two-dimensional projections, and the drawing instruments used to trace curves, including the very topical linkwork. The club appears to have survived only until the following May; a high point was no doubt the day its wares were set out in an exhibition in the Cavendish Laboratory.[65]

In emphasizing the importance of such visual evidence for the teaching *and* learning of geometry, Cayley noted: "There is no exercise more profitable for a student than that of tracing a curve from its equation, or say rather that of so tracing a considerable number of curves. And he should make the equation for himself."[66] He was as good as his word, and he extended his "hands-on" approach to the design of curve-generating mechanisms. Through his own curiosity about mechanical apparatus, he devised a range of wooden and metal contraptions. These devices, consisting of moving disks driven by gear wheels, were popular and he entered into the spirit of the times. No source of geometric problems was ignored. The standard geometric chucks used in lathes for producing ornamented turnings were studied, and chucks were specially devised for drawing ellipses, pear-shaped curves, various ovals, and figures-of-eight.[67] The visual experience for students was extended by his artistic inclinations, and when, in the 1880s, the number of students attending his lectures amounted to a class of a dozen, he made a point of demonstrating geometrical theories with handmade drawings and watercolor sketches.

Though Cayley had the power of abstract thought to a high degree, his affinity for painting, carpentry, and architecture further illustrates the down-to-earth side of his character. The joining of tangible model making to mathematics, as a method, was further enhanced when a former student, James Stuart, was promoted from his lectureship to the newly created chair of mechanism and applied mathematics in 1876. This man centered his teaching on the use of mechanical models and in his engineering laboratory, known locally as "The Workshop," students were taught such things as the making of steam engines and scientific instruments. Given Cayley's practical bent, it is not surprising to find him at the workbench himself, taking part in its activities.

"Wee Things"

Cayley's production of papers throughout the decade continued unabated. Admittedly, some were minor productions but he valued them, as the young man who once asked Thomson to "please take care of the wee things." Many of his 150 papers published during the 1870s were short notes and, when originality could not be found, he filled in by adopting the custom of Cambridge dons of publishing their solutions to mathematical Tripos and Smith's Prize questions. Miscellaneous topics were either suggested by his main work, or the work of others, and he could draw on his vast stock of past work and the legions of incidental questions which it suggested. This was a period when he was taking his fair share of society administration, and mathematical work was tucked away until he had leisure to pursue it. G. B. Mathews judged that few of his papers in this period were of "paramount importance . . . Cayley is, as it were, [was] brought into unfavourable comparison with himself." Mathews's admiration was undimmed, for it would be "unreasonable to expect an artist to produce an uninterrupted succession of masterpieces."[68] In a curious way, some of these "minor" productions have achieved greater recognition, especially his work on the mathematics of chemistry, group theory, and logic.

Early in 1872, Cayley heard from Peter Guthrie Tait on matrix algebra. Though Cayley stressed the importance of the algebra of matrices, they had not gained prominence, and determinants were still regarded as more important.[69] His 1858 memoir effectively codified matrix algebra, but many essential ideas about matrices were independently rediscovered by the Berlin school of mathematicians in one form or another.[70] One little known paper of Cayley's, produced in 1872 in response to Tait, was his "Extraction of the Square Root of a Matrix of the Third Order." In this, he compared matrices and quaternions in dealing with the physical problem of describing a "strain." Even as a well-established professor in his own right, Tait stood in a pupil-master relation to Cayley, his senior examiner at Cambridge, and he was in awe of his mathematical reputation. His tendency to attack, if anything, was undiminished, but on this occasion, he wrote in supplicant mood: "Thomson and I wish to introduce this [quaternion approach to finding the square root of a strain] into the new edition of our first volume on Natural Philosophy—but he objects utterly to Quaternions, and neither of us can profess to more than a very slight acquaintance with modern algebra—so that we are afraid of publishing something which you and Sylvester would smile at as utterly antiquated if we gave our laborious solutions to these nine quadratic equations."[71]

Written in response to Tait's overtures, Cayley's paper on matrix algebra was inventive but it indicates his somewhat fragmentary knowledge of the subject—when his work is viewed from a modern perspective. To find the square-root L of a given matrix M, he applied the Cayley-Hamilton theorem to each matrix separately and solved the resulting equations from which he obtained a formula for the square root matrix L in terms of M. It is illustrative of Cayley's way of proceeding: a rudimentary result was obtained by ingenuity, and *verified*—a method of proof to which he remained dearly attached. The testing of results is a time-honored way of going about research, but as a mathematical "proof" it would not pass muster today. Questions of *uniqueness* or *existence* did not play any part in Cayley's treatment of matrix algebra. He never exploited the theory of matrices, and for him it remained a fringe subject, though Tait's letter stirred him briefly. He later touched on matrices in connection with quaternions, but this spurt of interest was unsustained.[72] The 1858 memoir remained his most concentrated study of matrices and provided his own reference point in the theory.

Around this time, Cayley investigating triple orthogonal systems of surfaces, three systems of surfaces that orthogonally intersect each other in pairs; the work emphasizes the "concrete" aspect of Cayley's mathematics. It had been established by French geometers that the parameter ζ, which indexed a family of surfaces $\zeta = f(x, y, z)$ must satisfy a condition for it to be an orthogonal family. The condition itself awaited further attention. As he conveyed it to Hirst: "I have just worked out an interesting result—the partial diff[erential]l equation of the 3rd order satisfied by the function $\zeta = f(x, y, z)$ when this is a family of surfaces forming part of a triple orthogonal system. It was known that ζ satisfies such an equation but the equation itself is new."[73] Around this time too, he resurrected work of the Polish mathematician Hoëné Wroński, largely forgotten by the mathematical community after his death in 1853. Wronski had spent time in Marseille and Paris at the beginning of the century, where he had championed a highly algorithmic method in the theory of functions. This found expression in his "loi supreme," $F(x) = \sum_{r=0}^{\infty} A_r \Omega_r(x)$, an expansion for a function of extensive generality in terms of generating functions $\Omega_r(x)$ and which contains, as very special cases, Taylor and Maclaurin expansions. The work was connected to Burmann's Law, another special case known to Cayley. In his paper, he treats expansions that have coefficients A_r as framed in terms of determinants and shows the equivalence of various expansions.[74]

Cayley's attention to logic was slight, but his note, in which he described four fundamental syllogisms was regarded as innovatory.[75] His contact with this subject began with his youthful correspondence with George Boole in the 1840s and with his reading of Boole's *Mathematical Analysis of Logic* (1847) and *An Investigation of the Laws of Thought on which are founded the Mathematical Theories of Logic and Probabilities* (1854). In maturity, he remembered his difference of view with Boole and spoke of the "difficulty in making out the precise meaning of the symbols, and the remarkable theory there developed has, it seems to me, passed out of notice, without being properly discussed."[76] In another direction, his contact with Boole's work was continued, where his papers on the singular solutions of differential equations of the first order filled in lacunae generated by Boole's successful *Treatise on Differential Equations* (1859).[77] In other contributions which could definitely be classed among the "wee things," Cayley joined in a discussion with W. S. B. Woolhouse and T. B. Sprague of the Institute of Actuaries on a probability problem that was being discussed on both sides of the Atlantic.[78] Both gentlemen were highly influential in the Institute of Actuaries and had contributed to mathematics over the years. As a young man, Woolhouse had been an editor of *Leybourn's Repository* when he was deputy superintendent of the National Almanac Office and he was later the editor of *The Lady's and Gentleman's Diary*.

For the most part, Cayley's shorter articles published in minor journals were "diamond dust from the lapidary's workshop" as G. B. Mathews put it, and "they will doubtless help to polish gems not yet extracted from the mine."[79] One gem which has been extracted and polished has been the Cayley-Moser lottery problem, a problem now known as the "Secretary Problem" and that is now part of dynamical programming. As stated by Cayley: "There are n tickets representing $a, b, c, \ldots$ pounds [Sterling]. A person draws once; looks at his ticket; and if he pleases, draws again (out of the remaining $n-1$ tickets); looks at his ticket, and if he pleases draws again (out of the remaining $n-2$ tickets); and so on, drawing in all not more than k times; and he receives the value of the last drawn ticket. Supposing that he regulates his drawings in the manner most advantageous to him according to the theory of probabilities, what is the value of his expectation?"[80] Contributions of this sort from Cayley are found in Glaisher's *Messenger of Mathematics* or W. J. C. Miller's *Mathematical Questions with their Solutions from the "Educational Times."* Miller's journal was an almost entirely a British affair with only a few European mathematicians, such as Hermite, taking part. Cayley contributed to Miller's first volume in 1862, kept up a correspondence with the editor, and contributed a constant stream of problems and solutions. During his involvement with the journal he dealt with no less than 131 questions.[81]

Cayley would routinely solve problems set by himself, Sylvester, Clifford, Thomson, Glaisher, and Miller, or any unknown who ventured onto the pages of the Mathematical Questions with a challenge, unlike Sylvester who would propose problems, solve his own, but rarely solve those set by others. Cayley was prepared to work at an agenda he had not set himself, and he saw it a duty to submit little notes to place conundrums in a more general mathematical setting. Of the range of problems which attracted him, most were geometrical but they were also on topics in probability, algebra, combinatorics, and the elementary parts of invariant theory.

Writing a Treatise

On 14 November 1873, Cayley submitted to the Royal Society a long paper on the transformation of elliptic functions, which involved the calculation of the "modular equation." It was yet another problem in which calculations succeeded for small values of n but was difficult to apply for higher values. Here at least, Cayley's patience and persistence was rewarded and he was pleasantly surprised: "The [calculatory] process is a laborious one although less so than perhaps might beforehand have been imagined."[82] This was research, but elliptic functions were part of the undergraduate curriculum as well. J. W. L. Glaisher thought elliptic functions so central that they should be introduced as early as possible in a student's career, and to this end he taught them in a Tripos course at Cambridge year in and year out. We can appreciate why they were so regarded and their place in the kernel of a mathematical training. Elliptic integrals and the later elliptic functions were extensively studied in the eighteenth and nineteenth centuries. They occur in many problems, the most direct being to problems associated with the ellipse as a shape more general than the circle. The circumference of a circle has length $2\pi r$ but the length of an ellipse is only expressible as an elliptic integral. They also arise in the expression for the exact period of a simple pendulum and in elasticity problems. Carl Friedrich Gauss, Niels-Henrik Abel and Carl Jacobi, and Karl Weierstrass, as well as Cayley himself, studied the subject intensively.

It was a central subject in Glaisher's research for it was bound with a central object of the British Association Tables project. Yet, apart from Glaisher's lectures, there was no accessible sources to the theory and even students aiming for high places in the Tripos found that the prince of coaches, E. J. Routh, could not help them in this subject. While there was no shortage of school textbooks being produced, there were relatively few treatises being produced for the teaching of high-level university mathematics. There was an obvious need for one on elliptic functions. Toward this end, Cayley joined his research work in elliptic functions of 1873 with the publication of his sole book *An Elementary Treatise on Elliptic Functions*, which was published in 1876.

On the day it was published, Cayley's "Elliptic Functions" was already old-fashioned, its starting and finishing position being Jacobi's *Fundamental Nova Theoriae Functionum Ellipticarum* (1829). It was Jacobi's point of view that had inspired Cayley to study the subject as a young man and it was Jacobi's algebraic approach that Cayley still championed. Cayley's tack was to draw on the analogy between the elliptic functions and the ordinary circular functions.[83] His intention was to treat elliptic functions by "pure algebra" and to fill in the gaps left by Jacobi.

In this treatment, Jacobi took the beginning of elliptic function theory as 1751 when Euler formulated the *addition* theorem. The addition theorem for ordinary trigonometric functions is $\sin(u + v) = \sin u \cos v + \cos u \sin v$. For the elliptic functions sn, cn, and dn, there are analogous addition formulae:[84]

$$\operatorname{sn}(u + v) = \frac{\operatorname{sn} u \operatorname{cn} v \operatorname{dn} v + \operatorname{cn} u \operatorname{sn} v \operatorname{dn} u}{1 - k^2 \operatorname{sn}^2 u \operatorname{sn}^2 v}.$$

Cayley's included six different proofs of the addition theorem. He reverted to the line adopted by Jacobi and Legendre's *Traité des functions elliptiques* and made little use of Abel's work and later developments. He also dealt with theta functions (Θ-functions), a closely connected theory, even more basic than elliptic functions since these can be defined as the ratio of theta functions. Theta functions can be defined as infinite series, for example, $\Theta(q) - \sum_{k=-\infty}^{k=\infty} q^{k^2} = 1 + 2q + 2q^4 + \cdots$. They have a close connection with the theory of numbers, which are dealt with by Cayley in his book and subsequent papers.[85]

Francesco Brioschi, who prepared an Italian translation of Cayley's book in 1880, wrote of the timeliness of Cayley's book and noted that it "showed different parts of this difficult theory with a rare clarity and great rigor of demonstration; for this reason it was very useful to those who wished to learn about this theory. In publishing his book, ten years before G. H. Halphen's great *Treatise*, Cayley set an excellent example."[86] It came out in a second (posthumous) edition in 1895 and has since been reprinted but, intended as an elementary treatise, its terse style did not find favor with Cambridge undergraduates.

Invariant Theory Rejoined

In April 1870 Cayley submitted his "Ninth Memoir on Quantics" to the *Philosophical Transactions*. It reconciled his earlier results of the 1850s with Gordan's achievement of proving the finiteness theorem. With this milestone, Cayley's dedication to invariant theory ameliorated, and, but for the fact that he would return to the subject repeatedly, it would have been a fitting end to the whole series on quantics, being a defining memoir on the subject of the binary quintic algebraic form. In it he cataloged the twenty-three irreducible invariants and covariants of that polynomial and effected a correspondence between his results expressed in his own style and those of Paul Gordan.[87]

During the 1870s, invariant theory attracted a number of adherents, and significant additions to the literature were made, particularly to the theory of binary forms. Just before his death in 1872, Clebsch completed his *Theorie der binären algebraischen Forman*. In a less well-known book *Über das Formensystem Binärer Formen* (1875), Gordan consolidated and extended his own work.[88] In France, Edmond Laguerre wrote on the subject briefly, and Camille Jordan joined in the quest with two masterful papers published in *Liouville's Journal*.[89] Jordan adopted the symbolic notation, and in acknowledging the work of Clebsch and Cayley (for the original development of a symbolic notation phrased in terms of the hyperdeterminant operator), he sought to make Gordan's theorem precise. He outlined some of the main properties of the symbolic notation for potential French authors who might be attracted to the subject.

At the beginning of 1870, Sylvester parted company with the Woolwich Military Academy, at which he had spent the previous fifteen years. Under an institutional rationalization, an "experiment in reorganisation" inspired by a Royal Commission, the office of professor of mathematics was combined with the professorship of mechanics and an age limit placed on the new appointment. Sylvester was edged out, and thus he was forced to retire from a position he never fitted, either from his own perspective or that of the military authorities. At first, there was no offer of a pension but Sylvester was a fighter and he mobilized the scientific community to support him. Letters to the *Times* and a leading article gave greater prominence to Sylvester's plight. Eventually Gladstone stepped in and an annual pension of £300 was granted, in which sum all was taken into account, even his rights of pasturage on Woolwich Common. He retired to London's West End not far from his beloved Athenaeum, and rid of officialdom, he filled his days less with mathematics than with poetry and the arts. He attended Penny Readings, took singing lessons from the celebrated composer Charles-François Gounod, who had been displaced from France by the Franco-Prussian War, and he addressed working men's meetings and other groups with a view to standing as a candidate for the London School Board. Without a structure to his life and any inclination for mathematics, Sylvester was at sea. He was aged fifty-five and with little chance of finding suitable employment in Britain.

Sylvester thought vaguely about an academic position in Australia that had cropped up through the death of William Wilson, holder of a chair in Melbourne. He asked Cayley to act as an intermediary: "I see from the newspapers that [John Couch] Adams is to appoint to the vacant Mathematical Professorship at Melbourne. I am more than half induced to go out to the Antipodes rather than remain unemployed and *living upon Capital* in England if he should think me a suitable man for the place—Perhaps you will not mind putting the question to him."[90] This came to nothing, but another possibility arose later in the year when Daniel Coit Gilman from Baltimore was on a trawl for European talent to fill appointments for the newly proposed university funded by the Quaker millionaire Johns Hopkins. Benjamin Peirce urged Gilman to do "what his [Sylvester's] own country has failed to do—place him where he belongs—and the time will come, when all the world will applaud the wisdom of your selection."[91]

Gilman thought Sylvester admirably suited and he was appointed to Johns Hopkins in February 1876. In America, he was once again renewed to mathematical evangelizing. Now he had a new career, and no longer feeling cast aside, he found a fresh lease on life in forming and guiding the infant research school in Baltimore. Although his first advanced class consisted of precisely one student, G. B. Halsted, this man proved an ardent disciple and encouraged Sylvester to embark on invariant theory once again. It was a relatively young subject and there was much to be done. Apart from his paper on Newton's Rule in 1864, Sylvester was returning to a field in which he had cooperated closely with Cayley twenty-five years previously, but over that period of time, the subject had proved an "on-off" love affair.

At Johns Hopkins University Sylvester remained loyal to his and Cayley's methods of treating invariant theory and, though he attempted to learn the German symbolic method, he harbored the same misgivings about its ultimate worth as did Cayley. He voiced these doubts to William Spottiswoode: "Those piratical Germans, Clebsch & Gordan who have so unscrupulously done their best to rob us English of all the credit belonging to the discoveries made in the New Algebra will now suffer it is to be hoped the due Nemesis of their misdeeds. Nothing in Clebsch & Gordan is *really* new but their Cumbrous method of *limiting* (not *determining*) the Invariants to any given form. This part of their work is now I think destined to be blotted out of existence."[92]

Sylvester's tendency to exaggeration was given fresh encouragement by thoughts of academic theft. The German semiabstract method was artificial, he declared, whereas for him an invariant was an algebraic form annihilated by a differential operator, with only plain arithmetic operations in support. A week later, he wrote to Spottiswoode in mood for battle: "I believe to reduce this subject to 'Annihilation' all that the school of Clebsch and Gordan by aid of methods borrowed by the Germans without acknowledgement from Cayley and myself, have attempted in this subject. Their *Symbolical* method is nothing but the method of Hyperdeterminants in disguise."[93] Sylvester illustrated his and Cayley's commitment to their own methods when he wistfully reflected on the partial differential equation as the "subtlest of all instruments for putting Nature and

Reason to the question" and in the same breath he might also have mentioned generating functions.[94]

Much calculation and close work was inevitably required as the invariant theory program enjoyed a resurgence. It still required the finding of long algebraic expressions, and this, in turn, depended on the handling of generating function expressions. About one long expansion, Sylvester wrote to Spottiswoode that "[i]t is an awful long computation to find this Generator [generating function of the binary form of order 7] requiring the use of ruled paper with upward of 100 columns but one of our 'Fellows' [Miss Christina Ladd] is assisting me in it and I hope by tomorrow to have obtained a confirmation of the figures." His continuation of this passage amply shows the importance both he and Cayley placed on the establishment of the generating function method as leading to abstract rational argument. It was the belief that instances should be known thoroughly, from which the general argument could then be inducted. "Once [the generating function is] firmly planted in a foundation of facts, it will not I think be difficult to obtain a strict demonstration of the ground of their [invariants and covariants] existence."[95]

Calculation was the order of the day, whereas what was really needed were firmer foundations to the mathematical edifice rising up at Johns Hopkins. Since his own personal high- water mark, as laid out in the "Ninth Memoir," Cayley's focus had been diverted. But now Sylvester was stirring the waters of invariant theory again, just as he had done at the beginning of the 1850s. Would Cayley be pulled into the whirl of activity that was bound to ensue, as he and Sylvester were rejoined in the invariant theory quest?

CHAPTER 14

March On with Step Sublime

Cambridge and London were the twin spheres of Cayley's mathematical life, joined by the frequently traveled "Great Northern Line" that linked the university town and the metropolis. In only two hours from leaving home, he could be taking part in Royal Society soirées or lectures, scientific at-homes in grand London houses, trips to Greenwich for meetings of the Board of Visitors, or regular meetings of the London Mathematical Society and the Royal Astronomical Society. The administrative duties performed on behalf of these societies and the editorship of the *Monthly Notices* all made for an active scientific existence. But, with the death of his mother, family connections with Blackheath were broken and Cambridge life became more important. His young family was growing up and he wrote to J. J. Sylvester in America that "[w]e are all flourishing: St. Leonards [on the South Coast] has suited my sister Henrietta wonderfully" and, what was pertinent, "Mary & Henry are getting on very fairly in mathematics."[1]

"Prospecting"

While Cayley was settled, Sylvester was experiencing a hyperactive rejuvenation in Baltimore. Pensioned off from Woolwich, his new position in America was as unexpected as it was gratefully received. Starting a research school at Johns Hopkins would have reminded him of the freshness of the early 1850s, when he had the world before him. In choosing invariant theory as a research focus for the school, he would have felt the reverberation of those times when he and Cayley were newly qualified barristers and had done so much to create the subject. With Cayley prospecting for new topics, Sylvester would endeavor to entice him back. Could they return to their youthful journey once more?

In his enthusiasm, Sylvester rushed ahead, perhaps too quickly, and many new results had little of the certainty expected of mathematics. The ideal is to advance with unassailable results logically established, but Sylvester could hardly supply this for invariant theory. Determined to press ahead, he was quite prepared to publish highly conjectural "work in progress," though eloquently expressed footnotes warned readers of the dangers of accepting absolute truth in some instances.

One significant instance of this ad hoc approach was the method Sylvester used to identify invariants and covariants, whereby he minutely examined the numerator and denominator of generating functions. He called his method "tamisage," but the close inspection required hinged on an assumption that became known as the "Fundamental Postulate." In the heat of discovery, he described the whole procedure in a letter to Cayley:

> I think I may now announce with moral certainty that my method [of Tamisage] completely solves the problem of finding the *grundformen* [irreducible invariants and covariants] for binary forms and systems of binary forms (without mixture of superfluous forms) in all cases—I have sent an account of the method to the Comptes Rendus—I ought to add that *anterior to all* verification, this method *could not* give superfluous forms—but it is metaphysically conceivable that it might give *too few* grundformen. The principle [Fundamental Postulate] I proceed upon is that in interpreting the generating function we are not to assume the existence of more syzygetic relations than those which are necessary to make it consistent with itself and with the fact that every combination of Concomitants is a Concomitant. Even if this had not been true my method would have given a sure means of proceeding *step by step* from the lowest forms to higher ones until all were exhausted.[2]

The generating function approach to invariant theory was fundamentally weak; it was a method of arithmetic being used to make discoveries in algebra. It had drawn Cayley into error as Gordan had shown, and many results dependent on it had not been proved rigorously. Sylvester continued the use of generating functions to obtain invariant and covariant expressions but he needed the unproven Fundamental Postulate to help him. This was an article of faith based on speculation in partnership with an intuition gained from established results—guesswork based on experience.[3] To take such a step is dangerous mathematically but Sylvester was undeterred.

The German symbolic method applied to invariant theory offered a surer path to the fruits of invariant theory, but while it was powerful, it was not a panacea. It had weaknesses Cayley and Sylvester both regarded as serious.[4] To obtain the *irreducible* invariants and covariants using the symbolic method, the generated ones had to be examined and the reducible ones discarded. Failure to identify them could lead to an overestimate of their number by including the reducible with the irreducible ones. Sylvester wrote to Cayley about results achieved on the German side: "Gordan writes to me that he has *proved* their correctness—but the proof has never been produced—he says that any *à priori* proof by his method is not to be looked for—I mean a proof that the supposed *grundformen* being actually indecomposable."[5] In distinction to this unfortunate characteristic of the symbolic method, Cayley could be fairly sure that the invariants and covariants he had managed to find were irreducible, but he could not be certain he had found them all. He worked from firmly established irreducible invariants and covariants to unknown ones, but, in contrast to the German method, his method might lead to an incomplete system.

From the British perspective, progress in invariant theory was seen to depend on finding the best form of the generating functions. The binary quintic was still the focus of interest and Sylvester managed to find the needed expression in this case: "I have looked into your paper in the Q.M.J. [*Quarterly Mathematical Journal*]," he wrote to Cayley, "and find which I had no idea of (after inspecting y[ou]r ninth Memoir—where also I found you had been beforehand with me in the particular shape given to the *Crude* Generating Function) that your process is *identical* with mine—The form for the Quintic you will find 'ere this reaches you (or ought to do so in the *Comptes Rendus*)."[6]

In the summer of 1877, Sylvester made plans for his first visit to England following his appointment in Baltimore. Spasms of loneliness had accompanied him to America and he missed his friends, the Athenaeum, his life in London, and the opportunity for frequent jaunts to the Continent. He hoped to renew his home life in the summer months. The work in invariant theory was progressing well and he began to formulate a research direction. He was keen to focus on its combinatorial aspect by way of partitions and generating functions, the feature of the subject which had seduced him in the mid 1850s. With his bags packed for the two-week steamer voyage and leaving for New York the next day, he announced his plans in a letter to Cayley: "It seems to me that the time is now come for forming extensive tables of partitions of numbers but the best form to be given to such tables for the purpose of application to the invariantive theory will require consideration. Perhaps we may be able to consult on this subject when I have the pleasure D.V. of meeting you in England."[7]

Sylvester stepped ashore in England on 16 June. He visited France to attend a mathematical meeting and reported to Cayley that he had "made a step in the 'Nebular Theory' [a subtheory of invariant theory] as I term it." During his visit in England, he went to Cambridge, but he evidently spent much time at his Athenaeum desk. Toward the end of his stay he wrote to Cayley: "I fear it will be difficult for me to revisit Cambridge but I should like to do or for a day or to see you here if possible before I leave—I have had very very hard work with the new Nebular theory (as yet with very little actual point) and it has cost me several weeks work to discover the practical method of forming the Allied Generating Functions."[8]

Cayley could not match Sylvester's new enthusiasm for invariant theory and worked in a desultory way on a theory of "color mixing" in the style of James Clerk Maxwell. In drafting a linear theory for combining pigments, this line of "prospecting" seemed not to produce anything worth publishing, which was, in itself, an unusual turn of events.[9] Instead, Cayley became a source of encouragement to Sylvester, who had now returned to Baltimore:

Best thanks for your two letters, you seem to be getting on grandly—but I hope you will soon commence a complete Memoir. I have been quite taken off from the subject [of invariant theory]—first, after I came back from Plymouth [at the British Association for the Advancement of Science meeting in 1877], by the Trinity Fellowship Dissertations which were long but really very good—a success for the first time. . . . Since then I have been "prospecting" with one thing and another, and I have got into the double theta functions again, where I now see my way quite clearly, and have only to finish working out some long pieces of differentiation. . . . I have my hands quite full of this at present, and propose to get thro[ugh]' with it before going on with the quintic covariants. . . .

P.S. I hope to hear of the realisation of your plan for the new Mathematical Journal [*American Journal of Mathematics*].[10]

Candidates for Trinity College fellowships in mathematics were invited to submit dissertations for the first time in 1872 instead of sitting the written examinations that had been in place for many years. In a letter which crossed with Cayley's, Sylvester announced success with one outstanding problem, of especial interest to the recipient:

I have been fortunate enough in the last day or two to discover a *rigorous proof* of the theorem [Cayley's Law] that the number of linearly independent differentiants [semi-invariants] (i.e. functions D which satisfy the equation $(a\delta_b + 2b\delta_c + 3c\delta_d + \ldots)$ $D = 0$) is equal to and *cannot be greater* than the difference between the two well known denumerants—in other words the Independence of the Equations given by your fundamental theorem in the second Memoir on Quantics [published 1856]. The Independence only comes in as an inference. I *disprove* the possibility of the number being greater than the difference of the Denumerants by a wonderfully beautiful method and have sent the proof to [mediator] Borchardt for insertion in his Journal.[11]

"Cayley's Law" for counting the number of linearly independent covariants of a binary form had been relied on for twenty years and Sylvester was successful in giving a (partial) proof. As we have seen, "Cayley's Law" was central to the British approach to invariant theory. The difficulty in proving it hinged on the linear independence of a set of linear equations, but in 1856, Cayley had merely made the unsubstantiated remark that: "there is no reason for doubting that these equations are independent." Before discovering his own proof, Sylvester claimed this lack, had "rendered it to my mind as certain as any fact in nature could be, but that to reduce it to an exact demonstration transcended, I thought, the powers of the human understanding."[12]

While welcome, the proof of "Cayley's Law" did not vindicate the "Fundamental Postulate" Sylvester frequently invoked. Akin to Occam's razor, this postulate fitted the simplicity that characterized British invariant theory and can be contrasted with the sophisticated and abstract calculus adopted in Germany. Sylvester argued: "The validity of the fundamental postulate which is in accord with the law of parsimony [never do with more what can be done with less] is verified by its conducting to results which have been proved to be accurate for single binary quantics up to the sixth order inclusive."[13] He remarked to Camille Jordan: "M. Cayley has no doubt that this method is correct. But there is still a sort of *Metaphysical* doubt as to the possibility of there being other grundformen in addition to those given by the method of generating functions and 'tamisage.'"[14] Sylvester's belief in the veracity of the Fundamental Postulate was strengthened by established facts. He would have been sunk by one single counterexample, but this "extrinsic evidence" had not materialized and he used its lack as an "exemplification of that very reasonable postulate."[15]

In the absence of cast-iron proof, the theory was surely built on sand, for there was always the possibility of a counterexample appearing when it was least expected. Cracks in the invariant theory fabric might appear and threaten the whole structure, yet Sylvester remained confident. In dealing with higher order binary forms, he boasted that they were "out of the narrows," but while grandiose schemes were afoot, surely the dinghy was leaky. The postulate agreed with observed facts in the case of lower order forms but had no basis in proof, and if such methods were bold, the prospect of serious failure was always a danger.

The day-to-day cataloging of invariants and covariants progressed but there were substantive matters to settle. What Sylvester most desired was a proof of Gordan's theorem, a solid argument owing to the power of plain algebra and ingenuity rather than the abstract algebraic machinery of the German symbolic method. The theorem was so basic in invariant theory that it was baffling to Cayley and Sylvester that a simpler and direct proof could not be obtained. In December, Sylvester wrote Cayley one of his optimistic letters: "Since I wrote yesterday my ideas on the proof of Gordan's theorem have assumed a more precise form and what was surrounded by a sort of haze or nimbus is now perfectly clear and well defined so that I hope in the course of this very day to draw out my proof for insertion in the first number of our Journal."[16] Nothing arose from Sylvester's moment of expectation but he remained critical of Gordan's symbolic method. What he and Cayley could not deny was the success of the German method.

Technicalities aside, a deep-seated conservatism seems to have been at work. Mathematicians may be exposed to attractive theories but instant conversions are rare, and the high intellectual investment of a lifetime could not be dropped easily. New methods appear newfangled and they usually demand a complete alteration in viewpoint. Cayley's reluctance to change and his steadfastness in his own approach to invariant theory is perhaps understandable.

On 12 June 1878, Cayley completed his series on quantics with his submission of the "Tenth Memoir on Quantics." He had been working on the "quintic covariants" since his youth, but Sylvester's new research in America prompted him onto a new line of attack. Though he had written to Sylvester, "I hardly know what mathematics I have been doing," he introduced new types of generating functions in his "Tenth Memoir" and listed the degrees and orders of the fundamental syzygies for the binary quintic form, and even a list of the syzygies themselves. This was truly an end to the series begun twenty-five years before. It also represented his completion of an important subtask he had set himself to when he first became engaged with hyperdeterminants as a youth—"to find all the derivatives [invariants] of any number of functions [algebraic forms]."[17]

Beyond the binary quintic, mathematicians tackled higher order binary forms. Sylvester's adoption of pragmatic methods to obtain results was the way forward. He and his team at Baltimore would comb the expansion of generating functions for the terms that identified the irreducible invariants and covariants, and he was not even opposed to discovering them by chance. He tried to involve Cayley with the program:

> I mentioned to you that it was my intention to complete my tables of the N.G.F's [Numerical Generating Functions] by calculating the 7^c and 10^c and certain combinations in addition to those *already* worked by me—and would have preferred to have carried out myself the work which I had commenced—but do not desire to preclude you from taking possession of the case of the 7^c if you are particularly desirous to do so. But why not undertake the 9^c? That is a gigantic labor which I would most willingly relinquish to you and which I know would yield certain new and interesting results in the form of the N.G.F. especially as regards the denominator.[18]

Carried out by hand, the calculations were formidable, even when applied to binary forms of low order. Calculation was the perceived problem but the investigators lacked machines equal to the task. The rejection of "Babbage's machine" by the British Association committee prevented further investment in this direction for a time, and mathematicians fell back on the resources of human computers. They were inevitably slow and labor-intensive, but Sylvester had partially solved the problem by having a young and enthusiastic group of students at Johns Hopkins.

In particular, Sylvester had Fabian Franklin at his side, a computer he informed Cayley, who was "a good Algebraist and as reliable as a steam-engine."[19] Moreover, "Franklin is such a Calculator as probably has no superior (indeed I am sure he has none) in the whole world (for I have had experience of calculators before) and it seems a pity not to utilize such a force when it is at hand."[20] A Russian émigré, Franklin began his career as a civil engineer before becoming a computational aide who Sylvester counted an extraordinary "piece of good fortune to have such a man at my elbow."[21] Callow youth in America working for higher degrees were matched by professional calculators in England, paid from grants made available by the British Association for the Advancement of Science. During the years 1879-82, an initial grant of £50 was used for calculating the "Fundamental Invariants of Algebraical Forms" for binary forms of orders 7, 8, 9, and 10.[22] The binary forms were shown the greatest attention, but forms with three variables had also been widely studied because of their relevance to the theory of curves.

The wealth of intricate detail was an unremarkable feature of mathematics to the nineteenth-century explorers, however tedious it appears to later generations of mathematicians. Cayley and Sylvester were at home with detail and never shirked it, and to achieve their results, they applied the Victorian zeal and conscientiousness to any piece of worthwhile scientific work. As men who saw themselves as part of science, the "scientific" nature of the invariant theory enterprise was a well-understood reality. Sylvester went further than Cayley and likened the fundamental forms to concepts found in physics, to "standard rays, as they might be termed in the Algebraic Spectrum of the Quantic to which they belong." The knowledge that the fundamental algebraic forms were finite in number, the result furnished by Paul Gordan, added to the appeal of the cataloging task and even made it possible. It was an imperative that the fundamental forms be found and analyzed. Sylvester continued in characteristic vein, uniting the scientist with the mathematician in their quest for knowledge: "[a]nd, as it is a leading pursuit of the Physicists of the present day to ascertain the fixed lines in the spectrum of every chemical substance, so it is the aim and object of a great school of mathematicians to make out the fundamental derived forms, the Covariants and Invariants, as they are called, of these Quantics."[23]

Such scientific imagery was not employed by Cayley (at least not in writing), but his overall goals were the same, even if his was a restrained enthusiasm for the expanding calculation industry taking place in Baltimore. He undertook much of the calculatory work himself, but in March 1879 he wrote to Sylvester the unhappy news: "Nothing has occurred to me in regard to the calculations to be made with your [invariant theory] grant—& I think you had better decide altogether for yourself. I suppose you will have the work done in America. Mr. W. B. Davis [Cayley's English calculator] is dead—since a few days only."[24]

Cayley had his own objective. The symbolic forms obtained by the German school did not meet his need since their list was not easily translatable into his preferred homogeneous forms. From his point of view, the extensive homogeneous expressions were more efficiently obtained *de novo*, though this required patient attention to detail, aware as he was that error-free calculations were the only acceptable ones. The care he took is evident from his advice on how one might go about the task of "writing down in pencil a batch of terms, and counting them to see that the right number has been obtained, then, at the same time verifying the derivations, mark these over in ink; and so on with another batch of terms, until the whole number of the partitions [associated with invariants and covariants] of any particular number is obtained."[25]

Cayley's industry is quite remarkable. The length of calculated results for invariant theory have been compared with the formulae describing the motion of the Moon, and it is significant that Cayley was involved with both types of work—he and Charles Delaunay were leading exponents of the "lunar calculations" industry as well. In the light of invariant theory calculations, E. T. Bell commented that the *American Journal of Mathematics* "began storing up sheaves of calculations against an imminent famine that has yet to arrive."[26] He was right about Cayley and Sylvester filling the silos of mathematics with invariants and covariants, but he was judging them with a mindset that had moved away from such activity. The Victorian mathematicians saw this work as necessary, for they believed that progress in mathematics derived from the gradual accumulation of knowledge, and this meant storing up mountains of mathematical facts.

Cayley attempted to apply group theory to invariant theory. With the ultimate intention of obtaining a new proof of Gordan's theorem, he failed to combine these two branches, which are now so inseparably fused together. In a highly speculative paper, he threw out ideas of how such a proof might be constructed: "The infinite system of terms X, rational and integral functions [quantics] of a finite set of letters $(a, b, c, \ldots)$ which remains unaltered by all the substitutions of a certain group $G(a, b, c, \ldots)$ of substitutions upon these letters, includes always a finite set of terms P such that every term X whatever is a rational and integral function of these terms P."[27] The twentieth-century version of invariant theory is formulated in terms of group theory, but it was a novelty for Cayley, and he made little headway in this direction.

New Departures

Though Cayley published papers of great length, many now regarded as somewhat arcane, they are articulations of great theories that have since been superseded. In a curious twist, his shorter notes capturing insights are now best remembered—a vindication of his belief that the best work of a mathematician was done in five minutes. An instance of where this brevity is marked are the four contributions he made to the volumes of the newly founded *American Journal of Mathematics*. Sylvester commenced this journal at the end of 1877 and it was natural to summon help from England. He reminded Cayley of his duty to "occasionally lighten up my banishment" and for the journal, to "[s]end us any trifle at your *disposal as a page Iambi,* for our first number. The more the better."[28]

Ever ready to help, Cayley put together contributions published under the title *Desiderata and Suggestions.* The first, on the theory of groups, contains Cayley's famous theorem in group theory, the second his almost equally well-known result is on the diagrammatic graphical representation of groups, and the third an extended the Newton-Fourier iterative procedure for finding the root of an equation to the complex plane, an observation which has presaged modern work on fractals. The fourth is a lesser-known note on link work, which describes the geometrical curves traced out by a system of jointed rods. Interleaved in the chronology of these notes are his short incisive remarks on the four-color problem in graph theory.

Cayley immediately sent Sylvester his short piece on the theory of groups. In it "Cayley's theorem" was given informally: "the general problem of finding all the groups of a given order n, is really identical with the apparently less general problem of finding all the groups of the same order n, which can be formed with the substitutions upon n letters."[29] He hurriedly responded to Sylvester's demand. Group theory had moved on and Cayley had taken note of Jordan's work; moving with the times, he was not in the habit of looking back to his own papers of the 1850s to note that his new results were at variance with the old. In his short note, Cayley's new definition of a group is shorn of any appeal to the calculus of operations, a subject it was tied to in his papers of the 1850s. For Cayley, group theory appeared more abstract: "a group is defined by means of the laws of combinations of its symbols."[30] This short sentence is frequently seen as encapsulating abstract group theory and as laying the foundation stone for the modern theory of groups—it would provide a motif for combinatorial group theory a few years later in Germany. But was Cayley hearing the echoes of Duncan Gregory, who had put forward symbolical algebra as a science determined by the "laws of combination to which they are subject"? There are never influences so powerful as those learned in youth and Cayley showed himself turning to these time and again (see chap. 3).

By the following February, Cayley was occupied in other avenues, not least his preoccupation with the three-person working party delegated to frame new statutes for Trinity College following the passing of the Oxford and Cambridge Act (1877). In addition, he was a member of the university syndicate formed to consider reform of the mathematical Tripos. In these circumstances, he wrote to Sylvester: "I have to thank you very much for several letters, tho[ugh]' to be candid I have not been able to put myself sufficiently into your covariant theory to follow them in any satisfactory manner. I am always looking forward to the complete memoir which you are to write on the subject—and am a little disappointed to find from the table of Contents of your Journal, that you throw off with only *Notes* on the subject . . . I hardly know what Mathematics I have been doing—certainly not much—& I have been quite hindered from looking at my paper on the Covariants of a quintic."[31]

In May, he read a paper to the London Mathematical Society about representing a group diagrammatically. He explained it in terms of a hemipolyhedron (hemihedron), a three-dimensional figure obtained from a polyhedron by removing some of its faces to obtain the group diagram, which this figure projected on the plane. As was his custom, Cayley took a specific example of a group to illustrate the general idea.[32] In the course of the week, he wrote to Felix Klein: "I have just written out for the L[ondon].M[athematical].S[ociety]. a short paper on Groups, containing an idea which seems to me promising. In a substitution group of the order n upon n letters, every substitution must be *regular:* Taking $n = 12$, and considering the 12 letters *abcdefghijkl* suppose that one substitution of the group is *abc.def.ghi.jkl.* (*abc* written for (*abc*) means of course the cyclical change *a* into *b*, *b* into *c*, *c* into *a*): and that the group contains another substitution of the like form." Cayley constructed the diagram "in the first instance, with the arrows but without the letters, which are then affixed *at pleasure;* viz. the *form of the group* is quite independent of the way in which this is done, though the group itself is of course dependent upon it."[33] The following day he sent a synopsis for publication in Sylvester's journal.[34]

Group theory struggled for existence in England and only possessed two mid-Victorian champions: Thomas Kirkman, submerged in his Lancashire parish, and Cayley. In France, Camille Jordan produced a landmark with his *Traité des substitutions et des equations algébriques,* in which Galois's theory was central, an aspect of group theory that is mostly absent from Cayley's rather schematic papers. In the preface, Jordan acknowledged Cayley's contribution to the subject, and for his part, Cayley admired the *Traité* and congratulated Jordan on its completion. The subject was gaining importance and, ever intent to present the best course of study at Cambridge, Cayley argued that group theory should be part of the curriculum. It is likely that a small part of the subject was taught in his lectures, for he asked a Galois-style question in the Smith's Prize paper of 1877: "Write down the substitutions which do not alter the function $ab + cd$; and explain the constitution of the group." It is a simple problem and not one of the type he would describe as "rather catchy," but his solution was prefixed with the critical remark: "Relates to a theory which is not, but ought to be, treated of in the text books of the University."[35] Around 1878 Cayley wrote a short note on the life of Évariste Galois and acknowledged Galois's researches in the theory of algebraic equations and their solution by radicals: "and it is not too much to say that they [groups] are the foundation of all that has since been done, or is doing, in the subject."[36] By this time, the importance of Galois's ideas had generally been recognized, but in Britain, the subject was fifty years distant from inclusion in the undergraduate curriculum.

Cayley moved quickly from topic to topic. At the next monthly meeting of the London Mathematical Society in June he asked a question that has reverberated for generations of mathematicians: "Has a solution been given of the statement that in colouring a map of a country, divided into counties, only four distinct colours are required, so that no two adjacent counties should be painted in the same colour?"[37] Perhaps he was feeling ready for a new challenge in London's West End as he arrived for the meeting with his friend Thomas Archer Hirst. Just the previous day he submitted his final "Tenth Memoir on Quantics" to the Royal Society, a memoir marking the end of a series which had been in progress for so long.

Appearing in *Nature*, Cayley's question about the four-color problem would have been seen by Francis Galton, who had been taught by Cayley on the trip to Aberfeldy in Scotland as an undergraduate when Cayley was supervising a Cambridge reading party. Galton had kept in touch and had lost none of his reverence for the mathematician he had once judged a "brick and most gentlemanly-minded man." Never one to let an opportunity slip, the four-color problem appealed to Galton for several reasons. Through his passions for maps, he was interested in the possibility of using color in the printing process, now a possibility with the new lithographic technology, and he saw the possibility of a fillip for the new series of the Royal Geographic Society's journal. The scientific man-about-town was the go-between, linking the worlds of mathematics and geography, when he wrote to the editor Henry Bates:

> I wrote to Prof. Cayley of Cambridge, asking if he could send a short paragraph suitable for the Proceedings explaining as far as feasible the very curious problem of the sufficiency of 4 colours to distinguish adjacent territories in maps, which has lately excited great interest among the higher mathematicians.
>
> The problem is especially associated with Prof. Cayley's name & simple as it may be thought at first sight, it still baffles the highest analysis though there is no doubting its truth. I enclose his reply. It is much longer & more technical than I had hoped, but it makes the character of the problem very clear & I am sure it would interest a certain small section of our society. It is also something to obtain a paper from the most eminent mathematician in England, and probably in the world, on a geographical topic.... You see that the author wishes to have a proof.[38]

Indeed, Cayley did wish he had a proof! At their monthly meeting in February 1879, members of the society heard about geographical exploration of Afghanistan and India but in their monthly report, there was also Cayley's paper "On the Colouring of Maps," the only one he ever published in their journal. It began: "The theorem that four colours are sufficient for any map, is mentioned somewhere [in *The Athenaeum*] by the late Professor De Morgan, who refers to it as a theorem known to map-makers." He discoursed on the precise meaning of terms and continued: "It is easy to see that four colours are wanted." The question which captivated mathematicians was to decide whether four colours were sufficient. On this he admitted defeat: "And in any tolerably simple case it can be seen that four colours are sufficient. But I have not succeeded in obtaining a general proof."[39]

The paper on the "four- colors" written, Cayley moved on. In February 1879, he wrote to Sir William Thomson about the Newton-Fourier method of finding the root of an equation.[40] It achieved a high degree of significance in his mind for, a few days later, he wrote to Hirst about the same problem: "I have a beautiful question which is bothering me—the extension of the Newton-Fourier method of approximation to imaginary values: it is very easy and pretty for a quadric equation, but I do not yet see how it comes out for a cubic. The general notion is that the plane must be divided into regions; such that starting with a point P in one of these say the A-region ... [his ellipsis], the whole series of derived points $P_1 P_2 P_3$, ... up is P_∞ (which will be the point A) lies in this [planar] region; ... and so for the B. and C. regions. But I do not yet see how to find the bounding curves [of these regions]."[41]

The question may have been stimulated by research on complex variables and their geometrical interpretation on which he had worked eighteen months previously—but it had older antecedents, for he had been impressed by Sylvester's work on the imaginary roots using Newton's Rule done during the 1860s.[42] This was aligned to a different set of questions, but had in common the "boundary layer" idea which separated real and complex roots. The idea of "throwing aside the limitations as to reality" was a familiar question for him to take on.[43] He presented the Newton-Fourier problem to the Cambridge Philosophical Society, and by the beginning of March 1879 had done enough to send a short paper. He submitted this on 3 March 1879 to Sylvester for inclusion in the *American Journal of Mathematics*:

In connexion herewith, throwing aside the restrictions as to reality, we have what I call the Newton-Fourier Imaginary Problem, as follows.

Take $f(u)$ a given rational and integral function [a polynomial] of u, with real or imaginary coefficients; ξ, a given real or imaginary value, and from this derive ξ_1 by the formula

$$\xi_1 = \xi - \frac{f(\xi)}{f'(\xi)} \, ;$$

and thence $\xi_1, \xi_2, \xi_3, \ldots$ each from the preceding one by the like formula. ... The solution is easy and elegant in the case of a quadric equation: but the next succeeding case of the cubic equation appears to present considerable difficulty.[44]

The actual possibilities opened by this question depends heavily on computing power, but this was unavailable, and in the third note he contributed to Sylvester's journal, he made little progress on this problem.

In June, Cayley received a letter from Alfred Bray Kempe. It was exactly a year since both had attended the London Society Meeting when Cayley had raised the four-color question. Kempe had been a student at Trinity College Cambridge and had graduated as twenty-second wrangler in 1872. Much to the consternation of his friend J. W. L. Glaisher, who had graduated the year before, Kempe's lowly place in the wrangler list was nothing short of an injustice. Indeed it may have contributed to him seeking a legal career rather than an academic position. He became a barrister specializing in ecclesiastical law, but he kept up with mathematics and contact with Cambridge through Glaisher.

But now, the young man believed he had solved the four-color problem and though Cayley had lost sight of it himself, he responded with encouragement:

> The solution seems to me quite correct, & I am delighted with it. I think it might very well go, in nearly its present form to the *Geographical Magazine*, and I trust you will send it—I will write to Mr Bates to say that I have seen it . . .
>
> Geometrically, I think the correlative [dual] form of the theorem is more simple (tho[ugh]' less adapted for the *Geographical Magazine*) viz. "Considering on a sphere any number of points joined together in any manner by lines which do not cross each other then it is possible with four letters only, to mark the points, so that in each of the joining lines, the two extremities shall be marked by *different* letters." Your proof I think applies directly, and even more easily to this form of the theorem. . . .
>
> Is the theorem true where the surface is a multiply connected surface such as the anchor ring—I think then that at each point there might be 6 or more lines: and if so the proof would fail.[45]

Kempe's announcement of a proof appeared in the pages of *Nature*, and encouraged by J. J. Sylvester submitted it for publication in the *American Journal of Mathematics*.[46] The American journal was more appropriate than the geography journal; it was a technical paper and it was likely that the surveyors and explorers of the Empire had had sufficient exposure to the problem with Cayley's paper to want further mathematics. Kempe's proof contributed to his scientific standing though his mathematical credentials had been established through work on mechanical linkages. With such support from the leading mathematicians as Cayley and Sylvester, he was clearly in line for elevation to the Royal Society. Sylvester recommended Henry Smith to lead the application: "I mention the name of Henry Smith in especial [*sic*] because he is a more practical man than Cayley, who I am sure would be among your particular well wishers and do all in his power to promote your election."[47] Cayley was only troubled by his acute sense of propriety, as he wrote to Kempe: "I shall have much pleasure in signing your certificate: please send it soon, as I ought to do so *before* I am (as I suppose I shall be) a member of the Council."[48] Kempe was elected to the Royal Society a few years later with Cayley one of his leading supporters; the citation cautiously declaring that his proof "solving, it is believed for the first time, the problem of map delimitation with four colours."[49] The proof published in the second volume of Sylvester's *American Journal* survived for ten years before it was shown to be erroneous (but not without merit) by P. J. Heawood.[50]

By the time Kempe had announced his result, Cayley was more interested in Kempe's expertise in link-work. The Newton-Fourier problem was put aside and mechanical linkage problems became his focus. Only ten days after praising Kempe's proof, he wrote again to him about mechanical linkage problems, knowing him to be expert on this topic. The outcome formed the substance of the fourth note Cayley sent to Sylvester. The subject of graphical statics, of analyzing forces in systems of jointed rods, was popular in the 1870s, and had been spurred on by a lecture Sylvester gave to the Royal Institution in 1874. The next year, the well-regarded English mathematician Samuel Roberts wrote a paper on three-bar motion, and Cayley, acknowledging Roberts's priority, closely followed with an extensive paper of his own. In this most simple of linkages, three bars AB, BC, CD are fixed at A, D and freely jointed at B, C. A fixed point E on the coupler bar BC traces out a curve of order 6 as B, C move in the plane.[51] To determine the lengths of the bars so that the curve approximates a straight line is the classic "Watt's Parallel Motion" problem but Roberts and Cayley were more interested in the curve itself. (See "Three-bar motion" on p. 19g of the gallery.)

Of these four notes, only the first is abstract while the others are geometrical, indicating the balance of Cayley's work generally. They also offer another sidelight on his methods: in the few years, 1877–79, Cayley exploited the use of "color," something not usually associated with mathematics. His watercolor sketches taken into the lecture room could show off polyhedra to advantage, but he found color useful in the investigative process itself. His attempt to construct a linear theory of color came to little, but the graphical group theory (colorgroup) and the four-color problem are of lasting significance. In the Newton-Fourier problem, he was in a monochrome phase, in which regions were labeled white, gray, and black, but during these years, he reconsidered a theory of Augustin-Louis Cauchy's on the roots of equations using a broader palette. Adopting a geometrical perspective, the complexity of the visual evidence made the use of color essential, and he spoke of red and blue curves and regions tinted sable, gules, argent, and azure.[52] Looking at old problems and theories in a new way was a "Cayley-strength," and, much more than a mere cosmetic device. "Color" could be used for making new discoveries.

University Rounds

With Sylvester active in America, Cayley divided his Cambridge life between research, lecturing, administrative duties, and, in particular, Tripos reform. The large committee appointed on 17 May 1877 to consider further reform in mathematical studies would require his constant attention. The eighteen-member syndicate was drawn widely and included representatives from all phases of mathematical study at the university. It met each week for a year, and reputedly it was the most important year in the whole history of the Tripos. Cayley played his part in the formulation of the new regulations that were to emerge.[53] He again found himself at odds with the view of mathematics as the essential plank in a liberal education but without regard for it as a living subject. He accepted there were two classes of students, the "professed" mathematicians who would spend their lives studying the subject and the vast majority who would quit as soon as they crossed the Tripos finishing line. But, with a strong current of opinion to support the traditional liberal education position and preservation of the Cambridge system, Cayley placed his argument on support of the subject itself by his arguments for the "sake of the science."[54] It was a familiar debate, he was on known territory, and he no doubt knew it was a debate he had little chance of winning.

In the autumn of 1878 Cayley gave his lectures on solid geometry, but this was a time when academic matters appeared of secondary importance. There were difficulties with the endowment of his professorship owing to the deepening of the British agricultural depression. This would materially affect his stipend as a proposed reduction in rents would threaten the income of the Sadleirian estates in Hampshire. Together with dividends from consolidated stocks, the rents funded the chairs held by George Gabriel Stokes and James Challis as well. The tenant farmer, Clement Booth, complained to the university land agent: "The price of wheat is so low and labor risen so much that it is perfectly impossible to make all ends meet. This land is not at all good, I steam [i.e. by steam plough] a good deal and manure well too, but it is disheartening work with wheat at this price." After a small reduction in rent was offered, Booth wrote of "4 real bad years" and threatened to resign his lease. He was exasperated: "I can stand it no longer."[55]

For the university, the average income from the Sadleirian estates over the past thirteen years had been £1127-4s-9d and Cayley's stipend had been half of this. With uncertainty in the air, Cayley wrote to Stokes, partially accepting the difference in remuneration of the Lucasian and Sadleirian chairs but fighting his corner on behalf of pure mathematics:

> In reference to the proposed Stipend of the Sadlerian Professor, I think that,—not much, but still some weight is due to the consideration that if the Professorship had been now to be established—what the [University] Commission is proposing would

have been—from the £1100 income available for the Sadlerian to *take away* £500 for the Lucasian & Plumian Professorships—and then to *further increase* these up to £750.

The point which I rely upon is of course not this; but, Pure Mathematics = Applied Mathematics; and it [Pure Mathematics] is a distinct subject: whatever ground there may be for an inequality between two Professorships on like subjects, this would not apply to the Sadlerian: it would be only spoiling a Sadlerian Professor to give him in exchange one of the other Professorships.[56]

On the farm, Booth really was at the end of his tether and duly left. Without a tenant farmer on the Hampshire estates, the university income from this source went into steep decline.

Supposing his friend in financial trouble, and with his newly found American wealth, Sylvester offered to help. "I hope that the failure of your tenant," he wrote to Cayley, "has not caused you any inconvenience. Please excuse me if I am taking a liberty in saying that if any amount up to £1000 would be of any use to you, I should be very glad to make the advance to you and you could pay me interest if you wished to do so on the amount to be advanced which I could spare without the very slightest inconvenience."[57] Personal finance was a private matter but Sylvester was anxious to help in any way he could without obliging his friend to accept charity. Cayley was fortunate in one respect, for in 1875 he had been transferred from an honorary Trinity fellowship to an ordinary fellowship, a rare change in status that allowed him to receive financial benefit from the College Chest once more. Several years later, recognizing that its two most eminent professors were experiencing financial loss compared with past years, the university transferred both Cayley and Stokes to a payment schedule that would guarantee them assured incomes. For Cayley, this would mean the doubling of his lecturing duties; instead of lecturing for one term, he would have to lecture for two terms in each academic year.

Cayley may have reflected on his academic position when he thought of his previous occupation as a barrister. Still, the life of a mathematical professor had its compensations, and though Francesco Brioschi described him as living in modest circumstances, by his own standards, Cayley did occupy a large professor's house set in its own grounds with a complement of household servants and a horse-drawn carriage for trips around the outlying districts of Cambridge. It was not the same life as a high-earning London barrister, but he was always careful with money and despite the agricultural difficulties, it was a secure existence for the *pater familias* living at the end of Little St. Mary's Lane. There he kept a fatherly eye on the academic progress of his young son and daughter. Henry's mathematics called for attention and his father hoped he might have inherited his own passion for the subject. "Henry is getting on fairly with his mathematics," he wrote to Sylvester, "but he has not yet reached the stage of working a thing over for his own delectation. It is his 8th birthday in a few days." A year later he joined in the well-known parental refrain: "Henry is really very intelligent with his mathematics only he wants enthusiasm."[58] Poor Henry rescued himself later and his father must have felt proud when the 14-year-old won the school algebra prize at Sherbourne.

It is a measure of the informality of those days that Cayley was able to conduct mathematics from his private home. The Board of Mathematical Studies met there, lectures to the first women students were offered there, and examinations took place in this Cambridge backwater. Candidates for the Smith's Prize sat examinations set by Cayley, Stokes, Maxwell, and Isaac Todhunter and were required to attend each of their homes in turn. When it was Stokes's turn, the "dining-room mahogany supported the elbows" of the examined before and after the lunch provided by the professor.[59]

Karl Pearson, third wrangler in the Tripos in 1879, remembered sitting a Smith's Prize examination at Cayley's home, one of the last students to compete for the prize by a formal written examination before they were awarded for written dissertations.[60] On entering Garden House, Cayley's first words were:

"Throw off your gowns, gentlemen, you will work more easily without them," and accordingly they were dropped in a heap in a corner of the room, and we set to work unencumbered. Of course I knew nothing of the topics of Cayley's paper. My chance of scoring marks in the Tripos had depended only on my applied mathematics, and my pure mathematics were but sufficient to help in the former branch. But I took things leisurely, as if nothing depended on speed, and worked as one might work in solving crossword puzzles on a train journey. Cayley did not appear at lunch; sandwiches, biscuits and other light refreshments were brought up on a tray, accompanied by a decanter of excellent port wine; Cayley had not spared his cellar. . . . Cayley evidently did not think good port at all incompatible with the discussion of invariants or higher algebra.[61]

In the next year, Pearson was elected to a fellowship but remained undecided about his future life and the course it might take. He traveled to Berlin; on his return, he read for the bar at Lincoln's Inn and took temporary lecturing jobs around London. He was always proud he had been taught by Cayley and, when a few years later he went to University College to fill in for a pure mathematician, he carried it off by "dishing up a secular reflex of the choice herbs of my high priest, Cayley!"[62]

The lecture subjects chosen by Cayley during the 1870s indicates no discernible pattern, and he carried on much as he had always done—taking as his topics, those allied to current research, with the lectures themselves delivered to tiny audiences. His lectures delivered in the first term for the years 1877–82 were on algebra (1877), solid geometry (1878), differential equations (1879), theory of equations (1880), and Abel's theorem and the theta functions (1881). Reporting on a set of these "somewhat remarkable" lectures J. J. Thomson (no relation to William) wrote: "I was the only undergraduate; Glaisher and R. T. Wright, both Masters of Arts, made up the audience. Cayley did not use the blackboard, but sat at the end of a long narrow table and wrote with a quill pen on sheets of large foolscap paper. As the seats next [to] the Professor were occupied by my seniors, I only saw the writing upside down." Thomson, who went to Cambridge in 1876 the year after Pearson, was regarded by his Cambridge friends as an outstanding intellect and, no doubt for this reason, was permitted to attend the lectures. The future "J. J." wrote of the professor's method: "[i]t was a most interesting, valuable and educational experience to see Cayley solve a problem. He did not seem to trouble much about choosing the best method, but took the first that came to his mind. This led to analytical expressions which seemed hopelessly complicated and uncouth. Cayley, however, never seemed disconcerted but went steadily on, and in a few lines had changed the shapeless mass of symbols into beautifully symmetrical expressions, and the problem was solved." Thomson took away something of permanent value from Cayley's classes. It was not the material itself, but his realization that "[a]s a lesson in teaching one not to be afraid of a crowd of symbols, it was most valuable." The brilliant young mathematician, as he was then regarded, produced his first paper on quaternions, which was published in the *Messenger of Mathematics*. He became a fellow of Trinity College in 1881 and a fellow of the Royal Society in 1884 with Cayley's backing. At the end of that year, he succeeded Lord Rayleigh as professor of experimental physics at the Cavendish Laboratory.[63]

Pearson was the future statistician, Thomson the future physicist, but it was Andrew Russell Forsyth who was to be the future mathematician. A Glaswegian by birth, he was educated in Liverpool. Afflicted by eye trouble but determined to succeed, this pudgy-faced adolescent came to Cambridge a year ahead of entry to prepare for the scholarship examination. Charles Pendlebury, who became a leading textbook writer, was a contemporary at Liverpool, but it was his brother Richard, more of a mathematician and an established mathematical coach, who prepared the boy for the examination: in October 1877, Forsyth entered Trinity College with a Major Entrance Scholarship. Because of their advanced character, few undergraduates went to Cayley's classes, but Forsyth summoned up courage to attend, despite the traditional warning from his coach, E. J. Routh, that the lectures would not be of any service in the Tripos.

The attendance at Cayley's lectures changed Forsyth's perspective on pure mathematics. The first lecture gave him a profound shock, for the theory being expounded by the master mathematician was of the "complex variable," a phrase "which plunged me into complete bewilderment."[64] There too Forsyth learned that "old notes were never used a second time" and that the lectures were always on Cayley's latest research and, invariably, last-moment productions.[65] The course was listed as "Differential Equations" but Forsyth attended a course which danced around such topics as conformal representation, polyhedral functions, and Karl Schwarz's research on hypergeometric series. Forsyth gradually began to understand that pure mathematics consisted of more than tools for dealing with mathematical physics issuing forth from the Cavendish Laboratory.

A vignette of what it meant to attend one of Cayley's lectures was recalled when Forsyth was an old man:

> They were in the Michaelmas Term [first term] 1879, nominally on differential equations: the subject was never mentioned after the first ten minutes: instead he discussed icosahedral functions, groups, covariants, and so on: as comprehensible by me at that stage as if he had been dealing with Chinese syntax. The audience was small: five of us sitting in a row on a form. There was no chalk: the single blackboard was used as a rest for Cayley's blue draft paper manuscript, held several feet away from the nearest pair of eyes. Cayley held the manuscript more or less in place with his left hand; his right forefinger would move along a formula which he spoke aloud. Once or twice, he broke off, in order to write on a sheet of paper some "very important" formula.

The members of this improbable class balancing on wooden benches, striving to learn from the doyen of mathematics as he roamed over the mathematical terrain, experienced an idiosyncratic performance from a natural mathematician. He was likely oblivious that his listeners found the material unfamiliar or that they found any difficulty in understanding it. Forsyth continued:

> At the furthest end of the row sat Glaisher, reading proof sheets (Messenger or Quarterly, or something equivalent): Then [Richard] Pendlebury, who could see nothing and could not hear much for even then he was a little deaf: then R. C. Rowe who mostly stood up in his place and looked over Cayley's shoulder while he tried to make notes on the scribbling paper at arms length on the desk in front of him: then myself, writing down whatever could be seen or heard, a disconnected jumble to be developed into something more or less coherent by much labor: finally, nearest the door, and looking slantwise across the distant blue paper on the board, J. D. H. Dickson, writing much, sometimes notes of his own, sometimes fragments of mine. Not much of a lecture, one would say: but the great man was at work, and we all believed in him.[66]

Richard Rowe, the earnest young man making every effort to take away a record of the great man's work, was third wrangler in 1877. One of the Cambridge Apostles (an exclusive Cambridge fraternity of intellectuals), his dissertation on Abel's theorem was inspired by Niels-Henrik Abel's pathbreaking paper of 1826. At Cayley's instigation, Rowe was encouraged to publish his work in the *Philosophical Transactions of the Royal Society* in 1881.[67] Cayley was so impressed by Rowe's work that he recommended him for the professorship of pure mathematics at University College, London, but, sadly, Rowe died in 1884 before he could fulfill his early promise. The other attendee, J. D. Hamilton Dickson, a fellow Glaswegian of Forsyth's, had assisted William Thomson in the laying of the Atlantic cable before coming to Cambridge, where he was the fifth wrangler in 1874. On the occasion of his attendance at Cayley's lecture, he was a tutor at Peterhouse in Cambridge. Later he became the "tame mathematician" aiding Galton in his exploration of the statistical concept of correlation.

An exception to the small number of students attending Cayley's lectures occurred in 1881. In that year there were twelve students, which caused him to remark on the "exceptional number."[68] In the group was Charlotte Angas Scott, who was to progress to an academic career in America. Cayley's course was on Abelian functions, and Scott noted Cayley's empathy with Abel who had died when only twenty-six years of age.[69] About the course itself, Forsyth wrote "somehow, Abelian functions had sprung into a semblance of popularity: by that date Cayley had begun to chalk on the blackboard, though never at ease, and he had many a half-furtive look at his watch long before the hour was up . . . Once Sir William Thomson came; and he sat for the whole hour, without a single interrupting question. But again, it was not the information given us that mattered: again, it was the great man at work."[70] Thomson was generally uninterested in arcane questions of pure mathematics but sitting at Cayley's feet in the 1880s there was reverence in the air mixed with nostalgia for their student days together, or perhaps he just knew when to remain silent. How different he acted while attending a lecture given by Stokes, where he "would butt in after almost every sentence with some idea which had just occurred to him."[71]

Cayley was a significant figure in the movement toward women's education. That women should remain in the background was widely assumed in Victorian Britain, but with a sense of what was fair and proper, Cayley advanced the women's cause in material ways at both a personal and institutional level. It was inevitably delivered with a degree of seriousness from his all-male world but there was a lighter side, too. With his wry sense of humor, he let slip that he was quite satisfied with Queen Victoria on the throne, but when the Duke of York was born in 1865, "he could hardly rejoice to see a third male heir to the throne come into the world," since he himself "preferred a woman sovereign."[72] In 1869, the leading liberal in the university, the philosopher Henry Sidgwick, set up "Lectures for Women," to build on the movement that had taken place in other parts of the country. He recruited others of like mind to give courses. In the Victorian world, where every one of note knew everyone else, Henry Sidgwick and Cayley knew each other well (they had attended the same primary school though twenty years apart) and were university "liberals." Other people of this persuasion, who would support the educational ideal of women's education, could be guessed easily. Thus people such as Henry and Millicent Fawcett, Alfred Marshall, F. D. Maurice, John Venn, T. G. Bonney, N. M. Ferrers, and W. K. Clifford supported the "Lectures Association" and many advertised special courses of lectures. Cayley came quickly to the cause and, in the Lent term of 1870, offered a course on the "Principles of Arithmetic" and one a few years later on "Algebra"—to take place two afternoons a week in Garden House.[73]

Sidgwick's "Lectures Association" was formally constituted as "The Association for the Promotion of the Higher Education of Women in Cambridge" and this was complemented by the separate "Newnham Hall Company, Ltd." with shareholders, who put up the money and organized a hall of residence.[74] By 1880 these two bodies had joined forces to become "The Newnham College Association for advancing Education and Learning among Women in Cambridge," a "not-for-profit" company. At the incorporation of the new body, of what is now Newnham College, Cayley became the first president and chair of the college council.

Charlotte Scott was Cayley's best-known female disciple. She was in residence at Girton College, the other women's college asserting its rights to existence. Brought up in a Non-Conformist environment, her father was the principal of the Lancashire College near Manchester, whose mission it was to prepare young men for the Congregational Ministry. At home, the seven children were stimulated by father's mathematical riddles while home-tutoring was provided by members of the College teaching staff. With a scholarship from London's Goldsmith's Company, she entered Hitchin College in 1876, prior to it moving to Cambridge and becoming Girton.[75] She was the earliest women's success in the Tripos contest, being placed equivalent to the eighth wrangler in the order of merit in January 1880 after having taken the examination for the mathematical Tripos on an informal basis. She became a regular attendee at Cayley's professorial lectures and, realizing her ability, he did all he could to help her in the mathematical career she had chosen.

Not being eligible for formal admittance to a Cambridge degree, Charlotte Scott gained her official academic credentials from London University, where she was awarded an honors degree in 1882. Following this, she was a lecturer at Girton until 1885, when she was awarded a London D.Sc. Cayley supported her application for an academic appointment and brought the weight of his academic eminence to bear by testifying to her potential for original work in the mathematical field, an area not generally accepted as a woman's subject. Appointed to Bryn Mawr, she left for a career in the United States, where she spent over forty years, before taking retirement back in Cambridge. Her mathematical interests in analytical geometry coincided with Cayley's, an expertise acquired in her formative years in Cambridge. She published a book on *Plane Analytical Geometry*, and if attribution had been required, she wrote, "it would have been alike my duty and my pleasure to write on every page the name of Professor Cayley."[76]

Charlotte's Scott's achievement in the unreformed Tripos had an immediate effect. There was a public demand that women students at the university be granted the same rights as men. There was widespread agitation supported by petitions. In the first step toward women's degrees at Cambridge, Cayley's was not a banner-waving performance, but behind the scenes his cautious influence was felt in the right places, and he was shrewd enough to realize that such an upheaval in the status quo could be achieved only at a measured pace.[77] The implication of their being admitted to degrees was a constitutional matter, as graduates had the right to share in the governance of the university by voting on major issues.

In a country which did not have women's suffrage for another forty years, it would not be surprising to learn that the admission of women to a degree was resisted. Cayley did not press for too much too soon, but he was a signatory to the "The Three Graces" proposal put to the University Senate in February 1881 to allow women a first step—to sit the Tripos examinations but without the possibility of rewards enjoyed by the men.[78]

To East and West

For three months in the summer of 1880, Cayley visited Felix Klein in Munich. There he met Alexander von Brill who was to provide a rigorous proof of the geometrical correspondence theorem for plane curves, which, in 1866, he had only managed to grasp intuitively. Another contact was Walther von Dyck, who gained his doctorate in 1879 at Munich as a student of Klein and who picked up Cayley's intimation of an "abstract" group, which had appeared in the first volume of the *American Journal of Mathematics*. It was the characterization that a "group is defined by means of the laws of combination of its symbols" that von Dyck used as a motto for his founding of combinatorial group theory.[79]

Also at Munich was the young Adolf Hurwitz, barely beyond his teens, who had been taught by H. C. Schubert, author of *Abzählende Geometrie*, but was now a student with Klein. It was Hurwitz who ultimately proved the "16 squares theorem," which, in 1880, Cayley was vigorously but unsuccessfully attempting to prove himself. The visit coincided with a pivotal move in Klein's career. In May he accepted an invitation to a new academic position, a move which took him from the Technische Hochschule to a chair in geometry at the ancient University of Leipzig.[80]

At Cambridge, the demands on Cayley's time for lecturing remained slight and he contemplated following up his treatise on "Elliptic Functions" with another textbook. He wrote to Sylvester on Christmas Eve: "I have not been doing much mathematically—my lectures, on the theory of equations—I always go on with them very much from hand to mouth—took more time than they ought to have done. I think I shall write a book on the subject."[81] Sylvester was happy and active at Baltimore and had established a vibrant research school in the embryonic university. After retirement from Woolwich, he was once again in full flow and feeling at one with the world. He had just been awarded the Copley Medal of the Royal Society, and Spottiswoode, then president, singled out his work on invariant theory for general approval. When he wrote to Cayley, Sylvester's pleasure was transparent: "I ought to be very grateful to my friends in England who thought of bestowing so great an honour as the award of the Copley Medal on an unfortunate exile. I am told that you had much to do with it—I do not even know for certain if you were on the Council which bestowed it—It was very unexpected and *relatively speaking* I consider would need special arguments for its justification."[82] Cayley was a council member and would have supplied "the special arguments" in the case of pure mathematics.

Sylvester hoped Cayley would convert his lectures into a more permanent form and commence the writing of a book on the theory of equations, one which might include some Galois theory but inevitably be most concerned with the application of the rival invariant theory to the subject.[83] Unfortunately the project did not materialize. The full import of Sylvester's letter of early 1881 was not this, however, but to invite Cayley to visit him at Johns Hopkins: "I wish you would write a book on the theory of Qns [Equations] as you propose. I wish two times over that you would come over here—where you could more than double your emoluments as Professor and where you could command a class of 10 or 20 enthusiastic hearers and followers and really found a School . . . Besides founding a school your social relations here could be very agreeable: They are a great church-going and very friendly people."[84]

In May 1881, Sylvester pressed the invitation: "I wish you could come and join us here. I could promise you a class of some 10 at least of most intelligent and sympathetic auditors for your lectures: such men as [Thomas] Craig, [Fabian] Franklin, [Oscar H.] Mitchell, [Miss Christina] Ladd (although she is not exactly a man) myself, [William E.] Story and several most promising young men who bid fair to keep the succession of the Craigs, Franklins and the rest. In fact you would have a Class of Glaishers for your auditors and the seed you might sow would fall upon a fertile soil. . . . If you would come and join us you would be doing a great work for the glory of Pure Mathematics among the English speaking race throughout the world." Sylvester continued with other temptations, money, convivial society, and, not least, the opportunity of taking long walks in the wooded areas around Baltimore. He gave him a vivid account of the life in the new university making its mark on the academic world. He was always a man to encourage the growth of new ideas and ambitious projects: "We have not an idle student among us and no single case calling for the application of discipline has ever yet occurred—We number about 200 at present but sooner or later I am sure that a Boom will spring up in our favour and carry our numbers to a far higher figure. Newcombe [*sic*], Hill and other mathematicians are in our immediate Vicinity at Washington and Anapolis [*sic*] and Craig and I propose to convert our so called 'Mathematical Seminarum' into the 'American Mathematical Society.'"[85]

When he visited England that summer, Sylvester again discussed the possibility of a visit and issued a formal invitation. This time, Cayley decided to go. It would be a break in the routine, though he would be without his children, who were then being educated in Cambridge. His provisional acceptance was dispatched to Gilman: "I accept—subject only to my obtaining leave of absence from my own university: I anticipate no difficulty as to this, but the formal permission cannot be given until October or November. But for the distance and the absence from home, I would say that I accept with great pleasure. Thanking you very much for the kind terms of your letter."[86] On hearing the news, Sylvester was keen to make arrangements and suggest possible subjects for Cayley's guest lectures. One suggestion was the "Sixth Memoir on Quantics," which had gained attention owing to its important place in Felix Klein's Erlangen Program. C. S. Peirce, who taught logic at Johns Hopkins, rated it an "immortal memoir" and judged it one of the greatest mathematical influences of his life.[87] Sylvester also remembered Cayley's *Reports on Dynamics* to the British Association in 1856 and 1862, but there was a surfeit of subjects from which to choose. His primary intention was to raise Cayley's confidence in lecturing to an unknown audience: "It has occurred to me that your sixth memoir on Quantics might be a very desirable subject for you to lecture upon—or possibly *also* the substance of your report on Mechanics to the British Association with such additions as may be required by what has been done since the date when it was written—I only throw out this as a suggestion my idea being that the more strongly you can impress your own personality on the subject-matter of your lectures so much the more impressive and valuable they are likely to be."[88]

Cayley's formal acceptance was dispatched to Gilman three months later: "I can therefore now accept finally the flattering proposal made to me; and I trust to be in Baltimore on two or three days after New Years day. My wife will accompany me. The title for my lectures may be 'Analytical geometry and the Abelian and Theta Functions' [and] I am making some progress in putting the course together."[89] Some duties in England would have to be relinquished and, in November 1881, he resigned his editorship of the *Monthly Notices of the Royal Astronomical Society,* a position which he had held for twenty years.

Through Sylvester's energy and boundless enthusiasm, mathematics was firmly planted at Johns Hopkins by the time Cayley arrived. At the beginning of February, the Trustees gave a reception for him in Hopkins Hall. A large number was present to welcome the English professor and a brief welcoming address was given by Sylvester. There was a high level of expectation aroused among the faculty—the presence of a mathematician with a high European reputation was an event and provided a model for many of the young ambitious tyros of America. C. S. Peirce provided an impression made by their visitor: "His physiognomy was most striking. The shape of the head was extraordinary— rather flat, and seeming to have a cornice above it; the intensity of the gray eyes, which seemed to be looking through whatever opaque thing might be near to him was unparalleled. The handsome but rather small nose and the perfectly unaffected and smilingly watchful mouth seemed boyish in their unsophisticated interest in things. But what a curious catlike intensity as if just pouncing on a truth!"[90] While in Baltimore, Cayley and his wife stayed at Mrs. DuBois Egerton's rooming house, the "leading *salon* of the South," popular with visitors and where "almost every dinner was an evening party."[91]

Cayley duly gave a course of lectures on Abelian and theta functions, a subject at the forefront of his research. He had written a long paper on it in 1880, one which proved to be the last he published in the *Philosophical Transactions*. Based on his lectures at Baltimore, he began to compile another substantial memoir on the subject.[92] His work referred to the Clebsch-Gordan theory of Abelian functions in a vital way, but he introduced many of his own constructions. Sylvester was to write to Kempe that Cayley, apart from providing short papers for the *American Journal of Mathematics*, "is going to give us his great one on Elliptic & Abelian Integrals and theta functions."[93] While in Baltimore, Cayley took the opportunity of attending the short course of six lectures on the Clebsch-Gordan theory given by Sylvester's second-in-command William Story.[94]

Cayley conveyed his impressions of the daily round at Johns Hopkins to his friend Thomas Hirst in England: "I have been getting on satisfactorily enough with my own work on the subject [Abelian and theta functions], making out to myself the very beautiful manner in which Clebsch and Gordan in their book [*Theorie der Abelschen Functionen*] arrive at the Multiple theta functions—and completing the geometrical theory of the double theta functions as derivable from a nodal quartic—on two other days I hear Sylvester, who began a course of 'three or at most four' lectures on Multiple Algebra, which he has gone on with since the beginning of the year—& they are not finished—nor likely to be." There was time for sight-seeing, and, continuing his letter: "we have had two pleasant expeditions to Washington—the last time for a Saturday & Sunday with Prof. Newcomb—and shall be going sometime—I think the spring is now really come & we must be thinking about it—to Niagra."[95]

In Baltimore, Cayley made time to both encourage Sylvester on matrix algebra and make contributions himself. In a lecture in April 1882, he considered the problem of listing double-algebras that are both commutative and associative, but he did not attempt to classify algebras using invariant theory, as was done later. He had briefly tackled this problem as a young man (see chap. 5) but the subject was again topical. Benjamin Peirce's work on classifying algebras was current and his *Linear Associative Algebra*, which had existed in a limited edition, had been published in its totality in the *American Journal of Mathematics* the previous year.[96]

In a different contribution to the life of Johns Hopkins, Cayley gave a seminar on "isomers and trees" with the encouragement of Professor Ira Remsen, who was giving a lecture course to the chemistry students on the compounds of carbon. Remsen's scientific reputation was primarily associated with the discovery of the organic compound saccharine, but to his students, he was the "prince of teachers." He had studied in Germany and had been on the staff of a German university—he would have appreciated the joint work of Cayley and Carl Schorlemmer.

Cayley's visit was short—barely five months—but he made a substantial impression judging from the success of his lecture course and other incidental lectures. At the farewell reception "[t]he platform was abundantly decorated with flowers, and Cayley, with his retiring manner, looked very uncomfortable while Sylvester was speaking. Referring to his modesty, Sylvester suddenly turned towards him: 'There he sits, like a victim decked with flowers!'"[97] He and his wife boarded the White Star *Britannic* for England in June.

Cayley looked back on the visit to Johns Hopkins with warmth, and, in his letter of thanks to Gilman, wrote:

> I look back with great satisfaction to my five months residence at Baltimore, and shall always feel proud of my connection with your university. My lectures on the Abelian and theta functions, to a Class such as I had there, were a great pleasure to me, and I trust that they may have afforded suggestions for original research to some of my auditors. I was much gratified at having the opportunity of hearing the lectures of Prof. Sylvester and the short courses by Mr Peirce, Dr Story, and Dr Craig. I found in the university, a great amount of interest in Mathematics, and of good mathematical work—shown by the attendance at lectures, the meetings of the Scientific Society and Mathematical Seminary, and Dr Craig's Friday evenings; and by the *American Mathematical Journal:* and I confidently anticipate, that going on as it does, the university will, in Mathematics especially, hold a very high rank along with older Institutions in America and Europe—and will exercise a powerful influence for the advancement of the Science.[98]

By the beginning of the 1880s Cayley had secured a position of great eminence in the mathematical and scientific world, and to his European reputation, he had added a successful visit to Baltimore in the new world. With his friend returned, Sylvester began to reflect on his own position:

> I am hardly in a state of mind to draw up reports or do any kind of work—it was a *great mistake* my not going over to England this summer [1882]. The climate has been too much for me—I congratulate you heartily on your success with your work—as usual you accomplish whatever you set your mind upon. You and Mrs Cayley have left behind you—here in Baltimore the most agreeable recollections—and I am much gratified to be told by you that you enjoyed your visit. I have *not* done anything more with Matrices. I have moreover been incapable so far of drawing up my projected

memoir on the subject even with the aid of the copious notes I have retained of my lectures—Would there were any opening for me at home in England![99]

Sylvester was suffering from various ailments. His eyes gave him trouble and he was frequently overcome by the heat of the Baltimore summer. In these moods, his thoughts turned toward England. In mathematical work, he kept up the momentum with invariant theory, but it is all too evident that he missed the company of his friend.

PART 5

Make One Music as Before, 1882–1895

The vast extent of modern mathematics [is] a tract of beautiful
country seen first at a distance, but which will bear to be rambled
through and studied in every detail of hillside and valley, stream,
rock, wood, and flower.
—Arthur Cayley's *British Association Address*, 1883

In the early 1880s, Cayley stood at the pinnacle of European mathematics. Returning from the United States, he was elected president of the British Association for the Advancement of Science. He put all his years of erudition into the preparation of a broadly based address summarizing mathematical progress to the year 1883. It took into account Greek mathematics as well as recent accomplishments while staking out areas for future exploration by mathematicians. His high-minded Platonist views on the nature of number and space were thrown into relief by an attack on John Stuart Mill's empiricist conceptions.

Cayley's cherished research on invariant theory was exposed to criticism in the wake of David Hilbert's pathbreaking existentialist approach to the subject in the late 1880s. The leading problems were regarded as "solved," and the calculatory way of working was cast aside by the younger generation. Undeterred, he pursued his own objectives. Physically weak during his last days, he continued with mathematics, in which former forays raised further questions. Ever up to the mark, he commented on the very latest topics of interest exercising the mathematical community.

CHAPTER 15

"A Tract of Beautiful Country"

O n arrival in England from the United States, Cayley learned he had been proposed for the Royal Society's highest honor, the Copley Medal. On 15 June 1882, Thomas Archer Hirst proposed him "for his numerous profound and comprehensive researches in pure Mathematics," and the proposal was sec-onded by President William Spottiswoode.[1] The family spent their summer holidays in Switzerland and visited the places he had last seen as a young man just clear of the Tripos examination and beginning his academic career. On his return in August, he wrote to J. J. Sylvester: "The Swiss journey was quite a success—we went to the Bernese Oberland—places which I have not seen since I was first there in 1843—Thun, Lauterbrunnen, The Gemmi, Grindelwald &c.—to Andermatt—& by the St. Gothard & the lake of Lucerne—including the rail-way ascent of the Riga to Lucerne, Berne & so home."[2]

The next stop was Southampton for the annual meeting of the British Association. There he familiarized himself with the role of the president, the mantle he would assume in the following year. It was no doubt a topic of conversation when he visited William Thomson on board *Lalla Rookh*, his oceangoing yacht moored in the harbor.

Return to the Fray

Presidencies aside, Cayley's immediate mathematical concern was with the Clebsch-Gordan theory of algebraic curves, on which he lectured in Baltimore. He told Sylvester: "[I] have not got fairly into it yet; what I did as to the quartic curve is quite right, and a great step, but as I found out before Switzerland, does not by any means at once give me the final form (of integral of the third kind for the quartic curve) which I was in search of—and I have not since my return got any further."[3] He was unable to complete it to his satisfaction and the work came to a halt, and it did not restart until a few years later. Instead, he began to think about his Presidential Address to the British Association to be delivered one year's hence.

England also brought him closer to Peter Guthrie Tait and incipient contro-versy. Tait was always sensitive to being cast as the physicist in comparison to the mathematician, who was unrestrained by the world, possessed the glorious overview, and, in the eyes of some, was regarded as the superior one. "To some men," he declared, "physics is an abomination, to others it is something too trivial for the human intellect to waste its energies on."[4] While Sylvester, with his overblown insistence on generality in mathematics, was a target, Tait also engaged Cayley. Perhaps he wanted to expose the inadequacy of the mathematician's understanding of physical principles. Tait argued that Newton's laws should be modified, for they were based on the concept of force (that which changes a body's state of rest or motion), which had no objective existence for the physicist.

On the contrary, matter, momentum, and energy, Tait argued, had objective existence and the true basis of mechanics should include the conservation of these quantities. Cayley admitted to Tait he could not understand this proposal. "I understand force—I do not understand energy," he wrote to Tait, "I am willing to believe that Newton's Action = Reaction potentially includes d'Alembert's principle [the external forces acting on a body in motion balance the inertial forces]—but I never saw my way with the former, and I do see my way with the latter—and I accept Virtual Velocities + d'Alembert's principle as the foundation of Mechanics."[5] Cayley was thus a Lagrangian disciple and had not altered his stance since he submitted his "Report on Dynamics" to the British Association in 1858. He had been brought up with this approach as a student and was unmoved by Tait's offer of redemption: "[n]o more the arrows of the Wrangler race, Piercing, shall wound you."[6]

The subject of quaternions, Tait's other preoccupation, came to the fore in their role as a mathematical method for describing physical laws. In seeking mathematical symmetry between force and rotation, Cayley wrote to Tait, the "*sum* of two quaternions, quà *force,* is the resultant force [and the] product of two quaternions, quà *rotation,* is the resultant of rotation. But is there any interpretation for the sum quà *rotation* or for the product quà *force?* It would be very nice if there were. We enjoyed our American expedition very much." Tait was too busy to treat Cayley's enquiry of "symmetry" in more than a cursory way, and, for his part, Cayley did not fully understand the interpretation of the rotation operator $(q + r)(\ )(q + r)^{-1}$ given by Sir William Rowan Hamilton's champion. Offering a red rag to a bull, Cayley wrote an innocent postscript about the rotation operator: "I believe it was I who first gave in the *Phil[osoph-ical] Mag[azine]* the formula $q^{-1}(ix + jy + kz)q$, showing it was identical with that of Rodrigues for the effect of a rotation—but Hamilton was doubtless acquainted with it." Ever vigilant about the accuracy of his own historical facts, Tait unequivocally asserted that Hamilton had anticipated Cayley on connecting quaternions with rotations. Hamilton was certainly aware of the quaternion as a rotation but its succinct expression was first given by Cayley in 1844 (see chap. 4). Showing discretion, Cayley merely disarmed Tait: "I am rather glad to find that the formula was first given by Hamilton."[7]

In Baltimore, Sylvester missed the company of his friend, not least the erudition and the sounding board for his own ideas: "I find it very hard work to learn out of books," he wrote. "Would that you were here—as in your conversation I have ever found an all sufficient library for my wants."[8] On Christmas Day, 1882, Sylvester wished the new year would bring him "back to Mrs Egertons whose door I never pass without indulging regrets for the unrenewable past."[9] Of Cambridge gossip, he approved of Cayley voting for James Stuart, the Liberal candidate for the Cambridge University seat in Parliament.[10] He encouraged Cayley to continue with the material presented in his Baltimore lectures on Abelian functions: "I congratulate you on your success with the Clebsch Theory so far as attained and doubt not that you will eventually obtain all that you are in search of—as I have never known you to fail in any research you had once set your mind on."[11]

In invariant theory, Cayley continued with his probing of the binary quintic form and other binary forms of low order. The tedium of calculation continued, the generating function method employed still bolstered by the Fundamental Postulate. The postulate was a useful expedient in circumventing problems with generating functions used in invariant theory, but as the extensive tables of invariants and covariants were being assembled, nagging doubts over its veracity resurfaced. Sylvester had gone so far as announcing a prize of 1,500 French francs for proving or disproving the soundness of the Fundamental Postulate. He set a deadline of New Year's Day. True to form, he pitched in as a competitor and even stipulated that the prize would only be awarded if he had not already proved it himself.[12]

After Cayley's return to England in mid-1882, the postulate was found to be false by James Hammond.[13] A Cambridge graduate, Hammond had a facility for patient and accurate calculation, which enabled him to find an example which proved the postulate false. Cayley immediately reacted to the news of Hammond's counterexample with a hastily written paper, "[t]he extreme importance of Mr Hammond's result, as regards the entire subject of Covariants, leads me to reproduce his investigation in the notation of my Memoirs on Quantics."[14] Sylvester responded with a letter to Cayley that "Hammond has sent me a disproof of the Postulate: too late I ween for the prize." It was too late, because Sylvester had himself found the postulate false, and the caveat of the challenge invoked, he was spared the 1,500 francs expense. Commentating on the counterexample, Sylvester continued: "It [Hammond's counterexample] is quite different from mine and more explicit. He finds for instance a covariant of deg[ree]-order 5.13 to the 7^c [binary form of order 7] which is a ground form but which the method of Tamisage [Sylvester's method of examining the generating function] does not serve to disclose and which consequently will not be found in my table of groundforms for the 7^c!!!"[15]

As Cayley indicated, this was more than a storm in a tea cup. For the binary form of order 7, the application of "Cayley's Law" showed there were four linearly independent covariants of degree 5 and order 13.[16] Using the generating function method it was predicted that exactly four irreducible covariants existed; so, by invoking the Fundamental Postulate there were no syzygies between them and it was concluded all irreducible covariants had been found. Hammond rather spoiled this by displaying a syzygy and a new irreducible covariant that had not appeared before.

Cayley was hopeful that Hammond's counterexample would explain difficulties he had experienced with the binary form of order 7, a form which was proving as intractable as the binary form of order 5 (in general, binary forms of prime order were found to be more difficult to handle than those of composite order). The failure of the postulate meant that tables would have to be corrected. On this point, Cayley wrote a note to Sylvester: "I have just this moment, as I am writing, received the A.M.J. [*American Journal of Mathematics*]— and shall perhaps find that you have been before me, in what I was go[ing] to say—the disproof of the fundamental postulate will at any rate necessitate a revision of your table of the irreducible covariants of the seventhic, and I am in hopes that it may turn out that there is an irreducible invariant of the degree 20."[17] Sylvester followed Cayley's note several months later with a fulsome postscript:

In some previous note [above] you referred to the 7^c and the necessity of finding whether or not there is a 20.0 (i.e. invariant of the degree 20 attached to it) I had thought of the Canonical determinants method for ascertaining how this is but as you say of your own efforts without success—The 8^c is by the great labours of [August] Von Gall and my own, saved from the general wreck [resulting from the failure of the Fundamental Postulate] there was only one doubtful ground-form the 10.4 [covariant of degree 10, order 4] and that I may recall to you I have proved in the A.M.J. [*American Journal of Mathematics*] does not exist.[18]

The Fundamental Postulate had failed for the binary form of order 7, but Sylvester was under the impression that the binary form of order 7 was an isolated special case. He desperately wanted to buttress his methods and save the postulate from the "general wreck." He was defiant and Hammond's counterexample did not persuade him of the postulate's dubious character; only reluctantly did he admit that it was partially dubious. But he could offer only empirical evidence in its support: "Has it ever occurred to you," he wrote to Cayley years later, "to consider why my method in spite of a possible error [of arithmetic] in the result does as a matter of fact give *all* and not only *some* of the semi-invariants in all the cases to which it has been applied, viz. 5*cs, 6cs, 8cs* as shown by comparison with Clebsch and Gordans and as regards the 8*c* by Von Gall's calculations. . . . No case of discrepancy has yet been observed between the results of this method and the results obtained by others—or rather I should say whenever my results have differed from the German, mine have been *proved* to be correct and the German erroneous."[19]

The adoption of the postulate had been a bold step, but it had outlived its usefulness. At the root of the dilemma was the "problem of the syzygies." This did not exist for the cases of binary forms of low orders, but, for higher order forms, its presence caused the generating function methods to become unreliable. By the end of the century, P. A. MacMahon remarked that the method of generating functions "had become indeed a fruitful source of error."[20] In effect, Hammond's counterexample curtailed the use of generating functions for dealing with binary forms of order 7 and higher.[21] In both England and America invariant theory was running into difficulties.

On 9 February 1883, H. J. S. Smith, the holder of the Savilian chair at Oxford and a great figure of British mathematics died quite suddenly. Generally admired for his championing of such social causes as the extension of the franchise, Britain also lost a world-class number theorist. Cayley wrote to Sylvester that "[y]ou will have heard of the grievous loss we have had by the death of Prof. Smith: he was here at Cambridge not three weeks ago for the election of the Plumian Professsorship & the last accounts, [of] the evening before his death were quite favorable."[22]

Smith's death was a chance for aspirants both young and old. Sylvester signaled his intention of applying for the vacant chair in writing to Cayley: "I shall most probably offer myself as a Candidate for the succession to his professorship—but await the result of inquiries before coming to an absolute decision. Do you think I am likely to be appointed? If the chances are considerably against me—it would be impolitic to offer—but perhaps even impolitic it would be right on my part to do so by way of testing what I consider—although you may not perhaps agree with me—an important principle." The condition on applicants for the Savilian Chair that they be from the Christian World (*ex quacunque Natione Orbis Christiani*) was an obstacle Sylvester wanted to test, though it was not an absolutely insurmountable one. Sylvester continued his letter "[t]he salary I understand is very much less than what I receive here —which is equivalent to about £1250 English money [four US dollars to the £ sterling]: but the appointment would suit me."[23] Ever cautious, Cayley would not have wanted to raise Sylvester's hopes.

There was also the possibility of Felix Klein coming to England, though this was remote. Cayley evidently thought he would be an ideal candidate for the Savilian professorship. His stay with him in Munich three years earlier had been beneficial to both and he thought highly of Klein's ability. Cayley naturally applauded the Erlangen Program in geometry, which was based, in part, on his own discoveries. In return, Klein had especial regard for Cayley as a first-class mathematician and the originator of the scheme whereby non-Euclidean geometries could be classified; Klein saw Cayley as an algorithmic mathematician who excelled "in the given formal treatment of a given question, in devising for it an 'algorithm.'"[24]

As an elector for the Savilian Chair himself, Cayley dutifully informed Klein of the vacancy. While he was of the opinion that no Oxford mathematician was suitable qualified to fill the prestigious chair, a complicatory factor in outright support for Klein was Sylvester's possible candidature. It is possible he tried to dissuade Klein, for he warned him "there is I am afraid the serious objection of the want of interest in Mathematics at Oxford." This would have caused the leading German mathematician of his generation to think again. "I should think it a capital thing for Oxford," Cayley continued, and ever practical, "but as regards yourself, I do not venture at all to urge it upon you: the difference of income would probably be counter-balanced by the greater expence of living here [in England]."[25] There were now several possibilities: Klein to Oxford or Johns Hopkins; Sylvester to Oxford or full retirement.

Cayley was himself was in a settled position. Apart from the sojourn in Baltimore in 1882 and the fourteen years at the bar, he had spent twenty years at Cambridge. He had filled the Sadleirian Chair with distinction and was the recipient of a stream of honorary degrees, but he was not one to rest on his laurels or content himself to sit on the mathematical sidelines. There was presidency of the British Association at hand.

A View from the Summit

Perhaps the most conspicuous reminder of Cayley's place in the Victorian age is the address he gave to the British Association on assuming the presidency in September 1883. To be singled out as its president was a prominent honor. Scientists were invited, as well as politicians, but it was comparatively rare for a mathematician to be asked. During the previous fifty years, few mathematicians had served in the position of British Association president—George Peacock, in 1844, was the previous one whose scientific reputation rested solely on mathematics. William Spottiswoode had been president in 1878, but he had been active in science generally and was better known for his researches on the polarization of light and experiments on the electric discharges in gases than for his work in mathematics. Other past presidents who contributed to mathematics in a subsidiary way included John Herschel, George Biddell Airy, William Hopkins, George Gabriel Stokes, and William Thomson. It was significant that the organizing committee chose Cayley as the 1883 figurehead for he had been a constant support to the association and his position almost demanded public recognition.

For a brief time it was touch-and-go whether Cayley would be acceptable. Originally, the meeting was scheduled for Oxford, and this raised an obstacle to Cayley's candidature: a group of Oxford dons wanted Lord Salisbury as president.[26] The third marquis of Salisbury, who was to become prime minister on three separate occasions, was leader of the Conservative Party in the House of Lords. The British Association periodically chose an aristocrat as a figurehead for their meetings and the meeting at Oxford more than suggested he would be chosen. He was the Oxford chancellor, an office he had held for more than thirty years, and had been a student there himself. To their credit, the association did not bend to Oxford pressure and chose Cayley, but the ensuing rift caused the venue to be switched away from Oxford. Cayley was left with the prospect of presiding over the meeting at a small northern coastal town and the association audience with an invitation to a mathematics lecture, an experience not threatened by Lord Salisbury, who had emerged from his undergraduate studies with an "Honorary Fourth" in mathematics.[27] Cayley seemed oblivious of the backstage politics when writing to Sylvester: "[y]ou will have heard the meeting much to my wife's and my own disappointment is to be—instead of Oxford—at Southport. I scarcely heard anything in explanation of the failure of Oxford."[28]

Cayley accepted the invitation to be president while visiting Sylvester in America, and immediately began to think of likely topics for his address. In a letter to Hirst, he had sketched out some possibilities:

> I am astonished at my own audacity [in accepting the presidency]. Glaisher wrote to me that you & Adams were going to send me letters of exhortation & persuasion, but I have not yet received the one from Adams. I think I shall make the Address on pure Mathematics—including of course geometry—and the various directions— imaginaries & imaginary space, hyperspace, complex & ideal numbers, *Abzählende Geometrie* [Enumerative Geometry due mainly to H. C. Schubert] etc, in which the science has in modern times extended itself. I think it will be possible to be fairly interesting to a largish part of a non-Mathematical Audience.[29]

On his return, he prepared a first draft and informed Sylvester of progress: "I hardly see yet what I shall be able to make of it; I shall have to go over [it] again, with a good deal of difference to the point of view, much that was given—very well indeed—by Spottiswoode in his [Presidential] Address at Dublin in 1878."[30] Sylvester was ready in his support: "I feel assured that your address will repay the world and yourself for the time and trouble it will have cost you."[31]

To give such an address would be an ordeal. Cayley was a man of few words, rarely volunteering speech and when called upon to reply invariably did so in "five words long." He became well known for his taciturn nature, as if frequently he was absorbed in his own thoughts; to accompany him on walks one had to be at ease with long silences. The reading of a technical paper before an audience was something he regarded as an unprofitable formality to be dispensed with as quickly as possible. Oral expositions were not given much credence and he admitted to not learning much from them—his judgment on Spottiswoode's address was based on reading, as he had not attended the Dublin British Association meeting when he was president. Like many lawyers, he learned from the printed word.[32] Indeed, contemporaries attributed the severity of his written style to training in the law, perhaps forgetting that this hallmark of his work existed before he entered that profession. But the quip attributed to him that "the object of law was to say a thing in the greatest number of words, and of mathematics to say it in the fewest" he strenuously repudiated.

Sylvester would be able to help his friend with the basic points of exposition, for he was a man who could take an audience into his confidence and readily communicate ideas; "I think it rather a good thing to try and throw a little warmth and glow of personal feeling and passion into a recital of one's scientific experiences," he wrote to the editor of *Nature* on one occasion.[33] Sylvester paid little heed to the convention that held most mathematical authors to the path of pure symbolism in their technical expositions. Indeed, he never missed an opportunity to embroider his mathematical creations with philosophical commentary and personal anecdote. He was, however, ever conscious of Cayley's stringency: "I hope you will not be too severe in your judgement on this departure from conventional rules," he would write. "I know that you do not in general approve of any deviation from established usage in dealing with mathematical subjects. So I shall understand if you do not allude to the matter and that silence means *dissent*."[34] Here was the challenge for Cayley: to make a lecture on mathematics interesting and intelligible to a general audience.

When Sylvester visited England in the summer Cayley had a draft of his lecture approaching completion. It was a difficult writing task, but Sylvester gave him warm encouragement:

> I think that taken as a whole the Address is exceedingly good—and abounds with points of interest and, even with what the French might call witty reflexions. You ought I think to read it just *half* as fast as you did on that memorable afternoon. Two thirds of it read at a rate to ensure distinctness would occupy I believe considerably more than an hour in the delivery—I do not think that the multitude will be greatly edified by it as spoken—but the *contre-coup* [repercussion] of the judgement following its perusal will I think make ample amends and tend to support the merit of the Association as a body seriously bent on the promotion of science.[35]

Cayley was too aware that mathematics, and *pure* mathematics at that, would not appeal to a typical British Association audience, but he thought it right that the president should speak on his own subject. He benefited from Sylvester's skills as a popularizer while his own ability in dealing with popular mathematics was, at best, untested. He could cope with the "circle-squarers" and other cranks, who were attracted by his eminence and authority, on a one-to-one basis. Occasionally called on to adjudicate on their fantastic claims, one incisive remark was enough, as when he advised the Cambridge vice-chancellor that papers sent him were "without any value whatever, the author finds for the 'exact' value of π, a number which is a less accurate approximation than the 22/7 of Archimedes."[36] A lecture to a large body of the populace was a different proposition.

To detail the mathematical developments of the day and touch on the leading philosophical issues concerned with mathematics was the overall plan Cayley adopted. His views on what mattered to an active mathematician were clearly set out in the address. The image he evoked was the mathematician who rambles through "a tract of beautiful country," a metaphor which harkened back to the Romantic movement of the early nineteenth century. Leaders in this artistic and literary movement included some of Cayley's favorites—William Wordsworth and Samuel Taylor Coleridge, but above all, Sir Walter Scott—poets and writers who were widely influential in his youth and who remained favorites all his life. His attachment to the countryside and the Lake District year-on-year makes the allusion to Nature consistent with his approach to mathematics.

The metaphor of the scientist turning the pages of Nature's book was a common one for Victorian scientists but is perhaps an unusual one when applied to mathematicians who, by the commonly accepted canons of the subject, should be more adapted to the bare bones of rational argument. Clearly, logic and proof plays an important part in Cayley's mathematics, but the metaphor of Nature rises above this. Wordsworth's love of mountains and rivers coursing their ways on the valley floor finds a parallel with the way he looked on the mathematical endeavor. As the nineteenth century progressed and Britain became industrialized, there was an insatiable clamor for progress; but when Cayley looked for progress in mathematics, he maintained an outlook more akin to a pastoral idyll than a cold ideal based on relentless logic.

In the address, Cayley had to make an abstruse subject palatable to an assembly of the middle and upper classes more used to popular accounts of the recent triumphs of science and technology. They could relate to the invention of the telephone as demonstrated at the Plymouth meeting of 1877 and identify with the latest application of electricity. But pure mathematics? Cayley had also to compete with Krakatoa, which had erupted the month before, and scientists were engrossed in a debate on the atmospheric effects of one of the most vigorous volcanic eruptions the world had ever seen. He had also to contend with the immediacy of the latest technological success and a scientific sensation. He was aware of the presidential performance the year before by the metallurgist and electrical engineer Charles Siemens with his "wondrous practical applications of science to electric lighting, telegraphy, the St. Gothard Tunnel and the

Suez canal, gun-cotton, and a host of other purposes, complete with a grand concluding speculation on the conservation of solar energy."

A British Association meeting was a gathering of the "broad church"; it was always a varied affair, and this year the meeting at Southport would follow this tradition. Southport is a small coastal town in the north of England but within range of Liverpool and Manchester—there was a long pier and its sand flats stretch for miles. Southport had horse-drawn tramcars and was sedate, "lively but not noisy." Even today it is quintessentially Victorian. The Winter Gardens housing the association meeting occupied a site of eight acres.

As befitting a British Association president, Cayley had his portrait painted in oils. Ten years prior the London portraitist Lowes Dickinson had depicted him wearing his academic gown and sitting at his desk with his quill poised. That portrait was of the enduring student, but now the local artist William H. Longmaid portrayed him wearing a black coat, brown trousers and white collar fastened with a blue necktie—the president of the British Association for the Advancement of Science at Southport.[37] The program of the meeting would follow the usual lines; there would be the evening soirées and, on the weekend distinguished preachers would mount the pulpits of the town. The individual sections would gain the latest news; for instance, the geography section this year would hear the latest letter from Henry Morton Stanley in the Congo. Cayley would preside over the Saturday evening popular lecture to the working classes in the circus, this year delivered by the eminent consulting engineer Sir Frederick Bramwell. A stalwart of the association, Bramwell, inventor and engineer, spoke on the practical uses and benefits of science.[38] This promised to be immensely popular but what effect would a lecture on mathematics have?

The *Times* expressed an ambivalence for having a mathematician as president: "Of the president of the year, Professor Cayley, even Senior Wranglers speak with bated breath and hopeless wonder. Only Professor Sylvester is believed to come anything near to fathoming the depth of Professor Cayley's mathematical attainments; it has therefore, been feared that the Southport presidential address would not be of a character to appeal to a popular audience."[39] That the mathematics would be unintelligible to the general public is not surprising, but it would be of only marginal interest to the practical scientist. Mathematicians were generally regarded by the wider scientific community as occupying some rarefied domain, invariably mountainous, to which there was no hope of entry for the nonmathematician. Mathematicians were fringe members of the scientific community.

Though Cayley had overcome the painful shyness of youth and had grown self-assured, he was not entirely comfortable at the prospect of his lecture, a feeling which his audience could match. On the eve of the address, the *Times* reported the "ominous rumour [which] has been whispered about the 'halls' of Burlington-house that the president insisted on a blackboard and supply of chalk being provided to enable him to give demonstrations illustrative of the abstruse mathematics which he intended to introduce into his lecture."[40] The seriousness of meetings was offset by the advice offered to British Association speakers by scientists themselves: "[y]ou must prattle 'pure' like Cayley and pour out 'mixed' like Rayleigh, and make them twice as dry."[41] By accounts such as these the stage was set. On the opening Wednesday evening an audience assembled in the Southport Pavilion in the Winter Gardens eager to be entertained. Siemens handed them over to "one of the greatest of living mathematicians (applause)," who would tell them about a "much higher subject" than his own.[42]

The Address

As a true Platonist, Cayley's intention was to lay before his audience the beauty of the eternal ideas of mathematics. Moreover, it was an optimism for the future of mathematics that he radiated when he rose to speak "I wish to speak to you to-night upon Mathematics," he began, barely audible, but "I am quite aware of the difficulty arising from the abstract nature of my subject." He acknowledged that association audiences in recent times had heard of the "fairy-tales of science," another allusion to Tennyson's (first) *Locksley Hall,* a poetic work he knew intimately, and which served to make a link with his audience. Recalling former presidents with their practical results and the comparison with himself, he was just too diffident: "if, I say, recalling these or any earlier Addresses, you should wish that you were now about to have, from a different President, a discourse on a different subject, I can very well sympathise with you in the feeling." He started in classic Darwinian mode: "So much the worse, it may be, for a particular meeting; but the meeting is the individual, which on evolution principles must be sacrificed for the development of the race."[43]

It was to be a wide-ranging performance in which mathematics was surveyed from a historical perspective. He perhaps began on his weakest foot, on some questions in the underlying philosophy of mathematics, a subject which led him to John Stuart Mill. Mill's philosophy of science was an essential line of thought in the late nineteenth century. Mill argued that science should be based on empiricism, that is, scientific theories were arrived at by inducting general statements from instances of experiential data. In his view, scientific theories were not derived a priori (before experience) but through a process of introspection based on experience.

One of Cayley's first targets was empiricism in mathematics, a subject which Mill included in his thesis. Mill opposed *all* a priori reasoning, even in mathematics, as a quotation from his autobiography indicates: "The notion that truths external to the mind may be known by intuition or consciousness, independently of observation and experience, is, I am persuaded, in these times, the great intellectual support of false doctrines and bad institutions. . . . There

never was such an instrument devised for consecrating all deep-seated preju-
dices."[44] These thoughts were also set out in his widely influential *System of
Logic, Ratiocinative and Inductive: Being a Connected View of the Principles of
Evidence and the Methods of Scientific Investigation.*[45] Mill chose mathematics as
his central target, as he believed it was the most difficult case to argue: "the 'System
of Logic,'" he wrote, "met the intuitive philosophers on ground on which they
had previously been deemed unassailable."[46] This great Victorian, ten years
dead, influenced the whole century with philosophical theories and political
writings, but he was no mathematician. Cayley made no claim to be a philosopher either, but in choosing mathematics to put forward his general theory of
induction, Mill lay himself open to attack. Cayley took the Southport opportunity
as a suitable occasion to rebut his ideas and mount this attack.

In an altogether different direction, Cayley dissented from William Rowan
Hamilton's opinion that the idea of number was based on instances of "time."[47]
For himself, time was not an empirical concept; but as he expressed it at Southport, time was "a necessary representation lying at the foundation of all intu-
itions." If he rejected the notion of time as bound up with number, still less did
he entertain arguments which placed complex numbers (or as he called them,
imaginary numbers) on a temporal foundation: "we do not have in Mathematics
the notion of time until we bring it there," he resolutely concluded. He took
as his starting point for the basis for number the Greek idea of "plurality" of
indivisible units.[48]

Cayley believed that ordinal number was more fundamental than the concept of cardinality, or size of number[49]: "We think of, say, the letters, $a, b, c,$ &c.,
and thence in the case of a finite set—for instance a, b, c, d, e—we arrive at the
notion of number; co-ordinating them one by one with any other set of things,
or, suppose, with the words first, second, &c., we find that the last of them goes
with the word fifth, and we say that the number of things is = five: the notion
of cardinal number would thus appear to be derived from that of ordinal number." This understanding is at variance with the modern conventional view that
cardinal number logically precedes ordinal number in that it assumes *less* structure than ordinality and therefore precedes it. Cayley takes "logical precedence"
to mean subtracting of properties and so for him ordinality precedes cardinality.[50] In helping him with the preparation of the address, Sylvester reminded his
friend of their past conversations on this subject: "The observation of Ordinal
preceding Cardinal number in logical conception you got I believe from me—
but doubtless the parentage has escaped your recollection as it was a long time
ago when the subject was mentioned—the matter is of no importance but
when I saw my baby nursed in your arms it was impossible to restrain a cry of
natural affection and parental recognition."[51]

Although Cayley paid attention to these distinctions he did not question the
status of real numbers themselves (they were "quantity" or "magnitude and

were capable of continuous variation"). The Theory of Numbers itself was introduced at three different places in his address; he included a short exposition of Ernst Kummer's ideal numbers and saw "Fermat's Last Theorem" as an isolated result awaiting proof.[52] As for complex numbers, they were of the utmost importance in his work for they were the fundamental notion underlying geometry and analysis.[53] Though he rejected the interpretation of Duncan Gregory and William Walton on the meaning of imaginaries in geometry, the "principle of continuity" was Cayley's favored interpretation of their occurrence. This principle lay at the core of his mathematical thinking and was fundamental to his conception of modern geometry.

A substantial portion of the Address was devoted to geometry. Cayley's view on the basis for the truths of geometry expressed his Platonist beliefs and were an echo of the unchanging position he had expressed in his youth: he was influenced by William Whewell's philosophy of innate ideas and the existence of truths independent of any experience. These views were promulgated in Whewell's writings of the 1840s, which were Cayley's formative years at Cambridge.[54] His espousal of geometrical notions existing in a Platonic "realm," to which mathematicians can escape, identifies a major theme of his philosophical framework. For him space was an a priori notion and not unusually for his time, he understood *physical* space to be the unique three-dimensional Euclidean space and he never escaped from this Kantian notion.[55] The learning of Euclid as a schoolboy and a frame of reference prevalent in the Cambridge of his youth encouraged this viewpoint. Whewell, who played such a leading part in the shaping of Cambridge education, would have recognized Cayley's understanding on the nature of physical space as his own, an understanding expressed in his *Philosophy of the Inductive Sciences,* first published in 1840: "Space is not a General Notion collected by abstraction from particular cases; for we do not speak of *Spaces* in general, but of universal or absolute *Space*. Absolute Space is infinite. All special spaces are *in* absolute space and are parts of it."[56]

Whewell did not have Cayley's notion of projective geometry nor his sophisticated mathematical theories, but "absolute physical space" is present in the work of both. Each took positions in opposition to Mill, though Cayley did allow the philosopher some latitude: "I think it may be at once conceded that the truths of geometry are truths precisely because they relate to and express the properties of what Mill calls 'purely imaginary objects'; that these objects do not exist in Mill's sense, that they do not exist in nature, may also be granted; that they are 'not even possible,' if this means not possible in an existing nature, may also be granted." He took issue with Mill's argument involving the word "conceive." Many mathematicians would dismiss these arguments as no concern of theirs, but Cayley was remarkably patient. Again his Platonic view that there are a priori conceptions of such notions as "triangle" and "straightness" is underlined: "That we cannot 'conceive' them depends on the meaning which we attach to the word conceive. I would myself say that the purely imaginary objects are the only realities, the ουτως οντα [really real] in regard to which the corresponding physical objects are as the shadows in the cave; and it is only by means of them that we are able to deny the existence of a corresponding physical object; if there is no conception of straightness, then it is meaningless to deny the existence of a perfectly straight line."

As a young man, Cayley had been among the first to accept the notion of an n-dimensional mathematical space. By the 1880s this conception offered mathematics an obvious way forward. The idea of n-dimensional geometry was a fitting area for research, for so much was unknown, as he made clear in the address: "[i]n n-dimensional geometry, only isolated questions have been considered. The field is simply too wide; the comparison with each other of the two cases of plane geometry and solid geometry is enough to show how the complexity and difficulty of the theory would increase with each successive dimension." Cayley's expertise lay in projective geometry and the hallmark of his approach to this subject was *analytical* geometry. He was catholic in his geometrical interests and in his latter years studied differential geometry and read Bernhard Riemann's papers.[57] The pure geometer Olaus Henrici, president of the mathematics and physics section at Southport, drew his audience's attention to Christian von Staudt in his own lecture. In the 1880s Cayley became aware of the importance of von Staudt's foundational work and recognized the superiority of this treatment.

In company with such as William Clifford and Henri Poincaré, Cayley was one of the first to accept the legitimacy of non-Euclidean geometry. Often cast as a "conservative," he deserves to escape this label in connection with his thinking on non-Euclidean geometry—at least he was not so conservative as George Salmon, who offhandedly remarked that "he reserved such themes for the next world."[58] But though Cayley readily accepted the validity of models of non-Euclidean geometry, he could not see it as displacing the geometry of ordinary three-dimensional Euclidean space:

It is well known that Euclid's twelfth axiom, even in Playfair's form of it [through a point not on a given line there exists a *unique* line parallel to the given line], has been considered as needing demonstration; and that Lobatschewsky constructed a perfectly consistent theory, wherein this axiom was assumed not to hold good, or say a system of non-Euclidian plane geometry. There is a like system of non-Euclidian solid geometry. My own view is that Euclid's twelfth axiom in Playfair's form of it does not need demonstration, but is part of our notion of space, of the physical space of our experience—the space, that is, which we become acquainted with by experience, but which is the representation lying at the foundation of all external experience.

John Playfair's version of the twelfth axiom was published in his *Elements of Geometry* (1795), a work which included a presentation of Euclid's first six books. For Cayley, non-Euclidean geometries were "drawing-room geometries"—geometries that were mathematically consistent but had no relevance to physical space. He could not conceive of physical space in which Playfair's axiom was violated. Non-Euclidean geometries existed as part of Euclidean space but had no physical existence of their own: "[i]n regarding the physical space of our experience as possibly non-Euclidian, Riemann's idea seems to be that of modifying the notion of distance, not that of treating it as a locus in four-dimensional space. I have just come to speak of four-dimensional space. What meaning do we attach to it? Or can we attach any meaning? It may be at once admitted that we cannot conceive of a fourth dimension of space; that space as we conceive of it, and the physical space of our experience, are alike three-dimensional."

A few years later, when Thomson sent him a copy of *Popular Lectures and Addresses,* Cayley took a neutral position when confronted with the more revolutionary views on the nature of physical space. He wrote to Thomson: "In the lecture on the Wave Theory, you parenthetically ignore the notion of the curvature of space—Clifford would say that, going far enough, you might come—not to an end—but to the point at which you started. I have never been able to see whether this does or does not assume a four-dimensional space as the locus-in-quo of your re-entrant & therefore finite space."[59] The geometer D. M. Y. Sommerville judged that Cayley "never quite arrived at a just appreciation of the science [of geometry]. For Cayley, non-Euclidean geometry scarcely attained to an independent existence. It was always either the geometry upon a certain class of curved surfaces, like spherical geometry, or a mode of representing certain projective relations in Euclidean geometry."[60] Cayley held firm to the standpoint that non-Euclidean geometry existed *within* the framework of Euclidean geometry but did not have an objective physical existence in its own right. The step required to escape this thinking would take a few decades longer.

Cayley gave his Southport audience a wide view of modern mathematics. He touched on sophisticated theories, including details about subjects far removed from daily life: elliptic functions and hyperelliptic functions, the theory of equations, the theory of numbers, algebra and Galois's theory—what had been done in mathematics and what remained to be done in the future.[61] Toward the end of the evening, he turned his attention to the mathematics he regarded as *outside* ordinary mathematics—the extraordinary. Here his mode of expression is suggestive of modern axiomatic mathematics. A sense of the evolution of mathematical ideas being absorbed into a mainstream is indicated by his view on the placement of the calculus of operations as *within* the body of mathematical knowledge. In his youth this had been a well-defined subject of its own, an independent body of knowledge recognized as such by a group of practitioners. In the address, he reviewed the status of the "symbols of operation," and, noting that they do not obey the commutative law, he stated in a forthright manner: "it could hardly be said that any development whatever of the theory of such symbols of operation did not belong to ordinary algebra." Cayley separated multiple algebra, as encapsulated by Benjamin Peirce's *Linear Associative Algebra,* outside of ordinary mathematics. In Peirce's algebra, the symbols do not arise from considerations within ordinary mathematics but are defined externally (for example, an undefined symbol x may be assumed to satisfy $x^2 = x$).[62] He then explained, with a remarkably modern description, how the theory of multiple algebra might be presented:

> I would formulate a general theory as follows: . . . Two given entities are represented by two linear functions; the sum of these is a like linear function representing an entity of the same kind, which may be regarded as the sum of the two entities; and the product of them (taken in a determined order, when the order is material) is an entity of the same kind, which may be regarded as the product (in the same order) of the two entities. We thus establish by definition the notion of the sum of the two entities, and that of the product (in a determinate order, when the order is material) of the two entities.

Cayley saw this abstract framework as being suitable for the analysis of algebraic systems which had been developed separately during the course of the century, among them Hamilton's quaternions and the algebras contained in Hermann Grassmann's *Ausdehnungslehre.* When he was starting out in the 1840s and had discovered the quaternions in the writings of Hamilton, the quaternions were publicized as a system unlike that found in ordinary algebra but one which could be interpreted geometrically in ordinary space. While he readily accepted the idea of symbols being anticommutative, and had discovered the octonion system, which was both anticommutative *and* antiassociative, he had not seen these systems in a totally abstract framework as he did now.

Cayley also placed George Boole's *Laws of Thought* outside the boundaries of ordinary mathematics. In his Southport lecture, he recalled the difficulty of interpretation of probability (which he had raised with Boole thirty-five years previously) but noted the "remarkable theory" and lack of attention shown to the work, which, "it seems to me, passed out of notice, without having been properly discussed." He noted C. S. Peirce's work on systems of mathematical logic and the connection he made with logic and the linear associative algebras of his father Benjamin Peirce. These works, placed on the "borderland of logic and mathematics" were outside of ordinary mathematics. On another border lay Hermann Schubert's *Abzählende Geometrie.*[63]

By the end of his address, Cayley was led to speak of the "vast extent of modern mathematics." Never was the subject studied so intensively as at the present time, he declared—"the advances made have been enormous, the actual field is boundless, the future full of hope." The optimistic address combined broad discussion with technicalities; it had begun with philosophical enquiry but had rapidly ascended from the philosophical to the technical branches of mathematics. In his professional life as a mathematician, Cayley was not primarily concerned with the foundations of mathematics. He had a working mathematician's belief in the solid meanings of word and intuition and he trusted his own instinct. He was more concerned with widening the scope of mathematics and settling outstanding problems than in analyzing concepts into their ultimate elements. He was not an unreflective mathematician and would normally go some way to meet philosophers on their subject, but he would draw a line when, as he put it, the line of argument became "unprofitable."

Few others would attempt to cover Cayley's wide spectrum of activity. By the 1880s, scientists and mathematicians were moving toward an era of increased specialization, and a lecture of such breadth would become a rarity. But he had to admit his picture was incomplete. He lacked the omniscience Whewell had exercised in the 1840s: "My imperfect acquaintance as well with the mathematics as the physics prevents me from speaking of the benefits which the theory of Partial Differential Equations has received from the hydrodynamical theory of vortex motion, and from the great physical theories of heat, electricity, magnetism, and energy."

In the years ahead, mathematicians would become specialists: algebraists, geometers or analysts, hydrodynamicists, electrical theorists, and later still, even more specialized. In giving the address, his concern was for the integrity of mathematics, and at Southport, it was this unity he hoped to convey. "So much the worse, it may be, for a particular meeting," he had warned the audience at the beginning, "but the meeting is the individual, which on evolution principles must be sacrificed for the development of the race." He attempted to make his address understood by the audience, but concession was not part of his intellectual make-up.[64] Before the event, it was forecast that it would be an address of the "severely scientific class."[65] It had undoubtedly been one by the close of the opening Wednesday evening of the British Association meeting at Southport.

Applause

William Thomson proposed the vote of thanks and years later Susan Cayley wrote to him: "I still remember the delight with which I listened to your speech at Southport after the Presidential address had been given."[66] Cayley had said his piece but the audience of Victoria's England were probably bemused by their president expounding on his passion in that "odd, high-pitched voice" of his.[67] The immediate reaction might only be guessed. Indeed, it would have been miraculous if Cayley had been successful in reaching the popular ear. No doubt a ripple of polite applause was heard in the Pavilion at Southport, but the published reviews turned out to have more bite.

The occurrence of Cayley's address was reported in the local *Liverpool Daily Post.* His opposition to the dilution of Euclid was a newsworthy item and they reported his outlook: "there could be no possible harm to any person intending to go through the course of modern geometry to have studied Euclid, not only the ordinary course of the first four books, but also the fifth book, the omission of which he had always most deeply regretted."[68] Apart from news reporting, the kindest reviews judged that such an address was unsuitable to a typical British Association audience who were essentially "wonder loving" but nonscientific. The juxtaposition of "mathematical lecture" and "popular audience" presented the critics with too much ammunition and the fears Sylvester had expressed before the event, that "I do not think that the multitude will be greatly edified by it," were realized.

Richard Proctor, who had served as secretary to the Royal Astronomical Society during Cayley's tumultuous period as president, shrewdly assessed the impact of the esoteric address as one which "produced a somewhat singular impression. That not one in a hundred of those who heard or read it could form any opinion as to its value was generally admitted; yet hundreds expressed very strong opinions respecting its extreme value, its unusual profundity." Of its content, Proctor voiced a criticism which raised the ire of Cayley's friends, though they must have realized that his grasp on the subject matter was shaky. Proctor's inability to appreciate new mathematical thinking is palpable, when he applied "common sense" to demonstrate that Cayley's views on mathematics were inapplicable to the world. In particular, he seized on Cayley's exposition of "imaginaries" as a target for ridicule: "Mathematics in its prime, the mathematics of Newton and Lagrange and Laplace, advanced our knowledge like the mental work of a man in his prime; mathematics dealing with imaginary nonentities is like the unintelligible fancies of a dreaming dotard who *has been* learned and profound, but in his old age lets idle imaginations take possession of him."[69] Proctor rather spoiled his case by categorizing non-Euclidean geometry as dream-geometry and quadridimensional space as an absurdity. Conscious of Cayley's predicament, friends came to his defense. James Parker Smith, a Cambridge don (and son of Archibald Smith, founder of the *Cambridge Mathematical Journal*) wrote a sturdy rebuttal of Proctor's article.[70]

Cayley's decision to give the British Association a mathematics lesson also drew a stinging review from an anonymous reviewer in the *Athenaeum.* British Association audiences expected practical invention and the reviewer was critical of the lofty Platonic viewpoint Cayley had exercised: "Mathematics is to him a queen, who will not deign to recount the benefits she has conferred on her subjects, but is willing gracefully to acknowledge the services which they may have paid to her."[71] Most likely, this reviewer was Airy, now retired from his position as Astronomer Royal, but with his critical energies unquenched.

The upbeat address reflected Cayley's views on mathematics in terms of progress, and while it was submissive in tone, its message was one of self-confidence for the future. It was in keeping with the character of the late Victorian culture. Pointing to its self-assurance and alertness, G. M. Young caught its spirit: "[i]n its blend of intellectual adventure and moral conservatism, it is really Athenian. I doubt if any lines of Tennyson [from *In Memoriam*] were more often quoted by contemporaries than these:

> Let knowledge grow from more to more,
> But more of reverence in us dwell;
> That mind and soul, according well,
> May make one music as before,
> But vaster.

No words could express more perfectly the Victorian ideal of expansion about a central stability."[72]

CHAPTER 16

The Old Man of Mathematics

T he old man of mathematics still had the energy to attend meetings in London, though the sudden death of his younger brother in December 1883 loosened ties with the metropolis.[1] In the role of *éminence grise*, in which he was so obviously cast, mathematics took him there occasionally but it was less and less often. The Scot, Alexander MacFarlane, mathematician, logician, and historian, remembered meeting him on one of the occasions he was at the Albemarle Street rooms where a meeting of the London Mathematical Society was being conducted: "[t]he room was small, and some twelve mathematicians were assembled round a table, among whom was Prof. Cayley, as became evident to me from the proceedings. At the close of the meeting Cayley gave me a cordial handshake and referred in the kindest terms to my papers which he had read. He was then about sixty years old, considerably bent, and not filling his clothes. What was most remarkable about him was the active glance of his gray eyes and his peculiar boyish smile."[2]

In Cambridge, life quickened in the 1880s. Reform followed the Oxford and Cambridge Act (1877), and in 1882, Trinity College gained a new constitution, which Cayley had helped to frame. In this new world, governance was taken out of the hands of the master and seniority and put into the hands of an elected council. All Cambridge fellows were allowed to marry and still remain fellows, while students would sit for a newly structured mathematical Tripos quite different from the one that had remained substantially unaltered since its creation in the middle of the eighteenth century. Christopher Wordsworth wrote in his preface to *Scholae academicae* (1877) of "how completely we have been removed from Cambridge of half a century ago, or that we have lost almost the last glimpse of what our University, even forty years since, was like."[3]

Taking Stock

By the beginning of 1884, the chair of mathematics at the Johns Hopkins University, which James Joseph Sylvester had vacated, had not been filled. A continuation of the trawl for his replacement by its president must have caused Cayley some surprise, if not a little embarrassment. D. C. Gilman, who had dedicated himself to making the university a center for science, wrote to him: "We have this morning received a dispatch from Professor Klein saying that he declines our overtures,—and I now feel free to say that my preference from the first was *for you*, and I have reason to think that the entire board concurred with me; but we were prevented from saying so by the fact that Professor Sylvester had already enlisted your kind co-operation in correspondence with Professor Klein."[4]

The financial inducements were indeed favorable. Gilman believed the trustees would be willing to pay Cayley $6,000 in comparison with the $5,000 offered to Felix Klein, and the sum they had paid Sylvester. The reality of Gilman's offer, and what he had possibly overlooked, was that no amount of money could prize Cayley away from Cambridge. He was quite settled, for he had position and, what was unlikely to be duplicated elsewhere, he had the time to pursue mathematics without interruption, the only absolute condition he had ever required. He immediately cabled a reply: "Cannot accept will write." With characteristic politeness he expressed his admiration for the way Johns Hopkins was developing but though he was impressed by the university, felt bound to add: "But I am quite satisfied here [in Cambridge]: and as well for myself and my wife as on account of the children, cannot bring myself to the idea of abandoning England."[5]

Cayley's invariant theory work was now juxtaposed with his work on Abelian functions. Based on his lectures at Baltimore, he continued his substantial memoir on the Clebsch-Gordan theory.[6] A mathematical difficulty had caused a break in the work since his return from America, but the gap was bridged and he was able to write to Sylvester that his paper for the *American Journal of Mathematics* was ready.[7] Sylvester, with first-hand knowledge of the path of this research, wrote back: "Congratulations on the final success with your great labor which at one time you had abandoned in despair. I should like some day to know the history of its resuscitation."[8] In invariant theory Cayley still desired to prove Gordan's theorem using plain algebra and Sylvester too sought that elusive "natural method" of proof.[9] During his last days in America, Sylvester had written to Cayley of his unsuccessful attempts: "My supposed proof of Gordan's theorem was a *Delusion*—but I have considerable hopes of being able to found one upon the method of Deduction aided by the actual application of this method to the Quintic."[10] It was a problem which periodically obsessed them both.

In June 1884, further recognition for Cayley's work in invariant theory came with the award of the first De Morgan Medal by the London Mathematical Society.[11] The opening years of the 1880s represented a personal watershed in this subject. His long series of memoirs on quantics, begun in 1854, was now complete, and more than any of his other papers, these constituted his magnum opus. They effectively organized invariant theory and set out a program to lay bare the structure of algebraic forms. Cayley made inroads of substance to the objectives set out in the 1840s and had added to the *organic* part of algebra, for which he was awarded the Royal Society Medal in 1859. To make use of George Salmon's frank appraisal, invariant theory was "as different from what it was before Cayley's time as the knowledge of the human body possessed by one who has dissected it and knows its internal structure is different from that of one who has only seen it from the outside."[12] By comparison, Cayley's papers on matrix algebra and group theory were minor works. The scope of the series on quantics was much wider, concerning itself with interconnections of algebra and geometry. By the time he had been awarded the De Morgan Medal, the theory had gained a central place in mathematics, but when mathematicians honored him as founder of this silver pathway, he characteristically waived any claim to inaugurating the theory. He remembered the correspondence between himself and George Boole forty years previously and unequivocally assigned the grounding to his mentor.[13]

Cayley had made a flying start in his youth but now, in the 1880s, the leadership was shared with German mathematicians, who, with their efficient symbolic calculus, were able to make significant progress. While the "Anglo" group and the German group diverged in their method, they shared the common objective of cataloging and classifying invariants and covariants. Higher order binary forms were tackled by Sylvester and his disciples, and in Germany the plan was the same. In the 1880s, the German schoolmaster August von Gall calculated the invariants and covariants for the binary form of orders 7 and 8 using the symbolic method.[14] During the 1880s, invariant theory continued to occupy a central place in Cayley's research, and he became more involved with the minutiae of the subject. He conducted many of the long calculations himself, and as his collection of invariants and covariants expanded, so too did the length of the specimens to be stored away.

Primordial Germs

Cayley turned his attention from the "full-bodied" invariants and covariants to the more basic *semi*-invariants, a concept first discussed in the "Introductory Memoir" in 1854 and which became widely studied in the 1880s and 1890s. These are algebraic forms obtained from just *one* of Cayley's two differential operators (see chap. 8 for Cayley's discovery of semi-invariants). These were now adopted as the basic units of invariant theory—the "primordial germ," to use Sylvester's vibrant phrase.

The semi-invariant approach to invariant theory was spurred on by a discovery made by the relatively unknown Percy Alexander MacMahon. Born in 1842 into a military family based in Malta, MacMahon became an officer in the British Army after passing through the army school in Cheltenam and the Royal Military Academy at Woolwich. He was posted to India on the North-West frontier, but after his return to England in 1877, following a posting to Dover, he joined the staff at Woolwich in 1882 as an instructor in mathematics. There he was influenced by the former Cambridge don, George Greenhill, professor at mathematics at the Royal Artillery College. MacMahon was approaching thirty years of age and falling in line with a teaching career, but Greenhill convinced him he should consider research. Though he went to no university, this ex-Army major became a successor of Cayley in invariant theory and occupied that now discontinued vein of British mathematicians who maintained a keen interest in astronomy.[15]

MacMahon presented a result on elliptic functions at the 1882 British Association meeting at Southampton, which Cayley attended. His reference to Cayley's book on elliptic functions may have occasioned their meeting, but it was a result in invariant theory that had launched MacMahon's mathematical career. His "correspondence theorem," concerned with the binary form of infinite order, established another link between invariant theory and the algebra of symmetric functions and, through this, to the subject of partitions. It was a result which effectively reduced invariant theory to "arithmetic"—in principle.[16] Sylvester learned about it from Cayley but was not so impressed: "Have you not somewhat over-estimated the importance of MacMahon's discovery or theorem? . . . This is pretty (and likely to be useful) enough—but it does not seem to me to amount to a Revolution in the invariantive theory as it existed in the Pre-Mac-mahnic times."[17]

While Sylvester nursed doubts, Cayley had no hesitation in recognizing the theorem as a real step forward, and in response to Sylvester's reservation, wrote back immediately: "The great use of MacMahon's theory is in the means which it affords for making out the whole theory of the syzygies. It is a question of double partitions."[18] The connection with partitions had long been known, but Cayley informed Sylvester that double partitions held the key, and nothing else was required. He thought so highly of the new theory that he gave six special lectures on it at Cambridge in October 1883. Priority was briefly an issue as Sylvester made a claim for the result himself, but Cayley thought this unreasonable and poured cold water on Sylvester's claim.

MacMahon's discovery of the correspondence theorem set Cayley off on the tasks of constructing an algebra of symmetric functions and by this means, of discovering the previously unattainable syzygies. This work also led to a closer study of the binary form of infinite order and its irreducible semi-invariants, called perpetuants. These algebraic forms pitched forth some surprising facts not previously encountered. About one, Cayley wrote to Sylvester: "Would it surprise you to learn that the number of sextic perpetuants for any weight w whatever is $= 0$?" He could not be absolutely certain but was bolstered by there being "no à priori objection to this result." It was speculative but Cayley was intrigued: "Curious if true—but why not?"[19] In British hands, invariant theory was built on speculation, facts, and the occasional theorem: it is not surprising that the conjecture about perpetuants was later countermanded by MacMahon, the two of them working together at the cutting edge of the subject.

Cayley duly constructed an algorithm for the multiplication of nonunitary symmetric functions and the reduction of invariant theory to arithmetic became a real possibility.[20] By this means of attack, Cayley thought that a proof of Gordan's theorem would be in the offing. He sought one based on arithmetical principles, though this proved a false hope around 1885. He was involved with a vast amount of purely numerical work and, commentating on his "ingenious algorithm" for dealing with symmetric functions, MacMahon remarked: "It gave the requisite facility in dealing with combinations of forms represented in the notation of partitions. The great advance thus made will be apparent when it is stated that it became comparatively easy to deal with forms of as high a weight as forty or fifty and to assign the syzygies."[21] Not only was there much calculation, Cayley admitted it was troublesome calculation. He reminded MacMahon that a "single numerical error might be fatal" and tried to enlist his help in repeating the calculations as a check. In the middle of 1885, Cayley still had lingering doubts concerning the basic linear relationships that could exist between the covariants of the binary form of order 5—the syzygies of the binary quintic—and he earnestly hoped MacMahon would show some attention to the problem.[22] James Hammond, the young mathematician from Oxford, applied himself to it: he spent "every spare moment of [his] time" on the syzygy problem, and it was he who gave the final definitive final list of the 167 fundamental syzygies of the quintic.[23]

Sylvester Returns

At Johns Hopkins University, Sylvester had invigorated the study of mathematics and created a research environment on which the university could build. He was at his best when enthusing others at the beginning of projects—he was not one for the long haul, once the pioneering work was done. Mathematics at Johns Hopkins was now in a healthy state and he began to reflect on his future. He became unsettled and the feelings of restlessness to which he was prone were aggravated by health worries and the thought of further Baltimore summers in the heat. As he approached seventy years of age, his thoughts turned to England and the idea of applying for the Oxford chair left vacant by Henry Smith's death was acted upon. On 13 December 1883, Cayley cabled the glad news that Sylvester had been appointed and, in reply, heard from Sylvester: "I duly received your first telegram on Wednesday last at 12.30 P.M. announcing to me the gratifying and unexpected intelligence of my election to the Savilian Professorship."[24] Sylvester made plans to return on that "swift-courser of the deep," the *Arizona,* and he was in Oxford for the beginning of 1884.

Though invariant theory was Sylvester's principal research in Baltimore, this had latterly been replaced by an enthusiasm for matrix algebra, or as he termed it, the theory of multiple quantity. He was now too immersed in this subject to take part in Cayley's new investigation in invariant theory: "Would that I could do anything to assist you in your most interesting investigation concerning sextic Perpetuants! Unhappily I am out of that *dream* and out of the partition *dream* and have no present thought except for Multiple Quantity."[25] In the next month, Sylvester added: "What wonderfully beautiful work you & MacMahon appear to have been doing in Perpetuants! I must try and get back on the track."[26] As they had done so often in the past, Cayley and Sylvester were moving in different directions.

Cayley's attention to matrix algebra had been infrequent over the years but he was ever ready to help his friend. The thrust of Sylvester's work on the "new science of multiple quantity" was his treatment of equations in which the coefficients a, b, c, d and unknown value x were themselves matrices. The motivation for such questions came from basic algebra, but now the typical questions asked by Cayley and Sylvester were for the solution of equations in "multiple quantity." As usual, Sylvester tried the empirical approach, a way of working which left him immersed in a plethora of special cases and most usually concerned with matrices of small order. Cayley could only offer friendly advice: "I think you should take the bull by the horns and consider the general quadric equation $xax + bxc + d = 0$."[27]

Cayley tackled some questions himself: "Your formula as to $px = xq$ is of course quite right— I had begun working it out in the same way but made a mistake in the multiplication of the matrices—& I found some very pretty & consistent results belonging not to your problem at all but to the different one . . . which I was thus led to by a mere accident, [and] would belong as a very particular case of a theory more general than that of the functions of a single matrix—viz. to the theory of the functions of a matrix x and the transposed matrix $tr.x$."[28] Two months later there was more definite news from Cambridge: "I have solved $[px = xp^1]$ in a very compendious form," Cayley informed his friend.[29] Two papers resulted, identical in content, but one written in the language of matrices and the other using the quaternionic vocabulary, emphasizing both the separateness and identity of the two traditions that had grown up separately side-by-side.

Sylvester's return to England meant that he was able to resume his close links with his friend. In the summer, he visited Cayley in Grasmere in the English Lake District before going off to Paris, where he met their circle of mathematical associates. "I just missed meeting Hermite in Paris," he reported to Cayley, "but saw Picard, Darboux, Jordan, Jonquières, Mannheim and Bertrand. The latter is about to be made one of the Immortals i.e. a member of the Académie Française which is regarded as an immense honor."[30] On return he was in good humor with the prospect of a new start at Oxford—the new Heraclitus. An immediate task before him was the duty of giving an Inaugural Lecture, the "dreadful incubus of a public Lecture."[31] "Not having command of erudition sufficient to make it profound," he wrote to Cayley, "I shall try and make it amusing."[32] He was downbeat about the whole thing when he wrote to Rudolf Lipschitz of the inaugural being a combination of "a sort of autobiographical sketch with the announcement of the New Theory of Reciprocants."[33] Though settled in Oxford, he was left to himself more than he liked and predictably began to miss the vibrant atmosphere of Johns Hopkins and even began to complain about the English weather. He did have the company of Hammond, who acted as his amanuensis and general mathematical servant. On balance, he was satisfied with Oxford, a transition owing in no small measure to Cayley. "You ought to feel pleased in having taken so important a part in bringing me back to my native country," he wrote, "and enabling me to pass the evening of my days in so delightful an environment!"[34]

At the beginning of 1885, Cayley's ongoing work on Abelian functions came to a temporary standstill. As was his custom, he cast about for topics, or, as he would say, he would "go prospecting." His habit of extracting a work or writing an expository paper kept up the working pace during fallow periods. He wrote to Thomas Craig, the editor of the *American Journal of Mathematics*:

> I doubt whether I shall be able to go on at present with the Abelian paper: in fact I do not see what to do next. I have been studying and very much interested in Hermite's paper of 1855, On the transformation of the Abelian functions:[35] and intend to make a reproduction of this, altering the arrangement a good deal, and supplying the proofs which are only indicated: but as being substantially a mere reproduction, I would rather keep this distinct, and in fact I have announced it for the *Q[uarterly] M[athematical] J[ournal]*. I have also been working on the further development of the Semi-invariant theory and have found some interesting results: but there is a fundamental question which I have in vain attempted to solve, that of the correspondence of initial and final terms referred to in the paper *A[merican] M[athematical] J[ournal]*. I have been writing about it to MacMahon, but we have not yet got out anything satisfactory.[36]

There were still difficulties, as Cayley wrote to Craig six months later. It was summer and Cayley and family were holidaying in North Wales, in the walking country of Betws-y-coed and "very much delighted with the place." But the fresh air and gentle hikes were not having the desired mathematical effect, though he saw that the subject of semi-invariants was pivotal in his investigation: "I have *nothing* actually in hand, so please not to delay the No. [of the journal] on my account. I am quite stopped by a question in Semi-invariants partially solved by MacMahon & to which I have no doubt he refers in the paper of his which you have—but we are neither of us at present able to make the next step: if I succeed in doing so, I should be rather inclined to undertake a treatise on the subject; but I do not at all see my way."[37]

MacMahon's reputation was established among pure mathematicians and even Sylvester, whose cryptic note about the state of invariant theory in "pre-MacMahnic times," was impressed by the newcomer. He wrote to Cayley "I am very glad that you will give your support to MacMahon for a [Royal] medal. What a *Magnificent* formula is his representation of the lowest-weighted per-petuant of the nth degree! It almost takes ones breath away from its grandeur and simplicity." The award of a Royal Medal to MacMahon so quickly would have indeed been remarkable if it had been made, but he had to wait for the medal and, also, for membership of the Royal Society.[38]

Cayley had a brief encounter with the theory of reciprocants, a parallel "invariant theory" involving differential invariants, on which Sylvester was writing copiously.[39] There was a clear invitation for Cayley to join in: "Am very glad you take an interest in my new functions—provisionally we may call them Reciprocants . . . You will see that the whole of the game so to say of invariants has to be played out over again on a new field and subject to new laws but giving rise to a parallel theory of groundforms of perpetuants and syzygies and revolving on the same order of ideas." Cayley took sufficient interest in the subject to write a survey article but he did not play a significant part in this theory. Sylvester gave a course of lectures at Oxford and published over two hundred pages on the subject, a venture he believed to be a signifi-cant extension of traditional algebraic invariant theory. Just as he had done at Johns Hopkins, he gathered around him a young impressionable group of mathematicians to help him. The theory was not so novel as first supposed and when the pages had been written and published, it was later realized that the theory was properly subsumed under Sophus Lie's theory of continuous transformation groups.[40]

At this point, Cayley increasingly believed that Gordan's theorem could be approached through semi-invariants: "It appears a truism, and it might be thought that it would be, if not easy, at least practicable, to show for a quantic of any finite order n, that [using semi-invariants] . . . all the covariants be expressible as rational and integral functions of a finite number of irreducible covariants [Gordan's theorem]." He qualified the chances of success with his now familiar caveat in regard to higher order binary forms. Starting with the binary form of order 5, he noted: "the large number, 23, of the covariants of the quintic is enough to show that the proof, even if it could be carried out, would involve algebraical operations of great complexity."[41] MacMahon observed that Cayley's review of binary forms of infinite order and semi-invariants, "led him to desire a purely algebraic proof of Gordan's theorem concerning the finality of the covariants of quantics of finite order."[42] Sylvester also attempted to turn repeated failure into success: "In my off moments I have been thinking again of

Gordan's theorem and verily believe that I have found the proof of the finitude of the fundamental invariants and the fundamental reciprocants by a general method . . . Hammond is settled here and we meet for several hours daily. He will check me if I am under any delusion as to the Gordanic business." Sylvester hoped to use a particular class of semi-invariants, which he called proto-morphs, to achieve the result. Two weeks later, Cayley received a note from him born of frustration: "I nourish the undying hope that through the Protomorphs we shall be able to prove the finitude of the ground-forms of Invariants and Reciprocants by some simple process of reasoning."[43]

Sylvester's lack of sympathy with the semiabstract symbolic version of invariant theory advocated by the German mathematicians had not ameliorated. However, the deep-lying algebraic principle he sought was proving elusive, and, as his failure to discover it became only too obvious, his contempt for the German method increased. He had written to John Tyndall in 1882: "[h]is [Gordan's] own demonstration is so long and complicated and so artificial a structure that it requires a very long study to master and probably there is not *one person* in Great Britain who *has* mastered it."[44] Perhaps so, but Cayley had reservations about the entire symbolic approach, and he now took refuge in the belief that semi-invariants and partitions held the key to the riddle. Before going off to the Lake District for the summer, he wrote to Klein in May 1886: "I am putting together some results in regards to semi-invariants, but they will I am afraid take me some time to finish."[45] The following year Gordan attempted to give a simpler proof of his theorem in the second volume of his *Vorlesungen über Invariantentheorie,* but the work would not have appealed. Cayley wanted a nonsymbolic proof of a kind that had resisted all his efforts of finding, but such a proof was not yet available.[46] With MacMahon his nearest collaborator, he worked on in the belief he would eventually secure one.

Dusty Hedgerows

The new constitution of the mathematical Tripos, introduced in 1882, divided the student competition into two parts. The order of merit remained sacrosanct but its rankings were awarded on the basis of the first part of the Tripos, which still required the acquisition of traditional "tripos technique" in preparation for a once-and-for-all examination thus made accessible to the "quick and ready" men. This was as before, but a higher part (Part III) was introduced, and though this divided students into classes, they were not listed in merit order. For the whole process, Cayley's presence in the university remained irrelevant. Even for students studying the specialized higher Part III, designed for the education of future mathematicians, he could exercise only limited influence; as pure mathe-matics was in a subsidiary position within the range of the mathematics courses being offered. Students were given the impression that the sole purpose of pure mathematics in the Tripos was to provide tools for application in the physical sciences, an impression reinforced by E. W. Hobson, a later holder of the Sadleirian chair. Writing thirty years later, Hobson described it as a "a prejudice which did much to retard the due development of Pure Mathematics in this country during the nineteenth century, [and] is by no means extinct."[47]

Many teaching dons were unappreciative of mathematics as an end in itself but treated it in the old way as an essential component in a liberal education. Even those who valued mathematics per se, thought the higher reaches of mathematics should be out of bounds for undergraduates. As a professor, Cayley had little experience with teaching but some members of the university thought this was right. A professor's job was to carry out original research and not to lecture to undergraduates. Thus, potential disciples slipped through the net. In 1884, for instance, the Australian Senior Wrangler William Sheppard, who sailed halfway round the world for a Cambridge education, would not have automatically attended his classes. He became a respected statistician but appears not to have been taught by Cayley.[48] In the same year, Sheppard's friend, the future Nobel Prize–winning physicist William Henry Bragg had a daily routine, directed by his coach E. J. Routh, which consisted of nothing but mathematics: he "worked at the mathematics all the morning, from about 5–7 in the afternoon, and an hour or so every evening," but there is no sign of Cayley's influence.[49]

A new professor on the scene in 1883, G. H. Darwin, the second son of Charles Darwin, and a leading authority on the theory of gravitation, lamented that such men as Cayley had been without classes for years. He also referred to the detrimental effect of the Tripos mill and blamed the textbooks for infiltrating the view that mathematical knowledge is a closed book from which there is much to be learned but nothing to be added. "They impress the student in the same way that a high road must appear to a horse with blinkers," he wrote, "The road stretches before him all finished and macadamised, having existed for all he knows from all eternity, and he sees nothing of byways and foot-paths." He made the doleful observation: "I think it is not too much to say that there is no vitality [in mathematics] here," and he questioned: "It surely cannot be that Stokes, Adams and Cayley have nothing to say worth hearing by students of mathematics"; furthermore, he laid the blame for this state of affairs at the door of the mathematical Tripos.[50] Andrew Forsyth recognized Cayley as a voice crying in the wilderness, and, when Karl Pearson returned from Germany and suggested that a Mathematical Society might be founded in Cambridge along Continental lines, he only expected the "dusty hedgerows of Cambridge apathy."[51]

It was not a story of complete gloom. A few mathematicians from the younger generation would make their mark in mathematics. The young F. S. Macaulay graduated in 1883 and became a schoolmaster but proved himself a first-rate researcher as well, taking as his first subject Cayley's theorem on the intersection of curves. A few students who studied advanced subjects in the same era went on to make mathematics their life's occupation. Another one was Macaulay's younger brother Alexander. He went to Australia, taking with him all the missionary zeal of a fervent quaternionist, and spent much of his time involved with the International Association for Promoting the Study of Quaternions and Allied Systems of Mathematics. He wrote numerous books and papers on quaternions and octaves, evidently influenced by Cayley's Cambridge lectures on quaternions, though his essay on the subject submitted for the Smith's Prize in 1888 did not find favor with the professor.[52] In 1884, Frank Carey, from the same year, wrote his fellowship dissertation on number theory, and it was worthy enough for Cayley to make a comment in its favor.[53]

Forsyth, who had attended Cayley's lectures while an undergraduate, had become the Senior Wrangler in the unreformed Tripos of 1881 and a lecturer in the same year. He was Cayley's most successful student. While Cayley's lectures were inspirational for the few, it was young men like Forsyth who gave lectures to undergraduates. Forsyth published three papers while an undergraduate, one on the motion of a viscous incompressible fluid and the other two in pure mathematics. With extraordinary energy and diligence, he wrote his Trinity fellowship dissertation in barely three months on one of Cayley's subjects, the double theta functions, with this choice having been made after a brief flirtation with the kinetic theory of gases as a possible topic. Like Cayley himself, Forsyth was an expert in the manipulative processes of algebra, and perhaps seeing this, Cayley secured publication of the dissertation in the *Philosophical Transactions,* a sure recognition of the young man's talent. As both friend and pupil, Forsyth adopted invariant theory and became imbued with it; as one of his first duties, he gave a set of lectures based largely on Cayley's work on binary forms. In his neat handwriting, lecture by lecture, he charted Cayley's discoveries from their inception.[54] A steady stream of mathematical publications followed his appointment as lecturer, but strangely, he made no real impression on the literature of invariant theory for several years. In 1889 he made good this deficit by publishing no fewer than five papers of great length.

Forsyth acquired qualities Cayley would recognize as not abnormal: "the indomitable mastery of algebra and the almost incredible powers of production."[55] His output for his first decade established his mathematical reputation beyond question, and by the end of the decade he was recognized as "the most brilliant mathematician of the British Empire."[56] He became a man whose knowledge of mathematics was comparable to Cayley's own and through his drive to read widely in French and German and absorb mathematical ideas to the point of saturation, he did much to bring Continental pure mathematics to Cambridge.

Forsyth was a Cambridge mathematician par excellence, but a less obvious one, who socialized with him, was Thomas L. Heath. He started his intellectual career in the "R. L. Ellis mold" whereas Forsyth began with undiluted mathematics. Slow and methodical, Heath combined his classics and mathematics training when he submitted a fellowship dissertation in 1885 on Diophantus of Alexandria. By this time, he was a Treasury Civil Servant and in this capacity he was to remain while regularly publishing a string of scholarly work on Greek mathematics, which included his famous edition on Euclid. In recommending his *Diophantus* to the Pitt Press, Cayley played a critical role at the beginning of Heath's career.[57]

Cayley empathized with young people taking their first steps in research, though it is regrettable there were so few. A. N. Whitehead obtained a Trinity fellowship, and though he went on to other things, Cayley's influence on his mathematical activities is discernible. It is possible, in terms of subject matter that they collaborated, though I have found no evidence they worked together, as Cayley did with Forsyth, for example; by some accounts, Whitehead was a withdrawn student and kept to himself, so the influence may have been through a distant presence, as had occurred with George Chrystal's residence in the university. In 1887 Cayley published an expository paper on multiple algebra and included a section on Hermann Grassmann's *Ausdehnungslehre*.[58] Two years afterward, Whitehead, whose attention had shifted away from applied mathematics toward pure mathematics, gave Cambridge lectures on Grassmann's work and did so again the following year. His early work in multiple algebra and his willingness to offer a Cambridge course on non-Euclidean geometry, both "Cayley subjects," indicates that he was not left untouched by Cayley's presence in the university; he knew Cayley socially but may not have attended his classes.[59] In January 1891 he began work on *A Treatise on Universal Algebra* based on his Cambridge lectures and it was completed in 1898. A slim but more tangible link with Cayley came later with his paper on the algebra of symbolic logic, published in 1901, in which he endeavored to link logic with algebraic invariants and group theory.[60]

A more likely connection was with G. B. Mathews, who in 1883 was the Senior Wrangler to Whitehead's fourth wrangler position. A talented mathematician and mathematical author, it is likely he attended Cayley's lectures, especially in the year he had the "exceptional number" of a dozen auditors. Mathews reviewed Cayley's mathematical papers and in 1894 was involved in a discussion with him on Hamilton numbers.[61] A man who did attend Cayley's lectures was the geometer A. C. Dixon, the Senior Wrangler of 1886. His juvenilia in algebraic geometry was in the Cayley style but, in a period when the German and Italian schools were leading the way in algebraic geometry, his work failed to take note of their discoveries.[62] At a distance, Eliakim H. Moore at Yale College in the United States wrote a Ph.D. thesis entitled "Extensions of

Certain Theorems of Clifford and Cayley in the Geometry of n Dimensions," in which he included an extension of Cayley's theory of quadric surfaces in five-dimensional space.[63]

Born in Cambridge, H. F. Baker went to school with young Henry Cayley and remembered doing Latin prose with him and the "various mathematical diversions we had together."[64] Becoming a fellow of St. John's in 1888, Baker went on to a position of influence in the Cambridge mathematical world and became the first Cayley lecturer in 1903. His postgraduate work was carried out under Cayley's guidance and one substantial paper on invariant theory dealt with the enumeration of invariants and covariants in the style of his supervisor.[65] His mathematical technique and his basic approach to mathematics owed much to Cayley's formative influence. Following spells in Göttingen with Klein, he embraced the study of algebraic geometry from a modern viewpoint.

So much poorer were the chances for women. Charlotte Scott had benefited from Cayley's teaching and support, but another was critical. Looking back, Grace Chisholm Young, who became a student at Girton College in April 1889, was censorious; writing of her feelings as a young person eager to learn mathematics, she was of the opinion:

> Mathematical science had reached the acme of perfection. Through the long future ages, no new ideas, no new methods, no new subjects were to appear. The edifice of mathematical science was complete, roof on and everything. All that remained to be done was to consolidate and repair the masonry, and add to and correct the ornamentation.
>
> This was the view in those days, and the atmosphere was stifling to the young mathematician. Cayley, unconscious himself of the effect he was having on his entourage, sat, like a figure of Buddha on its pedestal, dead-weight on the mathematical school of Cambridge.[66]

On the lighter side, a publicly expressed squib, which lampooned Cayley's small classes, his "imaginary contact" with his audience and his ability to combine the "practical with the transcendental" was placed in the *The Cambridge Review:* "You have served your generation faithfully, and have done work of lasting value for posterity. You are a sort, of asymptote to the curve of progress, which the curve will approach more and more nearly as it goes off to infinity; but you are an asymptote with several finite points of contact as well. Or if we may compare the world to a huge n-ary quantic of high order, in which the coefficients are the facts of nature and of intellect and the variables are common men, you are the sort of a covariant function of the variables and coefficients."[67]

Through the provision of higher level courses for mathematical specialists, it was hoped that Cambridge would successfully emulate the Göttingen research school. In anticipation of this, Cayley's teaching duties increased in the academic year of 1886 when he was called on to give two courses of lectures each year.[68] The higher part of the Tripos gradually became the domain of student specialists and professors and it was out of the range of the private tutor. At this level, Cayley could interact with students, but there was never created at Cambridge anything comparable to Felix Klein's environment in Göttingen.

Being a Nonmathematician

If Cayley's influence in the undergraduate sphere was ultimately limited, he led a full life in Trinity College and in the university. In his period as professor, he brought a wealth of legal experience to his committee work, his rapid response to a letter, his immediate concern with the matter in hand, using his "common sense and native shrewdness" in the service of the university.[69] His influence was definitely felt on the Board of Mathematical Studies, but he also made important contributions to the university's web of committees: the Library Syndicate, the Cambridge University Press Syndicate, the Workshops Syndicate, and the Medical Syndicate. He was also a member of *Sex Viri*, the "six wise men," who acted as judges in the court of law with jurisdiction in the University of Cambridge. It used to amuse his former legal colleagues in the Inns of Court to remember his administrative skills and "to trace to their source at Lincoln's Inn those business qualities, which rendered the Professor hardly less esteemed in the Council chamber than in the professorial chair."[70] Ordinary matters of the Council of the Senate were allowed to progress smoothly but he did not let the business deflect him from mathematics.

There was also his presidency of Newnham College, and though the council formally met only four times annually, he was involved with advice and support throughout the year. Anne Jemima Clough, the principal, relied on his guidance, as did other officers concerned with the well-being of the college as it expanded. Eleanor Sidgwick, the treasurer (and a future principal) wrote to him: "We shall have very soon to decide whether to make an offer to St Johns [College] for the Newnham Land . . . I should be very much obliged to you if you would—if it would not give you a great deal of trouble—tell me what you think would be a reasonable offer to make, for as treasurer I ought to be thinking the matter over. . . . I trouble you with this question because I believe you know about such matters."[71] The year 1887 was one of great activity on the issue of higher education for women. In the middle of the year, J. J. Thomson wrote to a friend of the "great agitation going on to admit women to all the University privileges that men have," and the issue that "divided the friends of the women nearly as much as Home Rule has divided the Liberals." By the end of the year, he wrote that "we are just at the commencement of a great attack which is being made by the supporters of women to secure for them full admission to all the privileges of the university."[72] Thomson was on the same side as Cayley in pressing for the higher education of women, but Cayley was not in favor of asking for too much too soon. He was cautious and perhaps saw the danger of losing the gains that had already been made.[73]

If there was time to spare with all this activity, Cayley was kept busy by William Robertson Smith. Half-physicist and half-classicist and a former assistant to P. G. Tait, he became the University librarian in Cambridge who took on the responsibility of seeing the "Great Ninth Edition" of the *Encyclopaedia Britannica* through to publication.[74] Cayley's tally of articles written for this publishing venture during in the 1880s included ones on the British eighteenth-century mathematician John Landen (who famously discovered an important transformation of elliptic functions), Gaspard Monge, John Wallis, and the various technical subjects, "Locus," "Theory of Numbers," "Series," and "Surface."[75]

Cayley's experience in publishing his work was especially useful. When Isaac Todhunter died in 1884, Cayley, as a member of the University Press Syndicate, was left with the task of dealing with the publication of the unfinished manuscript of his ambitious work on elasticity. Karl Perarson was commissioned as editor and the first volume appeared in 1886 and the second in 1893; the monumental work became a classic in this field. It is a measure of Cayley's sensitivity that he could steer a diplomatic course between the concerns of the vigilant editor seeing the gaps in Todhunter's unfinished manuscript and the wishes of the Todhunter family concerned with his posthumous reputation. Cayley's role as negotiator was acceptable to Todhunter's widow, "he being so great a mathematician & also as a friend of my husband's." At the same time, he could advise Pearson on the editor's protocols and assure the young man that he would get credit for this worthy task, for "the work would always be 'Todhunter & Pearson.'"[76]

With ability across the divide between skillful administration and mathematical research, Cayley managed to avoid high office. Indeed, he appeared to lack the ambition for it and seemed content with a mathematical life. There was speculation that he might succeed to the mastership of Trinity, an appointment in the gift of the Crown. Henry Sidgwick, who had friends in high places, noted in 1886: "We think Lord Salisbury will want to appoint a cleric if he can: but that as there is no clerical candidate whose appointment will not be open to strong objections, he may acquiesce in a Conservative layman: and that in this case it will be either Cayley or Rayleigh."[77] Salisbury briefly considered Sir Henry Maine and Lord Rayleigh but, acknowledging a large postbag in support of a clerical appointment, settled on Henry Montagu Butler who assumed office in December 1886.[78]

An ex-headmaster of Harrow, and in his student days a Cambridge Apostle, it was hoped Butler would revive the fortunes of a moribund Trinity suffering the effects of continuous reform under its previous master and bring it back to the former glory that had existed under Whewell. The new master set out to improve the social side of college life and bring together undergraduate and don. He was only ten years younger than Cayley but wrote to a friend: "I am inviting the sixteen newly-elected Scholars to meet the Vice-Master, the Tutors, and some of the older Fellows, as Professor Cayley, H. Sidgwick, [Richard] Jebb etc."[79] With Butler's appointment, Cayley was seen in a different light by the incoming new order—an old man living on the fringes of college life.

Euclid's Champion

At school level, the vexed question of Euclid's *Elements* continued to simmer and had done so since the late 1860s. The AIGT (the Association for the Improvement of Geometrical Teaching) founded in 1870 had produced an alternative school syllabus to the one based on Euclid and a textbook of its own in 1884. In continuing to campaign, the AIGT petitioned for a reduction of the emphasis on Euclid's text in examinations. It was a very modest request. In January 1887, it sent a memorial to the Cambridge Board of Mathematical Studies signed by its members and prominent supporters, including T. A. Hirst and the disputatious Scot, P. G. Tait. The opening lines of the memorial indicate the thrust of their argument:

> We, the undersigned Members of the Association for the Improvement of Geometrical Teaching, and others interested in the teaching of Elementary Geometry, believing in the teaching of that subject than is consistent with a rigid adherence to the letter of Euclid's *Elements* is highly desirable, would welcome such a change in the Examinations in Elementary Geometry conducted by the Universities and other Examining Bodies, as would admit of the subject being studied from Text-books, other than editions of Euclid, without the student being thereby placed at a disadvantage in those examinations.[80]

At the meeting of the board in March, it was suggested they receive a "deputation from the Schoolmasters," and in April they met R. B. Hayward, E. M. Langley, and Rawdon Levett. Hayward had been on the British Association "Euclid" Committee appointed in 1869 and now he was the president of the AIGT. Levett had been a Cambridge undergraduate when Cayley was first elected as professor, and, like Cayley, he was deeply affected by the beauty of mathematics but, unlike him, vigorously opposed to the use of Euclid in schoolteaching—to the extent of forming an "Anti-Euclid Association," the forerunner of the AIGT. He remembered the meeting which Cayley dominated and reported on his preposterous view that "the proper way to learn geometry is to start with the geometry of n-dimensions and then come down to the particular cases of 2 and 3 dimensions."[81] But then, Cayley held unrealistic expectations of his own students, thinking less of their needs than of his subject; indeed, he recommended "a complete knowledge of invariants and covariants of ternary forms ought to be presupposed in the teaching of Higher Plane Curves."[82]

A man so completely out of touch might indeed have curious views on the teaching of Euclid. When one of the schoolteacher deputation pointed out that the authorized version of Euclid was Robert Simson's edition, Cayley suggested striking out Simson's additions and keeping strictly to the original treatise. Such a proposed course of action provoked a senior member of the university to whisper that to study Euclid in the original Greek would be better still.[83] To Cayley's mind, Euclid's treatment of geometry was superior to both the algebraic versions or the various simplified texts in books produced for the educational marketplace. His excessive admiration for Euclid's original text and his lofty approach only underscored a lack of appreciation for the teacher's task of teaching geometry in schools. Neither was he swayed by Henry Smith's witty observation that Euclid's work was written for men while in England it was administered to children.[84]

The appeal of Euclid, pure and unadulterated, reflected his own education at King's College, London, and Cambridge. Euclid had been a firm fixture on the King's curriculum when he was a pupil there, and, at Cambridge, a central plank of the undergraduate course, where George Peacock taught that Euclid's *Elements* approached perfection. In bringing together his thoughts on a Cambridge education, William Whewell had written in his *Principles of English University Education* (1837) that "Euclid has never been superseded, and never will be so without detriment to education." Cayley connected with the timelessness of Euclid as did the erstwhile Oxford tutor Charles L. Dodgson, who wrote: "neither thirty years, nor thirty centuries, affect the clearness, or the charm, of Geometrical truths." The whole weight of tradition, both personally and institutionally, was therefore against change.[85]

Cayley held an unwavering position and he was the most powerful member of the board. Though his was a minority view, he proved an uncompromising and formidable opponent. After the meeting, a member of the board confessed to a schoolmaster delegate that "[w]e cannot go against Cayley."[86] The message of the chairman conveyed to the AIGT was that Euclid should not be tampered with despite the archaic appearance of the text and the difficulty this put in the way of school pupils. In a subsequent account of the meeting, it was reported:

> He [Cayley] was absolutely irreconcilable. In his view Euclid was a thoroughly good book. Euclid had a certain number of idiosyncrasies: for instance, in his second proposition he would not allow a pair of compasses to be taken to measure off a certain length: but no more than that could be said against him. They [the pupils] had the advantage of a thoroughly good book, and the further advantage of one single book instead of a number of books with slight variations. The sort of book the Association desired would really differ very little from Euclid; and while, if it were

an open question, some parts of it might in themselves be improvements, they were unessential, and the gain would be far from compensating for the loss of the advantage of one book, and of studying geometry as it was studied by the Greeks. For the purpose of a liberal education it was best to study it in the form in which it had been studied from the first: as it stood, the form was part of the education.[87]

In Cayley's defense, it should be noted, that he always admitted his little contact with the teaching of elementary mathematics. Cayley the master mathematician and schoolchildren with "Euclid got by heart and not understood" were separated by a gulf dividing two blinkered viewpoints.[88] When he was once asked about the parts of elementary mathematics in which advanced students were most deficient, he answered with plain honesty: "My classes are very small and as I give lectures only I naturally assume in my hearers a sufficient acquaintance with the lower subjects."[89]

Had Cayley fully appreciated that geometry in schools was often reduced to memorization, he might have relaxed his opposition and admitted that rote learning was a poor substitute for real understanding. The way he tackled geometrical problems through the drawing of curves and the making of models was quite different from blindly memorizing known propositions and theorems. His own investigations required a spirit of discovery in the vein described by Hobson, who saw that "[t]he actual evolution of mathematical theories proceeds by a process of induction strictly analogous to the method of induction employed in building up the physical sciences; observations, comparison, classification, trials, and generalization are essential in both cases."[90] Given time, he might have been cajoled into accepting an "algebraic" Euclid. After all, he was the champion of *analytical* geometry, in which the central belief was that algebra offered the true basis for geometry in distinction to the ways of ancient geometry.[91]

The Great and the Good

In June 1887 the British Isles were immersed in the Queen's Jubilee Year and celebrating her fifty years on the throne. Education was to be one of the principal beneficiaries. On the strength of a popular monarchy, an Imperial Institute with a vague objective made a stuttering start. The women of England presented the Queen with a "thank offering" of £70,000 and many hoped she would use the money to establish a women's college. Instead, she erected a monument to her beloved Prince Albert at Frogmore near Windsor Castle. Cayley's reaction to those who demanded the return of their money conveys his wry sense of humor, for they "ought to have known what the Queen would do," he said. "England is peppered all over with monuments to the Prince Consort, but there was none at Frogmore."[92]

In the same month, Cayley attended John Tyndall's celebratory dinner on his retirement from the Royal Institution. He listened to Tyndall extolling the revolution brought about by landmarks in the progress of science: "the air is fresher than before; it fills our lungs and purifies our blood, and science in its Jubilee offering to the Queen, is able to add to the law of Conservatism [and] the principle of Evolution."[93] Darwin's principle of evolution rightfully puts other scientific revolutions in the shade, and the principle of conservation of energy became important for physicists in the 1860s, but there was not the merest mention of progress in mathematics. There was no fanfare for the silent revolution caused by the discovery of non-Euclidean geometry, the quaternions, or, indeed, of invariant theory. These innovations were significant within mathematics, particularly in British mathematics, but public acclaim was entirely absent.

There was, however, continued recognition for Cayley's work within the academic community. In November, he heard that his mathematical papers were to be edited and printed by the Cambridge University Press. The editorship would occupy him for years to come and he was eager to begin. "The Pitt Press are going to bring out a collected Edition of my mathematical papers," he wrote to Sylvester, "—4° [quarto], at about 2 volumes per annum. I shall have a good deal of work—which however will be interesting—in adding references to other papers."[94] He was sixty-six and looking forward to a review of a half-century of unremitting mathematical activity. He could summon the energy to retrace his steps, but the youthful vigor of the initial campaign was now missing. Signs of the passing years now pressed in and Sylvester was concerned for his friend's health. The year was dogged by minor health concerns and Cayley commiserated with Sylvester in a similar position: "I had myself a fall but escaped hurting myself coming out in the dark from a meeting among the stones &c of the New-Library buildings. My wife is very much better—but does not get well—we are having bright weather again, which is in her favor—Mary is quite recovered—& Henry flourishing—he will be going in for the Scholarship examination in about 3 weeks."[95]

Cayley's lecture for the autumn term were planned to fit around the title of "Quaternions and other Non-commutative Algebras."[96] In his class, he had Thomas S. Fiske, a 22-year-old American student from Columbia College who was to play a leading position in the establishment of the American Mathematical Society. Fiske recalled: "I had attended only a few lectures by Cayley on 'The calculus of the extraordinaries' when, slipping on the ice, he suffered a fracture of the leg, which brought the lectures to an end."[97] During his six months in England, Fiske was admitted into the circle of senior British mathematicians and attended lectures of the London Mathematical Society. In the next term, Oskar Bolza attended Cayley's Cambridge lectures on analytical geometry, which this time included a helping of geodesics and minimal surfaces. Then a specialist in invariant theory, the newly minted Dr. Bolza was making use of the Klein-Cayley-Newcomb transit route to the new world, where he too would become an influential figure in the founding of American mathematics.[98]

Two years later it was the turn of James Oliver to attend Cayley's lectures. As a youth, Oliver was influenced by Benjamin Peirce at Harvard, but now, as an elder on the faculty at Cornell University, his objective in traveling to Cambridge and Europe was to survey the scene in postgraduate teaching and research in order to transplant the best ideas to American soil. It was natural for him to call on Cayley, as he had been brought up in the culture of Cayley's mathematics. In his courses at Cornell, when he taught on such topics as quantics, quaternions, and analytical geometry, he consulted English texts. In his course in invariant theory, he used Salmon's *Modern Higher Algebra,* the text which transmitted Cayley's invariant theory work to a wider world, in both its objectives and style of working. Oliver was aware of the importance of encouraging the Cornell faculty to publish their work and may have seen the English mathematician as a model in this respect. But Cambridge turned out to be a way station, and after a very short visit, he went on to visit Klein in Göttingen.[99]

The high point for Cayley in the summer of 1888 was recognition from his home university. He had been showered with honorary degrees and foreign memberships—recently an honorary degree from Heidelberg as part of its 500-year Jubilee celebrations—but in Cambridge, he had always been either Mr. Cayley or Professor Cayley. In June he was awarded an honorary Sc.D., a distinction he received on the same day as his long-standing colleagues, G. G. Stokes, J. C. Adams, and Lord Rayleigh. They were joined by the upper layer of the late Victorian ("Unionist") Establishment, who came to Cambridge on 9 June for their investiture; the principal guest at the ceremony in the Fitzwilliam Museum was Prince Albert Victor (Prince "Eddy"), the eldest grandson of Queen Victoria, who received an honorary Doctor of Laws. Cambridge traditionally gave such awards to royalty, politicians, and to foreigners of distinction, but Cayley qualified in the category of being a British subject of conspicuous merit and one who served the university. The oration recalled his Platonic vision of mathematics:

> I have come at last to our distinguished Sadleirian Professor, a man who is held high in esteem among the experts not just in the subject of Algebra but, it is said, in the whole vast Kingdom of mathematical studies. This great man, although he could well have attained the highest honors from his skill in the law, preferred to devote all his energies to that science which strives to poetry the one nature of things with the minimum of words rather than in the most discursive. How enormously then did the Academy benefit from the sagacity of this man, and on more than one occasion the meeting of the entire Senate and the College bore witness to this. His sources of

inspiration stretched out not just as far as the banks of the Cam but even into Europe herself and to others across the Atlantic Ocean. Indeed, as if he were a second Socrates, once he had published that the theories he studied were only those which could be seen as if they were objects against a clear sky, he dedicated himself to the very same beauty and integrity that we must look upon with our own eyes. Moreover his theories were physical actualities rather than the filmy shadows of Plato's Cave. It is that very beauty which can be perceived but not explained by everyone. How clearly he held sway not on any infertile plain but in most beautiful country which can be investigated in each and every individual part, whose hills, valleys, brooks and rocks, flowers and woods can be examined at close quarters with immense delight. For many years, among hallowed groves, our reverend professor fruitfully traveled through that happy province.[100]

The orator summoned the memory of Cayley's 1883 address to the British Association. The quotation of his regarding the truths of geometry was repeated, and justifiably so since it identified the core of Cayley's philosophy: "I would myself say that the purely imaginary objects are the only realities, the 'ουτως οντα' in regard to which the corresponding physical objects are as the shadows in the cave." For him the imaginary theoretical objects of pure mathematics were the only truths. The orator coupled this statement with the other principal thread that runs through Cayley's mathematics. Taking part in research was an adventure, and it was inspired by the Romantic Age: Mathematics was a "tract of beautiful country seen first at a distance, but which will bear to be rambled through and studied in every detail of hillside and valley, stream, rock, wood, and flower."[101] In the afternoon of this day of public recognition, dressed in his new scarlet academics robes, he and his wife presided at the opening of a new hall of residence at Newnham College.[102]

In July the family capped the fruitful summer by holidaying in Switzerland and, at the end of the year, the local sculptor Henry Wiles completed a marble bust of the mathematician. Funded anonymously, it took its place in the Wren Library among the greats of the college, an honor for a living person. It was the more usual British practice for effigies to take their public place only when the originals had departed this life. Only two other Trinity alumni had ever achieved this distinction before—Alfred Tennyson and Adam Sedgwick.

Mathematics continued unabated. Contact with P. G. Tait in Edinburgh was maintained by some "wee things," which might be described as applied mathematics. One was on hydrodynamics but in another, Tait appealed for help with "Milner's Lamp." The problem, which had been raised by Augustus De Morgan, originated with Isaac Milner, the eighteenth-century Lucasian professor at Cambridge known for his "magic lanthorn." Cayley examined its construction and provided a solution through a differential equation, showing how one might be practically constructed from a semicircular base.[103]

Then Cayley resumed mathematics without any applied gloss. To Alfred Kempe's lengthy "memoir introductory to a general theory of mathematical form" (composed of 426 sections) he was lukewarm generally but was enthusiastic about the group theory section: "I attach considerable importance to the theories considered . . . in particular to the enumeration of the groups containing from one to twelve units—this is I think a really valuable contribution to the theory of groups."[104] Cayley worked on group theory himself in 1889. There was more calculatory work of his own on the subject of permutation groups in 1890, aided by E. H. Askwith, the Trinity College chaplain. In passing, he noticed the interrelationship between his group tables and Latin squares. The completion of a two-part paper on the transformation of elliptic functions for the *American Journal of Mathematics,* sundry notes on systems of three circles, and some additional thoughts on the relations that exist between four arbitrarily placed points on a circle (returning to his "juvenilia problem" once again) were published.[105] The way Cayley chose his topics might appear random but they were intimately linked with each other and with what others were doing. The link between symmetric functions and the all-pervasive invariant theory, connected with the emergent graph theory, and, as a result, it is not surprising that Cayley should turn his attention to graphical trees once again. In 1889 he gave a limited proof of a result of his own devising (that the number of labeled graphical trees which can be formed with n knots is n^{n-2}), which is now known as Cayley's theorem in graph theory.[106]

There was another point of contact with Tait. The Scotsman courted controversy and was about to ruffle some more academic feathers, for whenever his obsessions became topical, he sprang into action. The "strenuous controversialist" was about to act out his role once again. Quaternions were dear to Tait's heart, as was "priority of discovery," but when these two ingredients were combined, the barrage from Edinburgh was deafening. It was Tait's blinkered utterances that caused most antagonism within the academic body. In August 1888 Tait was preparing a third edition of his *An Elementary Treatise on Quaternions* and asked Cayley's advice. This given, he invited Cayley to write a chapter "On the Analytical View of Quaternions," a request in the same vein as he was asked to write whole chapters of Salmon's textbooks.[107] In enlisting the support of a well-known champion of the "coordinate method," as opposed to his own quaternionic bias, Tait could make the claim of being evenhanded in his treatment of the geometric material in his new edition.

The pure mathematicians were an obvious target for Tait's sorties over the years. His tongue-in-cheek remark that "Quaternions have one grand and fatal defect," presaged another broadside. As a physicist, he wrote that quaternions, "cannot be applied to space of n-dimensions, they are contented to deal with those poor three dimensions in which mere mortals are doomed to dwell, but which cannot bound the limitless aspirations of a Cayley or a Sylvester."[108]

In preparing the new edition of his book, Tait's intention for the inclusion of Cayley's chapter was far from flattering. He planned that Cayley's chapter should contain not much more than mathematical odds and ends, "[t]herein will naturally assemble all the disaffected or lob-sided members, which are not capable of pure quaternionic treatment but which are nevertheless valuable, like the occipital ribs and the anencephalous heads in an anatomical museum."[109] This hardly appealed to Cayley's sensibilities and he firmly requested Tait not to "compare [the chapter] with the Chamber of Horrors at Madame Tussaud's . . . I need not say anything as to the difference between our points of view; we are irreconcilable and shall remain so: but is it necessary to express (in the book) all your feelings in regard to coordinates?"[110]

Cayley staunchly defended coordinates as the way to study geometry. The subject of quaternions was a beautiful theory that could be interpreted geometrically and they constituted a theory with important connections in the sphere of pure mathematics (with matrix algebra, for example), but he saw it as a specialized theory and certainly not as a substitute for coordinates as a means of studying geometry. (Still less would he have agreed with the claim that quaternions were a discovery to compare with Newton's Calculus.)[111] Tait was a practical man and did not study the quaternions as an abstract system.

Sylvester had tried to explain the link between quaternions and binary matrices but he informed Cayley that the physicist "is still walking in outer darkness."[112] Tait was more concerned with quaternions only as a tool for physics and the rift with Cayley was the result of his strongly held but rather restricted view. He declared that he could "see pretty clearly in the real world, with its simple Euclidean space, by means of the quaternion telescope."[113] Cayley objected to the short form of a quaternionic formula and argued that it had to be unpacked to be understood. He was fond of explaining this in allegorical terms: "I compare a quaternion formula to a pocket-map—a capital thing to put in one's pocket, but which for use must be unfolded: the formula, to be understood, must be translated into coordinates."[114]

A New Maestro

On 6 September 1888 the 26-year-old David Hilbert dispatched his famous article on invariant theory to the *Göttinger Nachrichten*.[115] Cayley was inspired by Hilbert's success in proving Gordan's finiteness theorem for binary forms, and he wrote to Klein in early 1889: "I have read with great interest Hilbert's paper in the *Gött[inger] Nachr[ichten]*—which however I do not understand—and that in the last No. of the *Math[ematische] Ann[alen]*. It seems to me that if instead of applying this to the invariants of the binary function, we apply it to the covariants, or what is the same thing the seminvariants [*sic*], we have a very simple & beautiful proof of the finite number of the covariants—and I have written this out and send it to you herewith."[116]

Hilbert immediately saw the error in Cayley's reasoning but was reluctant to point it out.[117] A week later Cayley seemed to have his own doubts and wrote directly to Hilbert: "My difficulty was an *a priori* one, I thought that the like process should be applicable to semi-invariants, which it seems it is not; and now I quite see . . . I think you have found the solution of a great problem."[118] By the time Klein informed Cayley that there was an error in his reasoning, Cayley's doubts had evaporated and he wrote once again to Klein: "Thanks very much for your letter: I cannot see that there is any doubt as to the proof which I sent you—it depends only on the leading coefficient of a covariant [i.e. a semi-invariant] being a function of the differences of the roots and seems to me perfectly general. I shall be very much obliged if you will publish it—and of course any objections to it can afterwards be published."[119]

Klein discussed the matter with Hilbert: "With the stubbornness of an old gentleman Cayley has chosen the second alternative [to have the paper published]."[120] Hilbert was sensitive to Cayley's feelings and reluctant to refute his claim to have found a proof. He knew the mathematical argument was in error and informed Klein: "After much reflection I have convinced myself that it is better that I leave the refutation of Cayley's note to someone else, all the more so as Cayley does not attack me in the least; indeed, in the four detailed letters he has written to me, he shows me the greatest kindness. . . . Cayley's publication has at least the merit that it shows the attentive reader why every step in my proof is necessary"[121] By this time, Cayley would have been aware of the mounting criticism. His supposed proof of Gordan's theorem was an illusion, but he appeared unconcerned about the possibility of being contradicted.[122]

The question of Gordan's theorem and Cayley's "proof" would continue but the geometrical aspect of invariant theory was being actively discussed as well. Cayley's work in invariant theory was intimately tied to his expansive notation. The full homogeneous expression of an algebraic form was paramount and conveyed its meaning immediately, in contrast to the contracted German symbolic notation. His attitude parallels his view of the succinct quaternionic notation in geometry and his "pocket-map that had to be unfolded" for this notation to be understood. In the same way, the symbolic invariants and covariants had to be unfolded, whereas to him, they were easily understood if expressed in coordinates. This had been his view when he began his pioneering work in the 1840s and his constant view since that time. As he affirmed in his 1883 British Association address, "Descartes's method of co-ordinates is a possession for ever." He was at pains to point out the way by which coordinates were to be considered: "I regard the trilinear or quadriplanar coordinates as the appropriate forms including as particular cases the rectangular coordinates *x, y* or *x, y, z*—and bringing the theory into connexion with that of homogeneous forms of quantics—, . . . it is only in regard to these [homogeneous coordinates] that the notion of an invariant has its full significance."[123]

In the theory of non-Euclidean geometry, the logical difficulty to be circumvented was that the projective notion of distance, which Cayley had defined in terms of the inverse cosine in his celebrated "Sixth Memoir on Quantics" (1859), appeared circular; it was defined in terms of the ratios of coordinates, suggesting that projective distance was defined in terms of the ratios of distances. Since the appearance of the "Sixth Memoir," Klein had systematized non-Euclidean geometry using Cayley's work as a basis, thus bringing the memoir into a more prominent place in geometrical theory. Cayley recognized Klein's definition of distance (defined with the logarithmic function replacing Cayley's inverse cosine) as superior to his own but the "circularity" present in the definition still remained a puzzle. With several works on non-Euclidian geometry behind him, he addressed the problem, and he brought it to the fore in connection with the "Sixth Memoir": "As to my memoir," he wrote, "the point of view [in 1859] was that I regarded 'coordinates' not as distances or ratios of distances, but as an assumed fundamental notion, not requiring or admitting of explanation." In the 1880s, he had another suggestion: "It recently occurred to me that they might be regarded as mere numerical values, attached arbitrarily to the point, in such wise that for any given point the ratio *x:y* has a determinate numerical value, and that to any given numerical value of *x:y* there corresponds a single point." Thus he argued that coordinates were simply labels and did not entail a connotation of length.[124]

In 1889, after he had submitted a paper on geometry to the *Mathematische Annalen,* Klein drew his attention to Christian von Staudt's work. This mathematician had been long neglected but when discovered was heralded as the "Euclid of the nineteenth century." He had shown that the construction of a projective ratio (a ratio of ratios) could be made through pure geometry, thus avoiding the use of coordinates and circumventing Cayley's apparent circularity. Cayley wrote back to Klein: "I have looked at Staudt's *Beiträge zur Geometrie der Lage*—the sections on the Addition and Multiplication of 'Würfe' [Throws, or equivalently, cross-ratios] which leave nothing to be desired and the point of view is a better one than mine. If you will kindly return me the paper, I will not trouble you with it again, but I may perhaps rewrite it, explaining what Staudt has done, and send it to the *Messenger of Mathematics.*"[125] While seeing von Staudt's work as outstanding he did not regard it as settling the "circularity" conundrum. On writing out a summary of the problem, he appeared unconvinced: "It must however be admitted that, in applying this theory of Staudt's to the theory of distance, there is at least the appearance of arguing in a circle" since Staudt's construction was equivalent to the assumption that distance must satisfy $dist.PQ + dist.QR = dist.PR$, and, in effect, assuming a metric property.

In the summer of 1889, Klein visited England and discussed the matter of Gordan's theorem with him. From the first, Cayley had been appreciative of Hilbert's result and its importance and regretted he had not succeeded in finding a proof himself—as he openly acknowledged.[126] By this time he was beginning to suffer the effects of his long illness and Klein, seeing his old friend finding it difficult to walk any distance, became despondent. Cayley was insistent that Klein should publish his paper containing the errant proof of Gordan's theorem in the *Annalen* and Klein reported his conversation to the Danish mathematician Julius Petersen.[127] Klein advised him: "he assured me repeatedly that it would not be 'against his feelings' if anything were published in the *Annalen* to counter the arguments in his paper."[128] Petersen had used Hilbert's invariant theory papers as a departure point for founding a theory of graphs and had independently recognized that the proof was wrong and why it was wrong.[129]

Sylvester heard of Cayley's published attempt to prove Gordan's theorem through Petersen, as the two had corresponded on the theory of graphs at the end of 1889. Sylvester was taken aback to learn that Cayley had presented a proof that was in error, but accommodated the aberration by appealing to the mysteriousness of mathematics. He could not bring himself to believe that his hero of forty years was human after all. He thanked Petersen for opening his eyes "to the wonderful fallacy into which so great a Genius and so practised a Veteran as Cayley had allowed himself to be betrayed as if he had been some novice for the first time grappling with an arduous question."[130] The alleged proof depended on the supposed fact that a semi-invariant could be expressed as a symmetric function, but Petersen pointed out that this was only true for *certain* semi-invariants.[131] Sylvester was obliged to be publicly shocked by such a mistake but his deep-seated hero-worship was unmoved. "How very unfortunate that Cayley should have made such a mistake, and committed it to print!" he wrote to Klein. "It is wonderful and inexplicable to me that so piercing a genius and of such exactness of thought could have fallen into and persisted in such an error. He must have forgotten one of the most elementary theorems in symmetric functions when he fell into such a trap."[132]

Writing in 1889, MacMahon indicated a hope that Gordan's theorem could at last be proved using nonsymbolic methods. A possible lead to a simpler proof of the theorem might lie in the link between invariant theory with the long-established theory of symmetric functions, as had also been surmised by Cayley: "It is to be hoped that some of these facts [about symmetric functions] may help to forward the algebraical (as distinct from the symbolical) treatment; as yet, however, a purely algebraical demonstration of Gordan's great theorem concerning the finality of the ground covariants seems as far distant as ever."[133]

Cayley worked on applying his considerable skill in calculation—as MacMahon said of him: "No mathematician of Cayley's time possessed comparable powers over long expressions and tiresome calculations." In one of his last projects, Cayley attempted to establish a criterion for the reducibility of covariants. How can it be decided whether a given covariant is reducible? Reporting on this work, MacMahon noted that the subject "bristled with difficulties and exceptional cases," and he concluded that the work was only partially successful. It was a matter of "keen regret" to Cayley that he did not succeed but promised himself that he would return to the problem. His very direct approach, coupled with an inductive spirit of enquiry, was pitted against a problem which perhaps demanded more sophisticated techniques.[134]

CHAPTER 17

Last Years

The outside world was changing, though Cambridge remained the small market town dominated by a university as it had been for centuries. Within, the university was being transformed from the one Cayley had known as both man and boy to an institution without the old sureties and reference points. Modernity was nibbling at its traditions and new subjects were asserting their right to be included in the curriculum, though mathematics and classics held on tenaciously as the central core of a Cambridge education.[1] The university took positive steps to promote an engineering school, moves which had failed a few years before but which in the early 1890s were passed by the Senate without controversy.

At the same time, the efforts made to appoint a syndicate to inquire into the maintenance of Greek as an entrance requirement met stern resistance. The public schools such as Eton and Harrow provided courses in Greek, but those pupils from secondary schools without Greek were automatically debarred from Cambridge since the "Little-Go" entrance examination demanded some proficiency in the language. Many dons appreciated that the absence of an alternative as an entrance qualification was generally "distasteful to our middle classes since their sons were prevented from taking a Cambridge degree."[2] The cause had bubbled on for ten years, but in 1891 it provoked great excitement in the university, and though the matter now seems small, at the time the stakes were high, for if Greek were abandoned it would weaken classics as a cornerstone of Cambridge studies.

Set against the proposal to even inquire into a change of regulation were arraigned the full forces of conservatism, most prominently the archbishops of Canterbury and York, supported by a row of the lesser bishops. Regarded in the university as a leading liberal, Cayley voted for the establishment of an inquiry and, rarely for him, allowed his name to be used in propaganda by the modernizers. It was to no avail and the shock-troops of clergymen M.A.'s came to Cambridge for the day of decision and voted the proposal down by a huge majority.[3] Cayley thought an inquiry would be beneficial to the university but he was not taking sides in any future choice. He was not anti-Classics—far from it. On the afternoon of the vote, Cayley was at home reading the *Gorgias* from Plato's *Dialogues*.

At Home

Near to the Fitzwilliam Museum, Garden House provided an island of old-world charm where visitors remembered its peaceful atmosphere. The large house by the riverside, covered in ivy and Virginia creeper, maintained the measured pace of a former age. In his own home, Cayley was part of that age, a courteous Victorian gentleman remote from the bustle of the town. At social gatherings, he was ever the passive host, as one remembered him: "listening, attentive and watchful."[4] In his cultural kinship, he was no brash mid-Victorian, but a cultivated man. As Andrew Forsyth observed, he had an aversion to Charles Dickens, but was fond of Jane Austen and authors and poets of the Romantic Age, including Walter Scott and William Wordsworth. Scott was always a favorite and he never tired of John Gibson Lockhart's *Life*.

Indeed, many of Cayley's viewpoints were shaped in the afterglow of the European Romantic movement, the years when he was a young man and Scott was at the height of his popularity. He enjoyed Shakespeare and developed a taste for reading him in German, and he regularly reread the classics. Yet his tastes were catholic and he indulged in popular novel reading, an addiction he acquired in youth. In these latter years, he read the French writers Paul Féval and Xavier de Montépin, who had written such titles as *Les Mysterès de Londres* and *Les enfers de Paris*. His love of architecture was combined with watercolor painting. The Italian painting masters were admired on an early visit to Italy and one can appreciate his attachment to George Eliot's *Romola* (1863) set in fifteenth-century Florence.

If one glanced around his bookshelves in Garden House one would have sensed his broadly based affinities. First there was his splendid mathematical library. He had unbroken runs of the great mathematical journals, hundreds of collected offprints of his own and those produced by those from his wide circle of correspondents. Gaston Darboux gave him the two volume *Oeuvres de Fourier* (1888–90), which he edited. Cayley owned such tomes as Carl G. J. Jacobi's *Canon Arithmeticus* (1839) and the *Vorlesungen über Dynamik* (1844), Christian von Staudt's *Geometrie der Lage* (1847), P. A. Hansen's *Tables du Soleil* (1853), J. A. Serret's *Cours d'Algèbre Superiure* (1854), Charles Dodgson's *An Elementary Treatise on Determinants* (1867), the two volumes of Paul Gordan's *Vorlesungen über Invariantentheorie* (1885, 1887), and Charles Babbage's posthumous *Calculating Machines* (1889). He owned virtually all of George Salmon's mathematical treatises in their many editions.

On his shelves too, Cayley had placed such classics as William Bligh's *Narrative of the Mutiny on Board His Majesty's Ship the Bounty* (1790), Lawrence Sterne's *Tristam Shandy* (1797), and six volumes of Miguel de Cervantes' *Don Quixote de la Mancha* (an edition published in 1794). There were also books redolent of his youth, such as his copy of *Fifty Games at Chess* (1832) by the chess writer

William Lewis. He kept his law manuals detailing the intricacies of wills, conveyancing, partnership, and taxation. His interest in languages is apparent with an *English-and-Swedish Pocket Dictionary* (1832), and, as well, there were books in the common European languages and their dialects, plus introductions to Sanskrit and Icelandic. A booklet entitled *The Lathe and Its Uses; or, Instructions in the Art of Turning Wood and Metal* (1871), given its date, suggests a clear link to activities in James Stuart's mechanism laboratory.

Last to mention from his vast library is his copy of *The Book of Common Prayer, with Marginal References to Texts in the Holy Scriptures* (1840). Edmund Venables declared Cayley a devout Christian of the old-fashioned sort who based their faith on the Bible and church attendance. It was a side of his life he seldom spoke about, regarding it as a strictly private matter. A regular churchgoer, Cayley responded to questions of his religious belief curtly. He invariably replied in one sentence that he had settled such doubts long ago and the tone of that answer ended any discussion.[5] Faced with views at odds with his own, Forsyth remembered Cayley's silencing sentence, "I do not think so" *sotto voce*, but made without a trace of hostility.[6]

Mathematics sustained him. At the beginning of 1890, he received his copy of P. G. Tait's third edition of *An Elementary Treatise on Quaternions*, to which he had contributed a chapter. Clamped between the book's brick-red covers, the physicist and mathematician formed a curious pairing. Joint authors usually have some measure of agreement, but the composer of chapter 6 took exception to Tait's dogmatic view that the natural language of geometry was the quaternion and not the method of coordinates. In its preparation, he wrote to Tait: "Of course I receive under protest ALL your utterances in regard to coordinates," and in countering Tait's claim on the compactness of the quaternion notation, "Really, I might as well say, in analytical geometry we represent the equation of a surface of the second order by $U = 0$ compare this with the cumbrous and highly artificial [quaternion] notation $S\rho\phi\rho = -1$." It was an ongoing controversy. Tait offered a sop for Cayley's benefit "that we have to thank Cartesian processes for the *idea* of an Invariant." But this only ushered in his argument for the superiority of quaternions: "[i]n pure quaternions," Tait wrote, "you have them always, so that they present no feature for remark. ρ [length] itself is an Invariant just as much as ∇ is. But what do you say to my little three term formula which is equivalent to 189 Cartesian terms?"[7]

"Talking about Polyacra"

In May 1890, Grace Chisholm, the young aspiring mathematician at Girton College was a visitor to Garden House. She recorded the occasion when she and a friend sought permission to attend the professor's lectures. They were shown into the drawing room:

The room was furnished in a tasteful but old-fashioned manner. Books were in plenty, and old china and flowers. It was clearly a woman's haunt and nothing could have assorted more incongruously with it than the little bent old man with smooth-shaven wrinkled face and bent head who came in. The fire was gone that they say had once gleamed from the eyes of the great mathematician, the deep searching look that people have described was gone. Palest blue were the eyes, and they were for the most part fixed on the earth or directed sideways at some person or object. But the face was unmistakeable to anyone who had seen a portrait of the celebrated Cambridge professor.

Permission was readily obtained, for as we have seen, Cayley was an enthusiastic supporter of education for women. Grace continued:

> He came in and held out his hand, speaking hurriedly and in a high key. "You want to come to my lecture? Certainly, certainly. Do you know where it is?" Grace confessed that she did not. "You can follow me," said the professor. "I am going there now, you are just in time."
>
> He darted from the room and reappeared at the door in his cap and gown. Muttering he led the way into Mill Lane, and made off like an arrow down the passage opposite the front door.
>
> Grace and Isabel [Maddison] darted after him. It was a most amusing race. They were too afraid of losing sight of the swift moving professor to look at one another, or laugh. The flapping black gown sped ahead, across Trumpington Street, round a corner, in through a wicket gate and across a court, round another corner and, as the two girls hurried after, they beheld the tail of a gown whisking up a flight of stairs in a large building on the right. Up the stairs they flew and passed through an open door into a small lecture room. Professor Cayley, having already divested himself of his cap was fumbling about the blackboard with the chalk and sponge. At the long table Mr. Berry was seated with another man in academicals, and the mathematical dons of Girton and Newnham were present in their everyday attire with hats.[8]

Cayley's lectures in the Lent term in 1890 were nominally on the Theory of Equations but again he exercised "flexibility." One of those in academicals was Arthur Berry, the Senior Wrangler in 1885, who had studied in Germany and was a supporter of the University Extension Lecture Syndicate, in which he served as secretary. In the lecture, talking about the polyacra (polyhedra with triangular-shaped wedges at each of their vertices), Cayley's erudition was in full-flow:

> "I was talking about polyacra," said Professor Cayley.
> It was the beginning of a flow of words only to be likened to the flight of the great little man from Mill Lane to the lecture room. Grace's pen flew over the paper. Polyhedra

with vertices constantly springing from triangular faces, like crystals growing in a solution, trees with branches forking in all directions succeeded one another without intermission, twining this way and that round the professorial head, or emerging from under his flapping sleeves as he stood with his back to the listeners chalking and talking at the same time at the blackboard. The lecture came to an end as suddenly as it had begun. The little man came to a period, gathered up his papers under his arm, caught up his cap and, bowing deeply to his audience, made off as fast as he could down the staircase.[9]

Grace Chisholm's account gives us a glimpse of the great man preserved intact by a young, impressionable, but critical listener. In future research, Cambridge could not help her and she went on to become one of Felix Klein's doctoral students, the first female recipient of a doctoral degree in mathematics and, indeed, in any subject.

Change Afoot

Both Oxford and Cambridge were in the process of being transformed from clerical institutions to modern universities and an essential component of this was the founding of a research ethos in mathematics. Steps in the vitalization of mathematics could be traced back to 1872 when candidates for mathematical fellowships at Trinity College were invited to submit dissertations in preference to sitting yet another battery of examinations following graduation. They were to be specialists and the selection of fellows was made the basis of research ability rather than inculcated examination skills.

Cambridge enjoyed a reputation as the "mathematical" university based on the fame of the mathematical Tripos, but it did not escape criticism for complacency and it was compared unfavorably with Continental institutions. Reorientation to a new world, in which pursuance of research was a priority along with the establishment of research schools, was inevitably slow. From his survey of the Cambridge scene in 1891, J. W. L. Glaisher conceded: "I am afraid that the old saying that we have generals without armies is as true as ever."[10] Insofar as Cayley helped students, it was usually on an individual basis. When there were moves to establish postgraduate study in 1890, he was instrumental in drawing up regulations for *The Isaac Newton Studentships* designed primarily for postgraduate study in astronomy and physical optics. One brake on the establishment of postgraduate study was the refusal of the university to admit graduates from other universities bearing equivalent qualifications. In 1895 this embargo ended and "advanced students" from Britain and other parts of the world were allowed to enter Cambridge for graduate work.

In the latter years, Cayley continued to be driven by invariant theory though, from a modern perspective, he spent too much time dwelling on calculatory work. He was practically working alone and, not by nature gregarious, a learner rather than a teacher, he could not easily encourage the enthusiasms and ambitions of the young. All but a few students would have seen him as a remote professor and would have known him only by repute as the fount of all mathematical knowledge. He was doubtless encyclopedic, as Salmon's musing indicates: if the mathematical Tripos was thrown open to all-comers, Salmon observed that others might score higher on the problems but that Cayley would be far ahead in the bookwork.[11] Cayley was constantly in demand by students of all nationalities, Salmon whispered mischievously, who "desired to interest him in their investigations, and be assured by him that no unscrupulous predecessor had plagiarized their discoveries."[12] However erudite, Cayley was fully aware of the tidal wave of new mathematics being created by young full-time mathematical professionals. Thanking Klein for his second volume on elliptic modular functions, he acknowledged that "[t]he magnitude of the subject seems to have become somewhat colossal."[13]

While Forsyth, Henry Baker, and several others were students of Cayley during the 1880s, G. T. Bennett graduated Senior Wrangler in 1890 and was winner of the First Smith's Prize for a paper on number theory. Cayley was so impressed, he communicated the work to the Royal Society. Later, Bennett became expert on the geometry of mechanisms and link work while his particular attention to the subjects of Cayley's own geometrical interests is pronounced. Bennett was the *official* male Senior Wrangler, but it was Phillipa Fawcett's ascendancy that electrified the student population when she graduated "above the Senior Wrangler." The procession of males in the order of merit had been topped by this Newnham scholar. "Hail the triumph of the corset, Hail the fair Phillipa Fawcett," as one piece of doggerel celebrated, and which continued,

> *May she increase in knowledge daily*
> *Till the great Professor Cayley*
> *Owns himself surpassed.[14]*

It would have been a source of pride for Cayley to see Newnham's Phillipa triumph in the university town where twenty years previously he had offered the women a lowly course on the "Principles of Arithmetic."

In Phillipa Fawcett's year, Henry Cayley graduated twenty-fourth in the order of merit. As a young student in his freshman year, he had dutifully attended his father's lectures but his Tripos result was said to have disappointed his father. Henry was deterred from seeking a career in mathematics because of his father's eminence, which he felt he could never match. Instead, he authored a monograph on housing conditions existing in Cambridge and went on to a successful career as an architect.[15] Though daughter Mary passed the entrance examination for Cambridge and could have become one of the first women students, her father thought the place should go to a person less well-connected. The Victorian gentleman had a real sense of fairness, which he applied rigorously.

If students in England wished to further their mathematical career, they were obliged, like their American counterparts, to go to Germany. This career path was popular in other sciences—especially chemistry—and postgraduate students flocked to Germany in search of the German Ph.D. Realizing that there was little for promising students in Cambridge, Cayley was in the habit of recommending the best to take this course of action. About Ada Johnson, he wrote in warm support: "[s]he is of Newnham College and one of our most distinguished women students, being in the examination of June last [1894] alone (of women and men) in the first division of the first class; and she will I am sure derive pleasure and profit from your instruction."[16]

Other factors mitigated against the founding of an effective school of mathematics. Cayley's retiring disposition did not help. He was never the great expositor as were Klein or J. J. Sylvester, both of whom acted as magnets for the young. Even when Sylvester returned to Oxford, seventy-two years young, he invigorated mathematics there. In Cayley's case, this did not happen, yet Grace Chisholm's "figure of Buddha" suffocating mathematical growth seems too harsh, given the institutional factors working against him.

Old Worlds and New

The "Cayley, Sylvester, Salmon" triumvirate that had flourished in the middle of the century was now dispersed. Of the once close-knit group that studied invariant theory, it was Cayley who was the mainstay, the one who was not deflected from its study, and the one who was the lynchpin of their early partnership.

During his inaugural lecture at Oxford in 1884, Sylvester cast his mind back and remembered "Cayley, who, though younger than myself, is my spiritual progenitor—who first opened my eyes and purged them of dross so that they could see and accept the higher mysteries of our common Mathematical faith."[17] Being the rather proper Victorian, Cayley did not openly express his feelings as did Sylvester. In their "working together," there was that Victorian reserve Sylvester could not penetrate, though he tried. "My dear Sir" had become "Dear Cayley" early on but there was always Cayley's air of formality in the relationship. The two were strong individualists though they influenced one another and, although unacknowledged in the byline of a mathematical paper, they were each part of the other's work in a real way. Their common focus in invariant theory was to the same end and they shared the methods to achieve it. Through countless conversations and intensive letter writing, not withstanding the occasional contretemps, they had sought each other's opinions and advice for a half-century.

In the 1890s, Cayley tried to enlist Sylvester's support with the "semi-invariant approach" to invariant theory but he found it "a little intricate" and needed more explanation.[18] Sylvester continued at Oxford for a while but in February 1892 suffered cataracts in his eyes and this combined with the depression to which he was prone. He gained a year's exemption from teaching duties and finally retired to London's Mayfair in 1893. Invariant theory may have been too intricate, but in his last years he kept his mathematical interests alive with problems in number theory.

Salmon had effectively given up mathematics in the mid-1860s, but his publishing career continued unabated with ten years of contributions to the *Dictionary of Christian Biography.* Turning to prevailing theological matters, he played an active part in the reconstruction of the Irish Church after disestablishment. He argued for religious faith based on individual conscience and for many years, in lecture room and pulpit, attacked the Roman Catholic claim of papal infallibility. On the political stage, he staunchly opposed Home Rule for Ireland. Salmon was appointed provost of Trinity College, Dublin, by Lord Salisbury in 1888 to succeed John Jellett, but his scientific reputation remained sufficiently afloat for the Royal Society to award him the Copley Medal in 1889.

Over the years, Salmon became an archconservative and, to cap it all, he trenchantly opposed the admittance of women to Trinity College, Dublin, in his later years. "Over my dead body" would women be admitted, he is reputed to have declared. Though he had given up serious mathematics many years before, he remained addicted to number theory, and the man who had taken on the calculation of the most extensive invariants with such gusto was in old age "an ungainly and rather untidy old clergyman, scribbling hastily and without interruption on little scraps of paper . . . looking on arithmetic, searching for primes or finding the periods in their reciprocals."[19] Number theory had been the subject of his final paper in mathematics published twenty years previously and, like Sylvester, he treated it as a refuge. He once deemed the habit of finding the periods of reciprocals of primes to be a pernicious habit and promised himself that when he got to primes near 20,000 he would stop—but this vow fell by the wayside and he still continued.[20]

In connection with the quintic polynomial, which had for so long been the object of invariant theory, Cayley recognized the contribution of the American Emory McClintock as significant. In the same connection, he commented on the work of George P. Young, a philosopher transplanted to Canada from Scotland, who had turned to mathematics.[21] In his latter years, Cayley persevered with invariant theory on his own. Percy MacMahon observed that he "worked largely on his own initiative, although well acquainted with contemporary work on the Continent and in the United States of America."[22] His own work was no longer in the mainstream and the idea of laboriously calculating, cataloging, and preparing a taxonomic system was outmoded. MacMahon reported a view current at the end of the nineteenth century that undermined the attempt to set down a detailed map of systems of invariants and covariants. In the modern thinking worthy of a new century, the determination of an invariant system explicitly was only of sentimental interest—a quaint thing to while away the time but to no theoretical advantage.

Looking back over his achievements in invariant theory, Cayley must have felt his youthful objective of 1845: "To find all the derivatives [invariants] of any number of functions, which have the property of preserving their form unaltered after any linear transformation," excessively optimistic. It is true that many notable theorems were established and the theory had grown into a vast body of knowledge, but even in its own terms, the calculatory accomplishments were actually quite modest. The failure to establish the complete system of invariants and covariants for binary forms for all but the lowest order must have caused regret. At the culmination of his famous memoirs on quantics he had established the invariants and covariants of the binary quintic form and the fundamental syzygies. But what could be said about the next binary form along the scale, the binary form of order 6? Although a great deal was known in this case, the list of fundamental invariants and covariants was never completed by Cayley. A partial list of seventeen of the less lengthy covariants from the twenty-six known to exist was published by him in 1894, but he had to admit "their great length."[23] Paul Gordan had given the symbolic expressions for the system of the binary sextic in 1868, but Cayley never completed this part of the program on his own terms.

Cayley's attention to invariant theory spanned a period of fifty years and he had seen it grow from a few disconnected ideas to a place in mathematics that rivaled even the differential calculus in importance. It was indeed a subject that attracted the leading mathematicians of the late nineteenth century; through David Hilbert's success, it became more abstract. It was seen as applicable to a wide range of mathematical activity: the theory of algebraic equations, to functional analysis, to differential equations. By the end of the century, Cayley's successor at Cambridge, A. R. Forsyth, described the place and status of invariant theory in imperial terms: "It has invaded the domain of geometry, and has almost re-created the analytical theory; but it has done more than this, for the investigations of Cayley have required a full reconsideration of the very foundations of geometry. It has exercised a profound influence upon the theory of algebraical equations; it has made its way into the theory of differential equations; and the generalisation of its ideas is opening out new regions of the most advanced and profound functional analysis." Continuing, he was equally sanguine about its future: "And so far from its course being completed, its questions fully answered, or its interest extinct, there is no reason to suppose that a term can be assigned to its growth and its influence."[24]

Writing at the beginning of the twentieth century, MacMahon put the case for invariant theory more strongly: "It is the *idea* of invariance that pervades to-day all branches of mathematics."[25] Though Hilbert eventually abandoned it, others continued to work on the theory. Before leaving the theory for other fields, Hilbert paused to acknowledge the contribution Cayley had made to the subject. In a course of lectures at Göttingen in 1897, he recommended the constructive method as the best approach to invariant theory—in effect the historical pathway traced out by Cayley.[26]

The Central Luminary

Cayley was indeed fortunate. Throughout his life, he had been recognized for his mathematical accomplishment and once the Senior Wranglership was gained in 1842, his star had steadily risen. In England, he was without peer. Sylvester's estimate of him, driven by friendship and admiration, was not wide of the mark: "Our Cayley, the central luminary, the Darwin of the English School of mathematicians."[27] In old age, Cayley personified mathematical wisdom and knowledge to the mathematical community. To British eyes, he was the visible symbol of what it meant to be a mathematician, in the same way that Sir John Herschel, from an earlier generation, symbolized what it meant to be a scientist.

With Cayley it was almost always *pure* mathematics and though he contributed to the application of mathematics, he made little attempt to enter into the physical problems themselves. He was a pure mathematician. Unlike his contemporaries, George Gabriel Stokes, J. C. Adams, and William Thomson (Lord Kelvin), who also graduated from Cambridge in the 1840s, Cayley made little input to experimental science, and, whereas they were attracted to mathematics as an agent for expressing theories of the world, he was more inclined to use the world as a source for ideas of mathematics, itself. As Stokes had playfully reminded him once: "How horrible you would think it to prove, even in one's own mind, a proposition in pure mathematics by means of physics." Within his works there are papers on the mathematical aspects of optics and dynamics and, of course, astronomy, but they were chosen for their mathematical content. He had, in his student days, promised George Boole to "make a feint attempt to read some physical optics," but was forced to admit, "I found myself, getting back always to my favorite subjects—linear transformations and analytical geometry, and gave it up in despair."[28]

Toward the end of the 1880s, Cayley's health declined further. By March 1889, he began to weaken: "I have been rather unwell myself," he commiserated with Thomas Hirst. "I can only take very moderate walks and am afraid of a railway journey."[29] In May of 1890, he indicated to the Annual General Meeting of the Newnham Council that he wished to stand down as president, citing as the reason the "university claims on his time."[30] The truth was that all his activities (Newnham, university, societies, lecturing two terms in the year, editing his

Collected Papers) were placing a burden on his health. In July, he went to Oban in Scotland for a family holiday but was disappointed to be unable to join his children on their walking excursions. Later on that year, he felt much better but it was a temporary respite.[31]

The next summer he went to Barmouth in Wales and in the following year, 1891, Sylvester tried to tempt him to visit France for the *Association pour l'Avancement des Sciences* meeting to be held in Marseilles: "Shall you go to the meeting at Marseilles? There will be I believe, and [Éduoard] Lucas affirms, a special reception there by the Princely Merchants of the city and excursions arranged to visit Tunis and Algeria after the meeting."[32] It was a forlorn hope; Sylvester himself was succumbing to blindness. A fortnight later, there was a chance to meet a member of the rising generation when Henri Poincaré visited England. Writing to Cayley, he referred to the visit of the young man who had already impressed the scientific world:

> I expect Poincaré tomorrow and he will have rooms in College—I rather dread the encounter as there is so little in the way of Mathematics upon which I can hope to talk to him. His visit is to see our Universities—his staying with me is an accident due to our meeting in Paris. Of course I shall devote myself wholly to taking him about as long as he stays with me. Had he informed me a week earlier I could have procured for him a D.C.L. honorary degree—but the list was already made out and [Bartholomew] Price informed me that it was impossible to make any addition to it so late. Where shall you be in the course of the next week? How much *you* will have to say to him![33]

As the new year of 1892 dawned, Cayley's general health deteriorated and the problem with his walking had grew more severe. He could not walk far and railway journeys or even drives in his carriage were no longer possible.[34] His increasing ill-health was very likely the long-term effect of cancer.[35] Yet he continued with mathematics. He was then working on the fifth volume of his *Collected Papers*, sure in the knowledge that he had accomplished much. His friends estimated there would be another five quarto volumes to come but perhaps he knew better; there were a further eight volumes, the whole collection amounting to thirteen volumes, comprising 967 items. This editorial task would be enough to occupy most men in the evening of their lives, but as Cayley sat as his desk leafing through his previous papers, ideas were rekindled and he continued to publish new thoughts.

During his protracted illness, there were forty papers to come in the remaining three years. These present a microcosm of his work and reflected the way he had always written. With his insatiable appetite for mathematics, the titles embraced his whole spectrum of interests, virtually the whole field of pure mathematics. From his youthful days, there was a new paper on Julius Plücker's equations and one on Sylvester's method of elimination. He corresponded with Heinrich Weber (the mathematician) on work connected with elliptic functions, taking due note of the progress made since his own paper of fifty years previously.[36] His last report to the British Association was an update on Degen's tables for Pell's equation in number theory.[37]

Cayley did not dwell on the past but was ever up to the minute and conveyed messages of the latest research as in a note on Sylow's theorems on the theory of groups. If there was one regret, it was that the custom of English mathematicians to write only individual papers effectively placing a block on cooperative work. Not one of Cayley's papers is a joint paper, not even with Sylvester. He should have liked to cooperate with Samuel Roberts, who shared interests in link work and the eight-square problem in algebra, but papers were not written this way during his time. His choice of an ideal co-worker in Roberts is significant, for like himself, the erstwhile solicitor pursued mathematics for the love of the subject.

From 1892 onward, Cayley was almost completely confined to Garden House. He never regained his former strength and his friends were concerned for his well-being. At the beginning of 1893, his fortunes appeared to change and he was well enough for the doctor to pronounce him convalescent.[38] During his last summer, he wrote a short monograph on the *Principles of Book-keeping by Double Entry*. It might be thought a dull topic for a mathematician used to more esoteric titles, but even in a matter of arithmetic, it was typical that he saw something worthwhile, for in double-entry bookkeeping he saw that "negative magnitudes were used in a very refined manner."[39]

An eminently practical man, Cayley advised that double-entry accounting was "a theory which is mathematically by no means uninteresting: it is in fact like Euclid's theory of ratios an absolutely perfect one." The "perfection" in this piece of accounting procedure for Cayley lay mostly in its economy of design to meet utilitarian ends, but, as he remarked, "it is only its extreme simplicity which prevents it being as interesting as it would otherwise be."[40] It required little mathematical preparation before the system could be understood, yet it was a practical system and its perfection rendered it a beautiful one in his eyes. In perhaps his last communication to Sylvester, he wrote: "I send herewith a semi-mathematical production, the Principles of Bookkeeping by double entry— which I have been writing, & am rather pleased with. How very long it is since we have heard of you directly—I fear there is no good account to give. I keep very much of an invalid, getting up at 11 and going to bed at ½ past 7, but I am in [a] fairly permanent state—I hope to get on with my lectures [on Analytical Geometry] which should begin in about three weeks from now—and I correct proofs &c."[41]

He was poorly but not enough to keep him from mathematics. The steady stream of investigations continued. In 1894, in one of his last letters, he wrote to

Felix Klein: "I am too much of a permanent invalid for us to have visitors."[42] He sent his mathematical medals and the recently awarded Legion of Honour medal to the Fitzwilliam Museum for safekeeping, where they still remain today; among them are the prestigious Huygens and Copley medals but also the Silver medal that won the chemistry prize at King's College, London, in 1838.[43]

During his illness, Cayley took solace in reading. The Greek classics never failed to capture his imagination; his reading of the *Gorgias,* one of Plato's dialogues has been noted. It is named after the sophist Gorgias of Leontini, and in the dialogue Socrates exposes rhetoric as a sham. The questioning of the elderly Gorgias fails to discover a satisfactory definition of rhetoric, but Socrates summed up "in the manner of the geometricians" that "cookery is to medicine, so is rhetoric to justice."[44] Not least would Cayley have delighted in Plato's instruction on teaching children how to calculate. He would have been inclined to endorse all that had been said in ancient Greece. From his schooldays and through university, he had grown up with the Dialogues. The Gorgias selection had a particular resonance: his own subject of mathematics and the way he presented it had little to do with the practices of the rhetorician.

In lighter mood, Cayley favored sentimental stories concocted about country life in Scotland. Late in 1894 he enjoyed Ian Maclaren's enormously popular *Beside the Bonnie Brier Bush,* a simple tale derived from the Kailyard school of writing in which the Scottish vernacular is combined with a happy ending, a requirement of a story he shared with Charles Darwin, who had always insisted on such.[45]

Sir Walter Scott's *Guy Mannering* was also read, and Cayley would have been amused by Scott's picture of his startling character Domine Sampson "chewing the cud of a mathematical demonstration" or "stalking about with a mathematical problem in his head."[46] He would have appreciated Scott's treatment of the pompous Sir Robert Hazelwood, who spoke long sentences loaded with "triads and quaternions," words used by Scott well before the apparition of either subject in mathematics.[47] Guy Mannering's daughter could well have given him the phrases he needed to express the allegory of the mathematical researcher as the lover of nature. It was not incidental that Cayley had expressed the outline of mathematics as a "tract of beautiful country" that can "be rambled through and studied in every detail." Although he lived through the Victorian industrial period, his study of mathematics was colored by Scott's imagery: "sounding cataracts—hills which rear up their scathed heads to the sky—lakes, that, winding up the shadowy valleys, lead at every turn to yet more romantic recesses—rocks which catch the clouds of heaven."[48]

One of Cayley's last papers revived the Cartesian-quaternion controversy with Tait, his old protagonist. "I do not know what has made me write it just now," he wrote in the summer of 1894, "but it puts on record the views which I have held for many years past and which have not been before published."[49]

To the last, he could not bring himself to accept quaternions as a method, in the same way that coordinates constituted a method—they were for him a theory—though a very beautiful theory: "I consider the full moon far more beautiful than any moonlit view, so I regard the notion of a quaternion as far more beautiful than any of its applications."[50] He was a Platonist to the last.

Hearing of his illness, Edmund Venables, his friend of fifty years standing, the continental traveling companion of his youth, and the Anglican minister who had officiated at his wedding, visited him in December. In the same month, another close friend of his undergraduate days, William Thomson, now Lord Kelvin, consulted him about a question in polyhedra. Cayley duly obliged and, even in severe pain, could not constrain his feelings about mathematics. It was not the logic of the subject that had entranced him, but its form, its structure, and its beauty. He sketched out a representation of the geometrical polyhedron for his friend, informing him simply that the "symmetry is *a very wonderful one*."[51] Two days later, unable to put the problem aside, his wavering hand traced out an alternative representation of that symmetry.[52]

With his family around him, Arthur Cayley died at 6:00 P.M. on Saturday 26 January 1895. He was in his seventy-fourth year. At the regular Cambridge Philosophical Society meeting on the following Monday, George Gabriel Stokes proposed a motion of condolence which expressed its loss for a man "whose eminence conferred honour on the Society" and which made reference to his "simple and earnest character." J. J. Thomson, who was due to read a scientific paper, adjourned the meeting as a mark of respect.[53]

The funeral on Friday 1 February was held on a bitterly cold day heavy with snow. It was a local occasion attended by his friends of the university but in recognition of his stature, it was also a formal event attended by representatives of the German and United States embassies. "Those who were present in Trinity College on Friday last," wrote Glaisher, "will never forget the aspect of the great Court on that wintry day."[54] There were more than three hundred mourners including his family and close friends. Prominent members of the scientific establishment attended, including J. J. Thomson, R. S. Ball, Michael Foster, Stokes, Kelvin, and Rayleigh. The mathematicians included A. N. Whitehead, W. W. Rouse Ball, P. A. MacMahon, A. B. Kempe, E. B. Elliott, O. M. Henrici, and E. J. Routh, but the eighty-year-old Sylvester was evidently too ill to make the journey.

Cayley was held in the highest regard by his fellow mathematicians and the Cambridge academic community. Among them, veneration was not too strong a word, this passed down by the few who knew and understood the work contained in the thousands of pages of mathematics he had written. He had led a quiet, unpretentious life and never actively sought attention. The bond of academic friendships and support formed the social structures of his life and particularly so during this last decade at Cambridge. "Certainly Arthur was happy in the love of his friends," wrote Susan Cayley to Lord Kelvin after her husband's death.[55] A moving personal tribute was paid by Kelvin himself:

> In Cayley we have lost one of the makers of mathematics, a poet in the true sense of the word, who made real for the world the ideas which his ever fertile imagination created for himself. He was the Senior Wrangler of my freshman's year at Cambridge and I well remember to this day the admiration and awe with which, before the end of my first term, just fifty-four years ago, I had learned to regard his mathematical powers. When a little later I attained to the honour of knowing him personally, the awe was evaporated by the sunshine of his genial kindness; the admiration has remained unabated to this day, and his friendship has been one of the valued possessions of my life.[56]

Among men of the world, Cayley's life presented a puzzle. Here was a man whose obvious intellectual capacity could have guaranteed a successful career in the law, one which would have proceeded upward as a matter of course. Yet he rejected this to devote himself to mathematics. The *Times* noted he was "never was a public man, never had any 'career' in the ordinary sense of the term, was always the student, and almost a recluse."[57] It was sometimes said that he "lived in the Fourth Dimension" away from the concerns of the real world, and like many a mathematician, he did not escape being lampooned for the mathematicians' ineptitude in dealing with simple arithmetic.

It is a truism, but mathematicians are not well understood by the world at large; yet, at least, Cayley was a source of pride to members of his college, even if mathematical appreciations take curious turns. Standing in front of his portrait at Trinity, a member of the college staff was heard to tell a group of visitors: "he's that exact, he could take the earth in his hand and tell you its weight to a pound."[58] It was convenient for the *Times* to reinforce the stereotype of mathematicians in writing that "he was unable to count the change for a shilling."[59] In fact, Cayley had no need to count the change of a shilling for, always careful with money, he died a wealthy man.[60]

In the previous month, on the other side of the world, Robert Louis Stevenson had died in Tahiti. He was midway through writing *The Weir of Hermiston,* in the eyes of many, the finest of all his novels. This is hardly a work to unite with thirteen 600-page volumes of mathematics, but Cayley had led a life which Stevenson's Judge Advocate would have doubly appreciated. In front of his fire on a winter's evening, Adam Weir reflected on intellectual pleasures: "To be wholly devoted to some intellectual exercise is to have succeeded in life; and perhaps only in law and the higher mathematics may this devotion be maintained, suffice to itself without reaction, and find continual rewards without excitement."[61]

Inevitably, such a life spent by Cayley pondering erudite mathematical problems is uneventful. They might be solved out in the hills and mountains, and in the odd moments of everyday living, but much of the work is done at a desk. As has been remarked before, the more of life spent there, the more uneventful it is. When the *International Congress of Mathematicians* was held in Cambridge in 1912, a group of mathematicians was moved to lay a wreath on Cayley's grave. Sadly, the headstone no longer exists, and so he has achieved in Mill Road cemetery the anonymity of the motto used by Salmon to express an essential element of Cayley's existence. *Esse quam videri:* to see rather than be seen.[62]

Postscript

What are we to make of the life lived? Lowes Dickinson's oil portrait of Cayley sitting at his desk in his middle years conveys the lifelong student, probing, searching, and impatient for progress. On the surface, his life progressed smoothly, but the appearance of a contemplative life may camouflage the breakneck speed of its passage. The torrent of publications on a whole range of mathematics is overwhelming. Only a few have matched his vast output or the breadth of his work: the *Collected Mathematical Papers* runs to thirteen large volumes and comprises items on a multitude of mathematical topics. By any measure, Cayley was an extraordinary mathematician.

Speed of publication was a characteristic Cayley shared with other prolific Victorians, and he was merely repeating a Victorian adage in acknowledging that "one's best things are done in five minutes."[63] The quill pen poised over the barrister's blue draft paper was ever ready for the completion of another memoir to be dispatched that day. "I hate spinning out a tour" and "not finishing when one has done," he wrote on a walking tour in Scotland, sentiments which could equally apply to his finishing of a mathematical paper.[64] He was an opportunist mathematician with a competitive edge who felt bound to publish his work immediately and comment on the work of others. In his youth, he challenged men with established reputations, men like Sir William Rowan Hamilton, C. G. J. Jacobi, and Otto Hesse, and, when he became the *éminence grise* himself, what young author would not be flattered to read his *Addition* to their own efforts. C. S. Peirce's assessment of him as a "mathematical encyclopaedia" is not wide of the mark.[65]

Cayley was an example of a mathematician who reached the pinnacle of his subject rapidly and remained there throughout his life. But, of all the talented pure mathematicians produced by the Cambridge system throughout the nineteenth century, what so clearly separated Cayley from all the others? Given that success in the mathematical Tripos implied a high degree of technical expertise, no other British pure mathematician came close to his achievements.

Most who distinguished themselves in the Tripos examination system then passed on to a career in the church, the law, or commercial life. Mathematics, if retained at all, was regarded as a diversion, a serious avocation in some instances, but purely voluntary. For most, mathematics was simply an examination hurdle and the attention paid to it quickly terminated with the conferment of a Cambridge halo, gained by a high place in the order of merit. Cayley was a rarity, driven by the beauty of mathematics to the point of obsession. He was compulsive and he never stopped his mathematical activity. In letters written to friends on Christmas Day, jottings on postcards from the summit of Snowdon in Wales, and from isolated huts on the lakes of Norway, ideas were transmitted, and manuscripts were dispatched from the more remote parts of northern Scotland to editors awaiting his regular contributions. Passion is not a word that easily sits with a man of his reticence but it is not out of place. Without this inner drive, he could not have risen above mediocrity. At the same time, Cayley did not live in a vacuum; he was fortunate to enter university in company with a remarkable group of young scientists and an atmosphere of mathematical and scientific adventure. He found himself in a Cambridge that included men like Duncan Gregory, Stokes, and Thomson, and a new vehicle for the transmission of ideas, the *Cambridge Mathematical Journal.* These undergraduate years in the 1840s shaped his mathematical thinking and led to deeper studies during later decades. Ten years on, as the young men of the early 1840s moved into their full scientific lives, these cultural ingredients were lacking, and the environment that had sustained them in youth had evaporated. The successor to the *Cambridge Mathematical Journal* had ceased to be an undergraduate endeavor, and the examination course Cayley had followed had been watered down by reforms of the Tripos.

Infused with his quest for mathematics, and fortified by early accomplishment, Cayley set about establishing a national and international reputation. In France and Germany, and later in Italy, he quickly made links with serious mathematicians and published in their journals. He made contact in unlikely corners of Britain—in Lincoln where the impoverished Boole taught in a small school, in the village of Croft in Lancashire where Kirkman tended his parishioners, and in Dublin where Salmon was beginning a church career interlaced with one in mathematics. It has been suggested that Cayley was an isolated figure. Yet, while his mathematical life was solitary to some extent, his intellectual world was rich with widespread mathematical contacts. In London, he met Sylvester, who became a close confidante and provided frisson to an otherwise level existence. It was hardly a partnership in the ordinary sense but the collaboration was productive. Sylvester approached mathematics from an intensely competitive standpoint while Cayley, though keen to give credit where it was due, calmly absorbed new results into his own work.

In a rigid class-based society, Cayley's common denominator for choosing contacts was the commonality of their shared interest in the furtherance in mathematics. His correspondents found him congenial, helpful, and all recognized his exceptional attachment to his subject. He was a confident young man, and, as he progressed from short youthful gains to long definitive memoirs, he became an indelible member of the new Victorian scientific order that replaced the "gentleman of science" of an earlier era. His position at the center of Victorian mathematics was cemented by the scientific administration he undertook.

Cayley is one whose influence still pervades modern mathematics, a claim which cannot be made for many who graced the mathematical stage in Britain during the nineteenth century. He might be compared with his teacher George Peacock. A major figure in the transition of algebra, Peacock has left little discernible imprint on mathematics, and present-day mathematicians do not look on him as the instigator of ideas in the way they do of Cayley. Peacock's "principle of permanence of equivalent forms," a philosophical prop used to justify arguments in symbolical algebra, was pivotal in the 1830s and 1840s, but it finds no resonance in mathematical talk today and has long since disappeared from textbooks on algebra. On the contrary, mementos of Cayley's work are plentiful, as a glance at the "Glossary of Mathematical Terms" in this book clearly shows.

In a country of gentleman amateurs, Cayley introduced modern professional attitudes into the study of mathematics. He was out of step with the group that came before him, the men of science who took all of science as their playground. For William Whewell, "omniscience" was his foible, but Cayley did not attempt such a broad sweep. A contemporary comparison could be made with R. L. Ellis, who, four years older, was a competent mathematician, but one who took a serious interest in such topics as Chinese dictionaries and Roman aqueducts. When he graduated at Cambridge, his teacher regarded him as "a senior wrangler among senior wranglers" but even if spared illness, he would still have been the learned scholar with diffuse interests rather than the dedicated mathematician who was not diverted.

Cayley was more focused and out of step with those concerned with adding accretions to a broad band of knowledge. Still less could he become the regular "don of all work" with a few minor mathematical papers and the occasional slim volume on rarefied points of religious orthodoxy, produced during a life in college or in a country rectory. Mathematics was his theater and he hardly looked outside it for his creative work. He would not be deflected; indeed, many of his ventures in "mixed mathematics," the applied mathematics of the day, was but a thinly disguised excuse for pursuing a purely mathematical problem. It is symptomatic of his dedication that he should stray into "applied" subjects such as mechanics, astronomy, electricity, and organic chemistry when mathematical principles were at issue. Cayley substituted Whewell's omniscience in all departments of knowledge with an omniscience in pure mathematics.

The way Cayley shifted from topic to topic throughout a career of fifty years might now be construed as a weakness. It is tempting to surmise that with *more* focus (say on his major study, invariant theory) he would have achieved more. His reality was that he was contributing to one subject: mathematics. The life which began in the dying embers of the Romantic movement was not left untouched by its insistence on the unity of knowledge. To suggest that his darting about, a magpie picking up this bauble and that, would dilute his effort would be a reflection on the way mathematics is now perceived, in terms of ultraspecialization. Cayley saw the subject as one cloth on which to embroider and make connections, and he could not have done this by reducing his gaze. But when he came to the late 1880s, moving about this vast tapestry looked as out of place as Whewell's panorama had earlier. The adoption of Cayley's breadth no longer conformed to a modernist's specialized agenda.

Cayley's marked tendency to the organization of mathematics shows the administrator at work. This meant that the setting up of systems to accommodate an expanding field, and, in this, the morphology of mathematics was important. Geometry had a classificatory aspect to it, while the new symbolical algebra was seen as an organic subject, living and developing. With a broad knowledge of the geography of mathematics, Cayley could see connections between seemingly disparate areas: between quaternions and matrix algebra, quaternions and octonions, invariant theory and its applications to the solutions of polynomial equation and to curves and surfaces in geometry. This ability to make links across a wide field singled out Cayley as one of the most influential mathematicians of his generation, so that by the end of the nineteenth century he was generally regarded as the mathematician whose work was most frequently referred to by the entire mathematical community.

Remembering their close association in the 1850s, when both worked on the embryonic invariant theory, Charles Hermite remarked that Cayley's talent was backed up with a great capacity for hard work. This resulted in works of clarity and elegance, and he further noted that these works made people regard Cayley as comparable to the illustrious Augustin-Louis Cauchy. Where he differed from Cauchy was in the matter of calculation. There is no doubt that Cayley's time in his latter years were spent with long calculatory work, but, for him, it was an intrinsic part of his research program. Though human calculators were employed, he did not regard calculation as drudge's work. His calculatory approach—"a power of calculation that shunned no labour"—was central to the development of his intuition and the testing of results.[66]

The whole invariant theory program, conceived of by Cayley as algorithmic, was reminiscent of the work of his own hero, Carl Gustav Jacobi. But it was MacMahon's remark that reveals the kind of mathematician Cayley really was: "a great intuitionist, but his reputation in this respect may have suffered from his extraordinary skill as a formalist."[67] This intuitive approach to mathematics counterbalanced his lack of familiarity with a deductive style of proving results. The geometer Max Noether identified this deep intuitive understanding in his work. After Alfred Clebsch's early death in 1872, Noether assumed the mantle of Germany's leading algebraic geometer. He saw Cayley's strength as the way he was able to proceed from the concrete to general theories: "He is master of the *empirical* use of material: how he unites it in an abstract thought, generalises it and tests it through calculation, how then from the newly gained information the comprehensive idea appears at a single stroke, to which the subsequent crucial trial of verifying calculation will be applied over years of work. So is Cayley the *Natural Scientist* amongst mathematicians."[68] In this capacity as "Natural Scientist," Cayley displayed "old-fashioned" mathematical virtues, in which the manipulation of symbols was one of the essential skills. Francis Galton recalled that he "moved his symbols in battalions, along broad roads, careless of short cuts, and he managed them with the easy command of a great general."[69]

From this intuitive base, the printed word became highly formal. Writing in a somewhat opaque style (for example, the legalistically framed "An Introductory Memoir on Quantics"), Cayley's papers were apt to be forbidding. But if he ever heard whispers to this effect, they made no difference, for making concessions to his readers was not his way. In his flights of other-worldliness, he held the naive belief that others had his capacity to write and understand as he did. There is no doubt a less severe style would have been welcomed by some of his readers. While admiring the results, "their present value would be increased a hundred fold," wrote Philip Kelland, "were some fragments from them chiselled and made fit for ordinary hands."[70] Salmon spoke directly from one who struggled with Cayley's writings, which, he surmised, proceeded from the author's point of view and not the reader's and as a consequence of Cayley's habit of beginning at the point where his researches ended.

In his selection of topics, Cayley exercised free rein, and, like musicians or writers slipping fragments into their desk drawer to rework them from time to time, he would revisit old themes, sometimes after long intervals. Revising the central idea, he would quickly publish the result and move on, feeling no need to immediately polish and repolish his work before publication, as Salmon found to his cost. There was no attempt to take readers along the course of his own thinking or to gently guide them to the discovery of his results, was an opinion Salmon ventured.

Some believed that this severe style was cultivated deliberately. G. B. Mathews, not one to be cowed by difficult mathematics, inclined to the view that Cayley's restraint was due to "excess of sensitiveness, which made him follow an ideal of classic severity and shrink from any open expression of emotion."[71] Certainly this ideal was encouraged by an education based on classics, in which students were trained in the process of peeling away the unnecessary argument and acquired the literary skill of condensing. A piece of mathematics could likewise be boiled down to its bare essentials, in which extraneous words would be regarded as obscuring the argument and dismissed as verbosity. Uncompromising, Cayley relied on readers being familiar with the field of study. This mellowed in the latter years and from the 1880s he began to list the works of others with accompanying historical remarks, but when his readership saw his oft-repeated phrase "it is hardly necessary to remark," they knew there was work to be done.

Cayley was driven by mathematical novelty and, in discovering it, he was often (but not always) satisfied by confirmation of generally stated results in specific cases. In this, he ran the risk of constructing theories on shifting sand, but he was often saved from error by a sharply honed insight. If his method of verification resulted in a successful test, he was often satisfied, whereupon he boldly declared the result generally true and moved on. His admission that "the mere getting out [of] a simple result is a strong presumption of its correctness" is indicative of his research attitude.[72]

Forthright in his papers, he could be equally frank in his advice, as with one researcher seeking fruitful areas in astronomy: "[t]he theory of Mercury was very fully gone into by Leverrier in the *Annales de l'Obser.* & I do not imagine you could add anything to what he has done—or indeed that the existing observations furnish the means of doing so," he wrote. "My candid opinion is that the attempt would be a mere waste of time."[73] As a referee, he was impatient to get to the mathematical content. In considering work by Georges Halphen, he wrote: "I don't like Halphen's paper [on systems of conics]," adding, "I do not think so much talk is wanted before coming to the question: and the examples made *more special, and worked out more fully* [his emphasis]."[74]

Through his wide experience, he became a standard bearer for what was permissible to publish. Like many referees, he was influenced by an author's reputation and the author's position. He had doubts whether the Chief Baron's paper was worthy but he let it pass. Occasionally, he saw something he disliked, and then he could be blunt. Pity the hopeful author who submitted a memoir to the Royal Society, in which a cubic curve through nine points had actually been drawn, and which reaped Cayley's remark: "[i]t seems to me rather like taking five towns in England (forming a convex figure) and calculating the ellipse which can be drawn thro[ugh] them."[75] For authors who might be tempted to pass on trivia, he wrote of "results which assuming one cannot find them in a text book, one works out for oneself when one requires them."[76] If some errors were published, a short note from him would suffice to correct it. In his last summer, he pointed out an "unwarranted assumption" in another putative proof of Euclid's parallel axiom.[77] While he had exacting standards, he could also encourage. If he approved of their work, young mathematicians of promise could count on his support for publication in the prestigious Royal Society journals.

Cayley's technical ability may be taken for granted, but his knowledge of mathematics as an entire subject ran deep. When his erstwhile neighbor at Trinity, Robert Potts, sent him a copy of his *Elementary Algebra*, Cayley replied, "I have just looked at it enough to see that I shall find a great deal to interest me in the historical part."[78] Rouse Ball sent him a copy of his *Short Account of the History of Mathematics,* and he approved of this too. On the history of mathematics, his point of view was that of a working mathematician, who used his "power of commenting upon and developing the work of his predecessors."[79]

It is difficult to overestimate the influence of the Greeks on Cayley's thinking. He could read them directly through his knowledge of the classics and his familiarity with Latin and Greek. More generally, the classical world of the Greeks was central to the modes of thought of the Victorians.[80] The classical world has the quality of timelessness and Cayley, imbued with a study of world of algebraic forms likewise timeless, found in it a fit subject for study and exploration. Indeed the mathematical world of the Victorians conveys the feeling of a static, permanent world in which artifacts can be collected, displayed, and examined. The geometrical models fashioned by Cayley and his contemporaries and lovingly constructed from paper mâché and wire-frames are static, classical, and Greek.

At the end of the nineteenth century, his student J. W. L. Glaisher predicted: "A century hence Cayley's writings will be as familiar to the mathematicians of that time as Newton's, or Euler's, or Gauss's are to us."[81] It is perhaps true that Cayley's name is as familiar as these three, but his reputation is not as high as Glaisher's forecast would suggest. Being nearer to our own time, Cayley can best be seen as bridging two eras. He is a modern mathematician in his concern for structure in pure mathematics and he would give approval for the ideals for the axiomatic method. Nonetheless, readers who look for the modern "definition, theorem, proof" in his work will be sadly disappointed. In its place is a strong tendency to *classify,* and thus motivated, Cayley had more in common with the nineteenth-century classifying scientist than post-Victorian mathematicians. He often failed to provide "proof," the essential quality of their subject demanded by the mathematician of today. Instead, he assembled tables of algebraic invariants for page after page, an activity in keeping with the Victorian ideal that saw progress as the *gradual accumulation of knowledge.*

In attempting to address the entire field of mathematics, Cayley set himself an impossible task but the impossibility of it was only brought home to him in the 1880s. He was blessed with a prodigious memory—for, according to Galton, he "could repeat poetry by the yard"—and in the 1850s had even doubted the need for a *Catalogue of Scientific Papers,* since one would automatically know everything in one's own subject. But in the 1880s, he was forced to admit that knowing one's own subject entirely was not possible.[82]

During his lifetime, Cayley was admired and fêted in the mathematical community, but after his death, his mathematical reputation languished. The drive for his type of mathematics weakened and H. W. Turnbull found it necessary to reaffirm his hero's greatness. Invariant theory, the subject that made his reputation had reigned supreme, but this loyal disciple researched a subject that had dropped from its imperial position. No longer could it be said that all mathematical roads lead to the Holy Grail that Cayley had created. Mathematicians adopted a new abstract methodology through the axiomatic method vigorously launched by David Hilbert and his entourage in Germany. In its wake, the concrete algorithmic approach Cayley espoused was in retreat. His papers on group theory and matrices have been scrutinized, but his longitudinal studies involving much calculatory work are unread today.

However, as the twentieth century advanced, the vast extent of his work, as well as its constructive style, became appreciated once more, and the pioneering nature of his contributions recognized. Now, at the beginning of a new century ourselves, we have the cool assessment of one of the most influential of pure mathematicians, Bartels van der Waerden, the emissary of the new abstract algebra. The Dutchman's mature judgment of Cayley, as one who was "extremely influential," appropriately removes the nationalistic reservation from Turnbull's bluff defense of Cayley being the "greatest English mathematician." Cayley was an internationalist who, even as a youth, fully recognized that progress depended on cooperation across national boundaries.

How was Cayley viewed by his peers and immediate successors? Apart from the positive estimates made by Turnbull and van der Waerden, and, of course, the laudatory ones by Sylvester, various others have been made from time to time. The American astronomer Simon Newcomb, a personal friend, judged Cayley as a selfless pursuer of knowledge: "one of the creators of modern mathematics, he never had any ambition beyond the prosecution of his favorite science . . . his life was that of a man moved to investigation by an uncontrollable impulse; the only sort of man whose work is destined to be imperishable."[83] Another American, G. B. Halsted, who met Cayley when he visited the United States, said that he "must ever rank as one of the greatest mathematicians of all time."[84] A colleague of Halsted at Johns Hopkins, C. S. Peirce, was a great admirer, and stated simply that "all mathematicians ranked him as first among their living number."[85] Francesco Brioschi, an exact contemporary was better placed to appreciate the quality of Cayley's mathematics in terms of subject expertise. He judged him to be the first mathematician of Europe— compared with his lesser regard for Charles Hermite and Leopold Kronecker, whom he deemed the first mathematicians of their own countries.[86]

There are also contrary positions.[87] Though admitting that the unidimensional scale measuring "greatness" is of limited use, mathematicians have found such estimates alluring. G. H. Hardy, a later successor to Cayley's Cambridge chair, had decided views, as Lawrence Young recorded: "I once asked Hardy whether he thought Cayley was a great mathematician: Hardy simply glared at me." Hardy was perhaps influenced by the mood for debunking the Victorians, which existed after the World War; he was after all, a friend and contemporary of Lytton Strachey, author of *Eminent Victorians* (1918). Young also reported the views of Hardy's collaborator J. E. Littlewood, who "in his public lectures at Wisconsin, takes for granted the greatness of Cayley and Stokes in Cambridge."[88] Popular writers, like James Newman, who took the pulse of the mathematical profession in the 1950s, concluded that Cayley "brought mathematical glory to Cambridge, second only to that of Newton"—moreover, Newman added, it was the "fertility of his suggestions" which were especially valuable for the future growth of mathematics.[89]

Last Words

The tragedy of Arthur Cayley is that in the 1860s and 1870s, he was unable to pass on his erudition to the next generation and create a research school of mathematics in any way comparable to those in some German universities. When the chance did come in the 1880s, and a research attitude eventually stirred in Cambridge, it was too late. Of the few students who did benefit, one was Charlotte Angas Scott, whose mathematical career owed so much to Cayley's willingness to help and inspire. In her admiration of him, she noted that it was "his creative power that excites our admiration."[90] Happily there were a few others. Forsyth was steeped in Cayley's ideas and his earliest works show that he could engage with his mentor on equal terms.

When Forsyth inherited the Sadleirian chair, it was clear that he could no longer continue with Cayley's style of research. In the 1890s, the mathematical coinage was changing and by the beginning of the new century, Cayley's brand of invariant theory had passed out of fashion.[91] The skill and mastery of dealing with extensive algebraic forms was no longer wanted, and the cumbrous nature of the subject became an embarrassment to the young mathematicians. The appearance of Grace and Young's *Algebra of Invariants* in 1903, written in the concise German symbolic notation, was more important for the rising generation of invariant theorists in England. With its streamlined approach, one young man observed that a "new era dawned for the teaching and progress of higher algebra."[92] The German influence was felt more generally: "Much east wind has blown across the North Sea on our insular mathematics," declared another.[93]

Invariant theory in Cayley's tradition went into swift decline, and, falling from its imperial position, the descent led to sanctuary within the theory of groups. Since then, the classically shaped invariant theory has long been absorbed by a succession of abstract syntheses.[94] It would be rash, however, to announce its demise. The idea of invariance is so fundamental to mathematics and science that, though it occasionally fades, it is certain to be a recurring theme. To be completely ignorant of invariant theory is to be "simply illiterate," declared a leading geometer in the 1920s.[95] The subject was also beginning to acquire its powerful mythology. Hermann Weyl wrote that "the *theory of invariants* came into existence about the middle of the nineteenth century somewhat like Minerva: a grown-up virgin, mailed in the shining armor of algebra, she sprang forth from Cayley's Jovian head."[96]

Referring to a bygone age, a young Cambridge mathematician wrote in 1923: "I believe it is fashionable to try and despise Cayley's downright sledgehammer methods of attack; but he had a wonderful power of reaching his objective. He was a mathematical tank in some of his methods, but he knew where he was going."[97] It was this power and indomitable industry which took away the breath of the younger generation of mathematicians when they saw the thirteen volumes of his *Collected Papers* on the library shelves. G. T. Bennett was so reminded of his geometry teacher that he exclaimed: "what a worker the great man was!"[98]

Cayley was an explorer of an adventurous kind, and, at the time of his death, this species of bold Victorian was almost extinct. He was hardly a consolidator of acquired territory, content with tidying-up or making rigorous what was already known. He was a trailblazer who went tramping off into the wilderness in the great Victorian tradition. In this fashion, he took many mathematical paths and has left us his vast notebook for our perusal. His methods suited his own mathematical problems, but a hundred years later, many of his discoveries are still significant.

When Cayley was starting out on his mathematical journey in 1838, the future poet laureate Alfred Lord Tennyson had recently written in the widely read *Locksley Hall*:

> For I dipt into the future, far as human eye could see,
> Saw the Vision of the world, and all the wonder that would be.

Cayley had indeed "dipt into the future," and, after half a century and more, he observed in his presidential address to the British Association at Southport, "The advances made have been enormous, the actual field is boundless, the future full of hope."[99] With a keen perception for the future prospects of mathematics, the subject he cared so passionately about all his life, he embraced a couplet from the same poem, a few lines on:

> Yet I doubt not thro' the ages one increasing purpose runs,
> And the thoughts of men are widen'd with the process of the suns.

Arthur Cayley's Social Circle

Boldface entries indicate intimates in Cayley's larger community of scholars and friends. Ages are as of 1863, the year Cayley became the foundation Sadleirian Professor of Pure Mathematics at Cambridge.

Name	Age	Dates	Link with Cayley
Abel, Niels-Henrik	d.	1802–1829	Norwegian mathematician, theory of functions
Adams, John Couch	**44**	**1819–1892**	**Cambridge colleague, astronomer**
Airy, Hubert	25	1838–1903	Cambridge student of 1850s
Airy, Osmund	18	1845–1928	Cambridge student of 1860s
Airy, Sir George Biddell	**62**	**1801–1892**	**Astronomer Royal, educational reformer**
Airy, Wilfrid	27	1836–1925	Cambridge student of 1850s
Albert Victor, Prince		1864–1892	Eldest son of Prince of Wales, recipient of Cambridge Honorary Degree
Allen, John	53	1810–1886	Chaplain at King's College, London, teacher of mathematics
Allman, George J.	51	1812–1898	Irish botanist, Indian Civil Service examiner
Arbogast, L. F. A.	d.	1759–1803	French mathematician, calculus of operations
Aronhold, Siegfried H.	44	1819–1884	German mathematician, invariant theory
Askwith, Edward H.	——	1864–1946	Churchman, group theory
Atkinson, Michael A.	50	c. 1813–1890	Trinity College, Cambridge, don in 1840s, teacher of mathematics

Name	Age	Dates	Link with Cayley
Austin, William	43	1820–1909	Contemporary student, 1840s
Babbage, Charles	71	1792–1871	English mathematician, scientist, computer pioneer
Baker, Henry F.	——	1866–1956	English mathematician, Cambridge student, 1880s, geometry
Ball, Sir Robert S.	23	1840–1913	Irish mathematician, astronomy
Ball, W. W. Rouse	13	1850–1925	Trinity College, Cambridge, don, mathematician and colleague
Baltzer, Richard	45	1818–1887	German mathematician, determinants
Barstow, Thomas I.	45	1818–1889	Contemporary student, 1840s
Bates, Henry W.	38	1825–1892	Royal Geographical Society, editor
Bell, Eric T.	——	1883–1960	Biographer of mathematicians
Bennett, Geoffrey T.	——	1868–1943	English mathematician, Cambridge student, 1880s; link work
Bentham, George	63	1800–1884	English botanist
Berry, Arthur	1	1862–1929	Cambridge don, English mathematician, colleague
Bertrand, J. L. F.	41	1822–1900	French mathematician, dynamics
Bessel, Friedrich W.	79	1784–1864	German astronomer
Betti, Enrico	40	1823–1892	Italian mathematician, algebra
Bézout, Étienne	d.	1730–1783	French mathematician, algebra
Bickmore, Charles E.	15	c. 1848–1901	Oxford mathematician, calculator
Bienaymé, Irénée-Jules	67	1796–1878	Belgian mathematician, probability
Binet, Jacques P. M.	d.	1786–1856	French mathematician, determinants
Bird, Christopher	d.	c. 1820–1846	Contemporary student, 1840s
Blackall, Samuel	47	1816–1890	Cambridge don, teacher of mathematics, 1840s

Name	Age	Dates	Link with Cayley
Blackburn, Hugh	40	1823–1909	Contemporary student, Scottish mathematician
Bonney, T. G.	30	1833–1923	Cambridge don, women's education
Boole, George	**48**	**1815–1864**	**Friend, mathematician, and logician**
Booth, James	57	1806–1878	Irish mathematician, Royal Society colleague
Borchardt, Carl W.	46	1817–1880	German mathematician, editor of *Crelle's Journal*
Bragg, William H.	1	1862–1942	Cambridge student, 1880s; British physicist
Bramwell, Sir Frederick J.	45	1818–1903	BAAS meeting, Southport (1883)
Brewster, Sir David	82	1781–1868	Royal Society sponsor
Brill, Alexander von	21	1842–1935	German mathematician, geometry
Brioschi, Francesco	**39**	**1824–1897**	**Italian mathematician, invariant theory**
Bristed, Charles	43	1820–1874	American contemporary student at Cambridge, 1840s
Brodie, Sir Benjamin (elder)	d.	1783–1862	President of Royal Society, surgeon
Brodie, Sir Benjamin (younger)	46	1817–1880	Professor of chemistry, Oxford, calculus of operations
Bronwin, Brice	77	c. 1786–1869	English mathematician
Brooke, Charles	59	1804–1879	Surgeon and sometime mathematician
Brown, Alexander Crum	25	1838–1922	Scottish "mathematical" organic chemist
Browne, Robert W.	54	1809–1895	Professor of classical literature at King's College, London
Brummel, Edward	48	c. 1815–1901	Cambridge contemporary of J. J. Sylvester
Brunyate, William E.	——	1867–1943	Cambridge student, 1880s, educationalist
Bryan, George H.	——	1864–1928	Cambridge student, 1880s

Name	Age	Dates	Link with Cayley
Bryan, Reginald (Guy)	43	c. 1820–1912	Contemporary student, 1840s
Buchheim, Arthur	4	1859–1888	English mathematician, matrix algebra
Bunsen, Robert W.	52	1811–1899	Rival for Royal Medal, 1853
Burnside, William	11	1852–1927	British mathematician; Cambridge student, 1870s; group theory
Busk, George	56	1807–1886	X Club, surgeon
Butler, Henry M.	30	1833–1918	Master of Trinity College, Cambridge, 1880s and later
Buxton, Charles	41	1822–1871	Private pupil, 1840s
Buxton, Thomas F.	41	1822–1908	Private pupil, 1840s
Cantor, Georg	18	1845–1918	German mathematician, set theory
Carey, Frank S.	3	1860–1928	English mathematician; Cambridge student, 1880s; number theory
Casey, John	**43**	**1820–1891**	**Irish mathematician, geometry**
Catalan, Eugène C.	49	1814–1894	Belgian mathematician, algebra
Cauchy, Augustin-Louis	d.	1789–1857	French mathematician
Cayley, Arthur	**41**	**1821–1895**	**Subject of this biography**
Cayley, Arthur	d.	c. 1776–1848	Distant relative, fourth wrangler (1796)
Cayley, Charles Bagot	**40**	**1823–1883**	**Brother, author, translator**
Cayley, Cornelius	d.	1729–1780?	Great-uncle, religious author
Cayley, Cornelius	d.	1692–1779	Great-grandfather, Recorder of Kingston-upon-Hull
Cayley, Cornelius	d.	1762–1836	Uncle, brother of Henry
Cayley, Edward S., Jr.	39	1824–1884	Cousin; contemporary Cambridge student, 1840s
Cayley, George J.	37	1826–1878	Cousin; contemporary Cambridge student, 1840s
Cayley, Sir George	d.	1773–1857	Fourth cousin, English aeronautical pioneer

Name	Age	Dates	Link with Cayley
Cayley, Henrietta Caroline	**35**	**1828–1886**	**Younger sister**
Cayley, Henry	**d.**	**1768–1850**	**Father, Russia trade and City merchant**
Cayley, Henry	——	**1870–1949**	**Son, architect**
Cayley, John	**d.**	**1730–1795**	**Paternal grandfather, Russia trade and diplomat**
Cayley, John	d.	1761–1831	Uncle, brother of Henry
Cayley, Maria Antonia	69	1794–1875	Mother, née Doughty
Cayley, Mary	——	**1872–1950**	**Daughter**
Cayley, Sophia	**47**	**1816–1889**	**Elder sister**
Cayley, William	d.	1766–1803	Uncle, brother of Henry
Cayley, Sir William	d.	c. 1610–1681	First Cayley baronet
Cayley, William Henry	d.	1818–1819	Elder brother (died in infancy)
Challis, James	60	1803–1882	Cambridge colleague, astronomer
Chasles, Michel	**70**	**1793–1880**	**French mathematician, geometry**
Chisholm, Grace. *See* Young, Grace Chisholm			
Christie, Jonathan H.	**70**	**c. 1793 1876**	**Employer, conveyancing barrister**
Christie, Samuel H.	79	1784–1865	Secretary of Royal Society
Christoffel, Elwin B.	34	1829–1900	German mathematician, invariant theory
Chrystal, George	12	1851–1911	Scottish mathematician, algebra
Clebsch, (R. F.) Alfred	**30**	**1833–1872**	**German mathematician, geometry, invariant theory**
Clifford, William K.	**18**	**1845–1879**	**Cambridge student, 1860s; English mathematician; philosopher**
Clough, Anne J.	43	1820–1892	Newnham College principal, 1871–1892
Cobbold, Thomas Spencer	35	1828–1886	Western University of Great Britain, parasitologist
Cockle, Sir James	**44**	**1819–1895**	**English mathematician, algebra**

Name	Age	Dates	Link with Cayley
Collins, Wilkie	39	1824–1889	Contemporary law student, Lincoln's Inn; author
Coolidge, Julian L.	——	1873–1954	American mathematician, geometry
Cope, Edward M.	45	1818–1873	Elected fellow Trinity College, Cambridge, 1842
Cox, Homersham (Jr.)	6	1857–1918	Cambridge student, 1870s; non-Euclidean geometry
Coxwell, Henry T.	44	1819–1900	English balloonist
Cozens, Sarah	d.	1732–1803	Paternal grandmother
Craig, Thomas	8	1855–1900	American mathematician, Johns Hopkins University
Cramer, Gabriel	d.	1704–1752	Swiss mathematician, algebra
Crelle, August L.	d.	1780–1855	German mathematician, editor of *Crelle's Journal*
Cremona, (Antonio) Luigi	33	1830–1903	Italian mathematician, geometry
Crofton, Morgan W.	37	1826–1915	Irish mathematician, theory of curves, geometric probability, RMC Woolwich
Cubitt, Thomas	d.	c. 1819–1841	Contemporary undergraduate, 1840s
Cumming, James	d.	1777–1861	Cambridge professor, chemistry
Daniell, John Frederic	d.	1790–1845	Professor of chemistry, King's College, London
Darboux, Jean Gaston	21	1842–1917	French mathematician, geometry
Darwin, Charles R.	54	1809–1882	Rival for Royal Medal, 1853; scientist
Darwin, George H.	18	1845–1912	Cambridge colleague, English applied mathematician
Davidson, Charles	53	1810–1893	Contemporary law student, Lincoln's Inn
Davies, Emily	33	1830–1921	Women's education at Cambridge
Davies, John L.	37	1826–1916	Churchman, contemporary alpinist

Name	Age	Dates	Link with Cayley
Davies, Thomas S.	d.	1794–1851	English mathematician, RMC Woolwich
D'Avigdor, Elim Henry	22	1841–1895	Correspondent
Davis, William B.	47	c. 1816–1879	Calculator, invariant theory
Dedekind, (J. W.) Richard	32	1831–1916	German mathematician, algebra
De la Rue, Warren	48	1815–1889	Royal Astronomical Society Council, astronomer
Delaunay, Charles E.	47	1816–1872	French astronomer, lunar theory
De Morgan, Augustus	57	1806–1871	English mathematician, *Royal Society Catalogue*
Denman, George	44	1819–1896	Contemporary undergraduate at Cambridge, 1840s
Dew-Smith, Albert G.	15	1848–1903	Cambridge photographer
Dickens, Charles	51	1812–1870	Disliked author
Dickinson, G. Lowes	1	1862–1932	Cambridge don
Dickinson, Lowes Cato	44	1819–1908	Portrait painter
Dickson, James D. H.	14	1849–1931	Cambridge student, 1870s
Dirichlet, Johann P. G. Lejeune-	d.	1805–1859	German mathematician, algebra
Disraeli, Benjamin	59	1804–1881	Politician, pupil at Potticary's school
Dixon, Alfred C.	——	1865–1936	Cambridge student of 1880s
Dodgson, Charles L.	31	1832–1898	English mathematician and author, determinants
Donkin, William F.	49	1814–1869	Oxford mathematician
Du Chaillu, Paul Belloni	32	1831–1903	Explorer of Africa
Duhamel, Jean M. C.	66	1797–1872	French mathematician, differential equations
Dyck, Walther von	7	1856–1934	German mathematician, group theory
Earnshaw, Sammuel	58	1805–1888	Examiner for Smith's Prize, 1842
Eddis, Arthur Shelly	46	1817–1893	Contemporary fellow Trinity College, Cambridge; barrister

Name	Age	Dates	Link with Cayley
Edleston, Joseph	47	c. 1816–1895	Trinity College, Cambridge, don in 1840s; taught mathematics
Einstein, Albert	——	1879–1955	Physicist
Eisenstein, Ferdinand G. M.	d.	1823–1852	German mathematician, invariant theory
Elliott, Edwin B.	12	1851–1937	English mathematician, invariant theory
Ellis, Alexander J.	49	1814–1890	English scientist, foundations of mathematics
Ellis, Robert Leslie	**d.**	**1817–1859**	**Friend, English mathematician**
Everett, William	24	1839–1910	American visitor to Cambridge, 1860s
Ewing, James A.	8	1855–1935	Cambridge colleague, professor of mechanism and applied mathematics
Faà di Bruno, Francesco	38	1825–1888	Italian mathematician, invariant theory
Faraday, Michael	72	1791–1867	Scientist
Fawcett, Henry	30	1833–1884	Contemporary professor at Cambridge, economics
Fawcett, Millicent G.	16	1847–1929	Suffragist
Fawcett, Phillipa	——	1868–1948	"Above" the Senior Wrangler, 1890
Fenn, Joseph F.	43	1820–1884	Contemporary student, 1840s
Fenwick, John	48	1815–1889	Tripos competitor, 1842
Ferrers, Noman Macleod	34	1829–1903	Rival for Sadleirian Chair, 1863
Féval, Paul	46	1817–1887	Favored author
Finch, Gerald B.	28	1835–1913	Conveyancer, Lincoln's Inn
Fischer, Frederick W. L.	49	c. 1814–1890	Cambridge student, 1840s
Fiske, Thomas S.	——	1865–1944	American mathematician, visitor to Cambridge, 1880s
Forbes, James D.	54	1809–1968	Scottish scientist
Forsyth, Andrew Russell	**5**	**1858–1942**	**Friend, successor to Sadleirian Chair**

Name	Age	Dates	Link with Cayley
Foster, Michael	27	1836–1907	English physiologist
Frankland, Edward	38	1825–1899	English chemist
Franklin, Fabian	10	1853–1939	American mathematician, invariant theory
Frobenius, Georg F.	14	1849–1917	German mathematician, algebra
Frost, Andrew H.	44	c. 1819–1907	Tripos competitor, 1842
Frost, Percival	46	1817–1898	Competitor for Sadleirian Chair, 1863
Fuller, Frederick	44	1819–1909	Contemporary student, 1840s
Gall, August von	17	1846–1899	German mathematician, invariant theory
Galois, Évariste	d.	1811–1832	French mathematician, group theory
Galton, Francis	**41**	**1822–1911**	**Private pupil, English scientist**
Garnett, William	13	1850–1932	Member Cambridge Modelling Club
Gaskin, Thomas	54	c. 1809–1887	Competitor for Sadleirian Chair, 1863
Gauss, Carl F.	d.	1777–1855	German mathematician
Gell, Frederick	43	1820–1902	Undergraduate contemporary at Trinity College, Cambridge
Gilman, Daniel Coit	32	1831–1908	American university administrator, Johns Hopkins University
Gladstone, William Ewart	54	1809–1898	Politician
Glaisher, James (the elder)	54	1809–1903	Amateur mathematician, meteorologist
Glaisher, James W. L.	**15**	**1848–1928**	**English mathematician, colleague at Cambridge**
Glover, John	40	c. 1823–1884	Undergraduate taught by Cayley 1840s
Gordan, Paul A.	**26**	**1837–1912**	**German mathematician, invariant theory**
Goulburn, Henry	d.	1813–1843	Trinity College, Cambridge, don 1840s; teacher of mathematics

Name	Age	Dates	Link with Cayley
Grace, John H.	——	1873–1958	English mathematician, invariant theory
Grant, Robert	49	1814–1892	Scottish astronomer
Grassmann, Hermann G.	54	1809–1877	German mathematician, algebra and geometry
Gravatt, William	57	1806–1866	Engineer, promoter of Scheutz calculation engine
Graves, Charles	51	1812–1899	Irish mathematician, algebra
Graves, John T.	57	1806–1870	Irish mathematician, algebra, geometry
Gray, Benjamin	43	1820–1886	Contemporary student, 1840s; legal career
Gray, John E.	63	1800–1875	Royal Society colleague
Green, George	d.	1793–1841	Contemporary student of James Joseph Sylvester, theory of integration
Greenhill, Alfred G.	16	1847–1927	Cambridge don, 1870s
Gregory, Duncan F.	**d.**	**1813–1844**	**Friend; Trinity College, Cambridge; Scottish mathematician**
Gregory, Olinthus G.	d.	1774–1841	Teacher of mathematics at RMC Woolwich
Griffin, William N.	48	1815–1892	Contemporary student with James Joseph Sylvester
Grove, Sir William R.	52	1811–1896	Royal Society reformer, 1840s
Gudermann, Christolph	d.	1798–1852	German mathematician, elliptic functions
Hall, Thomas G.	**60**	**1803–1881**	**Professor of mathematics at King's College, London**
Halphen, Georges H.	19	1844–1889	French mathematician, geometry
Halsted, George B.	10	1853–1922	Student of James Joseph Sylvester, Johns Hopkins University, geometry
Hamilton, Henry Parr	69	1794–1880	Mathematical author
Hamilton, Sir William R.	**58**	**1805–1865**	**Irish mathematician and scientist, quaternions**

Name	Age	Dates	Link with Cayley
Hammond, James	13	1850–1930	English mathematician, invariant theory
Hansen, Peter A.	68	1795–1874	Danish astronomer
Hardy, G. H.	——	1877–1947	English mathematician
Hargreave, Charles J.	**43**	**1820–1866**	**English mathematician, legal career**
Harley, Robert	35	1828–1910	English mathematician, algebra
Hart, Andrew S.	52	1811–1890	Irish mathematician, geometry
Hawkshaw, John	52	1811–1891	British civil engineer
Hayward, Robert B.	34	1829–1903	Schoolmaster, geometry reformer
Heath, John M.	55	1808–1882	Trinity College, Cambridge, tutor in 1840s
Heath, Thomas L.	2	1861–1940	Cambridge student, 1880s
Helmholtz, Hermann L. F. von	42	1821–1894	German physicist
Hemery, James	d.	1814–1849	Trinity College don, 1840s; teacher of mathematics
Hemming, George W	42	1821–1905	Contemporary student; Senior Wrangler, 1844
Henrici, Olaus M.	23	1840–1918	Danish mathematician, resident in England; pure geometry
Hensley, Lewis	39	1824–1905	Contemporary student; Senior Wrangler, 1846
Hermite, Charles	**41**	**1822–1901**	**French mathematician, invariant theory, elliptic functions**
Herschel, John F. W.	71	1792–1871	English mathematician, calculus of operations
Hesse, (Ludwig) Otto	52	1811–1874	German mathematician, analytical geometry
Hicks, John W.	23	1840–1899	Member Cambridge Modelling Club
Hicks, William M.	13	1850–1934	Member Cambridge Modelling Club

Name	Age	Dates	Link with Cayley
Hilbert, David	1	1862–1943	German mathematician, invariant theory
Hill, Edwin	20	1843–1933	Member of Cambridge Modelling Club
Hill, Micaiah J. M.	7	1856–1929	Cambridge student, 1870s
Hirsch, Meier	d.	1765–1854	Mathematician, symmetric functions
Hirst, Thomas Archer	**33**	**1830–1892**	**Friend, English mathematician, geometry**
Hobson, Ernest W.	7	1856–1933	Third Sadleirian professor of pure mathematics
Hofmann, August Wilhelm von	45	1818–1892	German chemist
Hooker, Sir Joseph D.	46	1817–1911	English botanist
Hopkins, William	**70**	**1793–1866**	**Private tutor at Cambridge**
Hudson, R. W. H. T.	——	1876–1904	Cambridge mathematician, 1890s
Hudson, William H. H.	25	1838–1915	Cambridge don, 1870s; teacher of mathematics
Hurwitz, Adolf	4	1859–1919	German mathematician
Hutchinson, Charles E.	69	c. 1794–1870	Churchman, married to Cayley's cousin
Huxley, Thomas H.	38	1825–1895	English scientist
Hymers, John	60	1803–1887	Mathematical author
Ivory, James	d.	1765–1842	Scottish mathematician
Jacobi, Carl G. J.	**d.**	**1804–1851**	**German mathematician, algebra, elliptic functions**
Jebb, Richard C.	22	1841–1905	Contemporary Cambridge professor
Jellett, John H.	46	1817–1888	Irish mathematician
Jerrard, George B.	59	1804–1863	English mathematician
Joachimsthal, Ferdinand	d.	1818–1861	German mathematician, geometry
Jones, Richard	d.	1790–1855	Professor of political economy at King's College, London

Name	Age	Dates	Link with Cayley
Jonquières, Jean-Philippe F. de	43	1820–1901	French mathematician, algebraic geometry
Jordan, (M-E) Camille	25	1838–1922	French mathematician, group theory
Kay, Edward E	41	1822–1897	Private pupil, lawyer
Kay, Joseph	42	1821–1878	Private pupil, economist
Kelland, Philip	55	1808–1879	Scottish mathematician
Kelvin, Lord. *See* Thomson, William			
Kempe, Alfred B.	14	1849–1922	Cambridge student, 1870s; graph theory, group theory
King, Charles W.	45	1818–1888	Elected fellow Trinity College, Cambridge, 1842
King, Joshua	d.	1798–1857	Lucasian Professor of mathematics in 1840s
Kingsley, Charles	44	1819–1875	Contemporary student at King's College, London, 1830s, and at Cambridge, 1840s
Kirchoff, Gustav R.	39	1824–1887	German scientist
Kirkman, Thomas P.	**57**	**1806–1895**	**English mathematician, algebra**
Klein, (Christian) Felix	**14**	**1849–1925**	**German mathematician, non-Euclidean geometry**
Kronecker, Leopold	40	1823–1891	German mathematician
Kummer, Ernst E.	53	1810–1893	German mathematician
Lacroix, Sylvestre F.	d.	1765–1843	French mathematician
Ladd, Christina	16	1847–1930	Amcrican mathematician, Johns Hopkins University
Lagrange, Joseph-Louis	d.	1736–1813	French mathematician
Laguerre, Edmond N.	29	1834–1886	French mathematician, projective geometry, matrix algebra
Lamb, Sir Horace	14	1849–1934	Cambridge mathematician, 1870s
Lambert, Carlton J.	18	c. 1845–1921	Member Cambridge Modelling Club
Lamé, Gabriel	68	1795–1870	French mathematician

Name	Age	Dates	Link with Cayley
Langley, Edward Mann	12	1851–1933	Schoolmaster, geometry reformer
Laplace, Pierre Simon	d.	1749–1827	French mathematician
Latham, Henry	42	1821–1902	Cambridge friend
Lattimer, Thomas	9	1854–1912	Member Cambridge Modelling Club
Lebesque, Victor A.	72	1791–1875	French mathematician
Lejeune-Dirichlet, J. P. G. *See* Dirichlet, Johann P. G. Lejeune-			
Le Verrier, Urbain J. J.	52	1811–1877	French astronomer
Levett, Rawdon	20	1843–1923	Schoolmaster, geometry reformer
Lewis, William	76	1787–1870	Chess writer
Lie, (Marius) Sophus	21	1842–1899	Norwegian mathematician, group theory
Lindley, John	64	1799–1865	Rival for Royal Medal, 1853; English botanist
Liouville, Joseph	54	1809–1882	French mathematician, editor of *Liouville's Journal*
Listing, Johann B.	55	1808–1882	German mathematician, topology
Liveing, George D.	36	1827–1924	English scientist, civil service examiner
Lobachevsky, Nikolai Ivanovich	d.	1792–1856	Russian mathematician, non-Euclidean geometry
Locke, John	d.	1632–1704	Philosopher read by Cayley
Lockhart, John Gibson	d.	1794–1854	Friend of J. H. Christie, literature
Lockyer, (Joseph) Norman	27	1836–1920	English astronomer, editor of *Nature*
Lodge, Oliver J.	12	1851–1940	English physicist
Lubbock, Sir John W.	60	1803–1865	English scientist and astronomer
Lucas, François É. A.	21	1842–1891	French mathematician, number theory
Lyell, Charles	66	1797–1875	Professor of geology, King's College, London
Macaulay, Alexander	——	1863–1931	Student at Cambridge, 1880s; quaternions

Name	Age	Dates	Link with Cayley
Macaulay, Francis S.	1	1862–1937	Student at Cambridge, 1880s; algebra
MacCullagh, James	d.	1809–1847	Irish mathematician
MacFarlane, Alexander	12	1851–1913	Scottish mathematician, logic
Mackie, John H.	10	1853–1915	Member Cambridge Modelling Club
Maclaren, Ian (pen name of John Watson)	13	1850–1907	Favorite author
MacMahon, Percy A.	**9**	**1854–1929**	**Friend, English mathematician, invariant theory**
Maddison, Isabel	——	1869–1950	Student at Cambridge, 1880s
Maine, Henry J. S.	41	1822–1888	Candidate for mastership of Trinity College, Cambridge
Malfatti, Gian Francesco	d.	1731–1807	Italian mathematician, geometry
Manners, George John	43	1820–1874	Contemporary student at Trinity College, Cambridge, in 1840s
Mannheim, Victor M. A.	32	1831–1906	French mathematician
Marsh, Henry A.	46	1817–1901	Elected fellow Trinity College, Cambridge, 1842
Marshall, Alfred	21	1842–1924	Supporter of Newnham College
Mathews, George B.	2	1861–1922	English mathematician, algebra
Maurice Frederick D.	58	1805–1872	Member of "Lectures Association"
Maxwell, James Clerk	**32**	**1831–1879**	**Scottish physicist, member Cambridge Modelling Club**
Mayor, Robert B.	43	1820–1898	Tripos competitor, 1842
McClintock, Emory	23	1840–1916	American mathematician, algebra
Mill, John Stuart	57	1806–1873	Philosopher
Miller, Robert K.	21	1842–1889	Cambridge student, 1860s
Miller, William Hallowes	62	1801–1880	Contemporary professor at Cambridge

Name	Age	Dates	Link with Cayley
Miller, William J. C.	31	1832–1903	Editor of *Mathematical Questions for the "Educational Times"*
Mitchell, Oscar H.	6	1857–1889	Student at Johns Hopkins University
Mittag Leffler, G.	17	1846–1927	Swedish mathematician
Moberly, George	60	1803–1885	Relative, headmaster of Winchester, bishop of Salisbury
Möbius, August F.	73	1790–1868	German mathematician, geometry
Mohs, Friedrich	d.	1773–1839	Scientist
Moline, Mary	**71**	**c. 1792–1868**	**Mother of Susan Moline**
Moline, Robert	**73**	**c. 1790–1866**	**Father of Susan Moline**
Moline, Susan	**32**	**c. 1831–1923**	**Wife**
Monge, Gaspard	d.	1746–1818	French mathematician, geometry
Montépin, Xavier de	40	1823–1902	Favored author
Moon, Robert	46	1817–1889	English mathematician, contemporary student of law, 1840s (Inner Temple)
Moor, Allen P.	39	c. 1824–1904	Undergraduate taught by Cayley, 1840s
Moore, Eliakim H.	1	1862–1932	American mathematician
Morley, Frank	3	1860–1937	Cambridge student, 1880s, algebraic geometry
Moseley, Henry	62	1801–1872	Professor of natural and experimental philosophy and astronomy at King's College, London
Mould, James George	45	1818–1902	Competitor for Sadleirian Chair, 1863
Muir, Sir Thomas	19	1844–1934	Scottish mathematician, determinants
Munro, Hugh A. J.	44	1819–1885	Contemporary student and professor at Cambridge
Murphy, Robert	d.	1806–1843	Irish mathematician, calculus of operations

Name	Age	Dates	Link with Cayley
Neale, John Mason	45	1818–1866	Cofounder of Camden Society
Neumann, Carl G.	31	1832–1925	German mathematician, editor of *Mathematische Annalen*
Neville, Eric H.	——	1889–1961	English mathematician
Newcomb, Simon	28	1835–1909	American astronomer
Nichol, John Pringle	d.	1804–1859	Scottish astronomer
Noether, Max	19	1844–1921	German mathematician, algebraic geometry
Norris, John P.	40	1823–1891	Friend; churchman; and student contemporary at Cambridge, 1840s
O'Brien, Matthew	d.	1814–1855	Irish mathematician, author, and teacher
Oliver, James E.	34	1829–1895	American mathematician
Ostrogradsky, Mikhail	d.	1801–1862	Russian mathematician, dynamics, calculus
Otter, William	d.	1768–1840	Principal of King's College, London
Park, John James	d.	1795–1833	Professor of law and jurisprudence, King's College, London
Parkinson, Stephen	40	1823–1889	Contemporary Cambridge student; Senior Wrangler, 1845
Parnell, Hugh	44	1819–1906	Tripos competitor, 1842
Patmore, Coventry	40	1823–1896	Author
Peacock, George	**d.**	**1791–1858**	**English mathematician, tutor for Trinity College, Cambridge (1838)**
Pearson, Karl	6	1857–1936	Cambridge student in 1870s
Peirce, Benjamin	54	1809–1880	American mathematician, linear algebra
Peirce, Charles S.	24	1839–1914	Lecturer Johns Hopkins University; American philosopher, mathematician, logician

Name	Age	Dates	Link with Cayley
Pell, Albert	43	1820–1907	Contemporary student, 1840s
Pendlebury, Richard	16	1847–1902	Cambridge student, 1860s
Perigal, Henry	62	1801–1898	English mathematician, member of Royal Astronomical Society
Petersen, Julius	24	1839–1910	Danish mathematician, graph theory
Pfaff, Johann Friedrich	d.	1765–1825	German mathematician, differential equations
Philpott, Henry	56	1807–1892	Examiner for Smith's Prize, 1842
Picard, Charles E.	7	1856–1941	French mathematician
Pieri, Mario	3	1860–1913	Italian mathematician
Playfair, John	d.	1748–1819	Scottish mathematician, editor of Euclid's *Elements* (1795)
Plücker, Julius	**62**	**1801–1868**	**German scientist and mathematician, analytical geometer**
Poincaré, Henri	9	1854–1912	French mathematician, non-Euclidean geometry, automorphic functions
Poisson, Siméon-Denis	d.	1781–1840	French mathematician
Poncelet, Jean V.	75	1788–1867	French mathematician, geometry
Pontécoulant, Phillippe de	68	1795–1874	French astronomer
Potter, Richard	64	1799–1886	Examiner for mathematical Tripos 1842
Potticary, George B. F.	67	1796–1891	Proprietor of Eliot Place School, Blackheath
Potticary, John	d.	1763–1820	Former proprietor of Eliot Place School, Blackheath
Potts, Robert	58	1805–1885	Cambridge private tutor, editor of Euclid's works
Preston, Theodore	45	c. 1818–1873	Elected fellow Trinity College, Cambridge, 1842
Price, Bartholomew	45	1818–1898	English mathematician, Oxford

Name	Age	Dates	Link with Cayley
Proctor, Richard Anthony	26	1837–1888	English astronomer, Secretary of Royal Astronomical Society, author
Ramanujan, Srinivasa A.	——	1887–1920	Indian mathematician
Rankine, William J. M.	43	1820–1872	Physicist and civil engineer
Ranyard, A. C.	18	1845–1894	Member Royal Astronomical Society
Remsen, Ira	17	1846–1927	Organic chemist, Johns Hopkins University
Ricci-Curbastro, Gregorio	10	1853–1925	Italian mathematician, tensor calculus
Richmond, Herbert W.	——	1863–1948	British mathematician, geometer
Rickett, Mary E.	2	1861–1925	Mathematics lecturer at Newnham College
Riemann, (Georg F.) Bernhard	37	1826–1866	German mathematician
Roberts, Michael	46	1817–1882	Irish mathematician, invariant theory
Roberts, Samuel	36	1827–1913	English mathematician, linkwork
Roberts, William	46	1817–1883	Irish mathematician
Rodrigues, Olinde	d.	1794–1851	French mathematician, quaternions and rotation
Romilly, Joseph	72	1791–1864	University of Cambridge diarist
Roscoe, Henry E.	30	1833–1915	English chemist
Rose, Hugh James	**d.**	**1795–1838**	**Principal of King's College, London**
Rossetti, Christina	33	1830–1894	Friend of Charles Cayley, poet
Rossetti, Dante Gabriel	35	1828–1882	Leader of Pre-Raphaelite Circle
Rossetti, Gabriele	d.	1783–1854	Professor of Italian at King's College, London
Rossetti, William M.	34	1829–1919	Pre-Raphaelite circle
Rothschild, Mayer A. de	45	1818–1874	Student contemporary at Cambridge, 1840s

Name	Age	Dates	Link with Cayley
Rothschild, Nathan Meyer	d.	1777–1836	Colleague of Henry Cayley, Sr.
Routh, Edward J.	32	1831–1907	Competitor for Sadleirian chair, 1863
Rowe, Richard C.	10	1853–1884	English mathematician; student at Cambridge, 1870s
Ruskin, John	44	1819–1900	Art critic; contemporary student at King's College, London
Russell, William H. L.	40	1823–1891	English mathematician; student at Cambridge, 1840s
Salisbury, Third Marquis of	33	1830–1903	Rival for BAAS presidency, 1883; prime minister
Salmon, George	**44**	**1819–1904**	**Friend, Irish mathematician, churchman, provost of Trinity College, Dublin**
Scheutz, Georg	78	1785–1873	Swedish engineer, calculating machines
Schläfli, Ludwig	**49**	**1814–1895**	**Swiss mathematician, geometry**
Schorlemmer, Carl	29	1834–1892	Organic chemist
Schröter, H. E.	34	1829–1892	German mathematician, synthetic geometry
Schubert, Hermann C. H.	15	1848–1911	German mathematician, enumerative geometry
Schwarz, K. H. A.	20	1843–1921	German mathematician
Scott, Charlotte A.	5	1858–1931	Student at Cambridge, 1870s, 1880s; analytical geometry
Scott, George G.	52	1811–1878	English architect
Scott, Robert F.	14	1849–1933	Member Cambridge Modelling Club
Segar, Hugh W	——	1868–1954	Cambridge student, 1880s
Serret, Joseph A.	44	1819–1885	French mathematician, algebra
Sévigné, Marquise de	d.	1626–1696	Favored author
Shaw, Benjamin	44	1819–1877	Contemporary student
Shaw-Lefevre, John G.	66	1797–1879	Royal Society colleague

Name	Age	Dates	Link with Cayley
Sheppard, William F.	——	1863–1936	Cambridge student, 1880s; statistician
Shortland, Peter F.	48	1815–1888	Tripos competitor, 1842
Sidgwick, Eleanor M.	18	1845–1936	Principal of Newnham, 1892–1910
Sidgwick, Henry	25	1838–1900	Cambridge women's education, philosopher
Simpson, Charles T.	44	1819–1902	Contemporary student, 1840s
Smith, Archibald	50	1813–1872	Founder of *Cambridge Mathematical Journal*
Smith, Benjamin F.	**44**	**1819–1900**	**Contemporary student, 1840s**
Smith, Henry J. S.	37	1826–1883	English mathematician, geometry
Smith, James P.	9	1854–1929	Cambridge student, 1870s
Smith, Dr. Robert	d.	1689–1768	Founder of Smith's Prizes
Smith, William R.	17	1846–1894	Cambridge University librarian, 1886–1889
Sohncke, Ludwig A.	d.	1807–1853	German mathematician
Sommerville, Duncan M. Y.	——	1879–1934	Scottish mathematician, non-Euclidean geometry
Spencer, Herbert	43	1820–1903	English philosopher
Spottiswoode, William	**38**	**1825–1883**	**English scientist and mathematician, invariant theory**
Staudt, Christian von	65	1798–1867	German mathematician, geometry
Steele, William J.	d.	1831–1855	Cambridge private tutor, 1850s
Steiner, Jacob	67	1796–1863	Swiss mathematician, geometry
Stephen, Leslie	31	1832–1904	Cambridge don, 1850s
Stephen, Sir James	d.	1789–1859	Civil service examiner
Sterne, Laurence	d.	1713–1768	Favored author
Stevenson, Richard	d.	c. 1812–1837	Trinity College, Cambridge, don and mathematician
Stevenson, Robert L.	13	1850–1894	Scottish author

Name	Age	Dates	Link with Cayley
Stokes, George Gabriel	**44**	**1819–1903**	**Scientist, colleague at Cambridge**
Story, William E.	13	1850–1930	American mathematician, Johns Hopkins University
Strange, Alexander	45	1818–1876	Foreign Secretary of Royal Astronomical Society
Street, George E.	29	1834–1881	Favored architectural writer
Strutt, John W. (Lord Rayleigh)	21	1842–1919	English physicist
Stuart, James	20	1843–1913	Friend, contemporary professor at Cambridge
Sturm, Jacques C. F.	d.	1803–1855	Swiss mathematician, algebra
Sylvester, James Joseph	**49**	**1814–1897**	**Friend, English mathematician, invariant theory**
Tait, Peter Guthrie	**32**	**1831–1901**	**Scottish physicist**
Talbot, William H. Fox-	63	1800–1877	English mathematician, pioneer of photography
Taylor, Charles	23	1840–1908	Cambridge don, geometry
Taylor, Tom	46	1817–1880	Elected fellow Trinity College, Cambridge, 1842
Temple, Frederick	42	1821–1902	Churchman, civil service examiner
Tennyson, Alfred Lord	54	1809–1892	Poet Laureate, dramatist
Terquem, Olry	66	1797–1887	French mathematician and geologist
Thacker, Arthur	d.	c. 1814–1857	Trinity College, Cambridge, don, 1840s; teacher of mathematics
Thackeray, William M.	52	1811–1863	Author, English literature
Thomson, James (elder)	d.	1786–1849	Professor of mathematics, Glasgow University
Thomson, James (younger)	41	1822–1892	Professor of civil engineering at Belfast University
Thomson, Joseph John	7	1856–1940	British physicist
Thomson, William (Lord Kelvin)	**39**	**1824–1907**	**Friend, Scottish physicist, editor of *Cambridge and Dublin Mathematical Journal***

Name	Age	Dates	Link with Cayley
Thorp, Thomas	66	1797–1877	Tutor Trinity College, Cambridge, 1840s
Thurtell, Alexander	57	c. 1806–1884	Examiner for Tripos in 1842
Todhunter, Isaac	43	1820–1884	Competitor for Sadleirian Chair, 1863
Tooke, Thomas	d.	1774–1858	Contemporary of Henry Cayley, economist
Tooke, William	d.	1744–1820	Chaplain of British Factory in St. Petersburg
Townsend, Richard	42	1821–1884	Irish mathematician, geometry
Trotter, Coutts	26	1837–1887	Physicist, University administrator
Turnbull, Herbert W.	——	1885–1961	English mathematician, invariant theory
Turnbull, William P.	22	1841–1917	English mathematician, geometry
Tyndall, John	43	1820–1893	English physicist
Vandermonde, Alexandre T.	d.	1735–1796	French mathematician
Venables, Edmund	**44**	**1819–1895**	**Friend; churchman; student contemporary at Cambridge, 1840s**
Venn, John	29	1834–1923	Cambridge don, logician
Waley, Jacob	45	1818–1873	Editor of legal works
Wallis, A. J.	7	1856–1913	Cambridge student, 1870s
Walton, William	50	1813–1901	Cambridge private tutor, geometry
Watson, John	13	1850–1907	Favored author
Watt, Robert	46	1817–1890	Elected fellow Trinity College, Cambridge, 1842
Webb, Benjamin	44	1819–1885	Cofounder of Camden Society
Weber, Heinrich	21	1842–1913	German mathematician, algebra
Weddle, Thomas	d.	1817–1853	English mathematician, geometry
Wehnert, Edward	50	1813–1868	Western University of Great Britain

Name	Age	Dates	Link with Cayley
Weierstrass, Karl W. T.	48	1815–1897	German mathematician
Whatley, Richard	76	1787–1863	Oxford logician
Wheatstone, Charles	61	1802–1875	Professor at King's College, London
Whewell, William	69	1794–1866	Master of Trinity College, Cambridge; philosopher
Whitehead, Alfred North	2	1861–1947	English mathematician, algebra
Wilbraham, Henry	38	1825–1883	English mathematician, legal career
Williams-Ellis, John C.	30	1833–1913	Competitor for Sadleirian Chair, 1863
Williamson, Alexander W.	39	1824–1904	English chemist
Willis, Robert	63	1800–1875	Contemporary professor at Cambridge
Wilson, James M.	27	1836–1931	Cambridge student in 1850s, geometry reformer
Wilson, William P.	37	1826–1874	Senior Wrangler, 1847
Wollaston, Charles B.	47	c. 1816–1887	Western University of Great Britain
Wolstenholme, Joseph	34	1829–1891	English mathematician
Wood, Philip W.	——	1880–1956	English mathematician, invariant theory
Woolhouse, W. S. B.	54	1809–1893	Mathematician and astronomer; editor
Wordsworth, Christopher	d.	1774–1846	Master of Trinity College, Cambridge
Wright, Joseph E.	——	1878–1910	Cambridge student, 1890s; geometry
Wright, Richard T.	17	1846–1931	Cambridge student, 1860s; educational reformer
Wroński, Hoëné	d.	c. 1776–1853	Polish mathematician
Yeoman, Constantine B.	40	1823–1889	Private pupil
Young, Alfred	——	1873–1940	English mathematician, invariant theory
Young, George P.	44	1819–1889	Canadian philosopher and mathematician

Name	Age	Dates	Link with Cayley
Young, Grace Chisholm	——	1868–1944	Student at Cambridge in 1880s and 1890s
Young, William H.	——	1863–1942	English mathematician, set theory
Zeuthen, Hieronymus G.	24	1839–1920	Danish mathematician, algebraic geometry, history of mathematics

Glossary of Mathematical Terms

Cayley and Sylvester were prolific in their introduction of new terms into mathematics. As they found their mathematical world expanding, many old terms became inadequate for their needs. Theirs was not a unique experience—the language of mathematics undergoes constant change, but to the modern reader their choices often present a barrier.

To act as a checklist of technical terms that appear in this biography, a small selection is given below. No attempt has been made to be technically exact; indeed, such a goal would be out of place in a biography. (Many dictionaries of mathematics now exist and three have been listed in the bibliography: *Historical Dictionary of Mathematical Words, Elsevier Dictionary of Computer Science and Mathematics,* and the *Encyclopedic Dictionary of Mathematics.*) A fuller description of many terms may be obtained from the glossaries of the "Invariant Trinity" written when the terms were seen as new.[†]

Abel's theorem   A famous result in algebraic geometry submitted by Niels-Henrik Abel in 1826 but published posthumously in 1841. It involves the application of hyperelliptic functions (abelian functions) to the study of curves and surfaces. *See* chapter 14. *See also* **hyperelliptic functions.**

algebraic form (or **quantic**)   An expression composed of the addition of the products of variables. An example is the algebraic form $ax^2 + 2bxy + cy^2$ with two variables and "facients" a, b, c. Cayley's "quantic" was intended to displace the older "rational and integral function" (which refers to its form as a rational function, not to its coefficients being rational numbers). *See* chapter 8.

algorithm   A procedure in the form of a set of instructions—a "mathematical recipe."

analytical geometry   Geometry studied through coordinates and equations. This approach effectively reduces geometry to algebra and was thus opposed by the advocates of "pure" geometry who relied on the inspection of diagrams and arguments free of algebraic considerations.

associative property   If the combination of three symbols a, b, c is independent, of which two are combined first, then this operation is said to have the associative property. For example, ordinary addition has the associative property since $(a + b) + c = a + (b + c)$, but subtraction does not, since $(a - b) - c$ and $a - (b - c)$ are not the same in general.

[†]J. J. Sylvester wrote "Glossary of new or unusual terms, or of terms used in a new or unusual sense, in the preceding memoir: On a theory of the syzygetic relations of two algebraical functions" = [SP1 (1853), 580–86], and Cayley contributed "Recent terminology in mathematics" = [CP4 (1860), 594–608]. Salmon gave a comprehensive synopsis of terms that includes an early history of invariant theory (Salmon 1866, 266–82).

The term "associative" is attributed to W. R. Hamilton. For an important example, *see* **Cayley number**.

asyzygetic   Equivalent to the modern "linearly independent." *See* chapter 8.

bicenter   (of a **Cayley tree**) The *two* central vertices, c, of a graphical tree (*see* chapter 13):

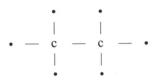

binary form (or **binary quantic**)   An algebraic form with *two* variables, for example, $ax^2 + 2bxy + cy^2$ (the binary quadratic), and $ax^3 + 3bx^2y + 3cxy^2 + dy^3$ (the binary cubic).

binary quantic   *See* **binary form**.

binary quintic   A binary form of order 5: $ax^5 + 5bx^4y + 10cx^3y^2 + 10dx^2y^3 + 5exy^4 + fy^5$. This algebraic form was a central interest to Cayley, as it was for many nineteenth-century mathematicians. *See* chapters 4, 8.

binary sextic   A binary form of order 6. *See* **binary form**.

calculus of operations   A branch of algebra in which symbols represent operators. Pioneered by L. F. A. Arbogast at the beginning of the nineteenth century, it was promoted by Duncan Gregory, George Boole, Cayley, and many others.

canonical form   A streamlined algebraic form that is in some sense the "simplest" attainable—for example, one consisting of "squared" terms only. A change of variable allows an algebraic form to be "reduced" to a canonical form, thus enabling it to be handled more easily.

catalecticant   An invariant of a binary form of even order that takes the form of a determinant with cyclically arranged rows.

caustic   A curve generated by light rays incident on a surface. An example is the curve with a cusp generated by light rays striking the interior surface of a coffee mug. *See* chapter 8.

Cayleyan[‡] (also known as the "**pippian**")   A curve dual to a given cubic curve (it is the curve "enveloped" by the lines joining pairs of corresponding pairs of points on the cubic curve $Hf = 0$). *See* chapter 4. *See also* **Hessian curve**.

Cayley-Bacharach theorem   A theorem concerned with the number of intersections of two algebraic curves. *See* chapter 3.

Cayley-Darboux-Lévy equation   The partial differential equation that must be satisfied by the curvilinear coordinates (p, q, r) of a point P if the surfaces $p = f(x, y, z)$, $q = g(x, y, z)$, $r = h(x, y, z)$ are to intersect orthogonally at the point P. Gaston Darboux and Maurice Lévy wrote on this subject. *See* chapter 13.

[‡]The references to Cayley's contributions listed here were either made by mathematicians during his lifetime or shortly after his death. I do not assert that Cayley was the "first" to discover any particular result, but in all cases Cayley played a part in the development—the ascriptions made here certainly illustrate the range of his work. I have not included items in which his name appears in a purely honorific capacity (for example, the group theory software CAYLEY, or the lunar crater in the uplands between Mare Vaporum and Mare Tranquillitatis christened by astronomers as the "Cayley lunar crater").

Cayley-Hamilton theorem An important theorem of linear algebra independently discovered by W. R. Hamilton. Cayley actually discovered a more general form of the theorem than the one usually given (*see* chapter 9). This version of the theorem was consequent on Cayley's preference for *homogeneous* polynomials over ordinary polynomials as they are understood today. For matrices P, Q, the "homogeneous characteristic polynomial" is the binary form $f(\mu, \lambda) = det(\mu P - \lambda Q)$. The theorem discovered by Cayley asserts that $f(Q, P) = 0$, if $PQ = QP$.

Cayley-Hermite problem Determination of the specific linear transformations that leaves a given bilinear form invariant. *See* chapter 9.

Cayley-Klein metric A metric (distance function) derived from Cayley's absolute conic. By choosing different conics, a variety of non-Euclidean geometries may be obtained. *See* chapters 9, 10. *See also* **Cayley's absolute**.

Cayley lines In connection with the Pascal's Hexagon theorem, Cayley found (1851) that the sixty "Kirkman points" lie in threes on twenty "Cayley lines." *See* **geometry of position**.

Cayley-Menger determinant A special determinant that expresses a geometrical relationship between points in space. This was the subject of Cayley's very first paper, written when he was an undergraduate. *See* chapter 2.

Cayley-Moser lottery problem A decision problem posed by Cayley that may now be formulated as a dynamic programming problem. Also known as the "Secretary problem." *See* chapter 13.

Cayley number A "number" with eight base units which extends the family of algebras composed of the complex numbers (two base units) and quaternions (four base units). A Cayley number is of the form $w = a + bi + cj + dk + ep + fq + gr + hs$, in which $i^2 = j^2 = , \ldots , h^2 = -1$. The Cayley numbers are "almost associative," but $(jp)q = k$ and $j(pq) = -k$. for instance. *See* chapter 5. *See also* **associative property**.

Cayley-Plücker coordinates In the 1860s, both Cayley and J. Plücker independently devised systems of six homogeneous coordinates to describe lines in space. *See* chapter 10.

Cayley-Plücker equations Equations that link together the singularities of skew curves. *See* chapter 5. *See also* **Plücker's equations** (for plane curves).

Cayley table The "multiplication table" that defines a group. Invented by Cayley around October 1853, the Cayley table illustrates the structure and pattern of a group—although it is impractical for dealing with large groups. *See* chapter 8.

Cayley tree A graph structure of vertices and edges that has no closed paths (cycles). Cayley applied their theory to various chemical structures, for example, to the family of *alcohols* (those with the formula $C_nH_{2n+1}OH$). *See* chapters 9, 13.

Cayley-Zeuthen equations Equations that link together singularities of surfaces—analogous to the Cayley-Plücker equations for skew curves. *See* chapter 11. *See also* **Cayley-Plücker equations**.

Cayley's absolute A conic or quadric surface that defines a distance in projective space. In the case of a degenerate conic (consisting of just two ideal points—"imaginary points at infinity"), the distance function is the ordinary Euclidean distance. Thus Cayley demonstrated that Euclidean geometry was a special case of projective geometry and not the other way around as had always been tacitly assumed. Cayley's absolute was seen by Felix Klein as the key to non-Euclidean geometry. *See* chapters 9, 10. *See also* **Cayley-Klein metric**.

Cayley's circular coordinates Points in the plane denoted by pairs $(z, \bar{z})$ of complex numbers. In this coordinate system, the unit circle has the equation $|z|^2 + |\bar{z}|^2 = 2$. These coordinates are now called minimal coordinates.

Cayley's colorgroup   (or Cayley diagram) A diagrammatic way of representing a group. *See* chapter 14.

Cayley's cubic surface   The surface described by an algebraic equation of order 3, which has four nodes, the maximum possible for a cubic surface. *See* chapter 4.

Cayley's formula   (in invariant theory) A general formula for an invariant or covariant of a binary form. *See* chapter 8.

Cayley's law   (in invariant theory) For a binary form, this is the formula $W = (w; \theta, n) - (w - 1; \theta, n)$ for counting the number of linearly independent invariants or covariants of a given degree and order. *See* **partitions, weight**. *See also* chapter 8.

Cayley's partial differential operators   Two specific operators X, Y used to characterize invariants; Sylvester would say that invariants are "annihilated" by X and Y. Semi-invariants are algebraic forms annihilated by *one* of these operators. The operators X, Y may be extended so as to characterize covariants. *See* chapter 7.

Cayley's sextic   A cardioid-like curve that has an Cartesian equation of order 6. *See* chapter 8.

Cayley's theorem   (for algebraic forms) A skew-symmetric determinant of even order is the square of a Pfaffian form. *See* chapter 6. *See also* **Pfaffian form**.

Cayley's theorem   (in elliptic functions) Related to a cubic transformation of an elliptic function. *See* chapter 9.

Cayley's theorem   (in graph theory) The number of distinct labeled trees with n vertices is n^{n-2}. *See* chapter 16.

Cayley's theorem   (in group theory) A pivotal result in this subject: every group may be represented as a permutation group. *See* chapter 14.

Cayley's transform   The orthogonal transformation that solves a variant of the Cayley-Hermite problem. Of use in quantum mechanics and other applications. *See* chapter 9.

Cayley's Ω-process   Method for finding invariants and covariants using the hyperdeterminant operator. This has the form of a determinant with partial differentiation operators as elements. *See* chapters 5, 12. *See also* **hyperdeterminant operator**.

center (of a **Cayley tree**)   The central vertex, c, of a graphical tree (*see* chapter 13, *compare* **bicenter**):

characteristic polynomial   The (standard) polynomial $f(\lambda) = \det(A - \lambda I)$, or, in homogeneous form the binary form $f(\mu, \lambda) = \det(\mu P - \lambda Q)$, where A, P, Q, are $n \times n$ matrices. *See* **Cayley-Hamilton theorem**.

Chasles-Cayley-Brill formula   Under a correspondence between points on a curve, the formula counts the number of self-corresponding (united) points. Chasles's version of the theorem applied only to unicursal curves (those with deficiency $D = 0$) but Cayley's version applied to curves of any deficiency. *See* chapter 11. *See also* **deficiency of a curve**.

Clebsch and Gordan theory   (in algebraic geometry) An important theory of the 1860s which associates abelian functions with curves and surfaces. *See* chapters 11, 14.

close   A region in the plane bounded by a closed curve. *See* chapters 10, 14.

combinant (in invariant theory)   An invariant or covariant of *several* quantics.

commutant   A generalization of the ordinary determinant. *See* **permutant**.

complex numbers   Numbers of the form $a + bi$ in which $i^2 = -1$. Cayley called them "imaginary quantities." *See* **quaternions, Cayley numbers**.

concomitant   A generic term used for an algebraic form that has the property of invariance.

conic   The classic curves obtained as cross-sections of a cone: circle, ellipse, hyperbola, parabola, a pair of straight lines. In analytical geometry, conics are synonymous with second-order equations.

convertible   Symbols a and b are convertible if $ab = ba$. In modern parlance, the operation is "commutative."

covariant   An algebraic form $C(a_1, a_2, \ldots, a_p; x_1, \ldots, x_n)$, that remains unchanged (except for a multiplicative constant) after a given algebraic form is transformed by a linear transformation. Unlike an invariant, a covariant involves variables $x_1, x_2, \ldots, x_n$ (although some authors use the term covariant to include invariant as a special case). The Hessian is an example of a covariant; the Hessian of the binary cubic form, for example, is the covariant Cayley would write as $Hu = (ac - b^2)x^2 + (ad - bc)xy + (bd - c^2)y^2$. *See also* **invariant, Hessian**.

cubic curve   A (plane) curve defined by a third-order algebraic equation $f(x, y, z) = 0$ in three homogeneous variables x, y, z. By setting $z = 1$ the curve in the "finite" plane is obtained and since this involves ten constants, a cubic curve is specified by nine constants (on division by one of them in $f(x, y, 1) = 0$). *See* chapter 11.

cubic surface   A surface defined by a third-order algebraic equation in four homogeneous variables $f(x, y, z, w) = 0$. There are forty-five triple tangent planes to a cubic surface and twenty-seven lines that lie in the surface. *See* **Cayley's cubic surface, triple tangent plane**.

curve of double curvature   *See* **skew curve**.

cusp   An elementary singularity on a curve. Two sections of the curve meet at a cusp and have a common tangent at the cusp point. The cubic curve $x^2 z - y^3 = 0$ has a cusp at the origin—at the point with homogeneous coordinates $(0, 0, 1)$.

deficiency of a curve   The maximum number of singularities that an nth order curve may have is $s = (1/2)(n - 1)(n - 2)$. If it does not have this number, the curve is said to be deficient with deficiency $D = s - \delta - \kappa$, where δ is the number of double points and κ is the number of cusps. Curves with $D = 0$ are called unicursal (for example, an ellipse, cubic curves with a single cusp) and enjoy special properties. Riemann's genus and deficiency coincide for curves with *simple* singularities such as double-points and cusps.

determinant   A number calculated from a matrix. In the usual expansion of a determinant, successive terms alternate in sign. Cayley stressed the difference between a determinant and a matrix. *See* **permanent**.

déterminant gauche   A skew-symmetric square array of symbols.

developable surface   (Cayley's torse) A surface that may be unfolded onto the plane without stretching or shrinking (for example, a cone).

discriminant   An important invariant Δ_n of a of a binary form of order n. It can be used to establish equality between the roots of a polynomial equation.

double point   A simple singularity on a curve. At a double point the curve crosses itself and at the double point the curves has two tangents. *Contrast* **cusp**.

double-sixer   Two disjoint sets of six-lines lying in a cubic surface which have interesting intersection properties.

edge of regression   (or cuspidal edge) A curve that divides a surface into two parts so that planes orthogonal to tangents of the curve intersect the surface in a cusp.

eliminant   *See* **resultant**.

elliptic function   Functions based on the ellipse. They are an extension of trigonometric functions based on the circle and analogous to them. The analogy is reinforced by the adopted notation for the basic elliptic functions of sn u and cn u, which
satisfy the "Pythagorean identity" $\text{sn}^2 u + \text{cn}^2 u = 1$. Whereas trigonometric functions have one period in which they repeat their values, elliptic functions have two
periods. Unlike trigonometric functions they also depend on a parameter k (called
the modulus).

An extensive branch of research in the nineteenth century, elliptic functions can be
defined from elliptic integrals or as the ratio of theta-functions. *See also* **elliptic integral, modular equation, theta function.**

elliptic integral   Typically an integral of the form $u = \int_0^y 1/\sqrt{P(x)}\, dx$, where $P(x)$ is a
polynomial of order 3 or 4.

Fundamental Postulate   A working hypothesis in invariant theory that limited the
number of syzygies of a given degree and order. It turned out to be false in general. *See*
chapters 14, 15.

generating function   Generating functions are the traditional way of solving intricate
counting problems and take the form of a function expandable as a power series. The
method usually involves a great deal of calculation—as was the case in invariant theory.

geometry of position   The branch of geometry concerned with incidence properties
of the lines drawn between points on curves. Typically the lines join points on conics,
from which setup many spectacular theorems arise. Cayley and his generation fully
explored this branch of geometry. *See* **Cayley lines.**

graph   A configuration of vertices linked by edges. They are widely applicable; for
example, to "mathematize" geographical maps in which towns (the vertices) are
joined by roads (the edges). The "graph of a function" frequently met with in mathematics is a different use of the word "graph." *See* **Cayley tree.**

group   A basic algebraic system in which certain rules are observed: elements of the
system are combined to produce elements of the same system, and it is further
assumed that there is an identity element, all elements have inverses, and the associative property holds. *See* chapters 8, 14. *See also* **associative property.**

Gudermannian   A special case of the elliptic function when the modulus $k = 1$. Studied by Christoph Gudermann, the teacher of Karl Weierstrass. *See* chapter 13. *See also*
elliptic function.

hemihedron   A word coined by Cayley to mean a geometrical object obtained from a
polyhedron by removing some of its faces. *See* chapter 14.

Hermite's law of reciprocity   (in invariant theory) It asserts that there is a one-to-one
correspondence between the invariants (*respectively,* semi-invariants, covariants) of
degree θ of a binary form of order n and the invariants of degree n of a binary form
of order θ. *See* chapter 8.

Hessian   A covariant Hf formed from the second partial differential coefficients of an
algebraic form f. The Hessian (so named by J. J. Sylvester) has widespread applications
in mathematics. *See* **Hessian curve.**

Hessian curve   For an nth order curve $f = 0$, the Hessian curve $Hf = 0$ is a curve of
order $3(n - 2)$. It intersects the curve $f = 0$ at its points of inflection. *See* **Hessian.**

homogeneous coordinates   Points in n-dimensional space are given $(n + 1)$ homogeneous coordinates. With these coordinates, all equations are homogeneous. For
instance, the equation of the unit circle is $x^2 + y^2 = 1$ in the ordinary Cartesian coordinates system, but becomes $x^2 + y^2 = z^2$ using three homogeneous coordinates.

hyperdeterminant   The term introduced by Cayley to denote those functions (as they
were then called) with the property of invariance. *See* chapter 4. *See also* **invariant.**

hyperdeterminant operator   An operator framed in terms of partial derivative operators. It can be used to generate invariants. It was written by Cayley as $\overline{12}^k\,uv$. *See* **transvection**.

hyperelliptic functions   (also called Abelian function) The inverse function $v = g(u)$ of the function defined by the hyperelliptic integral. Used in the study of analytical geometry. *See* **Abel's theorem, hyperelliptic integral**.

hyperelliptic integral   Typically an integral of the form $u = \int_0^v 1/\sqrt{P(x)}\,dx$, where $P(x)$ is a polynomial of order *greater* than 4.

invariant   (introduced to replace "hyperdeterminant")   An algebraic form $I(a_1, a_2, \ldots, a_p)$ that remains unchanged (except for a multiplicative constant) after an algebraic form is transformed by a linear transformation. For the algebraic form $ax^2 + 2bxy + cy^2$, $I(a, b, c) = ac - b^2$ is an invariant. Such a form I was said to have the property of invariance. *See* chapter 4. *Contrast* **covariant**.

Jacobian   An invariant or covariant of n algebraic forms expressed as a determinant in which the elements are partial derivatives. This is an example of a **combinant**.

Jacobi-Cayley resolvent   The resolvent equation of order 6 associated with the quintic polynomial equation. It was calculated first by C. Jacobi and then, twenty-five years later, by Cayley. *See* chapter 10.

Lagrange's theorem   (in the calculus) Taylor's theorem is a special case. *See* chapter 3. *See also* **calculus of operations**.

latin square   A square array of n rows and n columns, in which n symbols can be exactly once in each row and each column. Used in the design of experiments. *See* chapter 16. The **Cayley table** is an example of a latin square.

link-work   *See* **plane linkages**.

Mathews-Cayley-Hamilton table   An integer sequence 2, 3, 5, 11, 47, 923, . . . associated with the solution of polynomials, with W. R. Hamilton, and the British mathematician G. B. Mathews. *See* chapter 17.

matrix   A rectangular array of symbols fundamental in linear algebra. Their development began around the mid-nineteenth-century. *See* chapter 8.

modular equation   This expresses the relationship between k, the modulus of an elliptic function and λ, the modulus of its transformed elliptic function. The computation of a modular equation (dependent on the chosen transformation), was usually a long calculation. *See* chapter 13.

node   In analytical geometry, the term used to describe a singularity of a curve or surface. *See* **cusp, double point**.

non-Euclidean geometry   Different types of geometry obtained by varying the "parallel postulate." *See* **Cayley's absolute**.

octave, octonion   *See* **Cayley number**.

partitions   The partition of an integer is the dissection of an integer into its constituent parts. For example, one partition of 11 is $1 + 1 + 3 + 3 + 3$, while another is $3 + 3 + 5$. The total number of *different* partitions of w into θ or fewer integers selected from the range 1 to n is denoted by $(w; \theta, n)$. For example, $(11; 5, 7) = 30$.

permanent   Like a determinant but where each term has a positive sign. *See* **determinant**.

permutant   The generic term for the ultimate generalization of the common determinant. One subclass consists of Pfaffians, another of commutants and intermutants. *See* **commutant, Pfaffian form**.

perpetuant   An irreducible semi-invariant of the binary form of infinite order. Named for their property of "appearing and reappearing" as the leading term in expressions for covariants.

Pfaffian form An algebraic form named after the German mathematician J. F. Pfaff. They are also known as half-determinants. For example:

$$\begin{vmatrix} a & b & c \\ d & e \\ & f \end{vmatrix} = af - be + cd.$$

pippian *See* **Cayleyan.**

plane linkages A system of jointed rods that can move in the plane thereby generating curves. The need to understand mechanical devices in the eighteenth and nineteenth centuries contributed to the topicality of this study. *See* **three-bar motion.**

Plücker's equations Equations that connect the order and class of a plane curve with the number of its simple singularities. *See* chapter 5.

pole and polar The old term for duality in geometry. To a point (pole) in the plane there corresponds a line (polar), which joins the two points where the tangents from the pole meet a given conic. The interchange (the duality) of point and line results in corresponding sets of statements (for example, "two lines determine a point," and "two points determine a line"). "Pole and polar" can be extended to three dimensions, where the duality is between points and planes, and instead of conics, quadric surfaces are substituted.

polyacra Polyhedra whose vertices subtend triangular shaped faces. *See* chapter 17.

polyzomal curves Curves defined by equations involving the square roots of algebraic functions. They consist of lozenge shapes (called zomes) and are typically generated by plane linkage mechanisms. *See* chapter 12.

quantic *See* **algebraic form.**

quartic surface A surface defined by a fourth-order algebraic equation $f(x, y, z, w) = 0$ in four homogeneous variables. For example, $x^2y^2 + x^2z^2 + y^2z^2 + 2xyzw = 0$ is the equation that describes the Steiner surface (also called the "Roman surface"). Other examples are "Kummer's quartic surface" (with sixteen nodes, the maximum possible), "Weddle's surface" (the locus of the vertices of the quadric cones which pass through six points in space), and the "Wave surface" (which arises in Fresnel's wave theory of light).

quaternion A "number" of the form $w = a + bi + cj + dk$ in which $i^2 = j^2 = k^2 = -1$, and $ij = k$. Algebraic in character, they could be used to express rotations in space. After W. R. Hamilton discovered them in 1843, quaternions became a focus for a great deal of research—mostly by Hamilton himself. *See* chapter 4.

rational and integral function *See* **algebraic form.**

reciprocant An analog of invariant and covariant but also involving differential coefficients. An example is the Schwarzian derivative. *See* chapter 16.

regulus *See* **ruled surface.**

resultant (or **eliminant**) The resultant of eliminating variables from a set of algebraic equations—typically expressed as a determinant.

ruled surface (or **regulus**) A surface swept out by the motion of a line in space. The two types of ruled surface are (1) a developable surface (Cayley's torse), and (2) skew surface (Cayley's scroll). *See* **developable surface, scroll.**

scroll Cayley's generic name for a skew surface, the surface swept out by a straight line that continuously changes the plane of its motion. Examples are a hyperboloid of one sheet, a spiral staircase, a helix, the surface of a wood skew. *See* **ruled surface.**

secular acceleration of the moon The nonperiodic (and miniscule) increase in the angular velocity of the Moon's motion around the Earth. *See* chapter 10.

semi-invariant   *See* **Cayley's partial differential operators**.

skew curve   Curves in space, also known as a curve of double curvature, space curve, twisted curve, tortuous curve.

symbolic notation   (in invariant theory) A representation of a binary form by a symbolic expression $(\alpha_1 x_1 + \alpha_2 x_2)^n$. Pioneered by the German mathematicians Paul Gordan and Alfred Clebsch, algebraic forms were expressed in a "condensed" notation and an effective calculus constructed. *See* chapter 12. *See also* **transvection**.

symmetric function   A formal algebraic expression involving the roots of a polynomial equation. For example, $\Sigma\alpha$, the sum of roots. Symmetric functions have been studied by Isaac Newton and legions of mathematicians since the seventeenth century (including Cayley).

syzygy   A linear relation between covariants of a specified degree and order, which are thus "yoked together." The term originated in astronomy, in which three planets are in syzygy when they are in line—typically the moon, earth, and sun in conjunction.

tactic   A term owing to Sylvester to describe the branch of mathematics now known as combinatorics. Derivable from the Greek *tassō* "arrange."

tactions   The "problem of tactions" was described by Cayley as the problem to construct a circle touching three given circles. Derivable from the Latin *tactus* "touch."

Talbot's curve   A species of curve of order 6. Named after the pioneer of photography W. H. Fox-Talbot, a respected Victorian mathematician. *See* chapter 9.

tantipartite   Equivalent to the modern "multilinear."

ternary-quadratic form   An example of a two-part designation of algebraic forms—here, one with three variables (ternary) of order 2 (quadratic).

theta function   Infinite series involving exponential functions. *See* chapters 13, 14. *See also* **elliptic function**.

three-bar motion   A basic link-work mechanism considered by James Watt. Linked at two fixed points, the middle bar (the "coupler bar") is free to move. Points placed at different places on the coupler bar generates different curves, including the "approximate straight line" and the lemniscate. In general, curves of order 6 (sextic curves) are generated. *See* chapter 14.

torse   *See* **developable surface**.

transvection   A way of combining algebraic forms u, v to produce a third $(u\,v)^k$, called the kth transvectant (*Übereinanderschiebung*) of u and v. It was the key to P. Gordan's method of generating invariants and covariants. For example, the Hessian covariant of binary form u was generated as $(u\,u)^2$. Transvection is equivalent to the application of Cayley's hyperdeterminant operator $\overline{12}^k uv$. *See* **Hessian**, **symbolic notation**.

tree   *See* **Cayley tree**.

triple tangent plane   A plane that intersects a cubic surface in three of the twenty-seven lines that lie in the surface. *See* chapter 6.

weight   The weight w of a term in an algebraic form is found by assigning integer values to coefficients and variables. There are several conventions available, but in them the terms of invariants and covariants have the same weight (the "isobaric property").

Abbreviations

A system of abbreviations has been used as a matter of economy and for the reader's ease. There are five types of abbreviations listed here: (I) initials of correspondents for the many letters providing insight into the lives and work of this intellectual circle; (II) abbreviated titles for the collected papers of the various eminent authors cited within this work; (III) initials for the various libraries and archival manuscript collections researched for this study; (IV) general sources referenced frequently; and (V) abbreviated titles for journals cited frequently. Each section below provides a key to the abbreviations and any necessary explanation of how the reader may best understand the abbreviations while reading the Notes.

I. Correspondents

Correspondents are listed in alphabetical order according to their surname, except that Arthur Cayley heads the list.

C	Arthur Cayley
JCA	John Couch Adams
GBA	George Biddell Airy
SA	Siegfried Heinrich Aronhold
EA	E. Atkinson
EHA	Elim Henry d'Avigdor
HFB	Henry Frederick Baker
WWRB	Walter William Rouse Ball
GTB	Geoffrey Thomas Bennett
HB	Hugh Blackburn
GB	George Boole
AB	Arthur Buchheim
JC	John Casey
TC	Thomas Craig
LAC	[Antonio] Luigi G. G. Cremona
CD	Charles Darwin
CED	Charles E. Delaunay
ADM	Augustus De Morgan
PGLD	P. G. Lejeune Dirichlet
WFD	William Fishburn Donkin
RLE	Robert Leslie Ellis
ARF	Andrew Russell Forsyth
FF	Frederick Fuller

FG	Francis Galton
DCG	Daniel Coit Gilman
JWLG	James Whitbread Lee Glaisher
JTG	John Thomas Graves
WRH	William Rowan Hamilton
HH	Hermann L. F. von Helmholtz
JFWH	John Frederick William Herschel
LOH	Ludwig Otto Hesse
DH	David Hilbert
TAH	Thomas Archer Hirst
JDH	Joseph D. Hooker
THH	Thomas Henry Huxley
CJ	Camille Jordan
ABK	Alfred Bray Kempe
TPK	Thomas Penyngton Kirkman
FK	Felix Klein
RL	Rudolf Lipschitz
JNL	J. [Norman] Lockyer
PAM	Percy Alexander MacMahon
JCM	James Clerk Maxwell
RBM	Robert B. Mayor
WJCM	William J. C. Miller
MN	Max Noether
CSP	Charles Sanders Peirce
JP	Julius Petersen
RP	Robert Potts
CGR	Christina Georgina Rossetti
BR	Bertrand Russell
GS	George Salmon
LS	Ludwig Schläfli
DES	David Eugene Smith
HJSS	Henry John Stephen Smith
WS	William Spottiswoode
GGS	George Gabriel Stokes
JJS	James Joseph Sylvester
PGT	Peter Guthrie Tait
JJT	Joseph John Thomson
JT	James Thomson
WT	William Thomson
IT	Isaac Todhunter
RT	Richard Townsend
HWT	Herbert Westren Turnbull
WPT	William Peveril Turnbull
JOT	John Tyndall
JSV	J. S. Vaizey
EV	Edmund Venables
WW	William Whewell
CTW	C. T. Whitmell
PWW	Philip Worsley Wood

II. Collected Works

References to editions of collected papers of mathematicians and scientists are signaled in the Notes by the symbol *P* and in the Bibliography by (P) before the publication date. References to Cayley's papers are cited in the Notes in this form:

"A Memoir on the Theory of Matrices" = [*CP2* (1858): 475–96].

This means that the mathematical paper "Memoir on the Theory of Matrices" published in 1858 can be found in Cayley's *Collected Mathematical Papers* in volume 2, pages 475–96. If required, details of the original journal, volume number, and pagination can be found in the table of contents of the *Collected Mathematical Papers* (but several papers not printed in this collection are listed separately in the bibliography). For another example, James Joseph Sylvester's work is prominent in this biography and the style of citation is similar: "On a Remarkable Discovery in the Theory of Canonical Forms and of Hyperdeterminants" = [*SP1* (1851): 265–83], for which the elements in the citation have the corresponding meaning explained here. The same system applies to the collected works of other scientists and mathematicians. Also, the bibliography entry for a frequently cited collected work includes an annotation indicating to the reader what abbreviated form is used in the Notes.

III. Manuscript Collections

Manuscripts are located in the United Kingdom, unless noted otherwise. Entries are listed here in alphabetical order by abbreviation.

BUMI	Mathematische Institut der Universität, Bonn, Germany
CPS	Cambridge Philosophical Society
CRO	Cornwall Record Office, Truro, Cornwall
CUL	Cambridge University Library, Cambridge
CUNY	Columbia University (Rare Book and Manuscript Library), New York, USA
EPP	Archives of the École Polytechnique, Paris, France
EUL	Archives of the University of Exeter, Exeter
FRC	Family Records Centre, Islington, London
GLMCL	Guildhall Library Manuscripts City of London, City of London
GUL	Glasgow University Library, Glasgow
ICL	Imperial College, London
JHU	Johns Hopkins University, Baltimore, Maryland, USA
LC	Library of Congress, Washington, DC, USA
KCL	King's College, London, Archives
LIN	Lincoln's Inn Library
LMS	London Mathematical Society
LUBL	Brotherton Library, University of Leeds, Yorkshire
NCC	Newnham College, Cambridge
NPG	National Portrait Gallery
NSUG	Niedersächsische Staats und Universitätsbibliothek, Göttingen, Germany
RAS	Royal Astronomical Society, London
RIGB	Royal Institution of Great Britain, London
RGS	Royal Geographical Society, London
RSL	Royal Society of London
RUIM	Instituto Matematico, Università di Roma, Rome, Italy
RUL	Reading University Library, Reading, Berkshire

SJC	St. John's College, Cambridge
SNSP	Scuola Normale Superiore, Pisa, Italy
SPKB	Staatsbibliothek Preussischer Kulturbesitz, Berlin, Germany
St.BPL	St. Bride Printing Library, London
TCC	Trinity College, Cambridge
TCD	Trinity College, Dublin
UCC	The Boole Library, University College of Cork, Ireland
UCL	University College London Manuscripts, London
WSRO	West Sussex Record Office, Chichester, Sussex

IV. General Sources

Reference to the *Dictionary of National Biography* (22 vols., 1885–1901, London: Smith, Elder & Co.) edited by Leslie Stephen and Sidney Lee, its Supplements and additional volumes, will be made in the style *DNB*. Similarly, reference to the *Dictionary of Scientific Biography* (16 vols., 1970–80, New York: Scibner) will be cited using the abbreviation *DSB*. Reference to the *Encyclopaedia Britannica* is abbreviated as *Enc. Brit.*; in studying Cayley's life, the ninth edition is especially useful (24 vols., 1870–89, London: *The Times*).

V. Journals

AE	*Annals of Eugenics*
AES	*Archive of European Sociology*
AHC	*Annals of the History of Computing*
AHES	*Archive for the History of Exact Sciences*
AIHS	*Archive Internationale d'Histoire des Sciences*
AJM	*American Journal of Mathematics*
AM	*Annals of Mathematics*
AMM	*American Mathematical Monthly*
AMS	*Annals of Mathematical Statistics*
AS	*Annals of Science*
BAMS	*Bulletin of the American Mathematical Society*
BAuMS	*Bulletin of the Australian Mathematical Society*
BIMA	*Bulletin of the Institute of Mathematics and its Applications*
BJHS	*British Journal for the History of Science*
BJPS	*British Journal for the Philosophy of Science*
BLMS	*Bulletin of the London Mathematical Society*
BM	*Bibliotheca Mathematica*
BMS	*Bulletin des Sciences Mathématiques*
CDMJ	*Cambridge and Dublin Mathematical Journal*
CMJ	*Cambridge Mathematical Journal*
Coll.MJ	*College Mathematics Journal*
CR	*Cambridge Review*
CRend	*Comptes Rendus*
CUR	*Cambridge University Reporter*
DM	*Discrete Mathematics*
EHR	*Economic History Review*
HE	*History of Education*
HM	*Historia Mathematica*
HPL	*History and Philosophy of Logic*

HS	History of Science
HSPS	Historical Studies in the Physical Sciences
ILN	The Illustrated London News
J. rei. ang. Math.	Journal für die reine und angewandte Mathematik (Crelle)
JGMS	Journal of the Glasgow Mathematical Society
JHA	Journal for the History of Astronomy
JHUC	Johns Hopkins University Circulars
JIA	Journal of the Institute of Actuaries
JLMS	Journal of the London Mathematical Society
MA	Mathematische Annalen
MG	Mathematical Gazette
MI	Mathematical Intelligencer
MM	Messenger of Mathematics
MMLPS	Memoirs and Proceedings of the Manchester Literary and Philosophical Society
MNRAS	Monthly Notices of the Royal Astronomical Society
MQET	Mathematical Questions with their solutions from the "Educational Times"
NRRSL	Notes and Records of the Royal Society of London
ONFRSL	Obituary Notices of Fellows of the Royal Society
PLMS	Proceedings of the London Mathematical Society
PM	Philosophical Magazine
PMLPS	Proceedings of the Manchester Literary and Philosophical Society
PRIA	Proceedings of the Royal Irish Academy
PRSE	Proceedings of the Royal Society of Edinburgh
PRSL	Proceedings of the Royal Society of London
PTRSL	Philosophical Transactions of the Royal Society of London
QJPAM	Quarterly Journal for Pure and Applied Mathematics
QJRAS	Quarterly Journal of the Royal Astronomical Society
Rep. BAAS	Report of the British Association for the Advancement of Science
SEER	Slavonic and East European Review
SM	Scripta Mathematica
TAPS	Transactions of the American Philosophical Society
TCPS	Transactions of the Cambridge Philosophical Society
TPS	Transactions of the Philological Society
TRIA	Transactions of the Royal Irish Academy
TRSE	Transactions of the Royal Society of Edinburgh
VS	Victorian Studies

Notes

Introduction

Epigraphs: *Nature* 33 (7 Jan. 1886): 223 fn; HWT to WWRB, 2 Oct. 1923, TCC, 0.6.6.12. At this time, H. W. Turnbull was professor of mathematics in St. Andrew's, where he set up a research school specializing in invariant theory. A member of a "mathematical dynasty," he went to Trinity College, Cambridge, in 1903 to study mathematics. In a varied career, which included a spell as a lecturer in the University of Hong Kong, at the age of sixty-five he took on the onerous task of editing Newton's correspondence, a project he did not live to complete (Aitken 1962); van der Waerden 1985, 141.

1. Salmon 1883, 481. The basic source for Cayley's life is the magnificent obituary written by his student A. R. Forsyth ("Arthur Cayley" = [*CP8* (1895): ix–xliv]; Forsyth 1895). Cayley's mathematical work is outlined in other obituaries (Roberts 1895; Noether 1895), but these were written before the completion of Cayley's *Collected Mathematical Papers* (Cayley (P) [1889–98] 1963). A modern summary of Cayley's work is North 1971. Popular accounts of his life and work include MacFarlane 1916; Prasad 1933; Bell [1937] 1965. Prasad's account quotes extracts from twenty-nine of Cayley's papers published before he became Sadleirian professor at Cambridge in 1863. The unsigned obituary by C. S. Peirce is inaccurate in parts—Cayley's mother was not Russian and he was married once with two children rather than twice with no children (C. S. Peirce 1895, reprinted in C. S. Peirce 1976, 2: 642–46).

2. Halsted 1897, 164.

3. GB to WT, [15] Sept. 1846, CUL, Add Ms 7432.B155.

4. Leonhard Euler apart, Cayley was as productive as Augustin-Louis Cauchy and the recently deceased Paul Erdös.

5. C. G. J. Jacobi published 113 papers in *Crelle's Journal* whereas overall he published 135 papers (as listed in the *Royal Society Catalogue*).

6. *HilbertP2* (1893): 383.

7. Sylvester 1869, 4; "Address to the Mathematical and Physical Section of the British Association [for the Advancement of Science]" = [*SP2* (1869): 655].

8. Scott 1895, 133.

9. RLE to Lady Affleck (his sister), [1854], Whewell Papers, TCC, Add Ms a. 81.178(1). During the period of 1290 to 1294, Dante Alighieri published thirty-one poems in the *Vita Nuova*, most of which concerned his love for Beatrice.

10. Glaisher 1895b, 174.

11. The mathematician Solomon Lefschetz stayed at Garden House when it became a hotel after Susan Cayley died in 1923. On the occasion, he wrote to Oswald Veblen that "the old boy had pretty good taste you may believe me. It is alongside the Cam. at the Cam. dam falls, with a lovely view, grounds . . ." (Lefschetz to Veblen, 15 May 1931, LAC, Veblen Archive). I am grateful to D. Rogers for this reference.

12. Expressed by Michael Holroyd (Interview, Radio 4 [UK], 1 May 2001).

Part 1. Growing Up, 1821–1843

Epigraph: Salmon 1883, 481. The anonymous teacher who made this observation is possibly Elizabeth May Potticary (c. 1790–1871), the first wife of G. B. F. Potticary, the proprietor of the school in Eliot Place where Cayley went to school for four years. After Potticary gave up the school in 1850, he became the rector of Girton Village Church in Cambridge and may have met George Salmon, who was a frequenter of Cambridge and sermonized there. (In researching this book, my special historical oddity occurred in 1992 when speaking with J. E. Pollard [b. 1907], the *grandson* of G. B. F. Potticary, thus I experienced a living connection with the eighteenth century, when G. B. F. Potticary was born.)

Chapter 1. Early Years

1. Sir William Cayley was on the Royalist side during the English Civil War, but he was saved harassment from Parliamentary supporters by his wife's connections: Dorothy St. Quintin was the eldest daughter of the high-ranking Parliamentarian Sir William St. Quintin of Harpham, East Yorkshire. Far from suffering for being on the "wrong side" during the Civil War, he actually expanded his estate during the twenty-year interregnum (Roebuck 1980, 21).

2. Foster 1874–75, vol. 3; Gibbs-Smith 1962; *Times,* 18 Dec. 1857, p. 7, col. 6; Wood 1965, 17–24.

3. The older motto, "*Nullus quam unus. Per lucem ac tenebras mea sidera sanguine surgent* (Only one. By blood my stars rise through light and darkness)," seems to have been abandoned.

4. Thorpe 1908.

5. "The Reformed House of Commons," by Sir George Hayter, a large-scale painting depicting Parliament in session, which took ten years to complete, is now housed in the National Portrait Gallery in London.

6. Hudson 1974, 23–24, 178, 223–24. E. S. Cayley and G. J. Cayley were descendants of Edward Stillingfleet, Bishop of Worcester. Hugh Cayley, a son of G. J. Cayley, was the subject of a famous portrait (1867) by John Everett Millais.

7. GLMCL, MS11, 194/1 Part 1. Family records of the British in Russia are kept in microfilm at the Guildhall Library, London.

8. This layer of the family's genealogy is confused with the succeeding generation in Forsyth's obituary of Cayley (Forsyth 1895, ix).

9. John Cayley's surviving children were Elizabeth, John, Cornelius, George, Sarah, William, and Henry (the father of Arthur Cayley).

10. A. G. Cross 1997, 29, 63, 84–88. Cross treats John Cayley's leadership of the British community in St. Petersburg and gives a detailed account of his will and its beneficiaries. He establishes the existence of Freemasonry in Russia and analyzes its influence.

11. Cracraft 1969, 226 fn. 16, 229.

12. Storch 1801; A. G. Cross 1973.

13. Storch 1801, 9–10.

14. Grazley 1955.

15. Storch 1801, 8.

16. Cracraft 1969, 219–44. Much of the firsthand information on John Cayley is taken from this remarkable paper.

17. Ibid., 227, 228, 232.

18. Ibid., 231, 235.

19. Ibid., 232.

20. [Anon.] 1795. *Gentleman's Magazine* 65 (Part II): 705.

21. William Huskisson to John Cumming (copy), 20 May 1807, Public Record Office at Kew, London (Granville Papers), PRO.30/29/13/5/174. Henry Cayley to Lord Leveson-Gower, 1st Earl Granville, July 1807, PRO.30/29/13/5/175–78. Huskisson was the member of Parliament who came to a sticky end at the opening of the Liverpool to Manchester railway in 1830.

22. Edward Moberly married Sarah Cayley in St. Petersburg in 1785, and during periods spent in St. Petersburg, he followed his father-in-law John Cayley in becoming British consul. The Moberlys had nine sons and four daughters, some dying in infancy. Their eighth son, George Moberly, was a notable headmaster of Winchester and bishop of Salisbury.

23. Grattan-Guinness 1971, 122–24.

24. Cross 1993, 222–44.

25. *Minutes of the General Court [of the London Assurance Corporation].* GLMCL, MS8730/4, 43–44.

26. Drew 1928, 82.

27. *Kelley's London Guides, Post Office Directories, Robson's London Directories.*

28. Barty-King 1977, 86.

29. Forsyth 1895, x.

30. Simmons 1971, 28.

31. Evans [1845] 1852, 136–37.

32. *Times,* 17 June 1846, p. 6, col. d.

33. Mrs. Susan Cayley's albums were donated to the Fitzwilliam Museum (C to M. R. James, 30 Apr. 1894, CUL, Add Ms 7481.C18).

34. Dru-Drury 1973. The old house at 9 Eliot Place has been replaced by 9 and 9a Eliot Place (1973). Its large garden once extended to the existing Aberdeen Terrace with open fields in the area of today's Granville Park. Descriptions of this old part of Blackheath can be found in Rhind 1976.

35. Dickens [1850] 1996, chap. 5, p. 76.

36. J. P. Ashburnham to his sister, 17 Sept. 1834, quoted in Briggs 1983, 20.

37. The school is still in existence—as St. Pirans at Maidenhead, a private coeducational preparatory school catering for ages 3–13. Its fascinating history, with details of the Potticary era, has been written (Briggs 1983). Many of the details here are drawn from this book.

38. *Greenwich, Woolwich, Deptford Gazette and West Kent Advertiser,* 22 Mar. 1834, p. 1. Wilsher was able to take up to twelve pupils in his establishment.

39. *Greenwich, Woolwich, Deptford Gazette and West Kent Advertiser,* 4 Jan. 1834, p. 1.

40. Salmon 1883, 481.

41. Hilts 1975, 43.

42. *Nominations to the Senior Department* [of King's College, London], KCL, Archives, vol. 2: cert. 71 (8 Oct. 1835), cert. 74 (17 Oct. 1835); *Records of Council of King's College, London.* KCL, Archives, KA/E/A3.

43. For a historical survey of London school provision, see Bryant 1986.

44. Hearnshaw 1929, 40.

45. *King's College [London] Calendar 1835–1836,* KCL, Archives.

46. The birth of the modern monolithic University of London predated the foundation of King's. It started as a rather fragile Joint Stock company "London University" and converted to the "University of London" ten years later as a nonteaching examining body whose function was the award of degrees. This provided an essential route for

Dissenters and non-Anglicans to gain degrees and it became popular with students from University College (Rice 1997, 36–37, 155–62).

47. Barnard [1947], 1961, 84.

48. Kingsley [1850] 1983, chap. 2.

49. John Ruskin to John James Ruskin (father), 10 Mar. 1836, quoted in Hunt 1982, 75.

50. *King's College [London] Calendar, 1835–1836*, KCL, Archives.

51. Smith 1982; Platts [1894] 1909.

52. Hearnshaw 1929, 106.

53. *King's College [London] Calendar 1839–1840*, KCL, Archives.

54. Bowers 1975, 206.

55. *Nominations to the Senior Department* [of King's College, London], KCL, Archives, vol. 3: cert. 77 (1 Oct. 1838); *Records of Council of King's College London*, KCL, Archives, KA/E/A3.

56. Kingsley [1876] 1880, 1: 20. In this reference, "Archdeacon Brown" should read "Archdeacon Browne."

57. Browne 1851, 2: 235, 237.

58. *King's College [London] Calendar 1835–1836*, KCL, Archives.

59. *King's College [London] Calendars, 1833–1834, 1839–1840*, KCL, Archives.

60. Hall 1837. Hall's *Differential and Integral Calculus* (1837) was wide-ranging and covered advanced topics such as Taylor's theorem (for functions of both one and two variables), the application of calculus to the study of curves, curvature, and the singularities of curves. Integration was introduced as the inverse process to differentiation and differential equations (including second-order linear differential equations) were also treated. Following Lagrange, Hall introduced *derivatives* as the coefficients appearing in Taylor's expansion of a function—a point of view Cayley defended throughout his life. Cayley's understanding of a derivative remained an "algebraic" one though he was an avid reader of Cauchy's work.

61. Allen 1922.

62. Siddons 1936, 11; Brock 1975, 22.

63. A comparison between the mathematics taught by A. De Morgan and T. G. Hall is contained in a survey of post-school mathematics teaching in Victorian London made by Adrian Rice (Rice 1996).

64. "Presidential Address to the British Association, Southport, Sept. 1883" = [*CP*11 (1883): 429–59].

65. Hearnshaw 1929, 127.

66. Rigg [1897] 1909.

67. *King's College [London] Calendar 1837–1838*, KCL, Archives.

68. Rossetti 1906, 175.

69. *King's College [London] Calendar, 1840*, KCL, Archives.

70. The namesake and distant relative was the son of the owner of the Archangel trading firm Arthur Cayley and Co., to whom Henry Cayley's brother William was apprenticed (Cracraft 1969, 227 fn. 21).

71. *Guardian* (6 Feb. 1895): 201.

Chapter 2. A Cambridge Prodigy

1. Hudson 1842, 8.

2. Duncan F. Gregory to GB, 29 Mar. 1840, in MacHale 1985, 53.

3. *University Residents 1824–1913*, TCC. The large number (145) of students who entered Trinity College in 1838 can be compared with the total number of students (435)

formally admitted throughout the entire university (Ball and Venn 1911–16, 7). Of the seventeen constituent colleges of the university, Trinity and St. John's Colleges were the largest colleges by far. For a history of Cambridge University 1750 to 1870, see Searby 1997.

4. C. Wordsworth was the younger brother of the poet William Wordsworth.

5. Teaching at Cambridge during Cayley's undergraduate years was entirely college-based. Until the late 1860s, there were no intercollegiate lectures in the university and, apart from teaching by the system of private coaches, all classes and tutorials took place in the individual colleges.

6. Rouse-Ball 1921, 35. In Whewell's correspondence with his friend Hugh James Rose, the principal of King's College, London, there seems to be no reference to Cayley, the young prodigy about to begin his studies at Trinity College. Entrants to Cambridge from King's College, London, were generally strong mathematically; it was the leading school with 25 of the 220 who appeared in the wrangler class during 1830–60 (Becher 1984, 101).

7. Clark [1895] 1909.

8. Shipley 1913, 37.

9. Thorp 1841, 4.

10. Ibid., 6–7, 8.

11. The rule that required candidates for the classics degree to first acquit themselves in mathematics by appearing on the order of merit prevented several fine classics students from gaining an honors degree at all. Sons of the nobility were exempt from the requirement and the rule itself was relaxed in 1849 by only stipulating a first class in the ordinary degree as a qualification. Prequalification in mathematics was completely abandoned in 1854, and from 1857 onward students could proceed to the classics degree directly (Tanner 1917, 353).

12. Kingsley [1876] 1880, 1: 32 fn.

13. The high reputation of Trinity College in classics at the beginning of the nineteenth century gradually changed to a leadership in science by the end of the century, in step with the college's transition from a monastic institution to a secular one (Trevelyan 1946).

14. Pycior 1981. George Peacock was elected to the Lowndean chair in 1837 and appointed Dean of Ely in 1839 (Clark [1895] 1909, 583).

15. Peacock 1842, in the dedication.

16. Winstanley 1955, 409.

17. *Testimonials in Favor of Hugh Blackburn*, 1849, GUL.

18. Galton 1908, 75. King's Court was built at the beginning of George IV's reign as an overflow to Trinity's Great Court and Nevile's Court.

19. RLE Diary, 3 Dec. 1838, TCC, Add Ms a.219.1; RLE Diary, 29 May 1839, TCC, Add Ms a.82.1.

20. Joseph Romilly was a close friend of George Peacock. He graduated fourth wrangler in 1813 and was appointed registrary of the university in 1832.

21. Bury 1967, 192. B. Shaw and J. F. Fenn had been academic prizewinners at King's College, London, Fenn in mathematics and Shaw in classics. Cayley's election to a Trinity College fellowship in 1842 was followed by their own elections, Shaw in 1843 and Fenn in 1844. Shaw was called to the bar at Lincoln's Inn, practiced as a special pleader, 1847–60, and became a noted ecclesiastical lawyer, while Fenn followed a clerical career. The notable Latin scholar H. A. J. Munro was awarded a Trinity College scholarship in the same tranche as Cayley.

22. Trevelyan 1946, 99.

23. Robson and Cannon 1964; Garland 1980; Fisch and Schaffer 1991.

24. *Trinity College Examinations Book, 1836–1861,* TCC. Cayley was far in advance of the other students at Trinity College *in mathematics* as is indicated by the eventual positions of his closest competitors in the Tripos order of merit. In 1842 Reginald [Guy] Bryan was 23d wrangler, J. F. Fenn was 18th wrangler, B. Shaw gained a first-class degree in classics but was not a top wrangler, B. F. Smith was 17th wrangler. T. Cubitt, son of the builder and civil engineer William Cubitt, died before the Tripos examination took place.

25. Hilts 1975, 43.

26. RBM to Charlotte Mayor (mother), 26 Mar. 1840, TCC, Mayor B/8/195.

27. S. P. Thompson 1910, 1: 34.

28. If a Cambridge graduate went into schoolteaching, the honor of his ranking in the order of merit passed down to his school pupils. They would recall with pride that they had been taught by, say, a "fourteenth wrangler at Cambridge," and the school itself enjoyed reflected kudos.

29. Neville 1942, 239.

30. Kingsley [1876] 1880, 1: 36.

31. Kingsley [1850] 1983, chap. 13. These passages were excised from *Alton Locke* (1850) when C. Kingsley became professor at Cambridge in 1860.

32. Ball 1912, 320; Anderson [1891] 1908; Smyth 1867.

33. Bury and Pickles 2000, 167.

34. Hopkins 1854, 28, 30.

35. *Gentleman's Magazine,* Nov. 1866, p. 706.

36. FG to S. T. Galton (father), 11 Nov. 1841, in Pearson 1914–30, 1: 163.

37. Hopkins 1841, 24.

38. *TaitP1* (1876): 263.

39. Hopkins 1841, 10.

40. "On the Properties of a Certain Symbolical Expression" = [*CP1* (1841): 5–12]; "On Certain Definite Integrals" = [*CP1* (1841): 13–18]; "Notes and References" = [*CP1* (1889): 581–82].

41. RBM to Charlotte Mayor (mother), fragment, 15 Oct. 1841, TCC, Mayor B/8/204.

42. *Cambridge Advertiser and Free Press,* Extraordinary Edition, 21 Jan. 1842, no. 161.

43. Examination questions were divided between problem questions and bookwork questions. Of the thirty-three hours of examination in Cayley's Tripos examination, twenty-four and one-half were allotted to "propositions from books" and eight and one-half to "problems." The problems were regarded as the most testing part of the examinations. The printed examination paper presented candidates with a list of questions of varying length (between nine and twenty-four questions) with no limit on the number of questions which could be answered. The custom of allowing mathematics students in examinations to answer as many questions as they wished, and thus amass marks (rather than gain a percentage score), was still current in some British universities in the 1960s.

44. Roscoe 1906, 29.

45. J. J. 1844, 83.

46. Forsyth 1895, xxvi.

47. In pure mathematics, for example, the topics in the Tripos examination which proved signposts for Cayley's research included the singular solutions of differential equations, the calculus of operations, Bernoulli numbers, and Euler's theorem for polyhedra ($F + V - E = 2$).

48. Whewell 1833, 5.

49. Whewell adopted the description "Laplace's coefficients" from his reading of the *Mécanique céleste* (see Grattan-Guinness 1990, 1: 333).

50. Pell 1908, 71–72.

51. Roth 1971, 228. There seems to be some truth in the "Cayley story" that he finished an examination paper in less than the time allotted, though the details are uncertain. In another version, Cayley looked up from his footbath and remarked: "Likely enough he [C. T. Simpson] did. I floored it myself in two hours and a half" (Halsted 1895, 103).

52. Glaisher 1895, 174. According to W. W. Rouse Ball, an authority on the mathematical Tripos, *no vivâ voce* examination took place in the period 1833–48 (Ball 1921, 296–97). I am inclined to believe, however, that one was held in 1842, since Cayley himself told Glaisher the story of his own *vivâ voce* examination.

53. Salmon 1883, 481. Salmon noted that this story of Cayley's sangfroid was "best authenticated."

54. Journal of John Couch Adams, 21 Jan. 1842, SJC, Adams Papers, Box 25.

55. *Times,* 22 Jan. 1842, p. 5, col. 2.

56. Halsted, 1895, 103; RBM to his parents, 22 Jan. 1842, TCC, Mayor B/7/41.

57. FG to S. T. Galton (father), 21 Jan. 1842, in Pearson 1914–30, 1: 164. Like Cayley, C. T. Simpson became an Equity draftsman and conveyancer. Unlike Cayley, he made no attempt to carry on with mathematics "lest its spell should divert his attention from his professional work" (*The Eagle* [St John's College, Cambridge] 1901, 23: 359–60).

58. Journal of John Couch Adams, 22 Jan. 1842, SJC, Adams Papers, Box 25.

59. WT to JT (father), [Feb./Mar.] 1842, CUL, Add Ms 7342.T201.

60. Arnold 1873, 195–96.

61. Bristed 1852, 95.

62. WW to J. C. Hare, 13 Mar. 1842, in Whewell 1881, 264.

63. The annual competition for the Smith's Prizes involved the most successful students from the mathematical Tripos examination sitting another set of written examinations taking place each day for a week. The format of the "Professor's examination" was virtually the same as the Tripos, normally twenty-four questions on each examination paper, but they were more wide-ranging. The top student and the runner-up in these examinations were designated First Smith's Prizeman and Second Smith's Prizeman of their year.

64. Ross 1994, 468. Dr. Robert Smith was master of Trinity College from 1742 to 1768. Smith's own scientific reputation rested on his *Harmonics* (1749); Barrow-Green 1999.

65. Clark 1904, 95. Also, the *Cambridge Advertiser and Free Press,* Extraordinary Edition, 25 Feb. 1842, no. 166.

66. Bury and Pickles 2000, 196 fn. 5.

67. ADM to WW, 25 Jan. 1842, TCC, Add Ms. a.202.99.

68. Peacock 1834, 316.

69. The Ellis-Stokes-Cayley-Adams Senior Wrangler quartet was followed by the less-distinguished one of Hemming-Parkinson-Hensley-Wilson. G. W. Hemming pursued a law career, S. Parkinson became a college don and wrote textbooks, L. Hensley followed a church career and became a noted writer of hymns, and W. P. Wilson became professor of mathematics at the University of Melbourne, Australia.

70. CUL, University Archives, Min.V.7, p. 3; Becher 1995, 425.

71. "On a Theorem in the Geometry of Position" = [*CP*1 (1841): 1–4].

72. Although the paper is on "determinants," nowhere does Cayley use the specific term.

73. Coolidge [1940] 1963, 166. A. Cayley, "On a Theorem in the Geometry of Position" = [*CP*1 (1841): 1–4]; "Notes and References" = [*CP*1 (1889): 581].

74. In the case of four points lying on a circle, Cayley's determinant can be factored into the four factors $\overline{12.34} \pm \overline{14.23} \pm \overline{13.24}$, so that his result contains Ptolemy's quadrilateral theorem as a special case (Pritchard 2003, 42–49).

75. Thomas Muir, who devoted much of his energy to the study of determinants, wrote enthusiastically about Cayley's undergraduate paper: "we have for the first time in the notation of determinants the pair of upright lines so familiar in all the later work. The introduction of them marks an epoch in the history, so important to the mathematician is this apparently trivial matter of notation" (Muir 1906–23, 2: 6).

76. Aitken 1950, 72.

77. In his solution of the problem of expressing the relations which exist between points in space, Cayley made use of the double-subscript notation. This was unusual for him, and as a matter of principle he generally avoided the use of subscripts in later work. This aversion turned out to be a severe limitation and resulted in an inadequate notation for dealing with n-dimensional geometry effectively.

78. "Examples of the Dialytic Method of Elimination as Applied to Ternary Systems of Equations" = [$SP1$ (1841): 61–65].

79. "On Staudt's Theorems Concerning the Contents of Polygons and Polyhedrons with a Note on a New and Resembling Class of Theorems" = [$SP1$ (1852): 382–91]. Drawn to reconsider this classic determinant in old age, Sylvester and Cayley saw new interpretations for it (JJS to C, 7 May 1885, SJC, Sylvester Papers, Box 12).

80. "On the Rationalisation of Certain Algebraical Equations" = [$CP2$ (1853): 40–44].

81. "Note on the Value of Certain Determinants the Terms of Which Are the Squared Distances of Points in a Plane or in Space" = [$CP4$ (1860): 460–62]; Salmon 1860.

82. "On Some Formulae Relating to the Distances of a Point from the Vertices of a Triangle and to the Problem of Tactions" = [$CP4$ (1862): 510–12]. The problem of tactions is one of constructing a circle *touching* given systems of circles.

83. *Educational Times*, vol. 8, pp. 86–87 = [$CP7$ (1868): 585–87].

84. "Note on the Two Relations Connecting the Distances of Four Points on a Circle" = [$CP12$ (1888): 576–77]; "Note on the Relation between the Distance of Five Points in Space" = [$CP12$ (1889): 581–83].

85. In the only textbook he wrote, Albert Einstein used the example of "five points in space" to argue that it was not possible to select points in space *arbitrarily* (Einstein [1922] 1976, 7–8).

86. Sturmfels 1993, 17. Cayley's determinant expressing the relationship between five points in space is a Euclidean invariant treated in distance geometry (Blumenthal 1953 and 1961). It is referred in the twentieth-century literature as the Cayley-Menger determinant because of its use in Karl Menger's characterization of the Euclidean metric (Blumenthal and Gillam 1943).

Chapter 3. Coming of Age

1. "On the Properties of a Certain Symbolical Expression" = [$CP1$ (1841): 5–12]; "On Certain Definite Integrals" = [$CP1$ (1841): 13–18]; "Notes and References" = [$CP1$ (1889): 581–82].

2. CPS *General Meetings Minutes*, 1842. The Cambridge Philosophical Society was incorporated by Royal Charter (1832) and, in 1847, Prince Albert agreed to be its patron.

3. Laurence Sterne was greatly influenced by John Locke the author of the *Essay concerning Human Understanding* (1690).

4. Cayley was assistant tutor from 1843 to 1846. The assistant tutor was appointed and paid by the senior tutors of their "side" of Trinity College. They eased the senior tutor's

workload, took an active part in the life of the college, and occupied a recognized position in the college hierarchy (Ball and Venn 1911–16, 1: 23–24). The duties of an assistant tutor could be light and Forsyth suggests they were even nominal in Cayley's case; it is very likely that he did not lecture regularly in the college (Forsyth 1895, xi).

5. Galton 1908, 72.

6. E. Venables, *Guardian* (6 Feb. 1895): 200.

7. Courtney [1899] 1909. During a period spent as minister on the Isle of Wight, Edmund Venables was on friendly terms with Alfred Tennyson, the other Laureate of the Victorian Age.

8. The reading party in Scotland included C. B. Yeoman, J. Kay (the future economist), and E. E. Kay (a lawyer), younger brothers of Sir J. Kay Shuttleworth (the noted Victorian educational reformer), T. F. Buxton, and his brother C. Buxton, who became a well-known liberal MP and hunting friend of Anthony Trollope. All except J. Kay were Cayley's neighbors in Trinity's King's Court.

9. FG to S. T. Galton (father), 19 June 1842, in Pearson 1914, 1: 168. Being a "brick" (1840, *Oxford English Dictionary*) signaled approval for having such sound qualities as openness and sincerity.

10. *Guardian* (6 Feb. 1895): 201.

11. EV to GGS, 7 July 1842, CUL, Add Ms 7656.V14.

12. Ibid. In adopting the symbol Δ and the "topsy-turvy" ∇, Cayley was following Laplace and the French school involved with the calculus of operations. For a detailed outline of the use of these symbols, see Panteki (1991, 24–46). In his early papers, Cayley used ∇ in several different ways, and he also experimented with "topsy-turvy" J's and F's.

13. Pearson 1914, 1: 169.

14. Galton 1908, 72.

15. R. L. Stevenson 1886, *Kidnapped*, chap. 22.

16. The practice of electing fellows of Trinity on the basis of written examinations was continued until 1873. After this, some fellowships were awarded to candidates for dissertations or on the basis of published work, a move aimed at fostering research.

17. Winstanley [1940] 1955, 424.

18. The existence of fellowships as positions of "great respectability" that carried no obligation to produce work or give lectures was noted by V. A. Huber as one which had no counterpart in the German university system (Huber 1843, 2: 325–29).

19. Cayley was elected a Minor Fellow of Trinity College (3 Oct. 1842) and a Major Fellow (2 July 1845).

20. Of fellows elected in the same batch in 1842, T. Preston became professor of Arabic; E. M. Cope, a lifelong tutor at Trinity, was almost elected professor of Greek; C. W. King resided at Trinity and in a withdrawn existence became absorbed in his collection of antique gems, which were eventually bought by the Metropolitan Museum of Art in New York; R. Watt eventually became a clergyman; H. A. Marsh, a clergyman, became master of Corpus Christi College; R. P. Mate followed a legal career; and Tom Taylor became a notable dramatist of the theater and editor of *Punch*.

21. Bronwin 1843a.

22. Cayley 1843a, 197. This was Cayley's first published paper on elliptic functions, a subject on which he was to publish extensively. Unlike his "new subjects" (invariant theory, quaternions), elliptic integrals and elliptic functions had been studied since the mid-eighteenth century (Ayoub 1984).

23. Bronwin 1843b, 262, dated 5 Jan. 1843.

24. There was no room for political correctness in scientific communication in the early Victorian journals—as present-day readers quickly recognize.

25. Cayley 1843b, 360, dated 13 Apr. 1843.

26. Bronwin, 1844; Cayley 1845.

27. Brice Bronwin to Sir John Lubbock, 19 Feb. 1849, RSL B.475. I am grateful to Maria Panteki for this reference.

28. RLE to WT, [Feb. 1845], CUL, Add Ms 7342.E52.

29. "On the Transformation of Elliptic Functions" = [CP1 (1845): 132]. Boole criticized Brice Bronwin's work on differential equations as being less than "complete" though he could not fault it as being incorrect in general (Smith 1984a, 63–64, 71–74).

30. Forsyth 1995; "Arthur Cayley" = [CP8 (1895): xxvii].

31. "On the Intersection of Curves" = [CP1 (1843): 25].

32. By modern standards, the theorem on the intersection of curves was not rigorously proven by Cayley. Without justification, he assumed that an arbitrary curve passing through the eight points of intersection had a specific form (see Noether's theorem, chap. 13).

33. Thalberg 1927. Plücker's theorem is the case $n = l = m$, and Plücker-Jacobi's theorem deals with the case $n = l > m$.

34. The theorem stated by Cayley on the intersection of plane curves, is widely known as the "Cayley-Bacharach theorem" and has given rise to a vast body of knowledge (for example, Macaulay 1895; Semple and Roth 1949, 97–99; Baker 1933, 5: 99–107; Eisenbud, Green, and Harris 1996).

35. "On the Intersection of Curves" = [CP1 (1843): 27]; "Notes and References" = [CP1 (1889): 583]; "On the Intersection of Curves" = [CP12 (1887): 500–504]. Cayley's theorem needs $n \geq l$, m and $n < l + m$.

36. From a modern perspective, Cayley made both unwarranted and unstated assumptions. As with his approach to other questions in algebra and geometry, the theorem is only "generically true," that is, true in "almost all" cases; for example, it fails in the case of two quartic curves in which the remaining three points are collinear. Isaac Bacharach dealt with exceptional cases to the theorem forty years later (Bacharach 1886). Cayley first studied Bacharach's paper around March 1887 but maintained that the work was an addition to his statement of the theorem rather than a correction. To escape from cases that posed counterexamples, Cayley would occasionally use the phrase that curves were to be "taken at random" in order to *bar* such counterexamples to use a "Lakatosian" word. Cayley regarded such counterexamples as rare and therefore not worth considering. A polynomial equation of degree n "taken at random" would be assumed to have n distinct roots, for example. Similarly, matrices were assumed to be invertible "in general." In hearing lectures given by Cayley and Sylvester at Johns Hopkins University in the early 1880s, C. S. Peirce was critical of this mode of reasoning ("The Logic of Quantity" = [PeirceP4 (1881): 119]). This "generic reasoning" pervades Cayley's work; an analysis of his work in matrix algebra in the light of generic reasoning has been made by T. Hawkins (Hawkins 1977a).

37. "On the Motion of Rotation of a Solid Body" = [CP1 (1843): 28–35].

38. CPS General Minutes, 1843.

39. C to GB, 13 June 1844, TCC, R.2.88.1. A succinct account of n-dimensional determinants is given in Oldenburger 1940.

40. "On the Theory of Linear Transformations" = [CP1 (1845): 80–94].

41. Klein [1908] 1939, 143.

42. "Demonstration of Pascal's Theorem" = [CP1 (1843): 43–45]; Baker 1922, 2: 219–36.

43. Panteki 1987, 122. A neglected contributor to the calculus of operations is the Irish mathematician who graduated at Cambridge, Robert Murphy [Murphy 1836].

44. WT to JT (father), 5 Nov. 1841, CUL, Add Ms 7342.T183.

45. WT to JT (father), [Feb/Mar.] 1842, CUL, Add Ms 7342.T201. Quoted in Smith and Wise, 1989, 102.

46. Thomson 1874, iii.

47. Salmon 1883, 481.

48. CUL, Add Ms 7342.NB29.

49. Ellis 1844.

50. Gregory 1840; Ewald [1996] 1999, 323.

51. Gregory 1841, 1.

52. Thompson 1910, 1:31.

53. "On Linear Transformations" = [CP1 (1846): 95].

54. The step from ordinary algebra to symbolical algebra was argued along the lines: If the expression $(a + b)^n$, where a, b are symbols denoting "pure quantity," could be expanded by the binomial theorem, then it should be possible to apply the same reasoning to $(D + b)^n$, where D is the symbol of differential operation, which, by the "separation of symbols" can be separated from the function on which it operates (Grattan-Guinness 1990, 1: 211–19).

55. Gregory 1845.

56. Archibald Smith was one of the mathematicians to set this objective of "symmetry" in the early numbers of the *Cambridge Mathematical Journal.* Not only Smith and Gregory but other writers of articles in this journal sought the symmetry in their work on analytic geometry, a characteristic probably learned from earlier French writers. The presentation of such equations as the straight line in a balanced form (their equations involved only ordinary Cartesian coordinates and direction cosines) was the symmetry they desired.

57. Rawnsley 1896, 60.

58. RLE to C, [1846] CUNY, D. E. Smith Papers.

59. Ellis 1863, 272.

60. Glaisher 1895b.

61. ADM to JFWH, 28 May 1845, in De Morgan 1882, 151.

62. Grattan-Guinness 1990, 1: 185–268; Koppelman 1971; Panteki 1992.

63. "On Lagrange's Theorem" = [CP1 (1843): 40–42].

64. Peacock 1834, 312.

65. This occurred in the binomial expansion $(a + b)^n$, in which the value of n was not a natural number ("On a Theorem for the Development of a Factorial" = [CP2 (1853): 101]); Richards 1987, 14–15.

66. Peacock 1834, 283.

67. Clark and Hughes 1890, 2: 37.

68. *WhewellP*15 2001, 153.

69. WW to JFWH: 20 Aug. 1845, in *WhewellP*16 2001, 328.

70. As a Trinity College scholar, Cayley was entitled to reside in the college after graduation. He moved from King's Court into Room 4, staircase Z in Nevile's Court.

71. Eperson 2000, 9.

72. Babbage 1830, 1.

73. It is significant that much of the mathematical work of De Morgan, Peacock, and J. F. W. Herschel has been forgotten, yet many of the concepts and problems which concerned Cayley are still common currency in modern mathematics.

74. Kauvar and Sorenson 1969.

75. Hilts 1975, 43.

76. Montagnier 1918–19, 44.

77. John Ruskin to his parents, 1845, in Hunt 1982, 152.

78. Montagnier 1918–19, 51.

79. Ruskin 1903, 2: 382.

80. Another traveling party comprising C. G. J. Jacobi, J. P. G. Lejeune-Dirichlet, C. W. Borchardt, and L. Schläfli arrived in Rome in mid-November, 1843. J. Steiner was also residing there at this time. I have found no evidence that Cayley met any of these mathematicians on this occasion, but it is intriguing to believe there was a chance of his doing so.

81. Street [1855] 1874, 1; G. E. Street, a pupil and disciple of George Gilbert Scott (England's leading Gothic architect, influenced by the activities of the Camden Society) wrote *Brick and Marble in the Middle Ages,* which remained one of Cayley's favorite books.

Part 2. New Vistas, 1844–1849
Epigraph: C to GB, 3 Dec. [1845], TCC, R.2.88.22.

Chapter 4. A Mathematical Medley

1. "On the Theory of Algebraic Curves" = [*CP*1 (1844): 46–54]. This paper focused on the intersection of those curves with a certain number of asymptotes and Cayley treated the relatively simple case in which this number equals the order of the curve. Plücker 1837. Cayley traced the historical development of the geometrical theory of involution from Euler (1748) through to the then current analytical approach of J. Plücker and L. Hesse. He thought the *precision* of Plücker's results in one paper needed further attention ("On the Theory of Involution in Geometry" = [*CP*1 (1847): 259 fn.]).

2. "Chapters in the Analytical Geometry of (n) Dimensions" = [*CP*1 (1844): 55–62], quote on p. 55. Erhard Scholz makes the point that Cayley's geometrical language of n-dimensional geometry was limited to the title of the paper (Scholz 1999, 25).

3. "On Quaternions; or a New System of Imaginaries in Algebra" = [*Hamilton*P3 (1844): 108]. "On Certain Results Relating to Quaternions" = [*CP*1 (1845): 123].

4. "On a Class of Differential Equations, and on the Lines of Curvature of an Ellipsoid" = [*CP*1 (1843): 37].

5. Green 1835.

6. In August 1844, Hermann Grassmann published his *Ausdehnungslehre,* which contains the seeds of important ideas in linear algebra. I have found no specific evidence to suggest Cayley was aware of the importance of this book until it was generally recognized by British mathematicians in the 1870s. Cayley mentioned it in his British Association Address in 1883 (see chap. 15) and a subsequent (unfinished) survey: "On Multiple Algebra" = [*CP*12 (1887): 459–89].

7. "Sur quelques Théorèmes de la Géométrie de Position" = [*CP*1 (1846): 321].

8. Kline 1972, 858–59.

9. "On a Theorem in the Geometry of Position" = [*CP*1 (1841): 2]. "Note on a System of Imaginaries" = [*CP*1 (1847): 301]. "Sur quelques Propriétés des Déterminants Gauches" = [*CP*1 (1846): 332–36].

10. FF to WT, 16 Feb. 1845, CUL, Add Ms 7342.F290. F. Fuller's allusion to "$n + \frac{1}{2}$" is in reference to material published in Cayley's "Sur quelques integrals multiples" = [*CP*1 (1845): 195–203] and "Addition a la Note sur quelques integrals multiples" = [*CP*1 (1845): 204–6].

This work on potential theory, derived from Cayley's second and third papers written while he was an undergraduate.

11. Brock and Meadows 1984, 87.

12. WT to JT, 2 June 1844, CUL, Add Ms 7342.T264; Thompson 1910, 1: 79.

13. "Mémoire sur les Courbes du Troisième Ordre" = [CP1 (1844): 183–89]. "Notes and References" = [CP1 (1889): 586]; Baker 1933, 3: 214.

14. Cayley treats the cubic surface with the maximum number of four double points (nodes). Through these four nodes lie six lines which lie wholly within the surface. The six lines (each counted four times) plus three extra lines make the 27 lines which lie wholly in the surface. The equation of this surface may be written as an algebraic form in four homogenous variables of order 3, $\alpha yzw + \beta zwx + \gamma wxy + \delta xyz = 0$, and in the language of invariant theory the algebraic form is described as a "quaternary-cubic." Thus invariant theory informs analytical geometry. This particular surface (Cayley's cubic surface) is the reciprocal surface of J. Steiner's Roman (quartic) surface, which Cayley later modeled (see chap. 13).

15. GB to Joseph Hill, 5 May 1840, UCC, BP1/221/7(2).

16. Taylor 1956, 50.

17. Boole 1841.

18. C to GB, 13 June 1844, TCC, R.2.88.1.

19. Stevenson 1837. Richard Stevenson was third wrangler in 1834 and elected a fellow of Trinity College in the following year. He died of consumption in Beeston near Nottingham on 27 September 1837 aged 25 years (Gooden 2003; Crilly 2004).

20. The notion of Galilean invariance in classical mechanics is discussed and clarified in McKinsey and Suppes 1955. For a historical overview of the role of invariance in modern physics, see North 1965, 61–63.

21. Browne 1851, 1: 237, 242.

22. In the case of quadratic $ax^2 + 2bx + c$, for instance, the invariant $ac - b^2$ is linked to the "squared differences of the roots." That is, if α, β are the roots of $ax^2 + 2bx + c = 0$, or of any linearly transformed equation, then $ac - b^2 = (-a^2/4)(\alpha - \beta)^2$.

23. Peirce 1976, 2: 642.

24. A masterful survey of the history of *algebraic forms* over the course of the nineteenth century is given in MacMahon 1910a.

25. Numerical examples of this are $3x^2 + 4x + 20$ and $3x^2 + 22x + 59$ (in the first $a = 3, b = 2, c = 20$ and in the second $a = 3, b = 11, c = 9$). The remarkable fact is that the value of the expression $ac - b^2$ is the same in both—the common value being 56 and the reason for this is that $3x^2 + 22x + 59$ is obtained from $3x^2 + 4x + 20$ by replacing x by $x + 3$. Thus $ac - b^2$ is called an *invariant* since it does change when the variables are transformed in this way. The reader is warned that Cayley would not have considered numerical values. For him, a, b, c were part of symbolical algebra and were thus kept quite general.

26. "Allgemeine Auflösung der Gleichungen von den ersten vier Graden [General solution of equations of the first four degrees]" = [EisensteinP1 (1844): 7–9].

27. "Über eine merkwürdige identische Gleichung [On a remarkable identical equation]" = [EisensteinP1 (1844): 26–27]; "Note sur deux Formules données par M. M. Eisenstein et Hesse" = [CP1 (1845): 113–16].

28. "On the Theory of Linear Transformations" = [CP1 (1845): 93].

29. "On the Developable Derived from an Equation of the Fifth Order" = [CP1 (1850): 501]. The discriminant Δ_5 was calculated by George Salmon as a contribution to Cayley's research.

30. George Salmon calculated Δ_6 in the 1860s (Salmon 1866, 205–7).

31. In the 1840s, Cayley used a variety of terms for invariants: *hyperdeterminant, transforming function, derivative*. The terms *invariant* and *covariant* were introduced in 1851 by J. J. Sylvester.

32. Murphy 1837. R. Murphy was born in Mallow, Ireland, where at the age of eleven he was hailed as a "second Newton." He studied at Cambridge, where he was third wrangler in 1829. In 1834 he was elected F.R.S. He wrote the *Elementary Principles of the Theories of Electricity Heat, and Molecular Actions* (1833) and *Treatise on the Theory of Algebraical Equations* (1839). Drawn into drinking and gambling, he was forced into a life of poverty and died in London in 1843 at the age of thirty-seven.

33. "Solubility" was one question asked of the quintic equation, but a leading question which interested British mathematicians centered on the reality of its roots and their location.

34. C to GB, 13 June 1844, TCC, R.2.88.1.

35. Boole 1844.

36. Boole actually considered *orthogonal* linear transformations. He obtained the discriminant of a binary form by elimination of x, y from $\partial u/\partial x = 0$; $\partial u/\partial y = 0$. Cayley's corresponding differential equations were obtained for the *multilinear* function U. He chose not to use the now-standard notation for partial derivatives, which had been used by C. G. J. Jacobi in 1841, but instead, he wrote the equations in Cauchy's notation, which was then in favor at Cambridge:

$$d_{x_1}U = 0, d_{y_1}U = 0, \ d_{z_1}U = 0, \ d_{x_2}U = 0, \ d_{y_2}U = 0, \ d_{z_2}U = 0.$$

(See "On the Theory of Linear Transformations" = [*CP*1 (1845): 85]).

37. "On the Theory of Linear Transformations" = [*CP*1 (1845): 80].

38. For a homogeneous function with n sets of m variables, Cayley specialized to the case $n = 3$ and $m = 2$ to obtain the example described in his letter to Boole. Whereas Boole was concerned with the binary cubic form, Cayley concerned himself with a multilinear form U in three *sets* of two variables: $(x_1, x_2), (y_1, y_2), (z_1, z_2)$ The six partial differential equations (which the multilinear hyperdeterminant θU satisfies), were looked upon from the algebraic standpoint and not with a view to their analytic properties.

39. C to GB, 13 June 1844, TCC, R.2.88.1. Notice that a, b, c, are symbols that can be differentiated and subjected to the calculus of operations.

40. C to GB, 13 June 1844, TCC, R.2.88.1.

41. Identification of variables in multinomial algebraic forms reduces them to homogeneous algebraic forms, as in the case of the multilinear form (with $n = 3$ and $m = 2$), the identifications $x = x_1 = y_1 = z_1, y = x_2 = y_2 = z_2$ are made, effectively reducing the form to the binary cubic form. By this process, multilinear invariants are reduced to invariants of ordinary algebraic forms. For instance, the multilinear invariant θU, which Cayley showed Boole in his letter of 13 June 1844, reduces to the discriminant Δ_3.

42. The invariant $v = ae - 4bd + 3c^2$ was present in Cauchy's work on determinants of 1815 ("Notes and References" = [*CP*1 (1889): 584]).

43. C to GB, 23 Aug. 1844, TCC, R.2.88.2. The two multilinear invariants of degree 2 and 6 in the case $n = 4$ and $m = 2$ are given in Cayley's first invariant theory paper. The expression for the invariant of the sixth order contains 328 terms ("On the Theory of Linear Transformations" = [*CP*1 (1845): 91–92]).

44. C to GB, 7 Sept. 1844, TCC, R.2.88.3.

45. For a synopsis of the evolution of algebra leading to the study of linear associative algebra, see Pycior 1979.

46. For instance, complex numbers satisfy the commutative condition $zw = wz$ and the associative condition $(xy)z = x(yz)$ for *all* complex numbers x, y, and z.

47. WRH to Archibald Hamilton (son), 5 Aug. 1865, in R. P. Graves 1882–89, 2: 434; Hankins 1980, 291.

48. "Quaternions" = [*HamiltonP3* (1843): 103].

49. WRH to PGT, 15 Oct. 1858, in R. P. Graves 1882–89, 2: 435.

50. Hamilton 1844, 10.

51. Hamilton's note in the *Philosophical Magazine* announcing his discovery of the quaternions contained the definition of the modulus of a quaternion, $|Q| = w^2 + x^2 + y^2 + z^2$ and he noted that the "law of moduli" held true for quaternions: $|QQ'| = |Q||Q'|$ (Hamilton 1844, 10).

52. C to GB, 7 Sept. 1844, TCC, R.2.88.3.

53. Hamilton 1844, 241–46.

54. De Morgan 1844, quote from p. 241. De Morgan's paper was read 28 Oct. 1844, with an "Addition" read 17 Dec. 1844, *CPS General Minutes*. De Morgan published four papers on the "Foundations of Algebra" between 1839 and 1844.

55. De Morgan 1844, 242.

56. "On Certain Results Relating to Quaternions" = [*CP1* (1845): 123–24].

57. Here $q = 1 + \lambda i + \mu j + \nu k$ in $q^{-1}(xi + yj + zk)q$ defines the rotation of (x, y, z) ("On Certain Results Relating to Quaternions" = [*CP1* (1845): 123–24]).

58. "On the Motion of Rotation of a Solid Body" = [*CP1* (1843): 28–35]; Gray 1980.

59. "On the Geometrical Representation of the Motion of a Solid Body" = [*CP1* (1846): 234–36]; "On the Rotation of a Solid Body Round a Fixed Point" = [*CP1* (1846): 237–52].

60. Graves 1882–91, 3: 196. Cayley returned to the subject of rotations a few years later when he considered the composition of rotations ("On the Application of Quaternions to the Theory of Rotation" = [*CP1* (1848): 405–9]).

61. WRH to JTG, 17 Oct. 1843, Letter to [J. T.] Graves on quaternions; or on a new system of imaginaries in algebra = [*HamiltonP3* (1844): 106–10].

62. C to GB, 7 Sept. 1844, TCC, R.2.88.3.

63. A determinant (ordinarily defined) with quaternion entries vanishes if it has identical rows but not necessarily if it has identical columns. In seeking results for these determinants, Cayley stated the obvious result:

$$\begin{vmatrix} a & b \\ c & d \end{vmatrix} + \begin{vmatrix} b & a \\ d & c \end{vmatrix} = \begin{vmatrix} a & a \\ d & d \end{vmatrix} - \begin{vmatrix} c & c \\ b & b \end{vmatrix},$$

and declared it true for "determinants of any order" ("On Certain Results Relating to Quaternions" = [*CP1* (1845): 125]). Cayley briefly turned to determinants with unspecified noncommutative entries at the time he wrote his 1858 work on matrices (C to JJS, 19 Nov. 1857, SJC, Sylvester Papers, Box 2). Determinants with noncommutative entries were shown little attention in the nineteenth century (Muir 1906–23, 4: 490).

64. Cannell 1993; Cross J. J. 1985.

65. C to GB, 7 Sept. 1844, TCC, R.2.88.3; Green's paper on "Attractions" sent by G. Boole to Cayley is most likely Green 1835.

66. C to GB, 14 Sept. [1844], TCC, R.2.88.4.

67. C to GB, 24 Sept. 1844, TCC, R.2.88.5.

68. "Investigation of the Transformation of Certain Elliptic Functions" = [*CP1* (1844): 120–22].

69. C to GB, 11 Nov. 1844, TCC, R.2.88.9. Cayley first uses the term *dependent* hyper-determinant to mean a hyperdeterminant expressible as an *algebraic* function of others.

70. C to GB, 11 Nov. 1844, TCC, R.2.88.9. Here Cayley means 700–800 individual terms of the determinant formed by eliminating variables from the three quadratic equations. He partially computes this invariant in "On the Theory of Elimination" = [*CP*1 (1848): 372–74]. Cayley refers to Otto Hesse's work in "On the Theory of Involution in Geometry" = [*CP*1 (1847): 259 fn.].

71. C to GB, 11 Nov. 1844, TCC, R.2.88.9.

72. C to GB, 27 Nov. 1844, TCC, R.2.88.10. This particular speculation proved to be a false trail. For the invariants discussed in his early letters to Boole, Cayley used *degree* and *order* interchangeably. In the 1850s, when the study was widened to include covariants, *degree* became generally accepted terminology when considering the coefficients of forms and *order* when dealing with their variables.

73. "On Linear Transformations" = [*CP*1 (1846), 108]. Cayley made a check on his calculation of invariants by showing that the sum of the coefficients of the individual terms was zero (but this is only a *necessary* condition).

74. C to GB, 27 Nov. 1844, TCC, R.2.88.10.

75. In their correspondence, G. Boole determined there was no invariant of degree 6 for the binary quintic form. Initially, Cayley was unconvinced but within a few days had changed his mind. He also discovered that an invariant of degree 3 could not exist for the binary form of order 6. In invariant theory, the binary sextic form (of order 6) is only secondary in importance to the binary quintic. For the binary sextic form, there are invariants of degrees 6, 10, and 15, but these were unknown to Cayley at this stage.

76. C to GB, 11 Dec. 1844, TCC, R.2.88.11.

77. Ibid.

Chapter 5. From a Fenland Base

1. C to GB, 17 Jan. 1845, TCC, R.2.88.13.

2. "On Jacobi's Elliptic Functions, in Reply to the Rev. B[rice] Bronwin, and [post-script] on Quaternions" = [*CP*1 (1845): 127]. Cayley's work on the octaves is not mentioned in the principal obituaries (Forsyth 1895; Noether 1895).

3. Kleinfield 1963; van der Blij 1961.

4. WRH to JTG, 17 Oct. 1843, "Letter to [John] Graves on Quaternions; Or on a New System of Imaginaries in Algebra" = [*HamiltonP*3 (1844): 106–10]. As a youth in Dublin, J. T. Graves had investigated the logarithms of complex numbers (Rice 2000).

5. Dorling 1976.

6. R. P. Graves [1890] 1908.

7. JTG to WRH, 18 Jan. 1844 = [*HamiltonP*3 (1844): 649].

8. Hamilton 1853, 61.

9. Hamilton 1848.

10. "On Jacobi's Elliptic Functions, in Reply to the Rev. B[rice] Bronwin, and [post-script] on Quaternions" = [*CP*1 (1845): 127].

11. JTG to WRH, 3 Mar. 1845, in Hankins 1980, 304.

12. J. T. Graves 1845, 320.

13. JTG to WRH, 4 Dec. 1852, = [*HamiltonP*3 (1852): 655]; van der Blij 1961, 111–12 (see chap. 6).

14. The Cayley numbers are "almost associative." Cayley observed that the octave imaginaries *i, j, k, p, q, r, s* obey the distributive (associative) law *within* the triplets {*i, j, k*}, {*i, p, q*}, {*i, r, s*}, {*j, p, r*}, {*j, q, s*}, {*k, p, s*}, {*k, q, r*}, but when symbols belong to different

triplets, an antiassociative law holds—for example, $(ij)p = -i(jp)$. (The triplets identified by Cayley define lines in the finite [Fano] projective plane with 7 points and 7 lines.)

15. "Note on a System of Imaginaries" = [CP_1 (1847): 301]. Cayley found the expression
$\Lambda^{-1}X\Lambda$, in which Λ and X are octaves does not correspond to an orthogonal transformation, in contrast to the same expression $q^{-1}Xq$, in which q and X are quaternions.

16. C to GB, 19 Feb. 1845, TCC, R.2.88.17.

17. GB to ADM, 24 Feb. 1845, in G. C. Smith 1982, 15. I have not found the corresponding letter written from G. Boole to Cayley.

18. GB to WT, 6 Aug. 1845, CUL, Add Ms 7342.B145.

19. "On Algebraical Couples" = [CP_1 (1845): 128–31].

20. C to GB, 17 Jan. 1845, TCC, R.2.88.13.

21. Ibid. "Über die Wendepunkte der Kurven dritter Ordnung [On the inflection
points of cubic curves]" = [$HesseP$ (1844): 123–35].

22. C to WT, [4] Dec. 1846, CUL, Add Ms 7342.C43. The intersection of two cylinders
with their axes at right angles results in a space curve with two components. Problems
associated with the intersection of surfaces were an absorbing interest with Cayley. As he
pointed out to W. Thomson: "There is of course a great deal remaining to be done [on
the intersection of surfaces]; the theory of the *cusped* lines of a surface—and what is
probably more difficult the case where the double (or cusped) line is *not the complete
intersection of any two surfaces whatever*, e.g. when it is a curve of double curvature of the
third order" (C to WT, [4] Dec. 1846, CUL, Add Ms 7342.C43. Cayley's emphasis). Cayley
returned to this problem in 1859 (see chap. 10).

23. "Nouvelles remarques sur les Courbes du Troisième Ordre" = [CP_1 (1845): 190–94].

24. "Sur quelques Intégrales Multiples" = [CP_1 (1845): 195–203, 581–82].

25. C to GB, 21 Jan. 1845, TCC, R.2.88.14.

26. C to GB, 3 Feb. 1845, TCC, R.2.88.15.

27. C to WT, 14 Sept. 1846, CUL, Add Ms 7342.C37.

28. C to GB, 17 Jan. 1845, TCC, R.2.88.13.

29. WW to GBA, 20 Jan. 1845, TCC, O.15.97.8.

30. WT to JT, 10 Feb. 1845, CUL, Add Ms 7342.T295. In Thompson 1910, 1: 117.

31. J. M. Neale and B. Webb were the cofounders of the Camden Society in Cambridge.
It had an initial membership of 39 in 1839, and by 1843 there were 700 members consist-
ing of members from all layers of the Church, including the Archbishop of Canterbury
(Pevsner 1972, 123–29).

32. The announcement at the meeting held on 13 Feb. 1845 that the Camden Society
should be disbanded took the meeting by surprise since only three members had
resigned. In 1846, the society reemerged in London as the "Ecclesiological Late Cambridge
Camden Society" in Christ Church, Albany Street. The Rossettis attended this church
where the Anglican sisterhood of the Catholic Revival was established.

33. C to WT, 17 Feb. 1845, CUL, Add Ms 7342.C33; Bury and Pickles 1994, 132 fn. 41.

34. C to WT, 17 Feb. 1845, CUL, Add Ms 7342.C33.

35. C to WT, 4 Nov. [1847], CUL, Add Ms 7342.C56.

36. Michel Chasles became the first foreign member of the London Mathematical
Society founded in 1865.

37. RLE to C. B. Marlay, undated, incomplete. University of Nottingham Hallward
Library, My2603/1–2.

38. "Mémoire sur les Courbes à double Courbure et les Surfaces développables" =
[CP_1 (1845): 207–11]; Salmon and Rowe 1928, 1: 335–43. Plücker gave the equations that
connected the six parameters, which could be used to classify planar curves: order (n)

and class (*m*), and the number of nodes (δ), of cusps (κ), of points of inflection (ι), and of bitangents (τ) (Plücker 1839). Cayley's corresponding set of equations for space curves involved nine parameters.

39. "Note sur deux Formules données par M. M. Eisenstein et Hesse" = [*CP*1 (1845): 113–16].

40. Ibid.; "On the Theory of Linear Transformations" = [*CP*1 (1845): 80–94]; "Mémoire sur les Hyperdéterminants" [*CP*1 (1846): 117]; "On Linear Transformations" = [*CP*1 (1846): 95–112].

41. C to GB,11 Nov. 1844, TCC, R.2.88.9

42. C to GB, 5 Mar. 1845, TCC, R.2.88.18.

43. "On Linear Transformations" = [*CP*1 (1846): 95].

44. Using Cayley's hyperdeterminant operator, the calculations involved with the hyperdeterminant derivative are lengthy and cumbersome even in calculating the invariant $\Delta_2 = ac - b^2$. A similar mechanism can be use to calculate covariants but identification of variables after differentiation must take place. German mathematicians recognized the importance of the hyperdeterminant operator in the 1860s, but by this time Cayley had abandoned it.

45. C to GB, 5 Mar. 1845, TCC, R.2.88.18.

46. "On Linear Transformations" = [*CP*1 (1846): 95].

47. Turnbull 1926.

48. Olver 1999, 40.

49. "On Linear Transformations" = [*CP*1 (1846): 95].

50. C to GB, 15 Apr. 1845, TCC, R.2.88.20.

51. C to GB, 5 May 1845, TCC, R.2.88.21.

52. GB to WT, 18 July 1845, CUL, Add Ms 7432.B143.

53. "Note on the Calculus of Forms" = [*SP*1 (1853): 403]; "On the Calculus of Forms, Otherwise the Theory of Invariants" = [*SP*1 (1853): 411–22]. The technical terms invariant and covariant were introduced in the early 1850s by Sylvester, and Cayley began to use them immediately ("On the General Theory of Associated Algebraical Forms" = [*SP*1 (1851): 200]); in "Notes and References" = [*CP*1 (1889): 589]; "Note sur la Théorie des Hyperdéterminants" = [*CP*1 (1851): 577–79].

54. *Fifteenth meeting of the British Association for the Advancement of Science, List of members 1845*, 1st Suppl., pp. 2, 8. CUL, Rare Books, Cam.c.845.38. Cayley appeared to stay in Cambridge during the period the York meeting was taking place (C to GB, 24 Sept. 1844, TCC, R.2.88.5; C to GB, 15 Oct. 1844, TCC, R.2.88.6).

55. JCA to George Adams, 10 July 1845, CRO, AM 323.

56. MacHale 1985, 67, 176.

57. Herschel 1846, xxviii, xxix.

58. Bury and Pickles 1994, 136.

59. Report of Section A. *Report BAAS (1845)*, p. 3.

60. Graves 1847.

61. For the initial years, Cayley's annual dividend payment was £80 (1844), £85 (1845), £130-9s-8d (1846), but as a major fellow his share of the College Chest increased. Over the entire period 1844–53 of being "on the foundation," he averaged £240 per annum for his share of the dividend. This sum was approximately half the stipend paid to a university professor at Cambridge.

62. WT to GB, 17 July 1845, GUL, Kelvin Papers, B10.

63. WT to GB, 2 Sept. 1845, GUL, Kelvin Papers, B13.

64. C to GB, 3 Dec. [1845], TCC, R.2.88.22. See also "Recherches sur l'élimination et sur la théorie des courbes" = [*CP*1 (1847): 340].

65. "Sur quelques Théorèmes de la Géométrie de Position" = [CP_1 (1846): 317–28, 588]; "Problème de Géométrie Analytique" = [CP_1 (1846): 329–31].

66. "Mémoire sur les Fonctions doublement périodiques" = [CP_1 (1845): 156–82].

67. Cayley studied Abel's doubly-infinite products

$$u(x) = x \prod_m \prod_n \left(1 + \frac{x}{mw + nvi} \right)$$

and the two "most transcendental functions" $H(x)$, $\Theta(x)$ (as he described them) in terms of definite integrals—functions treated in Jacobi's *Fundamenta Nova*. He later observed that the work was superseded by the "beautiful theory of Weierstrass," but his own work in 1845 was published prior to Eisenstein's research ("Notes and References" = [CP_1 (1889): 586]); Scott 1895, 135.

68. G. C. Smith 1984b, 4. In the period 1845–47, Cayley briefly alluded to Cauchy's theory of complex integration. Augustus De Morgan had introduced this theory to English audiences in his substantial *Differential and Integral Calculus* (1842) though Cayley did not cite this source.

69. Cayley's work on skew determinants leads to the Cayley-Hermite problem (of describing the linear transformations which leaves the sum of squares on n variables unaltered). Though important, this problem is not in the mainstream of Cayley's invariant theory program but generalizes a problem couched in terms of quaternions ("Sur quelques Propriétés des Déterminants Gauches" = [CP_1 (1846): 332–36]). T. Hawkins has identified the Cayley-Hermite problem as prompting Cayley to consider the *addition* of matrices ten years later (Hawkins 1977a).

70. Students present at Cayley's supervision class on 14 Feb. 1845 were A. P. Moor and J. Glover. Both graduated in 1846, Moor with an ordinary B A degree, and Glover the eighteenth wrangler, and both progressed to clerical careers. Six pages of notes were written out by Cayley during the class and two (pages 2, 6) have survived. In the course of one solution, he showed students how to calculate $\int \tan^4 \theta d\theta$ (Wellcome Library for the History and Understanding of Medicine, WMS/ALS Cayley [Box 8]). That small audiences attended Cayley's lectures when he was professor has been documented, but he had the same experience as a young man.

71. WT to JT (father), 5 Oct. 1845, CUL, Add Ms 7342.T318.

72. The numbers of elections to Trinity minor fellowships in the decade 1840–50 were given by H. M. Innes as: 8, 5, 7, 8, 4, 6, 3, 3, 5, 5, 6 (Innes 1941). In 1846, when Cayley's associate Hugh Blackburn was made a Trinity fellow, only three new fellows were elected (Bury 1967).

73. RLE to C, [1846], CUNY, D. E. Smith Papers.

74. FF to WT, 11 Aug. 1850, CUL Add Ms 7342.F305.

75. Brooke's level of mathematical incompetence was discovered by his pupils during a short spell of teaching at University College, London (Rice 1997, 210–13).

Chapter 6. The Pupil Barrister

1. J. Smith [pseud. for J. Delaware Lewis] 1849, 7.

2. Heyck 1982, 72; Winstanley [1940] 1955, 370–71.

3. *Guardian* (6 Feb. 1895): 201.

4. The Cayleys had moved to Blackheath around 1843. They first occupied 59 Lee Road (demolished in 1961) and then moved to Cambridge House near "The Point" (destroyed by fire in June 1881).

5. J. G. Lockhart and J. H. Christie were friends for forty years (curiously, Lockhart's *Memoirs of the Life of Sir Walter Scott* was a favorite with Cayley). Peter George Patmore

was the father of the poet and essayist Coventry Patmore. See also Vaizey 1895–1907, 74–83; Millingen 1841, 2: 244–52.

6. Holborn 1999, 37.

7. Vaizey 1895–1907, 78. This story was reported by M. G. Davidson, a member of the Institute of Conveyancers.

8. Ibid., 81.

9. "Arthur Cayley" = [CP8 (1895): xiv].

10. Cunningham 1897, 76–77.

11. Robinson 1951, 48.

12. Denman 1897, 74.

13. Goldsmith 1843; Hudson 1842; Duman 1983, 78–86.

14. Joseph Liouville to WT [copy made by WT, mourning paper], 29 July 1847, GUL, Kelvin Papers, L17.

15. Hilts 1975, 43.

16. "On the Reduction of [integral], When U is a Function of the Fourth Order" = [CP1 (1846): 224–27]. "On Homogeneous Functions of the Third Order with Three Variables" = [CP1 (1846): 230–33]. "On Geometrical Reciprocity" = [CP1 (1848): 377]. "Sur la Surface des Ondes" = [CP1 (1846): 302–5]; "Note on a Geometrical Theorem Contained in a Paper by William Thomson" = [CP1 (1846): 253–54]; Hudson, R. W. T. [1905] 1990, 89–91.

17. "Recherches sur l'Élimination, et sur la Théorie des Courbes" = [CP1 (1847): 345].

18. "Note sur les Hyperdéterminants" = [CP1 (1847): 353].

19. C to WT, 7 Aug. 1846, CUL, Add Ms 7342.C35.

20. William Eccles to WT, 14 Nov. 1846, CUL, Add Ms 7342.E.16.

21. C to GB, 8 July 1846, TCC, R.2.88.24

22. GB to Joseph Hill, 4 June 1846, UCC, BP1/221/(28).

23. Cayley's testimonial, 19 June 1846, CUL, Ms 7342.Tm 4.

24. C to WT, 7 Aug. 1846, CUL, Add Ms 7342.C35.

25. Topham 1998, 364–69. The other "half" of Cauchy's work not liked by Cayley might have been his papers on topics that were *not* pure mathematics—Cauchy's publications on astronomy, optics, and mechanics. However, he could have been making the remark of not liking the "other half" in an ironical vein—an attempt at a humorous remark, to which he was prone when writing to William Thomson.

26. GB to WT, 17 Aug. 1846, CUL, Add Ms 7432.B151. Quoted in G. D. Smith 1984a, 12.

27. 1 Sept. 1846, UCC, BP/1/291.

28. C to GB, 1 Sept. 1846, TCC, R.2.88.29.

29. C to WT, 8 Sept. 1846, CUL, Add Ms 7342.C36.

30. GB to WT, [15] Sept. 1846, CUL, Add Ms 7432.B155; "On a Multiple Integral Connected with the Theory of Attractions" = [CP1 (1847): 285–89].

31. "Sur quelques Intégrales Multiples" = [CP1 (1845): 195–203, 204–6, 586]; "On Certain Formulae for Differentiation with Applications to the Evaluation of Definite Integrals" = [CP1 (1847): 267–75]; "On a Multiple Integral Connected with the Theory of Attractions" = [CP1 (1847): 285–89, 587]; "Démonstration d'un Théorème de M. Boole concernant des Intégrales Multiples" = [CP1 (1848): 384–87, 588]. He was to write to Liouville in 1857 pointing out the existence of this work when Liouville was working along similar lines ("Sur l'Intégrale [beta function]" = [CP4 (1857): 28–29]; Lützen 1990, 190).

32. C to GB, 14 Dec. 1846, TCC, R.2.88.23.

33. "On the Caustic by Reflection at a Circle" = [CP1 (1847): 273–75]. C to WT, [4] Dec. 1846, CUL, Add Ms 7342.C43.

34. C to WT, 6 Feb. 1847, CUL, Add Ms 7342.C45.

35. "Sur les Déterminants Gauches" = [*CP*1 (1848): 410–13, 589].

36. WT to GB, 21 Feb. 1847, GUL, Kelvin.B19. "On the Differential Equations which Occur in Dynamical Problems" = [*CP*1 (1847): 276–84]. "Demonstration of a Geometrical Theorem of Jacobi's" = [*CP*1 (1848): 362–63]. WT to GB, 27 Mar. 1847, CUL, Add Ms 7432.B20.

37. C to WT, 17 May 1847, CUL, Add Ms 7342.C48. The ratio of the doubly infinite products $H(u)$, $\Theta(u)$ is a way of defining an elliptic function such as sn u. This is the subject of the papers "On the Theory of Elliptic Functions" = [*CP*1 (1847): 290–300]; "Notes and References" = [*CP*1 (1889): 587]; Forsyth 1895, xli. Cayley's "wee things" are likely the follow-up papers: "On the Theory of Elliptic Functions" = [*CP*1 (1848): 364–65], "Notes on Abelian Integrals—Jacobi's System of Differential Equations" = [*CP*1 (1848): 366–69], "On the Theory of Elimination" = [*CP*1 (1848): 370–74]. In the last-mentioned, Cayley effectively gives a definition of "exactness," which is used in conjunction with exact sequences of groups in modern algebraic topology, though this is not to say that Cayley was doing anything remotely like algebraic topology (Gelfand and Kapranov and A. V. Zelevinsky 1994, 104).

38. WT to C, 28 June 1847, mourning paper, CUNY, D. E. Smith Papers. "On the Differential Equations which Occur in Dynamical Problems" = [*CP*1 (1847): 276–84, sub. 21 Feb. 1847].

39. C to WT, 5 June 1847, CUL, Add Ms 7342.C50.

40. C to WT, 5 June 1847, CUL, Add Ms 7342.C50; C to WT, 17 Aug. 1847, CUL, Add Ms 7342.C55.

41. C to WT, 17 Aug. 1847, CUL, Add Ms 7342.C55.

42. Ibid. This submission to Crelle's Journal is likely his paper "Note sur les Fonctions Elliptiques" = [*CP*1 (1848): 402–4], a continuation of his work on the solution of Jacobi's differential equation.

43. Cayley 1889b; MacMahon 1898; Archibald 1946; Elliott and Matheson 1909; Feuer 1987. The most substantial study of Sylvester to date is Parshall 1998a.

44. Halsted 1897, 160.

45. Joseph Romilly's diary entry 21 Jan. 1837 (Bury 1967).

46. Charles C. Atkinson (Secretary), 18 Nov. 1837, UCL, College Correspondence, AM/7.

47. "Analytical Development of Fresnel's Optical Theory of Crystals" = [*SP*1 (1837), 1–27; "On the Motion and Rest of Fluids" = [*SP*1 (1838), 28–32]; "On the Motion and Rest of Rigid Bodies" = [*SP*1 (1839), 33–35]. Robert Leslie Ellis read Sylvester's "motion of fluids" papers and pronounced them "clever, but eccentric" (Diary of Robert Leslie Ellis, 18 Dec. 1838, TCC, Add Ms a.219.1).

48. Royal Society, 25 Apr. 1839, vol. 8, cert. 288. A map of Sylvester's contributions to mechanics and mathematical physics is contained in Grattan-Guinness 2001.

49. ADM to TAH, 29 June 1865, in Collingwood 1966.

50. JJS to WT, 18 Nov. 1845, CUL, Add Ms 7342.S594.

51. Ibid.

52. Chasles gave Sylvester *La Géometrie* in Feb. 1847 (Cajori 1922, 340).

53. Venn 1954, 6: 526–27; Chapman 1985.

54. Simmonds 1948, 16–17. The Institute of Actuaries was founded on 8 July 1848 and on the following 14 October Sylvester was appointed one of its four vice-presidents.

55. C to WT, 4 Nov. 1847, CUL, Add Ms 7342.C56.

56. JJS to C, morning 24 Nov. 1847, SJC, Sylvester Papers, Box 10; Parshall 1998, 18–19.

57. JJS to C, afternoon 24 Nov. 1847, SJC, Sylvester Papers, Box 10. "The equation $Ax^3 + By^3 + Cz^3 = Dxyz$" = [$SP_1$ (1847): 107–9, 110–13, 114–18].

58. "Recherches sur l'Élimination, et sur la Théorie des Courbes" = [CP_1 (1847): 337]; "Nouvelles Recherches sur les Fonctions de M. Sturm" = [CP_1 (1848): 392]; Muir 1909.

59. White 1898, 85.

60. WT to GB, 5 Dec. 1847, GUL, Kelvin Papers, B21.

61. GB to C, 8 Dec. 1847, RSL, Mss782, E-13; Boole 1997, 192.

62. C to GB, 7 Dec. 1847, RSL, Mss782, E-13; Boole 1997, 191.

63. Playfair 1778. Cayley was utterly opposed to the geometrical interpretation of $\sqrt{-1}$, in which the equation of a circle $x^2 + y^2 = 1$ written in the form $x^2 - (\sqrt{-1}y)^2 = 1$ was interpreted as a hyperbola in the plane formed by the x-axis and the "imaginary" axis $(\sqrt{-1})y$ and which is orthogonal to the plane with standard x and y axes. See also Gregory 1839; Walton 1852.

64. The phrase "ουτωζ οντα" ("really real," or "how it really is") was a favorite expression and used on many occasions throughout his life. Its repetition during the course of this biography indicates the stability of Cayley's mathematical outlook and his understanding of the "reality" of three-dimensional real space. C to GB, 7 Dec. 1847, RSL, Mss782, E-13; Boole 1997, 191.

65. Ibid.

66. Ref. Rep., 25 Apr. 1860, RSL, RR.4.79.

67. "On the Applications of Quaternions to the Theory of Rotations" = [CP_1 (1848): 405–9, sub. 1 July 1848]. See Boole 1848, 278.

68. C to WT, 8 Feb. 1847, CUL, Add Ms 7342.C45. The problem tackled by T. Kirkman was derived from a prize question set in the *Lady's and Gentleman's Diary* (1844) by the editor W. S. B. Woolhouse.

69. Biggs 1981, 98–101.

70. C to TPK, [1848], in Kirkman 1848, 447.

71. The "n-squares problem" is to determine the natural number n for which the product of sums of n-squares is the sum of n-squares. Cayley gave a criterion for the existence of a formula of 2^n squares in 1852 ("Demonstration of a Theorem Relating to the Products of Sums of Squares" = [CP_2 (1852): 49–52]; Dickson 1919, 165–66). The rigorous confirmation that the permissible values of n were 2, 4, and 8 had to wait for another half-century. A. Hurwitz proved this in 1898 using matrix algebra ("Über die Komposition der quadratischen Formen von beliebig vielen Variablen" [On the composition of quadratic forms of an arbitrary number of variables] = [$HurwitzP_2$ (1898): 565–71]; van der Waerden 1985, 184–85). The matrix algebra tool was unavailable in the 1840s and, even in the 1850s, when the theory of matrices was being talked about, it was too primitive to be of use in proving this sophisticated result. Given Cayley's interest in the problem, it is appropriate that Hurwitz found Cayley's 1858 paper on matrices useful in formulating his proof.

72. "On the Triadic Arrangements of Seven and Fifteen Things" = [CP_1 (1850): 481–84, 589]. Poem published in *Mathematical Questions and Their Solutions from the Educational Times,* 1870, vol. 13, p. 79.

73. Ball [1892] 1949, 292.

74. "Note sur quelques Théorèmes de la Géométrie de Position" = [CP_1 (1851): 550–56]; Salmon 1869, 360–63.

75. HB to WT, 13 Apr. 1848, CUL, Add Ms 7342.B92.

76. Woodham-Smith [1962] 1991, 344.

77. The lectures were held on 21, 23, 26, and 28 June 1848 and they formed the basis for Hamilton's *Lectures on Quaternions* (1853). See Graves 1882–91, 2: 605.

78. WRH to WT, 9 Mar. 1846, CUL, Add Ms 7342.H13.

79. Salmon 1883, 482.

80. RT to WT, 1 Mar. 1847, CUL Add Ms 7342.T585.

81. "Sur la généralisation d'une Théorème de M. Jellett qui se rapporte aux Attractions" = [CP1 (1848): 388–91]; "Sur quelques Transmutations des Lignes Courbes" = [CP1 (1849): 471 fn.].

82. Ossory 1912; Mollan, Davis, and Finucane 1990, 34.

83. Townsend 1849, 251.

84. RT to WT, 10 Feb. 1846, CUL, Add Ms 7342.T580.

85. C to WT, [4] Dec. 1846, CUL, Add Ms 7342.C43.

86. Salmon 1847, 65; C to WT, [4] Dec. 1846, CUL, Add Ms 7342.C43.

87. "On the Triple Tangent Planes of Surfaces of the Third Order" = [CP1 (1849): 445–56]; "Notes and References" = [CP1 (1889): 589]; Salmon 1849. Quotes from "On the Triple Tangent Planes of Surfaces of the Third Order" = [CP1 (1849): 453].

88. Quote: "On the Triple Tangent Planes of Surfaces of the Third Order" = [CP1 (1849): 453]. Henderson 1911. In the 1860s, the "twenty-seven lines on a cubic surface" was studied using the embryonic group theory but this was not Cayley's way. Tantalizingly there are connections with this geometry of the "twenty-seven lines" and the algebra of octaves, but this was not discovered until the 1950s (van der Blij 1961, 111).

89. Lanczos 1970, 307.

90. Neville 1948.

91. C to WT, 28 Apr. 1847, CUL, Add Ms 7342.C47.

92. A pendulum with a double suspension like Blackburn's was developed by the American Nathaniel Bowditch. See HB to WT, 11 Mar. 1849, CUL, Add Ms 7342.B102.

93. *Testimonials in Favor of Hugh Blackburn*, 1849, GUL.

94. C to RLE, 24 Apr. [1849], TCC, Add Ms.c.65.28. Cayley was a Trinity College junior examiner in 1848 and senior examiner in 1849 and 1850.

95. C to GB, 15 Aug. 1849, TCC, R.2.88.30.

96. WT to GB, 5 Feb. 1847, GUL, Kelvin Papers, B18.

97. "Note sur l'Addition des Fonctions Elliptiques" = [CP1(151): 540–49]; "Note sur quelques Théorèmes de la Géométrie de Position" = [CP1 (1851): 550–56]; "Mémoire sur les Coniques inscrites dans une même Surface du Second Ordre" = [CP1 (1851): 557–63].

98. Dickens [1852–53], chap. 19, p. 258.

99. C to WT, 28 Apr. 1847, CUL, Add Ms 7342.C47. "On a Multiple Integral Connected with the Theory of Attractions" = [CP1 (1847): 289].

Part 3. A Rising Star, 1850–1862

Epigraph: Ref. Rep., 27 May 1861, RSL, RR.4.45. Philip Kelland was referee for the paper "On the Porism of the In-and-circumscribed Polygon" = [CP4 (1861): 292–308].

Chapter 7. Barrister-at-Law

1. Stone Buildings was built in the years 1775–85. It was severely hit by enemy action during the Second World War but has now been restored to its original form. As a young lawyer, Cayley saw an expanding Lincoln's Inn. The Great Hall was built 1843–45 to include a new library formerly housed in 2 Stone Buildings.

2. Rossetti 1906, 175.

3. Hilts 1975, 43.

4. Best [1971] 1973, 21–25.

5. Duman 1983, 19.

6. Trollope [1875–76] 1994, 28.

7. Vaizey 1895–1907, 166.

8. Park 1832, 69. John James Park held the chair of English Law and Jurisprudence at King's College, London. He was a brilliant barrister but a poor lecturer though the college hierarchy appreciated his conspicuous opposition to Jeremy Bentham (Hearnshaw 1929, 90).

9. Bury and Pickles 2000, 78, 98–99; N. M. Ferrers was called to the bar in 1855 but returned to Gonville and Caius College in Cambridge the following year, where he taught mathematics. He became master in 1880 (Routh 1905; Roberts and Gross 1912, 150–56; *The Caian*, 1902–3, 12: 85–92).

10. THH to Elizabeth Huxley, 3 May 1852, in Huxley 1900, 1: 100.

11. Hargreave 1849.

12. Hargreave 1866.

13. Charles S. C. Bowen to Austen Leigh, 21 Apr. [1857], in Cunningham 1897, 80–81. The other Senior Wrangler mentioned by Bowen may have been George Hemming (Senior Wrangler in 1844).

14. Vaizey 1895–1907, 80.

15. C to J. Vaizey, [1894], in Vaizey 1895–1907, 79.

16. A fellow law pupil during Cayley's training for the bar, T. C. Wright, was also involved with the production of Davidson's *Precedents* (Davidson, Waley, and Key 1873). Jacob Waley, a student of De Morgan's at University College, was the first student to be awarded an honors degree in mathematics in the University of London in June 1839.

17. Davidson, Waley, and Key 1873, 3 (Part 2): 1067–122.

18. Dickens [1852–53] 1998, chap. 1, p. 2.

19. Newcomb 1903, 281; a variant of this story is in Halsted 1899, 65.

20. C to WT, 6 Aug. 1850, CUL, Add Ms 7342.C59; "On Certain Multiple Integrals Connected with the Theory of Attractions" = [*CP*2 (1852): 35–40, sub. 6 Aug. 1850].

21. Only the wealthiest section of the population were liable for Death Duties in the early 1850s, which amounted to about 6 percent of the total population. Henry Cayley's estate was liable to a sum sworn under £12,000 (FRC, I.R. 26/1863/ff.595–96).

22. The Cayley family lived at Cambridge House, The Grove, Blackheath until around 1855. CGR to W. M. Rossetti, 25 Oct. 1861, in Christina G. Rossetti 1997, 1: 151; Jones 1991, 108).

23. The two servants present in the Cayley household on Census day, 1851, were Charlotte Wurn from Norwich and Amelia Willows from London (1851 British Census, FRC).

24. CGR to W. M. Rossetti, 25 Oct. 1861, in Christina G. Rossetti 1997, 1: 151; Jones 1991, 108.

25. W. Rosetti 1883, 776, 816–17.

26. Sawtell 1955, 129.

27. Hirst *Journal*, 11 Dec. 1860, RIGB, 3: 1563.

28. UCL, Galton Papers, GP.152/2; Burbridge 1994, 455.

29. W. Rossetti 1883, 776, 816–17.

30. C. B. Cayley 1846.

31. Charles Cayley's translation of Dante's *Divine Comedy* is compared with others of the period in Milbank 1998. His translation came out in the sequence *The Vision of Hell* (1851), *The Purgatory* (1853), *The Paradise* (1854), and *Notes* (which contains a critique of previous translations, 1855). Charles Cayley's poetry is discussed in Marsh 1994.

32. J. H. Röhrs to JCA, 11 Jan. [1855], SJC, Adams Papers 13/4/1. Röhrs was a mathematical physicist who wrote on such topics as precession and nutation and the mechanism

of artillery. He presented several of his papers to the Cambridge Philosophical Society.

33. RLE to Everina Frances, Lady Affleck (his sister), [1854], TCC, Whewell Papers, Add Ms a. 81.178(1).

34. Murray 1884, 502–5.

35. Rossetti 1906, 2: 312.

36. Sawtell 1955, 131.

37. "Observations on a New Theory of Multiplicity" = [SP1 (1852): 376].

38. JJS to C, 24 May 1850, SJC, Sylvester Papers, Box 9.

39. "On the Relation between the Minor Determinants of Linearly Equivalent Quadratic Functions" = [SP1 (1851): 246].

40. Nature 33 (7 Jan. 1886): 230.

41. Blake 1970, 13.

42. "On the Triadic Arrangements of Seven and Fifteen Things" = [CP1 (1850): 481–84, sub. 14 June 1850].

43. "On the Theory Of Skew Surfaces" = [CP2 (1852): 33–34]; Hill 1897, 141. "Sur le Problème des Contacts" = [CP1 (1850): 522–31, 519–21]; "On the Intersections, Contacts, and Other Correlations of Two Conics Expressed by Indeterminate Coordinates" = [SP1 (1850): 119–37]; "Notes and References" = [CP2 (1889): 598–601].

44. "Note on Quadratic Functions and Hyperdeterminants" = [SP1 (1851): 251]. Sylvester's research involved the family of conics $\alpha U + \beta V = 0$ passing through the intersection of conics $U = 0$, $V = 0$. For example, he investigated the discriminant $\Delta(\lambda) = discrt.(U + \lambda V)$, which Cayley would couch in the language of matrices a few years later.

45. JJS to C, 26 June 1850, SJC, Sylvester Papers, Box 9.

46. Galton 1908, 71.

47. "On Aronhold's Invariants" = [SP1 (1853): 607].

48. "Note on Quadratic Functions and Hyperdeterminants" = [SP1 (1851): 251].

49. JJS to C, 29 Sept. 1850, SJC, Sylvester Papers, Box 9. "Additions to the Articles 'On a New Class of Theorems,' and 'On Pascal's Theorem'" = [SP1 (1850): 145–51].

50. "Sketch of a Memoir on Elimination, Transformation, and Canonical Forms" = [SP1 (1851): 185 and 191]. Here the problem of "canonical forms" was one of writing a binary form of order $2n + 1$ as the sum of $n + 1$ powers of linear functions of x, y.

51. "An Essay on Canonical Forms, Supplement to a Sketch of a Memoir on Elimination, Transformation and Canonical Forms" = [SP1 (1851): 203 fn.]

52. JJS to C, 20 May 1851, SJC, Sylvester Papers, Box 9; Parshall 1998, 33–35.

53. "On the Theory of Permutants" = [CP2 (1852): 16–26]. Quote on p. 26.

54. "On the Theory of Permutants [post-script]" = [CP2 (1852): 26]; "On the Principles of the Calculus of Forms" = [SP1 (1852): 317].

55. Hugh MacColl to WJCM, 25 Apr. 1866, CUNY, D. E. Smith Papers, quoted in Astroh, Grattan-Guinness, Read 2000, 92.

56. Cayley 1889b, 217–19.

57. JJS to C, 18 June 1851, SJC, Sylvester Papers, Box 9.

58. A binary quadratic form is reducible to one squared term if and only if the discriminant $\Delta_2 = 0$. The necessary and sufficient condition for a binary form of order $2m$ to be reduced to the sum of m powers by linear substitutions is that the invariant known as the catalecticant (see the Glossary) vanishes (Elliott 1895, 267–90). Sylvester sets this theory out in "Sketch of a Memoir on Elimination, Transformation, and Canonical Forms" = [SP1 (1851): 190].

59. "Additional Notes to Prof. Sylvester's Exeter British Association Address" = [SP2 (1869–70): 714], the address first published in Sylvester 1870. The general problem of reducing a binary form to the sum of squared terms (the "canonical form") was partially solved by Sylvester in 1851 and completed by E. K. Wakeford. Sylvester wrote up his "remarkable result" as a postscript to a paper ("On the General Theory of Associated Algebraical Forms" = [SP1 (1851): 202.]; Elliott [1913] 1964, 284–85).

60. "On the Principles of the Calculus of Forms" = [SP1 (1852): 284–369]; "Note on the Calculus of Forms" = [SP1 (1853): 402]; "On the Calculus of Forms" = [SP2 (1854): 27].

61. "Note on the Calculus of Forms" = [SP1 (1853): 403].

62. "On a Remarkable Discovery in the Theory of Canonical Forms and Hyperdeterminants" = [SP1 (1851): 273, 280]. Here Sylvester introduced the term "discriminant," which we have used anachronistically until now.

63. "An Inquiry into Newton's Rule for the Discovery of Imaginary Roots" = [SP2 (1864): 380 fn.].

64. "Note sur la Théorie des Hyperdéterminants" = [CP1 (1851): sub. 21 Nov. 1851, 577].

65. C to JJS, 5 Dec. 1851, SJC, Sylvester Papers, Box 2; Parshall 1998, 37.

66. "Notes and References" = [CP2 (1889): 600–601].

67. "On the Principles of the Calculus of Forms" = [SP1 (1852): 352]. Sylvester demonstrated the importance of defining invariants as solutions to differential equations many years later and he then introduced the term "annihilator" to designate Cayley's partial differential operators.

68. The two differential equations given in Cayley's far-reaching letter are not of equal status. The second equation is merely a consequence of Euler's theorem for homogeneous functions. The important equations are the first equation and the equation obtained by taking the coefficients of the binary form in reverse order:

$$\left(k\frac{d}{dj} + 2j\frac{d}{di} + 3i\frac{d}{dh} + \cdots + nb\frac{d}{da} \right) U = 0.$$

Cayley was an admirer of Jacobi, but he failed to adopt Jacobi's notation for the partial derivative "∂-form," now commonly accepted, preferring to use the ordinary notation "d-form" in handwritten and printed work. He did, however, make extensive use of the "curly d" to denote a partial derivative as in ∂_x.

69. "Nouvelles recherches sur les covariants" = [CP2 (1854): 164–78].

70. Sylvester's inaugural lecture at Oxford. "On the Method of Reciprocants as Containing an Exhaustive Theory of the Singularities of Curves" = [SP4 (1886): 294].

71. "Notes and References" = [CP2 (1889): 600]. In the expression for the linear operator $a\partial_b + 2b\partial_c$ operating on $ac - b^2$, the symbols ∂_b and ∂_c are partial derivative operators. The individual calculations $(a\partial_b + 2b\partial_c)ac = 2ba$ and $(a\partial_b + 2b\partial_c)b^2 = a2b$, yields the net result $(a\partial_b + 2b\partial_c)(ac - b^2) = 0$.

72. "On the Theory of Linear Transformations" = [CP1 (1845): 84].

73. There are scattered clues suggesting a plausible reason for Cayley's abandonment of the hyperdeterminant derivative. I believe he found it relatively inefficient as an *agent for calculation*. He did not drop it through any suspicion of theoretical weakness, for he knew the hyperdeterminant derivative could be used to find any invariant or covariant in theory but he noted: "the application of it becomes difficult when the degree [of an covariant] exceeds 4" ("A Fourth Memoir on Quantics" = [CP2 (1858): 517]).

74. MacMahon 1896.

75. "On a Theorem in Covariants" = [*CP*8 (1872): 404–8]; "Note on a Hyperdetermi-nant Identity" = [*CP*13 (1892): 210]. "On the Theory of Semi-invariants" = [*CP*12 (1886): 344–57].

76. "Nouvelles Recherches sur les Covariants" = [*CP*2 (1854): 167].

77. "On the Principles of the Calculus of Forms" = [SP1 (1852): 356].

78. Sylvester wrote: "Suppose that [a putative invariant] C . . . alters neither when x receives such infinitesimal increment, y and z remaining constant, nor when y and z sep-arately receive corresponding increments z, x and x, y in the respective cases remaining constant . . . C will remain constant for any concurrent linear transformation of x, y, z . . ." ("On the Principles of the Calculus of Forms" = [SP1 (1852): 326]).

79. Cayley 1889b, 218.

80. Others claim that Cayley, Sylvester, and Hermite were the "Invariant Trinity," but Salmon has a greater claim than Hermite to be included (Broadbent 1964, 375).

81. GS to JJS, 14 Apr. 1852, SJC, Sylvester Papers, Box 3; Parshall 1998, 45–46.

82. "On Aronhold's Invariants" = [SP1 (1853): 606].

83. JJS to C, 1 Mar. 1850, CUL, Add Ms 7342.S597.

84. JJS to Lord Brougham, 20 June 1861, quoted in Parshall 1998, 104–5. As Karen Par-shall has pointed out, Sylvester gave an interesting account of the prehistory of invariant theory, one which dismissed Boole's contribution as of little value.

85. Salmon 1854, 19–20.

86. "The Aconic Function" = [*HamiltonP*3 (1853): 423]. Cayley had given his proof of Pascal's theorem ten years before: "Demonstration of Pascal's Theorem" = [*CP*1 (1843): 43–45].

87. "On Professor MacCullagh's Theorem of the Polar Plane" = [*CP*4 (1858): 12–20]; C to WRH, 3 Apr. 1865, in R. P. Graves 1882–89, 3: 199.

88. WRH to ADM, 31 Jan. 1852, in Hamilton 1882–91, 3: 331.

89. "On a System of Equations Connected with Malfatti's Problem, and, On Another Algebraical System" = [*CP*1 (1849): 468]; "On the Solution of a System of Equations in which Three Homogeneous Quadratic Functions of Three Unknown Quantities are Respectively Equated to Numerical Multiples of a Fourth Non-homogeneous Function of the Same" = [SP1 (1850): 152–56]. "Analytical Researches Connected with Steiner's Extension of Malfatti's Problem" = [*CP*2 (1852): 57–86]; "Notes and References" = [*CP*2 (1889): 593].

90. B. Price was elected professor of natural philosophy at Oxford in 1853; Bakewell 1996. The annual Royal Society fee of £14 a year could not have been a barrier to Cayley joining earlier.

91. T. H. Huxley, 3 May 1852, in Huxley 1900, 1: 80.

92. Royal Society, 3 June 1852, vol. 9, cert. 290.

93. Ref. Rep., 21 June 1852, RSL, RR.2.41. It is noteworthy that Peacock should think in terms of blocking pure mathematics given his own background in the subject.

94. Davies 1850, 41.

95. WT to GGS, 8 June 1852, CUL, Add Ms 7656.K58; in D. B. Wilson 1990, 1: 131.

96. A few days before their "incessant" conversation, Sylvester sent Joseph Liouville a copy of his printed work through Irenée-Jules Bienaymé. He also sent offprints to M. Chasles, C. Hermite, O. Terquem, E. Catalan, J. A. Serret, J. L. Bertrand, C. Borchardt, and A. L. Cauchy (Parshall and Seneta 1997, 213). WT to GGS, 29 June 1852, CUL, Add Ms 7656.K59.

97. J. Booth was educated at Trinity College, Dublin, and ordained in Bristol in 1842. His "tangential coordinates" (anticipated by Plücker) were known as "Boothian coordinates."

The author of educational, religious, and mathematical works, he was elected an F.R.S. in 1846 (Glaisher 1879; Foden 1989). See *Minutes of Council of the Royal Society* for quote re Cayley. The furor over Tyndall's award of the Royal Medal in 1853 for his work on magnetism indicates the competitive atmosphere among Royal Society scientists. After the award had been decided in favor of Tyndall, mutterings in the Royal Society Council suggested Plücker had priority, and Tyndall refused to accept the award (Eve and Creasey 1945, 46–49).

Chapter 8. A Grand Design

1. "Note on a Question in the Theory of Probabilities" = [*CP*2 (1853): 103–4]; "Notes and References" = [*CP*2 (1889): 594–98].

2. Cayley paraphrased a simplified version of the Challenge Problem Boole had set in 1851: "Given the probability α that a cause A will act, and the probability p that A acting the effect will happen; also the probability β that a cause B will act, and the probability q that B acting the effect will happen; required the total probability of the effect" ("Note on a Question in the Theory of Probabilities" = [*CP*2 (1853): 103]). Cayley's proposed solution, presented in the case $n = 2$, can be justified if three further assumptions are made (Hailperin 1986, 359–70). The Challenge Problem and Cayley's work on it is treated in (Dale [1991] 1999, 378–87).

3. "Note on the Porism of the In-and-circumscribed Polygon [and Corrections]" = [*CP*2 (1853): 87–92]. As Jacobi had done, Cayley attacked the problem using elliptic functions, but J. A. Todd observed that this is unnecessary and showed it can be obtained using invariant theory (Todd 1948). In its various forms and generalities, Cayley worked on this problem from 1853 through 1871. Its history has been discussed in the light of modern developments (Bos, Kers, Oort, and Raven 1987).

4. Cayley observed that the quadratic equation $ax^2 + bx + c = 0$ can be solved in ordinary algebra, yet he noted the substantially different question when the coefficients a, b, c, and the unknown x were regarded as symbols of operation. He rightfully suggested the problem of finding the unknown x in this case is far from clear.

5. Brock 1967, 101.

6. For example, Cayley regarded the Taylor expansion of $f(x + h)$ extendible to $f(D + h)$, where D represents the differential operator. Occasionally it was necessary to be precise and distinguish between scalars and operators, for example, between the ordinary *scalar* 1 and the identity *operator* 1. When Cayley wanted a symbol x to represent "ordinary quantity" he would use the notational device $\tilde{x}$. To solve a quadratic equation in ordinary algebra would mean solving the equation $a\tilde{x}^2 + b\tilde{x} + c = 0$.

7. Cayley put the operator identity $QP = QxP + Q(P)$ to extensive use in invariant theory. This was skillfully used throughout the century by British mathematicians and Cayley used it to the last ("On Reciprocants and Differential Invariants" = [*CP*13 (1893): 400]). The nature of operators were usually distinct from operands, but when Cayley found the need to consider operators acting on operators this had to be explained. For example, the idea of considering the differentiation symbol $P = A\partial_x + B\partial_y + \cdots$ as an operator *and* as an operand (as in $Q(P)$, above) was novel enough to warrant Cayley giving it the new name *Operandator*. This sensitivity stems from the same concern for distinguishing between x and ordinary quantity $\tilde{x}$. In passing, note that Cayley's reliance on unscripted notation prevented him from adequately describing P as a *finite* expression.

8. C to WRH, 23 Sept. 1853, TCD, Ms 7767.1239. I am grateful to Rod Gow for informing me of this significant letter and for sending me a copy of it. He noted that Hamilton sent out 86 presentation bound copies of the *Lectures on Quaternions* and in receiving the 31st on the list, Cayley was the first outside Hamilton's immediate circle in Ireland to receive

a copy. It is apparent from the letter and from his published papers that Cayley was inspired to study groups from geometrical considerations, though he chose to formalize the system of "functional symbols" 1, α, β, γ, δ, ε through the calculus of operations. Between sending his letter to Hamilton and submitting his group theory paper (during October 1853), he found that the composition of symbols was best seen through a "multiplication table" (now known as a "Cayley table"), a spatial array which vividly demonstrates the structure of a group. It is quite possible that Cayley's letter describing the symmetric group on three letters S_3, and Cayley's subsequent papers published in the *Philosophical Magazine*, encouraged Hamilton to construct the Icosian Calculus (1856), in which he defined a group in terms of generators and relations.

9. "On the Theory of Groups as Depending on the Symbolical Equation $\theta^n = 1$" = [*CP*2 (1854): 123–30, 131–32], [*CP*4 (1859): 88–91].

10. Cayley's papers on the wave surface and Sturm's functions appeared in the *Journal de Mathématiques Pures et Appliqués* in July 1846 ("Sur la Surface des Ondes" = [*CP*1 (1846): 302–5], "Note sur les Fonctions de M. Sturm" = [*CP*1 (1846): 306–8]). Galois's theory appeared later in the year, however, Cayley may not have seen this when the volume was current, for, on a walking tour in the new year he wrote to William Thomson: "*Liouville* I have not seen for an age, nor how my wave surface paper looks in print" (C to WT, 6 Feb. 1847, CUL, Add Ms 7342.C45).

11. Nicholson 1993.

12. Kiernan 1971, 101.

13. Equation = [*CP*11 (1878): 520]; "Sulla risoluzione delle equazioni algebriche" = [*BettiP*1 (1852): 31–80]. A full commentary on Galois's work appeared in the third edition of J. A. Serret's *Cours d'algèbre supérieure* (1866)—the first appearance of his theory in a textbook and a widely influential one (Kiernan 1971, 110–14). "On a Discovery in the Partition of Numbers" = [*SP*2 (1857): 87 fn.].

14. Brock 1967, 101.

15. "On the Theory of Groups as Depending on the Symbolical Equation $\theta^n = 1$" = [*CP*2 (1854): 123].

16. For the symmetric group on three letters, Cayley noted that the subgroup of order 3 was a "submultiple" (a normal subgroup) of the symmetric group but that no subgroup of order 2 could be a "submultiple." Cayley's placement of $\beta\alpha$ in the group table is guided by the convention for composing operators. Cayley tables are impractical for describing large groups but for the groups of small order, the groups which Cayley dealt with, they are ideal. In the second paper of the series, Cayley further explored the ideas of a "symmetric holder" (akin to a modern coset), "submultiple," and products of groups.

17. Cayley did not go as far in the direction of abstraction as the little known polymath A. J. Ellis, who graduated sixth wrangler at Cambridge in 1837, the same year as did Sylvester. Duncan Gregory was a friend and, from him, Ellis learned the importance of operators in symbolical algebra. He attempted to put the calculus of operations on a proper foundation "without any metaphysical or *a priori* reasoning" (Ellis 1859–60).

18. "On a Property of the Caustic by Refraction of a Circle" = [*CP*2 (1853): 118–22]. Cayley followed this by a survey of various cardioid-like curves set out in a lengthy memoir and supplement, "A Memoir on Caustics" = [*CP*2 (1857): 336–80], "A Supplementary Memoir on Caustics" = [*CP*5 (1867): 454–64]. This work may have inspired R. C. Archibald to name the curve with polar equation $r = 4\cos^3(\theta/3)$ and Cartesian equation of the sixth order $27(x^2 + y^2)^2 = 4(x^2 + y^2 - x)^3$, as "Cayley's sextic" (see especially [*CP*2 (1857): 368–69]).

19. Boole 1847.

20. "On the Homographic Transformation of a Surface of the Second Order into Itself" = [*CP*2 (1853): 105–12]; Moran 1975, 300–301; Klein, [1908] 1932, 65–75.

21. "On the Theory of Groups as Depending on the Symbolical Equation $\theta^n = 1$" = [*CP*2 (1854): 123–30]; van der Waerden 1985.

22. Sylvester 1854, 66.

23. "Notes and References" = [*CP*2 (1889): 598–601].

24. "Über die Resultante eines Systemes mehrerer algebraischen Gleichungen [On the resultant of a system of several algebraic equations]" = [*SchläfliP*2 (1850): 9–112]; "Sur un Théorème de M. Schläfli" = [*CP*2 (1855): 181–84].

25. C to JJS, 12 Oct. 1854, SJC, Sylvester Papers, Box 2; Parshall 1998, 73–75. "A Second Memoir upon Quantics" = [*CP*2 (1856): 251].

26. Skew invariants do not exist for binary forms of order 2, 3, and 4 but for $n = 5$, the binary quintic, there is exactly one: Hermite's skew-invariant (called skew because it changes sign when the coefficients of the binary quintic are reversed). Cayley was surprised at its existence since he had previously believed the degree of an invariant of the quintic to be a multiple of four ("evenly even").

27. "An Introductory Memoir on Quantics" = [*CP*2 (1854): 233 fn.]. "Tables of Covariants M to W of the Binary Quintic" = [*CP*2 (1889): 299–303].

28. "Sur la théorie des fonctions homogènes à deux indéterminées" = [*HermiteP*1 (1854): 312–13].

29. The condition $W = 0$ allows a binary quintic to be written as a product of linear factors and thus the corresponding polynomial (by setting $y = 1$) is soluble in terms of rational numbers (Elliott 1895, 308–9).

30. Bottazzini 1980; "Sulla teorica degli invarianti" = [*BrioschiP*1 (1854): 111–14]; "Intorno ad una proprietà degli invarianti" = [*BrioschiP*1 (1858): 151–56]; "La teorica dei covarianti e degli invarianti delle forme binaire e le sue principali applicazioni" = [*BrioschiP*1 (1858): 349–414].

31. "Sopra i covarianti delle forme binaire" = [*BettiP*1 (1858): 157–62].

32. Francesco Faà di Bruno calculated the invariant u (of degree 12) for the binary quintic form (Faà di Bruno 1857). See also Faà di Bruno 1859 and 1876. Inspired by Faà di Bruno, Giuseppe Peano began his career with invariant theory in 1881, but his interest was short-lived, and in the following year he dropped it to concentrate on calculus and the foundations of mathematics. Cayley carried out a small and intermittent correspondence with Faà di Bruno, mostly concerned with the subjects of transformation of elliptic functions, modular equations, symmetric functions, and differential equations (Faà di Bruno 2004).

33. Scientific Worthies XXI.—William Spottiswoode. *Nature* 27 (26 Apr. 1883): 597–601, esp. p. 597. By morphology in mathematics B. Price meant results derivable from the *form* of expressions and equations.

34. Spottiswoode 1851 and 1856.

35. WRH to John Kells Ingram, in McConnell 1944, 88.

36. GS to WRH, 24 June 1857, in R. P. Graves 1882–89, 3: 88.

37. Boole 1851a, 1851b.

38. GB to Dr. John Bury, 24 Mar. 1851, UCC, BP1/169. After George Boole's *Investigation of the Laws of Thought* (1854), Boole did return to his "mathematical public" but chose differential equations as his subject.

39. Boole 1852, 1853; Smith 1984a, 86–88, 92–93.

40. C to GB, 6 Feb. 1854, TCC, R.2.88.31.

41. Cannon 1978.

42. Dingle 1951a, 305.

43. Whitehead 1927, 37.

44. Bentham and Hooker 1862–83.

45. Dingle 1951b, 86, 98.

46. "Address to the Mathematical and Physical Section of the British Association" = [SP2 (1869): 656].

47. Mandelstam 1994, 227.

48. Cayley introduced a new notation $(a, b, c \ldots \langle x, y)^n$ for the binary quantic written incorporating binomial coefficients:

$$a\binom{n}{0}x^n + b\binom{n}{1}x^{n-1}y + \cdots + l\binom{n}{n}y^n.$$

This notation became universal among those espousing Cayley's approach to invariant theory. The avoidance of subscripts illustrates the weakness of the notation. The coefficients $a, b, c, \ldots$ are *literal* elements—Cayley called them "facients." They were regarded on a par with variables since they were required to be subject to the action of differential operators. Cayley again used an arrow device (to represent binary forms *without* binary coefficients).

49. "A Fourth Memoir on Quantics" = [CP2 (1858): 526].

50. "On a Remarkable Discovery in The Theory of Canonical Forms and of Hyperdeterminants" = [SP1 (1851): 280].

51. A Midsummer Night's Dream, Act Five, II. 16–17; "Algebraical Researches Containing a Disquisition on Newton's Rule . . ." = [SP2 (1864): 380].

52. JJS to C, 28 Oct. 1852, SJC, Sylvester Papers, Box 9.

53. JJS to [Leo Königsberger], [1885], SPKB Darmstadt H5 1923.54.

54. Crosland 1962.

55. Hirst Journal, 18 Nov. 1857, RIGB, 3: 1327.

56. "On Recent Discoveries in Mechanical Conversion of Motion" = [SP3 (1873): 8 fn.].

57. GS to JJS, 24 Apr. 1852, SJC, Sylvester Papers, Box 3. "Glossary of New or Unusual Terms, or of Terms Used in a New or Unusual Sense, in the Preceding Memoir: On A Theory of the Syzygetic Relations of Two Algebraical Functions" = [SP1 (1853): 580–86]; "Recent Terminology in Mathematics" = [CP4 (1860): 594–608]; Salmon gave a comprehensive synopsis of terms in his sketch of the early history of invariant theory (Salmon 1866, 266–82).

58. "A Seventh Memoir on Quantics" = [CP4 (1861): 335].

59. Ref. Rep., 25 Mar. 1861, RSL, RR.4.48.

60. "On the Equation for the Product of the Differences of All but One of the Roots of a Given Equation" = [CP4 (1861): 276–91]. "On the Developable Derived from an Equation of the Fifth Order" = [CP1 (1850): 500–506]. "A Memoir on Caustics" = [CP2 (1857): 365]. "On the Equation of Differences for an Equation of any Order and in Particular for the Equations of the Orders Two, Three, Four, and Five" = [CP4 (1860): 240–61].

61. MacMahon 1914–15.

62. "Tables of the Development of Functions in the Theory of Elliptic Motion" = [CP3 (1861): 360–474]; "On Tschirnhausen's Transformation" = [CP4 (1861): 376].

63. W. B. Davis was a graduate of the University of London (1855). Apart from the calculatory feats performed for Cayley, his own work included a check on the prime numbers which lie between 8 and 9 million (Mathematical Questions from the Educational Times, 1868, 8: 31). His historical interests came to light in papers to the British Association (Davis 1869; H. J. S. Smith 1867).

64. Babbage [1915], 7.

65. JJS to C, 20 Nov. 1862, SJC, Sylvester Papers, Box 9.

66. "An Introductory Memoir on Quantics" = [*CP*2 (1854): 221–34].

67. GB to ADM, 3 Jan. 1855, UCL, LMS Archives. Ref. Rep., 13 May 1854, RSL, RR.2.42. "An Introductory Memoir on Quantics" = [*CP*2 (1854): 232].

68. "Remarques sur la Notation des Fonctions Algébriques" = [*CP*2 (1855): 185–88]. In the printing of Cayley's paper, specifically written about notation, only the vertical bars were printed, so the distinction between matrices and determinants Cayley intended was completely missed.

69. C to JJS, [May, 1854], SJC, Sylvester Papers, Box 2. In this letter and the following one (C to JJS, 27 May 1854, SJC, Sylvester Papers, Box 2), Cayley is implicitly dealing with *irreducible* covariants.

70. C to JJS, 27 May 1854, SJC, Sylvester Papers, Box 2; Parshall 1998, 68.

71. In a postscript to his paper, Cayley introduced the notion of a *semi-invariant* and its importance was noted ("An Introductory Memoir on Quantics" = [*CP*2 (1854): 234]; "On the Equation of Differences for an Equation of any Order and in Particular for the Equations of the Orders Two, Three, Four, and Five" = [*CP*4 (1860): 241]; "On the Number of Covariants of a Binary Quantic" = [*CP*8 (1871): 566]).

72. Cayley's proof of "Cayley's Formula" demonstrated a mastery of the manipulation of operators X and Y ("A Second Memoir upon Quantics" = [*CP*2 (1856): 254–56]; Salmon 1866, 271–73). The differential operator $XY - YX$ is now known as a Lie product of differential operators and the leading semi-invariant A is a characteristic vector of this operator, but Sophus Lie and this terminology lay in the future. "Cayley's Law," which gives the combinatorial rule for the number of linearly independent covariants, is treated in Elliott [1895] 1913, 147–48. Thomas Hawkins refers to this problem as "Cayley's Counting Problem" and examines it historically in the light of modern developments in invariant theory (Hawkins 1986). See also "Postscript (7 Oct. 1854)" = [*CP*2 (1854): 234]; "Nouvelles Recherches sur les Covariants" = [*CP*2 (1854): 167]; and C to JJS, [1854/1855], SJC, Sylvester Papers, Box 2.

73. Cayley's employment of his algorithm for discovering the irreducible covariants of the binary quintic is illustrated in "An Eighth Memoir on Quantics" = [*CP*6 (1867): 151].

74. C to JJS, [1854/1855], SJC, Sylvester Papers, Box 2.

75. C to WT, 22 Sept. 1856, CUL, Add Ms 7342.C59D.

76. Ref. Rep., 23 Oct. 1855, RSL, RR.3.45.

77. JJS to C, [Thurs. 1856], SJC, Sylvester Papers, Box 9. Sylvester made a distinction between *exoscopic* and *endoscopic* methods in algebra, the difference between exterior and interior methods. For example, the formal manipulation of polynomials themselves was exoscopic while those methods dealing with the roots of a polynomials was endoscopic ("On a Theory of the Syzygetic Relations of Two Rational Integer Algebraical Functions Comprising as Application to the Theory of Sturm's Functions, and that of the Greatest Algebraical Common Measure" = [*SP*1 (1853): 431 fn., 580]; "Notes and References" = [*CP*1 (1889): 588]).

78. C to JJS, [1854/1855], SJC, Sylvester Papers, Box 2.

79. Additional notes to Prof. Sylvester's Exeter British Association Address = [*SP*2 (1869–70): 714]; "Note on the Calculus of Forms" = [*SP*1 (1853): 402–3].

80. JJS to C, 13 Aug. 1855, SJC, Sylvester Papers, Box 9.

81. Cayley also calculated sundry covariants for the binary form of order 6, the binary sextic (6 covariants), the binary quantic of order 7 (2 covariants), of order 8 (8 covariants), of order 9 (3 covariants), and order 12 (3 covariants).

82. MacMahon 1896, 17. "Researches on the Partition of Numbers" = [*CP2* (1855): rec. 14 and 20 Apr. 1855, 235–49]; "Supplementary Researches on the Partition of Numbers" = [*CP2* (1857): 506–512]; Dickson 1971, 2: 117–23; MacMahon 1896, 17–22.

83. Cayley's method of dealing with generating functions was to resolve them into partial fractions and analyze their expansions using "prime circulators." His solution of the Poncelet polygon problem involved the expansion of $\sqrt{discrt.(U + xV)}$ as a power series (see note 3). "On a Discovery in the Partition of Numbers" = [*SP2* (1857): 86–89].

84. JJS to C, 19 Apr. 1855, SJC, Sylvester Papers, Box 9.

85. JJS to C, 3 Oct. 1858, SJC, Sylvester Papers, Box 9; also 16 Oct. 1858.

86. "On the Problem of the Virgins, and the General Theory of Compound Partitions" = *SP2* (1858): 113–17]. The study of multidimensional partition theory was a subject that continued to fascinate Sylvester, and in the last year of his life, virtually blind, he turned to the problem once more ("On the Goldbach-Euler Theorem Regarding Prime Numbers" = [*SP4* (1896–97): 736]).

87. C to GGS, 3 May 1861, CUL, Add Ms 7656.C233.

88. F. Clark 1855, xxvii–xxix.

89. The Indian Civil Service examinations were held at King's College, London, in July 1855. *Times*, 21 June 1855, p. 12, col. 6; Danvers 1894, 127. See also GB to ADM, 3 Jan. 1855, in G. C. Smith 1982a, 68–69.

90. C to WT, 22 Aug. 1855, CUL, Add Ms 7342.C59A.

91. JJS to C, 8 Sept. 1855, SJC, Sylvester Papers, Box 9.

92. Lord Brougham to Lord Panmure, 28 Aug. 1855, in Halsted 1897, 163. Sylvester had acted quickly as M. O'Brien had died on Jersey on the 22nd of August.

93. "Biographical Notice of J. J. Sylvester" = [*SP4* (1912): xxvii].

94. GGS to C, 24 Oct. 1855, in Stokes 1907, 1: 387.

95. C to GGS, [Jan. 1857], CUL, Add Ms 7656.C216.

96. GGS to C, 23 Oct. 1858, in Stokes 1907, 1: 393.

97. "On Hansen's Lunar Theory" = [*CP3* (1857): 13–24].

98. C to WFD, 17 May 1855, RUL, Ms 139/1/2; Hansen 1838; Cayley's continuation paper is "A Memoir on the Problem of Disturbed Elliptic Motion" = [*CP3* (1859): 270–92].

99. JCA to C, 12 Dec. 1855, SJC, Adams Papers, 16/6/1/1.

100. WT to HH, 31 July 1864, in Thompson 1910, 1: 433. The problem in invariant theory suggested by W. Thomson was concerned with finding the joint invariants and covariants of the *three* quadratic forms in three variables; the physicist and civil engineer W. J. M. Rankine was involved with similar problems (Rankine 1857). The focus of Cayley's invariant theory was on the single binary form so that this problem would have been peripheral at this stage. C to WT, 14 Dec. 1855, GUL, Kelvin Papers, C10; 17 Dec. 1855, GUL, Kelvin Papers, C11; 26 Dec. 1855, CUL, Add Ms 7342.C59B. The invariant $x + y + z$ is now recognizable as the trace of a matrix.

Chapter 9. Without Portfolio

1. During 1853–56, Cayley published ten papers per year on average, but in the period 1857–60 this rose to thirty papers per year.

2. Cayley had a paper on geometry turned down by the *Philosophical Transactions*. He was given to understand it was "more suited to a mathematical journal," though why this should be the case with a paper on surfaces generated by an ellipsoid and not on his other papers is difficult to fathom. Perhaps the Royal Society wished to avoid being inundated by Cayley's papers in his recent productive surge (C to [TAH], 1 Dec. 1858, UCL, LMS Archive). Instead Cayley sent it to the *Annali di matematica pura e applicata*

set up by Barnaba Tortolini in 1858, where it was published as "Sur la surface qui est l'enveloppe des plans conduits par les points d'un ellipsoide perpendiculairement aux rayons menés le centre" = [*CP*4 (1859): 123–34]. Cayley wrote several papers on algebra and geometry in support of this journal in its early years.

3. "Note on Burman's Law for the Inversion of the Independent Variable" = [*SP*2 (1854) 44–49]; "On Differential Transformation and the Reversion of Serieses" = [*SP*2 (1856): 50–54]; "On the Change of Systems of Independent Variables" = [*SP*2 (1857): 65–85].

4. The operator identity $QP = QxP + Q(P)$ employed by Cayley in his researches on invariant theory (see chap. 8) suggested other questions. The side question, of how the composition of three operators could be so expressed, led him to formulate a theory of "trees." He calculated the number of *distinct* trees with n branches by expanding a generating function as a power series $1 + A_1x + A_2x + A_3x^3 + \cdots$ ("On the Theory of the Analytical Forms called Trees" = [*CP*3 (1857): 242–46]); "On the Analytical Forms Called Trees. Second Part" = [*CP*4 (1859): 112–15]; Biggs, Lloyd, and Wilson 1976, 37–45.

5. "A Memoir on the Symmetric Functions of the Roots of an Equation" = [*CP*2 (1857): 417–39]; Hirsch [1804] 1853; Decker 1910. Symmetric functions are traditionally expressed in terms of the coefficients of polynomials. Cayley made the reverse calculations whereby coefficients, their powers and their products, are expressed in terms of symmetric functions. The so-called Cayley-Betti-Ferrers Law of conjugate partitions states that there is a one-to-one correspondence between partitions and conjugate partitions (Kimberling 1999); "Supplementary Researches on the Partition of Numbers" = [*CP*2 (1858): 506–12]; "Apropos of Partitions" = [*CP*3 (1857): 36]. "On a Problem of the Partition of Numbers" = [*CP*3 (1857): 247–49]).

6. "Lectures on the Theory of Reciprocants" = [*SP*4 (1886): 329].

7. JJS to C, 25 Aug. 1856, SJC, Sylvester Papers, Box 9; Parshall 1998, 91 fn. 29.

8. C to JJS, 31 Oct. 1856, SJC, Sylvester Papers, Box 2.

9. JJS to C, Wednesday, 10 Dec. 1856, SJC, Sylvester Papers, Box 9.

10. *Minutes of Council of the Royal Society,* 8 Nov. 1855, 2: 333, 374.

11. Quoted in Lindgren 1987, 171.

12. Lindgren 1987, 185–211; Hyman 1984, 239–40; a Scheutz machine was eventually used commercially to calculate life-tables in London in 1864.

13. *Minutes of Council of the Royal Society,* 26 June 1856, 2: 362.

14. Charles Babbage to King Oscar I, 7 Apr. 1856, quoted in Lindgren 1987, 199.

15. C to WT, 14 Feb. 1856, CUL, Add Ms 7342.C59C.

16. JJS to C, 22 Feb. 1856, SJC, Sylvester Papers, Box 9.

17. J. D. Forbes to JCM, 13 Feb. 1856, in Campbell and Garnett 1882, 250.

18. John Clerk Maxwell (father) to JCM, 22 Feb. 1856, in Campbell and Garnett 1882, 252.

19. JCM to GGS, 22 Feb. 1856, in Stokes 1907, 2: 4.

20. JCM to WT, 22 Feb. 1856, in *MaxwellP*1 (1990): 400.

21. JJS to C, 22 Feb. 1856, Sylvester Papers, Box 9.

22. *Maxwell* (P and Letters), 1: 392 n. 2.

23. *Maxwell* (P and Letters), 1: 414, n. 4.

24. C to LS, 19 Mar. 1856, in Schläfli 1905, 76.

25. C to LS, [Jan.] 1856, in Schläfli 1905, 70–75.

26. Henderson 1911, 2; Taylor 1894.

27. "An Attempt to Determine the Twenty-Seven Lines upon a Surface of the Third Order, and to Divide Such Surfaces into Species in Reference to the Reality of the Lines upon the Surface" = [*SchläfliP*2 (1858): 198–218].

28. TRK to JJS, 25 Nov. 1856, SJC, Sylvester Papers, Box 2.

29. Ref. Rep. RSL, 5 Jan. 1857, RR.3.167; Kirkman 1857b.

30. "Note on the Summation of a Certain Factorial Expression" = [*CP*3 (1857): 250–53]; "On the Partitions of a Polygon" = [*CP*13 (1891): 93–113]. Biggs 1981, 105.

31. Royal Society, 11 June 1857, vol. 9, cert. 392.

32. JJS to C, 26 Nov. 1856, SJC, Sylvester Papers, Box 9.

33. TPK to JJS, 19 Feb. 1858, SJC, Sylvester Papers, Box 2; Kirkman 1857a.

34. Royal Astronomical Society, 9 Apr. 1857, cert. 851. John Couch Adams was also a sponsor of Cayley's membership. He was proposed on 8 May 1857 and elected 10 July 1857. Cayley served on the Council of the Royal Astronomical Society from Feb. 1858 until Feb. 1893, with only one break (in 1885–86). He was vice-president on fourteen separate occasions and president for 1872–74. See chap. 13.

35. RAS [1858], RAS Miscellaneous Papers 45(4).

36. "A Memoir on the Problem of Disturbed Elliptic Motion" = [*CP*3 (1859): 270–92]; "Memoir on the Development of the Disturbing Function in the Lunar and Planetary Theories" = [*CP*3 (1860): 319–43; CP7 (1872): 511–27].

37. RAS, 13 May 1859, RAS Miscellaneous Papers 45(4).

38. "On Sir W. R. Hamilton's Method for the Problem of Three or More Bodies" = [*CP*3 (1858): 97–103]; "On Lagrange's Solution of the Problem of Two Fixed Centres" = [*CP*3 (1858): 104–10].

39. "Report on the Recent Progress of Theoretical Dynamics" = [*CP*3 (1857): 156–204].

40. *Rep. BAAS, 1858*, Section A, p. 2.

41. Thompson 1910, 2: 600.

42. ADM to WRH, 27 July 1858, in R. P. Graves 1882–89, 3: 533.

43. "Report on the Progress of the Solution of Certain Special Problems of Dynamics" = [*CP*4 (1862): 513–93]. Cayley referred to the "Pfaffian Problem" in correspondence with George Boole (GB to C, 9 Jan. 1863, CUNY, D. E. Smith Papers).

44. Report of the committee appointed for the purpose of endeavoring to procure reports on the progress of the chief branches of mathematics and physics (*Rep. BAAS, 1879*, p. 38).

45. Klein [1908] 1932, 74.

46. Mikhailov 1984.

47. "On a Class of Dynamical Problems" = [*CP*4 (1857): 7–11]; "A 'Smith's Prize' Paper [1869]; Solutions by Prof. Cayley" = [*CP*8 (1869): 445–46].

48. Cayley showed only a passing interest in problems that involved optimizing integrals: "On a Problem in the Calculus of Variations" = [*CP*7 (1869–71): 263]); *Nature* 3 (2 Mar. 1871): 358–59; Knott 1911, 99.

49. JJS to C, 13 May 1856, SJC, Sylvester Papers, Box 9.

50. The specific aim of the École Centrale des Arts et Manufactures was to produce industrial engineering graduates. There was a three-year course including such practical subjects as descriptive geometry, mechanics, and machine theory (Grattan-Guinness 1990, 2: 1113–15). See Williams 1966. This episode in the history of education in Britain is largely unknown.

51. *Times*, 10 Sept. 1857, p. 3, col. 1.

52. Williams 1966, 32.

53. *Illustrated London News*, 12 Sept. 1857, 31 (no. 878): 269–70. p. 270, col. 1.

54. *Times*, 11 Sept. 1857, p. 3, col. 1.

55. Williams 1966, 39.

56. JJS to C, 14 Sept. 1857, SJC, Sylvester Papers, Box 9.

57. C to JJS, 19 Nov. 1857, SJC, Sylvester Papers, Box 2; Parshall 1998, 95–96.

58. "A Memoir on the Theory of Matrices" = [CP2 (1858): 475–96 rec. 10 Dec. 1857, read 14 Jan. 1858]; "A Memoir on the Automorphic Linear Transformations of a Bipartite Quadric Function" = [CP2 (1858): 497–505].

59. "On a Theorem Relating to Reciprocal Triangles" = [CP3 (1857): 5–7].

60. F. G. Eisenstein introduced a succinct matrix notation for composing linear substitutions in his fruitful year of 1844 ("Untersuchungen über die cubischen Formen mit zwei Variabeln [Investigations on cubic forms with two variables]" = [EisensteinP1 (1843): 10–25]; T. Hawkins has discussed these block notations in the context of Cayley's work (Hawkins 1977a, 82–87). Eisenstein was aware that linear substitutions could be added and multiplied and that the multiplication of them was not commutative.

61. "Eisenstein's Geometrical Proof of the Fundamental Theorem for Quadratic Residues (Translated from the Original Memoir, J. rei. ang. Math 28 (1844), with an addition by A. Cayley)" = [CP3 (1857): 39–43, quote on 42]).

62. W. R. Hamilton may have discussed matrices with Eisenstein a few months prior to the discovery of the quaternions. Eisenstein visited Hamilton at the Dunsink Observatory in the summer of 1843 and they discussed algebra. Eisenstein had visited England in 1842, but I have found no evidence that he met Cayley then; they may have encountered one another when Cayley visited Berlin in 1845.

63. "Sur la Théorie de la Transformation des Fonctions Abéliennes" = [HermiteP1 (1855): 444–78].

64. "Démonstration nouvelle d'une proposition relative a la théorie des formes quadratiques" = [DirichletP2 (1857): 209–14]. A linear substitution was written by Cayley as $x = \alpha X + \beta Y + \gamma Z$, $y = \alpha'X + \beta'Y + \gamma'Z$, $z = \alpha''X + \beta''Y + \gamma''Z$. They were typically given for two or three variables. Matrices of small order were of especial interest to him and found a place in other parts of his research: binary forms, the transformation of theta functions, linear fractional substitutions, problems in hydrodynamics, all of which only required 2×2 matrices.

65. "On a Theorem of M. Lejeune-Dirichlet's" = [CP2 (1854): 47–48].

66. Algebraic researches received an impulse in Germany by P. J. L. Dirichlet, who went to Göttingen in 1855. He became a friend and teacher of R. Dedekind, who taught a course on Galois's theory in the winter semester of 1856–57 (Corry 1996, 66–136).

67. "Additions to the Articles, 'On a New Class of Theorems,' and 'On Pascal's Theorem'" = [SP1 (1850): 150]; "On the Principles of the Calculus of Forms" = [SP1 (1852): 294 fn.].

68. JJS to C, 21 Sept. 1852, SJC, Sylvester Papers, Box 9. "On a Theorem in the Geometry of Position" = [CP1 (1841): 1–4]. Sylvester used the term *matrix* as an array from which determinants may be extracted. In his letter he gave the multiplication of 3×3 "matrices" as a multiplication carried out row by row. Sylvester dealt with the expression for the inverse using determinants ("On a Theory of the Syzygetic Relations of Two Algebraical Function . . ." = [SP1 (1853): 583]). At the beginning of the 1850s, Cayley used the term *matrix* as an array from which to extract determinants and used *double* vertical bars to stress the difference between these and ordinary determinants (C to JJS, 1 Nov. 1851, SJC, Sylvester Papers, Box 9; "On Linear Transformations" = [CP1 (1846): 95–112, for example, pp. 103–4]).

69. Cayley's intimation that "matrix" may be used in a "more general sense" might be considered as suggesting the possibility of n-dimensional matrices comparable to his earlier n-dimensional determinants. But here I believe Cayley meant the more mundane interpretation of "more general sense" to mean the arrangement of terms in

various two-dimensional patterns ("Recent Terminology in Mathematics" = [*CP*4 (1891): 601]).

70. Ref. Rep., 29 Mar. 1858, RSL, RR.3.55.

71. The elements of a matrix of Cayley's matrices may be considered as *variables* in the same way that "facients," which appeared as coefficients in algebraic forms, could be treated as variables. Cayley does not consider matrices with numerical entries. Moreover, he regarded the symbol *M* as a symbol denoting a matrix in much the same way he regarded writing *q* for a quaternion. This was useful but no substitute for the full expression in terms of coordinates. Cayley regarded the use of symbols *M, q* as a *condensed* notation but they had to be "unfolded" to be "properly understood" ("On Tschirnhausen's Transformation" = [*CP*4 (1861): 392–93]).

72. "Recent Terminology in Mathematics" = [*CP*4 (1860): 594]. Cayley did not reinforce his view on the primacy of matrices over determinants. He actually lapsed into dealing with linear transformations without them and could occasionally be casual about the distinction, to the extent of using a single symbol *L* to represent both a matrix *and* its determinant ("A Seventh Memoir on Quantics" = [*CP*4 (1861): 331]). At other times, he used matrices in the older sense of an array from which determinants may be extracted ("On the Double Tangents of a Plane Curve" = [*CP*4 (1859): 192]).

73. WRH to GS, [1857], in R. P. Graves 1882–89, 3: 88. Cayley observed that certain 2 × 2 matrices *L, M, LM* satisfy the same relationships as the quaternionic symbols, *i, j, k.* ("A Memoir on the Theory of Matrices" = [*CP*2 (1858): 491]).

74. Cayley noted that "matrices comport themselves as single quantities." It was thus necessary for him to state whether he was thinking of *A* as denoting an array or a single quantity. In the appropriate circumstances, a matrix *A* considered as single quantity would be written $\bar{A}$. Thus Cayley had both the matrix unity *I* (the identity matrix) and the ordinary unity $\bar{I}$. He would then speak of the matrix *A* *involving the matrix unity,* namely, $\bar{A}I$. Cayley in effect identified $\bar{A}I$ with

$$\begin{pmatrix} A & 0 \\ 0 & A \end{pmatrix}$$

and used this ambiguous device in his presentation of the Cayley-Hamilton theorem.

75. Ref. Rep., 29 Mar. 1858, RSL, RR.3.55.

76. "Sketch of a Proof of the Theorem that Every Algebraic Equation Has a Root" = [*CP*4 (1859): 117]

77. "On the Simultaneous Transformation of Two Homogeneous Functions of the Second Order" = [*CP*1 (1849): 431].

78. Cayley concluded his letter to Sylvester with several conjectures: "I believe the theorem might be extended to inconvertible [noncommutative] matrices by modifying the def[inition] of a determ[inan]t viz.

$$\begin{vmatrix} a, & b \\ c, & d \end{vmatrix} = \frac{1}{2}(ad + da - bc - cb)$$

but I am not at all sure as to this" (Crilly 1978). The study of determinants and matrices over noncommutative systems lay in the distant future, but Cayley's attempt to broach the question is notable.

79. WRH to GS, [1857], in R. P. Graves 1882–89, 3: 88.

80. C to PGT, [21] June 1894, in Knott 1911, 164.

81. Cayley returned to matrix algebra in 1865 when he applied his method of transforming a bilinear form to a specific bilinear form ("A Supplementary Memoir on the Theory of Matrices" = [*CP*5 (1866): 438–48]). Matrix algebra is the subject most frequently linked with Cayley's name. Yet, statistically, Cayley's attention to matrix algebra is even slighter than his attention to group theory and is insignificant when compared to the large corpus he produced on invariant theory.

82. W. Spottiswoode adopted Cayley's matrix notation in a revision of his memoir on determinants published in 1856 (Spottiswoode 1856). H. J. S. Smith and Cayley were in communication by letter between June 1857 and January 1858 and Smith utilized Cayley's matrix notation in his work on quadratic forms in relation to the theory of numbers (HJSS to C, 13 Jan. 1858, CUNY, D. E. Smith Papers). Cayley recognized the importance of Smith's work on linear equations and congruences but did not appear to publish in this area (Ref. Rep., RSL, RR.4.242).

83. An analysis of Cayley's matrix theory papers in the history of matrix algebra in general is given in Hawkins 1975; 1977a; 1977b; 1978.

84. Submitted 11 Feb. 1858. "The Fourth Memoir on Quantics" = [*CP*2 (1858): 513–26]; "The Fifth Memoir on Quantics" = [*CP*2 (1858): 527–57].

85. Submitted on 5 Mar. 1858, the paper was "On the Cubic Transformation of an Elliptic Function" = [*CP*3 (1858): 266–67, sub. 5 Mar. 1858]. W. K. Clifford knew the theorem on the cubic transformation as "Cayley's theorem" (*CliffordP* [1882]: 221).

86. RSL, 15 June 1858, RR3, 216.

87. R. W. Clark 1981, 61–68; Cayley 1857–59. Cayley joined the Alpine Club in December 1860 and remained a member until 1864.

88. Ball 1859, 1: 195.

89. Himmelfarb 1968, 209. Mountaineering was a common pursuit amongst "Men of Science" in the nineteenth century, and climbing in the Alps affords the traditional link with Romanticism which colored their scientific activity. To a research mathematician, mathematics means mathematics and Cayley was stringent as any when it came to "philosophic chatter." However, even he was influenced by the spirit of the times in which he lived: mathematics was not a matter of unadorned logic but a search for Wordsworth's "intuitive truths." At the time Cayley was climbing Le Dom in Switzerland in 1858, he was the same age as Wordsworth on the poet's completion of *The Prelude* (republished 1850). One can easily imagine him reading this work beloved of the Victorians, especially Book 6 (Cambridge and the Alps): "The pleasure gathered from the rudiments / of geometrical science," leading on to "distant mountains and their snowy tops" (Wordsworth 1994, 676, 685).

90. Perhaps unknown to Cayley, Maxwell had included the mathematical rendition of these topographical ideas in his geometrical lectures at Cambridge in 1856. For connections between Cayley and Maxwell on this topological subject, see "British Association Paper on Hills and Dales" = [*Maxwell* (P and Letters) 2 (1870): 566–67].

91. "On Contour and Slope Lines" = [*CP*4 (1859): 108–11]; "Notes and References" = [*CP*4 (1891): 609].

92. *Rep. BAAS, 1858.* Mathematics and Physics Section, p. 3.

93. C to JFWH, 13 Nov. 1858, RSL, 5.226.

94. ADM to JFWH, 15 Nov. 1858, in De Morgan 1882, 297.

95. JCA to C, 17 Nov. 1858, and 20 Nov. 1858, SJC, Adams Papers, 16/6/2 and 16/6/3.

96. JJS to C, 22 Dec. 1858, SJC, Sylvester Papers, Box 9. Sylvester despaired of the Woolwich appointment sapping his creative energy to publicly hope for "tranquility of mind" to continue his research on number theory ("Note on the Algebraical Theory of Derivative Points of Curves of the Third Degree" = [*SP*2 (1858): 109]).

97. JJS to C, 22 Dec. 1858, SJC, Sylvester Papers, Box 9.
98. C to WT, 7 Aug. 1846, CUL, Add Ms 7342.C35.

Chapter 10. The Road to Academe

1. Salmon 1883, 483.
2. "An Introductory Memoir on Quantics" = [*CP*2 (1854): 222].
3. C. S. Peirce (4 Feb. 1895). Obituary for Arthur Cayley. *Evening Post* [New York] p. 7, col. 2., reprinted in Peirce 1976, 2: 647.
4. "A Sixth Memoir on Quantics" = [*CP*2 (1859): 592]; Macaulay 1896.
5. "Demonstration of Pascal's Theorem" = [*CP*1 (1843): 43–45]. He defined distance in terms of bilinear forms for spherical geometry ("On Some Analytical Formulae, and Their Application to the Theory of Spherical Coordinates" = [*CP*1 (1846): 219]).
6. "A Sixth Memoir on Quantics" = [*CP*2 (1859): 592].
7. The year of 1859 provided several other landmarks in English cultural life. Charles Dickens's *A Tale of Two Cities*, George Eliot's *Adam Bede*, Samuel Smiles's *Self-Help*, John Stuart Mill's *On Liberty*. Cayley cared little for Dickens, admired Eliot's writing, and opposed Mill's views on the foundations of mathematics.
8. "On Poinsot's Four New Regular Solids" = [*CP*4 (1859): 81–85, 86–87]. Cayley identified *D*, the *density* of a polyhedron, but did not give it this name (Cromwell 1997, 257–59). Cayley ventured into "topology" with his "On the Partitions of a Close" = [*CP*5 (1861): 62–65, 617]. His definition of a *close* as an enclosed space which is path-connected was intuitively defined: "An enclosed space such that no part of it is shut out from any other part of it, or, what is the same thing, such that any part can be joined with any other by a line not cutting the boundary, is term[ed] a close" (p. 63). Cayley's notion of a close presaged his attempt to prove the "four-color theorem," in which the concept was utilized (see chap. 14). His notion of a "path" was similarly defined, and while his definition would hardly pass any "Cantorian test of rigour," it illustrates Cayley's perspective as an intuitionist mathematician: "The words line and curve are used indifferently to denote any path which can be described *currente calamo* [with the pen running on; his italics] without lifting the pen from the paper." In two letters to Sylvester, Cayley explored the notion of path-connected regions and topology of the plane [1860], SJC, Sylvester Papers, Box 2. Cayley is now regarded as one of the first to recognize the topological character of the polyhedral formulae connecting vertices, edges, and faces (Burde and Zieschang 1999, 104).
9. HB to WT, 11 Mar. 1849, CUL, Add Ms 7342.B102.
10. E. H. D'Avigdor was the grandson of the financier Sir Isaac Lyon Goldsmid (1778–1859), who had been a supporter of the setting up of London University, and one of Sylvester's proposers in his election to the Royal Society in 1839. D'Avigdor graduated from the University of London in 1861 and was employed by John Hawkshaw's civil engineering firm. He had hoped to enter the corps of Royal Engineers but was disqualified through nearsightedness. Starting his engineering career in Hull, he worked on projects in China, Hungary, and Austria. (1895, [obit.] *Institution of Civil Engineering Minutes of Proceedings* 121: 340–41.)
11. EHA to C, 5 Apr. 1859, 7 Apr. 1859, SJC, Sylvester Papers, Box 9, 7a. It is tempting to speculate that Cayley's speculation was connected with the remarkable Victorian figure Henry Perigal. He knew Cayley and was his sponsor for membership of the Royal Astronomical Society. He was a keen geometrician and a man of independent thought, among his interests being the popular mechanical devices that interested Cayley. He paid particular attention to the kind that could produce beautifully regular patterns of parallel

curves known as Lissajous figures. The kind of devices used for copying engineering drawings would have caught his attention.

12. *Stone Building's Order Book 1843–1891*, Lincoln's Inn Library, E3a3.

13. HB to WT, 11 Mar. 1849, CUL, Add Ms 7342.B102.

14. "Plan of a Curve-tracing Apparatus" = [*CP*8 (1871–73): 179–80. Around 1870 Cayley investigated the mechanical description of cubic and quartic curves. Later he put forward an idea for the evaluation of line integrals ("Suggestion of a Mechanical Integrator for the Calculation of $\int(Xdx + Ydy)$ along an Arbitrary Path" = [*CP*11 (1877): 52–54]).

15. WT to GGS, 6 Oct. 1859, Add Ms 7656.K106. In Wilson 1990, 1: 248, 250.

16. WT to GGS, 6 Oct. 1859, Add Ms 7656.K106; see also WT to GGS, 6 Oct. 1859, Add Ms 7656.K107 in Wilson 1990, 1: 250.

17. *Times*, 27 Oct. 1859, p. 12, col. d.

18. *Royal Astronomical Society Minutes*, 9 Dec. 1859, 6: 70.

19. *AdamsP*1 (1896): xxxv–xxxix.

20. JCA to C, 12 Dec. 1855, SJC, Adams Papers, 16/6/1/1; C to JCA, 12 June 1860, SJC, Adams Papers, 6/4/1/1.

21. "On the Secular Acceleration of the Moon's Mean Motion" = [*CP*3 (1862): 522–61, 568]; "Note on Plana's Lunar Theory" = [*CP*7 (1862–63): 357]; Glaisher 1896, 192–93; 1896–1900, 1: xxxv–xxxix. The calculations were made from a series solution to the differential equations which determine the motion of the moon and the Adams's result hinged on the evaluation of a multiplying factor. The Italian astronomer Giovanni Plana obtained $(3/2)m^2 - (2187/128)m^4$ and Adams obtained $(3/2)m^2 - (3771/64)m^4$, which was verified by Cayley via an independent method ("Notes and References" = [*CP*3 (1890): 568]). Results on the motion of the moon were important in the astronomy of the nineteenth century. The problem of the "secular acceleration" was treated in popular histories (Clerke 1893, 332–35).

22. "Address delivered by the President, Professor Cayley, on Presenting the Gold Medal of the [Royal Astronomical] Society to Professor Simon Newcomb" = [*CP*9 (1873–74): 180].

23. "Notes and References" = [*CP*3 (1890): 568].

24. ADM to JFWH, 19 May 1845, in S. De Morgan 1882, 150.

25. JJS to C, 26 Apr. 1859, SJC, Sylvester Papers, Box 9. The lectures were held on 6th, 9th, 16th, 20th, and [30th?] of June as well as the 4th and 11th of July 1859. JJS to C, 26 May 1883, SJC, Sylvester Papers, Box 11; "Inaugural Lecture at Oxford on the Method of Reciprocants" = [*SP*4 (1886): 280]; JJS to R. F. Scott, 22 Oct. 1888, SJC, Sylvester Papers, Box 1. JJS to C, 29 June 1859, SJC, Sylvester Papers, Box 9.

26. "Outlines of Seven Lectures on the Partitions of Numbers" = [*SP*2 (1897): 119–75]. The text of these lectures was rescued many years later and published in the *Proceedings of the London Mathematical Society* ("Outlines of Seven Lectures on the Partition of Numbers" = [*SP*2 (1897): 119–75]).

27. MacMahon 1897, 29.

28. Hirst *Journal*, 9 Oct. 1859, RIGB, 3: 1501.

29. GS to JJS, 1 May 1861, in the *Eagle*, 1908, 29: 380–81. Reprinted in "Biographical Notice of J. J. Sylvester" = [*SP*4 (1897): xxvi].

30. "Outlines of Seven Lectures on the Partitions of Numbers" = [*SP*2 (1897): 175].

31. C to JJS, 11 Aug. 1860, SJC, Sylvester Papers, Box 2.

32. "On a Problem of Double Partitions" = [*CP*4 (1860): 166–70].

33. E. Venables, *Guardian* (6 Feb. 1895): 200.

34. Lützen 1990, 233–35. JJS to LOH, 15 Mar. 1863, SPKB Darmstadt H5 1926.10; Parshall 1998, 108–13.

35. Hirst *Journal,* 9 Nov. 1861, RIGB, 3: 1593.

36. JJS to C, [1861], SJC, Sylvester Papers, Box 10.

37. Allibone 1976.

38. Jensen 1970, 63.

39. Hirst *Journal,* 21 Nov. 1858, RIGB, 3: 1423, and then 16 July 1859, RIGB, 3: 1494. Cayley's pitted facial features may have been caused by smallpox, a fairly common affliction in the first half of the century before the advent of a successful immunization treatment (a mild form of smallpox may have been the illness he suffered during the latter half of 1854).

40. Hirst *Journal,* 23 Dec. 1859, RIGB, 3: 1520.

41. Rossetti 1906, 175.

42. Hirst *Journal,* 1860, RIGB, 3: 1548.

43. Eve and Creasey 1945, 275.

44. TAH to [LAC], 18 Apr. 1862, RIGB, Tyndall Papers 7.H3.

45. Gardner and Wilson 1993; Hirst *Journal,* 3 May 1862, RIGB, 3: 1610.

46. In 1854, Cayley was proposed by S. H. Christie, secretary of the Royal Society and professor of mathematics at Woolwich Military Academy. In 1858 he was proposed by A. Smith. George Gabriel Stokes was the proposer in 1859, when he was awarded the Royal Medal. The other Royal Medal in 1859 was awarded to the botanist G. Bentham (nephew of Jeremy Bentham).

47. Brock 1967.

48. 1859. *PRSL* 10: 174.

49. Salmon 1866.

50. Biggs 1981, 106. Kirkman acquired a deep understanding of permutation groups. His contributions to group theory occurred during 1859–63 in connection with his entry for the Grand Prix de Mathématiques of 1860 set by the Académie des Sciences in Paris.

51. "On the Theory of Groups as Depending on the Symbolical Equation $\theta^n = 1$. Third Part" = [*CP*4 (1859): 88]. Cayley enumerated the five distinct groups of order 8 in terms of generators and relations. For the particular one defined by 1, α, α^2, α^3, β, $\alpha\beta$, $\alpha\beta^2$, $\alpha\beta^3$, where ($\alpha^4 = 1$, $\beta^2 = \alpha^2$, $\alpha\beta = \beta\alpha^3$), he noted that the symbols α, β, $\gamma = \alpha\beta$ were analogous to the quaternion symbols i, j, k and the group identical to the group $1, i, j, k, -1, -i, -j, -k$.

52. TPK to William Francis, 6 May 1861, Taylor and Francis Archive, St.BPL.

53. "Mémoire sur les arrangements que l'on peut former avec des lettres données et sur les permutations ou substitutions a l'aide desquelles on passe d'un arrangement a un autre" = [*CauchyP*XIII (2d Series) (1844): 171–282]; Belhoste 1991; Dahan 1980.

54. C to JJS, 16 Aug. 1860, Sylvester Papers, SJC, Box 10. Parshall 1998, 97–101. Cayley's abiding interest in the choice of notation is evident here. Cayley suggested the use of the arrowhead on "(" of (*a, b, c, d*), which he said was "a good piece of notation to distinguish Cauchy's circular substitutions," but it does not appear to have entered the common stock. He suggested another improvement on Cauchy's notation for a substitution, which imitated "multiplication of fractions" and at the same time ensured that substitutions could not be combined in the wrong order (Crilly 2000, 12–13).

55. C to JJS, 18 Aug. 1860, SJC, Sylvester Papers, Box 2.

56. *CauchyP*XIII (2d Series) (1844): 280. See note 55.

57. C to JJS, 18 Aug. 1860, SJC, Sylvester Papers, Box 2. Richard Dedekind worked on his own form of this question during 1855–58 when he first studied group theory (Hawkins 1974, 221).

58. TPK to William Francis, 6 May 1861, Taylor & Francis Archive, StBPL.

59. Ibid.

60. Kirkman 1860–62; 1862. Biggs 1981, 109–10.

61. TPK to JFWH, [undated], RSL, RS.11.55.

62. TPK to C, 14 Jan. 1863, SJC, Sylvester Papers, Box 2; Kirkman 1863.

63. TPK to C, 16 Feb. 1863, SJC, Sylvester Papers, Box 2.

64. *Times*, 30 Apr. 1863, p. 11, col. e.

65. "On Tschirnhausen's Transformation" = [CP4 (1861): 375–94].

66. "On the Equation of Differences for an Equation of any Order and in Particular for the Equations of the Orders Two, Three, Four, and Five" = [CP4 (1860): 240–61]; "On the Equation for the Product of the Differences of all but One of the Roots of a Given Equation" = [CP4 (1861): 276–91]. In the latter, Cayley made use of the notion of a *semi-invariant* introduced in his "Introductory Memoir on Quantics" = [CP2 (1854): 221–34]; "On a New Auxiliary Equation in the Theory of Equations of the Fifth Order" = [CP4 (1861): 309–24].

67. Cayley was the principal proposer of both Cockle and Harley for membership of the Royal Society, Harley in 1863 and Cockle in 1865.

68. "On a New Auxiliary Equation in the Theory of Equations of the Fifth Order" = [CP4 (1862): 324]; "Notes and References" = [CP4 (1891): 609–16]. Cayley computed Jacobi's resolvent equation but gave no details of his calculations (Dickson 1925).

69. MacMahon 1894 and 1904.

70. G. N. Watson commentated on Cayley's use of ungrammatical English but noted his extreme clarity (Berndt, Spearman, Williams 2002, 26).

71. Cayley 1862, 290.

72. SA to C, 17 June 1861 = [CP4 (20 Sept. 1861): 326]. S. H. Aronhold was a student of Otto Hesse.

73. "A Seventh Memoir on Quantics" = [CP4 (1861): 325–41, sub. 28 Feb. 1861, read 14 Mar. 1861].

74. Crilly 1986. The detail of the second, third and fifth, eighth, ninth, tenth memoirs on quantics deal mainly with the binary form.

75. H. J. S. Smith 1862, 296.

76. "Theory of Numbers" = [CP11 (1884): 609].

77. "Note on the Porism of the In-and-circumscribed Polygon" = [CP2 (1853): 87–92]; Salmon 1869, 330.

78. Spottiswoode 1865, 5.

79. "On a New Analytical Representation of Curves in Space" = [CP4 (1860): sub. 2 June 1859, 446–55, CP4 (1862): sub. 30 Oct. 1862, 490–94]. A paper published in 1862 suggested that a curve in space could be represented as the intersection of a cone and a "monoid surface." ("Considérations générales sur les courbes en espace" = [CP5 (1862): 7–20]). Thus a straight line in space, the simplest space curve of all, could be represented as the intersection of a standard cone with a tangent plane. Cayley described a representation of the special case of the straight line by "six-coordinates," thus enabling a line to be seen as the basic spatial element and not as an assemblage of points ("On the Six Coordinates of a Line" = [CP7 (1869): 66–98]; "Notes and References" = [CP4 (1891): 616]; Baker 1933, 4: 40; Plücker 1865). For an estimate of Cayley's attempt to deal with the general problem of representing a curve in space and its eventual solution in 1937, see Semple and Kneebone 1959, 207–9.

80. C to LAC, 2 Jan. 1861, in Cremona 1992, 1: 108.

81. Bertini 1903–4, vii. In 1866, Cremona shared the Berlin Academy Steiner Prize for his work on the transformation of plane curves and cubic surfaces and he was to win it again in 1874.

82. "Sur les cones du second ordre qui passent par six points donnés" = [*CP*5 (1861): 4–6]. When Cayley comprehensively treated quartic surfaces, he noted the special properties of Weddle's surface ("A Memoir on Quartic Surfaces" = [*CP*7 (1869–71): 162, 178–79]).

83. *MNRAS* 14 (1954), 116–19.

84. "On the Construction of the Ninth Point of Intersection of the Cubics which Pass Through Eight Given Points" = [*CP*4 (1862): 447].

85. "On Skew Surfaces, Otherwise Scrolls" = [*CP*5 (1863): 168–200]; Hill 1897, 141.

86. JJS to C, 7 Apr. 1859, SJC, Sylvester Papers, Box 9. Evidently Cayley and Sylvester applied to examine the mathematical Tripos at Cambridge in 1860 but the young E. J. Routh was to gain the post of examiner for his first time in that year.

87. JJS to C, 15 June [1861], SJC, Sylvester Papers, Box 10.

88. Clark 1904, 268–69. In 1847, for example, Samuel Blackall's lectures on algebra given in the second term of the first year included such topics as the elements of continued fractions and the theory of probability (Ramsey 1917, 1).

89. Hopkins 1854, 36.

90. Crilly 1999; Winstanley [1940] 1955, 333.

91. GGS to Mary Stokes, 31 Dec. 1857, in Stokes 1907, 1: 73.

92. *Times*, 21 May 1863, p. 8, col. 5.

93. C to EA, 21 May 1863, CUL, Add Ms 6580.50; Cayley 1863; C to EA, 27 May 1863, CUL, Add Ms 6580.54.

94. J. C. Williams-Ellis was the father of Bertram Clough Williams-Ellis (1883–1978), the architect who modeled an entire village in North Wales on the Italian town of Portofino. He made the first ascent of the Finsteraarhorn in the same year Cayley scaled the Dom and became a celebrated mountaineer.

95. Cayley and Sylvester were the first two signatories on the certificate proposing Todhunter's candidature for membership of the Royal Society when he was elected (Royal Society, 5 June 1862, vol. 10, cert. 29).

96. IT to GB, 23 May 1863, in MacHale 1985, 233.

97. CUL, University Archives, Sadleirian Chair of Pure Mathematics, O.XIV.52.128.

98. Davidson 1873, 3 (Part 2): 1067 fn.

Part 4. The High Plateau, 1863–1882
Epigraph: Arthur Cayley = [*CP*8 (1895): xv].

Chapter 11. The Mathematician Laureate

1. The Sadleirian Chair in Pure Mathematics was established in 1863. It was formally approved on 7 March 1860 by the Queen in Council and replaced 17 college-based lectureships in algebra. Cayley and his immediate successor, A. R. Forsyth preferred the spelling "Sadlerian," though "Sadleirian" conforms to the name of the chair's benefactor. Chairs in scientific subjects were gradually established at Cambridge: in zoology and comparative anatomy (1866), experimental physics (1871) and a chair of mechanism and applied mathematics (1875).

2. For the first year, Cayley received a stipend of £509-12s-9d, without the supplement of a college fellowship. By 1882, when new statutes of the university were enacted, the stipend for the Sadleirian chair was increased to £850, and he was an ordinary fellow of Trinity College and thus entitled to a share of College income.

3. For a survey of middle-class income in the nineteenth century, see Musgrove 1959. Cayley gave up the potential of a much higher salary by leaving the law profession. His

own principal in the law, J. H. Christie, for example, left more than £70,000 in his will (*Times,* 19 May 1876, p. 10e).

4. Glaisher 1996a, 192.

5. Several sons followed in their father's footsteps and became City merchants. In Greenwich, the Molines lived at 18 Nelson Street and then in Blackheath at 4 Westcombe Park. Nelson Street was only a short walk down the steep hill from the Cayley family home in The Grove, Blackheath.

6. E. Venables, *Guardian* (6 Feb. 1895): 201; JJS to C, 25 Dec. 1882, SJC, Sylvester Papers, Box 11; Forsyth 1895, xvi.

7. MacLaurin 1982, 66. Colin MacLaurin was professor of mathematics at Aberdeen and Edinburgh in the eighteenth century. There were minimal duties attached to his chairs and he enjoyed more or less unfettered freedom to carry out the academic projects of his own choosing. Cayley was a "new professor" (Rothblatt 1968). That professorships "bestowed as a reward for past service" were coming to an end was noted in (Heitland 1929, 267).

8. M[ayor] 1884.

9. WW to ADM, 19 Oct. 1863, TCC, O.15.47.30.

10. CUL, University Archives, Min.V.7, p. 1.

11. This was the first lecture of a series delivered at noon on Tuesdays and Thursdays during the Michaelmas [autumn] term, 1863. The text of the lecture with commentary has been published in full (Crilly 1999).

12. Glaisher 1895b, 174–76.

13. "On the Intersection of Curves" = [*CP*1 (1843): 25]. The generalization Cayley achieved as a young man became known as the Cayley-Bacharach theorem (see chap. 3); "On the Construction of the Ninth Point of Intersection of the Cubics Which Pass Through Eight Given Points" = [*CP*4 (1862): 495–504]; C to TAH, 7 Nov. [1865], UCL, LMS Archive.

14. Students at Cambridge in 1863 who went on to academic careers in which mathematics played a major part included: (Freshmen) Charles Niven, W. K. Clifford, C. J. Lambert, Osborne Reynolds; (Second year) Robert Morton, W. D. Niven, James Stuart, George Pirie; (Third Year) Lord Rayleigh, Alfred Marshall, H. M. Taylor, H. N. Grimley; (Finalists) H. J. Purkiss, W. P. Turnbull, Robert Pearce. For further details, see Crilly 1999, 133.

15. "Analytical Metrics" = [*CliffordP* (1865): 84–85]. W. K. Clifford saw the logical "vicious circle" of defining the projective distance (which he called a graphometric function) in terms of the cross-ratio, which itself was defined in terms of length. He evidently took the pragmatic view that cross-ratio belonged to metric geometry because it was defined in terms of length but it was also part of projective geometry because it was unaltered by projection.

16. C to JCA, 4 Nov. [1866], SJC, Adams Papers 24/4/4/1, mourning paper. "On the Determination of the Orbit of a Planet from Three Observations" = [*CP*7 (1868–70): 384–86, 400–477].

17. J. M. Wilson 1932, 42–43. J. W. Strutt's (Lord Rayleigh) reading list is given in Rayleigh 1924, 29. Wilson gave a reading list similar to the one given by Strutt.

18. Coaches were free to choose the level of their fees but most stayed within the commonly accepted standard charges. The standard charge for coaching in 1860 was about seven guineas (£7-7s-0d = £7.35p) per term and twelve guineas (£12-12s-0d = £12.60p) for coaching over the long summer vacation (Jackson 1910, 449).

19. Ridley 1970, 509.

20. Everett 1866, 43.

21. De Morgan 1882, 283.

22. Glaisher 1886, 38.

23. Garden House, 23 Little St. Mary's Lane, Cambridge. After 1923, the house was used first as a boarding house, then as a hotel. It was destroyed by a "mystery fire" on 24 April 1972. On the site is the modern Garden House Hotel, a hotel of international standard.

24. Hirst *Journal,* 1 July 1866, RIGB, 4: 1788.

25. Botton 1999, 1172–73.

26. JOT to C, Saturday [1864], TCC, Adv. Album, vol. 10.

27. ADM to WW, 15 Oct. 1863, TCC, Add Ms.a.202/152. Quoted in Biggs 1981, 112.

28. WW to ADM, 19 Oct. 1863, TCC, O.15.47.30.

29. Kirkman 1871, 52.

30. JJS to William Francis, [1865], Taylor & Francis Archive, St.BPL.

31. Guggisberg 1902, 107. This anonymous cadet began his studies at Woolwich in July 1863.

32. "On Newton's Rule for the Discovery of Imaginary Roots" = [*SP2* (1864): 376–479].

33. *Times,* 28 June 1865, p. 12. Ref. Rep., 28 Apr. [1864], RSL, RR.5.267. Cayley tempered his praise for Sylvester's paper. He rarely complained in his role of academic referee but noted: "I cannot help remarking that the manuscript is not such a one as should have been presented to the [Royal] Society, the illegibility is such as not only to make the referee's task a very irksome one, but to throw undue difficulty in the way of his forming a correct judgement of the paper" (Ref. Rep., 2 Oct. 1864, RSL, RR.5.269).

34. Ref. Rep., 23 June 1868, RSL, RR.6.65; "On the Conditions for the Existence of Three Equal Roots or of Two Pairs of Equal Roots of a Binary Quartic or Quintic" = [*CP6* (1868): 300–311].

35. Cayley did not study Fourier series in any depth (C to LS, 11 Sept. 1871, in Schläfli 1905, 107).

36. WT to HH, 31 July 1864, in Thompson 1910, 1: 433.

37. C to WT, 14 Dec. [1855], GUL, Kelvin Papers, C.10.

38. IT to C, 30 May 1876, UCL, Pearson Papers, 623; C to IT, [1876], UCL, Pearson Papers, 623.

39. "Note on the Numerical Calculation of the Roots of Fluctuating Functions" = [*RayleighP1* (1874): 194–95].

40. "On an Analytical Theorem Relating to the Distribution of Electricity upon Spherical Surfaces" = [*CP4* (1859): 92–107]; "On the Distribution of Electricity on Two Spherical Surfaces" = [*CP11* (1878): 1–6].

41. Cannon 1978, 117. Parts of Cayley's work in pure mathematics have gradually filtered into modern applied mathematics and physics. Matrix algebra, the representation of a rotation by the quaternion product $q^{-1}xq$, and graph theory are just three examples where this has happened.

42. Some of Cayley's memoirs are indeed extensive, some reaching more than a hundred pages in length. Notable in respect are his studies of curves and surfaces. In the study of curves, there is the sequence of three memoirs on cubic curves, those determined by an algebraic equation of order 3 ([*CP5* (1866): 313–53, 354–400, 401–15]). The memoirs on curves, "On Curves which Satisfy Given Conditions" = [*CP6* (1868): 191–291], and "On Polyzomal Curves, otherwise the Curves $\sqrt{U} + \sqrt{V} +$ &c. $= 0$" = [*CP6* (1868): 470–576] are vast. In the study of surfaces, those which are determined by an algebraic equation of orders 3 and 4, his memoirs include, "A Memoir on Cubic Surfaces" = [*CP6* (1869):

359–455] and "A Memoir on Quartic Surfaces" = [*CP7* (1869–71): 133–81, 256–60, 264–302]. Cayley's close connection with classic Greek geometry found expression in his many memoirs on ellipses and ellipsoids, and in particular, "On the Determination of the Orbit of a Planet from Three Observations" = [*CP7* (1870): 400–478], and "On the Centro-surface of an Ellipsoid" = [*CP8* (1873): 316–65]. In the theory of functions, he completed equally lengthy works, such as "A Memoir on the Transformation of Elliptic Functions" = [*CP9* (1874): 113–75] and "A Memoir on Prepotentials" = [*CP9* (1875): 318–423].

43. Talbot 1860.

44. "On the Classification of Cubic Curves" = [*CP5* (1866): 354–400, dated 8 Feb. 1864, read 18 Apr. 1864]; "On Cubic Cones and Curves" = [*CP5* (1866): 401–15, read 18 Apr. 1864, dated 19 Feb. 1865]; Ball 1891.

45. For quartic curves, there were ten genera corresponding to different Plückerian characteristics. Using Plücker's scheme, genera were labeled *anautotonic* (nonsingular), *uninodal, unicuspidal, binodal, nodocuspidal, bicuspidal, trinodal, binodocuspidal, nodobicuspidal, tricuspidal* (Basset 1902).

46. Salmon 1879, 213.

47. "Curve" = [*CP11* (1878): 480].

48. Ordinary inflectional points are defined in terms of points of contact of lines that have multiplicity three. With the "higher inflections points" the conic replaces the line as the defining object and these have multiplicity six, and these 27 special points on the cubic curve are called *sextactic* points. ("On the Conic of Five-pointic Contact at Any Point of a Plane Curve" = [*CP4* (1859): 227–28], "On the Sextactic Points of a Plane Curve" = [*CP5* (1865): 221–57]; Baker 1933, 5: 21–22).

49. Richards 1988; "Note on Lobatchewsky's Imaginary Geometry" = [*CP5* (1865): 471–72]. Nine days before submitting this note on Lobachevsky's geometry, Cayley was visited at Blackheath by the geometer T. A. Hirst and the two may have discussed the contents of it. Three weeks afterward, on 13 Feb. 1865, Cayley gave a paper at the Cambridge Philosophical Society on "Abstract Geometry" (CPS *General Minutes*, 1865), a lecture which no doubt prepared the way for his notable "A Memoir on Abstract Geometry" = [*CP6* (1870): 456–69].

50. Halsted 1899, 64. It is likely that Cayley had known of Lobachevsky's paper for some time but may not have realized its importance. Cayley regularly published papers in *Crelle's Journal* in the 1840s and little published work escaped his eye. He may have been prompted by the connection between Lobatschewsky's paper and the Güdermannian function, which he had recently investigated "On the Transcendant $gd\ u = (1/i)\log_e \tan(\pi/4 + iu/2)$" = [*CP5* (1862): 86–88]).

51. Rowe 1989.

52. "Notes and References" = [*CP5* (1892): 620]. "Presidential Address to the British Association, Southport, Sept. 1883" = [*CP11* (1883): 429–59].

53. "On the Notion and Boundaries of Algebra" = [*CP5* (1864): 292–94].

54. Calculus came under the umbrella of analysis, in keeping with the ideals of the Analytical Society in the early years of the century. As a staunch supporter of Lagrange's algebraic approach to the calculus, Cayley upheld Lagrange's proof of Taylor's theorem and thought it entirely legitimate to define derivatives from the successive terms of Taylor's theorem though he was also aware of such functions as e^{-x^2} and had been known by British mathematicians in the 1830s (Rice 2001). His comments on this occur in "On an Analytical Theorem Connected with the Distribution of Electricity upon Two Spherical Surfaces. Second Part" = [*CP4* (1859) 105]; "Note on Lagrange's Demonstration of Taylor's Theorem" = [*CP8* (1872): 493–95, 519 524]; "A Smith's Prize Paper and Dissertation

[1874]; Solutions and Remarks" = [*CP*9 (1874): 225]). Cayley regarded questions of the divergence or convergence of a series as a matter of *arithmetic,* this being numerical and separate from algebraic considerations.

55. Panteki 1993; Whewell 1833, 21.

56. Cayley's division between tactic and logistic can be analyzed with reference to the task of obtaining the expression $(n/2)(n + 1)$ for the sum of the first n natural numbers. As an art, the result depends on tactic (the rearrangement of the numbers) and on logistic (by performing the addition of $1 + 2 + \cdots + n$ and $n + (n - 1) + \cdots + 1$). The result itself belongs to the scientific part of algebra as it predicts the answer. Cayley seems to have adapted the Greek *logistiké* (arithmetic operations) to obtain the new term logistic. For a scholarly discussion of these points, see J. Klein 1968, 17–25.

57. "On the Notion and Boundaries of Algebra" = [*CP*5 (1864): 294].

58. CED to JCA, 22 Dec. 1862, Adams Papers SJC, 6/65/3/4; 12 Aug. 1863, Adams Papers SJC, 6/66/4/1.

59. Hirst *Journal,* 25 Sept. 1864, 4: 1688. E. Betti was invited to attend the British Association meeting at Bath (C to E. Betti, 1 Aug. 1864, SNSP, 379, 1).

60. *Bath Chronicle* 1864, p. 145.

61. TAH to C, 10 Oct. 1864, UCL, LMS Archive; see Hirst *Journal,* RIGB, 4: 1694.

62. Jensen 1970; Brock, McMillan, and Mollan 1981; Barton 1990.

63. Brock and Macleod 1976.

64. Rice, Wilson, and Gardner 1995; Rice and Wilson 1998. The establishment of the London Mathematical Society (LMS) followed that of the Geological Society (1807), the Royal Astronomical Society (1820), the Zoological Society (1826), the Royal Geographical Society (1830), the Statistical Society (1834), the Royal Botanic Society (1839), the Chemical Society (1841), and the Meteorological Society (1850) but came before the Aeronautical Society (1866) and the society for physics, the Physical Society of London (1874). The LMS initially met at University College, London, and then migrated to the old Burlington House, where they were tenants of the Chemical Society. From 1870 until 1916, the society shared rooms with the Asiatic Society at nearby Albemarle Street in a house occupied by the British Association for the Advancement of Science (a former residence of Florence Nightingale). The society has recently moved to permanent quarters at De Morgan House, Russell Square, London WC1B 4HP.

65. De Morgan 1882, 283.

66. T. A. Hirst became an important figure in the founding of the London Mathematical Society and was its fifth president (1872–74) (Gardner and Wilson 1993). In 1867 he succeeded De Morgan as professor of pure and applied mathematics at University College. C to TAH 15 June 1865, UCL, LMS Archive for quote. I understand the assets of the society are now in excess of £14m.

67. Glaisher 1926, 61; also 1914, li–lii.

68. "On the Problem of the Determination of a Planet's Orbit from Three Observations" = [*CP*7 (1868): 384–86, *CP*7 (1870): 400–478]; "On the Geodesic Lines on an Ellipsoid" = [*CP*7 (1871): 34–35, *CP*7 (1872): 493–510]. "Presidential Address to the British Association, Southport, Sept. 1883" = [*CP*11 (1883): 444].

69. Cayley sat on the Royal Society Council for 1868, 1874, and 1880 during the period 1863–82. Cayley's name had been mooted in the elections for honorary membership of Royal Society of Edinburgh in November 1864 but was passed over in favor of George Gabriel Stokes. In choosing three English members, John Stuart Mill and Tennyson were also elected. Cayley was elected to honorary membership on 4 Dec. 1865 (PGT to WT, 7 Nov. 1864, GUL, Kelvin Papers, T46).

70. C to TAH, 15 June 1865, UCL, LMS Archive. The field for the Copley Medal in 1865 initially included J. V. Poncelet but was narrowed down to J. Plücker, M. Chasles, and the physical chemist Henri V. Regnault.

71. Hirst *Journal,* 2 Nov. 1865, RIGB, 4: 1757–58.

72. "On the Higher Singularities of a Plane Curve" = [*CP5* (1866): 520–28].

73. In March 1866, Chasles proved that a curve on which is established an (α, β) correspondence possesses $\alpha + \beta$ self-corresponding points (united points) if the curve is *unicursal* (a curve with deficiency $D = 0$, such as a conic or a cubic curve with a single cusp).

74. "Curve" = [*CP11* (1877): 485].

75. C to LS, 29 Mar. 1866, in Schläfli 1905, 101. Cayley considered curves with nonzero deficiency and stated the Chasles-Cayley-Brill theorem, that there are $a + a' + 2kD$ united points if the curve has deficiency D. For a discussion of this theorem, see Semple and Roth (1949, 367–75).

76. J. L. Coolidge noted that Cayley came forth "with nothing suggestive of a rigorous proof" but in relation to this result admiringly referred to him as "that ingenious mathematician" (Coolidge [1940] 1963, 189). "On the Correspondence of Two Points on a Curve" = [*CP6* (1865–66): 9–13]; *London Mathematical Society Minutes of Meetings;* C to TAH, 19 Feb. 1866, 23 Feb. [1866], 27 Feb. [1866], UCL, LMS Archive; "Note sur la correspondance de deux points sur une courbe" = [*CP5* (1866): 542–45]; "Second Memoir on the Curves which Satisfy Given Conditions; the Principle of Correspondence" = [*CP6* (1868): 263–68]. The Chasles-Cayley-Brill theorem was presented to a meeting of the London Mathematical Society on 16 Apr. 1866. Hirst *Journal,* 2 May 1866, RIGB, 4: 1784.

77. "A Supplementary Memoir on Caustics" = [*CP5* (1867): 454–64]. "Note sur quelque formulas de M. E. de *Jonquières,* relatives aux courbes qui satisfont à des conditions données" = [*CP7* (1866): 41–43].

78. C to LS, 4 Dec. 1866, In Schläfli 1905, 103. Cayley followed the work of de Jonquières closely. In particular, he noted the result by which de Jonquières had calculated the number of conics which touch a curve of order *n* in five distinct points, a problem related to the "sextactic points" problem ("Addition to the Note on Problems in Regard to a Conic Defined by Five Conditions of Intersection" = [*CP7* (1866): 550]; Simons 1945, 254; C to WJCM: 15 Mar. 1866, CUNY, D. E. Smith Papers).

79. See Scott 1894, 81–86, for an introduction to "contact problems." The work was carried out by H. G. Zeuthen and H. Schubert, who were able to make Cayley's and Salmon's intuitively founded results more rigorous (Lützen and Purkett 1993, 3: 12–15).

80. Undated fragment. Hon. Fellows Proposal Form, CPS, Members Box.

81. "Note sur l'algorithme des tangentes doubles d'une courbe du quatrième ordre" = [*CP7* (1868): 123–25].

82. Cayley was unable to attend the Royal Society Council meeting of 26 Apr. 1866 when B. Riemann was elected as foreign member (shortly before his death on 16 June at the age of 39). At this time, he was not disposed to propose A. Clebsch and P. Gordan as foreign members, possibly because of a controversy stemming from his and their conflicting results on algebraic curves. Cayley had come to the conclusion that a transformed curve depended on a number of parameters different from the number calculated by Riemann ("On the Transformation of Plane Curves" = [*CP6* (1865–66): 1–8]; "Notes and References" = [*CP6* (1893): 593]).

83. Royal Society, 4 June 1863, vol. 10, cert. 45.

84. "A Second Memoir on Skew Surfaces, Otherwise Scrolls" = [*CP5* (1864): 201–20].

85. The first edition of Salmon's *Lessons Introductory to Modern Higher Algebra* appeared in 1859 with the enlarged second edition appearing in 1866. For quote, see

Salmon 1866, 210 fn., 253–65. The invariant *E* was not reproduced in later editions of Salmon's *Lessons Introductory to Modern Higher Algebra.*

86. C to LS, 4 Dec. 1866, in Schläfli 1905, 103. Cayley owned a copy of the *Melanges de Géométrie Pure* (1856) by E. Jonquières.

87. Noether 1905, 17.

88. Ernst Eduard Kummer to Leopold Kronecker, 25 July 1862, in *KummerP*1 (1975): 100–101.

Chapter 12. Years of Challenge

1. *Times*, 28 June 1865, p. 12, col. 1.

2. C to TAH, 21 Jan. 1867, UCL, LMS Archive.

3. C to TAH, 25 Jan. 1867, UCL, LMS Archive.

4. JJS to C, 14 June 1867, SJC, Sylvester Papers, Box 10. W. K. Clifford was awarded a first-class degree in mathematics from London University in the same year (1867).

5. MacMahon 1896, 7.

6. JJS to C, 14 Mar. 1867, SJC, Sylvester Papers, Box 10.

7. Minute, 28 Nov. 1864, in *Royal Society Catalogue of Scientific Papers 1800–1863*, 1: vi; Lyons 1944, 284–85.

8. Cayley's first paper read to the Cambridge Philosophical Society ("On the Theory of Determinants" = [*CP*1 (1843): 63–79]) was his first substantial paper on determinants (see chap. 3). Cayley was vice-president of the Cambridge Philosophical Society in 1868, 1873, 1877, 1883, and 1893.

9. Hall 1969, 17. The meeting of the Cambridge Philosophical Society took place on 18 March 1867.

10. WPT to Peveril Turnbull, March 1867, in Turnbull, W. P. 1919, 24–27. The Leonids meteors were observed in Europe on 14 Nov. 1866.

11. CPS, *General Minutes*, 1867.

12. *AdamsP*1 (1896): xxxix–xl.

13. WPT to Peveril Turnbull, March 1867, in Turnbull 1919, 24–27.

14. The deceptively difficult problem of plotting geodesic lines on the surface of an ellipsoid involves hyperelliptic integrals. There are two types of ruled surface. There is (1) a developable surface, which is a surface deformable onto the plane without stretching and shrinking, for example, a cone or cylinder. Cayley examined this type as early as 1845 (chap. 5) and there is (2) a skew-surface (Cayley's scroll), which is generated by a twisting straight line. The discriminant Δ_4 of the binary quartic form defines a developable surface $\Delta_4 = 0$ of the sixth order, known as Cayley's sextic torse ("On Certain Developable Surfaces" = [*CP*5 (1864): 267–83], "On a Special Sextic Developable" = [*CP*5 (1866): 511–19], "On a Certain Sextic Developable and Sextic Surface Connected Therewith" = [*CP*6 (1868): 87–100], "Note on quelques torses sextiques" – [*CP*7 (1868): 116–17, 118–20]. In particular, the manuscripts of the two last mentioned notes, published in the *Annali di Matematica pura ed applicata* are held in Rome (RIUM, Cremona Papers, 2739, 2740).

15. C to JJS, 14 Mar. [1865], UCL, LMS Archive; Schröter 1865. Sylvester's study of the amphigenous surface arose from his research on the roots of polynomial equations.

16. Scott 1895, 135.

17. One set of Plücker's models of quartic surfaces was donated to the London Mathematical Society. Copies of these models are normally on display at the Science Museum, South Kensington, London. See also *Royal Society Council Minutes of Meetings* 3: 458, and JCM to John Clerk Maxwell, 20 Feb. 1856, in *Maxwell* (P and Letters) 1: 398. In

his teaching at Cambridge in the 1850s, Maxwell illustrated his lectures on geometry with stereoscopic pictures.

18. See C to GGS, 22 Oct. 1855, in Stokes 1907, 1: 385; quote from C to JJS: 31 Oct. 1856: SJC, Box 2; Halsted 1897, 160.

19. "Notices of Communications to the London Mathematical Society" = [*CP6* (1866–69): 19]; *London Mathematical Society Minutes of Meetings*, 13 Dec. 1866. "Note on Cones of the Third Order" = [*CP4* (1859): 120–22]. See *Maxwell* (P and Letters) 2: 476n., for correspondence between Maxwell and Cayley on the stereographic representation of curves and surfaces. "A Notation of the Points and Lines in Pascal's Theorem" = [*CP6* (1868): 122].

20. "Sur l'Equation du cinquième degré" = [*HermiteP2* (1865, 1866): 347–424]; "An Eighth Memoir on Quantics" = [*CP6* (1867): 170].

21. "An Eighth Memoir on Quantics" = [*CP6* (1867): 147]. Cayley wanted knowledge of the linear relations between covariants since these would have a bearing on the relationships between the roots of the quintic ("On the Conditions for the Existence of Three Equal Roots or of Two Pairs of Equal Roots of a Binary Quartic or Quintic" = [*CP6* (1868): 300–311]; Ref. Rep.: 23 June 1868, RSL. RR.6.65).

22. "On the Curves Which Satisfy Given Conditions" = [*CP6* (1868): 191–262]; "Second Memoir on the Curves Which Satisfy Given Conditions; the Principle of Correspondence" = [*CP6* (1868): rec. 18 Apr. 1867, read 2 May 1867, 263–91]; Coolidge [1931] 1959, 426; Loria 1902, 300 n. 2.

23. "Second Memoir on the Curves Which Satisfy Given Conditions; the Principle of Correspondence" = [*CP6* (1867): 263–64].

24. C to LAC, 1 Mar. 1867, RUIM, 2749, in Cremona 1992, 1: 115, and C to TAH, 6 Apr. [1867], mourning paper, UCL, LMS Archive.

25. JJS to C, 27 May 1867, Sylvester Papers, SJC, Box 10; Parshall 1998, 131–32. Sylvester was awarded the Royal Medal in 1861.

26. Susan Cayley to TAH, 29 May [1867], and 4 June [1867], UCL, LMS Archive.

27. Ref. Rep, July 1867, RSL, RR.6.58: 3.

28. JJS to C, 15 Nov. 1867, Sylvester Papers, SJC, Box 10. Cayley published a very short Note in *Crelle*: "Note sur une transformation géométrique" = [*CP7* (1867): 121–22].

29. JC to C, 30 Apr. 1867 ("On Polyzomal Curves." = [*CP6* (1868): 472]). Cremona also settled questions of a similar type and Cayley undertook to have the results published in the *Quarterly Journal of Pure and Applied Mathematics* (C to LAC, 1 Mar. 1867, RUIM, 2749, in Cremona 1992, 1: 115). Cayley published his own observations in the *Annali di Matematica pura ed applicata* ("Démonstration nouvelle du théorème de M. Casey par rapport aux cercles qui touchent à trois cercles donnés" = [*CP7* (1867): 115], "Investigations in connexion with Casey's Equation" = [*CP6* (1867): 65–71]).

30. C to GGS, 3 Apr. 1875, CUL, Add Ms 7656.C259.

31. "On Polyzomal Curves, Otherwise the Curves $\sqrt{U} + \sqrt{V} + \&c. = 0$" = [*CP6* (1868): 470–576]. Cayley's 107-page paper was published in the *Transactions of the Royal Society of Edinburgh*. The subject of polyzomal curves was an interest he shared with the Woolwich-based M. W. Crofton, better known for his work on geometric probability.

32. PGT to C, 18 Aug. 1868, in Knott 1911, 149. Also C to PGT, 21 Oct. 1868, in Knott 1911, 149. In this exchange of letters with Tait, Cayley is referring to his paper, "On the Rotation of a Solid Body Round a Fixed Point" = [*CP1* (1846): 237–52].

33. JCM to J. S. Balfour, 28 Nov. 1870, in Knott 1911, 150.

34. Forsyth 1928, xv.

35. Schläfli's classification of cubic surfaces into twenty-two cases divides into (1) one ruled surface and (2) surfaces with various types of singularity (twenty-one cases). Of

the "general" cubic surface (without singularities), there are five types depending on the reality of the lines and tangent planes. On the first of these types, for example, there exists twenty-seven real lines and forty-five triple tangent planes. Cayley translated Schläfli's paper ("On the Distribution of Surfaces of the Third Order into Species, in Reference to the Absence or Presence of Singular Points, and the Reality of Their Lines" = [*SchläfliP2* (1863): 304–62, communicated by Arthur Cayley, with additional remarks]). Schläfli and Cayley omitted a particular ruled surface from their classification (C to LS, 20 Oct. 1862; 4 May 1863, in Schläfli 1905, 95–96). In 1870, Ludwig Schläfli won the Steiner Prize for his classificatory work on cubic surfaces.

36. "A Memoir on Cubic Surfaces" = [*CP6* (1869): 359].

37. Salmon 1859b.

38. Cayley considered the double-sixer on several other occasions, unsuccessfully in 1868 and successfully in 1870 (Dixon 1909); Cayley's most concentrated work on quartic surfaces is contained in a three-part series: [*CP7* (1869–71), 133–81, 256–60, 264–97].

39. Gordan 1868.

40. A bridge between the two notations can be obtained by comparing Cayley's Cartesian expression $ax^2 + 2bxy + cy^2$ with the expanded symbolic notation $u = \alpha^2 x + 2\alpha\beta xy + \beta^2 y^2$. In this way, the coefficients a, b, c are represented symbolically (a by α^2, b by $\alpha\beta$, and c by β^2). This is not quite the whole story, for there are equivalent symbolic representations of the coefficients a, b, c that have to be employed to make the calculus work. The algebraic form $(u, v)^k$ is called the kth transvectant (the kth *Überschiebung*) of u and v.

41. C. S. Peirce, "Professor Arthur Cayley." *Evening Post* [New York] 28 Jan. 1895, p. 7, cols. 1–2; for an exposition of the German symbolic method, see Osgood 1892. When Cayley finally listed the complete system of invariants and covariants for the binary quintic form, he gave their expressions in symbolic notation alongside his own long homogeneous expressions. The inductive algorithm which Gordan used, and which Cayley singled out as of the greatest importance in invariant theory, stated that an invariant or covariant of degree n could be obtained by a transvection operation of the parent quantic with an invariant or covariant of degree $n - 1$.

42. "A Tenth Memoir on Quantics" = [*CP10* (1878); 378].

43. C to JJS, 5 Dec. 1851, SJC, Sylvester Papers, Box 2; Parshall 1998, 37.

44. JJS to C, 21 Jan. 1869, SJC, Sylvester Papers, Box 11.

45. Salmon's *Lessons Introductory to the Modern Higher Algebra*, 2d ed., promulgated Cayley's incorrect result (Salmon 1866, 273).

46. *London Mathematical Society Minutes of Meetings*, February 1869.

47. "On Recent Discoveries in Mechanical Conversion of Motion" = [*SP3* (1873): 11 fn.].

48. C to [Secretary of the RSL], 17 Feb. 1869, RSL, MC.8.320. Cayley calculated the covariants O, P and his human computer W. B. Davis calculated the covariants R, S, T, V. These fully calculated covariants for the binary quintic were finally put to print in 1889 ("Tables of the Covariants M to W of the Binary Quintic" = [*CP2* (1889): 282–309]). For an *index* of the complete system of the binary quintic, see "A Tenth Memoir on Quantics" = [*CP10* (1878): 344].

49. C to [Secretary of the RSL], 24 Nov. 1869, RSL, MC.8.456.

50. "The Ninth Memoir on Quantics" = [*CP7* (1871): 335]; Cayley 1870.

51. "The Ninth Memoir on Quantics" = [*CP7* (1871): 353]; Cayley resolved the discrepancy between his results for curves and those of B. Riemann, which A. Clebsch and P. Gordan had published in their *Theorie der Abelschen Functionen* ("Notes and References" = [*CP6* (1893): 593]; C to LAC, 28 June 1869, RUIM, 2749, in Cremona 1992, 1: 115). Cayley reconciled his result with Riemann's through a refinement in invariant theory.

52. Cayley and Sylvester regarded the German symbolic notation as an "umbral notation" of limited use and, to them, it appeared unnatural. Their preference was for algebra based on algebraic forms expressed in homogeneous coordinates and the ordinary processes of addition and multiplication.

53. Craik 2000, 232–33. Alex Craik describes Ivory's work on the problem and his bitter disputes with Airy in the 1820s and 1830s.

54. GBA to the vice-chancellor, 5 Dec. 1857, CUL, Cam. Papers G.F.109.

55. Ibid.

56. Craik 2002, 196–97.

57. Report Board of Mathematical Studies, 8 May 1867, CUL, University Archives, Min.V.7. Quoted in Winstanley 1947, 223–24.

58. Stuart 1912, 156.

59. GBA to C, 8 Nov. 1867, in Airy 1896, 273–74; GBA to C, 8 Nov. 1867, TCC, 0.6.6.16.

60. C to GBA, 6 Dec. 1867, in Airy 1896, 275–76.

61. Ibid., 276.

62. GBA to C, 9 Dec. 1867, in Airy 1896, 277, and C to GBA, 10 Dec. 1867, in Airy 1896, 278. As if not to allow such suggestions to be passed up, mathematicians have taken delight in investigating "newly shaped billiard tables." Perhaps the first was Charles L. Dodgson, who examined circular billiards (Carroll 1889).

63. C to GBA, 10 Dec. 1867, in Airy 1896, 278–79.

64. C to GBA, 10 Dec. 1867, in Airy 1896, 278; GBA to C, 17 Dec. 1867, in Airy 1896, 279.

65. W. P. Turnbull 1867; H. W. Turnbull 1919, 14.

66. C to GBA, 10 Dec. 1867, in Airy 1896, 278.

67. *Times*, 16 Mar. 1868, p. 7, col. 1, 23 Mar. 1868, p. 7., col. 2.

68. Clifford 1879, 104.

69. Airy 1896, 268, and Clifford 1879, 1: 104.

70. C to [J. C. Adams], 19 Dec. [1867], SJC, Adams Papers 6/8/5/9. Cayley is quoting from the poem "Tam Glen," by Robert Burns, in which the girl seeks the advice she most wants to hear:

> *Come, counsel, dear Tittie, don't tarry;*
> *I'll gie you my bonie black hen,*
> *Gif ye will advise me to marry*
> *The lad I loe dearly, Tam Glen.*

71. *Times*, 23 Mar 1868, p. 7, col. 2.

72. CUR 5 (16 Nov. 1870): 94.

73. Report to University Syndicate (3 Dec. 1868), CUR (16 Nov. 1870): 97.

74. Crowther 1974, 34.

75. Brock 1975.

76. Wilson 1920–21. Brock 1975; Price 1983 and 1994.

77. It is tenable that Cayley made no contributions to the mathematics and physics section of the British Association in 1865–69.

78. Howson 1982.

79. Roseveare 1903; Royal Society, 1 June 1876, vol. 10, cert. 286.

80. Siddons 1936, 10.

81. Heitland 1929, 255.

82. Sylvester 1869, 7.

83. BA Report 1869, p. 6. (*The Tempest*, Act V, line 56.)

84. "A Third Memoir on Skew Surfaces, Otherwise Scrolls" = [*CP6* (1869): 312–28].

85. "A Memoir on Abstract Geometry" = [*CP*6 (1870): 456]; "Presidential Address to the British Association, Southport, September 1883" = [*CP*11 (1883): 441]. Kirkman, for one, was not impressed by the prospect of *n*-dimensional geometry. He preferred more down-to-earth subjects and he was still rueful about his work on "polyedra" being ignored. It is reasonable to assume Cayley was the target of his barbs for he had recently published his memoir on abstract geometry when Kirkman wrote: "The truths of Nature in our common space of three dimensions may well be left to wait for a century or two, till our eager analysts have discussed the geometries of all the superior dimensions. Yet is it a silly conceit to fancy that the way to the science of molecular forces and combinations may lie, perhaps, through such humble matter of fact as the enumeration and symmetries of polyedra?" (Kirkman 1871).

86. Henry Smith used the phrase the "absolute monarchy of Euclid" to indicate the position the "conservatives" upheld and the "radicals" opposed (*Nature* 8 [25 Sept. 1873]: 450).

Chapter 13. A Representative Man

1. Spencer 1904, 2: 258.

2. Christina Rossetti may have dedicated her acclaimed collection of poems *Sing-Song* (1872) to the young Henry Cayley: "Rhymes dedicated without permission to the baby who suggested them."

3. Portrait Fund, 1874, Campbell and Garnett 1882, 636–37. Dickinson's full-length portrait of Cayley in academic gown is reproduced in the *Collected Mathematical Papers* (vol. 6). Dickinson is linked with the pre-Raphaelites through the Working Men's College, where he taught drawing at the same time John Ruskin and D. G. Rossetti were there. His son, Goldsworthy Lowes Dickinson, was an influential member of King's College, Cambridge, and the subject of a biography by E. M. Forster.

4. C to WJCM: 17 Mar. [18??], CUNY, D. E. Smith Papers; Simons 1945, 257. This problem of Clifford's is connected with his proof that every rational [polynomial] equation has a root [*CliffordP* (1870): 20].

5. Ref. Rep., [undated]. RSL, RR.7.442; "A Memoir on Prepotentials" = [*CP*9 (1875): 318–423]. By considering a generally expressed potential function, Cayley sought to unify George Green's potential theory with that of P. G. L. Dirichlet's, taking into account the work of C. F. Gauss and C. Jacobi. In effect, he was returning to his very earliest undergraduate papers on potential theory (see chap. 5).

6. Royal Society, 4 Mar. 1874, vol. 10, cert. 231. W. K. Clifford was elected to the Royal Society on 4 June 1874.

7. [Anon.]. Prof. Clifford. *The Athenaeum*, no. 2680 (8 Mar. 1879): 315, col. 3; Cayley, *Nature* 19 (20 Mar. 1879): 475.

8. Hunt 1996. James Glaisher, the elder, had sponsored Cayley for election to the Royal Society in 1852.

9. Ref. Rep., 9 Feb. 1870, RSL, RR.7.29; Glaisher 1870.

10. Royal Society, 6 June 1875, vol. 10, cert. 266; J. J. Thomson, 1929.

11. Cayley declined an invitation to give evidence before the Commission. (*Report of Royal Commission on Scientific Instruction, 1874, 2:* Appendix VII, p. 31.)

12. GBA to NMF, 25 Feb. 1875, CUL, Cam. Papers, GF 109; Barrow-Green 1999.

13. Arthur Cayley [Flysheet], 1 Mar. 1875: CUL, Cam. Papers, GF 109.

14. Airy 1896, 327.

15. Sutherland 1911–12, 479.

16. Around 1870, the London Mathematical Society had 111 members compared with 528 in the Royal Astronomical Society, 2,150 in the Royal Geographical Society, and 3,200

in the Society for the Encouragement of Art, Manufacture, and Commerce (Livi 1869; Hilts 1975, 14). See also Rice and Wilson 1998, 191.

17. Collingwood 1966, 584. Cayley was on cordial terms with the leading Continental mathematicians and fostered widespread contacts (*see* note 82, p. 530; C to GGS, [undated]: CUL, Add Ms 7656.C285); Hirst *Journal*, 26 Apr. 1866, *RIGB*, 4: 1784. Cayley contributed to the *Jahrbuch* for twenty-four years, beginning with vol. 2. Other regular British reviewers were J. W. L. Glaisher, O. Henrici, and J. Casey (Müller 1904).

18. Hirst *Journal*, 10 Oct. 1872, RIGB, 4: 1953.

19. My principal source for this episode in the life of the Royal Astronomical Society is Meadows 1972, 97–112.

20. Pang 1994, 260–66. A. C. Ranyard was a prime mover in the establishment of the London Mathematical Society in 1864, but his interest in pure mathematics was overtaken by astronomy.

21. 1872–73, R.A.S. Annual Report, *MNRAS* 33: 189.

22. "Address Delivered by the President, Professor Cayley, on Presenting the Gold Medal of the [Royal Astronomical] Society to Professor Simon Newcomb" = [*CP*9 (1873–74): 184, 181]; Woolley 1966, 29.

23. Newcomb 1903.

24. JDH to CD, 12 Jan. 1873, in L. Huxley 1918, 1: 132.

25. JJS to WJCM, 15 Feb. 1873, in Parshall 1998, 140–41.

26. JJS to GGS, 23 Feb. 1873, CUL, Add Ms 7656.S2610.

27. JOT to THH, 16 June 1870, RIGB, Tyndall Correspondence, vol. 9, p. 2933.

28. "On the Porism of the In-and-circumscribed Polygon, and the (2, 2) Correspondence of Points on a Conic" = [*CP*8 (1871): 14–21]; "On the Problem of the In-and-circumscribed Triangle" = [*CP*8 (1871): 212–57]; Mathews 1898, 57: 217–18. The problem of the in-and-circumscribed triangle generalized to one of finding the number of polygons with vertices that lie on given curves. In this problem, Cayley's extremely detailed work involved considering no fewer than fifty-two distinct cases.

29. "A Ninth Memoir on Quantics" = [*CP*7 (1871): 334–53].

30. C to TAH, 2 July [1872], UCL, LMS Archive.

31. Lodge 1931b, 135, 138.

32. R. S. Ball 1915. Encouraged by F. Klein's "Ueber die sogenannte Nicht-Euclidische Geometrie" (1871) (Stillwell 1996, 63–111), Cayley wrote a paper on hyperbolic plane geometry in May 1872 ("On the Non-Euclidian Geometry" = [*CP*8 (1872): 409–413]). This was followed by his treatment of the geometry of the pseudosphere ("On the Non-Euclidian Plane Geometry" = [*CP*12 (1884): 220–38]) and an exposition of elliptic geometry ("Non-Euclidian Geometry" = [*CP*13 (1894): 480–504]). In this last paper, Cayley's analytical treatment, interpreted elliptic space as the imaginary space which resides in (complex) projective three-dimensional space. He attributed the pure geometry of elliptic space to Clifford. To a paper closely connected with Cayley's absolute written by Clifford and published posthumously, Cayley appended analytical notes ("On the Theory of Distance" = [*CliffordP* (1869): 157–63).

33. Klein [1909] 1939, 134.

34. Lodge 1931b, 136. Cayley described J. B. Listing's generalization of Euler's "$S + F = E + 2$" theorem for polyhedra. ("On the Partitions of a Close" = [*CP*5 (1861): 62–65]; "Communication to the London Mathematical Society Meeting" (12 Nov. 1868) = [*CP*6 (1868): 22]; "On Listing's Theorem" = [*CP*8 (1873): 540–47]).

35. The Association for the Improvement of Geometrical Teaching (AIGT) was formed in 1870 and its opening conference held on 17 Jan. 1871 at University College,

London, under the chairmanship of Thomas Archer Hirst; Howson 1982; Price 1994. See also *CUR* 17 (29 Mar. 1871): 274.

36. *Rep. BAAS 1876*, p. 12; Brock 1975, 29.

37. A. De Morgan wrote *The connexion of number and magnitude, or an attempt to explain the fifth book of Euclid* (1836) to explain the theory of proportion in a less formidable way than that presented in Euclid's *Elements*. Charles Dodgson, another Euclid loyalist, published *Euclid, Book V, Proved Algebraically* (1874), a booklet which presaged his humorous *Euclid and His Modern Rivals* (1879).

38. *Nature* 6 (31 Oct. 1872): 541–42. This British Association Committee underlines the importance of the construction of mathematical tables. It became a vehicle for the young James Glaisher to make his mark in the mathematical world. The tabulation of theta-functions was reckoned to be "the largest piece of numerical computation, with general application throughout the whole of mathematics, that has been undertaken since the original calculation of the logarithms of numbers and trigonometrical functions of Henry Briggs and Adriaan Vlacq, 1620–1633." Of the Committee's initial report of 175 pages, only two pages were reproduced in Cayley's *Collected Mathematical Papers*. The promised centerpiece of the whole program, the calculations of elliptic function tables, was never printed (Campbell-Kelly and Croarken 2000).

39. "[Extract from a] Report on Mathematical Tables" = [*CP*9 (1873): 424–425]; *Rep. BAAS* 1873, 1–175, quote on 164–65; "Report [of the Committee] on Mathematical Tables" = [*CP*9 (1875): 461–99].

40. *Rep. BAAS 1878*, p. 101; Dubbey 1978, 214.

41. C. W. Merrifield 1879. "Report of the Committee . . . appointed to consider the advisability and to estimate the expense of constructing Mr. Babbage's Analytical Machine, and of printing tables by its means." In *Rep. BAAS, 1878*, pp. 92–102.

42. Cayley compiled a report of this work for the British Association held in Bristol in 1875 ("On the Analytical Forms called Trees, with Application to the Theory of Chemical Combinations" = [*CP*9 (1875): 427–460]; "On the Number of the Univalent Radicals $C_n H_{2n+1}$" = [*CP*9 (1877): 544]; *MMLPS* 1895, 9: 237; in 1881, Cayley published a simpler solution for counting unrooted trees ("On the Analytical Forms Called Trees" = [*CP*11 (1881): 365–67]; Biggs, Lloyd, and Wilson 1976, 49–51). Cayley was then considering the problem of counting "trees" as one in pure mathematics (see chap. 10).

43. Hartog 1885–90; Roscoe 1892–93; Evans 1982, 168; quote in Challenger 1950–51, 183; Royal Society, 8 June 1871, vol. 10, cert. 185.; Partington 1964, 4: 775; Russell 1988.

44. "On the Mathematical Theory of Isomers" = [*CP*9 (1874): 204]; Rouvray 1977. Another alcohol from the $n = 5$ family was known at this time—one with the structural formula $C-C-C-C-C-OH$, having been shown to exist in 1871 (Simmonds 1919, 116).

45. Cayley 1875; *Nature* 12 (23 Sept. 1875): 463. Cayley's figure of 799 unrooted tree structures of the specified type (in the case $n = 13$) was subsequently corrected to 802, but Cayley's calculations are remarkable for their accuracy in general. Since his time the calculations have extended as far as $n = 22$ (Rains and Sloane 1999).

46. Brock 1967. Sir Benjamin Collins Brodie (the younger) proposed that chemical combination could be explained by the calculus of operations—Cayley owned a copy of his book (Brodie 1866).

47. C to FK, 12 July 1880, *NSUG*, 370; typically, Cayley would use plain alphabetic notation for coefficients (as in hx^2y compared with the German preference $a_{112}x_1{}^2x_2$). Similar sentiments by Cayley on the use of subscripts were expressed in ("On a Theorem Relating to the Multiple Theta-functions" = [*CP*11 (1880): 242–49]; "Seminvariant Tables" = [*CP*12 (1885): 275]).

48. "Report on Mathematical Notation and Printing" in *Rep. BAAS, 1876*, 337–39; C to GGS, 30 Sept. 1880, in Stokes 1907, 1: 397. Volume 1 of Stokes's *Mathematical and Physical Papers* had just appeared; Salmon 1879, 287 fn.

49. C to LS, 11 Sept. [1870], in Schläfli 1905, 107. This letter is dated 1871, but it would appear to have been written the preceding year.

50. Cayley referred to Noether's theorem in "On the Intersection of Curves" = [*CP*12 (1887): 500–504], while a rounded mathematical synopsis of it can be found in Semple and Kneebone 1959, 350. Charlotte A. Scott provided a geometrically based proof of the theorem in Scott, 1899, which is described in Kenschaft 1987, 106–7.

51. C to MN, 23 May 1871, SPKB, Darmstadt, H6, 1918.98.

52. "On the Deficiency of Certain Surfaces" = [*CP*8 (1871): 394–97]. To make the claim that the deficiency of a plane curve $u = 0$ is non-negative, it must be assumed that u is irreducible (Hilton [1920] 1932, 144–46). Cayley effectively showed that the concept of *deficiency*, expressed in terms of singularities of a surface, was distinct from deficiency in terms of the integrals defined on the surface (Gray 1999, 61; Baker 1913, 8). He was led to an expression for the deficiency of an nth-order surface by a theorem of A. Clebsch ("A Memoir on the Theory of Reciprocal Surfaces" = [*CP*6 (1869): 356]).

53. Around this time, Cayley produced a number of memoirs on quartic surfaces, including "Sketch of Recent Researches upon Quartic and Quintic Surfaces" = [*CP*7 (1869–71): 244–52]; "A Memoir on Quartic Surfaces" = [*CP*7 (1869–71): 133–81]; "A Second Memoir on Quartic Surfaces" = [*CP*7 (1869–71): 256–60]; "A Third Memoir on Quartic Surfaces" = [*CP*7 (1869–71): 264–97]. A progress report is given in *Nature* 3 (24 Nov. 1870): 78–79.

54. Darboux 1904, 157.

55. "On the Quartic Surfaces $(*\langle U, V, W)^2 = 0$" = [*CP*7 (1870): 304–13], [*CP*8 (1871): 2–11, 25–28].

56. *London Mathematical Society Minutes of Meetings,* 8 June 1871; *Nature* 4 (22 June 1871): 154; "On Plücker's Models of Certain Quartic Surfaces" = [*CP*7 (1869–71): 298–302].

57. "A Memoir on the Theory of Reciprocal Surfaces" = [*CP*6 (1882): 353–55].

58. *CUR* 33 (7 Oct. 1873): 19.

59. "On Steiner's Surface" = [*CP*9 (1873–74): 1–12; *Nature* 9 (25 Dec. 1873): 154; Viewed analytically, Steiner's surface has the equation $x^2y^2 + x^2z^2 + y^2z^2 + 2xyzw = 0$ considered by Steiner in 1844. Cayley had studied it previously as many others had done (notably E. E. Kummer, K. Weierstrass, H. E. Schröter, and L. Cremona).

60. William H. H. Hudson, its first secretary, was a well-known mathematics teacher at St. John's College before becoming professor at King's College, London, in 1882. His son R. W. H. T. Hudson, Senior Wrangler at Cambridge in 1898, was a geometer whose career was cut short by a fatal fall on Glydwr Fawr, Snowdon. His daughters were highly placed in the order of merit (Venn 1940, Part 2, 3: 474–75); Sidgwick 1938, 67.

61. *CUR* 55 (17 Feb. 1874): 236.

62. "On Dr Wiener's Model of a Cubic Surface with 27 Real Lines; and on the Construction of a Double Sixer" = [*CP*8 (1873): 366–84]; Blythe 1905.

63. Mathews 1898a, 217.

64. "A 'Smith's Prize' Paper [1868]; Solutions by Prof. Cayley" = [*CP*8 (1868): 430–31]; "On the Double Sixers of a Cubic Surface" = [*CP*7 (1870): 316–29]; Henderson 1911, 61 n. 39.

65. H. J. S. Smith wrote a masterful prospectus for the geometry section at the opening of the South Kensington Museum (*SmithP*2 (1876): 698–710]. *CUR* 199 (8 May 1877): 414. The active members of the Cambridge Modeling Club included Cayley, W. H. H. Hudson, R. F. Scott, C. J. Lambert, E. Hill, J. C. Maxwell, J. W. Hicks, W. M. Hicks, J. H. Mackie, T. Lattimer, Percival Frost, W. Garnett, G. W. Waterhouse.

66. "Curve" = [CP_{11} (1877): 462 fn.].

67. "Plan of a Curve-tracing Apparatus" = [CP_8 (1871–73): 179–80]; "On a Bicyclic Chuck" = [CP_8 (1872): 209–11].

68. Mathews 1898a, 217.

69. Following his "Memoir on the Theory of Matrices" in 1858 Cayley, turned only briefly to matrix algebra in the 1860s ("A Supplementary Memoir on the Theory of Matrices" = [CP_5 (1866): 438–48]).

70. Karl Weierstrass, Leopold Kronecker, and Georg Frobenius outlined the main properties of matrices in the context of algebraic forms (Hawkins 1977b, 144, 148).

71. Cayley 1872. This paper on matrix algebra is not contained in the *Collected Mathematical Papers*; PGT to C, 28 Feb. 1872, in Knott 1911, 152. Thomson came to terms with matrices as a notational device and recognized their value (Thomson W. 1874, Appendix).

72. In 1877, Cayley was led back to (binary) matrices by a problem in fluid motion and related to Sylvester that he had "found a pretty solution of an old question." The problem was the recurrent one, addressed by Charles Babbage and others, of computing $\phi^k(x)$ where $\phi(x) = (ax + b)/(cx + d)$. The solution in "closed form" represented an improvement on the solution presented in the 1858 memoir but predates a modern solution in terms of eigenvectors and eigenvalues (C to JJS: 5 Nov. 1877, SJC, Sylvester Papers, Box 2; "On the Matrix [binary matrix given], and in Connection Therewith the Function [fractional transformation]" = [CP_{11} (1880): 252–57]).

73. C to TAH, 2 July [1872], UCL, LMS Archive; "Sur la condition pour qu'une famille de surfaces données puisse faire partie d'un système orthogonal" = [CP_8 (1872): 269–91]. "On Curvature and Orthogonal Surfaces" = [CP_8 (1873): 292–315].

74. "On Wroński's Theorem" = [CP_9 (1873): 96–102]. Wroński adopted an idiosyncratic approach to analysis (Grattan-Guinness 1990, 1: 219–23). Burmann's law was probably made known at Cambridge by R. Murphy and D. F. Gregory ("Note on Burman's Law for the Inversion of the Independent Variable" = [SP_2 (1854) 44–49]).

75. "Note on the Calculus of Logic" = [CP_8 (1871): 65–66]; Church 1968, 422.

76. "Presidential Address to the British Association, Southport, Sept. 1883" = [CP_{11} (1883): 459].

77. "On the Theory of the Singular Solutions of Differential Equations of the First Order" = [CP_8 (1873): 529–34], [CP_{10} (1877): 19–23]. Boole's first edition of the *Treatise* was followed by the second edition (1865) and third edition (1872).

78. The probability problem is partially reproduced in [CP_{10} (1878): 600]; Woolhouse 1878.

79. Mathews 1898, 58: 50.

80. Cayley's solution technique for the lottery problem involved long calculation and did not yield a general result ("Problem 4528 to the Educational Times" = [CP_{10} (1874): 587–88]). Also known as the "Secretary problem," it is closely related to a modern optimal stopping problem, in which an employer wishes to hire a secretary chosen from a pool of n candidates. They are interviewed in turn but a candidate cannot be recalled once rejected (Moser 1956; Ferguson 1989).

81. W. J. C. Miller was vice-principal and professor of mathematics at Huddersfield College, Yorkshire, before becoming registrar and secretary of the General Medical Council in 1876. Some of Cayley's problems were also published in the *Giornale di Matematiche*.

82. "A Memoir on the Transformation of Elliptic Functions" = [CP_9 (1874): 113–75, read 8 Jan. 1874, quote on 127]; Cayley [1876] 1895, 172–73. Cayley extended the calculation of the elliptic modular equations to the cases $n = 7$ and $n = 11$ (partially), previously

accomplished by C. Jacobi for $n = 3$ and $n = 5$ in the *Fundamenta Nova* (1829). Cayley's work complemented the calculations made by L. A. Sohncke, who had made his calculations in the 1830s. It has been found that Cayley's approach was "unduly complicated" for all but the simplest cases and his method unsatisfactory as a basis for the theory. His approach leads to a system of nonlinear equations, which are practically impossible to solve analytically (Borwein and Borwein 1998, 102–6).

83. Cayley introduced the now standard notation for the Gudermannian function, a function which occurs in the mathematical theory of cartography: [*CP5* (1862): 86–88]; Cayley [1876] 1895, 58; Archibald 1934; Fletcher, Miller, Rosenhead and Comrie 1962, 1: 258–62.

84. Cayley [1876] 1895; "Note sur l'addition des fonctions elliptiques" = [*CP1* (1851): 548].

85. $$\{\Theta(q)\}^4 = 1 + 8 \sum_{k=1}^{k=\infty} \frac{kq^k}{1 + (-1)^k q^k}.$$

The expression for $\{\Theta(q)\}^4$ leads to the theorems connected with the decomposition of a natural number n into at most four-square numbers ("On a Theorem of M. Lejeune-Dirichlet" = [*CP2* (1854): 47–48]; Dickson [1919–23] 1971, 2: 240); S. Ramanujan was to study identities of this type and G. H. Hardy surmised that one source of his knowledge might have been Cayley's *An Elementary Treatise on Elliptic Functions* (Hardy 1940, 42).

86. Brioschi 1895, 196; Halphen 1886.

87. Cayley observed if two binary forms are equivalent under linear transformations, then the ratios of invariants are equal but that this does not provide a sufficient condition ("Note on the Theory of Invariants" = [*CP8* (1871): 385–87]; "Notes and References" = [*CP6* (1893): 593]). Felix Klein noted the limitations of invariant theory when applied to the theory of algebraic curves (Klein [1926] 1967, 1: 311).

88. Camille Jordan obtained estimates for the maximum orders of irreducible covariants of binary forms; see also Young 1913, lii.

89. "Covariants doubles des formes binaires" = [*LaguerreP1* (1872): 268–72]; "Représentation des formes binaires" = [*LaguerreP1* (1872): 273–76]. Edmond Laguerre also wrote memoirs on linear algebra in which his "matrix" was called a "système linéare." "Mémoire sur les covariants des formes binaires" = [*JordanP3* (1876): 153–208; (1879): 213–46].

90. JJS to C, 15 Feb. 1875, RSL, MM.15.19; Parshall 1998, 142–46.

91. BP to DCG, 18 Sept. 1875, in H. Hawkins 1960, 34–35.

92. JJS to WS, 19 Nov. 1876, SJC, Sylvester Papers, Box 1; Parshall 1998, 172–75.

93. JJS to WS, 25 Nov. 1876, SJC, Sylvester Papers, Box 1.

94. "Address on Commemoration Day at Johns Hopkins University, 22 February 1877" = [*SP3* (1877): 86].

95. JJS to WS, 26 Nov. 1876, SJC, Sylvester Papers, Box 1.

Chapter 14. March On with Step Sublime

1. C to JJS, 14 Mar. 1877, SJC, Sylvester Papers, Box 2.

2. JJS to C, 23 Apr. 1877, SJC, Sylvester Papers, Box 11; Parshall 1998, 177–81.

3. The Fundamental Postulate stated that an irreducible covariant and a basic syzygy *of the same degree and order* could not coexist. For Sylvester its failure would imply his sifting process on the generating function expansions was incapable of finding all irreducible invariants and covariants.

4. There was no a priori procedure for deciding whether an invariant or covariant generated by the German transvection operation was actually irreducible. In the case of

the binary cubic u and the Hessian covariant H, for example, the transvection operation generates the irreducible covariant $(u,H)^1 = \Phi$, but $(u,\Phi)^1 = H^2$ is reducible, though there was no way of deciding these conclusions a priori.

5. JJS to C, 9 May 1877, SJC, Sylvester Papers, Box 11. Sylvester complained that the German methods were not straightforward (JJS to JNL, 8 Mar. 1877, EUL MS 110 Lockyer Research Papers).

6. JJS to C, 9 May 1877, SJC, Sylvester Papers, Box 11; "Sur une méthode algébriques . . ." = [SP3 (1877): 59]. The paper of Cayley's referred to by Sylvester is likely "On an Identical Equation Connected with the Theory of Invariants" = [CP9 (1873): 52–55].

7. JJS to C, 28 May 1877, SJC, Sylvester Papers, Box 11.

8. JJS to C, SJC, Sylvester Papers, Box 11: 30 Aug. 1877, and 4 Sept. 1877, respectively.

9. C to GGS, 3 Nov. 1877, CUL, Add Ms 7656.C261.

10. C to JJS, 5 Nov. 1877, SJC, Sylvester Papers, Box 2. A new American mathematical journal had been proposed in November 1876 (see Parshall 1998, 164–66, 181). On Trinity College fellowship dissertations, Cayley had written favorably of John Cox's dissertation of the previous year, which discussed Hamilton's "Systems of Rays" (C to JCM, 7 Oct. 1876, CUL, Add Ms 7655, II/120; in [Maxwell (P and Letters) 3: 400].

11. JJS to C, 6 Nov. 1877, SJC, Sylvester Papers, Box 11; Parshall 1998, 184–87. "Sur les actions mutuelles des formes invariantives dérivées" = [SP3 (1878): 218–40].

12. Sylvester proved this in a restricted case but it was shown to be generally true in 1892 ("Proof of the Hitherto Undemonstrated Fundamental Theorem of Invariants" = [SP3 (1878): 117–26, quote on p. 117]; "Lectures on the Theory of Reciprocants" = [SP4 (1886): 364]; Elliott 1892). See chap. 9 re the British approach to invariant theory. According to Sylvester, Cayley's law had been ignored by the German mathematicians ("A Demonstration of the Impossibility of the Binary Octavic Possessing Any Groundform of Degree-order 10.4" = [SP3 (1881): 524]). C to JJS, [1854/1855], SJC, Sylvester Papers, Box 2.

13. "Tables of the Generating Functions and Groundforms for the Binary Quantics of the First Ten Orders" = [SP3 (1879): 309]. See also, JJS to RL, 28 Oct. 1878, BUMI.

14. JJS to CJ, 4 July 1878, EPP. Camille Jordan Correspondence.

15. "A Demonstration of the Impossibility of the Binary Octavic Possessing Any Groundform of Degree-order 10.4" = [SP3 (1881): 509].

16. JJS to C, 21 Dec. 1877, SJC, Sylvester Papers, Box 11.

17. "A Tenth Memoir on Quantics" = [CP10 (1878): 339–400]; "On Linear Transformations" = [CP1 (1846): 95].

18. JJS to C, 15 July 1878, SJC, Sylvester Papers, Box 11.

19. JJS to C, 21 Aug. 1878, SJC, Sylvester Papers, Box 11.

20. JJS to C, 11 Jan. 1879, SJC, Sylvester Papers, Box 11; Archibald 1936, 142.

21. JJS to C, 20 Apr. 1880, SJC, Sylvester Papers, Box 11. F. Franklin gained his doctorate in 1880 and later turned to a career in journalism.

22. JJS to C, 21 Aug. 1878, SJC, Sylvester Papers, Box 11; Cayley, Sylvester, and Spottiswoode were in charge of allocating funds for the British Association "Invariant Theory" grant.

23. "Address on Commemoration Day at Johns Hopkins University, 22 February 1877" = [SP3 (1877): 76].

24. C to JJS, 3 Mar. 1879, SJC, Sylvester Papers, Box 2. Within a few days there was another blow when Cayley heard of W. K. Clifford's death in Madeira.

25. "Specimen of a Literal Table for Binary Quantics, Otherwise a Partition Table" = [CP11 (1881): 359–60].

26. Bell 1945, 428.

27. "On the Theorem of the Finite Number of the Covariants of a Binary Quantic" = [*CP*11 (1881): 274]).

28. JJS to C, 24 Nov. 1877, SJC, Sylvester Papers, Box 11.

29. "The Theory of Groups" = [*CP*10 (1878): sub. 26 Nov. 1877, 403]. A little-known theorem in combinatorial group theory published by Cayley as a postscript around this time is "Two noncommutative symbols α, β which are such that $\beta\alpha = \alpha^2\beta^2$ cannot give rise to a group made up of symbols of the form $\alpha^p\beta^q$" ("A Problem in Partitions" = [*CP*11 (1878): 61–62]; "Note on the Equation $s_1 s_2 = s_2{}^2 s_1{}^2$, s_1 and s_2 Being Operators of a Finite Group" = [*MillerP*3 (1910): 151–54]).

30. "The Theory of Groups" = [*CP*10 (1878): sub. 26 Nov. 1877, 402]. It is at this point that Cayley comes close to the accepted modern definition of an *abstract* group. Cayley's papers of the 1850s (see chap. 8) utilize the calculus of operations and the "separation of symbols" to *generalize* the concept of a group of permutations.

31. C to JJS, 7 Feb. 1878, SJC, Sylvester Papers, Box 2.

32. The meeting was on 9 May 1878 (*Nature* 18 [16 May 1878]: 83). At this meeting F. Brioschi, G. Darboux, P. Gordan, S. Lie, and V. M. A. Mannheim were elected to foreign membership of the London Mathematical Society. Cayley represented two permutations α and β (which together define the alternating group on four letters) in his diagram: four black triangles represent the substitution α and the four red (dotted lines) triangles represent β.

33. C to FK, 15 May 1878, NSUG, 365. "On the Theory of Groups" = [*CP*10 (1878): 324–30]. For Cayley, a *regular* substitution is one which breaks up into cycles of the same length. In Cayley's example, the cycles are of length three.

34. "Desiderata and Suggestions. No. 2. The Theory of Groups; Graphical Representation" = [*CP*10 (1878): sub. 16 May 1878, 403–5]. He later formalized the diagrammatic representation of groups and introduced the term "colourgroup" for a specific class of graphs with colored edges. He described the colorgroups for groups of small order ("On the Theory of Groups" = [*CP*12 (1889): 639–56]; Chandler and Magnus 1982, 22–28).

35. "A Smith's Prize Paper, 1877" = [*CP*10 (1877): 41].

36. "Galois" = [*CP*11 (1879): 543]; "Equation" = [*CP*11 (1878): 518–21].

37. *Nature* 18 (11 July 1878): 294.

38. FG to H. W. Bates, 24 Jan. 1879, RGS JMS 19/24. Cayley submitted *two* manuscripts to the Royal Geographical Society, the second one which was used for the printed paper was (slightly) less technical (T. Crilly, "Arthur Cayley and the Four Color Map Problem" [forthcoming]). The first manuscript, in which he uses the technical term "close" (enclosed region) makes it clear that the problem was linked to Cayley's work in topology, see "On the Partitions of a Close" = [*CP*5 (1861): 62–65, 617].

39. "On the Colouring of Maps" = [*CP*11 (1879): 7–8]. Though Cayley did not solve the problem by providing a general proof, he made several insightful points and indicated where the difficulties lie.

40. C to WT, 7 Feb. 1879, CUL, Add Ms 7342.C62.

41. C to TAH, 10 Feb. 1879, UCL, LMS Archive; Peitgen and Saupe and Haeseler 1984.

42. "On the Geometrical Representation of Imaginary Variables by a Real Correspondence of Two Planes" = [*CP*10 (1878): 316–23]. This paper, on conformable representation was submitted on 13 Dec. 1877. Its implicit connection with mapmaking may also have triggered Cayley's recollection of the four-color problem. Beginning in 1885 and taking a lead from K. H. A. Schwarz, Cayley wrote several papers on *orthomorphosis*, his term for a conformal transformation.

43. When dealing with differential equations, Cayley adopted the same strategy of widening the scope of an investigation to include consideration of complex numbers.

He asked P. G. Tait: "Do you know anything as to the solution when the limitations are rejected, and imaginary [i.e. complex number] solutions taken account of ?" (C to PGT, 25 Mar. 1874, in Knott 1911, 154). The presentation to the Cambridge Philosophical Society on the Newton-Fourier problem took place on 24 Feb. 1879; *Nature* 19 (13 Mar. 1879): 451.

44. "Desiderata and Suggestions. No. 3. The Newton-Fourier Imaginary Problem" = [*CP*10 (1879): sub. 3 Mar. 1879, 405–6].

45. C to ABK, 24 June [1879], WSRO, Kempe/53/6/1–2.

46. *Nature* 20 (17 July 1879): 275.

47. JJS to ABK, 16 July 1879, WSRO, and 7 Sept. 1879, respectively. WSRO, Kempe/13/25–26.

48. C to ABK, 22 Nov. 1879, WSRO, Kempe/13/15.

49. Royal Society, 2 June 1881, vol. 11, cert. 12.

50. Kempe 1879; Geikie 1923; R. J. Wilson 2002. It is likely that Sylvester went to the grave thinking Kempe's proof was correct. Eight months before his death, he wrote to Kempe: "The problem of the 4 colors solved will ever be a bright leaf in your laurel crown" (JJS to ABK, 5 June 1896, WSRO, Kempe/29/5).

51. Cayley produced a long paper in response to the paper by S. Roberts: "On Three-bar Motion" = [*CP*9 (1876): 551–80]. At Cayley's suggestion, James A. Ewing, professor of mechanism and applied mechanics, constructed a curve-tracing mechanism in the workshops of the engineering laboratories at Cambridge. Cayley's interest in practical machinery is clearly seen by his close association with the Cambridge laboratories when his former student James Stuart was appointed to the chair and succeeded by Ewing.

52. "On the Geometrical Representation of Cauchy's Theorem of Root-limitation" = *CP*9 (1877): 21–39; Königsberger 1877, 1: 18–24.

53. The changes to the Cambridge mathematical Tripos introduced in 1873 threatened the existence of the order of merit. Many students wanted to graduate in mathematics owing to the prestigious place of the mathematical Tripos in a liberal education, but the increase of new topics in 1873 meant that students taking different subjects could not be compared on common ground, as had been the case previously. In 1882, a new scheme for examination of the mathematical Tripos was adopted which allowed the order of merit to regain its former meaning (Rouse Ball 1880). Cayley did not live to see the day when the order of merit was abandoned in 1907.

54. *CUR* 245 (14 May 1878): 522–26.

55. Clement Booth to George King, 26 Sept. 1878, CUL, University Archives, Char.II.14. p. 15, and 21 Feb. 1879, CUL, University Archives, Char.II.14. p. 17.

56. C to GGS, 16 Dec. 1879, CUL, Add Ms 7656.C265.

57. JJS to C, 12 May 1881, SJC, Sylvester Papers, Box 11; Parshall 1998, 201–5.

58. C to JJS, 7 Feb. 1878, and 3 Mar. 1879, SJC, Sylvester Papers, Box 2.

59. Stokes 1907, 1: 11.

60. In 1879 the examiners were unable to separate the two candidates for the Smith's Prizes awarded usually to the top student and the runner-up. They were awarded jointly with the same status to M. J. M. Hill (afterward holder of professorships at Mason College, Birmingham, and University College, London) and A. J. Wallis (who became a Cambridge don). Cayley wrote out solutions for essay questions that appeared in the Smith's Prize examination papers during the 1860s and 1870s and published them in minor journals. For an overview and history of the Smith's Prize at Cambridge, see Barrow-Green 1999.

61. Pearson 1936, 32.

62. Filon 1936–38, 74.

63. J. J. Thomson, 1936, 47. R. T. Wright, a fellow at Trinity College, was fifth wrangler in 1869 and an enthusiastic supporter of women's education. He was able to offer legal advice in the first days of Newnham College and was the first paid secretary of the Cambridge University Press. Cayley was one of J. J. Thomson's proposers for membership of the Royal Society (Membership Certificate, Royal Society, 12 June 1884, vol. 11, cert. 75).

64. Forsyth 1935, 172.

65. "Arthur Cayley" = [CP8 (1895): p. xvii].

66. ARF to JJT, 1 July 1935, CUL, Add Ms 7654/F23; Richard Pendlebury was Senior Wrangler in 1870, a lifelong Cambridge mathematician, and acclaimed alpinist but one who failed to make the most of his mathematical talents.

67. R. C. Rowe 1880. "Addition to Mr Rowe's 'Memoir on Abel's theorem.'" = [CP11 (1881): 29–36]; "Sur un théorème d'Abel" = [CP4 (1857): 5–6]; Rice 1997, 300.

68. C to TAH, 31 Mar. 1882, UCL, LMS Archive.

69. Scott 1895, 134–35.

70. ARF to JJT, 1 July 1935, CUL, Add Ms 7654/F23.

71. Thomson 1936, 50–51.

72. Mary Cayley to Mary Augusta Scott [1895], in Eisele 1976, 2: 648.

73. The course offered in 1873, for example, was on Algebra I, to take place on Monday and Thursday at 2 P.M. The normal fee was one guinea for the course but whether they took place is not known—they would only take place if a certain unspecified number of students was reached (CUR 18 [22 Apr. 1873]: 20).

74. Clough 1897; Gardner 1921.

75. Kenschaft 1987; Series 1997–98; Macaulay 1932.

76. C. A. Scott 1894, vi.

77. Cayley had been appointed to the syndicate to consider proposals relating to the higher education of women. CR1 (26, Suppl.; 9 June 1880): xxxiii. CUR 340 (8 June 1880): 618.

78. Following considerable agitation, the passing of "The Three Graces" by the University Senate in February 1881 was a partial triumph for the campaigners. Women were left in a position of knowing their position in the order of merit but ineligible for admittance to a Cambridge degree with its privileges and voting rights (Phillips 1979, 16–17).

79. W. von Dyck followed Klein to Leipzig and while there published his major papers on group theory (Dyck 1882, 1883); van der Waerden 1985, 152–53; Chandler and Magnus 1982, 5. Cayley's trilogy of the 1850s had escaped von Dyck's notice at the time he published his first paper on group theory. Cayley also visited Munich in 1879.

80. In proving the impossibility of a "16 squares" formula, A. Hurwitz, who was admired for his deductive proofs, remarked on Cayley's and Roberts's approach: "Their extremely laboured observations, which were based on trial and error, did not provide any convincing proof because they were based on particular assumptions concerning bilinear forms $z_1, z_2, \ldots$ and these were totally unjustified" ("Ueber die Komposition der quadratischen Formen von beliebig vielen Variablen" = [HurwitzP2 (1898): 565–66 fn.]). See chap. 6. See also Parshall and Rowe [1994] 1997, 175–82.

81. C to JJS, 24 Dec. [1880], SJC, Sylvester Papers, Box 2.

82. JJS to C, 19 Jan. 1881, SJC, Sylvester Papers, Box 11; Galton 1909, 71–72.

83. Cayley's article on the theory of equations written for the Encyclopaedia Britannica contains an outline of Galois's theory ("Equation" = [CP11 (1878): 490–521].

84. JJS to C, 19 Jan. 1881, SJC, Sylvester Papers, Box 11.

85. JJS to C, 12 May 1881, SJC, Sylvester Papers, Box 11; Parshall 1998, 201–5.

86. C to DCG, 15 Aug. 1881, JHU, Daniel Coit Gilman Papers MS1.

87. CSP to BR, 24 Apr. 1909, in C. S. Peirce 1976 3(2): 983–84.

88. JJS to C, 28 June 1881, SJC, Sylvester Papers, Box 11.

89. C to DCG, 8 Oct. 1881, JHU, Daniel Coit Gilman Papers MS1.

90. C. S. Peirce, "Professor Arthur Cayley." *Evening Post* [New York] 28 Jan. 1895, p. 7, cols. 1–2.

91. French 1979, 77–78.

92. "Function" = [*CP*11 (1879): 522–42]. Analogous to the three types of elliptic functions, there are sixteen different types of hyperelliptic functions (also called Abelian functions) and they can be expressed as quotients of double theta functions. He gave another set of lectures at Cambridge on this subject in 1881. "A Memoir on the Single and Double Theta-functions" = [*CP*10 (1880): 463–565]. "A Memoir on the Abelian and Theta Functions" = [*CP*12 (1882, 1885): 109–48, 149–216]. The first part of this long paper is effectively a record of the lectures he gave during his visit to Johns Hopkins University in 1882.

93. JJS to ABK, 3 June 1882, WSRO, Kempe/15/84/2.

94. *JHUC* 16 (July 1882): 218; W. E. Story was appointed at Johns Hopkins shortly after Sylvester.

95. C to TAH, 31 Mar. 1882, UCL, LMS Archive; Clebsch and Gordan [1866] 1967.

96. Peirce 1881; Pycior 1979, 537 fn.; Grattan-Guinness 1997. In the republication of B. Peirce's "Linear Associative Algebra," his son C. S. Peirce proved the celebrated theorem that the real numbers, complex numbers, and quaternions are the only associative finite dimensional real-division algebras (Parshall and Rowe [1994] 1997, 94). Of topical interest, Cayley discussed algebras with eight units as a generalization of the earlier Cayley numbers ("On the 8-square Imaginaries" = [*CP*11 (1881): 368–71). He and Samuel Roberts attempted to prove the nonexistence of a "16 squares formula" in 1881 (Dickson 1919, 168–69). *See* note 80.

97. Franklin 1896–97, 307.

98. C to DCG, 1 July 1882, JHU, Daniel Coit Gilman Papers MS1.

99. JJS to C, 3 Aug. 1882, SJC, Sylvester Papers, Box 11.

Part 5. Make One Music as Before, 1882–1895

Epigraph: Cayley's metaphor for mathematics is "Wordsworthian" if not by Wordsworth himself. It may have been inspired by his "A Night-Piece" ("Chequering the ground—from rock, plant, tree, or tower") or by "Sonnet 36" ("Westminster Bridge") in praise of London: "In his first splendour, valley, rock, or hill." Cayley's expression identifies his feelings about mathematics as from a preindustrial age. As has been observed by others, he rarely expressed his feelings for mathematics so openly as here (Mathews 1898b).

Chapter 15. "A Tract of Beautiful Country"

1. *Minutes of Council of the Royal Society,* 15 June 1882, 5: 326, 332.

2. C to JJS, 6 Sept. [1882], SJC, Sylvester Papers. Box 2.

3. Ibid.

4. *TaitP*2 (1882): 76.

5. C to PGT, 26 Feb. 1883, in Knott 1911, 235. Cayley believed in the same principles of mechanics as set out by Whewell in his texts on mixed mathematics (Guicciardini 1989, 134).

6. Knott 1911, 254.

7. C to PGT, 3 and 6 Nov. 1882, in Knott 1911, 155, 157, 158.

8. JJS to C, 6 Oct. 1882, SJC, Sylvester Papers, Box 11; Parshall 1998, 216–17.

9. JJS to C, 25 Dec. 1882, SJC, Sylvester Papers, Box 11.

10. In the by-election for the Cambridge University parliamentary seat caused by the resignation of Spencer Horatio Walpole, the Conservative candidate Henry Cecil Raikes was elected. Cayley was chosen as an official at the by-election, not being a publicly declared supporter of either candidate.

11. JJS to C, 25 Dec. 1882, SJC, Sylvester Papers, Box 11.

12. *Nature* 27 (28 Dec. 1882): 194–95.

13. Hammond 1882. J. Hammond was the eldest son of nine children. He was educated at King's College, London, and graduated from Cambridge 35th wrangler in 1874 (Elliott 1931).

14. "Addition to Mr Hammond's Paper 'Note on an Exceptional Case in which the Fundamental Postulate of Professor Sylvester's Theory of Tamisage Fails.'" = [*CP*11 (1883): 409–10].

15. JJS to C, 25 Dec. 1882, SJC, Sylvester Papers, Box 11.

16. Cayley's Law in invariant theory (see chap. 8) required the difference of the partition numbers $(w:5,7) - (w-1:5,7)$ where the weight of the putative covariants $w = \frac{1}{2}(5.7 - 13) = 11$. Since $(11:5,7) = 30$ and $(10:5,7) = 26$, there are four linearly independent (asyzygetic) covariants of degree 5 and order 13.

17. C to JJS, 12 Feb. 1883, SJC, Sylvester Papers, Box 2. Cayley's supposition was confirmed in 1890 when J. Hammond showed an invariant of degree 20 to exist for the binary form of order 7.

18. JJS to C, 26 May 1883, SJC, Sylvester Papers, Box 11.

19. JJS to C, 11 May 1885, SJC, Sylvester Papers. Box 12.

20. MacMahon 1904.

21. Morley 1912, 47.

22. C to JJS, 12 Feb. 1883, SJC, Sylvester Papers, Box 2.

23. JJS to C, 16 Mar. 1883, SJC, Sylvester Papers, Box 11.

24. Birkhoff 1975, 503.

25. C to FK, 4 Apr. 1883, NSUG, no. 374.

26. Hannabuss 2000, 192; Brock and Curthoys 2000, 452; the group of dons who agitated for the British Association meeting to be held in Oxford did not include Henry Smith.

27. When the Marquis of Salisbury did become president at the British Association Oxford meeting in 1894, his address was on "Evolution."

28. C to JJS, 6 Sept. [1882], SJC, Sylvester Papers, Box 2.

29. C to TAH, 31 Mar. 1882, UCL, LMS Archive. H. C. H. Schubert published his book on enumerative geometry in 1879 (Schubert 1879).

30. C to JJS, 6 Sept. [1882], SJC, Sylvester Papers, Box 2. After H. J. S. Smith's demise in February, Spottiswoode's sudden death in July 1883 at the age of fifty-eight was a further blow to the scientific establishment.

31. JJS to C, 6 Oct. 1882, SJC, Sylvester Papers, Box 11; Parshall 1998, 216–17.

32. Glaisher 1914, lii; Glaisher 1895b, 174; *Athenaeum*, no. 3510 (2 Feb. 1895): 151.

33. JJS to JNL, 3 Apr. 1870, EUL, MS 110, Lockyer Research Papers.

34. JJS to C, 8 Nov. 1884, SJC, Sylvester Papers, Box 12; Parshall 1998, 255–56; *Nature* 31 (13 Nov. 1884): 35–36.

35. JJS to C, 3 Aug. 1883, SJC, Sylvester Papers, Box 11. The "memorable afternoon" referred to by J. J. Sylvester is either the occasion when Cayley read his first paper at the Royal Society in 1852 or (more likely) his inaugural lecture at Cambridge on 3 Nov. 1863.

36. C to EA, 27 Sept. 1878, CUL, Add Ms 6583.577.

37. William H. Longmaid's portrait of Cayley as British Association president of 1883 hangs in the dining room of the master's lodge at Trinity College, Cambridge. Cayley was disappointed that the association meeting was not held at Oxford, as originally planned. Southport's promoter, Dr. Barron, had employed modern marketing techniques and proved an effective lobbyist in Southport being chosen as the venue (Lodge 1931, 54–57).

38. *Illustrated London News,* 29 Sept. 1883, 83 (no. 2319), p. 307. col. 2.

39. *Times,* 1 Sept. 1883, p. 7, col. 1.

40. *Times,* 17 Sept. 1883, p. 4, col. 3.

41. MacAlister 1935, 88.

42. *Liverpool Daily Post,* 20 Sept. 1883, p. 6, cols. 1, 2.

43. "Presidential Address to the British Association, Southport, Sept. 1883" = [*CP*11 (1883): 429–59]. Quotations from Cayley's presidential address will be made without further reference for the remainder of this chapter.

44. Mill 1874, 225–26.

45. The *System of Logic* (1843) quickly achieved a wide influence in England and Germany (Munday 1998). Mill's views had been prevalent at Cambridge (as retold by Leslie Stephen): "the young men who graduated in 1850 and the following ten years found their philosophical teaching in Mill's Logic, and only a few daring heretics were beginning to pick holes in his system" (Newsome [1997] 1998, 60).

46. Mill 1874, 226.

47. Initially, W. R. Hamilton wanted to show that "time" as a basis for the algebra of "number" gave it the status of a science, the "science of Pure Time." He later accepted that algebra is possible as a "science of symbols" (Øhrstrøm 1985).

48. Cornford 1941, 235.

49. This is a repetition of his view expressed in "On the Notion and Boundaries of Algebra" = [*CP*5 (1864): 292–94]. See chap. 11. The debate was far from being arcane and the question of primitive notions of mathematics was one of lively interest. W. R. Hamilton had chosen to base the concept of number on "Pure Time" while A. De Morgan chose to avoid the word "time" altogether and regarded number as continuous succession of points on a line. Whewell expressed it slightly differently: "Number is a modification of the conception of Repetition, which belongs to the Idea of Time" (Whewell, [1840] 1847, Part 2, p. 446).

50. The idea of ordinal number taking precedence over cardinal number was also expressed by W. R. Hamilton in his *Lectures on Quaternions* (1853) and also by A. De Morgan in the 1830s.

51. JJS to C, 3 Aug. 1883, SJC, Sylvester Papers, Box 11.

52. "Fermat's Last Theorem" states that nonzero integer solutions to the equation $x^n + y^n = z^n$ do not exist for all natural numbers $n > 2$. H. J. S. Smith noted it had only been proved in the case where n is a "nonexceptional" prime number ("Report on the Theory of Numbers. (Part 2)" = [*Smith*P1 (1860): 131–37]). Cayley referred to the theorem as an isolated result in the theory of numbers and noted that the "general" proof presents "very great difficulty," that is, a proof had not been found ("Theory of Numbers" = [*CP*11 (1884): 616, para. 44]). The theorem has since been proved by Andrew Wiles.

53. Cayley took a great deal of interest in the problem of computing the pth roots of unity.

54. Cayley sympathized with Herbert Spencer, who expressed a belief in the existence of a priori physical "truths" (such as *force*) on a par with a priori mathematical concepts (Spencer 1904, 2: 258).

55. Russell 1902.

56. Whewell [1847] 1967, Part 2: 446.

57. "Note on Riemann's Paper Versuch einer allgemeinen Auffassung der Integration and Differentiation,' *Werke*, pp. 331–344 [Essay on a general concept of integration and differentiation]" = [*CP*11 (1880): 235]; Ross 1977, 81.

58. Ball 1903–4.

59. C to WT, 25 Mar. 1889, CUL, Add Ms 7342.C63.

60. Sommerville [1914] 1958, 158.

61. It was only around the 1880s that Cayley systematically referred to the works of others in his research papers. His paper on the Schwarzian derivative was one of his first to contain a bibliography: "On the Schwarzian Derivative, and the Polyhedral Functions" = [*CP*11 (1883): 148–216].

62. Grattan-Guinness 1997.

63. Enumerative geometry deals with such questions as: "How many conics are there which touch five given conics?" Cayley had published a paper on it before the British Association meeting ("On Schubert's Method for the Contacts of a Line with a Surface" = [*CP*11 (1881): 281–93]).

64. The printed record of the address contains more than Cayley's oration at Southport. The essential channel of communication for Cayley was the published scientific paper and the lecture merely the prelude or preparation for the printed work. The content of a lecture might evaporate, but the paper could be read by future generations.

65. Forsyth 1895, xxi.

66. Susan Cayley to WT, 2 Dec. 1895, CUL, Add Ms 7342.C69.

67. M. A. Scott, 1895, reprinted in C. S. Peirce 1976, 2: 648.

68. *Liverpool Daily Post*, 21 Sept. 1883, p. 6.

69. Proctor 1884, 35, 38.

70. S[mith] 1884. James Parker Smith later became a Member of Parliament and secretary to Joseph Chamberlain.

71. *Athenaeum*, no. 2917 (22 Sept. 1883): 371–73.

72. Young [1937] 1948, 248. Much has been written of the Victorian age; a fascinating general portrait of the period is the recent work by A. N. Wilson (2002).

Chapter 16. The Old Man of Mathematics

1. In his latter years, Charles Cayley had moved to a square near the British Museum Library where he spent his hours as an impecunious *rat de bibliothèque* translating and writing poetry and other scholarly books (D. Hudson 1974, 127, 297).

2. MacFarlane 1916, 67.

3. Wordsworth [1877] 1968, v.

4. DCG to C, 31 Jan. 1884, JHU, Daniel Coit Gilman Papers MS1.

5. Crilly 1979.

6. "A Memoir on the Abelian and Theta Functions" = [*CP*12 (1882, 1885): 109–48, 149–216].

7. C to JJS, 11 July [1884], SJC, Sylvester Papers, Box 12.

8. JJS to C, [12] July 1884, SJC, Sylvester Papers, Box 12.

9. "On Subinvariants, that is, Semi-invariants to Binary Quantics of an Unlimited Order" = [*SP*3 (1882): 572].

10. JJS to C, 6 Oct. 1882, SJC, Sylvester Papers, Box 11; Parshall 1998, 216–19.

11. J. W. L. Glaisher proposed Cayley for this award. T. A. Hirst proposed the younger H. C. Schubert as one of the "rising generation" and he was disappointed that Cayley did not decline the honor (Hirst's *Journal*, 8 May 1884, *RIGB* 4: 2145, 2148).

12. Salmon 1883, 484.

13. Henrici 1884.

14. von Gall 1880, 1888. A brief survey of nineteenth-century results for the binary form can be found in MacMahon 1910a, 634–36.

15. Cayley, A. De Morgan, J. W. L. Glaisher, and P. A. MacMahon were all presidents of the Royal Astronomical Society.

16. MacMahon's correspondence theorem established a one-to-one correspondence between semi-invariants and (nonunitary) symmetric functions of a binary form of infinite order. As the tabulation of symmetric functions is equivalent to the arithmetical problem of partitioning an integer, MacMahon's correspondence theorem offered a *potential* reduction of invariant theory to a problem of arithmetic (Forsyth 1930). In character, it is similar to Cayley's observation of a few years earlier in which the study of abstract groups is potentially reduced to the study of permutation groups. The practicalities of these potentialities was a different issue, as Cayley acknowledged. The semi-invariants of the binary form of infinite order were established by Cayley, MacMahon, Sylvester, and E. Stroh (Elliott [1895] 1913, 238–40]).

17. JJS to C, 9 Aug. 1883, SJC, Sylvester Papers, Box 11.

18. C to JJS, 11 Aug. 1883, SJC, Sylvester Papers, Box 2.

19. C to JJS, *JHUC* 3, no. 27 (Nov. 1883): 13.

20. "A Memoir on Seminvariants" = [*CP*12 (1885) 239–62]. In a following paper ("On the Theory of Seminvariants" = [*CP*12 (1886): 344–57].) Cayley returned to his earliest papers on hyperdeterminants and indicated how his theory might be advanced in the light of MacMahon's discovery of the correspondence theorem.

21. MacMahon 1896, 5. Cayley's algorithm was limited to the product of *two* symmetric functions. MacMahon was later able to deal with longer products using the differential operators pioneered by J. Hammond.

22. C to PAM, 22 Jan. 1884, SJC MacMahon Papers, and generally in 1885.

23. JH to JJS, 16 Apr. 1883, *JHUC*, no. 25 (Aug. 1883): 150; Hammond 1886.

24. JJS to C, 13 Dec. 1883, SJC, Sylvester Papers, Box 11; Parshall 1998, 229–34.

25. JJS to C, 3 Feb. 1884, SJC, Sylvester Papers, Box 12; Parshall 1998, 243–46.

26. JJS to C, 2 Mar. 1884, SJC, Sylvester Papers, Box 12.

27. C to JJS, 30 June [1884], SJC, Sylvester Papers, Box 12; Sylvester 1888. A. Buchheim corresponded with Sylvester on matrix algebra from June 1884 to December 1886. Through Sylvester, Cayley was made aware of Buchheim's work (AB to JJS, 27 July 1885, SJC, Sylvester Papers, Box 12). Buchheim was a student of H. J. S. Smith at Oxford, where he took a first-class degree in 1880, and he studied with Felix Klein at Leipzig in 1881. On his return, he took a post at Manchester Grammar School and continued with research, placing matrix algebra in a wider and more rigorous context through his reading of H. Grassmann and G. Frobenius (Sylvester 1888).

28. C to JJS, 11 July [1884], SJC, Sylvester Papers, Box 12.

29. C to JJS, 2 Sept. 1884, SJC, Sylvester Papers, Box 12. "On the Quaternion Equation $qQ - Qq' = 0$" = [*CP*12 (1885): 300–304]; "On the Matricial Equation $qQ - Qq' = 0$" = [*CP*12 (1885): 311–13].

30. JJS to C, 22 Oct. 1884, SJC, Sylvester Papers, Box 12. The mathematical contacts were C. É. Picard, G. Darboux, C. Jordan, J. de Jonquières, V. M. A. Mannheim, and Picard's uncle J. L. F. Bertrand.

31. JJS to C, 12 July 1884, SJC, Sylvester Papers, Box 12.

32. JJS to C, 15 Oct. 1884, SJC, Sylvester Papers, Box 12.

33. JJS to RL, 13 Jan. 1886, BUMI. Sylvester's inaugural lecture at Oxford, given on 12 Dec. 1885, was on differential invariants (invariant functions involving derivatives which he called reciprocants).

34. JJS to C, 8 Nov. 1884, SJC, Sylvester Papers, Box 12.

35. "Sur la Théorie de la Transformation des Fonctions Abéliennes" = [*HermiteP*1 (1855): 444–78].

36. C to TC, 20 Feb. 1885, JHU; "On the Transformation of the Double Theta-functions" = [*CP*12 (1886): 358–89]; "Seminvariant Tables" = [*CP*12 (1885): 275–89].

37. C to TC, 12 Aug. 1885, JHU, Daniel Coit Gilman Papers MS1.

38. JJS to C, 22 Apr. 1885, SJC, Sylvester Papers, Box 12. MacMahon showed that a sextic perpetuant (irreducible semi-invariant of degree 6 of the binary quantic of infinite order) did exist for weight 31 and that they exist for each degree θ and are of weight $2^{\theta-1} - 1$ (MacMahon 1885; MacMahon 1910a, 638). In proposing P. A. MacMahon for a Royal Medal in 1885, Sylvester felt himself outmaneuvered at the Royal Society council meeting by G. G. Stokes, who proposed P. G. Tait (JJS to C, 7 May 1885, SJC, Sylvester Papers, Box 12). Tait received the award in 1886, but Sylvester justified his suggestion of a medal for MacMahon as putting down a marker for the future. MacMahon was elected F.R.S. in 1890, proposed by Cayley (Royal Society, 5 June 1890, vol. 11, cert. 182). He was awarded the Royal Medal in 1900.

39. The simplest example of a reciprocant is the Schwarzian derivative:

$$S(y) = \frac{dx}{dy}\frac{d^3y}{dx^3} - \frac{3}{2}\left(\frac{dx}{dy}\frac{d^2y}{dx^2}\right)^2$$

("On the Method of Reciprocants" = [*SP*4 (1886): 284–85, 304–10]). Cayley's extensive ("On the Schwarzian Derivative, and the Polyhedral Functions" = [*CP*11 (1883): 148–216]) was judged by W. Burnside as completely dealing with the theory of groups of a finite number of different substitutions (Burnside 1891).

40. JJS to C, 24 Oct. [1885], SJC, Sylvester Papers, Box 12; "Inaugural Lecture at Oxford, On Reciprocants and Differential Invariants" = [*CP*13 (1893): 366–404]; "On the Method of Reciprocants" = [*SP*4 (1885): 290]; "Biographical Notice" = [*SP*4 (1912): xxxiii]. Cayley touched on the classical differential geometry briefly ("On Curvilinear Coordinates" = [*CP*12 (1883): 1–18]). J. E. Wright of Trinity College, the Senior Wrangler in 1900, was active in the theory of invariants of differential forms pioneered by C. F. Gauss, G. Lamé, E. B. Christoffel, and G. Ricci-Curbastro. In his outline of the history of invariants and differential forms, Wright placed group theory at its center (Wright 1908, 1–8).

41. "On the Theorem of the Finite Number of the Covariants of a Binary Quantic" = [*CP*11 (1881): 273]. Cayley wrote a number of papers on semi-invariants in the 1880s and 1890s: [*CP*12 (1883): 22–29], [*CP*12 (1885): 239–62, 273–89, 326–27], [*CP*12 (1886): 344–57], [*CP*12 (1889): 558, withdrawn], [*CP*13 (1892): 195–200, 265–332, 362–65].

42. MacMahon 1896, 6.

43. JJS to C, 1 Feb. 1886, SJC, Sylvester Papers, Box 12; Parshall 1998, 261–62; JJS to C, 18 Feb. 1886, SJC, Sylvester Papers, Box 12.

44. JJS to JOT, 14 Sept. 1882, RIGB, Tyndall Papers, 4: 1519.

45. C to FK, 13 May 1886, NSUG, no. 379 mourning paper.

46. In the next generation of English mathematicians, J. H. Grace and A. Young employed the symbolic method effectively and gave an exposition of it in their influential text (Grace and Young 1903). They retraced the steps of Gordan's original papers and elaborated Gordan's theorem for binary forms; White 1899, 161.

47. Hobson 1911, 512.

48. Sheppard 1937–38. W. F. Sheppard graduated in 1884, the same year as Sir William Bragg, the textbook writer W. P. Workman, and W. H. Young.

49. Caroe 1978, p. 23.

50. Inaugural Lecture (1883) = [G. H. *DarwinP*5 (1916): 4 and 2, respectively]. O. Henrici had made similar criticisms of the Tripos years before (*Nature* 8 [9 Oct. 1873]: 492).

51. Forsyth 1930a, x; *CR*1, no. 22 (19 May 1880): 86.

52. "I have satisfied myself against this," wrote Cayley of the essay submitted by A. Macaulay on quaternions (C to GGS, 8 May 1888, CUL, Add Ms 7656.C272).

53. "The Binomial Equation $x^p - 1 = 0$; Quinqisection. Second Part" = [*CP*12 (1885): 72–73].

54. ICL, Forsyth Papers, B1/Forsyth, A. R./1/3.

55. Neville 1942, 240.

56. Whittaker 1936–38a, 217.

57. Heath 1885; Thompson 1939–41, 415.

58. It is possible Cayley knew about the *Ausdehnungslehre* in the 1850s through Salmon. Hamilton had read it and discussed it with Salmon in 1857 (WRH to GS, 23 June 1857, in R. P. Graves 1882–89, 3: 88).

59. Lowe 1985.

60. Whitehead 1901.

61. Natural numbers in the sequence: 2, 3, 5, 11, 47, 923, 409619, 83763206255, . . ., which are related to properties of polynomials (C to GBM, 20 Aug. 1894, CUNY, D. E. Smith Papers; Ref. Rep. 20 June 1887, RSL, RR.10.88; Berwick 1923; Sloane 1973, 51).

62. Whittaker 1936-38b, 166. G. H. Bryan, the joint Smith's Prize winner with A. C. Dixon in 1888 was sponsored for membership of the Royal Society by Cayley (Royal Society, 13 June 1895, vol. 11, cert. 274).

63. "On the Superlines of A Quadric Surface in Five-Dimensional Space" = [*CP*9 (1873): 79–83]; Parshall and Rowe 1994, 373–74.

64. HFB to WWRB, 1 Oct. 1923, TCC, 0.6.6.1.

65. Baker 1889. In the late 1880s, students who gained Smith's Prizes for essays on invariant theory included A. Berry, H. F. Baker, W. E. Brunyate (vice-chancellor of the University of Hong Kong, 1921–24) and H. W. Segar (a lecturer at Aberystwyth later professor of mathematics in Auckland, New Zealand).

66. Grattan-Guinness 1972, 115.

67. "Letters to Lecturers, XVIII: To Professor Cayley," *CR* 11, no. 270 (6 Feb. 1890): 180. I am grateful to June Barrow-Green for this reference.

68. Cayley's two courses of mathematics a year was a mild increase compared with the heavy lecturing duties shouldered by the professors of the Cavendish Laboratory (Kim 1995, 202). Hardy's view on the state of affairs at this time was that the Tripos was "somewhere near its lowest ebb." (Hardy 1948, 138).

69. M. A. Scott, 4 Feb. 1895, *Evening Post* [New York] p. 7., reprinted in C. S. Peirce 1976, 2: 648.

70. Vaizey, 1895–1907, 160. Cayley served on the Council of the Senate of Cambridge University from 9 Nov. 1876 until his resignation on 25 Oct. 1892.

71. E. M. Sidgwick to C, 7 Jan. 1887, NCC College Archives.

72. JJT to R. Threlfall, 7 Aug. 1887, and 11 Dec. 1887, in Thomson 1936, 29.

73. In 1887, Emily Davies and her London committee pressed for the admission of women to degrees on the same basis as men. Cayley and a small group (H. Sidgwick, N. M. Ferrers, C. Trotter) urged caution lest the hard-won existing privileges be withdrawn. The result Emily Davies desired came about in 1947 when the Queen Mother became the first Cambridge female graduate (McWilliams-Tullberg 1975, 85–100).

74. Thompson 1939–41, 414.

75. In the 1870s, Cayley contributed articles on "Curve," "Equation," "Function," "Galois," "Gauss" to the *Encyclopaedia Britannica*. The articles are reprinted in [*CP*11 (1878–88): 460–641]

76. L. A. Todhunter to KP, 26 Feb. 1890, UCL, Pearson Papers, 873/3; C to KP, 7 Feb. [1890], UCL, Pearson Papers, 655/7.

77. Henry Sidgwick's Journal, 11 Oct. 1886, in Sidgwick 1906, 460.

78. R. H. Williams 1988, 163.

79. Henry Montagu Butler to B. F. Westcott, 9 May 1887, in Montagu Butler 1925, 27.

80. CUL, University Archives, Min.V.7.

81. Siddons 1936, 18.

82. C. A. Scott 1895, 141.

83. Siddons 1936, 18. E. M. Langley, another reformer, commented on Cayley's conservatism where Euclid was concerned (Price 1994, 30–31).

84. Smith 1873, 451.

85. Carroll 1888; *WhewellP*15 (1837): 99.

86. Siddons 1936, 18. At their May meeting of 1887 following this encounter with the schoolmaster delegation, the Mathematics Board softened its attitude and J. W. L. Glaisher proposed that the board should not insist on the exact reproduction of Euclid's proofs, provided that Euclid's order was maintained (carried by seven votes to three).

87. *CUR* 718 (31 Jan. 1888): 376.

88. Winstanley 1947, 146.

89. C to DES, 9 Oct. 1893, in Simons 1945, 255.

90. Hobson 1911, 520.

91. Unlike Cayley, A. R. Forsyth was progressive in educational reform, and when he succeeded to the Sadleirian chair in 1895, he would have nothing to do with any report which argued for the retention of Euclid's order of presentation. The schoolteacher's argument was gradually accepted and from the beginning of the twentieth century, Euclid had ceased to dominate English mathematical education.

92. M. A. Scott, 4 Feb. 1895, *Evening Post* [New York] p. 7, reprinted in Peirce 1976, 2: 648.

93. Eve and Creasey 1945, 252.

94. C to JJS, 16 Nov. 1887, SJC, Sylvester Papers, Box 2. Cayley's rate of editing the *Collected Mathematical Papers* started out at two volumes a year but settled down to one volume per year: vols. 1, 2 (1889), vol. 3 (1890), vol. 4 (1891), vol. 5 (1892), vol. 6 (1893), and vol. 7 (1894). After Cayley's death, A. R. Forsyth took over the editorship and finished the work quickly: vol. 8 (1895), vols. 9, 10, 11 (1896), vols. 12, 13 (1897).

95. C to JJS, 16 Nov. 1887, SJC, Sylvester Papers, Box 2.

96. Cayley published an unfinished survey paper: "On Multiple Algebra" = [*CP*12 (1887): 459–89].

97. Quoted in Archibald [1938] 1980, 3–4. T. Fiske became the editor of the *Bulletin of the New York Mathematical Society* when it began publication in 1891. Cayley became a member of this society, which was renamed the American Mathematical Society around 1894.

98. Parshall and Rowe [1994] 1997, 267.

99. Cochell 1998; Parshall and Rowe [1994] 1997, 269–71.

100. *CR* 9, no. 230 (14 June 1888): 388 (trans. Victoria Fox); *CUR* 739 (5 June 1888): 766–69.

101. *CR* 9, no. 230 (14 June 1888): 388 (trans. Victoria Fox).

102. Cayley made a financial contribution of £135 to the building of the new hall at Newnham College; W. E. Gladstone gave £100; most contributions were in the range of £20–£30.

103. "On a Differential Equation and the Construction of Milner's Lamp" = [*CP*13 (1887): 3–5]. (*See* "Isaac Milner's lamp" on p. 19g of the gallery.)

104. Kempe 1887, 37–43; Ref. Rep., 17 Sept. 1885, RSL, RR.9.287. Cayley used "enumeration" to mean listing the possibilities, not merely counting them. The review of A. B. Kempe's paper was a daunting prospect but Cayley did not flinch from it. Sylvester felt otherwise: "The Herculean task of reporting on this memoir has also been imposed on me: this and the [Oxford] lectures together are enough to crush all the life out of me and to absorb my whole time" (JJS to C, 4 July 1885, SJC, Sylvester Papers, Box 12).

105. "On the Theory of Groups" = [*CP*12 (1889): 639–55]. In this paper, Cayley introduced a *colorgroup*, a diagram which he used to graphically represent *all* permutations. "On the Substitution-groups of Two, Three, Four, Five, Six, Seven, and Eight Letters" = [*CP*13 (1891): 117–49]; Askwith 1890. For a commentary on these papers, see Berwick 1929. "On Latin Squares" = [*CP*13 (1890): 55–57]. "On the Transformation of Elliptic Functions" = [*CP*12 (1887, 1888): 505–55]; "Systems of Equations for Three Circles Which Cut Each Other at Given Angles" = [*CP*12 (1888): 559–61, 564–70]; "Note on the Two Relations Connecting the Distances of Four Points on a Circle" = [*CP*12 (1888): 576–77].

106. Cayley's proof of "Cayley's law in graph theory" consisted in showing the result true for the single value of $n = 5$ ("A Theorem on Trees" = [*CP*13 (1889): 28]). There are at least ten distinct proofs of this theorem of Cayley's (Moon 1967).

107. See Dickson 1912; Tait [1867, 1873] 1890; PGT to C, 28 Aug. 1888, in Knott 1911, 159. Cayley duly wrote Chapter Six of P. G. Tait's *An Elementary Treatise on Quaternions* (1890), entitled "Sketch of the Analytical View of Quaternions." He dealt with Hamilton's biquaternions and the possibility of the modulus of a biquaternion being zero.

108. Address to Section A of the British Association = [*TaitP*2 (1871): 167].

109. PGT to C, [Sept./Oct.] 1888, in Knott 1911, 159.

110. C to PGT, [Oct./Nov] 1888, in Knott 1911, 159.

111. Ball 1906–7, lxix.

112. JJS to C, 3 Sept. 1884, SJC, Sylvester Papers, Box 12.

113. PGT to C, 22 Oct. 1888, in Knott 1911, 159. An example of the application of quaternions to geometry is the condition for three points in space represented by [quaternion] vectors α,β,γ to be coplanar. In the quaternion form, this condition is $S(\alpha\beta\gamma) = 0$, where $S(\alpha,\beta,\gamma)$ is the scalar part of the quaternion product $\alpha\beta\gamma$. In terms of Cartesian coordinates, the condition can be expressed in terms of a vanishing determinant formed from the coordinates, the method favored by Cayley.

114. Cayley 1895. "Coordinates versus Quaternions" = [*CP*13 (1895): 541–44]. Cayley had used the "pocket map" metaphor in the 1870s when he wrote to Tait of one succinct quaternionic expression as a "grand example of the pocket-map" (C to PGT, 25 Mar. 1874, in Knott 1911, 154).

115. Further papers in 1888, 1889 by D. Hilbert established a finiteness theorem for algebraic forms of any number of variables. His researches on invariants for a single algebraic form were brought together in a comprehensive paper ("Über die Theorie der algebraischen Formen" = [*HilbertP*2 (1890): 199–257]).

116. C to FK, 24 Jan. 1889, NSUG, no. 383; Cayley 1889a.

117. Paul Gordan's blunt appraisal was that Cayley's paper was "worthless" (in Sabidussi 1992, 152 fn. 7.)

118. C to DH, 30 Jan. 1889, in Reid 1970, 33.

119. C to FK, 22 Feb. 1889, mourning paper, NSUG, no. 384.

120. FK to DH, 24 Feb. 1889, in Sabidussi 1992, 103.

121. DH to FK, 27 Feb. 1889, in Sabidussi 1992, 103. Hilbert recognized Cayley as the "father" of invariant theory in his 1897 Göttingen lectures (Hilbert 1993, 2).

122. C to PAM, 22 Apr. 1889, SJC MacMahon Papers.

123. C to PGT, 25 June 1894, in Knott 1911, 165.

124. "Notes and References" = [CP2 (1889): 605]. For modern commentaries on Cayley's non-Euclidean geometry and Cayley's projective metric, see Jammer ([1954] 1969, 158) and North (1965, 322–24).

125. C to FK, 23 July 1889, mourning paper, NSUG, no. 386. Cayley's explanation did not appear in the *Messenger* but in the notes of the *Collected Mathematical Papers,* which he was then editing ("Notes and References" = [CP2 (1889): 604–6]. The Italian mathematician M. Pieri completed the work of von Staudt on the foundations of projective geometry using rational number coordinates by treating projective geometry with real number coordinates.

126. Cayley observed that Hilbert's theorem "is not easy to prove" ("Notes and References" = [CP5 (1892): 614]) and, a year later, an admission that "he never succeeded in finding a proof of the theorem" ("Notes and References" = [CP6 (1893): 596]).

127. Lützen, Sabidussi, and Toft 1992; Biggs, Lloyd, and Wilson 1976, 187–97.

128. FK to JP, 27 Sept. 1889, in Sabidussi 1992, 105.

129. JP to FK, 20 [Sept.] 1889, in Sabidussi 1992, 104–5. As is pointed out by G. Sabidussi, this letter is dated October 1889 but seems to have been written in September.

130. JJS to JP, 3 Oct. 1889, in Sabidussi 1992, 107.

131. JP to FK, 20 [Sept.] 1889, in Sabidussi 1992, 104–5; Petersen 1890. Petersen visited Cambridge and would have been able to discuss the problem with Cayley before he published his correction (JJS to ABK, 7 Jan. [1890], WSRO, Kempe Mss 23/10).

132. JJS to FK, 15 Oct. 1889, in Sabidussi 1992, 111.

133. MacMahon 1889, 36.

134. MacMahon 1897, 7, 8.

Chapter 17. Last Years

1. Salzman [1959] 1967, 278–79.

2. *Times,* 27 Oct. 1891, p. 14, col. 2.

3. The "Greek division" took place 29 Oct. 1891, where the vote was 185 votes for setting up a syndicate but 525 votes against the proposal (*CUR* 906 [3 Nov. 1891]: 154); Sidgwick 1906, 509–12.

4. Forsyth 1895, xvi.

5. According to the High Church newspaper, *Guardian* (6 Feb. 1895): 200.

6. Forsyth 1895, xxiv.

7. C to PGT, 24 Jan. 1890, in Knott 1911, 161; PGT to C, 25 Jan. 1890, in Knott 1911, 162.

8. Grattan-Guinness 1972, 117. Isabel Maddison was the first Ph.D. student of Charlotte Angas Scott at Bryn Mawr (U.S.) and received her degree in 1896.

9. Grattan-Guinness 1972, 117–18. Cayley was revisiting work carried out in the 1860s: "On the Δ faced Polyacrons, in Reference to the Problem of the Enumeration of Polyhedra" = [CP5 (1862): 38–43].

10. Glaisher 1891, 724.

11. Salmon 1883, 482.

12. *Athenaeum,* no. 3510 (2 Feb. 1895): 151; Salmon 1883, 482.

13. C to FK, 13 Oct. 1892, NSUG, no. 389; Klein 1890–92.

14. Anon. c. 1890. Quoted in Phillips 1979, 34.

15. Henry Cayley 1904.

16. C to FK, 19 Nov. 1894: NSUG, no. 390. Miss Ada Maria Jane Elizabeth Johnson gained a Clothworkers' Scholarship to pay for her undergraduate studies at Cambridge. In the competitive Part II of the Tripos, in the 1893 order of merit, she was placed between the fifth and sixth wrangler. On graduating in 1894, she studied with Klein in Göttingen (1895–96) in the company of Grace Chisholm and the Americans Mary Frances Winston and Annie MacKinnon (Parshall and Rowe [1994] 1997, 245).

17. "Inaugural Lecture at Oxford 'On the Method of Reciprocants'" = [*SP4* (1886): 300]; *Nature* 33 (7 Jan. 1886): 230.

18. JJS to C, 27 May 1891, SJC, Sylvester Papers, Box 12.

19. Joly 1905.

20. Salmon 1873b; R. S. Ball 1903–4.

21. "Notes and References" = [*CP4* (1891): 609–16]. E. McClintock was president of the New York Mathematical Society at the time Cayley's note appeared. Cayley saw McClintock's discovery of a covariant resolvent as an extension of his own work of 1861 and in his note he explained the connection between McClintock's and his approach to the problem (see chap. 10). An overview of the history of the problem of solving polynomial equations and its relationship to invariant theory is in Foulkes 1932. Re George P. Young and "On a Soluble Quintic Equation" = [*CP13* (1891): 88–92]; Berndt, Spearman, Williams 2002, 25–26.

22. MacMahon 1896, 7.

23. "Tables of Covariants of the Binary Sextic" = [*CP11* (1894): 377–88]. Those covariants labeled X, Y and Z given in 1894 are lengthy and were calculated to have 1002, 2012, and 1636 terms, respectively. The twenty-six invariants and covariants were conveniently labeled by the letters of the alphabet, where Z is the great skew-invariant of degree 15 calculated by G. Salmon.

24. Forsyth 1897, 548.

25. MacMahon 1901, 526.

26. Hilbert 1993, 2.

27. "Address to the Mathematical and Physical Section of the British Association" = [*SP2* (1869): 655].

28. GGS to C, 29 Oct. 1849, in Stokes 1907, 1: 382; C to GB, 3 Dec. [1845], TCC, RR.2.88.22.

29. C to TAH, 24 [Mar.] 1889, mourning paper. UCL, LMS Archive.

30. Newnham College Council Minutes, 10 May 1890, p. 175. NCC College Archives.

31. C to TAH, 1 Nov. 1890, UCL, LMS Archive.

32. JJS to C, 27 May 1891, SJC, Sylvester Papers, Box 12.

33. JJS to C, 11 June 1891, SJC, Sylvester Papers, Box 12.

34. C to TAH, 28 Dec. 1891, UCL, LMS Archive.

35. Cayley's death certificate gave his cause of death as disease of the prostate gland and cystitis; *Cambridge Chronicle and University Journal*, no. 6907 (1 Feb. 1895): 8.

36. Weber 1896.

37. In compiling these tables, Cayley was ably assisted in the calculations by the Oxford-based C. E. Bickmore.

38. GGS to WT, 26 Jan. 1893, in D. Wilson 1990, 2: 609; After this bout of illness, Cayley was not much better by the spring (C to C. T. Whitmell [Her Majesty's Inspector], 19 Apr. 1993, LUBL. MS 297, p. 4), and he had to cancel his teaching for a whole term this year (Griffin and Lewis 1990, 61).

39. "Presidential Address to the British Association, Southport, September 1883" = [*CP11* (1883): 434].

40. Cayley 1894, Preface.

41. C to JJS, 1 Oct. 1994, John Hay Library, Brown University, Providence, RI: Box 1, Folder 2.

42. C to FK, 19 Nov. 1894, NSUG, no. 390.

43. C to M. R. James (vice-chancellor), 17 Apr. 1894, CUL, Add Ms 7481.C17; C to M. R. James, 30 Apr. 1894, CUL, Add Ms 7481.C18.

44. Jowett [1871] 1953, 2: 554.

45. Ian Maclaren was the pen name of the English author John Watson.

46. Sir Walter Scott 1815, chap. 20, 173; chap. 8, p. 75.

47. Sir Walter Scott 1815, chap. 42, p. 379.

48. Sir Walter Scott 1815, chap. 17, Fourth extract, p. 152.

49. C to PGT, June 1894, in Knott 1911, 164.

50. "Coordinates versus Quaternions" = [CP13 (1895): 541–42].

51. C to WT, 5 Dec. 1894, CUL, Add Ms 7342.C67 [Cayley's emphasis].

52. C to WT, 7 Dec. 1894, CUL, Add Ms 7342.C68. It is probable that Cayley wrote his last paper for the new *Bulletin of the American Mathematical Society:* "Note on a Memoir in Smith's Collected Papers" = [CP13 (1895): 558–59, sub. 18 Dec. 1894]

53. CPS *General Minutes,* 1895; *CUR* 1059 (29 Jan. 1895): 467.

54. Glaisher 1895, 176.

55. Susan Cayley to WT, 2 Dec. 1895, CUL, Add Ms 7342.C69.

56. Thompson 1910, 2: 950.

57. *Times,* "Death of Arthur Cayley," 28 Jan. 1895, p. 4, col. 6.

58. Forsyth 1935, 162–63.

59. *Times,* "Death of Arthur Cayley," 28 Jan. 1895, p. 4, col. 6.

60. On his death, Cayley left the sum of £27,670-14s-2d to his wife Susan Cayley under a will made 3 April 1877. Thus by the time of his death he was easily a millionaire in today's terms (as the result of a calculation based on the salary of a 21st-century professor of mathematics).

61. Stevenson 1897, 151.

62. The broken headstone of Cayley's grave existed in Mill Road cemetery in the 1980s but it has not survived.

63. "Arthur Cayley" = [CP8 (1895): xxvi].

64. C to WT, 17 Aug. 1847, CUL, Add Ms 7342.C55.

65. C. S. Peirce, "Professor Arthur Cayley." *Evening Post* [New York], 28 Jan. 1895, p. 7, cols. 1–2.

66. Young 1922, li.

67. MacMahon 1896, 7.

68. Noether 1895.

69. Galton 1908, 72.

70. Ref. Rep., 27 May 1861, RSL, RR.4.45.

71. Roberts 1882–83; Mathews 1898b.

72. C to WT, 8 Feb. 1847, CUL, Add Ms 7342.C45.

73. C to [JCA], 10 Dec. 1872, SJC, Adams Papers, Box 6, folder 4, no. 3. It is hard to imagine Cayley offering advice to J. C. Adams on astronomical matters, but there is no doubting its forthright style; it may be advice for a third party.

74. C to TAH, 19 June [1878], UCL, LMS Archive. Though Cayley commented unfavorably on G. H. Halphen's work on conics, he praised the clarity and brevity of H. Zeuthen's contribution ("Notes and References" = [CP7 (1894): 594]). Halphen shared the Steiner Prize with Max Noether in 1881 for research on space curves. This work did not appeal to Cayley though he considered it clever (Hirst *Journal,* June 1889,

RIGB, 5: 2566) and he made critical remarks on it while editing the *Collected Mathematical Papers* ("Notes and References" = [*CP*5 (1892): 613–17]).

75. Ref. Rep., 11 June 1887, RSL, RR.10.45.

76. Ref. Rep., 29 July 1889, RSL, RR.10.291.

77. Cullovin 1895a, 190; Cullovin 1895b. A more complete refutation of Cullovin's argument is provided in Love 1895. Cullovin's "proof" is identical to that of Al-Abhari (d. 1264) as is shown in Al-Dhahir 1958.

78. C to RP, 5 Oct. [1879], WEL Manuscripts, WMS/ALS, Cayley, Box 8.

79. Mathews 1898b.

80. Turner 1981. The author does not include the influence of the Greek heritage in *mathematics* in this book.

81. Glaisher 1895b, 174.

82. Galton 1908, 72.

83. Newcomb 1903.

84. Halsted 1899, 59.

85. Peirce 1976, 2: 642.

86. Gardner and Wilson 1993.

87. Alexander MacFarlane, who met Cayley briefly, admired him, but rated Sylvester higher and thought that T. Kirkman and W. R. Hamilton, as pure mathematicians, were superior to both (MacFarlane 1916, 121).

88. Young 1981, 183.

89. Newman 1956, 1: 163–64.

90. Scott 1895, 139.

91. At Oxford, E. B. Elliott pursued invariant theory in Cayley's style. Elliott was born in Oxford, went to the university there, and graduated in 1873. He worked mainly in invariant theory and became widely known through his *Introduction to the Algebra of Quantics* (1895), a book which carried Cayley's methods into the twentieth century. It included a resumé of the German symbolic calculus and the results achieved by Hilbert (Allot, [1895] 1913). The way of applying invariant theory to elementary projective geometry is exemplified in Carr [1886] 1970, 681–95.

92. Turnbull 1941.

93. Elliott 1908. Forsyth imported German and French mathematics into Cambridge in the 1890s (Whittaker 1936-38a, 218).

94. Baker 1914, 371. In Sophus Lie's theory of continuous groups, Cayley's invariants and covariants appear as "group invariants and covariants" of groups of linear transformations.

95. Coolidge 1929, ix.

96. Weyl 1939, 489. Invariant theory as "romance" has been noted in Rota 1999.

97. PWW to WWRB, 2 Oct. 1923, TCC, o.6.6.15. P. W. Wood, an invariant theorist at Cambridge, was a distant cousin of Cayley's.

98. GTB to WWRB, 17 Oct. 1923, TCC, o.6.6.2.

99. "Presidential Address to the British Association, Southport, September 1883" = [*CP*11 (1883): 459]. The couplets quoted are ll. 119–20 and 137–38 of "Locksley Hall," published in 1842. Couplets such as these were liberally quoted by scientists in their memoirs, their reports, and their lectures. They conveyed "progress" and publicized science in language familiar to the Victorian reading public.

Bibliography

Archival Sources

Cayley's Nachlass was administered by W. W. Rouse Ball in 1923 after the death of Susan Cayley, the wife of Arthur Cayley. In the dispersal, letters written by living mathematicians were returned to their authors. Many letters written by others have not survived. An exception are the letters written by James Joseph Sylvester to Cayley (more than 500 items), which were sent to St. John's College, Cambridge, to join a small set of letters and notes (about 60) from Cayley to Sylvester, which were deposited there on the death of J. J. Sylvester in 1897. No letters written by Cayley were dealt with by Rouse Ball.

A list of recipients of mementos of Cayley's work in manuscript form sent out by Rouse Ball is available (TCC, 0.6.6).

A bound copy of letters written to George Boole by Cayley are held in Trinity College, Cambridge (TCC, R.2.88). These were presented to the college in 1896 by Susan Cayley, as they had been returned to Cayley after Boole's death in 1864. Boole's side of the correspondence does not appear to be extant. For the most part, the Cayley archive is scattered.

In writing this biography, I have used administrative records held at Trinity College, Cambridge. These include *Catalogus Bibliothecae Bodleianae*, 4 vols. (1738), which, used in conjunction with the *Trinity College Library Borrowing Books* (for October 1833–June 1839, and June 1839–March 1846), have made it possible to identify Cayley's library borrowings when he was a student and a fellow of Trinity. I have also made use of the *Trinity College Examinations Book, 1836–1861*, the *Examination Papers, 1842–1843*, the *Trinity College Room Rents, 1824–1842, 1843–1871*, the *Senior Bursars Audit Book, 1841–1847*, the *Senior Bursars Audit Book, 1848–1854*, and the *University Residents 1824–1913*.

At various scientific societies in London, I have consulted the *Minutes of the Royal Astronomical Society*, the *London Mathematical Society Minutes of Meetings*, the *Minutes of Council of the Royal Society*, and the *Institution of Civil Engineering Minutes of Proceedings*.

Printed Sources

Nearly all of Cayley's mathematical papers were reprinted in *The Collected Mathematical Papers* (Cayley 1889-98). Papers not published there or reproduced only in part are listed separately in this bibliography. Collected mathematical works of mathematicians appearing in the following list are designated by "(P)" before the publication date.

Abhyankar, S. S. 1976. "Historical Ramblings in Algebraic Geometry and Related Algebra." *AMM* 83: 409–48.

Adams, J. C. (P) 1896–1900. *The Scientific Papers of John Couch Adams*. Ed. W. G. Adams, 2 vols. Cambridge: Cambridge University Press. Referred to as *AdamsP* plus volume number and relevant year in the Notes.

Airy, G. B. 1896. *Autobiography of Sir George Biddell Airy.* Ed. W. Airy. Cambridge: Cambridge University Press.

Aitken, A. C. 1950. "Thomas Muir." *JGMS* 1: 64–76.

———. 1962. "Herbert Westren Turnbull 1885–1961." *ONFRSL* 8: 149–58.

Al-Dhahir, M. W. 1958. "Concerning the Parallel Postulate." *Bulletin of the College of Arts and Sciences, Baghdad,* 3: 60–65.

Alighieri, Dante. 1851–55. *Dante's Divine Comedy.* 4 vols. Trans. C. B. Cayley. London: Longman, Brown, Green & Longmans.

Allen, A. O. 1922. *John Allen and His Friends.* London: Hodder and Stoughton.

Allibone, T. E. 1976. *The Royal Society and Its Dining Clubs.* Oxford: Pergamon.

Anderson, R. E. [1891] 1908. "William Hopkins (1793–1866)." *DNB* 9: 1233–34.

Annan, N. 1999. *The Dons.* London: Harper Collins.

Archibald, R. C. 1929. "Notes on Some Minor English Serials." *MG* 14: 379–400.

———. 1934. "Notes and Queries." *SM* 2 (4): 300.

———. 1936. "Unpublished Letters of James Joseph Sylvester and Other New Information Concerning His Life and Work." *Osiris* 1: 85–154.

———. [1938] 1980. *A Semicentennial History of the American Mathematical Society 1888–1938.* 2 vols. Reprint. New York: Arno Press.

———. 1946. "Material Concerning James Joseph Sylvester." In *Studies and Essays in the History of Science and Learning,* ed. M. F. Montagu, pp. 211–17. New York: Henry Schuman.

Arnold, F. 1873. *Oxford and Cambridge: Their Colleges, Memories, and Associations.* London: Religious Tract Society.

Askwith, E. H. 1890. "On the Possible Groups of Substitutions That Can Be Formed with Three, Four, Five, Six and Seven Letters Respectively." *QJPAM* 24: 111–66.

Aslaksen, H. 1996. "Quaternionic Determinants." *MI* 18 (3): 57–65.

Astroh, M., I. O. Grattan-Guinness, and S. Read. "2001. A Survey of the Life of Hugh MacColl, (1837–1909)." *History and Philosophy of Logic* 22: 81–98.

Ayoub, R. 1984. "The Lemniscate and Fagnano's Contributions to Elliptic Integrals." *AHES* 29 (2): 131–49.

Babbage, C. 1830. *Reflections on the Decline of Science In England, and Some of Its Causes.* London: B. Fellowes.

———. 1851. *The Exposition of 1851.* London: John Murray.

Babbage, H. P. [1915]. *Memoirs and Correspondence of Major-General H. P. Babbage.* London: William Clowes.

Bacharach, I. 1886. "Über den Cayley'schen Schnittpunktsatz." *MA* 26: 275–99.

Baker, H. F. 1889. "On the Full System of Concomitants of Three Ternary Quadrics." *TCPS* 15: 62–106.

———. 1908. "On the Invariants of a Binary Quintic and the Reality of Its Roots." *PLMS* Ser. 2 6: 122–40.

———. 1913. "On Some Recent Advances in the Theory of Algebraic Surfaces." *PLMS* 12: 1–40.

———. 1914. "Presidential Address Section A." *Rep. BAAS 1913,* 367–73.

———. 1922–33. *Principles of Geometry.* 6 vols. Cambridge: Cambridge University Press.

———. 1930. "Percy Alexander MacMahon." *JLMS* 5: 307–20.

———. 1944. "G. T. Bennett." *JLMS* 19: 107–28.

Bakewell, M. 1996. *Lewis Carroll.* London: Heinemann.

Ball, J. Ed. 1859. *Peaks, Passes and Glaciers by Members of the Alpine Club.* London: Longman, Brown, Green, Longmans, and Roberts.

Ball, Sir R. S. 1903–4. "George Salmon." *PLMS* Ser. 2 1: xxii–xxviii.

———. 1906–7. "Charles Jasper Joly." *PRIA* 78: lxvii–lxix.

———. 1915. *Reminiscences and Letters of Sir Robert Ball.* Ed. W. V. Ball. London: Cassell.

Ball, W. W. Rouse. 1880. "The Mathematical Tripos." *CR* 1: 2–3, 18–19, 34–35.

———. [1888] 1960. *A Short Account of the History of Mathematics.* New York: Dover.

———. 1889. *A History of the Study of Mathematics at Cambridge.* Cambridge: Cambridge University Press.

———. 1891. "Newton's Classification of Cubic Curves." *Bibliotheca Matematica* 5: 35–40.

———. [1892] 1949. *Mathematical Recreations and Essays.* 11th Ed. Revised by H. S. M. Coxeter. London: Macmillan.

———. 1912. "The Cambridge School of Mathematics." *MG* 6: 311–23.

———. 1921. *Cambridge Notes, Chiefly Concerning Trinity College and The University.* Cambridge: Heffer.

Ball, W. W. Rouse, and J. A. Venn, eds. 1911–16. *Admissions to Trinity College, Cambridge (1546–1900).* 5 vols. London: Macmillan.

Barnard, H. C. [1847] 1961. *A History of English Education.* London: University Press.

Barrow-Green, J. 1999. "'A Corrective to the Spirit of too Exclusively Pure Mathematics': Robert Smith (1689–1768) and His Prizes at Cambridge University." *AS* 56: 271–316.

Barton, R. 1990. "An Influential Set of Chaps: The X-Club and Royal Society Politics, 1864–1885." *BJHS* 23: 53–81.

Barty-King, H. 1977. *The Baltic Exchange: The History of a Unique Market.* London: Hutchinson Benham.

Basalla, G., W. Coleman, and R. H. Kargon, eds. 1970. *Victorian Science: A Self-Portrait from the Presidential Addresses of the British Association for the Advancement of Science.* Garden City, NY: Doubleday.

Basset, A. B. 1902. "Classification of Quartic Curves." *Nature* 67 (27 Nov.): 80.

Bath Chronicle. 1864 (September). Reports of British Association. Bath: T. D. Taylor.

Becher, H. W. 1980. "William Whewell and Cambridge Mathematics." *HSPS* 11: 1–48.

———. 1984. "The Social Origins and Post-Graduate Careers of a Cambridge Intellectual Elite, 1830–1860." *VS* 28: 97–127.

———. 1995. "Radicals, Whigs and Conservatives: The Middle and Lower Classes in the Analytical Revolution at Cambridge in the Age of Aristocracy." *BJHS* 28: 405–26.

Belhoste, B. 1991. *Augustin-Louis Cauchy.* Trans. F. Ragland. New York: Springer.

Bell, E. T. [1937] 1965. *Men of Mathematics.* 2 vols. New York: Penguin Reprint.

———. 1945. *The Development of Mathematics.* 2d ed. New York: McGraw-Hill.

Bentham, G., and J. D. Hooker. 1862–83. *Genera Plantarum.* 3 vols. London: Spottiswoode.

Berndt, B. C., B. K. Spearman, and K. S. Williams. "Commentary on an Unpublished Lecture by G. N. Watson on Solving the Quintic." *MI* 24 (4): 15–33.

Bertini, E. 1903–4. "Life and Works of L. Cremona." *PLMS* Ser. 2. 1: v–xviii.

Berwick, W. E. H. 1923. "George Ballard Mathews, 1861–1922." *PLMS* Ser. 2. 31: xlvi–l.

———. 1929. "On Soluble Sextic Equations." *PLMS* 29: 4–5.

Best, G. [1971] 1973. *Mid-Victorian Britain 1851–1875.* St. Albans: Panther.

Bettany, G. T. [1888–61]. "James Cumming (1777–1861)." *DNB* 5: 296–97.

Betti, E. (P) 1903–1913. *Opere Mathematische.* 2 vols. Milan: Ulrico Hoepli. Referred to as *BettiP* plus volume number and relevant year in the Notes.

Biggs, N. L. 1981. "T. P. Kirkman, Mathematician." *BLMS* 13: 97–120.

Biggs, N. L., E. K. Lloyd, and R. J. Wilson. [1976] 1998. *Graph Theory 1736–1936.* Reprint. Oxford: Clarendon Press.

Birkhoff, G. 1975. "Foundations of Mathematics." *HM* 2: 503–5.

Blake, W. 1970. *A Choice of Blake's Verse.* Ed. K. Raine. Glasgow: Faber and Faber.

Blumenthal, L. M. 1953. *Theory and Applications of Distance Geometry.* Oxford: Clarendon.

———. 1961. *A Modern View of Geometry.* San Francisco: W. H. Freeman.

Blumenthal, L. M., and B. E. Gillam. 1943. "Distribution of Points in *n*-Space." *AMM* 50: 181–85.

Blythe, W. H. 1905. *On Models of Cubic Surfaces.* Cambridge: Cambridge University Press.

Bollobas, B. 1984. "Some Trinity Mathematicians." *CR* 105 (30 Jan. 1984): 15–20.

Boole, G. 1841. "Exposition of a General Theory of Linear Transformations." *CMJ* 3: 1–20, 106–19.

———. 1844. "Notes on Linear Transformations." *CMJ* 4: 167–71.

———. 1847. *The Mathematical Analysis of Logic.* Cambridge: Macmillan, Barclay and Macmillan.

———. 1848. "Notes on Quaternions." *PM* 33: 278–80.

———. 1851a. "On the Theory of Linear Transformations." *CDMJ* 6: 87–106.

———. 1851b. "On the Reduction of the General Equation of the *n*th Degree." *CDMJ* 6: 106–13.

———. 1852. "On Reciprocal Methods in the Differential Calculus." *CDMJ* 7: 156–66.

———. 1853. "On Reciprocal Methods in the Differential Calculus, (continuation)." *CDMJ* 8: 1–24.

———. 1854. *An Investigation of the Laws of Thought, on Which Are Founded the Mathematical Theories of Logic and Probabilities.* London: Walton and Maberley.

———. 1997. *George Boole: Selected Manuscripts on Logic and Its Philosophy.* Ed. I. O. Grattan-Guinness and G. Bornet. Boston: Birkhäuser.

Borwein, J. M., and P. B. Borwein. 1998. *Pi and the AGM.* Canad. Math. Soc. Monograph, vol. 4. New York: Wiley.

Bos, H. J. M., C. Kers, F. Oort, and D. W. Raven. 1987. "Poncelet's Closure Theorem." *Expositiones Mathematicae* 5: 289–364.

Bottazzini, U. 1980. "Algebraische Untersuchungen in Italien 1850–1863." *HM* 7: 24–37.

Botton, A. de, ed. 1999. *The Essential Plato.* Trans. Benjamin Jowett (1871) with M. J. Knight. Book-of-the-Month Club, Inc.

Bowers, B. 1975. *Sir Charles Wheatstone FRS 1802–1875.* London: HMSO.

Boyer, C. B. 1956. *History of Analytic Geometry.* New York: Scripta Mathematica.

———. 1968. *A History of Mathematics.* New York: Wiley.

Brent, J. 1993. *Charles Sanders Peirce: A Life.* Bloomington: Indiana University Press.

Briggs, D. R. 1983. *The Millstone Race: A Study of Private Education.* Exeter: Shortrun Press.

Brioschi, F. 1895. "Notice sur Cayley." *BSM* 19: 189–200.

Bristed, C. A. 1852. *Five Years in an English University.* 2d ed. New York: Putnam.

Broadbent, T. A. A. 1964. "George Boole (1815–1864)." *MG* 48: 373–78.

Brock, M. G., and M. C. Curthoys, eds. 2000. *The History of the University of Oxford, Nineteenth Century Oxford* (vol. 7, Part 2). Oxford: Oxford University Press.

Brock, W. H., ed. 1967. *The Atomic Debates.* Leicester: Leicester University Press.

———. 1975. "Geometry and the Universities: Euclid and His Modern Rivals 1860–1901." *HE* 4: 21–35.

Brock, W. H., and R. M. MacLeod. 1976. "The 'Scientists' Declaration': Reflexions on Science and Belief in The Wake of *Essays and Reviews, 1864–1865." BJHS* 9: 39–66.

————, eds. 1980. *Natural Knowledge in Social Context: The Journal of Thomas Archer Hirst* F.R.S. London: Mansell.

Brock, W. H., N. D. McMillan, and R. C. Mollan, eds. 1981. *John Tyndall: Essays on a Natural Philosopher.* Dublin: Royal Dublin Society.

Brock, W. H., and A. J. Meadows. 1984. *The Lamp of Learning: Taylor and Francis and the Development of Science Publishing.* London: Taylor and Francis.

Brodie, Sir B. C. 1866. *The Calculus of Chemical Operations. Part 1: On the Construction of Chemical Symbols.* London: Taylor and Francis.

Bronwin, B. 1843a. "On Elliptic Functions." *CMJ* 3: 123–31.

————. 1843b. "On M. Jacobi's Theory of Elliptic Functions." *PM* 22: 258–62.

————. 1844. "Reply to Mr. Cayley's Remarks." *PM* 23: 89–91.

Brooke, C. N. L. 1993. *A History of the University of Cambridge. Vol. 4: 1870–1990.* Cambridge: Cambridge University Press.

Browne, R. W. 1851. *A History of Classical Literature.* 2 vols. London: Bentley.

Bryant, A. [1932] 1979. *Macaulay.* Reprint. London: Weidenfeld and Nicolson.

Bryant, M. 1986. *The London Experience of Secondary Education.* London: Athlone.

Buckland, A. R. 1909. "John Pilkington Norris." *DNB* 14: 582–83.

Burbridge, D. 1994. "Galton's 100: An Exploration of Francis Galton's Imagery Studies." *BJHS* 27: 443–63.

Burde, G., and H. Zieschang. 1999. "Development of the Concept of a Complex." In *History of Topology,* ed. I. M. James, pp. 103–10. Amsterdam: Elsevier.

Burkill, J. C. 1979. "John Edensor Littlewood." *BLMS* 11: 59–103.

Burnside, W. 1891. "On a Class of Automorphic Functions." *PLMS* 23: 49–88.

Bury, J. P. T., ed. 1967. *Romilly's Cambridge Diary 1832–1842.* Cambridge: Cambridge University Press.

Bury, M. E., and J. D. Pickles, eds. 1994. *Romilly's Cambridge Diary 1842–1847.* Cambridge Records Society, vol. 10. Cambridge.

————, eds. 2000. *Romilly's Cambridge Diary 1848–1864,* Cambridge Records Society, vol. 14. Cambridge.

Busemann, H., and P. J. Kelly. 1953. *Projective Geometry and Projective Metrics.* New York: Academic Press.

Cajori, F. 1922. "George Bruce Halsted." *AMM* 29: 338–40.

————. 1929. *A History of Mathematical Notations.* 2 vols. Chicago: Open Court.

Campbell, L., and W. Garnett. 1882. *The Life of James Clerk-Maxwell.* London: Macmillan.

Campbell-Kelly, M., and M. Croarken. 2000. "Beautiful Numbers: The Rise and Decline of the British Association Mathematical Tables Committee, 1871–1965." *IEE Annals of the History of Computing* 22 (4): 44–61.

Campion, W. M. 1862. "On the Course of Reading for the Mathematical Tripos." In *The Students Guide to the University of Cambridge,* ed. Sir J. R. Seeley, pp. 82–102. Cambridge: Deighton Bell.

Cannell, D. M. 1993. *George Green, Mathematician and Physicist 1793–1841.* London: Athlone.

Cannon, S. F. 1978. *Science in Culture: The Early Victorian period.* New York: Dawson.

Carlyle, E. I. 1909. "John Casey (1820–1891)." *DNB* Suppl. 22: 395–96.

Caroe, G. M. 1978. *William Henry Bragg 1862–1942: Man and Scientist.* Cambridge: Cambridge University Press.

Carr, G. S. [1886] 1970. *Formulas and Theorems in Pure Mathematics*. 2d ed. New York: Chelsea.

Carroll, L. [1890] 1996. *Circular Billiards for Two Players*. In *The Universe in a Hankerchief* by M. Gardner, pp. 149–50. New York: Copernicus.

Cauchy, A. L. (P) 1882–1974. *Oeuvres completes d'Augustin Cauchy*. 27 vols. Paris: Gauthier-Villars. Referred to as *CauchyP* plus volume number and relevant year in the Notes.

Cayley, A. 1843a. "Note on Mr. Bronwin's Paper on Elliptic Integrals." *CMJ* 3: 197–98.

———. 1843b. "Remarks on the Rev. B. Bronwin's Paper on M. Jacobi's Theory of Elliptic Functions." *PM* 22: 358–60.

———. 1845. "On Jacobi's Elliptic Functions, In Reply To the Rev. Brice Bronwin; and On Quaternions." *PM* 26: 208–11.

———. 1846. "Mémoire sur les Hyperdéterminants." *J. rei. ang. Math*. 30: 1–37.

———. 1854. "Abstract of Introductory Memoir on Quantics." *PM* 8: 69–70.

———. 1857–1859. "On Curves of the Third Order." *PRSL* 9: 333–34.

———. 1862. "Final Remarks on Mr. Jerrard's Theory of Equations of the Fifth Order." *PM* 24: 290.

———. 1863. *A. Cayley's Mathematical Works 1841–1862*. London: Stangeways & Walden [CUL, Cam.c.863.33].

———. 1870. "The Ninth Memoir on Quantics." *Nature* 2 (23 June): 152–53.

———. 1872. "On the Extraction of the Square Root of a Matrix of the Third Order." *PRSE* 7: 675–82.

———. 1875. "Ueber die analytischen Figuren, welche in der Mathematik Bäume genannt werden und ihre Anwendung auf die Theorie chemischer Verbindungen." *Bericht deutsch. chem. Ges*. 8: 1056–59.

———. [1876] 1895. *An Elementary Treatise on Elliptic Functions*. 2d ed. Reprint. New York: Dover.

———. 1879. "Remarks 'On a Question of Probabilities' by W. S. B. Woolhouse." *JIA* 21: 204–13.

———. 1889a. "On the Finite Number of the Covariants of a Binary Quantic." *MA* 34: 319–20.

———. 1889b. "Science Worthies No. 25—James Joseph Sylvester." *Nature* 39 (3 Jan.): 217–19.

———. 1894. *The Principles of Book-Keeping By Double Entry*. Cambridge.

———. (P) [1889–98] 1963. *The Collected Mathematical Papers of Arthur Cayley*. Ed. A. Cayley and A. R. Forsyth. 13 vols. + Supplement, Reprint. New York: Johnson. [Cayley edited vol. 1, 2 (1889); vol. 3 (1890); vol. 4 (1891); vol. 5 (1892); vol. 6 (1893); vol. 7 (1894); vol. 8 (1895), 1–38. Forsyth edited vol. 8, (1895); vols. 9, 10, 11 (1896); vols. 12, 13 (1897).] Referred to as *CP* plus volume number in the Notes.

Cayley, C. B. 1846. "Inquiries in the Elements of Phonetics." *Phil. Mag*. 3d Ser. 28: 47–48.

Cayley, H. 1904a. *The Housing of Cambridge*. London: Rivingtons. (Reprinted from *The Economic Review*, Oct. 1904.)

———. 1904b. *Holy Trinity Church, Rothwell, Northamptonshire: A Short Historical Account*. Kettering.

Challenger, F. 1950–51. "Frederic Stanley Kipping." *ONFRS* 7: 183–219.

Chandler, B., and W. Magnus. 1982. *The History of Combinatorial Group Theory: A Case Study in the History of Ideas*. New York: Springer.

Chapman, J. V. 1985. *Professional Roots: The College of Preceptors in British Society*. Epping: Theydon Bois Publications.

Chasles, M. 1852. *Traité de géométrie supérieure,* Paris: Bachelier.

Church, A. 1965. "The History of the Question of Existential Import of Categorical Propositions." In *Logic, Methodology and Philosophy of Science II: Proceedings of the 1964 International Congress for Logic, Methodology and Philosophy of Science.* Ed. Y. Bar-Hillel, pp. 417–24. Amsterdam: North-Holland.

Clark, F., compiler. 1855. *India Register.* 2d ed. London: H. Allen.

Clark, J. W. 1904. *Endowments of the University of Cambridge.* Cambridge: Cambridge University Press.

———. [1895] 1909. "George Peacock (1791–1858)." *DNB* 15: 583–85.

Clark, J. W., and T. M. Hughes. 1890. *The Life and Letters of the Reverend Adam Sedgwick.* 2 vols. Cambridge: Cambridge University Press.

Clark, R. W. 1981. "Tyndall as Mountaineer." In *John Tyndall: Essays on a Natural Philosopher.* Ed. W. H. Brock, N. D. McMillan, and R. C. Mollan, pp. 61–68. Dublin: Royal Dublin Society.

Clebsch, R. F. 1872. *Theorie der binänaren algebraischen Formen.* Leipzig: Teubner.

Clebsch, R. F., and P. Gordan. [1866] 1967. *Theorie der Abelschen Functionen.* Reprint. Würzburg: Physica.

Clerke, A. M. 1893. *A Popular History of Astronomy.* 3d ed. London: Adam and Charles Black.

———. 1909. "Robert Grant 1814–1892." *DNB* Suppl. 22: 769–70.

Clifford, W. K. 1879. *Lectures and Essays.* Ed. L. Stephen and F. Pollock. London: Macmillan.

———. (P) 1882. *Mathematical Papers.* Ed. R. Tucker. London: Macmillan. Referred to as *CliffordP* plus volume number and relevant year in the Notes.

Clough, B. A. 1897. *A Memoir of Anne Jemima Clough.* London: Edward Arnold.

Cochell, G. G. 1998. "The Early History of the Cornell Mathematics Department: A Case Study in the Emergence of the American Mathematical Research Community." *HM* 25: 133–53.

Cocks, R. 1983. *Foundations of the Modern Bar.* London: Sweet and Maxwell.

Collingwood, Sir. E. F. 1966. "A Century of the London Mathematical Society." *JLMS* 41: 577–94.

Cooke, R. 1994. "Elliptic Integrals and Functions." In *Companion Encyclopedia of the History and Philosophy of the Mathematical Sciences.* 2 vols. Ed. I. O. Grattan-Guinnness, 1: 529–39. London: Routledge.

Coolidge, J. L. [1931] 1959. *A Treatise on Algebraic Plane Curves.* Reprint. Dover: New York.

———. [1940] 1963. *A History of Geometrical Methods.* Reprint. New York: Dover.

———. [1945] 1968. *A History of the Conic Sections and Quadric Surfaces.* Reprint. New York: Dover.

Cooper, C. H. 1842–53. *Annals of Cambridge.* 5 vols. Cambridge: Warwick.

Cornford, F. M. 1941. *The Republic of Plato.* Oxford: Clarendon.

Corry, L. 1996. *Modern Algebra and the Rise of Mathematical Structures.* Basel: Birkhäuser.

Courtney, W. P. [1899] 1909. "Edmund Venables (1819–1895)." *DNB* 20: 202–3.

Cracraft, J. 1969. "James Brogden in Russia, 1787–1788." *SEER* 47: 219–44.

Craik, A. D. D. 2000. "James Ivory, F.R.S., Mathematician: 'The most Unlucky Person That Ever Existed.'" *NRRSL* 54 (2): 223–47.

———. 2002. "James Ivory's Last Papers on the 'Figure of the Earth' (with biographical additions)." *NRRSL* 56 (2): 187–204.

Cremona, L. 1992. *La corrispondenza di Luigi Cremona (1830–1903)*. Ed. L. Nurzia et al. Rome: Università di Roma "La Sapienza," Facoltà di Scienze Matematiche, Fisiche e Naturali.

Crilly, T. 1965. "On the Existence of Cayley-Stiefel Manifolds." M.Sc. thesis, University of Hull.

———. 1978. "Cayley's Anticipation of a Generalised Cayley-Hamilton Theorem." *HM* 5: 211–19.

———. 1979. "A Klein Footnote." *MI* 2: 4.

———. 1981. "The Mathematics of Arthur Cayley with Particular Reference to Linear Algebra." Ph.D thesis, Council for National Academic Awards.

———. 1986. "The Rise of Cayley's Invariant Theory (1841–1862)." *HM* 13: 241–54.

———. 1988. "The Decline of Cayley's Invariant Theory (1863–1895)." *HM* 15: 332–47.

———. 1992. "A Gemstone in Linear Algebra." *MG* 76: 182–88.

———. 1995. "A Victorian Mathematician." *MG* 79: 259–62.

———. 1997. "A Mathematician Extraordinary—James Joseph Sylvester (1814–1897)." *MG* 81: 7–11.

———. 1998. "The Young Arthur Cayley." *NRRSL* 52 (2): 267–82.

———. 1998. "Arthur Cayley: The Road Not Taken." *MI* 20 (4): 49–53.

———. 1999. "Arthur Cayley as Sadleirian Professor: A Glimpse of Mathematics Teaching at Nineteenth-Century Cambridge." *HM* 26: 125–60.

———. 2000. "The Appearance of Set Operators in Cayley's Group Theory." *Notices of the South African Mathematical Society* 31 (2): 9–22.

———. 2004. "The *Cambridge Mathematical Journal* and Its Descendants: The Linchpin of a Research Community in the Early and Mid Victorian Age." *HM* 31: 455–97.

Cromwell, P. R. 1997. *Polyhedra*. Cambridge: Cambridge University Press.

Crosland, M. P. 1962. *Historical Studies in the Language of Chemistry*. London: Heinemann.

Cross, A. G. 1973. "The British in Catherine's Russia: A Preliminary Survey." In *The Eighteenth Century in Russia*. Ed. John G. Garrard, pp. 233–63. Oxford: Clarendon Press.

———. 1993. *Anglo-Russica: Aspects of Culutural Relations Between Great Britain and Russia in the Eighteenth and Early Nineteenth Centuries. Selected Essays by Anthony Cross*. Oxford: Berg.

———. 1997. *"By the Banks of the Neva" Chapters From The Lives and Careers of the British in Eighteenth-Century Russia*. Cambridge: Cambridge University Press.

Cross, J. J. 1985. "Integral Theorems in Cambridge Mathematical Physics, 1830–1855." In *Wranglers and Physicists*. Ed. P. M. Harman, pp. 112–48. Manchester University Press.

Crowe, M. J. [1967] 1985. *A History of Vector Analysis: The Evolution of the Idea of a Vectorial System*. Reprint. New York: Dover.

Crowther, J. G. 1974. *The Cavendish Laboratory, 1874–1974*. London: Macmillan

Cullovin, T. 1895. "A Rigorously Euclidean Demonstration of the Theory of Parallel Straight Lines To Be Introduced Immediatey After Euclid 1, 26." *QJPAM* 27: 188–91.

———. 1895. "Note on my Proof of Euclid's Twelfth Axiom." *QJPAM*, 27: 225–27.

Cunningham, H. S. 1897. *Lord Bowen: A Biographical Sketch with A Selection from His Verses*. London: John Murray.

Dahan, A. 1980. "Les trauvaux de Cauchy sur les substitutions. Étude de son approche du concept de groupe." *AHES* 23 (4): 279–319.

Dale, A. I. [1991] 1999. *A History of Inverse Probability.* 2d ed. New York: Springer.

Danvers, F. C., et. al. 1894. *Memorials of Old Haileybury College.* Westminster: Archibald Constable.

Darboux, G. [1904] 2003. "The Development of Geometrical Methods." In *The Changing Shape of Geometry.* Ed. C Pritchard, pp. 150–71. Cambridge: Cambridge University Press.

Darwin, Sir G. H. (P) 1907–1911, 1916. *Scientific Papers.* 4 vols. and suppl. vol. Cambridge: Cambridge University Press.

Dauben, J. W. 1995. *Abraham Robinson.* Princeton, NJ: Princeton University Press.

Davenport, H. 1966. "Looking Back." *JLMS* 41: 1–10.

Davidson, C., J. Waley, and Thomas Key, eds. 1873. *Davidson's Precedents and Forms.* 3d ed. 3 vols. London: William Maxwell.

Davies, T. S. 1850. [untitled]. *The Mathematician* 3 (Suppl): 41.

Davis, W. B. 1869. "[Notification] A Historical Note on Lagrange's Theorem." *Rep. BAAS 1868.*

Decker, F. F. 1910. *The Symmetric Function Tables of the Fifteenthic.* Washington: Carnegie Institution.

De Morgan, A. 1842. *Differential and Integral Calculus.* London: Baldwin and Cradock.

———. 1844. "On the Foundations of Algebra IV-on Triple Algebra and Addition." *TCPS* 8: 241–54.

De Morgan, S. 1882. *Memoir of Augustus de Morgan.* London: Longmans Green.

Denman G. 1897. *Autobiographical Notes of the Rt. Hon. George Denman, 1819–1847* (private circulation). London: Chiswick Press.

Dennis, J. L. 1972. *A Catalog of Special Plane Curves.* New York: Dover.

Dickens, C. [1850] 1996. *David Copperfield.* London: Penguin Classics

———. [1852–53] 1998. *Bleak House.* Oxford: Oxford University Press.

Dickson, J. D. H. 1912. "Peter Guthrie Tait." *DNB, Second Suppl.* 3: 471–74.

Dickson, L. E. 1914. *Algebraic Invariants.* New York: JohnWiley.

———. 1919. "On Quaternions and Their Generalisation and the History of the Eight Square Theorem." *AM* 20: 155–71.

———. [1919–23] 1971. *History of the Theory of Numbers.* 3 vols. Reprint. New York: Chelsea.

———. 1925. "Resolvent Sextics of Quintic Equations." *BAMS* 31: 515–23.

Dingle, H., ed. 1951a. *A Century of Science 1851–1951.* London: Hutchinson.

———. 1951b. "The Scientific Outlook in 1851 and in 1951." *BJPS* 2: 85–104.

Dirichlet, P. G. L. (P) 1889–97. *Werke.* 2 vols. Berlin: G. Reimer.

Dixon, A. C. 1909. "An Elementary Discussion of Schlafli's Double Six." *QJPAM* 40: 381–84.

Dixon, F. W. 1895. "Arthur Cayley." *The Observatory* 18: 112–13.

Dodgson, C. L. 1860. *A Syllabus of Plane Algebraical Geometry, (Part 1).* Oxford: J. H. and P. Parker.

———. 1888. *Curiosa Mathematica.* London: Macmillan

Dorling, A. R. 1976. "The Graves Mathematical Collection in University College London." *AS* 33: 307–10.

Dressel, P. L. 1940. "Statistical Seminvariants and Their Estimates with Particular Emphasis on Their Relation to Algebraic Invariants." *AMS* 11: 33–57.

Drew, B. 1928. *The London Assurance: A Chronicle.* Oxford: Kemp Hall.

Dreyer, J. L. E., and H. H. Turner, eds. 1923. *History of the Royal Astronomical Society.* London: Royal Astronomical Society.

Dru-Drury, E. 1973. "Reminiscences of Blackheath." *Trans. of the Greenwich and Lewisham Antiquarian Society* 8: 47–52.

Dubbey, J. 1978. *The Mathematical Work of Charles Babbage*. Cambridge: Cambridge University Press.

Duman, D. 1983. *The English and Colonial Bars in the Nineteenth Century*. Croom Helm: Canberra.

Dyck, W. 1882. "Gruppentheoretische Studien I." *MA* 20: 1–44.

———. 1883. "Gruppentheoretische Studien II." *MA* 22: 70–108.

Einstein, A. [1922] 1976. *The Meaning of Relativity*. London: Chapman and Hall.

Eisenbud, D., M. Green, and J. Harris. 1996. "Cayley-Bacharach Theorems and Conjectures." *BAMS* 33: 295–324.

Eisenstein, F. G. M. (P) 1975. *Mathematische Werke*. 2 vols. New York: Chelsea. Referred to as *EisensteinP* plus volume number and pertinent year in the Notes.

Elliott, E. B. 1892. "A Proof of the Exactness of Cayley's Number of Semi-invariants of a Given Type." *PLMS* 23: 298–304.

———. [1895] 1913. *An Introduction to the Algebra of Quantics*. 2d ed. Reprint. New York: Chelsea.

———. 1898. "Some Secondary Needs and Opportunities of English Mathematicians." *PLMS* 30: 5–23.

———. 1908. "Review of Introduction to Higher Algebra (by Maxime Bôcher)." *MG* 4: 291–93.

———. 1931. "James Hammond 1850–1930." *JLMS* 6: 78–80.

Elliott, E. B., and P. E. Matheson, 1909. "James Joseph Sylvester." *DNB*, 19: 258–60.

Ellis, A. J. 1859–1860. "On the Laws of Operation, and the Systematization of Mathematics." *PRSL* 10: 85–94.

Ellis, R. L. 1844. "Memoir of the Late D. F. Gregory, MA, Fellow of Trinity College, Cambridge." *CMJ* 4: 145–52.

———. (P) 1863. *The Mathematical and Other Writings of Robert Leslie Ellis MA*. Ed. William Walton. Cambridge: Deighton Bell.

Elsevier's Dictionary of Computer Science and Mathematics: In English, German, French, and Russian. 1995. Compiled by K. Peeva and B. Delijska. Amsterdam: Elsevier.

Encyclopedic Dictionary of Mathematics. 1993. 2 vols. 2d ed. Cambridge, MA: Mathematical Society of Japan and MIT Press.

Eperson, D. B. 2000. "Lewis Carroll—Mathematician and Teacher of Children." *MG* 84: 9–13.

Evans, D. M. [1845] 1852. *The City or the Physiology of London Business*. London: Groombridge.

Evans, F., ed. 1982. *The Daughters of Karl Marx*. London: André Deutsch.

Eve, A. S., and C. H. Creasey. 1945. *Life and Work of John Tyndall*. London: Macmillan.

Everett, W. 1866. *On the Cam: Lectures on the University of Cambridge in England*. London: Bceton.

Ewald, W., ed. [1996] 1999. *From Kant to Hilbert*. 2 vols. Oxford: Oxford University Press.

Faà di Bruno, F. 1857. "An Invariant of the Twelfth Degree of the Quintic [Formula]." *QJPAM* 1: 361–63.

———. 1859. *Théorie générale de l'élimination*. Paris.

———. 1876. *Théorie des Formes Binaires*. Turin.

———. 2004. *Franceso Faà di Bruno: Ricerca Scientifica, Insegnamento e Divulgazione*. Ed. Livia Giacardi. Torino: Deputazione Subalpina di Storia Patria.

Fairgrieve, M. M'C. 1931. "J. D. Hamilton Dickson." *PRSE* 51: 205–6.

Fauvel, J., R. Flood, and R. J. Wilson, eds. 2000. *Oxford Figures: 800 Years of the Mathematical Sciences.* Oxford: Oxford University Press.

Ferguson, T. S. 1989. "Who Solved the Secretary Problem?" *Statistical Science* 4 (3): 282–96.

Feuer, L. S. 1987. "Sylvester in Virginia." *MI* 9: 13–19.

Filon, L. N. G. 1936–38. "Karl Pearson." *ONFRS* 2: 73–110.

Fisch, M., and S. Schaffer, eds. 1991. *William Whewell: A Composite Portrait.* Oxford: Clarendon.

Fischer, G. 1986. *Mathematical Models.* 2 vols. Braunschweig: Vieweg.

Fisher, C. S. 1966. "The Death of a Mathematical Theory: A Study in the Sociology of Knowledge." *AHES* 3: 137–59.

———. 1967. "The Last Invariant Theorists." *AES* 8: 216–44.

Fifteenth Meeting of the British Association for the Advancement of Science, List of members 1845. 1st Suppl. Cambridge: Cambridge University Press.

Fletcher, A. J., et al. [1946] 1962. *An Index of Mathematical Tables.* 2 vols. Oxford: Blackwell.

Fletcher, C. 2002. *Mathematics by the Sea.* Aberystwyth: University of Wales.

Foden, F. 1989. "The Examiner: James Booth and the Origins of Common Examinations." Thesis, University of Leeds, Leeds.

Forsyth, A. R. 1880. "On the Motion of a Viscous Incompressible Fluid." *MM* 9: 134–39.

———. 1881. "The Symmetric Functions of the Roots of an Equation." *MM* 10: 44–54.

———. 1883 (subm. 1881). "Memoir on the Theta Functions, Particularly Those of Two Variables." *PTRSL* 173: 783–862.

———. 1895. "Arthur Cayley." *PTRSL* 58: ix–xliv.

———. 1897. "Presidential Address—Section A." *Rep. BAAS,* 541–49.

———. 1928. "William Burnside (1852–1927)." *PRSL* A117: xi–xxv.

———. 1930a. "James Whitbread Lee Glaisher." *PRSL* A126. i–xi.

———. 1930b. "P. A. MacMahon." *Nature* 125, no. 3146: 243–45.

———. 1935. "Old Tripos Days at Cambridge." *MG* 19: 162–79.

Foster, J. 1874–75. *Pedigrees of the County Families of Yorkshire.* 3 vols. London: Wilfred Head.

Foulkes, H. O. 1932. "The Algebraic Solution of Equations." *Science Progress* 26 (2): 601–8.

Franklin, F. 1880. "On the Calculation of the Generating Functions and Tables of Groundforms for Binary Quantics." *AJM* 3: 128–53.

———. 1896–97. "James Joseph Sylvester." *BAMS* 3: 299–309.

———. 1910. *The Life of Daniel Coit Gilman.* New York: Dodd & Mead & Co.

French. J. C. 1979. *A History of the University Founded by Johns Hopkins.* New York: Arno Press.

G., R. F. 1895. "Arthur Cayley." *Memoirs of the Manchester Literary and Philosophical Society* 9, Ser. 4: 235–37.

Gall, A. von. 1880. "Das vollständige Formensystem einer binären Form achter Ordnung." *MA* 17: 31–51, 139–52, 456.

———. 1888. "Das vollständige Formensystem der binären Form 7ter. Ordnung." *MA* 31: 318–36.

Galton, F. 1908. *Memories of My life.* London: Methuen.

———. 1914–30. *The Life, Letters and Labours of Francis Galton.* 4 vols. Ed. Karl Pearson. Cambridge: Cambridge University Press.

Gani, J., ed. 1975. *Perspectives in Probability and Statistics.* New York: Academic Press.

Gardner, A. 1921. *A Short History of Newnham College.* Cambridge: Bowes and Bowes.

Gardner, J. H., and R. J. Wilson. 1993. "Thomas Archer Hirst-Mathematician Xtravagant IV: Queenswood, France and Italy." *AMM* 100: 723–31.

———. 1993. "Thomas Archer Hirst-Mathematician Xtravagant V. London in the 1860s." *AMM* 100: 827–34.

Garland, M. M. 1980. *Cambridge before Darwin: The Ideal of a Liberal Education.* Cambridge: Cambridge University Press.

Geikie, A. 1923. "Sir Alfred Bray Kempe." *PRSL* 94B: i–x.

Gelfand, I. M., M. M. Kapranov, and A. V. Zelevinsky. 1994. *Discriminants, Resultants, and Multidimensional Determinants.* Boston: Birkhäuser.

Gibbs-Smith, C. H. 1962. "Sir George Cayley." *NRRSL* 17: 36–56.

Glaisher, J. W. L. 1870. "On the Numerical Values of the Sine Integral, Cosine Integral, and Exponential Integral." *PTRSL* 160: 367–88.

———. 1879. "James Booth." *MNRAS* 39: 219–25.

———. 1886. "The Mathematical Tripos." *PLMS* 18: 4–38.

———. 1891. "Presidential Address-Section A." *Rep. BAAS 1890,* 719–27.

———. 1895. "Arthur Cayley." *CR* 29 (7 Feb. 1895): 174–76.

———. 1896. "Arthur Cayley." *MNRAS* 56: 191–97.

———. 1896–1900. "Biographical Notice [of J. C. Adams]." In *The Scientific Papers of John Couch Adams.* 2 vols. Ed. W. G. Adams, vol. 1, pp. xv–xlviii. Cambridge: Cambridge University Press.

———. 1914. "Samuel Roberts." *PLMS* 13: xlix–liii.

———. 1926. "Notes on the Early History of the [London Mathematical] Society." *JLMS* 1: 51–64.

Glazebrook, R. T. 1932–1935. "Sir Horace Lamb." *ONFRS* 1: 375–92.

Goldsmith, G. 1843. *The English Bar: or, Guide to the Inns of Court: Comprising an Historical Outline of All the Inns of Court, the Regulations for Admission, etc.,* London: Edmund Spettigue.

Gooden, A. C. 2003. *Cambridge in the 1830s: The letters of Alexander Chisholm Gooden, 1831–1841.* Ed. J. Smith and C. Stray. Woodbridge: Boydell.

Gordan, P. 1868. "Beweis, dass jede Covariante und Invariante einer binaeren Form eine ganze Function mit numerischen Coefficienten einer enlichen Anzahl solcher Formen ist." *J. rei. ang. Math.* 69: 323–54.

———. 1885–87. *Vorlesungen über Invariantentheorie.* 2 vols. Leipzig: Teubner.

Grace, J. H., and A. Young. 1903. *The Algebra of Invariants.* Cambridge: Cambridge University Press.

Grattan-Guinness, I. O. 1971. "The Correspondence between Georg Cantor and Philip Jourdain." *Jahresbericht der Deutschen Mathematiker-Vereinigung* 73 (3): 111–30.

———. 1972. "A Mathematical Union: William Henry and Grace Chisholm Young." *AS* 29 (2): 105–86.

———. 1990. *Convolutions in French Mathematics 1800–1840.* 3 vols. Basel: Birkhäuser.

———. 1992. "Charles Babbage as an Algorithmic Thinker." *AHC* 14 (3): 34–48.

———. 1994. "Beyond Categories: The Lives and Works of Charles Sanders Peirce." *AS* 51: 531–38.

———. 1997. "Benjamin Peirce's Linear Associative Algebra (1870): New Light on Its Preparation and Publication." *AS* 54: 597–606.

———. 2001. "The Contributions of J. J. Sylvester, F.R.S., to Mechanics and Mechanical Physics." *NRRSL* 55 (2): 253–65.

Graves, C. 1847. "On Algebraic Triplets." *PRIA* 3: 51–54, 57–64, 80–84, 105–8.

Graves, J. T. 1839. "On the Functional Symmetry Exhibited in The Notation of Certain Geometrical Porisms, When They Are Stated Merely with Reference to the Arrangement of Points." *PM* 15: 129–36.

———. 1845. "On a Connection between the General Theory of Normal Couples and the Theory of Complete Quadratic Functions of Two Variables." *PM* 26: 315–20.

Graves, R. P. 1882–91. *Life of Sir William Rowan Hamilton.* 3 vols. + Addendum. Dublin: Hodges, Figgis.

———. [1890] 1908. "John Thomas Graves (1806–1870)." *DNB* 8: 430–31.

Gray, J. 1980. "Olinde Rodrigues' Paper of 1840 on the Transformation Groups." *AHES* 21: 375–85.

———. [1986] 2000. *Linear Differential Equations and Group Theory from Riemann to Poincaré.* 2d ed. Boston: Birkhäuser.

———. 1995. "Arthur Cayley (1821–1895)." *MI* 17: 62–63.

———. 1999. "The Classification of Algebraic Surfaces by Castelnuovo and Enriques." *MI* 21: 59–66.

Grazley, J. 1955. "The Rev. Arthur Young 1769–1827: Traveller in Russia and Farmer in the Crimea." *Bulletin of the John Rylands Library* 37: 393–428.

Green, G. 1835. "On the Determination of the Exterior and Interior Attractions of Ellipsoids of Variable Densities." *TCPS* 5: 395–430.

Greenhill, Sir G. 1893. "Collaboration in Mathematics." *PLMS* 24: 5–16.

———. 1912. "Presidential Address of the London Branch of the Mathematical Association." *MG* 6: 104–8.

Gregory, D. F. 1839. "On the Existence of Branches of Curves in Several Planes." *CMJ* 1: 259–66.

———. 1840. "On the Real Nature of Symbolical Algebra." *TRSE* 14: 323–30.

———. 1841. "On the Elementary Principles of the Application of Algebraical Symbols to Geometry." *CMJ* 2: 1–9.

———. [1845] 1852 (completed by W. Walton). *A Treatise on the Application of Analysis to Solid Geometry.* 2d ed. Cambridge: Deighton Bell.

———. (P) 1865. *The Mathematical Writings of Duncan Farquharson Gregory.* Ed. W. Walton. Cambridge: Deighton Bell.

Griffin, N., and A. Lewis. 1990. "Bertrand Russell's Mathematical Education." *NRRSL* 44: 51–71.

Guggisberg, Sir F. G. [1900] 1902. *"The Shop": The Story of the Royal Military Academy.* 2d ed. London: Cassell & Co.

Guicciardini, N., 1989. *The Development of Newtonian Calculus in Britain 1700–1800.* Cambridge: Cambridge University Press.

Gunther, R. T. 1937. *Early Science at Cambridge.* Oxford: Oxford University Press.

Hailperin, T. 1986. *Boole's Logic and Probability.* 2d ed. Amsterdam: North Holland.

Hall, A. R. 1969. *The Cambridge Philosophical Society. A History 1819–1969.* Cambridge: Cambridge Philosophical Society.

Hall, M. B. 1984. *All Scientists Now: The Royal Society in the Nineteenth Century.* Cambridge: Cambridge University Press.

Hall, T. G. 1837. *A Treatise on the Differential and Integral Calculus.* 2d ed., enlarged. Cambridge: Cambridge University Press.

Halphen, G. H. 1886. *Traité des functions elliptiques et de leur applications.* Paris: Gauthier-Villars.

Halsted, G. B. 1895. "Arthur Cayley." *AMM* 2: 102–6.

———. 1897. "James Joseph Sylvester." *AMM* 4: 159–68.

———. 1899. "The Collected Mathematical Papers of Arthur Cayley." *AMM*. 6: 59–65.

Hamilton, Sir W. R. July 1844. "On Quaternions." *PM* 25: 10–13, 241–46.

———. 1848. "Note Concerning the Research of J. T. Graves." *TRIA* 21: 338–41.

———. 1853. *Lectures on Quaternions.* Dublin: Hodges and Smith.

———. (P) 1931–2000. *The Mathematical Papers of Sir William Rowan Hamilton.* 4 vols. Vol. 1 ed. A. W. Conway and J. L. Synge, 1931; vol. 2 ed. A. W. Conway and A. J. McConnell, 1940; vol. 3 ed. H. Halberstam and R. E. Ingram, 1967; vol. 4 ed. B. K. P. Scaife, 2000. Cambridge: Cambridge University Press. Referred to as *HamiltonP* plus volume number and pertinent year in the Notes.

Hammond, J. 1882. "Note on an Exceptional Case in Which the Fundamental Postulate of Prof. Sylvester's Theory of Tamisage Fails." *PLMS* 14: 85–88.

———. 1886. "Syzygy Tables for the Binary Quintic." *AJM* 8: 19–25.

Hankins, T. 1979. "In Defence of Biography: The Use of Biography in the History of Science." *HS* 17: 1–16.

———. 1980. *Sir William Rowan Hamilton.* Baltimore: Johns Hopkins University Press.

Hannabuss, K. 1983 (31 March). "The Mathematician [H. J. S. Smith] the World Forgot." *New Scientist* 97 (1351): 901–3.

———. 2000. "The Mid-Nineteenth Century." In *Oxford Figures,* ed. John Fauvel, Raymond Flood, and Robin Wilson, pp. 187–217. Oxford: Oxford University Press.

Hansen, P. A. 1838. *Fundamenta nova investigationis orbitae verae quam luna perlustrat, quibus annexa est solutio problematis quatuor corporum breviter exposita.* Gothae: in commissis apud Carolum Glaeser.

Hardy, G. H. 1926. "The Case against the Mathematical Tripos." *MG* 13: 64–65.

———. 1928–29. "Dr. Glaisher and the 'Messenger of Mathematics.'" *MM* 58: 159–60.

———. 1940. *Ramanujan.* Cambridge: Cambridge University Press.

———. 1948. "The Case against the Mathematical Tripos." *MG* 32: 134–45 [first published in 1926].

Hargreave, C. J. 1849. "Analytical Researches Concerning Numbers." *PM* 35: 36–53.

———. 1866. "An Essay on the Resolution of Algebraic Equations." Preface by G. Salmon. Dublin: Private circulation.

Harley, R. 1894–95. "Sir James Cockle." *MMLPS* 9: 215–28.

Harman, P. M., ed. 1985. *Wranglers and Physicists.* Manchester, UK: Manchester University Press.

Harrison, H. M. 1994. *Voyager in Time and Space: The Life of John Couch Adams.* Lewes: The Book Guild.

Hartog, P. J. 1885–90. "Carl Schorlemmer 1834–1892." *DNB* 17: 928–29.

Hassé, H. R. 1951. "My Fifty Years of Mathematics." *MG* 35: 153–64.

Hawkins, H. 1960. *Pioneer: A History of the Johns Hopkins University 1874–1889.* Ithaca, NY: Cornell University Press.

Hawkins, T. 1974. "New Light on Frobenius' Creation of the Theory of Group Characters." *AHES* 12 (3): 217–43.

———. 1975. "Cauchy and the Spectral Theory of Matrices." *HM* 2: 1–29.

———. 1977a. "Another Look at Cayley and the Theory of Matrices." *AIHS* 27: 82–112.

———. 1977b. "Weirstrass and the Theory of Matrices." *AHES* 17: 119–63.

———. 1978. "The Theory of Matrices in the 19th Century." In *Proceedings of the International Congress of Mathematicians (Vancouver, BC, 1974),* 2 vols. ed. R. D. James, 2: 561–70. Montreal: Canadian Mathematical Congress.

———. 1986. "Cayley's Counting Problem and the Representation of Lie Algebras." In *Proceedings of the International Congress of Mathematicians (Berkeley, California)*. 2 vols. Ed. A. M. Gleason. American Mathematical Society, 2: 1642–56.

Hearnshaw, F. J. C. 1929. *The Centenary History of King's College, London 1828–1929*. London: Harrap.

Heath, T. L. 1885. *Diophantus of Alexandria*. Cambridge: Cambridge University Press.

Heitland, W. E. 1929. "Cambridge in the Seventies." In *The Eighteen-Seventies*, ed. H. Granville-Barker, pp. 249–72. Cambridge: Cambridge University Press.

Henderson, A. 1911. *The Twenty-Seven Lines upon a Cubic Surface*. Cambridge Tracts Number 13. Cambridge: Cambridge University Press.

Henrici, O. M. F. 1884a. "Presidential Address." *PLMS* 16: 1–5.

———. 1884b. "Modern English Mathematics." *Nature* 31 (18 Dec.): 151–52.

Hermite C. 1895. "Notice sur M. Cayley." *CRend.* 120 (5): 233–34.

———. (P) 1905–17. *Oeuvres de Charles Hermite*. 4 vols. Ed. E. Picard. Paris: Gauthier-Villars. Referred to as *HermiteP* plus volume number and relevant year in the Notes.

Herschel, Sir J. W. F. 1846. "Presidential Address." *Rep. BAAS 1845*, xxvii–xliv.

Hesse, L. O. P. (P) [1897] 1972. *Gesammelte Werke*. Reprint. New York: Chelsea. Referred to as *HesseP* and pertinent year in the Notes.

Heyck, T. W. 1982. *The Transformation of Intellectual Life in Victorian England*. London: Croom Helm.

Hilbert, D. (P) 1932–35. *Gesammelte Abhandlungen*, 3 vols. Berlin: Springer. Referred to as *HilbertP* plus volume number and pertinent year in the Notes.

———. 1993. *Theory of Algebraic Invariants*. Trans. R. C. Laubenbacher. Cambridge: Cambridge University Press.

Hill, J. E. 1897. "Bibliography of Surfaces and Twisted Curves." *BAMS* 3: 133–46.

Hilton, [aka Simpson] H. [1920] 1932. *Plane Algebraic Curves*. 2d ed. Oxford: Oxford University Press.

Hilts, V. L. 1975. "A Guide to Francis Galton's *English Men of Science*." *TAPS* 65 (5): 3–85.

Himmelfarb, G. 1968. *Victorian Minds*. London: Weidenfeld and Nicholson.

Hirsch, M. [1804] 1853. *Sammlung von beispielen, formeln and aufgaben aus der buchstabenrechnung und algebra*. 8th ed. Berlin: Duncker und Humblot.

Historical Dictionary of Mathematical Words. http://members.aol.com/jeff570/mathword. html.

Hobson, E. W. 1911. "Presidential Address-Section A." *Rep. BAAS 1910*, 509–22.

Hodge, W. V. D. 1957. "Henry Frederick Baker." *JLMS* 32: 112–28.

Hoffman, P. 1998. *The Man Who Loved Numbers*. London: Fourth Estate.

Holborn, G. 1999. *Sources of Biographical Information on Past Lawyers*. British and Irish Association of Law Librarians.

Hopkins, W. 1841. *Remarks on Certain Proposed Regulations Respecting the Studies of the University and the Period of Conferring the Degree of B.A.* Cambridge: J & J.J. Deighton; London: J. W. Parker.

———. 1854. *Remarks on the Mathematical Teaching of the University of Cambridge*. Cambridge.

Hort, A. F. 1927. "John Llewelyn Davies (1826–1916)." *DNB 1912–1921*: 147–48.

Howson, A. G. 1982. *A History of Mathematics Education in England*. Cambridge: Cambridge University Press.

http://www.austega.com/familyhistory/cayley.htm. This site contains some details of Cayley family history, including the Moberly family.

Huber, V. A. 1843. *The English Universities.* 2 vols. [abr. trans.]. Ed. and trans. Francis W. Newman. London: Pickering.

Hudson, D. 1974. *Munby, Man of Two Worlds.* London: Sphere.

Hudson, J. C. 1842. *The Parents Hand-book; or Guide to the Choice of Professions, Employments, and Situations.* London: Longman, Brown, Green, & Longmans.

Hudson, R. W. T. [1905] 1990. *Kummer's Quartic Surface.* Cambridge: Cambridge University Press.

Hunt, J. D. 1982. *The Wider Sea. A Life of John Ruskin.* London: Dent.

Hunt, J. L. 1996. "J. W. L. Glaisher, FRS, ScD (1848–1928)." *QJRAS* 37: 743–57.

Hunter, M. 1999. "Robert Boyle (1627–1691): A Suitable Case for Treatment?" *BJHS* 32: 261–75.

Huxley, L. 1900. *Life and Letters of Thomas Henry Huxley.* 2 vols. London: Macmillan.

———. 1918. *Life and Letters of Sir J. D. Hooker.* 2 vols. London: John Murray.

Hyman, A. 1984. *Charles Babbage.* Oxford: Oxford University Press.

Innes, H. M., ed. 1941. *Fellows of Trinity College.* Cambridge: Cambridge University Press.

Jackson, H. 1910. "Cambridge Fifty Years Ago." *CR* 31: 449–51.

J. J. 1844. "Desultory Remarks on Academic and Non-Academic Mathematics and Mathematicians." *PM* 25: 81–93.

James, I. M. ed. 1999. *History of Topology.* Amsterdam: Elsevier.

Jammer, M. [1954]. 1969. *Concepts of Space.* 2d ed. Cambridge, MA: Harvard University Press.

Jensen, J. V. 1970. "The X-Club: Fraternity of Victorian Scientists." *BJHS* 5: 63–72.

Jolly, W. P. 1974. *Sir Oliver Lodge.* London: Constable.

Joly, C. J. 1905. "George Salmon (1819–1904)." *PRSL* 75: 347–55.

Jones, K. 1991. *Learning Not To Be First: The life of Christina Rossetti.* Gloucestershire: Windrush Press.

Jordan, C. 1870. *Traité des substitutions et des équations algébriques.* Paris: Gauthier-Villars.

———. (P) 1961–64. *Oeuvres.* 4 vols. Ed. G. Julia and J. Dieudonné. Paris: Gauthier-Villars. Referred to as *JordanP* plus volume number and relevant year in the Notes.

Jowett, B., ed. and trans. [1871] 1953. *The Dialogues of Plato.* 4 vols. 4th ed. Oxford: Oxford University Press.

Kanigel, R. 1991. *The Man Who Knew Infinity.* London: Scribners.

Kauvar, G. B., and G. C. Sorenson, eds. 1969. *The Victorian Mind.* London: Cassell.

Kempe, A. B. 1879. "On the Geographical Problem of the Four Colors." *AJM* 2: 193–200.

———. 1887. "A Memoir Introductory to a General Theory of Mathematical Form." *PTRSL* 177: 1–70.

Kenschaft, P. 1987. "Charlotte Angas Scott, 1858–1931." *Coll.MJ* 18 (2): 98–110.

Kiernan, B. M. 1971. "The Development of Galois Theory from Lagrange to Artin." *AHES* 8: 40–154.

Kim, D. W. 1995. "J. J. Thomson and the Emergence of the Cavendish School, 1885–1990." *BJHS* 28: 191–226.

Kimberling, C. 1999. "The Origin of Ferrers Graphs." *MG* 83: 194–98.

Kingsley, C. [1850] 1983. *Alton Locke, Tailor and Poet: An Autobiography.* Ed. E. A. Cripps. Oxford: Oxford University Press.

———. [1876] 1880. *Charles Kingsley: His Letters and Memories of His Life.* 2 vols. Ed. F. E. Kingsley, 8th ed. abr. London: Kegan Paul.

Kirkman, T. P. 1847. "On a Problem in Combinations." *CDMJ* 2: 197–204.

———. 1848. "On Pluquaternions, and Homoid Products of Sums of *n* Squares." *PM* 33: 447–59, 494–509.

———. 1857a. "On the 7-partitions of *X*." *MMLPS* 14: 137–50.

———. 1857b. "On the K-partitions of the *r*-gon and *r*-ace." *PTRSL* 147: 217–72.

———. 1860–62. "On the Theory of Polyedra." *PRSL* 11: 218–32 (rec. 10 May 1861).

———. 1862. "On the Theory of Polyedra." *PTRSL* 152: 121–65 (rec. 3 Jan. 1862).

———. 1863. "The Complete Theory of Groups, Being the Solution of the Mathematical Prize Question of the French Academy for 1860." *PMLPS* 3: 133–52, 161–62.

———. 1871. "Note on Question 3167." *MQET* 14: 49–52.

Klein, F. [1908] 1932. *Elementary Mathematics from an Advanced Standpoint: Arithmetic, Algebra, Analysis*. 3d ed. Trans. E. R. Hedrick and C. A. Noble. London: Macmillan.

———. [1909] 1939. *Elementary Mathematics from an Advanced Standpoint: Geometry*. 3d ed. Trans. E. R. Hedrick and C. A. Noble. London: Macmillan.

———. 1890–92. *Vorlesungen über die Theorie der Elliptischen Modulfunctionen*. 2 vols. Leipzig: Teubner.

———. [1926] 1967. *Vorlesungen Ueber die Entwicklung der Mathematik im 19. Jahrhundert*. 2 vols. Reprint. New York: Chelsea.

Klein, J. 1968. *Greek Mathematical Thought and the Origin of Algebra*. Cambridge, MA: MIT Press.

Kleinfield, E. A. 1963. "Characterization of the Cayley Numbers." In *Studies in Modern Algebra No. 2*, ed. A. A. Albert, pp. 126–43. Englewood Cliffs, NJ: Mathematical Association of America.

Kline, M. 1972. *Mathematical Thought from Ancient to Modern Times*. New York: Oxford University Press.

Knobloch, E., and D. E. Rowe, eds. 1993. *The History of Modern Mathematics*, vol. 3. Boston: Academic Press.

Knott, C. G. 1911. *Life and Scientific Work of Peter Guthrie Tait*. Cambridge: Cambridge University Press.

Königsberger, L. 1877. *Repertorium der literarischen Arbeiten aus dem Gebiete der reinen und angewandten Mathematik*. 2 vols. Leipzig.

Koppelman, E. 1971. "The Calculus of Operations and the Rise of Abstract Algebra." *AHES* 8 (3): 155–242.

Kummer E. E. (P) 1975. *Collected Papers*. 2 vols. Ed. A. Weil. Berlin: Springer. Referred to as *KummerP* plus volume number in the Notes.

Lacroix, S. F. 1816. *An Elementary Treatise on the Differential and Integral Calculus*. Trans. C. Babbage, G. Peacock, and J. W. F. Herschel. Cambridge: J. Deighton.

Laguerre, E. N. (P) 1898–1905. *Oeuvres de Laguerre*. 2 vols. Paris: Gauthier-Villars. Referred to as *LaguerreP* plus volume number and relevant year in the Notes.

Lanczos, C. 1970. *Space through the Ages*. London and New York: Academic Press.

Laugwitz, D. [1996] 1999. *Bernhard Riemann 1826–1866*. Trans. A. Shenitzer. Boston: Birkhäuser.

Lindgren, M. 1987. *Glory and Failure: The Difference Engines of Johann Müller, Charles Babbage and Georg and Edvard Scheutz*. Trans. C. G. McKay. Stockholm Papers in the History and Philosophy of Science 2017, Linköping: Linköping University.

Livi, L. 1869. "A Scientific Census." *Nature* 1: 99–100.

Lodge, Sir O. 1931a. *Advancing Science*. London: Ernest Benn.

———. 1931b. *Past Years: An Autobiography of Sir Oliver Lodge*. London: Hodder and Stoughton.

Loria, G. 1902. "L'oeuvre mathématique d'Ernest de Jonquières." *BM* 3d (ser. 3): 276–322.

Love, A. E. H. 1895. "Note of Mr. Cullovin's Demonstration of the Theory of Parallels." *QJPAM* 27: 353–56.

Lowe, V. 1985. *Alfred North Whitehead: The Man and His Work, 1861–1910*. Baltimore: Johns Hopkins University Press.

Lützen, J. 1990. *Joseph Liouville, 1809–1882*. New York: Springer.

Lützen, J., and W. Purkett. 1993. "Conflicting Tendencies in the Historiography of Mathematics: M. Cantor and H. G. Zeuthen." In *The History of Modern Mathematics*, ed. E. Knobloch and D. E. Rowe, 3: 1–42. Boston: Academic Press.

Lützen, J., G. Sabidussi, and B. Toft. 1992. "Julius Petersen 1839–1910, A Biography." *DM* 100: 9–82.

Lyons, Sir Henry. 1944. *The Royal Society 1660–1940*. Cambridge: Cambridge University Press.

MacAlister, E. F. B. 1935. *Sir Donald MacAlister of Tarbet*. London: Macmillan.

Macaulay, F. S. 1895. "Point Groups in Relation to Curves." *PLMS* 26: 495–544.

———. 1896. "Cayley's Theory of the Absolute." *MG* 1: 155–58.

———. 1932. "Charlotte Angas Scott." *JLMS* 7: 230–40.

Macfarlane, A. 1895. "Arthur Cayley." *AMM* 2: 99–102.

———. 1916. *Lectures on Ten British Mathematicians of the Nineteenth Century*. London: Chapman and Hall.

MacHale, D. 1985. *George Boole*. Dublin: Boole Press.

MacLaurin, C. 1982. *The Collected Letters of Colin Maclaurin*. Ed. S. Mills. Nantwich: Shiva.

MacLeod, R. M. 1970. "The X-Club: A Social Network of Science in Late-Victorian England." *NRRSL* 24: 305–22.

MacLeod, R. M., and P. Collins, eds. 1981. *The Parliament of Science: The BAAS 1831–1981*. Northwood, UK: Science Reviews.

MacMahon, P. A. 1885. "On Perpetuants." *AJM* 7: 26–46.

———. 1889. "Memoir on a New Theory of Symmetric Functions." *AJM* 11: 1–36.

———. 1894. "Cayley's Papers (Review of vols. 1–6)." *Nature* 49 (18 Jan.): iv–vi.

———. 1896. "Combinatory Analysis: A Review of the Present State of Knowledge." *PLMS* 28: 5–31.

———. 1897. "Presidential Address." *PLMS* 28: 5–32.

———. 1898. "James Joseph Sylvester." *PTRSL* 63: ix–xxv.

———. 1901. "Presidential Address-Section A." *Rep. BAAS 1901*, 519–28.

———. 1904. "Review of Grace, J. H., and A. Young 1903, The Algebra of Invariants." *MG* 3: 8–10.

———. 1910a. "Algebraic Forms." *Enc. Brit.* 1: 620–41.

———. 1910b. "Arthur Cayley (1821–1895)." *Enc. Brit.* 5: 589–90.

———. 1914. "Robert Harley, 1828–1910." *PRSL* 91A: i–v.

———. (P) 1978, 1986. *Collected Papers*. 2 vols. Ed. G. E. Andrews. Cambridge, MA: MIT Press.

McConnell, A. J. 1944–45. "The Dublin Mathematical School in the First Half of the 19th Century." *PRIA* 50: 75–88.

McKinsey, J. C. C., and P. Suppes. 1955. "On the Notion of Invariance in Classical Mechanics." *BJPS* 5: 290–302.

McWilliams-Tullberg, R. [1975] 1998. *Women at Cambridge*. London: Victor Gollancz.

Mandelstam, J. 1994. "Du Chaillu's Stuffed Gorillas and the Savants from the British Association." *NRRSL* 48 (2): 227–45.

Marsh, J. 1994. *Christina Rossetti—A Literary Biography.* London: Jonathan Cape.

Mathews, G. B. 1895. "Note on Hamilton's Numbers." *QJPAM* 27: 184–88.

———. 1898a. "Cayley's Papers (Review of vols. 8, 9)." *Nature* 57: 217–18.

———. 1898b. "Cayley's Mathematical Papers (Review of vols. 10, 11)." *Nature* 58: 50.

Maxwell, J. Clerk. (P and Letters) 1990, 1995, 2002. *The Scientific Letters and Papers of James Clerk Maxwell.* 3 vols. Ed. Peter M. Harman. Cambridge: Cambridge University Press. Referred to as *MaxwellP* plus volume number and pertinent year in the Notes.

M[ayor], J. E. B. 1884 (5 March). "In Memoriam: Dr. Todhunter." *CR* 5, 229.

Maz'ya, V., and T. O. Shaposhnikova, 1998. *Jacques Hadamard: A Universal Mathematician.* Providence, RI: American Mathematical Society; London: London Mathematical Society.

Meadows, A. J. 1972. *Science and Controversy: A Biography of Sir Norman Lockyer.* Cambridge, MA: MIT Press.

Meo, M. 2004. "The Mathematical Life of Cauchy's Group Theorem." *HM* 31: 196–221.

Merzbach, U. C. 1984. *Carl Friedrich Gauss: A Bibliography.* Wilmington, DE: Scholarly Resources.

Meyer, F. 1890–91. "Bericht über den gegenwärtigen Stand der Invariantentheorie." In *Jahresbericht der Deutschen Mathematiker-Vereinigung* 1: 79–288.

Mikhailov, G. K. 1984. "The Dynamics of Mechanical Systems with Variable Masses as Developed at Cambridge during the Second Half of the Nineteenth Century." *BIMA* 20: 13–19.

Milbank, A. 1998. *Dante and the Victorians.* Manchester: Manchester University Press.

Mill, J. S. 1843. *The System of Logic, Ratiocinative and Inductive: Being a Connected View of the Principles of Evidence and the Methods of Scientific Investigation.* London: Parker.

———. 1874. *Autobiography.* 3d ed. London: Longmans, Green, Reader, and Dyer.

Millingen, J. G. 1841. *The History of Duelling.* 2 vols. London.

Mills, S. 1980. "Thomas Kirkman—The Mathematical Cleric of Croft." *MMLPS* 120: 100–109.

Minkowski, H. (P) 1967. *Gesammelte Abhandlungen.* New York: Chelsea.

Moberly, C. A. E. 1911. *Dulce Domum.* London: John Murray.

Mollan, C., W. Davis, and B. Finucane, eds. 1990. *More People and Places in Irish Science and Technology.* Dublin: Royal Irish Academy.

Moon, J. W. 1967. "Various Proofs of Cayley's Formula for Counting Trees." In *A Seminar on Graph Theory,* ed. Frank Harary, pp. 70–78. New York: Holt.

Montagnier, H. F. 1918–19. "From Early Records of the Col de Théodule, the Weissthorn, the Adler, and other Passess of the Zermatt District." *The Alpine Journal* 32: 44, 51.

Montagu, M. F., ed. 1946. *Studies and Essays in the History of Science and Learning.* New York: Henry Schuman.

Montagu-Butler, J. R. M. 1925. *Henry Montagu Butler—Master of Trinity College Cambridge 1886–1918.* London: Longmans Green.

Moran, P. A. P. 1975. "Quaternions, Harr Measure and the Estimation of a Palaeomagnetic Rotation." In *Perspectives in Probability and Statistics,* ed. J. Gani, pp. 295–301. New York: Academic Press.

Morley, R. K. 1912. "On the Fundamental Postulate of Tamisage." *AJM* 34: 47–68.

Moser, L. 1956. "On a Problem of Cayley." *SM* 22: 289–92.

Moss, B. 1979. W. K. "Clifford: A Mathematician and His Heritage." Unpublished ms. read to the British Society for the History of Mathematics, 17 December 1979.

Mowat, R. B. 1939. *The Victorian Age*. London: George Harrap.

Muir, Sir T. 1909. "A Theorem of Cayley's Regarding Sturm's Functions." *MM* 38: 105–11.

———. 1906–23. *The Theory of Determinants in the Historical Order of Development*. 2d ed. 4 vols. London: Macmillan.

———. 1930. *Contributions to the History of Determinants, 1900–1920*. London: Blackie.

Müller, F. 1904. "Das Jahrbuch über die Fortschritte der Mathematik 1869–1904." *BM* Ser. 3, 5: 292–97.

Munday, P. 1998. "Politics by Other Means: Justus von Liebig and the German Translation of John Stuart Mill's *Logic*." *BJHS* 31: 403–18.

Murphy, R. 1837. "Analysis of the Roots of Equations." *PTRSL* 127: 161–78.

———. 1837. "First Memoir on the Theory of Analytical Operations." *PTRSL* 127: 179–210.

Murray, J. A. H. 1884 (16 May). "Charles Bagot Cayley." *TPS* 502–5.

Musgrove, F. 1959. "Middle-class Education and Employment in the Nineteenth Century." *EHR* 12: 99–111.

Neumann, P. M. 1999. "What Groups Were: A Study of the Development of the Axiomatics of Group Theory." *BAuMS* 60: 285–301.

Neville, E. H. 1942. "Andrew Russell Forsyth 1858–1942." *JLMS* 17: 237–56.

———. 1948. "Salmon." *MG* 32: 273.

Newcomb, S. 1903. *The Reminiscences of an Astronomer*. New York: Harper.

Newman, J. R. 1956. *The World of Mathematics*. 4 vols. New York: Simon and Schuster.

Newsome, D. [1997] 1998. *The Victorian World Picture*. London: Fontana.

Newton, Sir I. 1860. [*Enumeratio linearum tertii ordinis*] *Sir Isaac Newton's Enumeration of Lines of the Third Order, Generation of Curves by Shadows, Organic Description of Curves, and Construction of Equations by Curves*. Trans. C. R. M. Talbot. London: H. G. Bohn.

Nicholson, J. 1993. "The Development and Understanding of the Concept of Quotient Group." *HM* 20: 68–88.

Noether, M. 1895. "Arthur Cayley." *MA* 46: 462–80.

———. 1905. "George Salmon." *MA* 61: 1–19.

North, J. D. 1965. *The Measure of the Universe: A History of Modern Cosmology*. Oxford: Clarendon Press.

———. 1969. "Arthur Cayley (1821–1895)." In *Some Nineteenth Century British Scientists*, ed. R. Harré, pp. 31–64. New York: Pergamon Press.

———. 1971. "Arthur Cayley." *DSB* 3: 162–70.

Nový, L. 1973. *Origins of Modern Algebra*. Trans. Jaroslav Tauer. Prague: Academy of Sciences.

Oldenburger, R. 1940. "Higher Dimensional Determinants." *AMM* 47: 25–33.

Olver, P. J. 1999. *Classical Invariant Theory*. Cambridge: Cambridge: Cambridge University Press.

Øhrstrøm, P. 1985. "W. R. Hamilton's View of Algebra as the Science of Pure Time and His Revision of This View." *HM* 12: 45–55.

Osgood, W. F. 1892. "The Symbolic Notation of Aronhold and Clebsch." *AJM* 14: 251–61.

Ossory, J. 1912. "George Salmon." *DNB* Second Supplement, 3: 251–54.

Pang, A. S. K. 1994. "Victorian Observing Practices, Printing Technology, and Representations of the Solar Corona, (1): The 1860s and 1870s." *JHA* 25: 249–74.

Panteki, M. 1987. "William Wallace and the Introduction of Continental Calculus to Britain: A Letter to George Peacock." *HM* 14: 119–32.

———. 1991. "Relationships between Algebra, Differential Equations and Logic in England 1800–1860." Ph.D. thesis, Council for National Academic Awards.

———. 1993. "Thomas Solly (1816–1875): An Unknown Pioneer of the Mathematization of Logic in England, 1839." *HPL* 14: 133–69.

Park, J. J. 1832. *The Legal Observer*. Supplement. 5 (111): 69–77.

Parshall, K. H. 1988. "America's First School of Mathematical Research: James Joseph Sylvester at The Johns Hopkins University, 1876–1883." *AHES* 38: 153–96.

———. 1989. "Towards a History of Nineteenth-Century Invariant Theory." In *The History of Modern Mathematics*. 2 vols. Ed. D. E. Rowe and J. McCleary, 1: 157–206. Boston: Academic Press.

———. 1990. "The One-Hundredth Anniversary of the Death of Invariant Theory?" *MI* 12: 10–16.

———, ed. 1998a. *James Joseph Sylvester: Life and Work in Letters*. Oxford: Clarendon Press.

———. 1998b. "The Role of Community in the Life of and Work of J. J. Sylvester." *MI* 20 (3): 35–39.

———. 1999. "The Mathematical Legacy of James Joseph Sylvester." *Nieuw Archief Voor Wiskunde* 17 (2): 247–67.

Parshall, K. H., and E. Seneta. 1997. "Building an International Reputation: The Case of J. J. Sylvester (1814–1897)." *AMM* 104 (3): 210–22.

Parshall, K. H., and D. E. Rowe. [1994] 1997. *The Emergence of the American Mathematical Research Community, 1876–1900: J. J. Sylvester, Felix Klein, and E. H. Moore*. Reprint. Providence, RI: American Mathematical Society and London Mathematical Society.

Partington, J. R. 1961–1970. *A History of Chemistry*. London: Macmillan.

Peacock, G. 1834. "Report on the Recent Progress and Present State of Certain Branches of Analysis." *Rep. BAAS 1833,* 185–352.

———. 1842. *A Treatise on Algebra,* Arithmetical Algebra, vol. 1. Cambridge: Deighton.

———. 1845. *A Treatise on Algebra,* Symbolic Algebra, vol. 2. Cambridge: Deighton.

Pearson, K. 1936. "Old Tripos Days at Cambridge, as Seen from Another Viewpoint." *MG* 20: 27–36.

Peirce, B. 1870. "Linear Associative Algebra." Washington, DC, 1870. Lithographed.

———. 1881. "Linear Associative Algebra with Notes and Addenda, by C. S. Peirce." *AJM* 4: 97–229.

Peirce, C. S. (P) 1931–58. *Collected Papers*. 8 vols. Cambridge, MA: Harvard University Press. Referred to as *PeirceP* plus volume number and pertinent year in the Notes.

———. 1976. *The New Elements of Mathematics by C. S. Peirce*. 4 vols. Ed. Carolyn Eisele. The Hague, Paris: Mouton.

Peitgen, H. O., D. Saupe, and F. v. Haeseler. 1984. "Cayley's Problem and Julia Sets." *MI* 6 (2): 11–20.

Pell, A. 1908. *The Reminiscences of Albert Pell*. Ed. T. Mackay. London: John Murray.

Perkin, H. 1996. *The Origins of Modern English Society 1780–1880*. London: Routledge and Kegan Paul.

Petersen, J. 1890. "Ueber die Endlichkeit des Formensystems einer binären Grundform." *MA* 35: 110–12.

Pevsner, N. 1972. *Some Architectural Writers of the Nineteenth Century*. Oxford: Clarendon Press.

Phillips, A., ed. 1979. *A Newnham Anthology*. Cambridge: Cambridge University Press.

Platts, C. [1894] 1909. "Matthew O'Brien (1814–1855)." *DNB* 14: 764.

———. [1896] 1909. "Robert Potts (1805–1885)." *DNB* 16: 228.

Playfair, J. 1778. "On the Arithmetic of Impossible Quantities." *Phil. Trans. Royal Society of London* 68: 318–43.

Plücker, J. 1828–31. *Analytisch-geometrische Entwicklungen*. 2 vols. Essen: Baedeker.

———. 1835. *System der analytischen Geometrie*. Berlin: Duncker and Humblot.

———. 1837. "Théorèmes généraux concernant les equations à plusieurs variables, d'un degré quelconque entre un nombre quelconque d'inconnues." *J. rei. ang. Math.* 16: 47–57.

———. 1839. *Theorie der algebraischen Kurven*. Bonn: Adolf Marcus.

———. 1865. "On the New Geometry of Space." *PTRSL* 155: 725–91.

Potts, R. 1879. *Elementary Algebra, with Brief Notices of its History*. London: Longman.

Power, E. A. 1970. "Exeter's Mathematician—W. K. Clifford, F.R.S. 1845–1879." *Advancement of Science* 26: 318–28.

Prasad, G. 1933–34. "Cayley." In *Some Great Mathematicians of the Nineteenth Century, Their Lives and Works*. 2 vols. 2: 1–33. Benares City: Benares Mathematical Society.

Price, M. H. 1983. "Mathematics in English Education 1860–1914: Some Questions and Explanations in Curriculum History." *HE* 12 (4): 271–84.

———. 1994. *Mathematics for the Multitude?—A History of the Mathematical Association*. Leicester, UK: The Mathematical Association.

Pritchard, C. 1998a. "Tendril of the Hop and Tendril of the Vine: Peter Guthrie Tait and Quaternions, Part I." *MG* 82: 26–36.

———. 1998b. "Flaming Swords and Hermaphrodite Monsters: Peter Guthrie Tait and Quaternions, Part II." *MG* 82: 235–41.

———, ed. 2003. *The Changing Shape of Geometry*. Cambridge: Cambridge University Press.

Proctor, R. A. 1884. "Dream-Space." *Gentleman's Magazine* (January–June) 1884. 256: 35–46.

Pycior, H. M. 1979. "Benjamin Peirce's *Linear Associative Algebra*." *Isis* 70: 537–51.

———. 1981. "George Peacock and the British Origins of Symbolical Algebra." *HM* 8: 23–45.

Rains, E. M., and N. J. A. Sloane. 1999. "On Cayley's Enumeration of Alkanes (or 4-valent Trees)." *Journal of Integer Sequences*, vol. 2. (LANL math. CO/0207176).

Ramsey, A. S. 1917. "William Henry Besant. 2." *PLMS* 16 Ser. 2: l–liii.

Rankine, W. J. M. 1857. "On the Invariant Sum of the Products of the Coefficients of a Pair of Homogeneous Functions." *QJPAM* 1: 357–59.

Rawnsley, H. D. 1896. *Harvey Goodwin*. London: John Murray.

Rayleigh, Baron [R. J. Strutt]. 1924. *John William Strutt, Third Baron Rayleigh*. London: Edward Arnold.

———. (P) 1899–1920. *Scientific Papers*. 6 vols. Cambridge: Cambridge University Press. Referred to as *RayleighP* plus volume number and relevant year in the Notes.

Reid, C. 1970. *Hilbert*. London: Allen and Unwin.

Report of Royal Commission on Scientific Instruction and the Advancement of Science 1872– 1874. 3 vols. London: HMSO.

Rhind, N. 1976. *Blackheath and Environs, 1790–1970*. London: Bookshops, Blackheath.

Rice, A. C. 1996. "Mathematics in the Metropolis: A Survey of Victorian London." *HM* 23: 376–417.

———. 1997. "Augustus De Morgan and the Development of University Mathematics in London in the Nineteenth Century." Ph.D. thesis, Middlesex University.

———. 2001. "Inexplicable? The Status of Complex Numbers in Britain, 1750–1850." In *Around Caspar Wessel and the geometric representation of complex numbers*, ed.

J. Lützen, pp. 147–80. Copenhagen: The Royal Danish Academy of Sciences and Letters.

———. 2001. "A Gradual Innovation: The Introduction of Cauchian Calculus into Mid-Nineteenth-Century Britain." *Proceedings of the Canadian Society for the History and Philosophy of Mathematics* 13: 48–63.

Rice, A., R. J. Wilson, and J. H. Gardner. 1995. "From Student Club to National Society: The Founding of the London Mathematical Society in 1865." *HM* 22 (4): 402–21.

Rice, A., and R. J. Wilson. 1998. "From National to International Society: The London Mathematical Society, 1867–1900." *HM* 25: 185–217.

Richards, J. L. 1987. "Augustus De Morgan, the History of Mathematics, and the Foundations of Algebra." *Isis* 78: 7–30.

———. 1988. *Mathematical Visions—The Pursuit of Geometry in Victorian England.* London: Academic Press.

Ridley, J. 1970. *Lord Palmerston.* Constable.

Rigg, J. M. [1897] 1909. "Hugh James Rose (1795–1838)." *DNB* 17: 240–42.

Roberts, S. 1882–1883. "Remarks on Mathematical Terminology and the Philosophic Bearing of Recent Mathematical Speculations Concerning the Realities of Space." *PLMS* 14: 5–15.

———. 1895. "Arthur Cayley." *PLMS* 26: 546–51.

Roberts, E. S., and E. J. Gross, 1912. *Biographical History of Gonville and Caius College.* Vol. 4, 1899–1910. Cambridge: Cambridge University Press.

Robertson, D. 1977. "Mid-Victorians amongst the Alps." In *Nature and the Victorian Imagination,* ed. U. C. Knoepflmacher and G. B. Tennyson, pp. 113–36. Berkeley: University of California Press.

Robinson, K. 1951. *Wilkie Collins: A Biography.* London: Davis-Poynter.

Robson, R., and W. F. Cannon. 1964. "William Whewell (1794–1866)." *NRRSL* 19: 168–90.

Roebuck, P. 1980. *Yorkshire Baronets 1640–1760.* Hull University and Oxford University Press.

Roscoe, H. E. 1892–93. "Carl Schorlemmer." *ONFRS* 52: vii–ix.

———. 1906. *The Life and Experiences of Sir Henry Enfield Roscoe.* London: Macmillan.

Rosenfeld, B. A. 1988. *A History of Non-Euclidean Geometry.* New York: Springer.

Roseveare, W. N. 1903. "Robert Baldwin Hayward." *PLMS* 35: 466–69.

Ross, B. 1977. "The Development of Fractional Calculus 1695–1900." *HM* 4: 75–89.

Ross, S. 1994. "De Morgan Tussles with Smith's Harmonics in a Comic Poem." *BJHS* 27: 467–71.

Rossetti, C. G. 1872. *Sing-Song, A Nursery Rhyme Book.* Illus. Arthur Hughes. London: George Routledge & Sons.

———. 1997–99. *The Letters of Christina Rossetti.* 2 vols. Ed. Anthony Harrison. Charlottesville, VA: University Press of Virginia.

Rossetti, W. M. 1883. "Charles Bagot Cayley." *Athenaeum* 2: 776, 816–17.

———. 1906. *Some Reminiscences of William Michael Rossetti.* 2 vols. London: Brown, Langham.

Rota, G. C. 1999. "Two Turning Points in Invariant Theory." *MI* 21: 20–27.

Roth, L. 1971. "Old Cambridge Days." *AMM* 78: 223–36.

Rothblatt, S. 1968. *The Revolution of the Dons, Cambridge and Society in Victorian England.* London: Faber and Faber.

Routh, E. J. 1905. "Norman Macleod Ferrers 1829–1903." *PRSL* 75: 273–79.

Rouvray D. H. 1977. "Sir Arthur Cayley—Mathematician/chemist [*sic*]." *Chemistry in Britain* 13: 52–57.

Rowe, D. 1989. "The Early Geometrical Works of Sophus Lie and Felix Klein." In *The History of Modern Mathematics,* 2 vols., ed. D. E. Rowe and J. McCleary, 1: 209–73. Boston: Academic Press.

———. 2000. "Episodes in the Berlin-Göttingen Rivalry 1870–1930." *MI* 22 (1): 60–69.

Rowe, R. C. 1880. "Memoir on Abel's Theorem." *PRSL* 30: 515–19.

Ruskin, J. 1903. *The Works of John Ruskin.* 39 vols. Ed. E. T. Cook and A. Wedderburn. London: George Allen.

Russell, B. 1902. "Non-Euclidean Geometry." In *Enc.Brit.,* 10th ed., 28: 664–74.

Russell, C. A. 1971. *The History of Valency.* Leicester: Leicester University Press.

———. 1988. "'Rude and Disgraceful Beginnings': A View of History of Chemistry from the Nineteenth Century." *BJHS* 21: 273–94.

Sabidussi, G. 1992. "Correspondence between Sylvester, Petersen, Hilbert and Klein on Invariants and the Factorisation of Graphs 1889–1891." *DM* 100: 99–155.

Salmon, G. 1844. "On the Properties of Surfaces of the Second Degree Which Correspond to the Theorems of Pascal and Brianchon on Conic Sections." *PM* 24: 49–51.

———. 1847. "On the Degree of a Surface Reciprocal to a Given One." *CDMJ* 2: 65–73.

———. 1848. *A Treatise on Conic Sections Containing an Account of Some of the Most Important Modern Algebraic and Geometric Methods.* Dublin: Hodges, Smith.

———. 1849. "On the Triple Tangent Planes to a Surface of the Third Order." *CDMJ* 4: 252–60.

———. 1854. "Exercises in the Hyperdeterminant Calculus." *CDMJ* 9: 19–33.

———. 1859a. *Lessons Introductory to the Modern Higher Algebra.* Dublin: Hodges, Smith.

———. 1859b. [sub. 1855] "On the Degree of the Surface Reciprocal to a Given One." *TRIA* 23: 461–88.

———. 1860. "On the Relation Which Connects the Mutual Distances of Five Points in Space." *QJPAM* 3: 282–88.

———. 1866. *Lessons Introductory to the Modern Higher Algebra.* 2d ed. Dublin: Hodges, Smith.

———. 1869. *A Treatise on Conic Sections.* 5th ed. London: Longmans Green, Reader and Dyer.

———. 1873a. *A Treatise on the Higher Plane Curves.* 2d ed. Dublin: Hodges, Smith.

———. 1873b. "On the Periods of Reciprocals of Primes." *MM* 2: 49–51.

———. 1879. *A Treatise on Conic sections.* 6th ed. London: Longmans.

———. 1883. "Science Worthies, Number 22—Arthur Cayley." *Nature* 28 (20 Sept.): 481–85.

———. 1928. *A Treatise on the Analytic Geometry of Three Dimensions.* 7th ed. 2 vols. Ed. C. H. Rowe. London: Longmans, Green.

Salzman, L. F., ed. [1959] 1967. *The Victoria History of the County of Cambridgeshire and the Isle of Ely.* Reprint. London: Oxford University Press.

Sawtell, M. 1955. *Christina Rossetti: Her Life and Religion.* London: A. R. Mowbray.

Schläfli, L. 1905. Briefwechsel von Ludwig Schläfli mit Arthur Cayley mit dem Facsimile eines Briefes von A. Cayley. [Correspondence 1856–71]. Ed. J. H. Graf. *Mitteilungen der Naturforschenden Gesellschaft in Bern* Nr. 1591–1608: 70–107.

———. (P) 1950–56. *Gesammelte mathematische Abhandlungen.* 3 vols. Basel: Birkhauser. Referred to as *SchläfliP* plus volume number and pertinent in the Notes.

Scholz, E. 1999. "The Concept of Manifold, 1850–1950." In *History of Topology,* ed. I. M. James, pp. 25–64. Amsterdam: Elsevier.

Schröter, H. E. 1865. "Über die Steinersche Fläche vierten Grades." *Crelle* 64: 79–94.

Schubert, H. C. 1879. *Kalkül der abzählenden Geometrie.* Leipzig.

Scott, C. A. 1894. *An Introductory Account of Certain Modern Ideas and Methods in Plane Analytical Geometry.* London: Macmillan.

———. 1895. "Arthur Cayley." *BAMS* 2: 133–41.

———. 1899. "A Proof of Noether's Fundamental Theorem." *MA* 52: 592–97.

Scott, Sir W. [1815] 1987. *Guy Mannering, or, The Astrologer.* London: Soho.

Searby, P. 1997. *A History of the University of Cambridge. Vol. 3: 1750–1870.* General ed. C. N. L. Brooke. Cambridge: Cambridge University Press.

Sedgwick, W. F. [1896] 1909. "Michael Roberts (1817–1882)." *DNB* 16: 1275.

Segre, B. 1942. *The Non-singular Cubic Surface.* Oxford: Oxford University Press.

Semple, J. G., and L. Roth. 1949. *Introduction to Algebraic Geometry.* Oxford: Oxford University Press.

Semple, J. G., and G. Kneebone. 1959. *Algebraic Curves.* Oxford: Oxford University Press.

Series, C. 1997–98. "And What Became of the Women?" *Mathematical Spectrum* 30 (3): 49–52.

Sharlin, H. I. 1979. *Lord Kelvin, the Dynamic Victorian.* University Park: Pennsylvania State University Press.

Sheppard, N. F. 1937–38. "W. F. Sheppard." *AE* 8: 1–14.

Shenitzer, A. 1991. "The Cinderella Career of Projective Geometry." *MI* 13 (2): 50–55.

Shipley, A. E. 1913. *"J", A Memoir of John Willis Clark.* London: Smith, Elder, and Co.

Siddons, A. W. 1936. "Progress." *MG* 20: 7–26.

Sidgwick, E. 1938. *Mrs. Henry Sidgwick.* London: Sidgwick and Jackson.

Sidgwick, H. 1906. *A Memoir.* London: Macmillan.

Simmonds, R. C. 1948. *The Institute of Actuaries 1848–1948.* Cambridge: Cambridge University Press.

Simmons, J., ed. 1971. *The Birth of the Great Western Railway.* Bath: Adams and Dart.

Simons, L. G. 1945. "Among the Autograph Letters in the David Eugene Smith Collection." *SM* 11: 247–62.

Sloane, N. J. A. 1973. *A Handbook of Integer Sequences.* New York: Academic Press.

Smith, M. C., and N. Wise. 1989. *Energy and Empire.* Cambridge: Cambridge University Press.

Smith, G. C. 1981. "De Morgan and the Laws of Algebra." *Centaurus* 25: 50–70.

———, ed. 1982a. *The Boole-De Morgan Correspondence 1842–1864.* Oxford: Clarendon Press.

———. 1982b. "Matthew O'Brien's Anticipation of Vectorial Mathematics." *HM* 9: 172–190.

———. 1984a. *The Boole-Thomson letters.* Monash University History of Mathematics series (no. 32). Clayton, Victoria.

———. 1984b. "How Cauchy's Complex Function Theory Became Known in Britain." Monash University History of Mathematics Series (no. 30). Clayton, Victoria.

Smith, H. J. S. 1862. "Report on the Theory of Numbers.—Part III." *Rep. BAAS 1861,* 292–340.

———. 1867. "On Large Prime Numbers Calculated by Mr. Barrett Davis [Notification]." *Rep. BAAS 1866,* Section A: 6.

———. 1873. "Presidential Address to Section A, BAAS." *Nature* 8 (25 Sept.): 448–52.

———. (P) 1894. *Collected Mathematical Papers of Henry John Stephen Smith.* 2 vols. Oxford: Clarendon Press.

S[mith], J. P. 1884 (20 February). "Imaginary Science." *CR* 5: 194–96.

Smith, J. [pseud. for J. Delaware Lewis] 1849. *Sketches of Cantabs*. 2d ed. London: George Earle.

Smyth, W. W. 1867. "William Hopkins." *Quartely Journal of the Geological Society of London*, 23: xxix–xxxii.

Sommerville, D. M. Y. [1914] 1958. *The Elements of Non-Euclidean Geometry*. Reprint. New York: Dover.

Spencer, H. 1904. *An Autobiography*. 2 vols. London: Williams and Norgate.

Spottiswoode, W. 1851. *Elementary Theorems Relating to Determinants*. London: Longman, Brown, Green.

———. 1854–55. "Researches on the Theory of Invariants." *PRSL* 7: 204–7.

———. 1856. "Elementary Theorems Relating to Determinants." *J. rei. ang. Math.* 51: 209–71, 328–81.

———. 1865. "President's Address (Mathematics and Physics)." *Rep. BAAS 1865*.

———. 1883. "Sur les invariants et les covariants d'une fonction transformée par une substitution quadratique." *Roma. R. Acad. Lincei Trans.* 7: 218–23.

Stephen, L. 1865. *Sketches from Cambridge by a Don*. London: Macmillan.

———. 1885. *Life of Henry Fawcett*. London: Smith Elder.

Stevenson, R. 1937. "Solution of Two Problems in Analytical Geometry." *CMJ* 1: 32–35.

Stevenson, R. L. 1897. *The Weir of Hermiston and Other Fragments*. Edinburgh: Longmans, Green.

Stillwell, J. 1996. *Sources of Hyperbolic Geometry*. History of Mathematics Series, vol. 10. Providence, RI: American Mathematical Society; London: London Mathematical Society.

Stokes, Sir G. G. (P) 1880–1905. *Mathematical and Physical Papers*. 5 vols. Cambridge: Cambridge University Press.

———. 1907. *Memoir and Scientific Correspondence of Sir George Gabriel Stokes*. 2 vols. Ed. Joseph Larmor. Cambridge: Cambridge University Press.

Storch, H. 1801. *The Picture of Petersburg*. London: T. N. Longman & O. Rees.

Street, G. E. [1855] 1874. *Brick and Marble in the Middle Ages: Notes of a Tour in the North of Italy*. 2d ed. London: Murray.

Stuart, J. 1912. *Reminiscences*. London: Chiswick Press.

Sturmfels, B. 1993. *Algorithms in Invariant Theory*. New York: Springer.

Sutherland, J. 1911–12. "Professor George Chystal." *PRSE* 32: 477–503.

Sylvester, J. J. 1835. *A Collection of Examples on the Integral Calculus: In Which Every Operation of Each Example Is Completely Effected*. Cambridge: Deighton.

———. 1854. "Provisional Report on the Theory of Determinants." *Rep. BAAS 1853*, 66.

———. 1869. "Address to the Mathematics and Physics Section of the British Association." *Rep. BAAS 1869*, Mathematics and Physics Section, 1–9.

———. 1870. *The Laws of Verse*. London: Longmans, Green.

———. 1888. "The Late Arthur Buchheim." *Nature* 38 (27 Sept.): 515–16.

———. (P) 1904–12. *The Collected Mathematical Papers of James Joseph Sylvester*. 4 vols. Cambridge: Cambridge University Press. Corrected Reprint 1973; New York: Chelsea. Referred to as *SP* plus volume number and pertinent year in the Notes.

Tait, P. G. [1867, 1873] 1890. *An Elementary Treatise on Quaternions*. 3d ed. Cambridge: Cambridge University Press.

———. 1895. "On the Intrinsic Nature of the Quaternion Method." *PRSE* 20: 276–84.

———. (P) 1898–1900. *Scientific Papers*, 2 vols. Ed. C. G. Knott. Cambridge: Cambridge University Press. Referred to as *TaitP* plus volume number and pertinent year in the Notes.

Tanner, J. R., ed. 1917. *The Historical Register of the University of Cambridge Being a Supplement to the Calendar with a Record of University Offices Honours and Distinctions to the Year 1910.* Cambridge: Cambridge University Press.

Taton, R., ed. 1965. *Science in the Nineteenth Century.* Trans. A. J. Pomerans. London: Thames and Hudson.

Taylor, Sir G. 1956. "George Boole FRS 1815–1864." *NRRSL* 12: 44–52.

Taylor, H. M. 1894. "On the Special Form of the General Equation of a Cubic Surface and on a Diagram Representing the Twenty-Seven Lines on a Surface." *PTRSL* 185 (A): 37–69.

Thalberg, O. M. 1927. "On the Intersections of Plane Algebraic Curves." *MM* 57: 94–99.

Thompson, D. W. 1939–1941. "Sir Thomas Heath." *ONFRS* 3: 409–26.

Thompson, S. P. 1910. *The Life of William Thomson, Baron Kelvin of Largs.* 2 vols. London: Macmillan.

Thomson, J. J. 1878. "On the Resolution of the Product of Two Sums of Eight Squares Into the Sum of Eight Squares." *MM* 7: 73–74.

———. 1929 (25 January). "Dr. Glaisher." *CR* 50 (1228): 212–13.

———. 1936. *Recollections and Reflections.* London: G. Bell and Sons.

Thomson, Sir W. 1874. "Archibald Smith." *PTRSL* 22: i–xxiv.

———. 1889. *Popular Lectures and Addresses.* Vol. 1: Constitution of Matter. London: Macmillan.

Thorp, T. 1841. *A Few Words to Freshman.* Cambridge: Cambridge University Press.

Thorpe, J. H. 1908. "Cornelius Cayley (1729–1780?)." *DNB* 3: 1300.

Todd, J. A. 1948. "Poncelet's Poristic Polygons." *MG* 32: 274–80.

Tolstoy, Ivan. 1981. *James Clerk Maxwell.* Edinburgh: Cannongate.

Topham, J. R. 1998. "Two Centuries of Cambridge Publishing and Bookselling: A Brief History of Deighton, Bell and Co., 1778–1998, with a Checklist of the Archive." *Transactions of the Cambridge Bibliographical Society* 11: 350–403.

Townsend, R. 1849. "On the Surfaces of the Second Order Which Pass Through Nine Given Points." *CDMJ* 4: 241–52.

Trevelyan, G. M. 1946. *Trinity College—An Historical Sketch.* Cambridge: Cambridge University Press.

Trollope, A. [1875–76] 1994. *The Prime Minister.* London: Penguin.

Turnbull, H. W. 1919. *Some Memories of William Peveril Turnbull.* London: George Bell.

———. 1926. "Recent Developments in Invariant Theory." *MG* 13: 217–21.

———. 1941. "Alfred Young (1873–1940)." *ONFRSL* 3: 761–78.

Turnbull, W. P. 1867. *An Introduction to Analytical Plane Geometry.* Cambridge: Deighton Bell.

Turner, F. M. 1981. *The Greek Heritage in Victorian Britain.* New Haven, CT: Yale University Press.

Tvrdá, J. 1971. "On the Origin of the Theory of Matrices." *Acta historiae rerum naturalium necnon technicarum* (Special Issue 5) Czechoslovak Studies in the History of Science, Prague.

Vaizey, J. S. 1895–1907. *The Institute—A Club of Conveyancing Counsel. Memoirs of Former Members.* London: Lincoln's Inn.

van der Blij, F. 1961. "History of the Octaves." *Simon Stevin, Wis. en Natuurkundig Tijdschrift* 34 (3): 106–25.

van der Waerden, B. L. 1985. *A History of Algebra.* Berlin: Springer.

Venn, J. 1912. "Norman Macleod Ferrers (1829–1903)." *DNB* Second Suppl. 2: 20–21.

Venn, J. A. 1922–1954. *Alumni Cantabrigienses.* Cambridge: Cambridge University Press.

Walton, W. 1852. "On the Doctrine of Impossibles in Algebraic Geometry." *CDMJ* 7: 234–51.

Warwick, A. 2003. *Masters of Theory: Cambridge and the Rise of Mathematical Physics.* Chicago: University of Chicago Press.

Watson, E. C. 1939. "A Possible Portrait of Arthur Cayley." *SM* 6: 32–36.

———. 1939. "College Life at Cambridge in the Days of Stokes, Cayley, Adams, and Kelvin." *SM* 6: 101–6.

Weber, H. 1896. "Vier Briefe von Arthur Cayley über elliptische Modulfunctionen." *MA* 47: 1–5.

Weyl, H. 1939. "Invariants." *Duke Mathematical Journal* 5: 489–502.

Whewell, W. 1833. *Address* [to the British Association for the Advancement of Science]. Cambridge: Pitt Press.

———. [1847] 1967. *Philosophy of the Inductive Sciences, Founded upon Their History.* 2 parts. 2d ed. Reprint. London: Frank Cass.

———. 1881. *The Life of William Whewell. DD.* Ed. Mrs. Stair Douglas. London: Kegan Paul.

———. (P) 2001. *Collected Works of William Whewell.* 16 vols. Bristol: Thoemmes Press. Referred to as *WhewellP* plus volume number in the Notes.

White, H. S. 1899. "Report on the Theory of Projective Invariants: The Chief Contributions of a Decade." *BAMS* 5: 161–75.

White, W. 1898. *The Journals of Walter White.* London: Chapman and Hall.

Whitehead, A. N. 1901. "Memoir on the Algebra of Symbolic Logic." *AJM* 23: 139–65, 297–316.

———. 1927. *Science and the Modern World.* Cambridge: Cambridge University Press.

Whittaker, E. T. 1936–1938a. "Andrew Russell Forsyth." *ONFRS* 2: 209–27.

———. 1936–1938b. "Alfred Cardew Dixon." *ONFRS* 2: 65–174.

Williams, I. W. 1966. "The Western University of Great Britain." *Collegiate Faculty of Education Journal* (Univ. College, Swansea), 32–40.

Williams, R. H., ed. 1988. *Salisbury-Balfour Correspondence.* Ware: Hertfordshire Record Society.

Wilson, A. N. 2002. *The Victorians.* London: Hutchinson.

Wilson, D. B. 1985. "The Educational Matrix: Physics Education at Early-Victorian Cambridge, Edinburgh and Glasgow Universities." In *Wranglers and Physicists,* ed. P. M. Harman, pp. 12–48. Manchester: Manchester University Press.

———. 1987. *Kelvin and Stokes.* Bristol: Adam Hilger.

———, ed. 1990. *The Correspondence between Sir George Gabriel Stokes and Sir W. Thomson, Baron Kelvin of Largs.* 2 vols. Cambridge: Cambridge University Press.

Wilson, J. M. 1920–21. "The Early History of the [Mathematical] Association." *MG* 10: 239–44.

———. 1932. *James Maurice Wilson: An Autobiography 1836–1931.* London: Sidgwick and Jackson.

Wilson, R. J. 2002. *Four Colours Suffice.* Allen Lane

Wingfield-Stratford E. 1930. *The Victorian Tragedy.* London: Routledge.

Winstanley, D. A. [1940] 1955. *Early Victorian Cambridge.* Cambridge: Cambridge University Press.

———. 1947. *Later Victorian Cambridge.* Cambridge: Cambridge University Press.

Wood, E. M. 1965. *The History of the Polytechnic.* London: MacDonald.

Woodham-Smith, C. [1962] 1991. *The Great Hunger: Ireland 1845–1849.* New York: Penguin.

Woods, G. T. 1976. "Chemically, the Same or Different?" *MG* 60: 247–56.

Woolhouse, W. S. B. 1878. "On a Question of Probabilities." *JIA* 204–13.

Woolley, Sir R. 1966. "The Kinematical and Chemical History of the Galaxy." *JLMS* 44: 29–48.

Wordsworth, C. [1877] 1968. *Scholae academicae: Some Accounts of Studies at the English Universities in the Eighteenth Century.* London: Frank Cass.

Wordsworth, W. 1994. *The Works of William Wordsworth.* Hertfordshire, Eng.: Wordsworth Editions Limited.

Wright, J. E. 1908. *Invariants of Quadratic Differential Forms.* Cambridge Tracts in Mathematics and Mathematical Physics. Cambridge: Cambridge University Press.

Wussing, H. 1984. *The Genesis of the Abstract Group Concept.* Trans. A. Shenitzer. Cambridge, MA: MIT Press.

Yaglom, I. M. 1988. *Felix Klein and Sophus Lie.* Ed. H. Grant and A. Shenitzer; trans. S. Sossinsky. Boston: Birkhäuser.

Young, A. 1913. "Paul Gordan." *PLMS* 12 Ser. 2: li–liv.

Young, G. M. [1937] 1948. *Daylight and Champaign.* London: Rupert Hart-Davis.

Young, L. 1981. *Mathematicians and Their Times.* Amsterdam: North Holland.

Young, W. H. 1922. "Adolf Hurwitz." *PLMS* 20: xlviii–liv.

Index

Page numbers in *italics* followed by a *g* refer to illustrations in the galleries. "C." in subentries refers to Arthur Cayley.

Abel, Niels-Henrik, 53, 72, 89, 117, 160, 443

Abelian functions, 360, 373, 396, 474, 545n. 92

Abel's theorem, 70, 357, 359–60, 468

absolute, Cayley's. *See* geometry, Cayley's absolute in

Abzählende Geometrie (Schubert), 362, 377, 386

academic positions, C. attracted to: Cambridge University (Lowndean Chair), 233; Cambridge University (Sadleirian Chair), 261, 263; Glasgow (Regius Professor of Astronomy and Directorship of the Observatory), 238; Marischal College, Aberdeen (Professor of Natural Philosophy), 218–20; St. Andrews (Regius Professor of Mathematics), 233; Western University of Great Britain, South Wales (Professor of Mathematics), 223–26

Adams, John Couch, 443; and Cambridge lectures, 268–69, 339; at Cambridge Philosophical Society, 292; C. confides to, 308, 556n. 73; in competition with C., 233; and C. on secular acceleration of the moon, 239–40, 475; on C.'s Tripos success, 49–50; elector for Sadleirian chair, 255–56; encourages C. in astronomy, 212; impressions of, at British Association for the Advancement of Science (1845), 114; influential position of, 172, 180, 338; and Leverrier at British Association for the Advancement of Science, 132; and Neptune, 32

agricultural depression, 355–56

Airy, Sir George Biddell, *16g*, 302–8, 443; at British Association for the Advancement of Science (1845), 114; calculators employed by, 199; disputes with C., 302–8, 316–18; as President of the Royal Society, 321; ranked with C., 263

Airy, Hubert, 303, 443

Airy, Osmund, 303, 443

Airy, Wilfrid, 303, 443

Albert, Prince, 171, 217, 407, 490n. 2

Albert, Prince Victor, 409, 443

algebra: new, 160; notion and boundaries of, 279–80; organic part of, 392; organization of, 110, 173, 184, 188, 279–81, 529n. 56, 547n. 49; "purely idle," 317; symbolical, 53, 71–72, 132, 141, 184, 187. See also *Lessons Introductory to Modern Higher Algebra*

algebraic forms, 468; canonical form for, 171, 269, 507n. 50, 508nn. 59, 62; Cayley's theorem for, 131, 471; history of, 495n. 24. *See also* binary forms; quantics

algorithm, 110–11, 375–76, 394, 468

Allen, John, 23, 443

Allman, George J., 210, 443

Alpine Club, The, 230, 273, 520n. 87

American Journal of Mathematics, 344, 348, 366, 376, 391, 411

American Mathematical Society, 363, 556n. 52

Analyse des équations indeterminées (Fourier), 34

analytical geometry. *See* geometry, analytical

Analytisch Geometrische Entwicklungen (Plücker), 65

Annali di matematica pura ed applicata, 190, 295

Annali di Scienze mathematische e fisiche, 186, 190

annihilator. *See* differential operators, as annihilators

Aperçu historique sur l'origine et le développement des methods en géométrie (Chasles), 65

appearance, C.'s physical. *See* Cayley, Arthur, physical appearance of applied mathematics, C.'s influence on modern, 527n. 41

Arbogast, L. F. A., 71, 443

Archimedes, 61

Architectural Antiquities of Rome (Taylor and Cresy), 76

Aristotle, 61–62

Aronhold, Siegfried H., 443, 524n. 72; C. and, compared, 251; C. credits, 178; notation introduced by, 189, 294

Ashburnham, A. P., 16, 485n. 36

Askwith, Edward H., 411, 443

Association for the Improvement of Geometrical Teaching (AIGT), 405, 536n. 35

associative property, 102, 186, 468, 497n. 46, 498n. 14

astronomy: C. elected to Académie des Sciences in, 250; C.'s undergraduate studies in, 46, 52; physical vs. mathematical, 284; takes up, as study, 212. *See also* Royal Astronomical Society

astrophysics, 320

asyzygetic. *See* invariants, asyzygetic

Athenaeum Club, The, 243

Atkinson, Edward, 443; elector for Sadleirian chair, 255–56

Atkinson, Michael A., 443

Ausdehnungslehre, Die, 386, 401, 494n. 6, 551n. 58

Austin, William, 49, 50, 444

Babbage, Charles, 444; calculating machines of, 217, 325, 346, 418, 537n. 41; on England's decline in science, 75; a mathematical problem of, 539n. 72; as professor, 32

Babbage, H. P., 200

Baker, Henry F., 422, 444, 551n. 65

Ball, Robert S., 323, 444

Ball, W. W. Rouse, 438, 444

Baltic Coffee House, 14

Baltzer, Richard, 191, 444

Barstow, T. I., 35, 48, 50, 444

Bates, Henry W., 353, 444

Bateson, William, 255–56

Bath Power, Alexander, 224

Bell, Eric T., xiv, 444

Bennett, Geoffrey T., 422, 441, 444

Bentham, George, 194, 444, 523n. 46

Bernoulli numbers, 276; in undergraduate syllabus, 488n. 47

Berry, Arthur, 420, 444, 551n. 65

Bertrand, J. L. F., 395, 444, 509n. 96, 549n. 30

Beside the Bonnie Brier Bush (Maclaren), 429

Bessel, Friedrich W., 276, 444

Betti, Enrico, 444; C. invites, to British Association for the Advancement of Science, 529n. 59; and Euclid, 309; and group theory, 186; and invariant theory, 191; and London Mathematical Society, 318; and Sylvester, 242

Bézout, Etienne, 66, 444

bicenter. *See* Cayley trees, bicenter of

Bickmore, Charles E., 444, 555n. 37

Bienaymé, Irénée-Jules, 444, 509n. 96

binary forms, 88, 469; as content of memoirs on quantics, 524n. 74; era of, 111; higher order, 346–47, 392; notation for, 513n. 48; principal results for, 549n. 14. *See also* binary quintic form

binary quintic form, 88–89, 295, 373, 469, 533n. 41; campaign, 206–7, 373; elaborate structure of, 295; fundamental syzygies of the, 346, 394, 425; system of invariants and covariants for, *12g, 15g*, 98–99, 301, 346, 425, 498n. 72, 512n. 29, 514n. 73

binary sextic, 88, 469, 498n. 75, 514n. 81, 555n. 23

Binet, Jacques P. M., 55, 66, 444

Bird, C., 49, 444

Blackall, Samuel, 444, 525n. 88

Blackburn, Hugh, 445; at Cambridge, 69, 501n. 72; C. writes testimonial for, 151; at Glasgow, 218; at Lincoln's Inn, 145, 150; pendulum of, 505n. 92

Blackheath. *See* Cayley, Arthur, Blackheath homes of

Blanshard, Henry, 13

Board of Mathematical Studies, 73, 357, 403, 405, 534n. 57

Bolza, Oskar, 408

Bombelli, Rafael, 89

Bonney, T. G., 360, 445

Boole, George, *8g*, 445; abandons invariant theory, 192–93, 206; appointed at Cork, 152–53; C. assigns priority in invariant theory to, 112, 392; C. proposes for Royal Society, 221; C.'s correspondence with, 85–86, 88–92, 101–6, 498n. 75; C. writes testimonial for, 129–30; and differential equations, 512n. 38; on importance of calculation, 199; and matrix algebra, 228; and symbolical algebra, 71, 132; tempts C. to study optics, 131

Boole's Challenge Problem, 184, 246, 510n. 1

Booth, Clement, 355–56

Booth, James, 182, 204, 445, 509n. 97

Borchardt, Carl W., 117, 132, 445, 494n. 80, 509n. 96

Bowditch, Nathaniel, 505n. 92

Bowen, Charles S., 506n. 13

Bragg, William H., 399, 445, 550n. 48

Bramwell, Sir Frederick J., 380, 445

Brantschen, Jean Baptiste, 76

Brewster, Sir David, 114, 180, 445

Brill, Alexander von, 286, 362, 445

Brioschi, Francesco, 445; on C., 356, 439; and elliptic functions, 337; and Euclid, 309; and invariant theory, 190–91; and London Mathematical Society, 542n. 32

Bristed, Charles, 50–51, 445

British Association for the Advancement of Science: C. as vice-president at Bath, 281–82; C.'s presidential address to, 376–89; C. supports, 181, 217, 322–25, 328; grants awarded by, 200, 347, 541n. 22; past (mathematical) presidents of, 376. See also British Association for the Advancement of Science, annual meetings of

British Association for the Advancement of Science, annual meetings of: Bath (1864), 281–82; Bradford (1873), 322–25; Brighton (1872), 322; Bristol (1875), 328; Cambridge (1845), 113–15, 500n. 54; Dublin (1878), 325, 377; Edinburgh (1871), 322; Exeter (1869), 322; Glasgow (1855), 210; Hull (1853), 188; Leeds (1858), 232, 236; Liverpool (1870), 322; Oxford (1847 and 1860), 132–33; Plymouth (1877), 344; Southampton (1882), 371; Southport (1883), 264, 376–89, 494n. 6, 547n. 37, 548n. 64

Brodie, Sir Benjamin (elder), 246–47, 445

Brodie, Sir Benjamin (younger), 246, 328, 445, 537n. 46

Brogden, James, 8–9

Bromhead, Sir Edward ffrench, 85, 97

Bronwin, Brice, 63–64, 101, 445, 492n. 29

Brooke, Charles, 119, 445

Brougham, Peter (Lord Brougham), 210

Browne, R. W., 445; at King's College, London, 21–22, 87

Brummel, Edward, 136, 445

Brunel, I. K., 14

Brunyate, William E., 445, 551n. 65

Bryan, George H., 445, 551n. 62

Bryan, Reginald (Guy), 36, 446, 488n. 24

Buchheim, Arthur, 446, 549n. 27

Bunsen, Robert W., 182–83, 194, 446

Burnside, William S., 318, 446

Busk, George, 282, 446

Butler, Henry M., 404–5, 446

Buxton, Charles, 446, 491n. 8

Buxton, Thomas F., 446, 491n. 8

Calcul des dérivations (Arbogast), 71

calculus of operations, 469; applied to chemistry, 246, 537n. 46; and group theory, 510n. 8; Lagrange's theorem in, 23, 200, 474, 528n. 54; and matrix algebra, 228; and Robert Murphy, 493n. 43; "separation of symbols" in, 69, 186, 493n. 54; and symbolical algebra, 71, 184–85, 511n. 17; symbols of operation in, 491n. 12, 510n. 4; in undergraduate mathematics syllabus, 488n. 47

Cambridge Advertiser and Free Press, 43, 51

Cambridge Analytical Society, The, 23, 46, 528n. 54

Cambridge and Dublin Mathematical Journal, xv, 116, 128, 131–32, 146, 153, 181

Cambridge Camden Society. *See* Camden Society

Cambridge Mathematical Journal, 42–43, 55–59, 63–74, 83, 115, 137, 433

Cambridge Modelling Club. *See* Modelling Club, Cambridge

Cambridge Philosophical Society: C. at, 292–93, 332, 531n. 9; C. elected to, 57; C. elected to Council of, 117, 132; C. presents his first paper to, 66–67, 531n. 8; founding of, 490n. 2

Cambridge University, 215, 349, 355–56, 403, 409–10, 417

Cambridge University Act (1856), 215, 255

Camden Society, 59, 107–8, 499nn. 31, 32

canonical form. *See* algebraic forms, canonical form for

Canterbury, archbishop of, 18, 417, 499n. 31

Cantor, Georg, 11, 446

Cardano, Girolamo, 89

Carey, Frank S., 400, 446

Carnot, L., 48, 55

Casey, John, 297, 446, 536n. 17

Catalan, Eugène C., 446, 509n. 96

catalecticant. *See* invariants, catalecticants as

Cauchy, Augustin-Louis, 446; C. compared with, 435, 483n. 4; C. reads, generally, 34, 60, 124, 486n. 60; C. reads, on applications of mathematics, 131, 502n. 25; C. reads, on complex analysis, 117, 501n. 68; C. reads, on determinants, 66; C. reads, on equations, 354; C. reads, on group theory, 247–48; C. reads, on polyhedra, 237; cited in Peacock's report, 72; and overproduction, 129; and Sylvester, 509n. 96

caustic. *See* curves, caustics

Cavendish Laboratory, 304, 331–32

Cayley, Arthur: affection for, in academic community, 313; ancestry of, xxiii, 3–5; apprentice to George Boole, 86–100; award of honorary ScD to, 409–10; awarded Chemistry medal, 25, 326; and beauty of mathematics, xvi; birth of, 3, 11; Blackheath homes of *13g, 14g,* 165, 501n. 4, 506n. 22; as by-election official, 546n. 10; called to Bar, 152; as chess player, 108, 418; childhood in Russia, 11–12; as consultant to scientists, 211, 276; contemplates taking private pupils, 234, 254; Copley medal awarded to, 371; death of, 430; descendant of "St. Petersburg Cayleys," 5; and early taste for arithmetic, 17; elected to Cambridge Philosophical Society, 57; elected to London Mathematical Society, 283; elected to Royal Astronomical Society, 221–22; elected to Royal Society of London, 180; the *éminence grise,* 390, 426, 432; enters Cambridge University, 27; as examiner and moderator for mathematical Tripos, 159–60, 254; as examiner for Indian Civil Service, 209–10, 254, 505n. 94; as explorer of mathematics, xv, 99, 441; extensive memoirs by, 527n. 42; fellowship gained by, 62–63, 356, 491nn. 18, 19, 500n. 61; first publication of, 54–56; as First Smith's Prizeman, 51–53; in Garden House, xvii, *22g,* 271–72, 357, 418–19, 428, 483n. 11, 527n. 23; great geometrical period of, 277; headstone of, *24g;*

health of, 230, 320–21, 390, 408, 415, 426–27, 523n. 39, 555n. 38; holistic view of mathematics of, 284; humor of, 127, 133, 360, 365, 379, 407; inaugural lecture of, 263–68; individualism of, 72, 170; influence of Greeks on, 438; involved with adult education, 239; in Lake District, 37, 58, 204, 216, 220, 247, 281, 395, 398; lectures at Cambridge, 268–70, 419, 526n. 11; and lectures on physics, 308; on legal conveyancing, 163–65; liking for Italian architecture and art, 76–77, 418; London home of, *3g;* and "long-division," 16, 17; marriage of, 261–62; on mathematical education, 302–8; as a mathematical referee, 275–76, 379, 437, 527n. 33; mathematical style of, 64, 70, 278, 283–84, 378, 436, 524n. 70; Mathematician Laureate, xv, 238; meets J. J. Sylvester, 139–41; a member of *Sex Viri,* 403; membership of university syndicates, 403; the "new professor," 262, 526n. 7; and Newton, 301, 440; obituaries for, 483n. 1; the *pater familias,* 356; at the peak of production, 312; perseverance of, 296–97; physical appearance of, 157, 244–45, 320–21, 364–65, 523n. 39; and Platonist viewpoint, 381–83, 388; pleasure in research of, 282; portraits of, *1g, 6g, 10g, 17g, 23g;* practical skills of, 333; pragmatic attitude of, 295; as president of the British Association for the Advancement of Science, 371, 376–89; as president of the London Mathematical Society, 300–301; as president of the Royal Astronomical Society, 319–21; primary education of, 15–17, 485nn. 34, 37; publishing record of, *18g,* 483n. 4, 515nn. 1, 2; pupilage at Lincoln's Inn, 120–26, 150–53; quotes Robert Burns, 308, 534n. 70; quotes Tennyson's "Locksley Hall," 381, 441, 557n. 99; reading material of, 37, 418–19; renowned as a mathematician, 312–13; restlessness of, 126; and Royal Medal, 182–83, 246, 523n. 46; as Sadleirian professor, 261–63, 305, 525nn. 1, 2; scales Le Dom, 231, 520n. 89; secondary education of, 17–25; the Senior Wrangler, 49–51; sense of propriety of, 354; and short mathematical notes, 348–54; supports Sylvester, 395; on Sylvester, 173–74; on Sylvester's work, 275–76, 527n. 33; taciturn nature of, 282,

377; as tutor on reading parties, 58–60, 127; undergraduate education of, 27–56; in United States of America, 363–67; unknown to general public, xiv, 172; as vice president of British Association for the Advancement of Science, 281–82; walking exploits of, 37, 157, 177; will and testament of, 556n. 60; on "wills," 317; wins school prizes, 25. *See also* Trinity College, Cambridge

Cayleyan curves. *See* curves, Cayleyan

Cayley-Bacharach theorem. *See* geometry, Cayley-Bacharach theorem in

Cayley, Charles Bagot (younger brother), *14g*, 165–68, 245, 446, 548n. 1; and "algebraic" linguistics, 166–67; birth of, 11; and Christina Rossetti, 165; death of, 390; financial setback of, 209; influenced by Gabriele Rossetti, 21; mathematical skills of, 166; physical appearance of, 166; translator of *Dante*, 167, 506n. 31

Cayley, Cornelius (great grandfather), 122, 446

Cayley, G. J., 5

Cayley, Henrietta Caroline (sister), 12, 447

Cayley, Henry (father), 447; Russia merchant, 10–15, 26, 164–65, 485n. 21, 486n. 70, 506n. 21

Cayley, Henry (son), *23g*, 312, 341, 356–57, 422, 447, 535n. 2

Cayley, Hugh, 484n. 6

Cayley, John (grandfather), 5–10, 447, 484nn. 9, 10, 16

Cayley, Lucy (cousin), 26

Cayley, Maria Antonia (mother). *See* Doughty, Maria Antonia

Cayley, Mary (daughter), 312, 341, 422–23, 447, 544n. 72

Cayley, Sir George (cousin), 3–4, *4g*, 446

Cayley, Sir William (ancestor), 3, 447, 484n. 1

Cayley, Sophia (sister), 3, 165, 447

Cayley, Susan (wife). *See* Moline, Susan

Cayley, William Henry (elder brother), 447

Cayley-Hamilton theorem. *See* matrix algebra, Cayley-Hamilton theorem in

Cayley-Klein metric, 470

Cayley-Moser "Secretary Problem," 335, 470, 539n. 80

Cayley number, 101–5, 470, 498n. 14, 505n. 88, 545n. 96

Cayley's theorem (graph theory). *See* graph theory, Cayley's theorem in

Cayley's theorem (group theory). *See* group theory, Cayley's theorem in

Cayley's transform. *See* matrix algebra, Cayley's transform in

Cayley table. *See* group theory, Cayley table in

Cayley trees, 470; applications to Chemistry, 325–28, 365, 537nn. 42, 44, 45; bicenter of, 469; center of, 471; origins of, 214, 516n. 4

Challis, James, 447; at Cambridge, 31–32, 51, 218; at Cambridge Philosophical Society, 57, 292; elector for Sadleirian chair, 255–56; and Sadleirian estates, 355

characteristic polynomial, 471

Chasles, Michel, 447; *Aperçu* of, 65, 108; C. meets, 140; C.'s admiration for, 284–85; C. supports, for medal, 530n. 70; and London Mathematical Society, 318; "principle of correspondence" due to, 285–88; and Sylvester, 509n. 96; and Weddle's surface, 253

Chasles-Cayley-Brill formula, 286, 471, 530nn. 73, 75, 76

chemistry: analogous to invariant theory, 194, 196; C.'s education in, 21, 25. *See also* Cayley trees

Chisholm, Grace (Mrs. Grace Young), 402, 419–21, 423, 467, 555n. 16

Christie, Jonathan II., 447, 501n. 5, 525n. 3; colleague of C., 157, 161–62; and duel at Chalk Farm, 122–23; legal cases of, 162–64; pupil master of C., 123–25, 141

Christie, Samuel H., 447, 523n. 46

Christoffel, Elwin B., 447, 550n. 40

Chrystal, George, 318, 401, 447

Clairault, Alexis, 97

Clarendon and Taunton Royal Commissions, 309

classics degree, prequalification in mathematics for, 45, 487n. 11

classification in mathematics, 193–95

Clebsch, (R. F.) Alfred, 447; and geometry, 330; and invariant theory, 322, 338; and London Mathematical Society, 318; and Royal Society, 530n. 82

Clebsch and Gordan theory, 287, 365, 391, 471

Clifford, William K., *16g*, 447, 541n. 24; at British Association for the Advancement of Science, 309, 323; C. defends, 314, 535n. 4; C. describes work of, 314; C. inspires the young, 268; C. observes, during mathematical Tripos, 290–91; comes to C.'s aid, 307; compared with C., 315; on C.'s absolute, 526n. 15, 536n. 32; C. states view of, on space, 385; elected to Royal Society, 535n. 6; election of, to Cambridge Philosophical Society, 292; on non-Euclidean geometry, 536n. 32; as student, 256, 531n. 4; as teacher at Cambridge, 316, 360; values C.'s work on prepotentials, 314

close, 471, 542n. 38

Clough, Anne J., 403, 447

coaches at Cambridge, mathematical, 38–41, 272, 526n. 18

Cobbold, Thomas Spencer, 224, 447

Cockle, James, 447; and C., 251, 524n. 67; the lawyer, 150, 160–61; and Robert Harley, 199, 250

Colenso, J. W., 24, 135

Collected Mathematical Papers (Cayley), xv, xvii–xviii; production of, 408, 427, 432, 552n. 94

Collection of Examples (Peacock), 23

Collège de France, 242

Collins, Wilkie, 125, 448

color, C.'s use of, 354

colorgroup, Cayley's. *See* group theory, Cayley's colorgroup in

color mixing, 343

combinant, 198, 471

Commentaries on the Laws of England (Blackstone), 163

commutant, 242, 472

complex numbers, 471. *See also* imaginary numbers

compound permutation, method of, 172–73

convertible, 472, 497n. 46

conveyancing. *See* legal profession, conveyancing in

Cookson, Henry, 255–56

Coolidge, Julian L., 54, 448, 530n. 76

Cope, Edward M., 448, 491n. 20

couples, algebraic, 103–5

Cours d'algèbre supérieure (Serret), 186, 418, 511n. 13

covariants, 175–76, 472

Cox, Homersham (younger), 448

Cox, J., 541n. 10

Coxwell, Henry T., 281, 448

Cozens, Sarah (grandmother), 448

Craig, Thomas, 363, 366, 448

Cramer, Gabriel, 66, 124, 264, 448

Cramer's paradox. *See* geometry, Cramer's paradox in

cranks, mathematical, 379

Creedy, Charles, 200

Crelle, August L., xvi, 134, 448

Crelle's Journal. See Journal für die reine und angewandte Mathematik

Cremona, (Antonio) Luigi, 448; and "Cayleyan," 84; C. corresponds with, 253, 295; and London Mathematical Society, 318; and quartic surfaces, 330, 538n. 59; wins Steiner Prize, 524n. 81

Crichton-Stuart, John, 162

Crofton, Morgan W., 448, 532n. 31

Crookes, William, 194

cubic curves. *See* curves, cubic

cubic surfaces. *See* surfaces, cubic

Cubitt, Thomas, 36, 448, 488n. 24

Cumming, James, 68, 135, 448

curves: caustics, 185, 187, 199, 215, 286, 469; Cayleyan, 84, 197, 469; Cayley's sextic, 471, 511n. 18; cubic, 106, 278, 294, 472, 522n. 14, 527n. 42, 528n. 44; deficiency of, 285, 472; of double curvature (skew curves, space curves), 46, 97, 105, 147, 253, 472, 476, 499n. 22, 524n. 79; edge of regression, 52, 472; elliptic, 286; Hessian, 473, 540n. 4; intersection of, 65, 81, 492nn. 31–36, 494n. 1, 538n. 50; and invariant theory, 406; Plücker's theory of, 105; polyzomal, 297, 475; quartic, 277–78, 371, 522n. 14, 528n. 45; red and blue, 354; representation of, in space, 105, 253; sextactic points on, 278, 528n. 48, 530n. 78; singularities of, 278; Talbot's, 476; tracing of, 237–38, 294, 306, 332, 521n. 11, 14, 543n. 51; unicursal, 272

Daniell, John Frederic, 20–21, 69, 448

Darboux, Jean Gaston, 330, 418, 448, 542n. 32, 549n. 30

Darwin, Charles, 182–83, 448

Darwin, George H., 399, 448

Davidson, Charles, 163, 448
Davidson, M. G., 502n. 7
Davidson's Precedents and Forms, 163–64
Davies, Emily, 448, 551n. 73
Davies, John Llewelyn, 231, 448
Davies, Thomas S., 181, 449
D'Avigdor, Elim Henry, 237, 449, 521n. 10
Davis, William B., 200, 250, 301, 347, 449, 513n. 63, 533n. 48
Dedekind, (J. W.) Richard, 184, 449, 518n. 66, 523n. 57
deficiency. *See* curves, deficiency of; surfaces, deficiency of
De la Rue, Warren, 319–20, 449
Delaunay, Charles E., 240, 348, 449
De Morgan, Augustus, 449; on *Cambridge Mathematical Journal*, 71; commends C. for a chair, 233; as critic of mathematical Tripos, 52; on C.'s report on dynamics, 222; on examinations at Cambridge, 47, 52, 270; on Euclid, 537n. 37; on mathematical societies, 240, 283, 292; on mathematical topics, 184, 351, 501n. 68, 547nn. 49, 50; opposes Scientist's Declaration, 282; remarks on mathematicians, 128, 136, 273; sponsors C. for Royal Astronomical Society, 221; as teacher, 486n. 63, 493n. 73; on triple algebra, 94–95, 104, 115; at University College, 18, 24, 102, 256, 506n. 16, 529n. 66
De Morgan medal. *See* London Mathematical Society, C. awarded De Morgan medal of
Denman, George, 125, 449
Descartes, R., 55, 264, 317, 413
déterminant gauche, 117, 131, 472, 501n. 69
determinants, 472; Cayley-Menger, 470, 490n. 86; C.'s axiomatic approach to, 242–43; first paper on, 54–56, 489n. 72; importance of 67; *n*-dimensional, 66, 518n. 69; paper on, presented to Cambridge Philosophical Society, 66; and quaternions, 96, 497n. 63. See also *déterminant gauche*
developable surfaces. *See* surfaces, developable
Devonshire, Duke of, 316
Dew-Smith, Albert G., 449
Dickens, Charles, 418, 449, 521n. 7
Dickinson, G. Lowes, 449, 535n. 3
Dickinson, Lowes Cato, 312, 380, 432, 449, 535n. 3
Dickson, James D. H., 359–60, 449

Differential and Integral Calculus (Lacroix), 23
differential equations, 539n. 77; Brioschi and, 512n. 32; Cayley-Darboux-Lévy, 469; and dynamics, 132; for motion of moon, 522n. 21; partial, 334, 387; singular solutions of, 539n. 77; and Tait, 542n. 43; in undergraduate syllabus, 357, 488n. 47
differential operators, 471; as annihilators, 339; and Cayley trees, 214; C.'s partial, 175–77, 508nn. 67, 71, 510nn. 6, 7, 514n. 72; the *D*, 493n. 54; nature of, 513n. 48
Dirichlet, Johann P. G. Lejeune-, 449, 518n. 66; and C. on potential theory, 83, 105–6, 535n. 5; and matrix algebra, 227
discriminants, 88, 472; calculation of, 90, 199, 288; in geometry, 507n. 44, 531n. 14; of low degree, 110; obtained by elimination, 496n. 36
Disquisitiones Arithmeticae (Gauss), 66
Disraeli, Benjamin, 15, 449
Dixon, Alfred C., 401, 449, 551n. 62
Dodgson, Charles L. (Lewis Carroll), 449; and circular billiards, 534n. 62; and determinants, 418; on Euclid, 74, 406, 537n. 37; meets C., 281
Donkin, William F., 178, 191, 212, 240, 449
double-sixer, 220, 298, 472, 533n. 38
Doughty, Maria Antonia (Mrs. Maria Cayley), 11, 15, 447
Doulcet, P. G. *See* Pontécoulant, Phillipe de
duality, geometric. *See* geometry, pole and polar (reciprocal polars) in
Du Chaillu, Paul Belloni, 195–96, 449
Duhamel, Jean M. C., 243, 449
Durham University, 24
Dyck, Walter von, 449, 544n. 79
dynamics, 222–23; and Jacobi, 132. *See also* three-body problem

Earnshaw, Sammuel, 51–52, 449
Eccles, William, 127, 234
École Centrale des Arts et Manufactures, 223, 225, 517n. 50
Eddis, Arthur S., 120, 449
edge of regression. *See* curves, edge of regression
Edleston, Joseph, 450
Einstein, Albert, 187, 450, 490n. 85
Eisenstein, (Ferdinand) Gotthold, *8g*, 450, and invariant theory, 87–88, 91, 108, 178, 195; and matrix algebra, 226–27, 518nn. 60–62

Elementary Treatise on Elliptic Functions, An (Cayley), 337, 540n. 85

Elements de géométrie (Legendre), 33

Elements of Euclid: C.'s attachment to, 72–73, 388, 428; C.'s defense of, 323–24, 405–7; C.'s education in, 24, 270, 383, 406; in the Dock, 309–11; and non-Euclidean geometry, 384–85; Potts's edition of, 74; Simson's edition of, 406

Elements of Geometry & Plane Trigonometry (Leslie), 61

eliminant. *See* resultant

Elliott, Edwin B., 450, 557n. 91

elliptic functions, 473, 491n. 22; addition theorem in, 337; C.'s controversy with Bronwin on, 63–64; Cayley's theorem in, 471, 520n. 85; doubly infinite products and, 70, 97, 117, 189, 503n. 37; multiplication of, 128–29; and Poncelet's polygon problem, 510n. 3, 536n. 28; transformation and modular equations of, 336, 474, 512n. 32, 527n. 42, 539n. 82

elliptic integrals, 153, 304, 336, 473

Ellis, Alexander J., 186–87, 450, 511n. 17

Ellis, Robert Leslie, 450; career of, 53, 120; C. follows, as examiner, 151; on C.'s career plans, 118; on C.'s character, xvi; C.'s friendship with, 70; on Chasles, 108; in coach with C., 35; in comparison to C., 434; as editor of the *Cambridge Mathematical Journal*, 64; on Sylvester, 503n. 47

Ellis, William, 450

Emerson, R. W., 312

Encyclopaedia Britannica, C.'s contributions to, 404, 544n. 83, 552n. 75

Encyclopaedia Metropolitana, 25

Engels, F., 325

Equations Numériques (Lagrange), 314

Erdös, Paul, 483n. 4

Euclid. See *Elements* of Euclid

Euler, L., 61, 207–8, 494n. 1

Euler's theorem on homogeneous functions, 508n. 68

Eureka!, 204

Everett, William, 270, 450

Ewing, James A., 450, 543n. 51

Exercices de mathématiques (Cauchy), 34

extraordinary mathematics, 385–86

Faà di Bruno, Francesco, 191, 450, 512n. 32

Faraday, Michael, 69, 115, 217, 450

Fawcett, Henry, 269, 360, 450

Fawcett, Millicent G., 360, 450

Fawcett, Phillipa, 422, 450

Fenn, Joseph F., 35–36, 450, 487n. 21, 488n. 24

Fenwick, John, 49, 450

Fermat's Last Theorem, 286, 383, 547n. 52

Ferrers, Noman Macleod, 450, 506n. 9; editor of *Quarterly Journal for Pure and Applied Mathematics,* 255; and higher education of women, 551n. 73; and Sadleirian chair, 255, 263; as student, 160; as teacher at Cambridge, 316, 360

Féval, Paul, 418, 450

figure of the Earth, 47, 53, 303–4

Fischer, Frederick W. L., 69, 450

Fiske, Thomas S., 408, 450, 552n. 97

"five points in space." *See* space, "five points in"

Forbes, James D., 219, 450

Forsyth, Andrew Russell, *21g;* 358–59, 450; brilliant mathematician, 400–401, 525n. 1; on C., 262, 358–59, 399; on Euclid, 552nn. 91, 94; on invariant theory, 425; succeeds C., 440; on Tait, 298

Foster, Michael, 430, 451

four color problem, 351–54, 521n. 8, 542nn. 38, 39, 42, 543n. 50

Fourier, Joseph, 34, 527n. 35

Frankland, Edward, 182, 282, 325–26, 451

Franklin, Fabian, 346, 363, 451

Frobenius, Georg F., 451, 539n. 70

Frost, Andrew H., 49, 451

Frost, Percival, 256, 263, 451, 538n. 65; and Sadleirian chair, 256, 263

Fuller, Frederick, 49–50, 83, 118–19, 309, 451

Fundamenta Nova Theoriae Functionum Ellipticarum (Jacobi), 63, 132, 337, 501n. 67, 539n. 82

Fundamental Postulate, 473, 540n. 3; assumption of the, 342, 344–45; found to be false, 373–75

Gall, August von, 374–75, 392, 451

Galois, Évariste, 53, 185–86, 190, 350, 451

Galois's theory, 350, 511n. 10, 518n. 66, 544n. 83

Galton, Francis, 17, 34, 58–60, 157–58, 351, 451

Garden House. *See* Cayley, Arthur, in Garden House

Garnett, William, 451, 538n. 65
Gaskin, Thomas, 43–44, 48, 451; and
    Sadleirian chair, 256
Gauss, Carl F., 66, 72, 86, 451, 535n. 5, 550n. 40
Gell, Frederick, 34, 451
Genera Plantarum (Bentham, Hooker),
    194–95, 198
generating function, 340, 473
geometric models. *See* geometry, geometric
    models in
Géométrie de position (Carnot), 48
Géométrie descriptive (Monge), 33
geometry: abstract, 311, 535n. 85; analytical,
    45–46, 54, 65–66, 264, 468; Cayley-
    Bacharach theorem in, 65–66, 469, 526n. 13;
    Cayley lines in, 145, 148, 470; Cayley-
    Plücker coordinates in, 253, 470, 524n. 79;
    Cayley-Plücker equations in, 108, 470, 499n.
    38; Cayley's absolute in, 232, 235–36, 470,
    554n. 124; C.'s tactile approach to, 293–94;
    Cayley-Zeuthen equations in, 287, 470; con-
    ics in, 286, 472; Cramer's paradox in, 267;
    enumeration of curves in, 286–87; homoge-
    neous coordinates used in, 70, 414, 473;
    Kirkman points in, 145; Malfatti's problem
    in, 179, 161; *n*-dimensional, 82–83, 265, 293,
    311, 384, 402, 405, 494n. 2, 535n. 85;
    Noether's theorem in, 329–30; non-Euclidean,
    278–79, 384–85, 414, 474, 536n. 32, 554n.
    124; pole and polar (reciprocal polars) in,
    127, 147, 266, 268, 297, 475; of position, 54,
    144–45, 473; Poncelet's polygon problem in,
    184–85, 252, 322, 515n. 83; principle of con-
    tinuity in, 383; projective, 265, 384; stereo-
    graphic representations in, 531n. 17, 532n.
    19; symmetry of expression in, 69, 493n. 56.
    See also Modelling Club, Cambridge
Germany, Cayley in, 116–17, 362, 518n. 62,
    544n. 79
Gibbs, George Henry, 13
Gilbert, W. S., 332
Gilman, Daniel Coit, 339, 366, 390, 451
Girton College, Cambridge, 231, 361, 402
Gladstone, W. E., 291, 338, 451
Glaisher, James (elder), 211, 281–82, 451,
    535n. 8
Glaisher, James W. L. (younger), *17g*, 451; on
    C., xvii, 283–84, 261, 438; as colleague of C.,
    211, 271, 336, 421, 430, 536n. 17; compared

with C., 315; on Euclid, 552n. 86; and math-
    ematical tables, 537n. 38; proposes C. for
    medal, 548n. 11; as protégé of C., 315; and
    rare porcelain collection, 315
Glover, John, 451, 501n. 70
Gordan, Paul A., *15g*, 451; on invariant theory,
    298–302, 338, 342, 425, 533n. 41, 553n. 117;
    and London Mathematical Society, 542n.
    32; and Royal Society, 530n. 82
Gordan's theorem, 299, 345, 391, 394, 397,
    412–13, 415
Göttingen, mathematical school at, 402–3,
    426
Goulburn, Henry, 451
Grace, John H., 452, 550n. 46
gradual accumulation of knowledge, process
    of, 195, 438
Grant, Robert, 217, 221, 238–39, 452
graph theory, 473; Cayley's theorem in, 411, 471,
    553n. 106. *See also* Cayley trees, origins of
Grassmann, Hermann G., 452, 494n. 6
Gravatt, William, 217, 452
Graves, Charles, 115, 452
Graves, John T., 102–4, 115, 180, 452
Gray, Benjamin, 150, 452
Gray, David, 218
Gray, John E., 180, 194, 452
Great Exhibition (1851), 171
Great Northern Railway, 271, 341
Great Reform Act, 5, 158
Great Stink, 230
Great Western Railway, 13–14
Green, George, 82, 96–97, 105–7, 136, 452,
    535n. 5
Greenhill, Alfred G., 316, 452
Greenwich, Board of Visitors at, 320, 341
Gregory, Duncan F., 68–70, 120, 452; on alge-
    bra and analysis, 69, 115, 187, 511n. 17,
    539n. 74; and Boole, 85; and C.'s examina-
    tion for mathematical Tripos, 43–44; on
    geometry, 142
Gregory, Olinthus G., 135, 452
Griffin, William N., 136, 452
Grimley, H. N., 526n. 14
group theory, 473; abstract, 349, 511n. 17,
    520n. 81, 542n. 30; attention shown to, by
    C., 392; and calculus of operations, 510n. 8;
    Cauchy's theorem in, 248, 471; C. begins in,
    185–88; Cayley's colorgroup in, 471,

group theory (*continued*)
553n. 104; Cayley's theorem in, 349, 471; Cayley table in, 186, 510n. 8, 511n. 16, *11g*; diagramatic representation of groups in, *19g*, 350, 542n. 34; generalization in, 247–48; within invariant theory, 348, 557n. 94; and Kirkman, 247–49, 523n. 50; and partitions, 542n. 29; and quaternions, 523n. 51; subsequent connections with, 505n. 88; the term "group," 473

Grove, Sir William R., 179, 452

Gudermann, Christolph, 452

Gudermannian, 473, 540n. 83

Guy Mannering (Scott), 429

Hall, Thomas G., 452; as teacher at King's College London, 22, 35, 486nn. 60, 63; textbooks by, 23

Halphen, Georges H., 337, 437, 452, 556n. 74

Halsted, George B., 339, 439, 452

Hamilton, Henry Parr, 46, 452

Hamilton, Sir William Rowan, 452; at British Association for the Advancement of Science, 114, 132; on C., 145–46, 229; and Cayley numbers, 102–4; and matrix algebra, 227–29, 518n. 62; on "Pure Time," 279, 382, 547nn. 47, 49; and George Salmon, 192. *See also* quaternions

Hamilton numbers, 401

Hammond, James, 373–75, 394, 398, 453, 546nn. 13, 17, 549n. 21

Hansen, Peter A., 212, 453

Hardy, G. H., 439, 453, 540n. 85, 551n. 68

Hargreave, Charles J., 161, 180, 453

Harley, Robert, 199, 250, 282, 453, 524n. 67

Hart, Andrew S., 148, 453

Hawkshaw, John, 453, 521n. 10

Hayward, Robert B., 309–10, 453

Heath, John M., 33, 453

Heath, Thomas L., 401, 453

Helmholtz, Hermann L. F. von, 276, 453

Hemery, James, 453

hemihedron, 350, 473

Hemming, George W., 150, 453, 489n. 69, 506n. 13

Henrici, Olaus M. F. E., 245, 384, 453, 536n. 17

Henry, Joseph, 217

Hensley, Lewis, 193, 453, 489n. 69

Hermite, Charles, *13g*, 453; on C., 435; C. compared with, 439; compared, with Sylvester, on the binary quintic form, 294; and law of reciprocity, 189, 202; and London Mathematical Society, 318; and matrix algebra, 227; and potential for invariant theory, 190; and problem solving, 335; proves *e* is transcendental, 323; and skew invariant, 189–90, 205–6, 301, 512n. 26

Herschel, John F. W., 37, 114–15, 178, 233, 453, 493n. 73

Hertslet, Lewis, 224

Hesse, (Ludwig) Otto, 453; and analytical geometry, 494n. 1; and invariant theory, 91, 108, 178, 189, 524n. 72; and London Mathematical Society, 318

Hessian, 197–98, 473

Hexagon theorem, Pascal. *See* Pascal Hexagon theorem

Hicks, John W., 453, 538n. 65,

Hicks, William M., 453, 538n. 65,

higher education of women: C.'s course on "Algebra I" in, 544n. 73; C.'s course on "Principles of Arithmetic" in, 422; "The Three Graces," 362, 544nn. 77, 78. *See also* Newnham College, Cambridge

Hilbert, David, 454; and axiomatic method, 439; and invariant theory, xvi, 412–13, 425–26, 553n. 115, 554n. 121

Hill, Edwin, 454, 538n. 65

Hill, George William, 363, 454

Hill, Micaiah J. M., 454, 543n. 60

Hirsch, Meier, 214, 454

Hirst, Thomas Archer, *15g*, 454; on C., 166, 244–46, 281–82; in company of C., 242–43, 272, 321; corresponds with C., 285, 290; and Euclid, 309–10, 405, 536n. 35; and London Mathematical Society, 529n. 66; mathematical pedigree of, 243–44; proposes C. for medal, 371, 548n. 11; records views on C., 197; on Sylvester, 241; and X Club, 282

Hobson, Ernest W., 399, 407, 454

Hofmann, August Wilhelm von, 182, 454

homogeneous coordinates. *See* geometry, homogeneous coordinates used in

Hooker, Joseph D., 194, 282, 321, 454

Hopkins, William, *4g*, 39–42, 454; advocates medal for C., 182; and examinations, 51; guides and sponsors C., 57, 180; and

Marischal appointment, 219; and Sadleirian lectureships, 254
Hudson, R. W. H. T., 454, 538n. 60
Hudson, William H. H., 331, 454, 538n. 60
humor, C.'s sense of. *See* Cayley, Arthur, humor of
Hurwitz, Adolf, 362, 454, 504n. 71, 544n. 80
Hutchinson, Charles E., 26, 454
Hutton, Charles, 60
Huxley, Thomas H., 50, 193, 282, 454
hydrocarbons, enumeration of, 325–28
hydrostatics, 45
Hymers, John, 46, 454
hyperdeterminant operator, 109, 176, 474, 500n. 44, 508n. 73
hyperdeterminants, 67, 91, 126, 171–80, 473, 496n. 38. *See also* hyperdeterminant operator
hyperelliptic functions. *See* Abelian functions
hyperelliptic integral, 474, 531n. 14

imaginary numbers, "English interpretation" of, 141–42, 383, 504n. 63
inaugural lecture, C.'s, 263–68
Indian Civil Service, 20, 209–10, 515n. 89
Inns of Court, 121–22, 170
Institute of Actuaries, 503n. 54
integrals, multiple, 42, 130–31. *See also* potential theory
International Association for Promoting the Study of Quaternions, 400
International Congress of Mathematicians, 432
invariants, 474, 508n. 67; asyzygetic, 204, 469; catalecticants as, 198, 469; independence of, 97, 111; joint, 212, 515n. 100; semi-, 344, 392–94, 397–98, 412–13, 424, 476, 514n. 71; weight of, 176. *See also* covariants
invariant theory, 188–93; Boole as instigator, 112; Cayley's Formula in, 204, 471, 514n. 72; Cayley's Law in, 204, 344, 471, 514n. 72; Cayley's Ω-process for, 109–10; 176, 299–300, 339–40, 471; classification in, 193–96, 438; combinatorial problems in, 92, 300; concomitants in, 198, 472; C.'s first papers in, 109; C. as initiator of, 554n. 121; C.'s strategy in, 112; decline in C.'s approach to, 440; equivalent binary forms in, 540n. 87; *genera and species* in, 196; generating functions

and, 342–43; and group theory, 348, 557n. 94; Hermite's law of reciprocity in, 473; importance of, 425; invariance in, 86–87, 495n. 20; leadership in, shared with Germany, 392; likened to physics, 347; and links with science, 87; multilinear (tantipartite), 67, 90–91, 476, 496nn. 36–43; the name, 113; new foundation for, *11g*, 175–77; nomenclature in, 174–75, 196–98; notation in, 251, 513n. 48, 537n. 47; objectives for, 110–11, 252, 425, 541n. 12; origins of, 86; prehistory of, 87, 509n. 84; "primordial germs" in, 392–94; "problem of the syzygies" in, 111, 127, 190, 294, 301, 375; reciprocants in, 396–97; role of calculation in, 91, 99–100, 127, 176, 198–201, 288, 300, 340, 346–47, 416, 425, 435, 508n. 73; symbolic method in, 339, 342, 345, 392, 398, 413; and symmetric functions, 415, 476, 512n. 32; transvection operation in, 299, 476, 533n. 40, 540n. 4; as universal method, 126. *See also* invariants
Invariant Trinity, 177–79, 247, 423–24, 509n. 80
Investigation of the Laws of Thought (Boole), 192–93, 335, 386, 512n. 38
Ireland, Cayley in, 145–50
Isaac Newton Studentships, 421
isomerism, 326–27
Italy, Cayley in, 76–77
Ivory, James, 118, 454, 534n. 52

Jacobi, Carl G. J., *8g*, 454; auxiliary equation of, 250; C. reads, 131–32, 418; compared with C., xvi; on determinants, 66–68; and potential theory, 535n. 5
Jacobian, 68, 197–98, 474, 496n. 36
Jacobi-Cayley resolvent. *See* polynomial equations
Jahrbuch über die Fortschritte der Mathematik, 319, 536n. 17
Jarvis, Charles G., 200
Jebb, Richard C., 405, 454
Jellett, John H., 146, 424, 454
Jerrard, George B., 250–51, 454
Joachimsthal, Ferdinand, 116–17, 454
Johns Hopkins University, 363, 390–91
Johnson, A. M. J. E., 423, 555n. 16
Jones, Richard, 21, 454

Jonquières, Jean-Philippe F. de, 286, 455; on geometry, 295–96, 530n. 78, 531n. 86

Jordan, Camille, 338, 346, 350, 455, 540n. 88, 549n. 30

Journal de Mathématiques Pures et Appliquées (*Liouville's Journal*), 84, 118

Journal für die reine und angewandte Mathematik (*Crelle's Journal*), xvi, 84, 118, 200; C. submits block of seven papers to, 202–3

Kay, Edward E., 455, 491n. 8

Kay, Joseph, 455, 491n. 8

Kelland, Philip, 309–10, 455

Kelvin, Lord. *See* Thomson, William

Kempe, Alfred B., 353–54, 411, 455, 553n. 104

Kepler, Johannes, 237

Kepler-Poinsot polyhedra, 237, 521n. 8

King, Charles W., 455, 491n. 20

King, Joshua, 32, 455

King's College London: education of C. at, 17–26, 487n. 6, *1g*; prizegiving at, *3g*

Kingsley, Charles, 455; *Alton Locke* by, 19, 39; on Cambridge mathematics, 31, 39; and Plato at King's College, London, 21–22

Kirchoff, Gustav R., 276, 455

Kirkman, Thomas P., 455; on abstract geometry, 535n. 85; on combinatorial problems, 142–44, 504n. 68; on geometry of position, 144–45; on group theory, 247–48; isolated at Croft, 143, 273–74; on pluquaternions, 143; on polygons, 216, 220–21; on polyhedra, 249; and "schoolgirls problem," 144, 170

Klein, Felix, 455; admires C.'s work, 187, 223; at British Association for the Advancement of Science, 323; C. communicates with, 67, 412–15, 422, 429; C. visits, 362, 544n. 79; considers chair at Oxford, 375–76; and invariant theory, 540n. 87; suggests C. for Johns Hopkins, 391

Kronecker, Leopold, 289, 439, 455, 539n. 70

Kummer, Ernst E., 289, 330, 383, 455, 538n. 59

Lacroix, Sylvestre F., 23, 34, 455

Ladd, Christina, 340, 363, 455

Lady's and Gentleman's Diary, 199, 335

Lagrange, Joseph-Louis, 455; C. reads, 33–34, 42, 55, 61, 66–67; and invariant theory, 86; and mechanics, 372; and solution of equations, 250; and three-body problem, 306

Lagrange's theorem. *See* calculus of operations, Lagrange's theorem in

Laguerre, Edmond N., 338, 455

Lamb, Horace, 316, 455

Lambert, Carlton J., 455, 526n. 14, 538n. 65

Lamé, Gabriel, 455, 550n. 40

Lanczos, Cornelius, 149

Langley, E. M., 405, 456

Laplace, Pierre Simon, 34, 61, 66, 124, 456

Laplace's coefficients, 47, 53, 304

Latham, Henry, 456

Lathe and Its Uses, The, 419

latin squares, 411, 474

Lattimer, Thomas, 456, 538n. 65

law, Cayley's contribution to English, 163–65

Lebesque, Victor A., 66, 456

Leçons sur le calcul des functions (Lagrange), 34

Lectures Association, 360–61

Lectures on Quaternions (Hamilton), 185, 504n. 77, 547n. 50

Lefschetz, Solomon, 483n. 11

legal profession: C. as barrister in, 157–62; C. called to Bar, 150; C. enters, 120–24; C. leaves, 257; C. in training for, 124–26; C.'s frustration with, 164. *See also* law, Cayley's contribution to English

Legendre, A. M., 33

Lejeune-Dirichlet, Johann P. G. *See* Dirichlet, J. P. G. Lejeune-

Leslie, John, 61

Lessons Introductory to Modern Higher Algebra (Salmon), 247, 288, 409, 530n. 85, 533n. 45

Leverrier, Urbain J. J., 132, 456

Levett, Rawdon, 405, 456

Lewis, William, 419, 456

Lie, (Marius) Sophus, 397, 456, 514n. 72, 542n. 32,

Lincoln's Inn, Stone Buildings in, 123, 157–59, 505n. 1, *7g*. *See also* legal profession

Lindley, John, 182, 456

Linear Associative Algebra (Peirce), 365, 386, 545n. 96

linear transformations, 98, 109, 129, 131, 175–76, 201

linkages, plane. *See* link-work

link-work, 474–75, 543n. 51; *19g*; three-bar motion, 354, 476

Liouville, Joseph, 119, 126, 186, 197–98, 242, 456

Liouville's Journal. See *Journal de Mathématiques Pures et Appliquées*
Listing, Johann B., 456, 536n. 34
Littlewood, J. E., 440
Liveing, George D., 210, 456
Liverpool Daily Post, 388
Lobachevsky, Nikolai Ivanovich, 278, 384, 456, 528n. 50
Locke, John, 58, 61, 456, 490n. 3
Lockhart, John Gibson, 122, 456, 501n. 5
"Locksley Hall" (Tennyson), 381, 441, 557n. 99
Lockyer, (Joseph) Norman, 319–20, 456
Lodge, Oliver J., 322–23, 456
logic, 335, 539n. 75
London Assurance Corporation, 13, 14
London Mathematical Society: C. at meetings of, 318, 350–51, 390; C. awarded De Morgan medal of, 391–92, 548n. 11; C. introduced as member to, 283; C. as president of, 300–301, 319; C. reads a paper to, 285; as Council member of, 271; history of, 529n. 64, 535n. 16
Longmaid, William H., 380, 547n. 37
Lowdean professorship, 32, 233
Lubbock, John W., 240, 282, 456
Lucas, François É. A., 427, 456
lunar theory, 215, 230, 233, 239–40
Lyell, Charles, 20, 33, 281, 456

Macaulay, Alexander, 400, 456, 551n. 52
Macaulay, Francis S., 399–400, 457
MacCullagh, James, 118, 147, 457
MacFarlane, Alexander, 390, 457, 557n. 87
Mackie, John H., 457, 538n. 65
Maclaren, Ian, 429, 457, 556n. 45
Maclaurin, Colin, 84, 526n. 7
Maclaurin's theorem (geometry), 267
MacMahon, Percy A., *20g,* 457; on C., 176, 424; collaborates with C., 396–98; "correspondence theorem" due to, 392–94, 549nn. 16, 20; on Fundamental Postulate, 375; on Gordan's theorem, 415; on importance of invariant theory, 425; and Royal medal, 550n. 38, 554n. 8
Maddison, Isabel, 420, 457, 554n. 8
Maine, Henry J. S., 404, 457
Malfatti, Gian Francesco, 72, 179, 457
Manners, George John, 30, 457
Mannheim, Victor M. A., 457, 542n. 32, 549n. 30

Marsh, Henry A., 457, 491n. 20
Marshall, Alfred, 360, 457, 526n. 14
Maskelyne, Nevil, 59–60
Mate, R. P., 150, 491n. 20,
Mathematical Analysis of Logic (Boole), 141, 187, 335
Mathematical Questions with their Solutions from the Educational Times, 268, 273, 314, 335, 513n. 63
Mathematical Tables, British Association for the Advancement of Science report on, 324–25, 537n. 38
mathematical Tripos, *5g,* 34–35; C. and Airy debate topics for, 302–8; order of merit (1842) in, 48–49; reforms in, 53–54, 349, 355, 398, 543n. 53; Senate House examination for, 43–48
Mathematician, The, 181
mathematics, teaching of, in public schools, 23–24
Mathematische Annalen, 299
Mathews-Cayley-Hamilton table. *See* Hamilton numbers
Mathews, G. B., 333, 401, 436, 457
matrix algebra, 226–30, 474; attention shown to, by C., 392; Cayley-Hamilton theorem in, *13g,* 228–30, 334, 470; Cayley Hermite problem in, 230, 470, 501n. 69, 520n. 81; C. on, 226–30; C. returns to, 333–34, 520n. 81, 539nn. 69, 71; Cayley's transform in, 471, 501n. 69; and determinants, 227; equations in, 395; and group theory, 227, 392; history of, 520n. 83; multiple algebra and, 365, 386, 395; notation and basics of C.'s, 203, 514n. 68, 518n. 69, 519nn. 71, 72, 74; and quaternions, 412, 519nn. 73, 78; and Sylvester, 171, 518n. 68. *See also* linear transformations
Maurice F. D., 360, 457
Maxwell, James Clerk, 457; applies to Marischal College, with C., 219–20; at Cavendish Laboratory, 309; and electricity, 276; evaluates C.'s reports on dynamics, 222; and geometrical models, 293, 331, 538n. 65; on hills and dales, 520n. 90; on vision and the eye, 211; writes poem for C., 313
Mayor, Robert B., 37, 42, 49, 457
McClintock, Emory, 424, 457, 555n. 21
mechanics, C.'s view of energy and force in, 372

Méchanique analytique (Lagrange), 33, 57
mechanisms, 237–38, 332
memoir on quantics, 252, 425; Eighth, 294, 299; Fourth and Fifth, 230, 520n. 83; Introductory, 201–4; Ninth, 302, 337–38, 343; Second, 204–5, 299, 301, 344; Seventh, 251, 275, 294; Sixth, 232, 235–37, 266, 268, 278, 323, 364, 414; Tenth, 351; Third, 220
Messenger of Mathematics, 315, 335
Mill, John Stuart, 457, 529n. 69; philosophy of, 381–84; *System of Logic* by, 382, 547n. 45
Miller, Robert K., 290, 457
Miller, W. H., 119, 217, 292, 321, 457
Miller, William J. C., 458, 539n. 81
Milner, Isaac, *19g*, 410, 553n. 103
Mitchell, O. H., 363, 458
"mixed mathematics," undergraduate, 42–47, 52
Moberly, George, 10–11, 458, 485n. 22
Möbius, August F., 277; and his strip, 323, 458
Modelling Club, Cambridge, 331–32, 538n. 65
modular equations. *See* elliptic functions, transformation and modular equations of
Moline, Mary (mother in law). *See* Pritchard, Mary
Moline, Robert (father in law), 261–62, 458
Moline, Susan (Mrs. Susan Cayley), 261–63, 296, 387, 431, 458
Moline family, 211, 526n. 5
Monge, Gaspard, 33, 61, 458
Montépin, Xavier de, 418, 458
Monthly Notices. See Royal Astronomical Society, C. editor of *Monthly Notices*
Moon, Robert, 150, 458
Moor, Allen P., 458, 501n. 70
Moore, Eliakim H., 401, 458
Morley, Frank, 458
Morton, R., 526n. 14
Moseley, Henry, 21, 458
Mould, J. G., 256, 458
mountaineering, 76, 133, 231–32, 272–73, 330, 520n. 89
Muir, Thomas, 54, 458, 490n. 75
multiple algebra. *See* matrix algebra
Munro, Hugh A. J., 458, 487n. 21
Murphy, Robert, 89, 458, 493n. 43, 496n. 32, 539n. 74
Murray, J. A. H., 168

n-dimensional geometry. *See* geometry, *n*-dimensional
n-squares, problem of, 103–4, 144, 504n. 71
Neale, John Mason, 31, 459, 499n. 31
Neumann, Carl G., 299, 459
Neville, Eric H., 39, 459
Newcomb, Simon, 320–21, 363, 365, 439, 459
Newnham College, Cambridge, *21g*, 361, 403, 410, 423, 426, 544n. 63, 552n. 102
Newton, H. A., 292
Newton, Isaac: classification of cubic curves by, 277, 317; the *Principia*, of, 22, 45, 270
Newton-Fourier Problem, 352, 542n. 43
New York Mathematical Society, 555n. 21
Nichol, John Pringle, 238, 459
Niven, Charles, 526n. 14
Niven, W. D., 526n. 14
Noether, Max, 289, 435–36, 459, 556n. 74
Noether's theorem, 329–30, 538n. 50
non-Euclidean geometry, *See* geometry, non-Euclidean
Norris, John P., 107, 459
Norway, Cayley in, 116
notation, mathematical: abridged, 55; British Association for the Advancement of Science report on, 538n. 48; C.'s, for determinants, 54–55; C.'s, for groups, 523n. 54; C.'s attitude toward subscripts, 328–29, 537n. 47; in geometry, 148
notion and boundaries of algebra. *See* algebra, notion and boundaries of
Nouveaux Exercices d'Analyse et de Physique Mathématique (Cauchy), 247–49
number, ordinal and cardinal, 382, 547n. 50
numerical solutions, C.'s attitude toward, 200

O'Brien, Matthew, 20, 66, 206, 210, 459, 515n. 92
octave. *See* Cayley number
octonion. *See* Cayley number
Oliver, James E., 409, 459
optics: as application of group theory, 187; C. tempted to study, 131; as study for students, 41, 45, 52
order of merit (1842). *See* mathematical Tripos, order of merit in
Ostrogradsky, Mikhail, 222, 459
Otter, William, 25, 26, 459
Owen, Richard, 57, 180

Oxford and Cambridge Act (1877), 349, 390
Oxford University, 376–77

Palmerston, Lord, 270
Park, John James, 459, 506n. 8
Parkinson, Stephen, 106, 459, 489n. 69
Parnell, Hugh, 49, 459
partitions, 207–9, 474; Cayley-Betti-Ferrers
   Law on conjugate, 516n. 5; and invariant
   theory, 214–15; and Poncelet's polygon
   problem, 515n. 83; "Problem of Virgins" on,
   208–9, 515n. 86; Sylvester's lectures on,
   240–41, 522nn. 25, 30
Pascal Hexagon theorem, 65, 67, 144–48, 178,
   266–68, 294
Pascal's Hexagram. *See* Pascal Hexagon theorem
Patmore, Coventry, 167, 459, 501n. 5
Peacock, George, *4g*, 459; at British Association
   for the Advancement of Science (1845),
   114–15; at Cambridge Philosophical Society,
   66; on C. and Royal Society, 180–81; lectures
   given by, 31–32; links with King's College,
   London, 27; report on analysis to British
   Association for the Advancement of Science
   by, 89; as Smith's Prize examiner, 51–53,
   543n. 60; on symbolical algebra, 69, 71–72,
   386, 511n. 17; as teacher, 493n. 73; and *Treatise on Algebra* (1830), 31; and *Treatise on
   Algebra* (1842, 1845), 32; as tutor, 27–28,
   406
Peano, G., 512n. 32
Pearce, R., 526n. 14
Pearson, Karl, 357, 399, 404, 459
Peaucellier, C. N., 332
Peirce, Benjamin, 339, 386, 459
Peirce, Charles S., 459; on C., 364–65, 432, 439;
   on C.'s absolute, 236; "division algebra theorem" by, 545n. 96; generic reasoning criticized by, 492n. 36; on invariant theory, 87,
   299; on logic, 386; obituary for C. by, 483n. 1
Pell, Albert, 47–48, 460
Pendlebury, Richard, 316, 358–59, 460, 544n.
   66
Perigal, Henry, 460, 521n. 11
periodic variation of planet's orbit, 52. *See also*
   secular acceleration of moon's motion
"permanence amidst change," 87
permutant, 173, 474

perpetuant, 474, 550n. 38
Petersen, Julius, 460, 554n. 131
Pfaff, Johann Friedrich, 131, 460
Pfaffian form, 131, 475
Philosophical Club (Royal Society), 243
Philosophical Transactions of the Royal Society,
   201
philosophy of mathematics, 381–83
Philpott, Henry, 460
Picard, Charles E., 460, 549n. 30
Pieri, Mario, 460, 554n. 125
pippian. *See* curves, Cayleyan
Pirie, G., 526n. 14
Plana, Giovanni, 240, 522n. 21
Plato, 21–22, 58–61, 62, 87, 272, 383, 417, 429
Playfair, John, 60, 142, 384–85, 460
Plücker, Julius, 460; and analytical geometry,
   65, 494n. 1; C. advocates, for medal, 217,
   284–85, 530n. 70; migrates to physics,
   152–53; provides benchmark for C., 81–82;
   visits England with geometrical models,
   293, 499n. 38, 531n. 17
Plücker's equations, 108, 331, 475, 499n. 38
Poincaré, Henri, 427, 460
Poinsot, Louis, 237
Poisson, S. D., 97, 222, 460
polyacra. *See* polyhedra
polygon problem, Poncelet's. *See* geometry,
   Poncelet's polygon problem in
polyhedra, 215, 248–49, 521n. 8; Euler's formula for, 237, 488n. 47, 536n. 34; polyacra
   as type of, 419–20, 475
polynomial equations: Hamilton numbers,
   401, 474; Jacobi-Cayley resolvent, 474, 524n.
   68; solution of, 53, 250; symmetric func-
   tions in, 199, 214, 215, 516n. 5
polyzomal curves. *See* curves, polyzomal
Pompeii Illustrated with Picturesque Views
   (Donaldson), 76
Poncelet, Jean V., 184, 266, 460, 530n. 70
Poncelet's polygon problem. *See* geometry,
   Poncelet's polygon problem in
Pontécoulant, Phillippe de, 239–40, 460
potential theory, 96, 105–6 , 314, 527n. 42
Potter, Richard, 44, 460
Potticary, George B. F., 15, 211, 460, 484
Potticary, John, 15, 460
Potts, Robert, 74, 437, 460
pre-Raphaelite circle, 166

Preston, Theodore, 460, 491n. 20
previous examination ("Little Go") at Cambridge, 35, 417
Price, Bartholomew, 179, 191, 460, 509n. 90, 512n. 33
Principles of Book-keeping by Double Entry (Cayley), 15, 428
Principles of Geology (Lyell), 33, 57
Pritchard, Mary (Mrs. Mary Moline), 261, 458
"problem of the syzygies." *See* invariant theory, "problem of the syzygies"
Proctor, R. A., 319–20, 388, 461
proof, mathematical: absence of, 345; conjecture and analogy in, 89, 296; C.'s blend of induction and deduction in, 99, 305–6; C.'s use of Peacock's principle in, 493n. 65; of existence and uniqueness, 334; of general results, 492n. 36, 497n. 63; and intuition, 228–29; by verification, 334; vs. algorithms, 110. *See also* truths, search for
Purkiss, H. J., 526n. 14

quantics, 201, 475; quaternary cubic, 495n. 14; ternary-quadratic, 476. *See also* memoir on quantics
Quarterly Journal for Pure and Applied Mathematics, 200, 242, 255, 263, 315
quaternions, 475; C's attitude to, 96, 178; C's first papers on, 95; C.'s lectures on, 408; in Chamber of Horrors, 412; vs. coordinates, 412, 553nn. 113, 114; Hamilton and, 81–82, 115, 92–96, 145–49, 178, 187, 510n. 8; and rotations in space, 95, 142, 187, 297–98, 372, 497n. 60; and W. Thomson, 334. See also *Lectures on Quaternions* (Hamilton)
quartic surfaces. *See* surfaces, quartic
quintic form, binary. *See* binary quintic form
quintic equation, solution of, 189, 424, 512n. 29

Rabelais, F., 34
Ramanujan, Srinivasa A., 461, 540n. 85
Rankine, William J. M., 461, 515n. 100
Ranyard, A. C., 319–20, 461, 536n. 20
rational and integral function. *See* algebraic form
Rayleigh, Lord. *See* Strutt, J. W.
reciprocal polars. *See* geometry, pole and polar (reciprocal polars) in

reciprocant, 474, 549n. 33
Regnault, H. V., 530n. 70
regulus. *See* surfaces, ruled (regulus)
religion: C. on, 419; the Scientist's Declaration on, 282
Remsen, Ira, 365–66, 461
resultant, 189, 475
Reynolds, O., 526n. 14
Ricci-Curbastro, Gregorio, 461, 550n. 40
Richmond, Herbert W., 461
Rickett, Mary E., 461
Riemann, G. B. F., 285, 287, 384, 461, 530n. 82
Riemann surfaces, 332
Roberts, Michael, 146, 192, 461
Roberts, Samuel, 354, 428, 461
Roberts, William, 118, 146, 461
Robertson Smith, W. *See* Smith, William R.
Rodgers, J. E. D., 224
Rodrigues, Olinde, 66, 95, 372, 461
Röhrs, J. H., 167, 506n. 32
Romanticism, 410, 418, 520n. 89
Romilly, Joseph, 136, 461, 487n. 20
Roscoe, Henry E., 325, 461
Rose, Hugh James, 25, 461, 487n. 6
Rossetti, Christina, 165, 168, 461, 535n. 2
Rossetti, Dante Gabriel, 165, 461, 535n. 3
Rossetti, Gabriele, 21, 167, 461
Rossetti, William M., 25, 157, 165–66, 245, 461
rotations. *See* quaternions, and rotations in space
Rothschild, Meyer A. de, 30, 461
Rothschild, Nathan Meyer, 13, 30, 462
Routh, Edward J., 462; as coach for Tripos, 40, 263, 316, 336, 399; as examiner, 254, 525n. 86; performance of, and Maxwell in Tripos, 52; and Sadleirian chair, 255, 263
Rowe, Richard C., 359, 462
Royal Astronomical Society: C. as editor of *Monthly Notices*, 239–40, 284, 364; C. introduced to, 221–22, 517n. 34, 521n. 11; C. as president of, 319–21, 517n. 34, 536n. 19; past presidents of, 549n. 15; position of, as a national society, 529n. 64, 535n. 16
Royal Commission on Technical and Scientific Instruction, 316
Royal medal. *See* Royal Society of London, C. proposed for Royal medal of
Royal Society Catalogue of Scientific Papers. See Royal Society of London, *Royal Society Catalogue of Scientific Papers* project of

Royal Society Club, 243

Royal Society of Edinburgh, 529n. 69

Royal Society of London, *11g*; C. awarded Copley medal of, 371; C. as member of Council, 245, 354, 529n. 69; C. proposed for Royal medal of, 182–83, 246, 523n. 46; C.'s introduction to, *10g*, 189; C. submits first paper to, 179; grants awarded by, 200; *Royal Society Catalogue of Scientific Papers* project of, 218, 291

Ruffini, Paolo, 72

ruled surfaces. *See* surfaces, ruled

Ruskin, John, 19, 76, 167, 462, 535n. 3

Russell, William H. L., 462

Russia. *See* St. Petersburg

Sabine, Sir Edward, 217, 246

Sadleirian chair of pure mathematics: applicants for, 255–57; C. as holder of, 376; duty attached to, 263–64, 308; establishment of, 255, financial provision for, 355–56, 525n. 2

Sadleirian lecturers, 254

Salisbury, Third Marquis of, 376–77, 404, 462, 546n. 27

Salmon, George, *9g*, 146–50, 287–89, 462; calculates invariants for C., 199, 288; on C., 235, 392, 422, 521n. 1; corresponds with C. on cubic surfaces, 148–49; and difficulty reading C., 245; on geometry, 298, 309; and W. R. Hamilton, 192, 551n. 58; meets C. in Ireland, 146–50; and partitions, 241; on terminology in invariant theory, 198

Saunders, Trelawny, 224

Scheutz, Georg, 217, 462

Scheutz difference engine, 217, 516n. 12

Schläfli, Ludwig, 462; C. extends work of, 298; classification of cubic surfaces by, 532n. 35; corresponds with C., 189, 285–86; mathematical papers translated by C., 220; on Noether's theorem, 329–30

Schorlemmer, Carl, 325–26, 462

Schröter, H. E., 462, 538n. 59

Schubert, Hermann C. H., 362, 377, 462, 530n. 79, 546n. 27, 548nn. 11, 63

Schwarz, K. H. A., 330, 462, 542n. 42

Schwarzian derivative, 548n. 61, 550n. 39

Scotland, Cayley in, 58–60, 133–34, 210, 322, 427

Scott, Charlotte A., 360–61, 462; on C., xvi, 440; compares Plücker with C., 293; on

Noether's theorem, 330, 538n. 50; as research supervisor, 554n. 8

Scott, George Gilbert, 462, 494n. 81

Scott, R. F., 462, 538n. 65

Scott, Sir Walter, 379, 429

scroll. *See* surfaces, skew

Secretary Problem. *See* Cayley-Moser "Secretary problem"

secular acceleration of moon's motion, 239–40, 475, 522n. 21

Sedgwick, Adam, 57, 410

Segar, Hugh W., 462, 551n. 65

semi-invariant. *See* invariants, semi-

Senate House examinations. *See* mathematical Tripos, Senate House examinations for

Senior Wrangler, *6g*, 38, 49, 270

"separation of symbols." *See* calculus of operations, "separation of symbols" in

Serret, Joseph A., 119, 186, 462, 509n. 96

set operators, Cayley's, 186, 248

Sévigné , Marquise de, 48, 462

sextic, binary. *See* binary sextic

Shaw, Benjamin, 35–36, 165, 462, 487n. 21, 488n. 24

Shaw-Lefevre, Sir John G., 180, 462

Sheppard, William F., 399, 463, 550n. 48

Sherbourne school, 357

Shortland, Peter F., 49, 50, 463

Sidgwick, Eleanor M., 403, 463

Sidgwick, Henry, 360, 404–5, 463, 551n. 73

Siemens, Charles, 379

Simpson, Charles T., 37, 42, 47–49, 53, 463, 489nn. 51, 57

skew curves. *See* curves, of double curvature

Smith, Archibald, 68, 120–21, 150, 463, 493n. 56, 523n. 46

Smith, Benjamin F., 36–37, 48, 463, 488n. 24

Smith, Dr. Robert, 463, 489n. 64

Smith, H. J. S., 463; on algebraic forms, 252; and British Association for the Advancement of Science, 322, 546n. 26; A. Buchheim taught by, 549n. 27; with C. at Oxford, 281; death of, 375, 546n. 30; on Euclid, 309–10, 406, 535n. 86; and Fermat's Last Theorem, 547n. 52; and London Mathematical Society, 283; and matrix algebra, 230; more practical than C., 353; as proponent of pure geometry, 245; and South Kensington exhibition, 538n. 65

Smith, James P., 388, 463, 548n. 70

Smith, William R., 404, 463

Smith's Prizes at Cambridge, 51–54, 308, 316–18, 489nn. 63, 64

Sohncke, Ludwig A., 463, 539n. 82

Sommerville, Duncan M. Y., 463

Southey, Robert, 20

space: C. on physical, 235, 278; "five points in," 54–56, 490n. 74; reality of, 292–93; "really real," 142, 265, 384, 410, 504n. 64

space curves. *See* curves, of double curvature

Specimens of Ancient Sculpture (Dilettanti Society), 76

Speke, J. Hanning, 281

Spencer, Herbert, 282, 463, 547n. 54

Spitalfields, Mathematical Society of, 240, 242

Spottiswoode, William, 206, 463; and British Association for the Advancement of Science, 132; C. "at home" with, 182; and determinants, 188; on invariant theory, 191–92, 252; and matrix algebra, 230, 520n. 82; supports C. for medal, 371

Sprague, T. B., 335

St. Petersburg: Cayley family background in, 5–9; map of, 2g. *See also* Cayley, Henry (father)

Staudt, Christian von, 414, 418, 463, 554n. 125

Steele, William J., 160, 463

Steiner, Jacob, 117, 179, 463, 494n. 80

Stephen, Leslie, 270–71, 303, 463, 547n. 45

Stephen, Sir James, 210, 463

Sterne, Laurence, 58, 418, 463, 490n. 3

Stevenson, Richard, 463, 495n. 19

Stevenson, Robert Louis, 60, 431, 463

Stokes, George Gabriel, 464; C. advises, 59–60, 211–12, 293; C. and, as examiners, 209–10; C. and, on notation, 329; with C. on *Royal Society Catalogue* project, 217; on death of C., 430; elector for Sadleirian chair, 255–56; and Glasgow appointment, 238; proposes C. for Royal medal, 523n. 46; and Sadleirian estates, 355; teases C. on "proof," 426

Stone Buildings. *See* Lincoln's Inn, Stone Buildings in

storehouse principle, Victorian, 324, 389

Story, William E., 363, 365, 366, 464

Strachey, Lytton, 439

Strange, Alexander, 319–20, 464

Street, George E., 76, 464, 494n. 81

Stroh, E., 549n. 16

Strutt, John W. (Lord Rayleigh), 269, 276, 381, 404, 464, 526nn. 14, 17

Struve, Otto, 250

Stuart, James, 274, 464; on Airy, 304; C. in laboratory of, 419; and chair at Cambridge, 333, 543n. 51; Liberal candidate, 373

Sturm, Jacques C. F., 275, 464; theorem named after, 45

Sturm's functions, 100, 511n. 10,

surfaces: amphigenous, 531n. 15; Cayley's cubic, *18g*, 471; Cayley's sextic torse, 531n. 14; cubic, 149, 220, 253, 293, 298, 472, 495n. 14, 527n. 42, 532n. 35; deficiency of, 330, 538n. 52; developable, 105, 472, 513n. 60, 531n. 14; Fresnel's wave, 511n. 10; orthogonal systems of, 334; quartic, 298, 318, 330–31, 400, 475, 525n. 82, 527n. 42, 533n. 38, 538nn. 53–56, 59; quintic, 318; ruled (regulus), 293, 475, 531n. 14; singularities of, 330–31; skew, (scroll), 254, 288, 310, 475, 531n. 14; Steiner's, 293, 331, 538n. 59; Tetrahedroid, 127, 331; torse, 476; triple tangent plane on, 476; twenty-seven lines in the cubic, 148–49, 220, 293, 313, 332, 505n. 88, 516n. 27, 532n. 35

Switzerland, Cayley in, 76, 371

Sylvester, James Joseph, *9g*, 464; at British Association for the Advancement of Science (1845), 114; called to Bar, 168; C.'s friendship appreciated by, 169–70, 291, 396, 426; creativity of, in invariant theory, 169, 188; depressions of, 205–6, 295–97; on Euclid, 309–11; internationalist viewpoint of, 395, 509n. 96; joins Johns Hopkins University, 339; as lawyer-mathematician, 150; life of, prior to meeting C., 134–39; the "Mathematical Adam," 197; on matrices and determinants, 518n. 68; meets C., 139–41; merges invariants and gorillas, 195; on natural philosophy of, 503n. 47; at Oxford University, 394; partnership with C., *12g*; and "priority" disputes with C., 172–74, 296-97; supportive of C., 225–26, 234, 254, 356, 377–78, 415; and tracing of curves, 293–94; at University College, London, 136; at University of Virginia, 136-37; at Woolwich Military Academy, 210–11, 215–16, 274–75, 338

symbolic notation (invariant theory), 299, 476, 533nn. 40, 41 symmetric functions. *See* polynomial equations, symmetric functions in

System der analytischen Geometrie (Plücker), 65

System of Logic (Mill), 279

syzygy, 46, 198, 476, 532n. 21

tables, mathematical, 324–25

tactic, 280, 476

tactions, 476

Tait, Peter Guthrie, *16g*, 464; C. advises, on matrix algebra, 333; C. at variance with, on Euclid, 405; C. examines, 160; and C. on mechanics, 371–72; C. on rotation formula formulated by, 297–98, 532n. 32; C. writes chapter for, 411–12, 419; C. writes to, on quaternions, 429–30; views of, on matrix algebra, 229

Talbot, W. H. Fox, 464. *See also* Talbot's curve

Talbot's curve, 214, 476

tantipartite. *See* invariant theory, multilinear

Taylor, Charles, 323, 464

Taylor, H. M., 526n. 14

Taylor, Tom, 464, 491n. 20

Temple, Frederick, 210, 464

Tennyson, Alfred, 167, 410, 464, 491n. 7, 529n. 69

Terquem, Olry, 464, 509n. 96

Tetrahedroid. *See* surfaces, Tetrahedroid

Thacker, Arthur, 464

Thackeray, William M., 75, 464

Théorie analytiques des probabilités (Laplace), 33

Theorie der Abelschen Functionen (Clebsch and Gordan), 287, 371, 533n. 51

Theorie der algebraischen Kurven (Plücker), 65

theory of numbers, 215, 252, 383, 424, 428

theta function, 337, 357, 476

Thomson, J. J., 358, 403–4, 464

Thomson, James (elder), 150, 464

Thomson, James (younger), 238, 464

Thomson, William (Lord Kelvin), *9g*, 464; attends C.'s lecture, 360; at British Association for the Advancement of Science (1845), 115; C. on *Lalla Rookh* with, 371; C. writes testimonial for, 128; on C. and Sylvester, 182; on C.'s attention to *pure* mathematics, 276; C.'s last communication with, 430, *24g*; on C.'s prospects in Scotland, 218, 238;

consults C. on invariant theory, 212–13, 515n. 100; gives memorial tribute for C., 431; on matrix algebra, 539n. 71; in Paris, 106-8; praises C.'s report on dynamics, 222; professor at Glasgow, 129–34; on reform of mathematical Tripos, 308; supports Kirkman for Royal Society, 221; undergraduate with C., 38, 49, 50, 68–69, 84

Thorp, Thomas, 28, 29, 107, 465

three-bar motion. *See* link-work, three-bar motion

three-body problem, 215, 306

Thurtell, Alexander, 44, 465

Todhunter, Isaac, 465, 525n. 95; and C. on elasticity, 276, 404; and Sadleirian chair, 256, 263

Tooke, Thomas, 13, 465

Tooke, William, 9, 465

topology, 237, 503n. 37, 520n. 90, 521n. 8. *See also* close

Townsend, Richard, 146, 309, 465

Traité de méchaniques céleste (Laplace), 34

Traité des fonctions elliptiques (Legendre), 53, 337

Traité des substitutions et des equations algébriques (Jordan), 350

Traité du calcul différentielle et du integral (Lacroix), 34

Transactions of the Cambridge Philosophical Society, 66, 71

transvection. *See* invariant theory, transvection operation

Traveller's Book, 76

Treatise on Conic Sections (Salmon), 149

Treatise on Differential Equations (Boole), 335

Treatise on Quaternions, An Elementary (Tait), 411, 419, 553n. 107

Treatise on the Analytic Geometry of Three Dimensions (Salmon), 288

Treatise on the Differential and Integral Calculus (Hall), 23

Treatise on Universal Algebra, A (Whitehead), 401

trees. *See* Cayley trees

Trinity College, Cambridge: academic competition in, 34–38; annual dividend at, 215, 500n. 61; C. as assistant tutor at, 57–63, 490n. 4; C. assists the framing of new statutes of, 349, 390; C. as bachelor-scholar

Trinity College, Cambridge (*continued*)
at, 57–58; C. as freshman at, 27–34; C. begins research career in, 74–75; C. enters, 27–30; classics and science at, 487n. 13; C. leaves, 117–18; C.'s performance in examinations of, 36, 488n. 24; examiner at, 132, 151–52, 505n. 94; fellowship dissertations at, 344, 400, 421, 491n. 16, 541n. 10; fellowship elections at, 491nn. 19, 20, 501n. 72; fellowship examinations at, 60–63, 491n. 16; mementos of C. at, xiv, *24g*; ordinary fellow of, again, 356; student rooms in, 34, 493n. 70; tipped for mastership of, 404–5
triple algebra, 94–95, 104, 115
triple tangent plane. *See* surfaces, triple tangent plane on
Tripos, mathematical. *See* mathematical Tripos
Trotter, Coutts, 465, 551n. 73
truths, search for, 520n. 89
Tschirnhausen, E. W. von, 72, 250
Turnbull, H. W., 439, 465, 483
Turnbull, William P., 292, 307, 465, 526n. 14
Tyndall, John, 465; and mountains, 230, 272–73; at Royal Institution of Great Britain, 408; at Royal Society, 179, 182–83, 509n. 97; and X Club, 282

United States of America, Cayley in. *See* Cayley, Arthur, in United States of America
University College, London, 18, 24, 55
University of London, 24–25, 485n. 46

Vandermonde, A. T., 66, 124, 465
van der Waerden, B. L., 439
Venables, Edmund, 465, 491n. 7; C. supervises students with, 58–60; comforts C., 165, 430; officiates at C.'s marriage, 261; on tour with C., 76
Venn, John, 360, 465
verification, Cayley's method of, 334
Victoria, Queen, 255, 332, 360, 525n. 1, 407

Wales, Cayley in, 184, 396, 427
Waley, Jacob, 465, 506n. 16
Wallace, William, 68
Wallis, A. J., 465, 543n. 60
Walton, William, 69, 142, 150, 316, 465
Waring, Edward, 250
Watson, John, 465

Watt, Robert, 465, 491n. 20
Webb, Benjamin, 465, 499n. 31
Weber, Heinrich, 428, 465
Webster, William Bullock, 223–25
Weddle, Thomas, 253, 465, 525n. 82
Wehnert, Edward, 224, 465
Weierstrass, Karl W. T., 466, 501n. 67, 538n. 59, 539n. 70
weight. *See* invariants, weight of
Weir of Hermiston (Stevenson), 431
Wellington, Duke of, 4, 17
Weyl, Hermann, 440–41
Whatley, Richard, 466
Wheatstone, Charles, 21, 180, 217, 466
Whewell, William, 466; on art vs. science, 279; on astronomy, 46; on C., 263; C. influenced by, 383; on C.'s report on dynamics, 222; elector for Sadleirian chair, 255–56; on Euclid, 406; as fellowship examiner at Trinity College, 61–63; King's College, London, and, 25; and Kirkman, 273–74; as Master of Trinity College, 36, 118; on nature of space, 383; as Smith's Prize examiner, 51–52; on Thomson's Tripos result, 106; as tutor, 28
Whitehead, Alfred North, 194, 401, 466
Whitmell, C. T., 555n. 38
Whymper, Edward, 230
Wiener, Ludwig Christian, 332
Wilbraham, Henry, 150, 184, 466
Wiles, Henry, 410
Williams-Ellis, John C., 255, 466, 525n. 94; and Sadleirian chair, 255
Williamson, Alexander W., 196, 466
Willis, Robert, 217, 466
Wilson, Andrew, 224
Wilson, James M., 269, 309–10, 466, 526n. 17
Wilson, Richard, 135, 138
Wilson, William P., 338, 466, 489n. 69
Wollaston, Charles B., 224, 466
Wolstenholme, Joseph, 466
Wood, Philip W., 466, 557n. 97
Woodham-Smith, Cecil, 145
Woolhouse, W. S. B., 335, 466
Wordsworth, Christopher (Master of Trinity College), 31, 35, 73, 390, 466, 487n. 4
Wordsworth, William, 379, 520n. 89, 545
Workman, W. P., 550n. 48
Wright, Joseph E., 466, 550n. 40
Wright, Richard T., 466, 544n. 63

Wright, T. C., 506n. 16
Wroński, Hoëné, 334, 466, 539n. 74

X Club, 191, 282–83, 321–22

Yeoman, Constantine B., 466, 491n. 8
Yorkshire, Cayley in, 127–29
Young, Alfred, 466, 550n. 46

Young, G. M., 389
Young, George P., 424, 466
Young, Grace Chisholm. *See* Chisholm, Grace
Young, Lawrence, 439
Young, William H., 467, 550n. 48

Zeuthen, Hieronymus G., 287, 467, 530n. 79, 556n. 74

Acronyms used in the list below:

CA: Collection of the author
CUP: Cambridge University Press
EB: *Encyclopedia Britannica*
KCLA: King's College London Archives
LMS: London Mathematical Society
MFSJCC: Master and Fellows of St. Johns College, Cambridge
MFTCC: Master and Fellows of Trinity College Cambridge
NCL: Newnham College Library
NPG: National Portrait Gallery, London
RS: Royal Society
SCUL: Syndics of Cambridge University Library

Gallery following page 100

Page 1g. *Top*: MFTCC. *Bottom*: KCLA.
Page 2g. *Top*: CA. *Bottom*: London University Press.
Page 3g. *Top*: KCLA. *Bottom*: City of Westminster Archives Centre.
Page 4g. *Top*: NPG. *Middle* and *bottom*: MFTCC.
Page 5g. *Top*: SCUL. *Bottom*: CA.
Page 6g. *Top*: SCUL. *Bottom*: MFTCC.
Page 7g. *Top*: Treasurer and Benchers of Lincoln's Inn. *Bottom*: CA.
Page 8g. *Top*: LMS. *Middle*: G. Eisenstein, *Mathematische Werke* (New York: Chelsea, 1975). *Bottom*: NPG.

Page 9g. *Top*: A. Enthoven. *Middle*: MFTCC. *Bottom*: LMS.
Page 10g. *Top*: MFSJCC. *Bottom*: RS.
Page 11g. *Top*: RS. *Middle*: CUP. *Bottom*: MFSJCC.
Page 12g. *Top*: MFSJCC. *Bottom*: CA.

Gallery following page 234

Page 13g. *Top*: MFSJCC. *Middle*: LMS. *Bottom*: Lewisham Local Studies and Archives.
Page 14g. *Top* and *middle*: CA. *Bottom*: Harry Ransom Humanities Research Center, University of Texas at Austin.

Page 15g. *Top* and *middle*: LMS. *Bottom*: CA.
Page 16g. *Top*: CA. *Middle*: CUP. *Bottom*: LMS.
Page 17g. *Top*: LMS. *Bottom*: MFTCC.
Page 18g. *Top*: CA. *Bottom*: Mathematical Institute, University of Oxford.
Page 19g. *Top*: CUP. *Middle* and *bottom*: EB.
Page 20g. *Top*: MFSJCC. *Bottom*: LMS.
Page 21g. *Top* and *middle*: NCL (reprinted from *The Graphic*). *Bottom*: LMS.
Page 22g. *Top*: SCUL. *Bottom*: CA.
Page 23g. *Top*: CA. *Bottom*: MFTCC.
Page 24g. *Top*: SCUL. *Middle* and *bottom*: CA.